GLOBAL HISTORY OF TECHNIQUES

GLOBAL MATTERS

CULTURE MATÉRIELLES, HISTOIRE DES TECHNIQUES ET CIRCULATION GLOBALE DES SAVOIRS

TECHNE

KNOWLEDGE, TECHNIQUE, AND MATERIAL CULTURE

VOLUME 9

La série Global Matters fait partie de la collection TECHNE, dirigée par Dániel Margócsy et Koen Vermeir.

Global History of Techniques

(Nineteenth to Twenty-First Centuries)

Edited by
GUILLAUME CARNINO, LILIANE HILAIRE-PÉREZ AND
JÉRÔME LAMY

BREPOLS

Ouvrage publié avec le soutien des partenaires suivants :

Nous avons essayé de contacter toutes les personnes qui bénéficient de droits d'auteur sur les photos et les illustrations utilisées dans ce livre. Si des photos ou des illustrations ont été utilisées à l'insu des ayants droit, ces personnes peuvent s'adresser à rights@brepols.net.

D/2024/0095/99
ISBN 978-2-503-59151-3
eISBN 978-2-503-60054-3
DOI 10.1484/M.TECHNE-EB.5.129707
ISSN 2736-7452
eISSN 2736-7460

Printed in the EU on acid-free paper.

Table of Contents

We dedicate this book to Larissa Zakharova, whom we have missed throughout the completion of this project and whom we continue to miss.

GUILLAUME CARNINO, LILIANE HILAIRE-PÉREZ AND JÉRÔME LAMY

The Global History of Techniques and the Globalization of the History of Techniques

This book is the first overall study of the global history of techniques in the contemporary period. Building on work published in the last fifteen years (Misa, Schot, 2005; van der Vleuten, Kaijser, 2006; Krige, Barth, 2006; van der Vleuten, 2008; Schot, Scranton, 2013–2019; Kohlrausch, Trischler, 2014; Hilaire-Pérez, Zakharova, 2016; Hård, 2017; Pretel, Camprubi, 2018a; Krige, 2019), it provides comparative and connected studies of technology seen as 'techniques' on a global scale. While it puts particular emphasis on means of production and communication, the exploitation of natural resources, major technological systems, infrastructures, and networks, it also reflects upon overarching issues that both underpin and help renew our understanding of the field, whether it be the risk governance, repair and maintenance, conflict, everyday techniques and the time scales involved in different techniques. Far from painting a simplified picture of evolutions on a global scale, this book opens with a world tour of techniques in a bid to restore the complexity of regional historiographies and of the meanings given to technical activities in society. This approach seems particularly necessary considering the return of grand narratives which, in the name of global history, fall back on the mostly Eurocentric meta-narrative that long dominated this field of research (Berg, 2013; Bray, 2015; Bray, Hilaire-Pérez, 2016; Popplow, 2016).

Part and parcel of our decentred vision, is our use of the terms 'technique' and 'technology', While 'technology' has become ubiquitous on a global scale and somewhat emphatically refers to the advanced technological productions of industrial societies ('new technologies') as opposed to traditional techniques, this meaning is rather ambiguous, and the term actually polysemic. For a long time, the term 'technology' did signify the science of technique (*techno-logos*), the science of the arts and of manufacture; in 1728 Christian Wolff defined *technologia* as 'the science of the arts and works of art, the science of things that men produce by the work of organs of the body, mainly by the hands', and in 1806 Johann Beckmann stated the same in his *Entwurf der allgemeine Technologie* or 'General Project of Technology'

Global History of Techniques, ed. by Guillaume Carnino, Liliane Hilaire-Pérez and Jérôme Lamy, Global Matters Culture Matérielles, histoire des techniques et circulation globale des savoirs, 9 (Turnhout: Brepols, 2024), pp. 11–26
BREPOLS PUBLISHERS 10.1484/M.TECHNE-EB.5.136421

(Carnino, Hilaire-Pérez, Hoock, 2017). This sense of the concept of technology, which opened up the possibility of a science of human activity and was marked both by encyclopaedism and the political issues of different states' efficient management of resources, faded during the nineteenth century, when the science of engineers, process engineering, and the acceptance of technology as an applied science emerged (Schatzberg, 2006, 2018). Technology then no longer referred to the reasoned and comparative study of the means of production, but rather to the industrial operation of transforming materials into products. The former, more reflexive meaning was henceforth only used in the small, yet active, circles of collection curators, philosophers, anthropologists, and ethnologists who argue in favour have seeing 'technology as a human science' (Haudricourt, 1987) and contributed to its reappearance as a field of study in the social sciences (Iribarren, 2021; Loeve, Deldicque, 2018).

The way the field has developed raises a number of issues, including that of the term 'technique', which has almost disappeared from the English language in favour of the term 'technology'. While the history of technology is increasingly globalized, its research community is increasingly international. English being the *lingua franca* of modern globalization as well as the language of historians of globalization, the use of the term 'technology' has also increased, as John Krige (Krige, 2019: 15) has pointed out. This raises further questions and invites critical reflection, as the term 'technology' is associated with western techniques understood to be applications of science, or at best techno-scientific hybrids, that helped the economic growth that characterized this part of the globe from the eighteenth century onwards. As Mikaël Hård explains, historians generally follow 'the common understanding of technology as more akin to engineering than to material culture. For example, the hydropower plant, in its close association to the field of engineering, is more likely to be understood as an artefact of technology than, say, bows and arrows' (Hård, 2017). Faced with the limits of the term 'technology', which fails to capture techniques outside the scientific field and, as was the case for many centuries and many regions of the world, created the myth of a 'technological vacuum' in some places or time periods, historians have resorted to using other expressions, like 'useful knowledge' or 'useful and reliable knowledge', to restore cultures to their appropriate know-how, skills, and technological knowledge (Mokyr, 2002; Berg, 2007; O'Brien, 2019). Other historians have proposed an examination of the meaning given to techniques in non-western cultures by looking at words used in different societies to refer to processes or aspects of human activity. For instance looking at the word 'way' used among the vaShona hunters of Zimbabwe to refer to a means of action and displacement (Mavhunga, 2014); or 'fundi' used in Kiswahili manuals in Tanzania to refer to a group of tasks ranging from divination, carpentry, and tailoring to playing soccer (Grace, 2021); or the use 'gong', as both an accounting unit for the task and a quantity of work in Song China (Lamouroux, 2010b) helps enrich the very notion of technique (Mavhunga, 2017; Coupaye, 2021).

Encouraged as we are by a historiography of techniques that has largely benefited from the contributions of the social sciences and that does not reduce techniques to efficiency, it seems important for us to maintain, and even extend, the use of the word 'technique' to refer to human activities and know-how, whether these activities use tools or not. This position is in keeping with the modern emergence of the term 'technique', as an activity that is conceptually distinct. Socially and linguistically identified in France as '*la technique*' from the end of nineteenth century to the 1940s (Camolezi, 2021), it remained connected to other human activities and interests, hence philosophers and social scientists from Alfred Espinas to Marcel Mauss for instance understanding 'technical action as a social fact' (Iribarren, 2021). It is also in this sense that the term 'technic' is sometimes employed in recent historiography (Kumar, 2019). At the same time the words 'technique' and 'technical' have increasingly been used in English to describe method, skill, and know-how (Krige, 2019; Mateos, Suárez-Díaz, 2019). We can then define 'technique' as a set of actions and processes and their cumulative material manifestation in objects produced by human beings. Technique acts as a key mediation between man and the world: in this sense, it is controlled and embodied by actions (e.g. know-how necessary to the handling of an object) while objects are informed by these actions (e.g. tools used whose very form is shaped by their use and transformed by their wear and tear). This is what properly constitutes technique. The relationship between embodied knowledge and tools started when human beings first started using tools. It conditions and defines the relationship of humans to the world. It therefore deserves a dedicated term.

We think it is important to continue to be aware of 'technology' in the sense of the science of the arts, which was its meaning in French and in German in the nineteenth^th century. According to recent research, this meaning has persisted in different intellectual contexts which are now gradually being better identified. At the same time, there is currently research on the complexity of the long-term codification of techniques. This raises several issues, not least that of the interaction with tacit and localized knowledge including in recent times, despite the rejection of this tacit knowledge by forms of standardised knowledge and the automation of production (Le Roux, Guéritte, 2016; Carnino, Marquet, 2019). It is therefore also appropriate to consider 'technology as a human science' more closely, since our aim is to place technology back at the heart of intellectual history, an aspect that has long been neglected by hitorians who tend to reduce technology to instrumentation, applied or industrial science, and productivism. As a result all authors in this volume were free to use the word 'technology' in its different meanings, in keeping with the long history of the term. This book reflects the diverse uses of this term, the various historiographical traditions the term draws from, while chapters also contextualise the term within the specific fields and activities they examin.

Apart from this central linguistic question, this book's multifaceted perspective as it combines different levels of analysis without imposing any linear narrative to the technical transformations of the last two centuries is also explained in part by the present collection bringing together authors from different disciplines and countries. Conversely the book bears witness to the recent globalization of the history of techniques, the history of which should begin to be written. The authors and the research programmes that have given impetus to the global history of techniques are largely based in English-speaking countries, but this book forms part of a recent expansion of this globalized outlook (Jasanoff, Kim, 2015; Pretel, Camprubi, 2018a; Krige, 2019) by presenting a wider range of research, and in particular francophone research, which has hitherto rarely been included in historiographical accounts of the field. It thereby follows another collection of essays on the global history of techniques that was recently published in France (Carnino, Hilaire-Pérez, Kobiljski, 2016) and concluded that it was necessary to think about the global history of techniques in terms of — to use the phrase coined by the French ethnologist André–Georges Haudricourt (Haudricourt, 1987) — 'technology as a human science' (Carnino, Hilaire-Pérez, 2016). Guillaume Carnino and Liliane Hilaire-Pérez suggested that one might situate the global history of technique between the search for structuring principles on the one hand, and the collection of the infinite variety of techniques on the other, bringing into play two key notions: with on the one hand convergence — what André Leroi-Gourhan calls 'trends' (Leroi-Gourhan, 1943), what Gilbert Simondon calls 'concretization' (Simondon, 1958), and is sometimes referred to as 'codification of practice' by other authors (Schäfer, Popplow, 2015) — and on the other hand divergence, since according to Marcel Mauss, there is a myriad variations in techniques that make up the texture of human societies. These kaleidoscopic variations justify the existence of such a field as 'technological sociology', which is encyclopaedic in nature (Mauss, 1960 [1934]; Schlanger, 1991). Clearly the meaning assigned to 'divergence' and 'convergence' here is radically different from that used by economists, who consider the former as a significant break in trajectories, captured at macro level (Pomeranz, 2000), and the latter as a process of growth development leading to the inversion of economic trends (Riello, 2013: 292–94). By contrast, for us the notion of divergence refers to forms of local fragmentation (including failure and exclusion) that coexist with trends towards functional integration and convergence (van der Vleuten, 2008: 992).

Convergence and divergence thus refer to processes related to the very make up of technical objects, with the notions of concretization and deconcretization initially borrowed from the philosophy of techniques now being used by historians (Simondon, 1958: 19–23; Livet, 2005: 26–31), a binary reminiscent of the terms 'coupling' and 'decoupling' used in sociology (Flichy, 1995: 167; White, 1992; Grossetti, Bès, 2003). As Antoine Picon explains, 'it may be that technical progress does not inevitably proceed from a passage from the abstract to the concrete, in other words to the increasing integration of functions and forms, but to a successive moments concretization and deconcretization' (Picon, 1992: 387). Thus, another possible reading of the processes of concretization/convergence and deconcretization/divergence of techniques emerges, which should be considered in terms of cycles and alternations. These definitions stretch beyond the usual economic meaning of these terms, and also beyond the socio-cultural meaning of appropriation and agency, to think more specifically about techniques. Such an approach, which is inseparable from the development of the history of techniques in France in dialogue with the social sciences since the *Annales*, explains why our work combines history with sociology, anthropology, ethnology, and philosophy. Interdisciplinarity and reflexivity play an essential role in the critical investigation of today's teleology of growth, progress and innovation, three notions which have often kept us from properly understanding the 'mode of existence' of technical objects (Simondon, 1958;Bontems, 2019).

This book then bears the imprint of two rationales. On the one hand, it builds on research carried out internationally that rejects overarching diffusionist narratives in favour of an approach that stresses the complexity of how techniques circulate (Hilaire-Pérez, Verna, 2006; Krige, 2019), and more generally stresses the many forms of divergence that exist — what anthropologists call socialized technique and which economic historians are just beginning to consider (Davids, 2019). On the other hand, this book's chapters contribute to asking new questions because of their self critical approach (Lamy, 2018; Crépel, Lamy, Petitjean, 2017; Grunwald, 2018; Heßler, Weber, 2019; Böhn, Möser, 2010), their attention to terminology (Schatzberg, 2006, 2018; Carnino, Hilaire-Pérez, Hoock, 2017), to issues of scale, and because they highlight the complex role of various timescales on techniques (Grunwald, 2012; Benoit, 2020).

In line with current research, this book also adopts, a critical point of view on the history of techniques (Feenberg, 2004; Jarrige, 2014; Jasanoff, Kim, 2015; Hård, Jamison, 1998). As the authors of a dossier on infrastructures and their place in the ideologies of power and modernity put it,

> While the Silicon Valley's industrial, institutional and financial infrastructures are now a model that is spreading to many parts of the world, while so-called 'Smart Cities' are fashioning the deployment of new urban surveillance and control infrastructures, the humanities and social sciences cannot simply stand by and describe evolutions: they must also interrogate them and analyse the deeper ideological issues at work' (Jarrige, Le Courant, Paloque-Bergès, 2018: 8).

This is also the approach of some authors who outline the hidden and paradoxical issues at stake in European technologies, including their vulnerabilities (van der Vleuten, Högselius, Hommels, Kaijser, 2013), and others who focus on the role played by 'technocratic experts', from engineers to economists, in the establishment of post-colonial hierarchies via asymmetrical technical transfers (Pretel, Camprubi, 2018b). Our book reflects these current positions, investigating the central role played by innovation, growth, and techno-sciences in the history of techniques. This critical approach is particularly well represented in France in relation to the environmental impacts of technology (Bonneuil, Fressoz, 2016; Jarrige, Le Roux, 2017), to notions of risk and technical uncertainty (Fressoz, 2012; Fridenson, 2012; Le Roux, 2016), to those of breakdowns and accidents (Lambert, Raveux, 2019) and to workers' opposition to machinism (Bourdeau, Jarrige, Vincent, 2006; Biagini, Carnino, 2010; Jarrige, 2014).

At the heart of these approaches is the paradigm of the construction of technopolitical regimes, an idea that underpins the global history of modern techniques and has dominated the literature for a generation (Winner, 1980; Noble, 1984; Hecht, 1998, 2012; Mitchell, 2011; Kurban, Peña-Lopez, Haberer 2017; Krige, 2019), inviting us to re-read transnational political processes — such as European integration for instance — in the light of the history of techniques — with the ideas of a 'hidden integration' and a 'hidden fragmentation' (Misa, Schot, 2005: 3). This political dimension was the focus of Larissa Zakharova's work on the global history of techniques, and it has largely inspired this book (Zakharova, 2020; Hilaire-Pérez, Zakharova, 2016). Going against any nationalistic appropriation of techniques, a tendency as potent in France as in the USSR, Larissa Zakharova showed that supposedly national techniques resulted in fact from multiple encounters that resulted in an 'interdependence of territories' — whether the latter be desired or imposed, embraced or negated (Hilaire-Pérez, Zakharova, 2016: 25) — and led to various forms of adaptation rather than any homogeneity (Zakharova, 2016, 2020). This book, therefore, seeks to understand and contextualise these nationalistic narratives in the light of a political history of techniques at the same time as it to uncovers complex processes made up of hybridizations, multiple timescales, and territories, highlights polysemic notions and controversies. This calls for the adoption of a variety of points of view, methods, and conceptual tools.

What this book also does, beyond simply mapping out the field of the global history of techniques today, is to highlight the diversity of reflections and conceptual tools at work. Several choices made for this book bear witness to specific points that we feel it is important to make. One of these is the relationship between techniques and globalization. At the beginning of this book we emphasize the multiple pitfalls that threaten any attempt to reconstruct a history spanning more than two centuries of technological practices on a global scale. One major focus is the definition of globalization itself. The term is often used as an alternative to such terms as 'connected', or 'transnational'. This is not merely a linguistic issue. Each term is connected to a historiographical project, so that clarity is important. David Edgerton defines globalization as a means of studying 'all places that use technology, not just the small number of places where invention and innovation is concentrated' (Edgerton, 2006: xiii). Focusing on the analysis of how technologies are used rather than invented, Edgerton calls for 'a history of technology engaged with all the world's population, a population, which is mostly poor, non-white, and half female' (Edgerton, 2006: xiii). Globalization here thus refers to an actual situation. Technology considered as techniques, that is as a means of acting on the world is therefore part of his analysis. In a way, this global history of material practices is based on a naturalization of diffusionism. The study of centres of technical production and innovation and how technologies circulated from these centres was at the core of traditional anthropology and prehistoric studies (Leroi-Gourhan, 1943, 1945; Haudricourt, 1987). The fact it echoes with current reflections on the globalization of techniques explains its recent revival (Gazagnadou, 2008). Due to the prevalence of information infrastructures (Mattelart, 1992, 1999) and very large-scale energy flows (Gras, 2015) in the contemporary period, among other things, globalization has been taken for granted as a key factor for the circulation and adaptation of techniques.

In contrast with this vision, we first need to remind ourselves that long-distance communications were used prior to the nineteenth century. It is also important to stress that the supposed fluidity of globalization processes is by no means obvious (Didry et al., 2004): issues linked to the (in)compatibility of various national norms

(Méadel, 1994), path dependency (David, 1985; Gardey, 1998), various instances of resistance (Jarrige, 2014; Fischer, 2022), and state control of flows (Krige, 2019) all contradict such neat narratives of technologies smoothly spreading around the world.

The predominance of science (to the detriment of techniques) in the study of globalized forms of exchange and appropriation poses another difficulty that needs to be navigated. Asymmetry is not a new concept (Febvre, 1935), but it calls for increased vigilance here. In fact, recent attempts to develop a framework for the global history of science (which incidentally often includes techniques although this is not clearly stated) have led to the West being overvalued (Roberts, 2009) and techniques to be seen merely as components of scientific activity (Fan, 2012).

In this collection of essays we use the term globalization in the sense of a relational capacity of individuals to produce, transmit, appropriate, transform, contest, and even ignore techniques. The large cultural areas defined in the first part of the book thus exemplify congregated coherent technological structures as well as areas of more or less stable technical exchanges. These divisions make it possible to identify different levels of of appropriation, standardization, and circulation. These are by no means rigid structures limiting the possibilities of a multifaceted approach to the continuous mutation of technical appropriation.

Connected history, unlike comparative history, intends to focus on the social, political, and cultural processes where the 'local and specific' meets the 'supra-local' (Subrahmanyam, 1997: 745). One of the difficulties in discussing techniques lies in the material nature of the exchanges at stake. Roger Chartier asked with regard to this type of approach:

> How can we think about the relationship between appropriation and acculturation, between inventive reuse and cultural uprooting? How can we define the processes of "interaction" or "negotiation" (...) depending on whether they operate within relations of domination or in relations of exchange? (Chartier, 2001: 123).

The quest for an illusory 'global' overarching template — a grand narrative accounting for the supposed formalization of techniques — is another major pitfall for any attempt at dealing with the history of techniques on a global scale. The period spanning the nineteenth, twentieth and twenty-first centuries has too narrowly been studied through a purely economic lens: industrial revolutions, the transformations of modern capitalism, the massive monetization of human activities and endeavours and since the 1970s, the deployment of a neoliberal programme supported by international institutions (IMF, World Bank) have resulted in a focus on economic models (Pretel, Camprubi, 2018a). From this perspective, techniques have been considered as a variable that can explain any increase in trade, the speed of communication, increased productivity, etc. (Verley, 1997). Taken for granted techniques are included in a purely economic narrative. However some current studies attempt to disentangle the analysis of techniques from such a reductive economic understanding:

> In the face of the often simplistic discourse that the spectacular transformations of present-day Chinese society give rise to, it seemed essential for us to recall that

> these have a history, a history during which the activity and technical knowledge that bring about change have never depended solely on economic conditions, and techniques here cover the unequal capacity of different groups and individuals to act on their environment and to legitimize this activity (Lamouroux, 2010a: 161).

Generally speaking, in response to the questions of 'who invented what' and which techniques may have generated increases in productivity, historians increasingly tend to examine the multiple issues related to techniques, their fitness to purpose, and their transmission in specific societies before embarking on large-scale comparisons and grand narratives of globalization.

The search for homogeneity in the history of technology also applies to attempts at model building like Immanuel Wallerstein's world-systems (Wallerstein, 2009). The multiple criteria used (economic, political, cultural, ideological, etc.) obfuscate the specific characteristics of technical practices so that it becomes impossible to identify them. On the other end of the spectrum is the division of technical practices into a myriad of case studies (Douki, Minard, 2007: 18) that explain each process in all its finest detail, but do not attempt to define a more comprehensive intellectual framework that might account for the processes at work.

In order to overcome these various conundrums in the historiography of the globalization of techniques, we have chosen to focus on the processes of circulation, a position which means taking into account actors' situations and motivations, potential failures, intermediaries, while transport and communication infrastructures are put into context (Hilaire-Pérez, 2008). Even the most modern dematerialized data cannot be abstracted from stubborn materiality (Laumonier, 2019). Using circulation as a lens draws attention to a variety of operations: it allows for the debunking of apparently self evident notions such as innovation (Sainsaulieu, Saint-Martin, 2015) and for the exploration of various facets of human activities to identify forms of technical appropriation. Connections and exchanges are studied along instances of miscommunication and of techniques failing to spread: techniques thus emerge as infinitely varied practices that are inseparable from global processes, be they political, cultural, economic, or social.

Another contention of the volume is to challenge the traditionally event-driven approach to the history of techniques. In the same way as the *Annales* school was founded upon the rejection of the history of great kings and famous battles (Burguière, 2006), it seems important today to initiate a radical departure from the event-driven history of techniques. Instead, it is crucial to embrace a much more ambitious approach which forgoes the anecdotal and any simplistic focus on particular discoveries or inventions, to adopt a multiscalar perspective. French historiography in particular has developed the study of the temporality of techniques with specific emphasis on the *longue durée*. This particular historiographical development owes both to the notion of 'technical systems' and to Fernand Braudel's tripartite vision of time that has made it possible for historians to think about techniques unfolding over different timescales (Hilaire-Pérez, Jarrige, 2020). We also want to build on David Edgerton's critique of novelty (2006) while refraining from seeing practices as a cure-all go-to concept offering protection from the evils of an over-fascination for innovation. If attention to social developments once partly succeeded in debunking the grand narrative of progress (Bijker, Hughes, Pinch, 1987), today's received wisdom tends to downplay the role of techniques in favour of social factors. But in the name of avoiding technical determinism what is foregrounded is often an equally simplistic vision dominated by social determinism. If issues of the social uptake, circulation and appropriation of techniques is obviously important, the widespread deployment of a technique over an entire territory is also shaped, and in turns gives rise to, crucial phenomena that require analysis.

Insisting on the scale of technical implementations in particular eschews the twin issues on the one hand of the diffusionist paradigm (which we know to be a mere fiction) and on the other of the fragmentation of the narrative into a myriad factors making any causal analysis impossible. Having an emphasis on critical masses also makes it possible to move beyond the false debate centred on the notion of innovation, opposing rupture and continuity or, for the history of techniques, invention and imitation. Shunning the usual focus on the inventive, creative, and god-like moment of invention, and foregrounding instead the development of structured technical systems as Thomas P. Hughes (1983; 1989) suggested a while ago already, is a way of highlighting actors' ability to hybridize, build and negotiate between different systems and techniques. Leaving the event-driven history of techniques behind shifts the focus from inventors (who are no longer described as brilliant and misunderstood individuals: Hugues, 1989; Hilaire-Pérez, 2000; Carnino, 2015) to system builders, and at the same time unsettles the other false dichotomy between social determinism and technique, revealing crucial elements that are played out in diverse places and times in each phase of any technical deployment (van der Vleuten, Oldenziel, Davids, 2017). The emergence of a coherent technical universe then invariably contributes to the closing and opening of new worlds. If we pay attention to the various sizes and geographies of technical devices, we can put today's burning question of social inequalities back center stage at last. Factoring in technical issues when analysing colonization thus provides, for example, a way towards understanding certain formative events, such as the battle of Omdurman at the end of the Anglo-Egyptian conquest of Sudan. Winston Churchill recounts that on 2 September 1898, after five hours of fighting (during which, according to him, the infantry fired calmly, while the enemy was still far away), twenty English men, twenty of their Egyptian allies, and 11,000 Dervishes had been killed (Headrick, 1981). Clearly, the spread of techniques and technical diffusionism are two sides of the same colonial coin, they are part of the same power dynamics (Headrick, 1988; Pessis, 2019).

These methodological stances have informed the structure of the present work.

Our criticism of global reductionism has therefore prompted us to devote the first part of the book to a world tour of techniques, which aims to preserve the uniqueness of each region of the globe during the specified period and to highlight reappropriations as well as misunderstandings: in an unabashedly arbitrary fashion, and in order to curtail any ethnocentric bias, the chapters in this first section take the reader on an intellectual journey around the world from Oceania — which all too often gets short changed in the existing literature — just as it has done in power dynamics — to South America. Some chapters are more historiographical, others focus more on the specific contexts of some developments, but all provide a critical analysis of contemporary techniques at the crossroads of global and local dynamics.

The emphasis on the large scale deployment of techniques rather than supposedly revolutionary inventions has led us in the second part to focus on sectors whose importance seems critical to the constitution of the modern world although the relative importance of each sector in today's world is open to discussion naturally. We start with a chapter on heavy industries, and end with urban techniques, but this order should not be invested with more meaning than it carries.

Distinguishing between technique and technology is a means to avoid reducing phenomena to their materiality and makes it possible to think about significant trends. The third part of the book focuses on various structural issues that have dominated technique in the last three hundred years. The section starts with a discussion of technology and, with each chapter leading to the next, weaves its way from from labour to social conflicts, from the body to gender or from technology of the everyday to food.

Whatever the section, authors have been expressly instructed to cover all time periods, themes or historiographical debates that they felt were important for their topic. If the adoption of a comprehensive approach from a chronological, thematic, and historiographical point of view was a guiding principle, each author was then left free to choose a particular element or approach. Any work of this type also unavoidably has lacunae. Some topics — such as musical instruments, packaging or technical failures — unfortunately have had to be left out because the specialist authors who could have written on them were either unavailable at the time of the project or unwilling to conform to its format. These shortcomings are, of course, also caused by the constraints of an already very full but we hope rich volume.

References

Benoit, Serge, *D'eau et de feu: forges et énergie hydraulique, XVIIIe–XXe siècle. Une histoire singulière de l'industrialisation française* (Rennes: Presses Universitaires de Rennes, 2020).

Berg, Maxine, 'The Genesis of "Useful Knowledge"', *History of Science*, 45.2 (2007), 123–33.

———, ed., *Writing the History of the Global: Challenges for the 21st Century* (Oxford: Oxford University Press, 2013).

Bert, Jean-François, ed., *Lire 'les techniques du corps'. Relire Marcel Mauss* (Paris: Éditions de la Sorbonne, 2022).

Biagini, Cédric and Guillaume Carnino, *Les Luddites en France. Résistance à l'industrialisation et à l'informatisation* (Montreuil: Éditions L'échappée, coll. Frankenstein, 2010).

Bijker, Wiebe E., Thomas P. Hughes, and Trevor Pinch, eds, *The Social Constructions of Technological Systems* (Cambridge, MA: MIT Press, 1987).

Böhn, Andreas and Kurt Möser, eds, *Techniknostalgie und Retrotechnologie* (Karlsruhe: Karlsruher Institut für Technologie Scientific Publishing, 2010).

Bonneuil, Christophe and Jean-Baptiste Fressoz, *The Shock of the Anthropocene. The Earth, History and Us* (London: Verso, 2016).

Bontems, Vincent, 'La recherche au nom de l'innovation', habilitation thesis, Paris, ENS, 2019.

Bourdeau, Vincent, François Jarrige, and Julien Vincent, *Les Luddites. Bris de machine, économie politique et histoire* (Maisons-Alfort: Éditions Ère, 2006).

Bray, Francesca, 'Technological Transitions', in *The Cambridge World History*, vol. 6, The Construction of a Global Wordl, 1400-1800 CE, part 1 : Foundations, ed. by Jerry H. Bentley, Sanjay Subrahmanyam and Merry E. Wiesner-Hanks.

——— and Liliane Hilaire-Pérez,'Les techniques et l'histoire globale', in *Histoire des techniques. Mondes, sociétés, cultures (XVIe–XVIIIe siècle)*, ed. by Guillaume Carnino, Liliane Hilaire-Pérez, Aleksandra Kobiljski (Paris: Presses universitaires de France, Nouvelle Clio, 2016), pp. 7–22.

Burguière, André, *L'École des annales. Une histoire intellectuelle* (Paris: Odile Jacob, 2006).

Camolezi, Marcos, 'Technique, technics, Technik: les mots à travers les dictionnaires du XIXe siècle', *Artefact. Techniques, histoire et sciences humaines*, 15 (2021), 61–106.

Carnino, Guillaume, *L'Invention de la science. La nouvelle religion de l'âge industriel* (Paris: Seuil, 2015).

——— and Clément Marquet, 'Du mythe de l'automatisation au savoir-faire des petites mains: une histoire des datacenters par la panne', *Artefact. Techniques, histoire et sciences humaines*, 11 (2019), 161–88.

——— and Liliane Hilaire-Pérez, 'Les temps longs, l'histoire globale et la complexité en histoire des techniques', in *Histoire des techniques. Mondes, sociétés, cultures XVIe-XVIIIe siècles, XVIe–XVIIIe siècle)*, ed. by Guillaume Carnino, Liliane Hilaire-Pérez, Aleksandra Kobiljski (Paris: Presses universitaires de France, Nouvelle Clio, 2016), pp. 485–90.

———, ———, and Jochen Hoock, eds, *La Technologie générale. Johann Beckmann Entwurf der allgemeinen Technologie / Projet de technologie générale (1806)* (Rennes: Presses universitaires de Rennes, 2017).

———, ———, and Aleksandra Kobiljski, eds, *Histoire des techniques. Mondes, sociétés, cultures (xvie–xviiie siècle)* (Paris: Presses universitaires de France, Nouvelle Clio, 2016).

Chartier, Roger, 'La Conscience de la globalité', *Annales. Histoire, Sciences Sociales*, 56.1 (2001), 119–23.

Coupaye, Ludovic, 'Technology', in *Cambridge Handbook of Material Culture*, ed. by Lu Ann De Cunzo and Catharine Dann Roeber (Cambridge: Cambridge University Press, 2022), 436–68.

Crépel, Pierre, Jérôme Lamy, and Johann Petitjean, eds, *Histoire critique des sciences, xvie–xviiie siècle, Cahiers d'histoire. Revue d'histoire critique*, 136 (2017).

David, Paul A., 'Clio and the Economics of QWERTY', *The American Economic Review Papers*, 75.2 (1985), 332–37.

Davids, Karel, 'Toolkits, Creativity and Divergences: Technology in Global History', in *Global Economic History*, ed. by Giorgio Riello and Tirthankar Roy (London: Blooomsbury, 2019), pp. 67–82.

Didry, Claude, Patrick Dieuaide, Laurence Roulleau-Berger, Monique Selim, Richard Sobel, 'La Mondialisation n'existe pas: regards sur les expériences singulières du travail globalisé', *L'Homme et la société*, 152–53 (2004), 9–16.

Douki, Caroline and Philippe Minard, 'Histoire globale, histoires connectées: un changement d'échelle historiographique?', *Revue d'histoire moderne et contemporaine*, 54.4 bis (2007), 7–21.

Edgerton, David, *The Shock of the Old: Technology and Global History since 1900* (New York, NY: Oxford University Press, 2007).

Fan, Fa-ti, 'The Global Turn in the History of Science', *East Asian Science, Technology and Society: An International Journal*, 6 (2012), 249–58.

Febvre, Lucien, 'Réflexions sur l'histoire des techniques', *Annales d'histoire économique et sociale*, 7 (1935), 531–34.

Feenberg, Andrew, *(Re)penser la technique. Vers une technologie démocratique* (Paris: La Découverte/MAUSS, 2004).

Fischer, Daniel, *Dietrich, un entrepreneur des savoirs* (Paris: Presses des Mines, 2022).

Flichy, Patrice, *L'Innovation technique. Récents développements en sciences sociales. Pour une nouvelle théorie de l'innovation* (Paris: La Découverte, 1995).

Fressoz, Jean-Baptiste, *L'Apocalypse joyeuse. Une histoire du risque technologique* (Paris: Le Seuil, 2012).

Fridenson, Patrick, 'L'Histoire de l'incertitude technique et ses enjeux', *Revue d'histoire moderne et contemporaine*, 59.3 (2012), 7–18.

Gardey, Delphine, 'La Standardisation d'une pratique technique: la dactylographie (1883–1930)', *Réseaux*, 87 (1998), 75–103.

Gazagnadou, Didier, *La Diffusion des techniques et les cultures* (Paris: Éditions Kimé, 2008).

Grace, Joshua, 'Ufundi and Tekinolojia in Independent Tanzania: Kiswahili Lexicons of Specialization in Adult Education Manuals, 1960s to 1980s', *Artefact. Techniques, histoire et sciences humaines*, 15 (2021), 17–42.

Gras, Alain, *Oil. Petite anthropologie de l'or noir* (Paris: Éditions B2, 2015).

Grossetti, Michel and Marie-Pierre Bès, 'Dynamiques des réseaux et des cercles. Encastrements et découplages', *Revue d'économie industrielle*, 103 (2003), 43–58.

Grunwald, Armin, *Technikzukünfte als Medium von Zukunftsdebatten und Technikgestaltung* (Karlsruhe: Karlsruher Institut für Technologie Scientific Publishing, 2012).

———, 'Technik als Transformation von Möglichkeitsräumen. Technikphilosophie anders gedacht', in *Möglichkeiten der Reflexion. Festschrift für Christoph Hubig*, ed. by Philipp Richter, Jan Müller and Michael Nerurkar eds (Baden–Baden: Sigma, 2018), pp. 203–16.

Hård, Mikael, ERC 'A Global History of Technology, 1850–2000 GLOBAL–HOT' (2017) <https://www.academia.edu/37125306/A_Global_History_of_Technology_1850_2000_GLOBAL_HOT> [accessed 21 March 2022].

———, *Microhistories of Technology. Making the World* (Cham: Palgrave Macmillan, 2023).

——— and Andrew Jamison, eds, *The Intellectual Appropriation of Technology: Discourses on Modernity, 1900–1939* (Cambridge, MA: MIT Press, 1998).

Haudricourt, André-Georges, *La Technologie science humaine. Recherches d'histoire et d'ethnologie des techniques* (Paris: Éditions de la Maison des Sciences de l'Homme, 1987).

Headrick, Daniel, *The Tools of Empire: Technology and European Imperialism in the Nineteenth Century* (Oxford: Oxford University Press, 1981).

Headrick Daniel, *The Tentacles of Progress: Technology Transfer in the Age of Imperialism, 1850–1940* (Oxford: Oxford University Press, 1988).

Hecht, Gabrielle, *The Radiance of France. Nuclear Power and National Identity after World War II* (Cambridge, MA: The MIT Press, 1998).

———, *Being Nuclear. Africans and the Global Uranium Trade* (Cambridge, MA: The MIT Press, 2012).

Heßler, Martina and Heike Weber, eds, *Provokationen der Technikgeschichte. Zum Reflexionsdruck historischer Forschung* (Paderborn: Ferdinand Schöningh, 2019).

Hilaire-Pérez, Liliane, *L'Invention technique au siècle des Lumières* (Paris: Albin Michel, 2000).

——— and François Jarrige, 'Serge Benoit et la pluralité des temporalités techniques', in *D'eau et de feu: forges et énergie hydraulique, xviiie–xxe siècle. Une histoire singulière de l'industrialisation française*, Serge Benoit (Rennes: Presses Universitaires de Rennes, 2020), pp. 415–34.

Hilaire–Pérez, Liliane, and Catherine Verna, 'Dissemination of technical knowledge in the middle ages and the early modern history. New approaches and methodological issues', *Technology and Culture*, 47 (2006), 536–65 ('La circulation des savoirs techniques du Moyen Âge à l'époque moderne. Nouvelles approches et enjeux méthodologiques', *Tracés*, 16 (2009), 25–61).

Hilaire-Pérez, Liliane and Larissa Zakharova, 'Introduction', in *Les Techniques et la globalisation au xxe siècle*, ed. by Liliane Hilaire-Pérez and Larissa Zakharova (Rennes: Presses Universitaires de Rennes, 2016), pp. 23–39.

Hughes, Thomas P., *Networks of Power. Electrification in Western Society, 1880–1930* (Baltimore, MD: Johns Hopkins University Press, 1983).

———, *American Genesis. A Century of Invention and Technological Enthusiasm, 1870–1970* (New York, NY: Viking, 1989).

Iribarren, Leopoldo, 'La Technologie: paradigmes grecs d'une science sociale française', *Artefact. Techniques, histoire et sciences humaines*, 15 (2021), 175–200.

Jarrige, François, *Technocritiques. Du refus des machines à la contestation des technosciences* (Paris: La Découverte, 2014).

———, Stefan Le Courant, and Camille Paloque-Bergès, 'Infrastructures, techniques et politiques', *Tracés. Revue de Sciences humaines*, 35 (2018), 7–26.

——— and Thomas Le Roux, eds, *La Contamination du monde. Une histoire des pollutions à l'âge industriel* (Paris: Le Seuil, 2017).

Jasanoff, Sheila and Sang-Hyun Kim, eds, *Dreamscapes of Modernity. Socio-technical Imaginaries and the Fabrication of Power* (Chicago, IL: University of Chicago Press, 2015).

Kohlrausch, Martin and Helmuth Trischler, eds, *Building Europe on Expertise. Innovators, Organizers, Networkers* (Cham: Palgrave Macmillan, 2014).

Krige, John, 'Introduction: Writing the Transnational History of Science and Technology', in *How Knowledge Moves. Writing the Transnational History of Science and Technology*, ed. by John Krige (Chicago, IL: The University of Chicago Press, 2019), pp. 1–31.

——— and Kai-Henrik Barth, eds, *Global Power Knowledge: Science and Technology in International Affairs, Osiris*, 21 (2006).

Kumar, Prakash, 'Modalities of Modernization: American Technic in Colonial and Postcolonial India', in *How Knowledge Moves: Writing the Transnational History of Science and Technology*, ed. by John Krige (Chicago, IL: The University of Chicago Press, 2019), pp. 120–48.

Kurban, Can, Ismael Pena-Lopez, and Maria Haberer, 'What is Technopolitics? A Conceptual Scheme for Understanding Politics in the Digital Age', *IDP. Revista de Internet, Derecho y Política*, 24 (2017), 3–20.

Lambert, Guy and Olivier Raveux, eds, *Pannes et accidents (XIXe–XIXe siècle). Au cœur de l'économie, des techniques et de la société*, special issue, *Artefact. Techniques, histoire et sciences humaines*, 11 (2019).

Lamouroux, Christian, 'Présentation', in *Travail et savoirs techniques dans la Chine pré-moderne*, in special issue, *Revue de synthèse*, 131, ed.by Christian Lamouroux (2010a), 161–66.

———, 'Le Travail artisanal en Chine (X^e–XIIe siècles). Entre curiosité lettrée et expertise bureaucratique', *Revue de synthèse*, 131 (2010b), 167–92.

Lamy, Jérôme, *Faire de la sociologie historique des sciences et des techniques* (Paris: Hermann, 2018).

Laumonier, Alexandre, 4 (Brussels: Zones Sensibles, 2019).

Leroi-Gourhan, André, *Évolution et techniques*, vol 1: *L'Homme et la matière* (Paris: Albin Michel, 1943).

———, *Évolution et techniques*, vol 2: *Milieu et techniques* (Paris: Albin Michel, 1945).

Le Roux, Muriel and Françoise Guéritte, *Navelbine® and Taxotère®. Histories of Sciences* (Amsterdam: ISTE Press-Elsevier, 2016).

Le Roux, Thomas, ed., *Risques industriels. Savoirs, régulations, politiques d'assistance, fin XVIIe–début XXe siècle* (Rennes: Presses Universitaires de Rennes, 2016).

Livet, Pierre, *Qu'est-ce qu'une action?* (Paris: Vrin, 2005).

Loeve, Sacha and Timothée Deldicque, 'Les malentendus de la technologie', *Artefact. Techniques, histoire et sciences humaines*, 8 (2018), 215–54.

Mateos, Gisela and Edna Suárez-Díaz, 'Technical Assistance in Movement: Nuclear Knowledge Crosses Latin American Borders', in *How Knowledge Moves. Writing the Transnational History of Science and Technology*, ed. by John Krige (Chicago, IL: The University of Chicago Press, 2019), pp. 345–67.

Mattelart, Armand, *La Communication-monde* (Paris: La Découverte, 1992).

———, *Histoire de l'utopie planétaire. De la cité prophétique à la société globale* (Paris: La Découverte, 1999).

Mauss, Marcel, 'Les techniques du corps', in *Sociologie et anthropologie* (Paris: Presses universitaires de France, 1950), pp. 363–86 [*Journal de psychologie normale et pathologique*, 32.3–4 (1935), 271–93].

Mavhunga, Clapperton Chakatetsa, *Transient Workspaces: Technologies of Everyday Innovation in Zimbabwe* (Cambridge, MA: MIT Press, 2014).

———, *What Do Science, Technology, and Innovation Mean from Africa?* (Cambridge, MA: MIT Press, 2017).

Méadel, Cécile, 'Les Belles images de la télévision. Une histoire du D2Mac', *Annales des Mines. Gérer et comprendre*, 134 (1994), 18–29.

Misa, Thomas J. and Johan Schot, 'Inventing Europe: Technology and the Hidden Integration of Europe', *History and Technology*, 21.1 (2005), 1–19.

Mitchell, Timothy, *Petrocratia. La démocratie à l'âge du carbone* (Maisons-Alfort: Éditions Ère, 2011).

Mokyr, Joel, *The Gifts of Athena: Historical Origins of the Knowledge Economy* (Princeton and Oxford: Princeton University Press, 2002).

Noble, David F., *Forces of Production. A Social History of Industrial Automation* (New York, NY: Alfred A. Knopf, 1984).

O'Brien, Patrick, 'Useful and Reliable Knowledge in Europe and China', in *Global Economic History*, ed. by Giorgio Riello and Tirthankar Roy (London: Blooomsbury, 2019), pp. 54–66.

Pessis, Céline, 'Défendre la terre. Scientifiques critiques et mobilisations environnementales des années 1940 aux années 1970', doctoral thesis, EHESS Paris, 2019.

Picon, Antoine, *L'Invention de l'ingénieur moderne. L'École des Ponts et Chaussées 1747–1851* (Paris: Presses de l'École nationale des Ponts et Chaussées, 1992).

Pomeranz, Kenneth, *The Great Divergence: China, Europe and the Making of the Modern World Economy* (Princeton, NJ: Princeton University Press, 2000).

Popplow, Marcus, 'Technology and Technical Knowledge in the Debate on the "Great Divergence"', *Artefact. Techniques, histoire et sciences humaines*, 4 (2016), 275–85.

Pretel, David and Lino Camprubi, eds, *Technology and Globalisation. Networks of Experts in World History* (Cham: Palgrave Macmillan, 2018a).

——— and ———, 'Technological Encounters: Locating Experts in the History of Globalisation', in *Technology and Globalisation. Networks of Experts in World History*, ed. by David Pretel and Lino Camprubi (Cham: Palgrave Macmillan, 2018b), pp. 1–26.

Riello, Giorgio, *Cotton. The Fabric that Made the Modern World* (Cambridge: Cambridge University Press, 2013).

Roberts, Lissa R., 'Situating Science in Global History. Local Exchanges and Networks of Circulation', *Itinerario*, 33.1 (2009), 9–30.

Roy, Tirthankar and Giorgio Riello, eds, *Global Economic History* (London: Bloomsbury, 2019).

Sainsaulieu, Ivan and Arnaud Saint-Martin, *L'Innovation en eaux troubles. Sciences, techniques, idéologies* (Vulaines sur Seine: Éditions du Croquant, 2015).

Schäfer, Dagmar and Marcus Popplow. 'Technology and innovation within expanding webs of exchange', in *The Cambridge World History*, ed. by Benjamin Z. Kedar and Merry E. Wiesner-Hanks (Cambridge: Cambridge University Press, 2015), pp. 309–38.

Schatzberg, Eric, 'Technik Comes to America: Changing Meanings of Technology before 1930', *Technology and Culture*, 47.3 (2006), 486–512.

———, *Technology: Critical History of a Concept* (Chicago, IL: The University of Chicago Press, 2018).

Schot, Johan and Philip Scranton, eds, book series *Making Europe*, 6 vols (Basingstoke: Palgrave McMillan, 2013–2019).

Schlanger, Nathan, 'Le fait technique total. La raison pratique et les raisons de la pratique dans l'œuvre de Marcel Mauss', *Terrain*, 16 (1991), 114–30.

Simondon, Gilbert, *Du mode d'existence des objets techniques* (Paris: Aubier, 1958).

Subrahmanyam, Sanjay, 'Connected Histories: Notes Towards a Reconfiguration of Early Modern Eurasia', *Modern Asian Studies*, 31.3 (1997), 735–62

Tesnière, Valérie, 'Lucien Febvre lu par Maurice Daumas: comment écrire une histoire des techniques au XXe siècle », *Artefact*, 15 (2021), 281–97.

van der Vleuten, Erik, 'Toward a Transnational History of Technology: Meanings, Promises, Pitfalls', *Technology and Culture*, 49.4 (2008), 974–94.

———, Per Högselius, Anique Hommels, and Arne Kaijser, 'Europe's Critical Infrastructure and its Vulnerabilities: Promises, Problems, Paradoxes', in *The Making of Europe's Critical Infrastructure: Common Connections and Shared Vulnerabilities*, ed. by Per Högselius, Anique Hommels, Arne Kaijser and Erik van der Vleuten (Basingstoke: Palgrave Macmillan, 2013), pp. 3–19.

——— and Arne Kaijser, eds, *Networking Europe: Transnational Infrastructures and the Shaping of Europe, 1850–2000* (Sagamore Beach, MA: Science History Pub, 2006).

———, Ruth Oldenziel, and Mila Davids, *Engineering the Future, Understanding the Past: a Social History of Technology* (Amsterdam: Amsterdam University Press, 2017).

Verley, Patrick, *L'Échelle du monde. Essai sur l'industrialisation du monde* (Paris: Gallimard, 1997).

Wallerstein, Immanuel, *Comprendre le monde. Introduction à l'analyse des systèmes-mondes* (Paris: La Découverte, 2009).

White, Harrison C., *Identity and Control. A Structural Theory of Action* (Princeton, NJ: Princeton University Press, 1992).

Winner, Langdon, 'Do Artifacts Have Politics?', *Daedalus*, special issue, *Modern Technology: Problem or Opportunity?*, 109.1 (1980), 121–36.

Zakharova, Larissa, 'L'Insertion des techniques étrangères de communication dans l'environnement industriel soviétique: le cas des centraux téléphoniques Ericsson à Leningrad', in *Les Techniques et la globalisation au XXe siècle*, ed. by Liliane Hilaire-Pérez and Larissa Zakharova (Rennes: Presses Universitaires de Rennes, 2016), pp. 83–102.

———, *De Moscou aux terres les plus lointaines. Communications, politique et société en URSS* (Paris: Éditions de l'EHESS, 2020).

Part 1

Techniques Across the World

LILIANE HILAIRE-PÉREZ

Introduction

Going against Eurocentric approaches that have long underpinned the history of technology and persist under the guise of a so-called European (and British) exception — a marked nationalistic strain in the historiography (van der Vleuten, 2008: 982) — global economic history has highlighted what the Great Divergence owes to geopolitics and imperialist and colonial dynamics on the one hand (Pomeranz, 2000; O'Brien, 2009), and the appropriation of techniques and economic mechanisms from the rest of the world on the other. For Kenneth Pomeranz, insofar as China was still comparable to Europe regarding the size of its economy and the sophistication and volume of its productive sectors until the eighteenth century, how can we understand the transformation that took place in England and its neighbours which changed from a Smithian (i.e. mercantile) growth to a so-called 'modern' industrial boom? (Pomeranz, 2000) With Pomeranz's work, the comparison between China and England received much attention, with an emphasis on natural resources and forms of governance, counterbalancing discourses on the cultural predispositions of the two countries' populations. Pomeranz emphasized the crucial role of coal deposits and of the 'ghost acres' released by the plantation economy — and slavery — in English economic development. In recent times, scholars have stressed the changing geopolitics of resources, labour, and riches as factors for the Great Divergence (O'Brien, 2009). As Giorgio Riello explains, 'divergence was a process based on a changing interaction between different areas of the world' (Riello, 2013: 293). In any case, the linear and normative models based on the economics of development in the wake of Walt W. Rostow (1916–2003) and his theories which gave a central place to the notion of 'backwardness' (Krige, 2019b: 10), have now lost all scientific value (Calais, Mortier, 2020). The changes in trajectories that occurred in the nineteenth century are now being defined as the results of global developments, covering both transformations of equilibriums and processes of circulations on various scales. These remained obscured for a long time because of the weight of national historiographies in many regions of the world.

Although Pomeranz paid little attention to techniques, his perspective nevertheless was valuable in allowing new interpretations to emerge that ran against the traditional narrative of England's so-called inherent advantages in terms of inventiveness. However, techniques did feature when it came to the management of resources in

Global History of Techniques, ed. by Guillaume Carnino, Liliane Hilaire-Pérez and Jérôme Lamy, Global Matters Culture Matérielles, histoire des techniques et circulation globale des savoirs, 9 (Turnhout: Brepols, 2024), pp. 29–42
BREPOLS PUBLISHERS
10.1484/M.TECHNE-EB.5.136422

China, as the author later confirmed (Pomeranz, 2015), an age-old trend intrinsically linked to the mythology of power to this day (Schäfer, 2018). Such approaches remind us that techniques are a 'total social fact' and can no longer be reduced to growth factors or seen in purely quantitative terms (Mauss, 1923–24, Popplow, 2016). They cannot be utilized to simplistically account for the upturns and downturns of the global economy, or serve as mere props in a story of countries competing or catching-up with each other, whether it be in the race to industrialization or the intensification of underpaid work (Riello, 2013). The emblematic case of the steam engine has therefore been revisited by historians wishing to test the canonical account of 'English superiority' against the political, social, and cultural stakes of techniques in Qing China (Inkster, O'Brien, 2004). As we have noted previously,

> [a]s historians increasingly question how techniques have been able to generate increases in productivity, they increasingly prefer to question the multiple issues related to techniques, their meaning, and their transmission in given societies — before embarking on large-scale comparisons and accounts of globalization (Bray, Hilaire-Pérez, 2016: 20).[1]

This multidimensional nature of techniques also underlies the complexity of long-distance appropriation, which remains difficult to study (Hilaire-Pérez, Verna, 2009; Raj, 2013; Davids, 2019: 69) and requires collective research. This has been the case for silk and cotton technologies moving from Asia to Europe in the modern era (Schäfer, Riello, Molà, 2018; Riello, 2013) and for techniques in contemporary Europe involved in the vast programmes from which the *Making Europe* collection has emerged (Schot, Scranton, 2013–2019; Högselius, Kaijser, van der Vleuten, 2015: xx–xxi).

In the same vein, the centrality of techniques has also recently been recognized in the history of international relations (van der Vleuten, 2008; Krige, Barth, 2006). Now that their mobility or 'fluidity' is being increasingly recognized (Schot, Scranton, 2015), the history of techniques cannot be separated from the history of global political interactions (Krige, Barth, 2006), of the Cold War, of decolonization (Hecht, 2006), and of regional integration phenomena such as the European construction (Misa, Schot, 2005; van der Vleuten, Kaijser, 2006). As John Krige and Kai-Henrik Barth have suggested, contemporary globalization itself is based on a vast technical system, a 'global network of power', to use Thomas P. Hughes's expression (Krige, Barth, 2006: 20). In this process of globalization, the United States (and the English language) have been dominant (Krige, 2019a, 2019b; Daniels, 2019). Nevertheless, techniques never travel in a straight line, as shown by the adaptations of European and American techniques by the Soviet industry (Hilaire-Pérez, Zakharova, 2016), and also by instances of failure, resistance, and conflict (Pretel, Camprubi, 2018). Finally, the weight of national histories and processes of reterritorialization (what David Pretel and Lino Camprubi call 'de-globalization') in the techno-scientific field, not to mention the scale of state investment in research (Krige, 2019b), invite a

1 Unless otherwise stated all translations are the present author's own.

dialectical vision of the opening and closing of borders to technologies (Krige, Barth, 2006; Hilaire-Pérez, Zakharova, 2016; Pretel, Camprubi, 2018: 6; Krige, 2019b). Additionally, on the symbolic level, as we know technical legacies have often been utilized in techno-nationalist mythologies of power, and this in various contexts (Schäfer, 2018; Roche, 2019).

In a context not entirely dissimilar to that of the early modern period, the study of the processes of integration and technical convergence constitutes a goal which, apart from the particularly well-studied European case, allows us to raise methodological questions concerning notions (such as borders, networks, circulation, or transfer) as well as the modalities of encounter and hybridization in 'contact zones' (Krige, 2019b; Pretel, Camprubi, 2018). This means avoiding trying to come up with an overarching narrative about the evolution of technical lineages over the last two centuries — except when focusing on an object or a network, as we propose to do in the second part of the book. In this sense, too, the notion of convergence cannot be reduced to the acquisition of relative advantages by countries that were excluded from growth in the nineteenth century. A broadening of the focus beyond the economic stakes gives technological meaning to divergence and convergence. It allows for a joined examination of conflicting trends with on the one hand, the multiplication and opening up of possibilities (sometimes leading to failures) and, on the other, the increased functionality of technical achievements which has been amplified over the last two centuries by processes of transmission and the homogenization and standardization of knowledge, procedures, materials, artefacts, and networks.

Any discussion of techniques on a global scale today is necessarily contextualized and regionalized. This leads to highlighting divergence rather than convergence in the sense that the studies, above all, reveal the possibilities, i.e. the multiplicity of (often conflicting) technical trajectories that are usually forgotten in asymmetrical narratives nourished by ideals of modernity and development. These tend to erase the complexity of long-term developments, which are dismissed as 'tradition' (Calais, Mortier, 2020; Adas, 2006) and to obscure the potential for innovation of local societies (Durand, 2019). Finally, economic processes hitherto analysed in the light of global developments are now reassessed in their local components, such as the de-industrialization of African regions during colonization (Frederick, 2017, 2020). It seems to us that these critical approaches guard scholars against the use of categories forged by the great unifying, teleological narratives and by the simplistic attribution of techniques to different national entities (van der Vleuten, 2008). Recontexutalizing the global history of techniques in various regions is the goal of this chapter. Just as in our previous work, the order of the chapters is deliberately random, in contrast to publications of global history starting with Europe or America (Carnino, Hilaire-Pérez, Kobiljski, 2016). Also unlike existing works, we make room for spaces that are usually under represented if represented at all, such as Oceania, the Middle East, North Africa, Russia, and Central Europe.

Oceania, a region generally forgotten in global overviews, is obviously an area of interest, as Marie Durand demonstrates. Its general absence from global history studies reflects the absence of historicization of local developments and technical hybridizations, although they have long been part of 'Asian-Australian transnationalism' (Galliot, 2016). Local techniques that were studied by anthropologists in the nineteenth century were considered to be living traces of primitivism, particularly in terms of navigation and fishing. More recently, the revival of anthropological approaches, outside of evolutionary patterns, has been coupled with the history of imperialist domination (from plantations to nuclear testing). Durand focuses on three areas: transportation, resources, and telecommunications. The transformation of means of transport in the nineteenth century led to increased regional and international integration within colonial networks ensuring the exploitation of resources, as evidenced by the related development of gold mining and aviation in New Guinea. In terms of resource exploitation, plantation agriculture and mining (of, for instance, gold, nickel, copper, cobalt, zinc, and phosphate) were promoted by the colonial powers but also by some countries after independence, with deleterious effects on the environment. These countries have also taken over the means of telecommunications that were first installed at the end of the nineteenth century, and the region is currently equipped with a high-performance digital network that shapes everyday family interaction. Far from the impression given by established narratives, Oceania is characterized by the central role that innovation has in the lives of populations.

With her emphasis on the Middle East in the modern period and her stress on the absence of a comprehensive historical narrative, Elisabeth Mortier highlights the impact of imperialism. Long-established local industries (be it glass, soap, pearling or carpets) were displaced or transformed to meet the demands of world markets, at the same time as exports of raw materials such as cotton increased (even if this resulted in the reimportation of yarn from Manchester). Interest in a new resource, oil, is emblematic of European domination, which was also evident in urban planning and infrastructure, starting with the Suez Canal, but also involving the telegraph and railways. The role of European engineers as key players in globalization has been confirmed (Pretel, Camprubi, 2018). It continues today via consulting agencies (Cariou, 2020). As Mortier has shown in other works, European domination was accompanied by a highly ideological discourse extolling modernization even as traditional local techniques were sometimes appropropriated, as was the case in Palestine with irrigation (Mortier, 2020). This enthusiasm for technical modernity is also at play in the newly independent states, as evidenced by the rise of 'mining agriculture', which resulted from the combined effects of oil revenues, technical transfers such as drilling, and global investments (Calais, 2020; Cariou, 2020).

The chapter on the Maghreb region, written by Yamina Bettahar, is also a departure from traditional literature, as this space has so far remained on the margins of the history of globalization and, more broadly, of the modern history of techniques. This case study, despite the diversity the contexts covered (in Morocco, Algeria, and Tunisia), is nevertheless emblematic of the pressing questions raised by the technical choices linked to the colonial period and by developmental policies, including those of Algerian socialism at the height of the Cold War and the way it managed the country's hydrocarbon resources. The developmental rationale imposed by the geographical proximity and shared history of Europe and the indirect effects of internationalization, sometimes unbridled globalization, continue to have their detrimental effects. Bettahar thus proposes a reflection on the failure of a model of accelerated industrialization advocated by the rulers in the 1960s and 1970s, which she attributes to the weight of the colonial heritage on one hand, and to the lack of democracy on the other, a theme that is also addressed in other contributions. She also links this failure to the difficulties of appropriation posed by North-South technical transfers encouraged by international institutions such as the World Bank. Today, the strength of a growth model based on the high status of innovation, even in recently established technopoles with no link to the technical cultures of the country in question, continues to persist and demonstrates the durability of unequal technopolitical regimes.

The case of East Asia, presented by Francesca Bray and Aleksandra Kobiljski, is significantly different. The rich historiography on the topic has seen a shift away from the traditional questions of 'backwardness' or 'catching up'. The authors highlight local appropriations and the intra-regional circulation of techniques, particularly under the influence of Japan, and then South Korea and Taiwan, but also in the context of rivalry and the combination of interests between communist and capitalist countries. The case of China is emblematic: the anti-technocratic Maoist experience was followed by an accelerated techno-scientific industrialization during the 1980s and 2000s, with the support of Western, Japanese, South Korean, and Taiwanese firms. Techno-nationalism also characterized South Korea, which launched the construction of a road supposed to bring about reunification after the Cold War which was actually framed as a military operation aimed at defeating the North. The highlighting of national trajectories and intra-regional dynamics renders any Eurocentric narrative obsolete, confirming Ian Inkster's point of view on the need to take technical autonomy into account in these countries, which were never colonized by Europe (Inkster, 1988).

By contrast, on the chapter devoted to India, Dhruv Raina, Omprasad, and Nayani Sarma underline the strong Eurocentric bias of the history of technology in the country, linked to its use in studies focused on modernization and colonization. Pre-colonial techniques, which are rarely taken into account, have primarily served to tell stereotypical narratives about the Indian 'backwardness' (Roy, 2016). The authors reveal the tropes of the literature that links colonization and modernization: the technical tools of the colonial power, the transformation of local societies, the appropriation of imported techniques such as the railway (which was the source of new national cohesion), and also the Gandhian criticism of industrialization. The failure of technical transfers carried out in the name of development theories have encouraged historians' interest for local technical conditions, in connection with social studies'and global history's reassessment of the (essential) share that India and China have had in world production since the eighteenth century. The study of local dynamics, instead of only the colonizer-colonized relationship, have prevailed, for example, in the history of engineering and technical education in India. From the 1980s onwards, as India asserted itself as a nuclear, space, and computer power, the history of technology opened up to connected approaches and interlocking temporalities, redrawing boundaries between traditional and modern techniques, as for instance with the history communication. The history of technosciences has also focused on the links between international collaboration and the complexity of local paths to high-tech innovation.

Continuity is also important in the new historiography of the history of technology in Africa, which is presented by Joshua Grace. While the colonial hold on the one hand and the developmentalist ideology on the other denied Africans the possibility to appropriate techniques associated with western modernity and erased the history of pre-colonial techniques, creating the myth of a 'technological absence', Grace highlights the multiple ways in which an 'alternative' narrative has been affirmed. He stresses the important role played by independentists, of leaders like Julius Nyerere, and of anonymous users and technicians, but also of intellectuals and writers, and finally of the researchers who questioned the idea of colonization acting as a watershed moment. Two strong ideas guide Grace's argument. On the one hand, European techniques aimed at the exploitation of resources; they consisted of selective and not necessarily sustainable developments. On the other hand, if European techniques were able to establish themselves in Africa, this was because of the know-how of the populations who have made them 'work' by appropriating them despite the limitations that the Europeans placed on the technical autonomy of the inhabitants. The question of repairs, but also that of mobility and its place in technical cultures, is thus raised. The author emphasizes that this reversal of the perspective not only helps to distance oneself from dated historical narratives, but that it allows one to understand the meaning given to techniques in African cultures. He explains for example the term *nziya* ('way'), a means of action and displacement, used among the VaShona hunters of Zimbabwe (Mavhunga, 2014), which expands the very concept of technique.

Such a shift in focus makes it necessary to question the place to be given to Europe in the history of techniques. Far from seeing Europe as a key player of global history, the authors of the two chapters devoted to Europe distance themselves from eulogistic discourses of progress, in the East as well as in the West, and a critical reflection on these geographical entities that historiography invites us to decompartmentalize. The ideological dimension of the techniques is of prime importance. Jawad Daheur stresses the institutional creations in Central Europe in the 1950s (including committees for the history of science and technology, academies of science, and museums) and connects the development to socialist ideals and historical materialism. The impact was clear in terms of their subjects: the history of coal mining and metallurgy dominated studies in Czechoslovakia and Poland (where this production was massive) and benefitted from strong nationalist promotion. By contrast, from the 1970s onwards, research developed on the earlier periods and on a diversity of subjects, abandoning the recent history of technology in the socialist countries. If the fall of communism saw the continuation of these projects, analyses since the 2000s have changed the situation by studying techniques in socialist regimes and by promoting new themes, such as gendered history, and approaches open to social sciences and critical reflection, all against a backdrop of the internationalization of research. One sign of this is the inclusion of Central European researchers in the important project 'Tensions of Europe: Technology and the Making of Europe', which is devoted to the role of technology in a history which looks at both processes of fragmentation and integration. A European research network has thus developed through programmes and publications (van der Vleuten, Kaijser, 2006; Högselius, Hommels, Kaijser, van der Vleuten, 2013; Schot, Scranton, 2013–2019).

One of the promoters of this work, Erik van der Vleuten, is a specialist of infrastructures as creating both links and fragmentation (Högselius, Kaijser, van der Vleuten, 2015). He also proposes a reflective approach to techniques in Europe. His aim is to analyse the social and political construction of the Western Eurocentric discourse that has long underpinned the idea of universal progress (van der Vleuten, Oldenziel, Davids, 2017). Beginning with the Saint-Simonian utopia and the Universal Exhibitions, the author highlights the diversity of interpretations of the universalism of techniques in different contexts. The idea of the universalism of techniques serves colonial expansion, but also repressive urban policies or the opportunism of entrepreneurs, supported by engineers who are both project managers and mediators. This sometimes leads to the rise of alternatives of collective management, such as the rural wind turbines of Poul la Cour in Denmark, a case recently brought to attention (Bruyerre, 2020). The myth of European civilizational progress has broken down with wars, the Shoah, Hiroshima, colonial violence, and industrial capitalism, and given rise to technophobia. Two responses have emerged as a result: on the one hand, the technocratic movement, which aimed at reinforcing expertise as a counterweight to politics, and on the other hand the valorization of participative practices and of users (in the form of repair for example). For van der Vleuten, the multiplicity of discourses on techniques constitutes a challenge for global history, which fails to encompass the plurality of local issues within a connected narrative.

Russia is investigated by Elena Kochetkova and Julia Obertreis, who pay tribute in this chapter to its initial author, Larissa Zakharova. Russia is an emblematic example of the tension between dependence on the West, including in the midst of the Soviet period, and the recurrent affirmation of a national and even imperialist ambition. If Russian historiography has been influenced by the importance given to techniques in the Soviet propaganda, and its downplaying of technological transfers in particular (Gouzévitch, 2016), the authors also underline the recurrent theme of Russian technical backwardness in Western historiography. Without following this line, Elena Kochetkova and Julia Obertreis underline that, despite media successes, scientific and technical research has suffered from totalitarianism, which raises, as Larissa Zakharova expressed it, 'the general question of the conditions and possibilities of innovation in authoritarian regimes with a state-run economy' (Hilaire-Pérez, Zakharova, 2016, 30). In the nineteenth century, the tsarist regime embarked on the modernization of railways and the military through foreign collaborations. The Bolsheviks, and then the Stalinist regime, made technology a means and a symbol of modernization and invested their energy in electrification, aviation, and the training of engineers, by practising espionage, hiring Western workers, and establishing contracts with foreign companies, with sometimes unforeseen boomerang effects in the

West, as for instance with telecommunications (Zakharova, 2016). The tension between dependence and nationalism was brought to a head under Stalin. Technicians recruited and experts trained abroad who were accused of espionage were eliminated (Zakharova, 2015). American aid and Soviet technical mobilization were combined during the Second World War in a large-scale military production that continued during the Cold War. The Khrushchev era, which was marked by technological enthusiasm with the conquest of space and the boom in consumer goods (via the chemical industry) is misleading. In spite of an expansionist policy (in the Soviet republics, developing countries, and China), the USSR faced numerous hurdles at the same time as pollution and industrial risks increased. The current regime, despite the display of technological ambition, remains constrained by the lack of democratic institutions and is therefore dependent on foreign countries.

As in previous work Jérôme Baudry, continues his analysis of the American narratives of industrialization and of the hold of technology on daily life and national concerns (Baudry, 2016). Baudry highlights the place of the 'technologies of the frontier' (roads, canals, railways), which were the engine of the American '[c]ulture of improvement'. While the technical transfers with England are obvious, these infrastructures encourage the idea of America's civilizing mission carried by techniques. The same is true of industrialization. It was born from a 'transit of technology' but was quickly characterized by a distinctive feature, mass production via the 'American manufacturing system'. Later it was further bolstered by Fordism and Taylorism, which mapped out a specific path based on management techniques, the aura of the engineer, and the belief in the superior authority of scientific and technical expertise. The rise of technosciences through institutionalized research, both private and public, is expressed in the power of the 'military-industrial-university' complex, which allows for extraordinary achievements and what Thomas P. Hughes calls 'broad technical systems'. This results in large scale enthusiasm for progress, particularly for the civilian spin-offs of military innovations. However, the situation has changed since the 1990s, with the smallest military investment made by the state after a series of failures and opposition (during and after the Vietnam War in particular). The current climate is one of mistrust for the promises of techniques.

Latin America, as presented by David Pretel, was at the heart of technical, economic, and political issues for the United States and Europe in the nineteenth and twentieth centuries. Pretel cites the example of Porfirio Diaz's Mexico, between 1876 and 1911, which was the main destination of American investments in mining, metallurgy, and railways, leading to the lasting presence of industrialists and engineers. The same is true in the West Indies, where the sugar economy of Puerto Rico and Cuba depended on US imports of capital goods. Foreign hegemonies in Latin America were also the result of the Cold War, among other things through Soviet influence. Pretel emphasizes the role of international expert communities, an argument he recently developed elsewhere (Pretel, Camprubi, 2018). But far from concluding that Latin America was dependent, he rather highlights the complexity of interactions as experts form heterogeneous networks, leading to processes of globalization and 'de-globalization' and promoting internationalist values in the manner of 'technical diplomats' while responding to local priorities (Pretel, Camprubi, 2018, 7). For Pretel, the simplistic narrative of dependency and backwardness has marginalized the place of post-colonial Latin America in global history. By contrast, he argues that processes of autonomy (including through local expertise and innovation) and the geopolitics of techniques should be explored jointly. The benefit, in return, for global history is a renewed attention to the reciprocity of circulation and to the local and global co-production of techniques, to the point of questioning the distinction between international and domestic expertise.

All these chapters, whose rich diversity I have sketched briefly, suggest the necessary rehabilitation of so-called 'old' techniques (the ancient, the well established, and the indigenous; see Hård, 2017). These have usually been excluded from the history of techniques, which has long been understood as a narrative of power marked by tcolonial and developmental history and later by the Cold War. The chapters to follow invite a shift in focus, whereby the techniques which had been made invisible are restored as effective means of action, including with the purpose to make the colonial machine work. It is a new contribution to the necessary consideration of the *longue durée*, which we know is by no means limited to the long term and to the great narratives, but which presupposes the consideration of multiple temporalities, overlaps, and divergences to be analysed at different scales (Trivellato, 2015), which Heike Weber and Jonas van der Straeten perfectly analyse in this volume.

References

Adas, Michael, *Dominance by Design: Technological Imperatives and America's Civilizing Mission* (Cambridge, MA: The Belknap Press of Harvard University Press, 2006).

Baudry, Jérôme, 'Rencontres et politique de la technique en Amérique du Nord', in *Histoire des techniques. Mondes, sociétés, cultures XVIe–XVIIIe siècles*, ed. by Guillaume Carnino, Liliane Hilaire-Pérez, and Aleksandra Kobiljski (Paris: Presses universitaires de France, Nouvelle Clio, 2016), pp. 231–43.

Bray, Francesca and Liliane Hilaire-Pérez, 'Les Techniques et l'histoire globale', in *Histoire des techniques. Mondes, sociétés, cultures XVIe–XVIIIe siècles*, ed. by Guillaume Carnino, Liliane Hilaire-Pérez, and Aleksandra Kobiljski (Paris: Presses universitaires de France, Nouvelle Clio, 2016), pp. 7–22.

Bruyerre, Philippe, *La Puissance du vent. Des moulins à vent aux éoliennes modernes*, Histoire & Techniques (Toulouse: Presses universitaires du Midi, 2020).

Calais, Damien, 'Les Cultures hydroponiques aux Émirats arabes unis: un nouvel entrepreneuriat entre insertion locale et réseaux globaux', *Artefact. Techniques, histoire et sciences humaines*, 12 (2020), 177–202.

——— and Élisabeth Mortier, 'Pour une histoire des techniques d'irrigation au Moyen-Orient au XXe siècle', *Artefact. Techniques, histoire et sciences humaines*, 12 (2020), 167–76.

Cariou, Alain, 'Les Techniques d'irrigation dans la péninsule arabique: de la tradition oasienne millénaire à la révolution des périmètres irrigués', *Artefact. Techniques, histoire et sciences humaines*, 12 (2020), 203–30.

Carnino, Guillaume, Liliane Hilaire-Pérez, and Aleksandra Kobiljski, eds, *Histoire des techniques. Mondes, sociétés, cultures XVIe–XVIIIe siècles* (Paris: Presses universitaires de France, Nouvelle Clio, 2016).

Daniels, Mario, 'Restricting the Transnational Movement of "Knowledgeable Bodies": the Interplay of US Visa Restrictions and Export Controls in the Cold War 35', in *How Knowledge Moves. Writing the Transnational History of Science and Technology*, ed. by John Krige (Chicago, IL: University of Chicago Press, 2019), pp. 35–60.

Davids, Karel, 'Toolkits, Creativity and Divergences: Technology in Global History', in *Global Economic History*, ed. by Tirthankar Roy and Giorgio Riello (London: Bloomsbury, 2019), pp. 67–82.

Durand, Marie, 'Les Lieux de l'innovation: gestes techniques et intégration des valeurs sociales dans les îles Banks au Vanuatu', in *Valeurs et matérialités*, ed. by Frédéric Keck (Paris: Presses de l'ENS, 2019), pp. 15–28.

Frederick, Katherine, *Deindustrialization in East Africa: Textile Production in an Era of Globalization and Colonization, c. 1830–1940* (London: Palgrave Macmillan, 2020).

———, 'Global and Local Forces in Deindustrialization: the Case of Cotton Cloth in East Africa's Lower Shire Valley', *Journal of Eastern African Studies*, 11.2 (2017), 266–89.

Galliot, Sébastien, 'Les Techniques océaniennes: problèmes de sources et pistes de réflexion', in *Histoire des techniques. Mondes, sociétés, cultures XVIe–XVIIIe siècles*, ed. by Guillaume Carnino, Liliane Hilaire-Pérez, and Aleksandra Kobiljski (Paris: Presses universitaires de France, Nouvelle Clio, 2016), pp. 245–54.

Gouzévitch, Irina and Dmitri Gouzévitch, 'La Construction de l'État russe à l'épreuve du transfert technique', in *Histoire des techniques. Mondes, sociétés, cultures XVIe–XVIIIe siècles*, ed. by Guillaume Carnino, Liliane Hilaire-Pérez, and Aleksandra Kobiljski (Paris: Presses universitaires de France, Nouvelle Clio, 2016), pp. 143–61.

Hård, Mikael, ERC 'A Global History of Technology, 1850–2000 GLOBAL-HOT' (2017) <https://www.academia.edu/37125306/A_Global_History_of_Technology_1850_2000_GLOBAL_HOT> [accessed 12 June 2021].

Hecht, Gabrielle, 'Negotiating Global Nuclearities: Apartheid, Decolonization, and the Cold War in the Making of the IAEA', in *Global Power Knowledge: Science and Technology in International Affairs*, special issue, *Osiris*, 21.1 (2006), 25–48.

Hilaire-Pérez, Liliane and Catherine Verna, 'La circulation des savoirs techniques du Moyen Âge à l'époque moderne. Nouvelles approches et enjeux méthodologiques', *Tracés*, 16 (2009), 25–61.

——— and ———, 'Dissemination of Technical Knowledge in the Middle Ages and the Early Modern Era. New Approaches and Methodological issues', *Technology and Culture*, 47 (2006), 536–65.

——— and Larissa Zakharova, eds, *Les Techniques et la globalisation au XXe siècle* (Rennes: Presses Universitaires de Rennes, 2016).

Högselius, Per, Anique Hommels, Arne Kaijser, and Erik van der Vleuten, eds, *The Making of Europe's Critical Infrastructure: Common Connections and Shared Vulnerabilities* (Basingstoke: Palgrave Macmillan, 2013).

———, Arne Kaijser and Erik van der Vleuten, eds, *Europe's Infrastructure Transition: Economy, War, Nature*, vol. 3, *Making Europe: Technologies and Transformations 1850–2000* (Basingstoke: Palgrave Macmillan, 2015).

Inkster, Ian, 'Prometheus Bound: Technology and Industrialization in Japan, China and India Prior to 1914. A Political Economy Approach', *Annals of Science*, 45.4 (1988), 399–426.

Inskter, Ian and Patrick O'Brien, eds, *The Global History of the Steam Engine*, special issue, *History of Technology*, 25 (2004).

Krige, John, 'Export Controls as Instruments to Regulate Knowledge Acquisition in a Globalizing Economy', in *How Knowledge Moves. Writing the Transnational History of Science and Technology*, ed. by John Krige (Chicago, IL: University of Chicago Press, 2019a), pp. 60–92.

———, 'Introduction: Writing the Transnational History of Science and Technology', in *How Knowledge Moves. Writing the Transnational History of Science and Technology*, ed. by John Krige (Chicago, IL: University of Chicago Press, 2019b), pp. 1–32.

——— and Kai-Henrik Barth, 'Introduction: Science, Technology, and International Affairs', *Global Power Knowledge: Science and Technology in International Affairs*, special issue, *Osiris*, 21.1 (2006), pp. 1–21.

Mauss, Marcel, 'Essai sur le don. Forme et raison de l'échange dans les sociétés archaïques', L'Année sociologique, t. I (1923–1924), pp. 30–186.

Mavhunga, Clapperton Chakatetsa, *Transient Workspaces: Technologies of Everyday Innovation in Zimbabwe* (Cambridge, MA: MIT Press, 2014).

Misa, Thomas J. and Johan Schot, 'Inventing Europe: Technology and the Hidden Integration of Europe', *History and Technology*, 21.1 (2005), 1–19.

Mortier, Élisabeth, 'Les Circulations de savoir et les transferts de techniques d'irrigation entre populations arabe et juive en Palestine ottomane et mandataire (fin XIXe siècle–1948)', *Artefact. Techniques, histoire et sciences humaines*, 12 (2020), 231–53.

O'Brien, Patrick, 'The Geopolitics of a Global Industry: Eurasian Divergence and the Mechanization of Cotton Textile Production in England', in *The Spinning World: a Global History of Cotton Textiles, 1200–1850*, ed. by Giorgio Riello and Prasannan Pathasarathi (Oxford: Oxford University Press, 2009), pp. 351–65.

Pomeranz, Kenneth, 'Les Eaux de l'Himalaya: barrages géants et risques environnementaux en Asie contemporaine', *Revue d'histoire moderne et contemporaine*, 62.1 (2015), 7–47.

———, *The Great Divergence: China, Europe and the Making of the Modern World Economy* (Princeton, NJ: Princeton University Press, 2000).

Popplow, Marcus, 'Technology and Technical Knowledge in the Debate on the "Great Divergence"', *Artefact. Techniques, histoire et sciences humaines*, 4 (2016), 275–85.

Pretel, David and Lino Camprubi, 'Technological Encounters: Locating Experts in the History of Globalisation', in *Technology and Globalisation. Networks of Experts in World History*, ed. by David Pretel and Lino Camprubi (Cham: Palgrave Macmillan, 2018), pp. 1–26.

Raj, Kapil, 'Beyond Postcolonialism and Postpositivism. Circulation and the Global History of Science', *Isis*, 104 (2013), 337–47.

Riello, Giorgio, *Cotton: the Fabric that made the Modern World* (Cambridge: Cambridge University Press, 2013).

——— and Tirthankar Roy, eds, *How India Clothed the World: the World of South Asian Textiles, 1500–1850* (Leiden: Brill, 2009).

Roche, Sylvain, 'S'appuyer sur le passé pour construire l'avenir: l'exemple du projet nationaliste écossais dans le domaine des énergies renouvelables offshore', *Artefact. Techniques, histoire et sciences humaines*, 10 (2019), 143–74.

Roy, Tirthankar, 'Institutions et systèmes techniques en Inde', in *Histoire des techniques. Mondes, sociétés, cultures XVIe–XVIIIe siècles*, ed. by Guillaume Carnino, Liliane Hilaire-Pérez, and Aleksandra Kobiljski (Paris: Presses universitaires de France, Nouvelle Clio, 2016), pp. 53–78.

Schäfer, Dagmar, 'The Historical Roots of Modern Bridges: China's Engineers as Global Actors', in *Technology and Globalisation. Networks of Experts in World History*, ed. by David Pretel and Lino Camprubi (Cham, Palgrave Macmillan, 2018), pp. 27–39.

———, Giorgio Riello, and Luca Molà, eds, *Threads of Global Desire: Silk in the Pre-modern World* (Woodbridge, Rochester, and New York, NY: Boydell Press, 2018).

Schot, Johan and Philipp Scranton, '*Making Europe*: An Introduction to the Series', in *Europe's Infrastructure Transition: Economy, War, Nature*, ed. by Per Högselius, Arne Kaijser, and Erik van der Vleuten (New York, NY: Palgrave Macmillan, 2015), pp. xi–xvii.

——— and ———, eds, *Making Europe*, 6 vols (Basingstoke: Palgrave Macmillan, 2013–2019).

Trivellato, Francesca, 'Un nouveau combat pour l'histoire au XXIe siècle?', *Annales. Histoire, sciences sociales*, 70.2 (2015), 333–43.

van der Vleuten, Erik, 'Toward a Transnational History of Technology: Meanings, Promises, Pitfalls', *Technology and Culture*, 49.4 (2008), 974–94.

——— and Arne Kaijser, eds, *Networking Europe: Transnational Infrastructures and the Shaping of Europe, 1850–2000* (Sagamore Beach, MA: Science History Publications, 2006).

———, Ruth Oldenziel, and Mila Davids, *Engineering the Future, Understanding the Past: a Social History of Technology* (Amsterdam: Amsterdam University Press, 2017).

Zakharova, Larissa, 'L'Insertion des techniques étrangères de communication dans l'environnement industriel soviétique: le cas des centraux téléphoniques Ericsson à Leningrad', *in Les Techniques et la globalisation au xxe siècle*, ed. by Liliane Hilaire-Pérez and Larissa Zakharova (Rennes: Presses Universitaires de Rennes, 2016), pp. 83–102.

———, 'Quand la politique se mêle de technologie, ou le progrès technique en contexte répressif. L'Union soviétique de l'entre-deux-guerres', in *Les Savoirs-mondes: mobilité et circulations des savoirs depuis le Moyen Âge*, ed. by Pilar Gonzalez-Bernaldo and Liliane Hilaire-Pérez (Rennes: Presses Universitaires de Rennes, 2015), pp. 275–86.

MARIE DURAND

From the 'Stone Age' to the Digital Era

Techniques in Oceania: A Brief Historiography (Nineteenth to Twenty-First Centuries)

Any attempt to draft a global history of techniques in Oceania from the nineteenth to the twenty-first century presents several difficulties. Despite the growth of Pacific indigenous scholarship and historical narratives focussed on islanders' experiences, a significant amount of asymmetry persists in the available sources. In the nineteenth and early twentieth centuries, accounts were mostly produced by Westerners of various kinds: missionaries, traders, colonial administrators, art collectors, and scholars of different disciplines. This made it difficult to take indigenous voices, agency, and knowledge into account.[1] Following the 'island-oriented history' that developed after J. W. Davidson (1966), historians of the Pacific region, notably in the Australian National University at Canberra and at universities in New Zealand and Hawaii, tried to shift the focus away from the grand narratives of European-centred imperial histories (Merle, Naepels, 2003). Relying upon 'against-the-grain' readings of Western texts, oral histories, and alternative sources such as museum collections and visual documents, their attempts to refocus the scholarly gaze were mostly successful at correcting a prior notion of passive islanders as entirely submissive to rules imposed by foreign powers (Thomas, 1990). Although the transmission of knowledge was largely oral before conversion to Christianity, writing was quickly and willingly adopted, for instance, by the missionaries' converts in Polynesia at the beginning of the nineteenth century, on islands where the population already had direct and frequent access to trade goods (Parsonson, 1967). With regard to techniques, however, while variously advanced technologies were noted in Glen Barclay's 1978 *History of the Pacific from the Stone Age to the Present Day*, they were mentioned to explain how some populations came to dominate others, in a rather debatable, deterministic interpretation. Generally, techniques and technologies were still only considered peripherally in

1 This is in spite of the fact that invaluable work was done in both finding and digitizing sources by the Pacific Manuscript Bureau initiative (PAMBU) that was set up by the Department of Pacific History of the Australian National University. PAMBU started in 1968 as a non-profit organization sponsored by the National Library of Australia, the National Library of New Zealand/Te puna Mâtauranga o Aotearoa, the Mitchell Library (State Library of New South Wales), and the University of Hawaii, and continues today to preserve sources of historical interest and Pacific heritage.

Global History of Techniques, ed. by Guillaume Carnino, Liliane Hilaire-Pérez and Jérôme Lamy, Global Matters Culture Matérielles, histoire des techniques et circulation globale des savoirs, 9 (Turnhout: Brepols, 2024), pp. 43–68
BREPOLS PUBLISHERS 10.1484/M.TECHNE-EB.5.129781

more recent histories of the Pacific region, although these did promote a global approach (Matsuda, 2012; Salesa, 2012; Igler, 2013; however, see Armitage, Bashford, 2014). Overall, it must be acknowledged that technology seems to have elicited more work from archaeologists and anthropologists than from historians, at least until very recently.

The resulting, fragmented nature of information on techniques makes it difficult to give a coherent and sufficiently nuanced impression of technical patterns, changes, and forms of circulation from the nineteenth to the twenty-first centuries. This fragmentation must also be linked to the extremely heterogeneous histories, as well as the environmental and social contexts, of the area under consideration. For example, the social, economic, and political lives of atoll inhabitants, which often revolve around marine exploitation and sea-mediated exchange networks, may differ widely from those of the horticulturalist peoples living inland on large, high volcanic, or continental islands. With regard to chronology, in the nineteenth century, islands such as Guam in the Marianas already looked back upon over two centuries of dealing with a European presence (Rogers, 1995). The initial contact between Guam's Chamorros people and the Spanish occurred in 1521, during Magellan's first circumnavigation. By contrast, the inland mountainous valleys of the New Guinea Highlands had not yet been visited by 'white men' and would not receive them until the 1930s and 1940s (Moore, 2003; Nelson, 1976). The entire region thus reveals complex histories of uneven transformations over time, and yet despite this multiplicity of experiences, increased cultural contact between Euro-American foreigners and Pacific Islanders occurred across the entire area from the nineteenth century onwards, accompanied by what were likely changes happening at unprecedently fast pace.

In this context, one is faced with dichotomized academic approaches. Much of the available material about techniques and technologies is embedded in theoretical anthropological paradigms and agendas which, before the second half of the twentieth century, largely disregarded the contemporary transformations of societies. In the nineteenth and beginning of the twentieth centuries, however, these societies experienced rapid changes and creatively reproduced themselves as part of larger networks. This paper will therefore begin with a brief, general historiographical overview of the pertinent literature before focussing on three themes — transport, resource exploitation, and communication — to provide a glimpse into the dynamics of the circulation of techniques that have always been a characteristic pattern of Pacific societies.

Techniques in Oceania: a Brief Historiography

The early decades of the nineteenth century saw the last of the scientific expeditions organised in the 'Islands of the Great Ocean' and the entire region mapped according to Western spatial criteria and standards. Renaming the area was necessary to develop operational geographical knowledge as well as cultivating an image of these spaces as 'territories', that could thereby be opened up, to a certain extent, to further imperial and commercial activities (Blais, 2001). The terms 'Malaysia', 'Melanesia', 'Polynesia', and 'Micronesia', that had been proposed by Jules Sébastien César Dumont d'Urville (1832), were generally adopted, along with a conception of the islanders' physical features and social and material contexts that placed them on the newly defined evolutionary, linear path of progress. In a lecture delivered at the Royal Institution of Great Britain on 28 May 1875, Augustus Henry Lane Fox Pitt-Rivers linked material manifestations to the evolution of human beings' mental structures, asserting that techniques were better suited than languages for tracing the origins of human development. As materializations of evolutionary mental states, objects made it possible to glimpse evidence of prehistoric developments (Pitt-Rivers, 1906). Pacific societies were thus regarded as contemporary instances of the most 'primitive' human evolutionary stages, living exemplars of the so-called 'Stone Age'. Lewis Henry Morgan, for instance, defined 'ethnical periods' — savagery, barbarism, and civilization — based on the interaction between technical achievements, family and kinship institutions, and political organization. Australian Aborigines and most Polynesian peoples occupied the 'middle status of savagery', according to the fact that they knew how to use fire and lived on fish (Morgan, 1877: 27).

Among the things that most astonished Europeans were navigation and fishing techniques, which both attracted comment and prompted the composition of scholarly works. Surveys of the canoes of Oceania and Pacific peoples' maritime techniques appeared in the early literature (Pâris, 1841; Barron, 2019; Rieth, 2010). At the beginning of the twentieth century, anthropologists' interests in material culture gradually receded before studies which focused more on religion, social organization, or magic, according to the functionalist paradigm popularized by Bronislaw Malinowski (1922). However, marine technologies and the technical features of watercraft remained central to some diffusionist studies attempting to trace the historical movements of early Oceanian populations. The anthropologist Alfred C. Haddon and the zoologist James Hornell described the types of canoes they observed in the 1920s and 1930s and published between 1936 and 1938 a major technological overview covering the entire Oceanian region. They included the information they were able to gather about historic types of watercraft and proposed a model in which different technical features were equated with a variety of population movements from Indonesia and the West (Haddon, Hornell, 1975; see also Neyret, 1974; Di Piazza, 2014). Papuan peoples who used only rafts and simple dugout canoes, either without sails or masts or rigged with vertical oblong sails made of plaited palm leaves, are thought to have populated New Guinea and some parts of Melanesia. They were supposedly followed by migrants from Indonesia who introduced double outrigger canoes with booms and square sails rigged with a mast, yard, and boom. The subsequent wave of settlers is thought to have arrived before AD 450 on single outrigger and double sailing canoes to form the basis of the Western Polynesian population. The next migrants, corresponding to Rivers' kava-people (1914), used large double sailing canoes and other large sailing canoes with single outriggers that could be rendered more seaworthy by the addition of strakes. This ultimately led to the development of plank-built canoes, which were taken from the Moluccas to

Polynesia by the final immigrants, whom the authors equate with Rivers' betel-people (Rivers, 1914; Haddon, Hornell, 1975: 75–79). Other surveys with diffusionist goals and similarly concerned with techniques were authored by scholars using museum collections to complement their fieldwork (Schurig, 1930; Speiser, 1923; Doran, 1981; see also Rivers, 1914). Further, institutions like the Bernice Pauahi Bishop Museum of Honolulu or the Dominion Museum (from 1936 the National Art Gallery and Dominion Museum) in Wellington, New Zealand became important sources for publications on the archaeology and material culture of the Pacific, particularly the Polynesian and New Zealand Maori cultures.[2]

Nevertheless, before the Second World War, most anthropological scholarship describing the technologies of the Pacific inhabitants was concerned with cataloguing pre-colonial or pre-Christian sets of objects rather than with their actual manufacture or the technical systems that brought them into existence, or with the contemporary transformations of these systems (Lemonnier, 2014). By contrast, in the 1960s and 1970s, as the Marxist economic theory gained momentum and changes in productive means and forces returned to the forefront of research agendas, the Pacific appeared most promising, especially those areas that had only recently been entered by white colonists or were considered to be outside their influence for the time being. While anthropologists took a special interest in the introduction of steel to places where stone tools were used by Australian Aborigines (Sharp, 1952) or New Guinea highlanders (Salisbury, 1962; Townsend, 1969; Godelier, Garanger, 1973), historians looked at transformations in the economic sphere, in labour, and in the development of industry since the nineteenth century as well as the impact on islanders' travel, geographical reach, local experiences, and social organization (Scarr, 1967; Shineberg, 1967, 1999; Corris, 1973; Bennett, 1976, 1987; Maude, 1981). Many influential studies concerning ethno-archaeology and ethno-ecology in the Pacific context appeared from the 1960s onwards and investigated the local knowledge, classifications, and understanding of material environments (Best, 1942; Barrau, 1962; Haudricourt, 1964; Rappaport, 1968; Dwyer, 1989; Pétrequin, Pétrequin, 1993, 2006; Brunois, 2007, Mondragon, 2018) complementing other studies which were published in the form of monographs (Blackwood, 1950; Sillitoe, 1988). More recent processual anthropological works also have highlighted how technical and material practices are at the core of people's constant modification of their physical and social worlds and how they reorganize the environmental limits and the perceptual categories they consider operative (MacKenzie, 1991; Revolon, 2012; Lemonnier, 2012; Leach, 2002, Hermkens, 2013; Coupaye, 2013; Damon, 2017; Durand, 2019; Küchler, 2002; Küchler, Were, 2009; Were, 2013). These dynamic technical appropriations are intertwined with colonial and post-colonial histories

2 See http://hbs.bishopmuseum.org/pubs-online/bpbm-bulletins.html for a list of the Bishop Museum's *Bulletins* (1922-present), *Memoirs* (1899–1949), *Special Publications* (1892-present) and *Occasional Papers* (1899-present). Prominent writers for the Dominion Museum's *Bulletin* include the ethnographers Elsdon Best (see e.g. Best, 1912, 1916, 1925) and Te Rangi Hiroa (Sir Peter Buck), a medical doctor, anthropologist, and museum director (Buck, 1923–1924, 1930, 1944, 1957); both also made significant contributions to the *Journal of the Polynesian Society*, which was founded in 1892.

of exploitation and forms of domination and resistance in the region. They have been associated with the wide circulation of materials and products by large trading companies such as the German firms Godeffroy and Son and Hernsheim and Co., the Australian companies Burns Philp, and the French business Maison Ballande, all of which were established in the region (Mann Spoehr, 1963; Firth, 1973; Purcell, 1976; Bollard, 1981; Buckley, Klugman, 1981, 1983; Aldrich, 1989; Newbury, 1998; Thompson, 2000; Angleviel, 2016). Indeed, sandalwood, sea cucumbers, pearl shells, copra, and sugar cane, along with gold, copper, phosphate, nickel, and sand, have long been material mediators for globally connected histories of domination, influence, creative appropriation, and transfers at a technical and social level. On the other hand, another type of exploitation closely linked to colonization started after the Second World War, as the Pacific was adopted by several countries as a test site for nuclear and thermonuclear weapons. From 1945 to 1962, Americans conducted tests essentially in the Marshall Islands, while the British chose the Montebello Islands, the Australian desert, and the Kiribati Islands to try out their nuclear weapons in the atmosphere from 1952 to 1963 (Firth, 1986). The French government, which lost its Sahara test sites when Algeria gained its independence, chose two atolls in French Polynesia, Moruroa and Fangataufa. From 1966 to 1996, the Centre d'essai du Pacifique (CEP) conducted 193 tests on these atolls (Meltz, Vrignon, 2022). These choices seem to have had the general perception in common that these low-lying, dry lands were of no particular use, presented little resource potential, and were so remote that they presented ideal locations for these tests. In recent decades, historians have started to uncover the entwined histories of colonial domination, scientific and technical research, and dramatic, ongoing social and health issues related to nuclear weapons tests in the Pacific (Barker, 2004; Gusterson, 2004; Guérin, 1972; Regnault, 1993; Danielsson, 1993; Alexis-Martin, 2019; Meltz, Vrignon, 2022).There are no civil nuclear stations in the Pacific, except for a few research reactors in Australia (Argounès, 2013).

Finally, research agendas merging environmental and ecological histories with social, political, and economic issues are at the centre of recent interdisciplinary interests in infrastructures and networks (e.g. Duensing, 2009, 2015; Foster, Horst, 2018; Starosielski, 2015). They question divergent conceptions of space, temporal and environmental knowledge, actors' intersubjectivity, practices, and scales of action as well as intertwined patterns of power and dominance. Among these, transportation, exploitation of the local environment and resources, and communications and media are dynamic areas of scholarly enquiry. The following sections will focus on these themes as case studies.

'A Sea of Islands': Transportation Techniques

Canoe-making skills and navigational techniques are both important, albeit unevenly shared technologies for many Pacific Islanders (Hau'ofa, 1994). In the nineteenth century, a wide variety of watercraft was still in use in the region, ranging from the reed bundles of Tasmania, New Zealand, and the Chatham Islands, via the bark canoes of the Australian Aborigines, the rafts with two-boom triangular mat sails of Fiji, Santa-Cruz, and Mangareva, and dugout canoes with single or double outriggers, to the large double canoes of Fiji, Tonga, and the Society Islands (Haddon, Hornell, 1975). Whether the watercraft were hollowed out from a single log and with raised planks on the sides or made with irregular planks that are assembled and sewn together, in both cases stem and stern pieces were then added. Apparently, a similar five-part design has been found throughout the regions settled by populations speaking Austronesian languages (Horridge, 2008: 88–89).[3] Whereas double outrigger canoes corresponded to inbay usage, off-shore, sea-going canoes had a single outrigger on the windward side or a second hull, and their raised platforms made it possible to accommodate passengers and cargo for long voyages (Horridge, 2008: 90). Technical abilities and skills related to canoe building, including knowledge of materials and appropriate measurements, could be limited and reserved to groups of specialists, while larger canoes demanded the cooperation of entire communities and testified to their commissioners' power (Guiot, 2001, 2003; Hiquily et al., 2008; Feinberg, 1988, 1995). Canoe construction, repairs as well as sailing expeditions were accompanied by rituals and ceremonies (Malinowski, 1922; Best, 1925; Dodd, 1972; Lewis, 1972; Buck, 1957; Holmes, 1981). Therefore, canoes held some of the societies' fundamental values (Munn, 1977) and affected both the emotional and the aesthetic senses (Gell, 1992; Campbell, 2002). Navigation called for detailed knowledge of the winds, currents, and cosmological patterns, i.e. the ability to determine the position of the sun, the moon, and the stars at different times of the year, as well as knowledge of the shapes of clouds and swells or the behaviour of birds (Lewis, 1971, 1972; Finney, 1976, 1998).

Although 'pacific outrigger canoes were originally as homogeneous as the Austronesian people', they were 'later [...] influenced by [...] introduced technology' (Horridge, 2008: 85). In the early decades of the nineteenth century, these technical skills and a spatio-temporal knowledge of navigation and trade routes were met with the skills of Euro-American mariners and traders who arrived in those waters in increasing numbers. In the wake of the 1788 foundation of a British convict colony at Port Jackson (later Sydney) on the east coast of Australia, commercial ventures multiplied all over the Pacific (Tench, 2006): between the end of the eighteenth century and the 1860s, sealing (Hainsworth, 1967) and, to a much larger extent, whaling introduced hundreds of ships to the Pacific who were in search of the oil needed for America's and the Old Continent's developing city street lighting systems. In 1844, for example, 571 whalers operated in the Pacific (Bennett, 1987; Dodge, 1965; Forster, 1985; Ralston, 1977; Starbuck, 1964; Ward, 1966–1967). The sandalwood and sea cucumber trades also flourished in the region in the first half of

3 Austronesian-speaking populations are distributed from Madagascar to Easter Island. Although the precise history of their expansion and encounters with other populations in the Pacific are still being debated, it is generally acknowledged that they derive from a common origin in the region of Taiwan some 5000 or 4000 years ago (Kirch, 2017).

the nineteenth century. After the first discovery of sandalwood in Hawaii in 1789, the trade mainly developed from 1811. In the early years, Fiji was a major centre for this trade until its resources were exhausted. From 1825, New Caledonia and the New Hebrides (Vanuatu) became important places for sandalwood provision, but the trade gradually declined in the second half of the century due to the rapid depletion of resources. Sea cucumbers, which are used in Chinese gastronomy and medicine, were collected in lagoons at the depth of some metres. They were prepared by drying over fire and required working parties composed of Europeans and Islanders to establish stations ashore. This trade started in the 1820s in Fiji and, after peaking in around 1835, decreased in the 1850s. Both the sandalwood and the sea cucumber trade led to various and varying collaborations and conflicts with local leaders and groups. Numerous objects were exchanged, with some contributing to important political and social change; they included metal tools, firearms (Shineberg, 1971; Howe 1974), and materials which were locally most valued such as whale ivory in Fiji. All of this transformed the local economies (Thomas, 1991). These trades also helped to increase contact between island populations. In 1829, the British trader Samuel Henry took a group of Tongans to cut sandalwood on Erromango in the southern New Hebrides. In the following year, hundreds of Hawaiians, Rotumans, and Tongans came to the island, and these massive arrivals caused significant conflicts which ended in bloodshed. Consequently, the entire trade on Erromango came to a halt and was resumed only in the 1840s and the 1850s (Shineberg, 1967).

As a result of these numerous encounters, transportation changed and new paths and networks opened up. So did local techniques: canoes' mat sails, 'heavy, weak and soon rotted', were replaced by woven textile sails, while fibre lashings and plaited coconut sennit ropes were changed for hemp and sisal (and later nylon and polymer cordage) (Horridge, 2008: 94). In the nineteenth century, pulleys and halyards were more widely adopted along with masts. In Hawaii, square sails and topsails were added to older technical influences that had probably been appropriated from Spanish ships in Eastern Polynesia (Horridge, 2008: 97). Pacific Islanders, especially those who had a high social status locally, also acquired foreign ships in the nineteenth century. The historian Peter Mills notes that

> Pomare II of Tahiti purchased a ship in 1821, and Tamatoa of Raiatea followed suit in 1821. [...] By the 1850s, New Zealand Maori were purchasing and building schooners, and by 1858, there were 53 Maori vessels registered in Auckland of more than 14 tons [...]. By the early 1900s, even the natives of the small island of Tanna in Vanuatu had purchased their own vessel (Fetukai) so that they could sell their copra without middlemen (see Mills, 2003: 54 for further references).

Mills also suggests that at least eighty Western-style vessels were purchased or used by Hawaiian chiefs before 1830, and that a good number of them had been locally built by Hawaiian shipwrights (Mills, 2003: 60). On shore, foreign carpenters, blacksmiths, rope-makers, and sailmakers worked with Hawaiian craftspeople in dedicated buildings and areas, while 'hauling wood from the mountain to make masts was a ritualized event, with *kahuna* chanting as the work progressed, similar to what was done to procure timber for wooden idols and canoe hulls' (Mills, 2003: 61). However, after 1820, chiefs often chose to purchase their ships rather than build them, and they traded as far as the Pacific Northwest, California, Central Polynesia, Melanesia, and Canton. They extended their local authority by using these ships in warfare and to transport tributes (Mills, 2003: 64).

In the nineteenth century, technical achievements such as the screw propeller in the 1840s, the triple expansion engine in the 1870s, and 'the introduction in the 1880s of steel as a structural material' brought major changes in transportation patterns as steam navigation developed alongside the imperial powers' increasing importance (Steel, 2011: 4). Several companies shared the Pacific transportation business, including Burns Philp and Company, who were based in Sydney, the Australasian Steam Navigation Company, the German Norddeutscher Lloyd steamship company, the French Messageries Maritimes (Berneron-Couvenhes, 2007), and the Union Steam Ship Company of New Zealand (USSCo.). They connected the area with colonial metropolitan centres and contributed to the creation of regional centres. Focussing on the USSCo., the historian Frances Steel (2011) has highlighted its links with Australia, New Zealand, and several of the Pacific Islands, with Fiji as a central node. Operating on fixed schedules, large steamers served island ports that became administrative hubs. There, they connected with smaller ships belonging to or crewed by islanders, which serviced the numerous plantations and industries scattered among the islands and formed a major component of the island economies in the first half of the twentieth century (Young, 1993). Thus, local island services were mostly separate from the large steamers servicing intercolonial and trans-colonial networks. On board the latter, the labour force was largely of European origin, since at the end of the nineteenth century, Australian seamen's strikes in reaction to the perceived threat of cheaper Chinese and then Indian labour had ended in the banning of people of colour as steamer crews (Steel, 2011: 101). After the First World War, other transportation technologies emerged in the region, which were similarly linked to the (im)balances of imperial and, later, post-colonial powers.

As early as 1913, a committee of political figures and influential scientists was formed in Germany to discuss the possibility of building airship sheds in the Sepik River and Dutch New Guinea for dirigibles and other aircraft to be flown there, with costs shared by the German, Dutch, and Australian governments. Three years later, another proposal was made by Erik Mjöberg, a Swedish aviator, to explore the Australian-administered Territory of Papua by plane, but this was rejected by the Australian authorities, partly out of fear of the local populations' reactions to such an 'advanced' technology (Sinclair, 1978: 11; Wolfers, 1975). By the beginning of the 1920s, however, the first flights to the new Mandated Territory of Papua and New Guinea had been made by expeditions led by the photographer Frank Hurley (1921), the United States Navy (1923), the Dutch-American Expedition to Dutch New Guinea (1926), and by the Royal Australian Air Force, who flew from the Mandated Territory to the Solomon Islands (1926). The goals were mainly the exploration and documentation of the areas, and in the Royal Australian Air Force case, also included an evaluation of the possibility of a regular air mail service between Papua,

New Guinea, and north Queensland. Regular services finally started in the context of the Morobe gold rush of the 1920s. The discovery of gold in the area around 1910 attracted many prospectors, and mining activities soon flourished in the region. Papua New Guineans were employed by prospectors in large numbers, although the mountainous environment necessitated the transport of supplies by portage, at very expensive rates (Nelson, 1976). By the end of 1925, about fifty European prospectors and 1500 indigenous workers were active on the Bulolo and Watut rivers in the mountainous interior of Morobe Province (Sinclair, 1978: 25).

In 1927, therefore, the Australian Guinea Gold company started operating an air service between the coast and the goldfields, so that supplies could be transported more cheaply. The company was soon joined by other air services proposed by rival companies (first using a De Havilland 37 biplane with wings covered in fabric, then metal aircraft like the German Junkers). Technical details were instrumental in this competition, with companies looking for the best aircraft, engines, pilots, and mechanics for the purpose. Renamed 'Guinea Airways Limited', the Guinea Gold Air Services expanded in the following years to serve larger parts of the Territory (Sinclair, 1978). Altogether, aviation services played a crucial part in the economy of Papua New Guinea, providing support for Australia's colonial advance into parts that had not yet been explored (Leahy, 1991) and for missionary activities, as well as transporting supplies for commercial mining ventures (Hyndman, 1994; Nelson, 2016) and agroforestry businesses (Bell, 2010).

In 1973, when it became self-governing, Papua New Guinea created its own national airline, Air Niugini, which was a move towards national control of air services that had already been achieved or would also be made by most newly independent countries in the Pacific (Samoa became independent in 1962, Nauru in 1968, Fiji and Tuvalu in 1970, Papua New Guinea in 1975, the Solomon Islands in 1978, Kiribati in 1979, and Vanuatu in 1980.) For instance, although Australian interests were still initially present in Air Niugini, by 1976 the government of Papua New Guinea controlled 83% of capital investment (Guthrie, 2013: 300). Outside New Guinea, air transportation owed much to Melanesia's strategic position during the Second World War. Many ground installations, including airfields, resulted from the presence of American and Japanese troops on the islands. After the war, these installations were used by colonial powers who developed regular trans-oceanic services. At first these flights required refuelling stops on islands located along the way, and thereby ensured a connection of these places to Asia, the United States, and Australia. This was undermined by the new jet technology which made it possible to fly nonstop across the Pacific (Ward, 1989; Guthrie, 2013). Consequently, most Pacific countries remained dependent on larger international carriers such as the Australian Qantas, Air Niugini, or Air New Zealand, according to patterns that also reflected the post-colonial balance of powers and regionalism in the area (Guthrie, 2013).

In the final decades of the twentieth and in the early twenty-first centuries, when regional institutions such as the Pacific Island Forum emerged to discuss cultural, economic and political specificities of the area, canoe and navigation skills became a major aspect of indigenous identity claims.[4] Motivated by research on how Polynesians first managed to settle on distant islands, a Hawaiian team built a 20-metre long, double-hulled sailing canoe, based on the characteristics of ancient Polynesian voyaging craft but using modern materials. The *Hōkūle'a* proved the vitality and contemporary value of Polynesian seafarers' skills by sailing from Hawaii to Tahiti in 1976, and to other islands in the 1980s (Finney, 1992). It continues to sail around the world to promote environmental causes and commitments to the indigenous community.[5] Fleets of Polynesian canoes gathered on the occasion of the sixth Festival of Pacific Arts and Culture in Rarotonga (Cook Islands) in 1992 and in Raiatea in 1995. These events, as well as numerous contemporary canoe races involving Pacific nationals (Guiot, 2003: 36–38), ensure and demonstrate the dynamism of the transmission and re-creation of technical knowledge in the region.[6] In this respect transportation is only one of a number of topics that could have equally served as case studies for this paper.

Agroforestry, Fisheries, and Mining: the Exploitation of Resources

Over time, Pacific peoples have developed very elaborate resource management techniques. For example, Australian Aboriginal populations carefully control bush fires to maintain vegetation and encourage wildlife reproduction in certain areas (Whitehead et al., 2003; Russell-Smith, Whitehead, Cooke, 2009; Fache, 2019). Similarly, agricultural pits dug into the coral base of atolls, which provides access to the freshwater lens, have vegetable compost deposited within them to create a humus-rich soil and allow many Polynesian and Micronesian peoples to grow taro (*Colocasia* or *Cyrtosperma*) in otherwise difficult environments (Yamaguchi et al., 2005; Ward, Proctor, 1980; Lebot, 2009). Damon (2017) describes how peoples in the Trobriand islands use certain trees to renew soil fertility, preserving the trunks when they cut the bush to open a new patch of garden, and then concentrating crops around them so that the plants experience better growth. In Vanuatu, villagers accumulate plants in their gardens to grow the most diverse cultivars possible, which is evidence of the broad range of their social networks and supports biodiversity (Caillon, Lanouguère-Bruneau, 2005; Calandra, 2017). Techniques of forestry and marine resource management are also highly developed. All these techniques are embedded in relations based on exchange, and social networks. Additionally, they can be closely associated with power structures and the assertion of authority. Numerous taboos, often signalled by specific palms or bundles of leaves and plants that are

4 The South Pacific Forum was established in 1971 by Pacific Island leaders to discuss social and political issues, in a context where they felt their voices were not heard in discussions by other regional institutions like the South Pacific Commission. It was renamed the Pacific Island Forum in 2000.

5 See http://www.hokulea.com/.

6 See also the project on sustainable sea transportation in the Pacific using craft powered by the wind, the sun, and coconut biofuel, supported by the Okeanos Foundation for the Sea (https://okeanos-foundation.org/).

placed on restricted areas, prevent individuals from exploiting certain resources for a specific amount of time and help to channel them into future stocks that will serve the tabooer's interests. These taboos therefore identify the seat of local authority over resources, an authority which can be contested on those grounds (Bennett, 2000; Fache, Pauwels, 2016).

Encounters with Euro-Americans created new ways of relating to these resources. Whereas coconut was only a secondary food in the diet of Pacific Islanders, whose staple foods were starchy root crops such as taro or yams that provided complex carbohydrates combined with dietary fibre, vitamins, and minerals (Pollock, 1992, 2017), it soon became a major good for trade (Bennett, 2018). In the 1840s, new European and American techniques to manufacture soap and candles increased the use of coconut oil (Daumalin, Girard, Raveux, 2003) and prompted the development of trade in the Eastern and Central Pacific (Maude, Leeson, 1965). However, as Judith Bennett has pointed out, it was only in the 1860s that the German firm J.C. Godeffroy and Son developed a new method

> of drying the mature kernel as copra and exporting this to European mills to extract the oil, while the residue was used as a food for cattle and pigs, particularly in the northern winters. Thus, there was no chance of the oil going rancid on the two to three months sea voyage or lost through cask leakage (Bennett, 2018: 357).

Islanders' agricultural lives were changed as they integrated the production of copra as a new, lucrative activity that, in turn, allowed them to gain some access to the goods that planters and traders sold in their small stores and to pay the taxes imposed on them by colonial governments. At the other end of the commodity chain, coconut oil soon came to join other vegetable oils that were used to 'lubricate machinery, to produce explosives and by the turn of the nineteenth century to make margarine for human consumption' in Hamburg, London, Marseille, Tokyo, and San Francisco (Bennett, 2018: 362).

During the second half of the nineteenth century and in the twentieth century, plantations of cotton, sugar cane, and coconut palms relied on cheap and abundant labour that was usually imported from abroad, thus increasing mobility and work migration in the region. This labour trade, at times equated to slavery, is often referred to as 'blackbirding' (Moore, Leckie, Munro, 1990; Panoff, 1991; Lal, Munro, Beechert, 1993; Graves, 1993; Shineberg, 1999; Lal, 2008; Stead, Altman, 2019; Moore, 2019). The development of steam navigation in the Pacific in the course of the twentieth century served plantation agriculture well, but the large monocultures weakened the islands' ecological systems. Often situated on coastal land, the vast, cleared areas of coconut plantations that were devoid of any understory were less efficient in absorbing wave wash overs and high seas, while insects and pests welcomed these undiversified environments, so that entomologists reacted and introduced exotic biological pest controls (Bennett 2018: 364).

In the 1960s and the 1970s, plantation agriculture was locally re-organised into cooperatives run by Pacific Islanders. As such, plantations also played a part in forming economic infrastructures for the future independent nations of Oceania (Denoon, Snowdon, 1979; Ponter, 1983; Weightman, 1989; Bennett, 2018: 365).

Mining was not generally practised by Pacific peoples before contact with Euro-Americans who were settling in the region (Burke, 1994). By the early decades of the twentieth century, prospectors had discovered gold deposits in several parts of New Guinea, including the Louisiade and Woodlark archipelagoes as well as the Mambare, Gira, Yodda, Waria, and Lakekamu rivers, Milne Bay, and Morobe Province (Nelson, 1976). A significant part of the miners who searched for gold in New Guinea had already gained digging experience in the north Queensland fields of Australia. While at first miners worked mainly on small-scale, individually owned concessions, sifting through the rivers' alluvia using sluice boxes, this work was generally taken over and developed on a larger scale by companies at the beginning of the twentieth century. On Misima Island, for instance, Broken Hill Proprietary Limited dug tunnels and installed 'crushmills, cyaniding vats, sawmills, black-smiths' and carpenters' shops, electric lighting, staff accommodation, port facilities', and even a seven-mile tramway (Nelson, 1976: 41).

For New Caledonia, Yann Bencivengo has shown how the birth of the nickel industry in the 1870s was

> part of the second industrialization emerging then in Europe and North America, powered by new energies (petroleum, electricity) relying upon the new driving sectors of electric, chemical and automobile industries [Woronoff, 1994 : 348]. This start is also embedded in a movement of globalization linked to colonization, the growth of the North American economy and the progress of long-distance transportation and communication (Bencivengo, 2014: 137).

Tracing the colonial history of nickel mining, Bencivengo recalls how, for a short period at the end of the nineteenth century, this small Pacific archipelago and the French Société Le Nickel were at the centre of the global nickel trade. The Société's local and international success has been linked to several factors: the development of a reliable technical method to process nickel ore that was inspired by copper metallurgy; the company's management of the entire production chain with mines and factories in both New Caledonia and Europe; and the colonial exploitation of convicts as cheap labour in the nineteenth, as well as Japanese, Indonesian, and Vietnamese workers later in the twentieth century (Bencivengo, 2014). More recently, nickel mining, for example the Koniambo project, has also raised economic hopes for a Kanak-led development, supporting the idea of New Caledonia future independence (Horowitz, 2004).

In addition to gold and nickel, other minerals such as copper, cobalt, and zinc have been extracted from the ground of South Pacific countries, often with considerable environmental impact (Howard, 1991; Connell, Howitt, 1991). From early twentieth century to the 1960s, islands including Christmas Island, Nauru, and Banaba/Ocean Island in the British Empire, and Makatea in French Polynesia, were heavily mined for phosphate, which was used as a fertilizer in Australia, the United Kingdom, and New Zealand. On these islands the surface soil was largely stripped away, and the mines left only hard calcite pinnacles, i.e. sterile landscapes that were subject to erosion and could no longer support the local populations (Williams, MacDonald, 1985; Weeramantry, 1992; Pollock, 2014; see also Cushman, 2013). On Banaba Island, all inhabitants had to relocate to Fiji in 1945 (Teaiwa, 2015). Since the second half of the twentieth century, large-scale transnational mining ventures have developed, some of which are criticized for their environmental and social impacts. The Ok Tedi mine in the Western province of Papua New Guinea, for example, has been discharging waste into the Alice and Fly rivers, whose levels of pollution are acknowledged today, for several decades (Filer, Jenkins, 2017; Banks, Ballard, 2003), while the activities at the Panguna mine, on the island of Bougainville, are deemed partly responsible for the civil war that raged in the island from 1988 to 1998 (Quodling, 1991; Regan, 2017).

Finally, deep sea mineral resources have recently attracted the attention of exploration companies. Massive sulphide deposits on the sea floor (containing copper, lead, zinc, gold, and silver), manganese nodules, and ferromanganese crusts (containing nickel and cobalt, and sometimes rare earth elements) have been identified at several places in the South Pacific Ocean seabed, which creates new economic, environmental, and legal issues for the region (Baker, Beaudoin, 2013).

Communication Techniques

A global history of communication techniques in the South Pacific is one that requires looking underground and down at the ocean floor, as well as up into the air. Wired and wireless technologies have been complementing each other, as Nicole Starosielski (2015) has noted in her history of the infrastructure of trans-Pacific undersea cable networks. Starosielski shows that security has always been crucial to the technologies and routes chosen by telecommunication companies. This ensured that undersea cables continued to be a valued option even after the Second World War, in the age of radio and satellite.

Australia used telegraph lines to connect its urban settlements from 1858 onwards, but it was only in 1872 that a connection with the overseas network was established through a submarine telegraph cable linking Banjoewangie, Java to Darwin. In the same year, the Australian Overland Telegraph line connecting Darwin with Port Augusta in the south was completed (Moyal, 1984; Taylor, 1980). In the following years, undersea connections multiplied: the Eastern Telegraph Company installed a cable from Botany Bay in Australia to Nelson in New Zealand in 1876; in 1902, the British colonies were connected through the All-Red Line operated by the Pacific Cable Board, thereby inaugurating the first trans-Pacific cable network; and two years later, the United States established its own cable, reaching the Philippines and then China via Hawaii and Guam, while New Caledonia had been connected to Australia via a French cable in 1893 (Johnson, 1903; Boyce, 2000; Laborie, 2013; Starosielski, 2015).

Undersea cables were judged to be safer than terrestrial cables, and they also presented an economic advantage, as companies were able to avoid transit taxes if they went through extraterritorial waters. These first submarine telegraph cables roughly followed the maritime transportation routes, surfacing in ports where the smooth transition from sea to land had already been organised, and where colonial infrastructures could provide the necessary support for stations to be set up and secured. In the late nineteenth century, the British largely dominated the market for producing and laying submarine communication cables since they had mastered an insulation method using gutta-percha, which was obtained from Malaysian trees, and whose production was part of the British Empire's forestry industry (Starosielski, 2015). However, imperial powers should not be seen as the sole force driving the telegraph cable industry: the telegraph companies' economic and industrial interests often extended beyond the boundaries of colonial empires (Laborie, 2013; Winseck, Pike, 2007).

As Nicole Starosielski explains, telegraph cables opened up the way for later networks of analogue coaxial cables which became possible thanks to the invention of the submersible repeaters that were laid from the 1950s onwards, and then fibre-optic cables from the 1980s and 1990s onwards. These respectively catered for the transmission of telephone conversations, telegraph and telex messages, photo telegrams, and slow-scan television on the one hand, and for the boom of information traffic in the internet era on the other. Instead of opening multiple new routes, companies mostly tended to make use of existing infrastructures, thereby strengthening prior paths of connexion while nevertheless being influenced by the new geographies and polarities of the Cold War, and later neo-liberalism and global financial capitalism. Gradually, the technical problem of insulation from the ocean waters shifted, and the ocean depths were seen as providing the necessary insulation from the multiple social and political troubles on the surface of the earth. Increased interconnections would ensure diversified routes to prevent possible failure due to a problem at any point. Therefore, previous hubs were reinforced, and Fiji, for instance, became a strategic node (Starosielski, 2015).[7]

Shortwave radio communications had also developed in the 1920s and the 1930s, and after the Second World War, teleradio networks made scheduled transmissions for different groups possible. Administrations, companies, and missionaries all used scheduled transmissions to connect with remote rural outstations (Johnson 1988; Bolton, 1999; Handman, 2018). In the 1960s and the 1970s, specific radio programmes were developed that supported, as Lissant Bolton argues for Vanuatu, the emergence of national sentiment through the definition of shared concepts such as 'kastom' — from the English 'custom' (Bolton, 1999).

Finally, the entire South Pacific region has been the stage for what has widely been acknowledged as a telecommunication 'revolution' during the last thirty years (Cave, 2012: 2). The telephone appeared during the colonial era, when fixed landlines serviced urban settlements and were gradually set up even in remote rural villages, powered by solar panels or diesel generators. Following the independence, services

7 See also www.surfacing.in for a multi-layered and innovative historical visualization of the world's undersea networks.

were provided by monopolistic national companies such as Telikom PNG in Papua New Guinea, Solomon Telekom in the Solomon Islands, and Telekom Vanuatu Limited in Vanuatu (Foster, Horst, 2018). However, the liberalization of the market in the 1990s and development of mobile phone technologies contributed to a much more widespread coverage and a large increase in users. Heather Horst and Robert Foster indicate that 'average mobile coverage (2G) in Fiji, Samoa, the Solomon Islands, Tonga and Vanuatu jumped from less than half of the population in 2005 to 93% of the population in 2014 (Pacific Region Infrastructure Facility 2015). A 2015 Groupe Special Mobile Association (GSMA) report notes that 'unique subscriptions across the entire region increased from 2.3 million in 2009 to 4.1 million at the end of 2014 — a 12.6 per cent annual growth rate' (Foster, Horst, 2018: 1). Scholars point to the crucial turn caused by the arrival of Digicel Group Limited on the market: between 2006 and 2008, Digicel obtained licences and a significant share of the mobile communication markets of Samoa, Papua New Guinea, Fiji, Tonga, and Nauru (Cave, 2012). Sustained by a network of towers set up in remote rural areas, Digicel developed specific offers based on the values of family and reciprocity that are widely shared in the South Pacific region as well as targeting low-income individuals (Foster, Horst, 2018: 2–3). Receiving and sending small amounts of prepaid airtime to and from family and friends was facilitated by a large network of local retail vendors and various means of requesting and acquiring additional credit (Kraemer, 2018; Foster, Horst, 2018).

In the past decade, mobile phones followed by smartphones have become the major tool for digital connection throughout the Pacific, and they have been used creatively for text messages and phone calls as well as for watching, listening to, storing, and sharing photographs, videos, and music (Stern, 2014; Crowdy, 2015; Hobbis, 2017a). The use of mobile phones and internet access especially to social media including Facebook, Twitter, and Instagram have had vast effects on the social life of people. Mobile technologies now enable radical changes in geographical, temporal, and intimate personal boundaries, while the uses of phones by companies are also frequently diverted and transformed by people so that they talk to their deceased relatives or else (see e.g. Telban, Vávrová, 2014). Today, mobile phones are also major tools allowing governments to communicate the threats of environmental events such as cyclones or tsunamis (Calandra, 2017: 274; Ballard, McDonnell, Calandra, 2020: 5). For ceremony and ritual, expanding the range of family and kin who can be reached and who can contribute impacts the amount of goods exchanged at marriages and funerals as well as the social extent of the reciprocal exchanges expected to be performed. At funerals in northern Vanuatu, for example, the number of the deceased's classificatory siblings to whom his or her children must give ritual compensation in order to secure their rights regarding their parent's resources has risen sharply due to the availability of mobile phones and communication (Durand, 2015). As privately owned devices, mobile phones also transform customary boundaries of communication, since it is now possible to move into the private sphere what was expressed in public beforehand and that was therefore subject to social control. For women in particular, mobile technologies have provided both a means to escape from strict family control and a new means of being harassed by men (Servy, 2013; Taylor, 2015; Hobbis, 2017b). Finally, even the moral values and practices associated with the uses of vernacular or vehicular languages are being influenced by these new technologies (Vandeputte, 2013). Generally, Pacific peoples today creatively use digital tools and technologies for mediations which connect past practices with future representations in numerous ways (De Largy-Healy, 2016; Geismar, 2012).

As these three case studies show, a global history of techniques analysing external and internal forces, the circulation of ideas and practices, and the great diversity of creative practices and appropriations of the inhabitants, environments, and nations of the South Pacific is yet to be developed as a structured field of research. Nevertheless, scholars' interests in various disciplines help to show how dynamic the circulation of techniques is in the region. Indeed, communication, the arts, architecture, and agriculture, among many other fields, constantly testify to the fact that technical innovation is very important to the Oceanian peoples' lives.

References

Aldrich, Robert, 'French Economic Interests in the Pacific Colonies', in *The French Presence in the South Pacific*, ed. by Aldrich Robert (London: MacMillan, 1989), pp. 105–38.

Alexis-Martin, Becky, *Disarming Doomsday: the Human Impact of Nuclear Weapons since Hiroshima* (London: Pluto Press, 2019).

Angleviel, Frédéric, *Les Ballandes et l'appel du large: cinq générations d'aventuriers et d'entrepreneurs, 1838–2016* (Noumé: GRHOC and Groupe Ballande, 2016).

Argounès, Fabrice, 'Le Nucléaire dans le Pacifique', *Hermès*, 65.1 (2013), 67.

Armitage, David and Alison Bashford, eds, *Pacific Histories: Ocean, Land, People* (Basingstoke: Palgrave Macmillan, 2014).

Baker, Elaine and Yannick Beaudoin, eds, *Deep Sea Minerals: Deep Sea Minerals and the Green Economy*, vol. II ([s.l.]: Secretariat of the Pacific Community, 2013).

Ballard, Chris, Siobhan McDonnell, and Maëlle Calandra, 'Confronting the Naturalness of Disaster in the Pacific', *Anthropological Forum*, 30.1–2 (2020), 1–14.

Banks, Glenn and Chris Ballard, 'Resource Wars: The Anthropology of Mining', *Annual Review of Anthropology*, 32.1 (2003), 287–313.

Barclay, Glen, *A History of the Pacific from the Stone Age to the Present Day* (London: Sidgwick and Jackson, 1978).

Barker, Holly M., *Bravo for the Marshallese: Regaining Control in a Post-Nuclear, Post-Colonial World* (Wadsworth: Cengage Learning, 2004).

Barrau, Jacques, *Les Plantes alimentaires de l'Océanie. Origines, distributions et usages* (Marseille: Annales du Musée colonial de Marseille, 1962).

Barron, Géraldine, *Edmond Pâris et l'art naval: des pirogues aux cuirassés* (Toulouse: Presses universitaires du Midi, 2019).

Bell, Joshua, 'Sugar Plant Hunting by Airplane in New Guinea: a Cinematic Narrative of Scientific Triumph and Discovery in the "Remote Jungles"', *The Journal of Pacific History*, 45.1 (2010), 37–56.

Bencivengo, Yann, 'Naissance de l'industrie du nickel en Nouvelle-Calédonie et au-delà, à l'interface des trajectoires industrielles, impériales et coloniales (1875–1914)', *Journal de la Société des Océanistes*, 138–39 (2014), 137–49.

Bennett, Judith, 'Immigration, "Backbirding", Labour Recruiting? The Hawaiian Experience 1877–1887', *The Journal of Pacific History*, 11 (1976), 3–27.

———, 'Pacific Coconut: Comestible, Comfort and Commodity', *The Journal of Pacific History*, 53.4 (2018), 353–74.

———, *Pacific Forest: a History of Resource Control and Contest in Solomon Islands c. 1800–1997* (Cambridge: The White Horse Press, 2000).

———, *Wealth of the Solomons: a History of a Pacific Archipelago 1800–1978* (Honolulu, HI: University of Hawaii Press, 1987).

Berneron-Couvenhes, Françoise, *Les Messageries maritimes: l'essor d'une grande compagnie de navigation française, 1851–1894* (Paris: Presses Paris Sorbonne, 2007).

Best, Elsdon, *Fishing Methods and Devices of the Maori* special issue, *Dominion Museum Bulletin*, 12 (1929).

———, *Forest Lore of the Maori: With Methods of Snaring, Trapping, and Preserving Birds and Rats, Use of Berries, Roots, Fern-Root, and Forest Products, With Mythological Notes on Origins, Karakia Used etc.* (Wellington: Polynesian Society, 1942).

———, *The Maori Canoe*, special issue, *Dominion Museum Bulletin*, 7 (1925).

———, *Maori Storehouses and Kindred Structures*, special issue, *Dominion Museum Bulletin*, 5 (1916).

———, *The Stone Implements of the Maori*, special issue, *Dominion Museum Bulletin*, 4 (1912).

Blackwood, Beatrice, *The Technology of Modern Stone Age People in New Guinea* (Oxford: Pitt Rivers Museum, 1950).

Blais, Hélène, 'Apprendre à dire l'espace. L'invention du triangle polynésien dans les récits de circumnavigation (1817–1845)', *Genèses*, 45 (2001), 91–113.

Bollard, A. E., 'The Financial Adventures of J. C. Godeffroy and Son in the Pacific', *The Journal of Pacific History*, 16.1 (1981), 3–19.

Bolton, Lissant, 'Radio and the Redefinition of "Kastom" in Vanuatu', *The Contemporary Pacific*, 11.2 (1999), 336–60.

Boyce, Robert, 'Imperial Dreams and National Realities: Britain, Canada and the Struggle for a Pacific Telegraph Cable, 1879–1902', *English Historical Review*, 460 (2000), 39–70.

Brunois, Florence, *Le Jardin du casoar. Savoir-être et savoir-faire écologiques* (Paris: CNRS Éditions, Éditions de la Maison des Sciences de l'Homme, 2007).

Buckley, Kenneth and Kris Klugman, *The Australian Presence in the Pacific: Burns, Philp 1914–1946* (Sydney: George Allen and Unwin, 1983).

Buckley, Kenneth and Kris Klugman, *The History of Burns, Philp: the Australian Company in the South Pacific* (Sydney: Burns, Philp & Co. 1981).

Burke, Gill, 'Mining in the Pacific: Past History, Present Practice, Continuing Troubles', review of *Bougainville: the Mine and the People* by Paul Quodling; *Mining and Indigenous Peoples in Australasia* by John Connell and Richard Howitt; *Mining, Politics and Development in the South Pacific* by Michael C. Howard; *Nauru: Environmental Damage under International Trusteeship* by Christopher Weeramantry, *The Journal of Pacific History*, 29.1 (1994), 105–09.

Caillon, Sophie and Virginie Lanouguère-Bruneau, 'Gestion de l'agrobiodiversité dans un village de Vanua-Lava (Vanuatu): stratégies de sélection et enjeux sociaux', *Journal de la Société des Océanistes*, 120–121 (2005), 129–48.

Calandra, Maëlle, 'Jardins de terre, jardins de mer à Tongoa (Vanuatu). Une anthropologie de la nature domestique dans un milieu affecté par la catastrophe' (unpublished doctoral thesis, EHESS Paris, 2017).

Campbell, Shirley, *The Art of Kula* (Oxford and New York, NY: Berg, 2002).

Cave, Danielle, *Digital Islands: How the Pacific's ICT Revolution is Transforming the Region* (Sydney: Lowy Institute for International Policy, 2012).

Connell, John and Richard Howitt, eds, *Mining and Indigenous People in Australasia* (Sydney: Sydney University Press 1991).

Corris, Peter, *The South Sea Islanders and the Queensland Labour Trade* (Canberra: Australian University Press, 1973).

Coupaye, Ludovic, *Growing Artefacts, Displaying Relationships. Yams, Art and Technology amongst the Nyamikum Abelam of Papua New Guinea* (New York, NY and Oxford: Berghahn Books, 2013).

Crowdy, Denis, 'When Digital is Physical and Ethnomusicologists are File Sharers', *Journal of World Popular Music*, 2.1 (2015), 61–77.

Cushman, Gregory T., *Guano and the Opening of the Pacific World* (Cambridge: Cambridge University Press, 2013).

Damon, Frederic, *Trees, Knots and Outriggers: Environmental Knowledge in the Northeast Kula Ring* (New York, NY and Oxford: Berghahn Books, 2017).

Danielsson, Bengt, *Moruroa, notre bombe coloniale: histoire de la colonisation de la Polynésie française* (Paris: L'Harmattan, 1993).

———, *Poisoned Reign: French Nuclear Colonialism in the Pacific* (New York, NY: Penguin Books, 1986).

Daumalin, Xavier, Nicole Girard, and Olivier Raveux, *Du savon à la puce. L'industrie marseillaise du XVIIe siècle à nos jours* (Marseille: Éditions Jeanne Laffitte, 2003).

Davidson, James W., 'Problems of Pacific History', *The Journal of Pacific History*, 1 (1966), 5–21.

De Largy Healy, Jessica, 'Retours sur images. Nouveaux médias et transmission du secret dans les rituels du nord de l'Australie', *Ethnographiques.org*, 33 (2016) <https://www.ethnographiques.org/2016/De-Largy-Healy> [accessed 12 June 2021].

Denoon, Donald and Catherine Snowdon, *A Time to Plant and A Time to Uproot. A History of Agriculture in Papua New Guinea* (Port-Moresby: Institute of Papua New Guinea Studies, 1979).

Di Piazza, Anne, 'La pirogue monodrome à voile à livarde: l'outil de peuplement de la Polynésie occidentale', in *Les Sciences humaines et sociales dans le Pacifique sud. Terrains, questions et méthodes*, ed. by Laurent Dousset, Barbara Glowczewski, and Marie Salaün (Marseille: Pacific-Credo Publications, 2014), pp. 33–47.

——— and Erik Pearthree, eds, *Canoes of the Grand Ocean* (Oxford: BAR-Archeopress, 2008).

Dodd, Edwards, *Polynesian Seafaring. The Ring of Fire*, 2 vols (New York, NY: Dodd, Mead and Company, 1972).

Dodge, Ernest, *New England and the South Seas* (Cambridge, MA: Harvard University Press, 1965).

Doran, Edwin, *Wangka, Austronesian Canoe Origins* (College Station, TX: A&M University Press, 1981).

Duensing, Dawn E., *Hawai'i's Scenic Roads: Paving the Way for Tourism in the Islands* (Honolulu, HI: University of Hawaii Press, 2015).

———, 'Haleakalā Highway: Bringing the World to Maui', *The Journal of Pacific History* 44.3 (2009), 303–04.

Dumont d'Urville, Jules Sébastien César, 'Notice sur les îles du Grand Océan et sur l'origine des peuples qui les habitent', *Bulletin de la Société de Géographie*, 17 (1832), 1–21.

Durand, Marie, 'Les Lieux de l'innovation: gestes techniques et intégration des valeurs sociales dans les îles Banks au Vanuatu', in *Valeurs et matérialités*, ed. by Frédéric Keck (Paris: Presses de l'ENS, 2019), pp. 15–28.

———, 'Yams, Mobile Phones and Rice, About the Organisation of a Funerary Feast in the Banks Islands, Vanuatu' (unpublished paper presented at the 10th *ESfO Conference*, Brussels, 24–27 June 2015).

Dwyer, Peter, 'Etolo Traps: Techniques and Classification', *Memoirs of the Queensland Museum*, 27.2 (1989), 275–87.

Fache, Élodie, *Sur la piste des rangers aborigènes. Prendre soin de la savane tropicale australienne à l'ère néolibérale* (Marseille: Pacific-Credo Publications, 2019).

——— and Simonne Pauwels, *Fisheries in the Pacific. The Challenges of Governance and Sustainability* (Marseille: Pacific-Credo Publications, 2016).

Feinberg, Richard, *Polynesian Seafaring and Navigation: Ocean Travel in Anutan Culture and Society* (Kent, OH: Kent State University Press, 1988).

———, ed., *Seafaring in the Contemporary Pacific Islands* (DeKalb, Northern Illinois, IL: University Press, 1995).

Filer, Colin and Philippa Jenkins, 'Negotiating Community Support for Closure or Continuation of the Ok Tedi Mine in Papua New Guinea', in *Large-scale Mines and Local-level Politics. Between New Caledonia and Papua New Guinea*, ed. by Colin Filer and Pierre-Yves Le Meur (Canberra: ANU Press, 2017), pp. 229–59.

——— and Pierre-Yves Le Meur, eds, *Large-scale Mines and Local-level Politics. Between New Caledonia and Papua New Guinea* (Canberra: ANU Press, 2017).

Finney, Ben, *From Sea to Space* (Palmerston North: Massey University Press, 1992).

Finney, Ben, 'Nautical Cartography and Traditional Navigation in Oceania', in *The History of Cartography*, vol. II, book III, *Cartography in the Traditional African, American, Arctic, Australian, and Pacific Societies*ed, ed. by David Woodward and G. Malcolm Lewis (Chicago, IL and London: The University of Chicago Press, 1998), pp. 443–92.

Finney, Ben, ed., *Pacific Navigation and Voyaging* (Honolulu, HI: University of Hawaii Press, 1976).

Firth, Stewart, 'German Firms in the Western Pacific Islands, 1857–1914', *The Journal of Pacific History*, 8.1 (1973), 10–28.

———, 'The Nuclear Issue in the Pacific', *The Journal of Pacific History*, 21.4 (1986), 202–16.

Forster, Honore, *The South Sea Whaler:an Annotated Bibliography of Published Historical, Literary and Art Material Relating to Whaling in the Pacific Ocean in the Nineteenth Century* (Fairhaven, MA: The Kendall Whaling Museum and E. J. Lefkowicz, 1985).

Foster, Robert J. and Heather Horst, eds, *The Moral Economy of Mobile Phones* (Canberra: Australian National University Press, 2018).

Geismar, Haidy, 'Museum + Digital =?', in *Digital Anthropology*, ed. by Heather A. Horst and Daniel Miller (London: Berg, 2012), pp. 266–87.

Gell, Alfred, 'The Technology of Enchantment and the Enchantment of Technology', in *Anthropology, Art, and Aesthetics*, ed. by Anthony Coote and Anthony Shelton (Oxford: Oxford University Press, 1992), pp. 40–63.

Godelier, Maurice and José Garanger, 'Outils de pierre, outils d'acier chez les Baruya de Nouvelle-Guinée. Quelques données ethnographiques et quantitatives', *L'Homme*, 13.3 (1973), 187–220.

Graves, Adrian, *Cane and Labour: the Political Economy of the Sugar Industry 1862–1906* (Edinburgh: University of Edinburgh Press, 1993).

Guérin, Daniel, 'Tahiti malade de la bombe', *Les Temps Modernes*, 316 (1972), 788–811.

Guiot, Hélène, 'La Construction navale polynésienne traditionnelle. Dimension culturelle d'un processus technique', *Techniques & Culture*, 35–36 (2001), 445–78.

———, *Les Pirogues reflets de la Polynésie* (Paris: Société des Océanistes, 2003).

Gusterson, Hugh, *People of the Bomb: Portraits of America's Nuclear Complex* (Minneapolis, MN: University of Minnesota Press, 2004).

Guthrie, Karina, 'Aviation Regionalism in the Pacific', *The Journal of Pacific History*, 48.3 (2013), 29–308.

Haddon, Alfred C. and James Hornell, *Canoes of Oceania*, 3 vols (Honolulu, HI: Bernice P. Bishop Museum, 1975).

Hainsworth, David R., 'Exploiting the Pacific Frontier: the New South Wales Sealing Industry 1800–1821', *The Journal of Pacific History*, 2 (1967), 59–75.

Handman, Courtney, 'The Broad — and the Narrow — Cast: Christian Radio in the Making of Colonial Social Relations' (unpublished paper presented at the 12th *ESfO Conference*, Cambridge, 7–10 December 2018).

Haudricourt, André-Georges, 'Nature et culture dans la civilisation de l'igname: origine des clones et des clans', *L'Homme*, 4 (1964), 93–104.

Hau'ofa, Epeli, 'Our Sea of Island', *The Contemporary Pacific*, 6.1 (1994), 147–61.

Hermkens, Anna-Karina, *Engendering Objects: Dynamics of Barkcloth and Gender Among the Maisin of Papua New Guinea* (Leiden: Sidestone Press, 2013).

Hiquily, Tara *Va'a. La Pirogue polynésienne* (Pirae: Au vent des îles — Musée de Tahiti et des Îles – Te Fare Manaha, 2008).

Hobbis, Geoffrey, 'The MicroSDs of Solomon Islands: an Offline Remittance Economy of Digital Multi-Media', *Digital Culture & Society*, 3.2 (2017a), 21–40.

———, 'The Shifting Moralities of Mobile Phones in Lau Communicative Ecologies (Solomon Islands)', *Oceania*, 87.2 (2017b), 173–87.

Holmes, Tommy, *The Hawai'ian Canoe* (Hanalei, HI: Editions Limited, 1981).

Horowitz, Leah S., 'Toward a Viable Independence? The Koniambo Project and the Political Economy of Mining in New Caledonia', *The Contemporary Pacific*, 16.2 (2004), 287–319.

Horridge, Adrian, 'Origins and Relationships of Canoes and Rigs', in *Canoes of the Grand Ocean*, ed. by Anne Di Piazza and Erik Pearthree (Oxford: BAR-Archaeopress, 2008), pp. 85–105.

Howard, Michael, *Mining, Politics and Development in the South Pacific* (Boulder, CO: Westview Press, 1991).

Howe, Kerry R., 'Firearms and Indigenous Warfare. A Case Study'. *The Journal of Pacific History*, 9 (1974), 21–38.

Hyndman, David, 'A Sacred Mountain of Gold: the Creation of a Mining Resource Frontier in Papua New Guinea', *The Journal of Pacific History*, 29.2 (1994), 203–21.

Igler, David, *The Great Ocean: Pacific Worlds from Captain Cook to the Gold Rush* (Oxford and New York, NY: Oxford University Press, 2013).

Johnson, George, *The All Red Line: The Annals and Aims of the Pacific Cable Project* (Ottawa: James Hope and Sons, 1903).

Johnson, Lesley, *The Unseen Voice. A Cultural Study of Early Australian Radio* (London: Routledge, 1988).

Kirch, Patrick Vinton, *On the Road of the Winds. An Archaeological History of the Pacific Islands Before European Contact (Revised and Expanded Edition)* (Oakland, CA: University of California Press, 2017).

Kraemer, Daniela, '"Working the Mobile": Giving and Spending Phone Credit in Port Vila, Vanuatu', in *The Moral Economy of Mobile Phones*, ed. by Heather Horst and Robert Foster (Canberra: Australian National University Press, 2018), pp. 93–106.

Küchler, Susanne, *Malanggan: Art, Memory and Sacrifice* (Oxford: Berg, 2002).

——— and Graeme Were, 'Empathie avec la Matière', *Techniques & Cultures*, 52–53 (2009), 190–212.

Laborie, Léonard, '"Via Pacific". Télégraphie, impérialismes et mondialisation dans le Pacifique au XIXe siècle', *Hermès, La Revue*, 65.1 (2013), 132–36.

Lal, Padma Narsey, *Ganna: Portrait of the Fiji Sugar Industry* (Lautoka: Sugar Commission of Fiji, 2008).

Lal, Brij V., Doug Munro, and Edward D. Beechert, *Plantation Workers: Resistance and Accommodation* (Honolulu, HI: University of Hawaii Press, 1993).

Leach, James, 'Drum and Voice. Aesthetics and Social Process on the Rai Coast of Papua New Guinea', *Journal of the Royal Anthropological Institute*, 8 (2002), 713–34.

Leahy, Michael, *Explorations into Highland New Guinea, 1930–1935* (Tuscaloosa, AL: University of Alabama Press, 1991).

Lebot, Vincent, *Tropical Root and Tuber Crops* (Montpellier: CIRAD, 2009).

Lemonnier, Pierre, 'De la collecte d'objets à l'anthropologie des actions matérielles. Approches des objets et des techniques en Océanie', in *Les Sciences humaines et sociales dans le Pacifique Sud: terrains, questions et méthodes*, ed. by Laurent Dousset, Barbara Glowczewski, and Marie. Salaün (Marseille: PCP, 2014), pp. 89–109.

———, *Mundane Objects. Materiality and Non-verbal Communication* (Walnut Creek, CA: Left Coast Press, 2012).

Lewis, David, 'Expanding the Target in Indigenous Navigation', *The Journal of Pacific History*, 6 (1971), 83–95.

———, *We, The Navigators. The Ancient Art of Landfinding in the Pacific* (Honolulu, HI: University of Hawaii Press, 1972).

MacKenzie, Maureen, *Androgynous Objects: String Bags and Gender in Central New Guinea* (Amsterdam: Harwood Academic Publishers, 1991).

Malinowski, Bronislaw, *Argonauts of the Western Pacific* (London: Routledge & Kegan Paul, 1922).

Mann Spoehr, Florence, *White Falcon: the House of Godeffroy and Its Commercial and Scientific Role in the Pacific* (Palo Alto, CA: Pacific Books, 1963).

Matsuda, Matt K., *Pacific Worlds: A History of Seas, Peoples, and Cultures* (Cambridge: Cambridge University Press, 2012).

Maude, Henri E., *Slavers in Paradise: the Peruvian Slave Trade in Polynesia 1862–1864* (Stanford, CA: Stanford University Press, 1981).

——— and Ida Leeson, 'The Coconut Oil Trade of the Gilbert Islands', *The Journal of the Polynesian Society*, 74.4 (1965), 396–437.

Meltz, Renault and Alexis Vrignon (eds.), *Des bombes en Polynésie: les essais nucléaires français dans le Pacifique* (Paris: Vendémiaire, 2022).

Merle, Isabelle and Michel Naepels, 'Comme à la limite de la mer...', in *Les Rivages du temps. Histoire et Anthropologie dans le Pacifique*, ed. by Isabelle Merle and Michel Naepels (Paris: L'Harmattan, 2003), pp. 11–32.

Mills, Peter, 'Neo in Oceania: Foreign Vessels Owned by Hawaiian Chiefs before 1830', *The Journal of Pacific History*, 38.1 (2003), 53–67.

Mondragón, Carlos, 'Forest, Reef and Sea-Level Rise in North Vanuatu: Seasonal Environmental Practices and Climate Fluctuations in Island Melanesia', in *Indigenous Knowledge for Climate Change Assessment and Adaptation*, ed. by Douglas Nakashima, Igor Krupnik, and Jennifer T. Rubis (Cambridge: Cambridge University Press, 2018), pp. 23–40.

Moore, Clive, *Hardwork: Australian South Sea Islander Bibliography: With a Select Bibliography of the Sugar Industry, and the Pacific Labour Trade* (Pyrmont, NSW: Australian South Sea Islanders, 2019).

———, *New Guinea: Crossing Boundaries and History* (Honolulu, HI: University of Hawaii Press, 2003).

———, Jacqui Leckie, and Doug Munro, *Labour in the South Pacific* (Townsville: James Cook University, 1990).

Morgan, Lewis H., *Ancient Society* (Chicago, IL: Charles, Kerr & Company, 1877).

Moyal, Ann, *Clear Across Australia. A History of Telecommmunications* (Melbourne: Thomas Nelson, 1984).

Munn, Nancy, 'The Spatiotemporal Transformation of Gawa canoes', *Journal de la Société des Océaniste*, 33.54 (1977), 39–53.

Nelson, Hank, *Black, White and Gold: Goldmining in Papua New Guinea 1878–1930* (Canberra: Australian University Press, 2016).

Newbury, Colin, 'Mammon in Paradise: Economic Enterprise in Pacific Historiography', *The Journal of Imperial and Commonwealth History*, 26.2 (1998), 196–212.

Neyret, Jean, *Pirogues océaniennes*, 2 vols (Paris: Association des Amis du Musée de la Marine, 1974).

Panoff, Michel, 'The French Way in Plantation Systems', *The Journal of Pacific History*, 26.2 (1991), 206–12.

Pâris, François-Edmond, *Essai sur la construction navale des peuples extra-européens ou collection des navires et pirogues construits par les habitants de l'Asie, de la Malaisie, du Grand Océan et de l'Amérique...* (Paris: Arthus Bertrand, 1841).

Parsonson, Gordon S. 'The Literate Revolution in Polynesia', *The Journal of Pacific History*, 2 (1967), 39–57.

Pétrequin, Anne-Marie and Pierre Pétrequin, *Écologie d'un outil: la hache de pierre en Irian Jaya (Indonésie)* (Paris: CNRS Éditions, 1993).

———, and ———, *Objets et pouvoir en Nouvelle-Guinée. Catalogue de la donation Anne-Marie et Pierre Pétrequin* (Paris: Éditions de la Réunion des Musées Nationaux, 2006).

Pitt-Rivers, Augustus H. L. F., *The Evolution of Culture and Other Essays* (Oxford: Clarendon Press, 1906).

Pollock, Nancy, *These Roots Remain: Food Habits in Islands of the Central and Eastern Pacific Since Western Contact* (Laie: Institute for Polynesian Studies, 1992).

———, 'Nauru Phosphate History and the Resource Curse Narrative', *Journal de la Société des Océanistes*, 138–39. 1–2 (2014), 107–20.

———, 'Diversification of Foods and their Values: Pacific Foodscapes', in *Tides of Innovation in Oceania: Value, Materiality and Place*, ed. by Elisabeth Gnecchi-Ruscone and Anna Paini (Canberra: ANU Press, 2017), pp. 261–94.

Ponter, Brian, 'Co-operatives in Vanuatu, 1962–1980' (unpublished doctoral thesis, University of the South Pacific, 1983).

Purcell, David C., 'The Economics of Exploitation: the Japanese in the Mariana, Caroline and Marshall Islands 1915–1940', *The Journal of the Pacific History*, 11.3 (1976), pp. 189–211.

Quodling, Paul, *Bougainville, the Mine and the People* (Sydney: Centre for Independent Studies, 1991).

Ralston, Caroline, *Grass Huts and Warehouses. Pacific Beach Communities of the Nineteeth Century* (Canberra: Australian National University, 1977).

Rappaport, Roy A., *Pigs for the Ancestors: Ritual in the Ecology of a New Guinea People* (New Haven, CT: Yale University Press, 1968).

Regan, Anthony J., 'Bougainville: Origins of the Conflict, and Debating the Future of Large-scale Mining', in *Large-scale Mines and Local-level Politics. Between New Caledonia and Papua New Guinea*, ed. by Colin Filer and Pierre-Yves Le Meur (Canberra: ANU Press, 2017), pp. 353–414.

Regnault, Jean-Marc, *La Bombe française dans le Pacifique: l'implantation, 1957–1964* (Papeete: Scoop Ed., 1993).

Revolon, Sandra, 'L'Éclat des ombres. Contraste, iridescence et présence des morts aux îles Salomon'. *Techniques & Culture*, 58 (2012), 252–63.

Rieth, Éric, 'François-Edmond Pâris (1806–1893): aux origines de l'ethnographie nautique', in *Tous les bateaux du monde*, ed. by Eric Rieth (Grenoble: Chasse-marée, Glénat, Musée national de la Marine, 2010), pp. 10–23.

Rivers, William H., *The History of Melanesian Society* (Cambridge: Cambridge University Press, 1914).

Rogers, Robert F., *Destiny's Landfall: A History of Guam* (Honolulu, HI: University of Hawaii Press, 1995).

Russell-Smith, Jeremy, Peter Whitehead, and Peter Cooke, *Culture, Ecology and Economy of Fire Management in North Australian Savannas. Rekindling the Wurrk Tradition* (Collingwood: CSIRO Publishing, 2009).

Salesa, Damon, 'The World from Oceania', in *A Companion to World History*, ed. by Douglas Northrop (Chichester: Wiley-Blackwell 2012), pp. 392–404.

Salisbury, Richard F., *From Stone to Steel* (Melbourne: Melbourne University Press, 1962).

Scarr, Deryck, 'Recruits and Recruiters: A Portrait of the Pacific Islands labour Trade', *Journal of Pacific History*, 2.1 (1967), 5–24.

Schurig, Margarete, *Die Südseetöpferei* (Leipzig: Bruno Schindler, 1930).

Servy, Alice, '"As-tu un petit copain? Non je n'ai pas de téléphone". Moralité, progrès technique et sexualité en milieu urbain', *Hermès, la Revue*, 65.1 (2013), 137–43.

Sharp, Lauriston, 'Steel Axes for Stone-age Australians', *Human Organization*, 11.2 (1952), 17–22.

Shineberg, Dorothy, 'Guns and Men in Melanesia', *The Journal of Pacific History*, 6 (1971), 61–82.

———, *They Came for Sandalwood: a Study of the Sandalwood Trade in the South-west Pacific 1830–1865* (Melbourne: Melbourne University Press, 1967).

———, *The People Trade: Pacific Laborers and New Caledonia 1865–1930* (Honolulu, HI: University of Hawaii Press, 1999).

Sillitoe, Paul, *Made in Niugini. Technology in the Highlands of Papua New Guinea* (London: British Museum in association with theUniversity of Durham Publication Board, 1988).

Sinclair, James, *Wings of Gold. How the Airplane Developed New Guinea* (Bathurst: Pacific Publications, 1978).

Speiser, Felix, *Ethnographische Materialien aus den Neuen Hebriden und den Banks-Inseln* (Berlin: Kreidel Verlag, 1923).

Starbuck, A., *History of the American Whale Fishery From Its Earliest Inception to the Year 1876* (New York, NY: Argosy Antiquarian Ltd., 1964).

Starosielski, Nicole, *The Undersea Network* (Durham, NC: Duke University Press, 2015).

Stead, Victoria and Jon Altman, *Labour Lines and Colonial Power. Indigenous and Pacific Islander Mobility in Australia* (Canberra: ANU Press, 2019).

Steel, Frances, *Oceania under Steam: Sea Transports and the Cultures of Colonialism c. 1870–1914* (Manchester: Manchester University Press, 2011).

Stern, Monika, '"Mi wantem musik blong mi hemi blong evriwan" ["I want my music to be for everyone"]. Digital Developments, Copyright and Music Circulation in Port Vila, Vanuatu', *First Monday*, 19.10 (2014) <http://firstmonday.org/ojs/index.php/fm/article/view/5551/4130> [accessed 19 December 2020].

Taylor, John, 'Drinking Money and Pulling Women: Mobile Phone Talk, Gender and Agency in Vanuatu', *Anthropological Forum*, 25 (2015), 1–16.

Taylor, Peter, *An End to Silence. Building the Overland Telegraph Line from Adelaide to Darwin* (Sydney: Methuen of Australia, 1980).

Teaiwa, Katerina, 'Ruining Pacific Islands. Australia's Phosphate Imperialism', *Australian Historical Studies*, 46.3 (2015), 374–91.

Telban, Borut. and Daniela Vávrová, 'Ringing the Living and the Dead: Mobile Phones in a Sepik Society', *TAJA*, 25.2 (2014), 223–38.

Tench, Watkin, *Botany Bay. La fondation de l'Australie coloniale*, ed. by Isabelle Merle, trans. by Frédéric Cotton (Toulouse: Les éditions Anacharsis, 2006).

Te Rangi, Hiroa (Buck, Peter), *Maori Plaited Basketry and Plaitwork*, 2 vols (Wellington, NZ: Government Printer, 1923–1924).

———, *Samoan Material Culture*, special issue, *Bernice P. Bishop Museum Bulletin*, 75 (1930).

———, *Arts and Crafts of the Cook Islands*, special issue, *Bernice P. Bishop Museum Bulletin*, 179 (1944).

———, *Arts and Crafts of Hawaii*, special issue, *Bernice P. Bishop Museum Special Publication*, 45 (1957).

Thomas, Nicholas, *Entangled Objects. Exchange, Material Culture and Colonialism in the Pacific* (Cambridge, MA: Harvard University Press, 1991).

———, 'Partial Texts. Representation, Colonialism and Agency in Pacific History', *Journal of Pacific History*, 25.2 (1990), 139–58.

Thompson, Anne-Gabrielle, *John Higginson: un spéculateur aventurier à l'assaut du Pacifique Nouvelle-Calédonie, Nouvelles-Hébrides* (Paris: L'Harmattan, 2000).

Townsend, William H., 'Stone and Steel Tool Use in a New Guinea Society', *Ethnology*, 8.2 (1969), 199–205.

Vandeputte, Leslie, 'New Technologies and Language Shifting in Vanuatu', *Pragmatics*, 23.1 (2013), 171–81.

Ward, Gerard, *American Activities in the Central Pacific 1790–1870, 8 vols.* (Ridgewood, NJ: Gregg Press, 1966–1967).

———, 'Earth's Empty Quarter? The Pacific Islands in a Pacific Century', *Geographical Journal*, 155.2 (1989), 235–46.

——— and Andrew Proctor, *South Pacific Agriculture: Choices and Constraints. South Pacific Agricultural Survey 1979* (Canberra: The Australian National University and The Asian Development Bank, 1980).

Weeramantry, Christopher, *Nauru, Environmental Damage under International Trusteeship* (Oxford: Oxford University Press, 1992).

Weightman, Barry, *Agriculture in Vanuatu. A Historical Review* (Port-Vila: The British Friends of Vanuatu, 1989).

Were, Graeme, 'On the Materials of Mats: Thinking Through Design in a Melanesian Society', *Journal of the Royal Anthropological Institute*, 19 (2013), 581–99.

Whitehead, Peter J., David M.J.S. Bowman, Noel Preece, Fiona Fraser and Peter Cooke, 'Customary Use of Fire by Indigenous Peoples in Northern Australia: Its Contemporary Role in Savanna Management', *International Journal of Wildland Fire*, 12.4 (2003), 415–25.

Williams, Maslyn and Barrie MacDonald, *The Phosphateers: A History of the British Phosphate Commissionners and the Christmas Island Phosphate Commission* (Melbourne: Melbourne University Press, 1985).

Winseck, Dwayne and Robert Pike, *Communication and Empire: Media, Markets, and Globalization, 1860–1930* (Durham, NC: Duke University Press, 2007).

Wolfers, Edward, *Race Relations and Colonial Rule in Papua New Guinea* (Sydney: Australia and New Zealand Book Co., 1975).

Woronoff, Denis, *Histoire de l'industrie en France du XVIe siècle à nos jours* (Paris: Seuil, 1994).

Yamaguchi, Toru Kayanne, Hajime,Yamano, Hiroya, Najima, Yayoi, Chikamori, Masashi, and Hiromune Yokoki, 'Excavation of Pit Agriculture Landscape on Majuro Atoll, Marshall Islands, and Its Implication', *Global Environmental Research*, 9.1 (2005), 27–36.

Young, John, 'Sailing to Levuka: the Cultural Significance of the Island Schooners in the Late 19th Century', *The Journal of Pacific History*, 28.1 (1993), 36–52.

ELISABETH MORTIER

The Middle East in the Nineteenth and Twentieth Centuries

At the Crossroads of Technological Continuity and Sudden Western Modernity

Today, the Middle East includes many countries with different religions and ethnic groups: Egypt, Turkey, Iraq, Syria, Lebanon, Israel and the West Bank, Iran, and every country of the Arabian Peninsula. The expression 'Middle East' was created at the beginning of the twentieth century, by Alfred Mahan, a naval officer from the United States, and popularized by the British journalist Valentine Chirol (Crouzet, 2016). They used this new expression to refer to the area located to the west of British India, which they perceived only through the lens of British strategic interests. The populations of the nineteenth-century Middle East lived under the rule of two Empires: the Ottoman Empire and the Persian Empire. Due to the fall of the Ottoman Empire at the end of the First World War and to the major upheavals caused by the Second World War, the political and social map of the Middle East changed completely throughout the twentieth century. During the first half of the twentieth century, new countries and societies emerged, such as Protectorates over Palestine, Syria, Lebanon, and Iraq during the inter-war period, the Republic of Turkey in 1923, and the Kingdom of Saudi Arabia in 1932. Since the end of the Second World War in 1945 and the foundation of the State of Israel in 1948, Middle Eastern societies have been transformed by several wars and revolutions.

During the nineteenth and twentieth centuries, the history of technology in this part of the world was deeply connected to its political and social contexts. The Middle East was a place of technological encounters and transfers between the western world and multi-confessional and multi-ethnic countries. Technology and know-how relating to, among other things, transport, architecture, city planning, and craft industries, were developed over several centuries. Throughout the nineteenth century, the political, social, and technological influence of European countries grew stronger in both Empires. According to Quataert, the Ottoman Empire entered 'the processes of westernization and modernization' in order to gain the ability to compete against Europe (Quataert, 1992). European technology infiltrated every aspect of daily life and public affairs, including railways, printing, the telegraph, the army, power production, the industry, agriculture, and architecture. In the Ottoman

Global History of Techniques, ed. by Guillaume Carnino, Liliane Hilaire-Pérez and Jérôme Lamy, Global Matters Culture Matérielles, histoire des techniques et circulation globale des savoirs, 9 (Turnhout: Brepols, 2024), pp. 69–82
BREPOLS PUBLISHERS 10.1484/M.TECHNE-EB.5.129801

Empire, these technological transformations took place during the Tanzimat, i.e. the period from 1839 to 1876, which experienced radical reforms that led to the modernization of the Ottoman society based on a European model (Mantran, 2010). The transfer of Western technology to the Persian Empire did not follow the same path and pace as it had done in the neighbouring empire. The Shahs of the Qajar dynasty, who ruled Iran until the 1920s, sought advice from European experts for the modernization of the Empire's administration and army (Djalili, Kellner, 2017). However, the spread of European technology was unevenly paced until the reign of Reza Shah, who seized power in 1925 (Cronin, 2003).

The rise of European imperialism from the second half of the nineteenth century onwards and the establishment of colonial rule in the Middle East abruptly reshaped the economic structures and production processes by targeting a foreign market. The technological history of the Middle East in the nineteenth and twentieth centuries fluctuated between the continuation of local techniques, their adjustment to new economic and social contexts, and bursts of European technological modernity. These technological transformations impacted the working structures and daily lives of the region's inhabitants. For instance, at the turn of the twentieth century, the introduction of cigarette rolling machines to Egypt led to social struggles in this industry (Beinin, 2001). The hydraulic works built in Egypt under British colonial rule — most notably the Aswan Low Dam on the Nile, which was inaugurated in 1902 — forced Egyptian farmers to use and rely on chemical fertilizers for growing their crops. During the Second World War, the depletion of chemical fertilizers led to severe food shortages that affected both the British troops and the Egyptian population (Mitchell, 2002).

The intrusion / introduction of European technology in the Middle East must be analysed on a more global scale, i.e., in the context of the expansion of imperialism on every continent (Asia, Africa, America, and Oceania) in the nineteenth and twentieth centuries (Headrick, 1988). In the Middle East, it is necessary to define the different ways, pace, and forms taken by the implementation of European technology. Moreover, the local responses and adjustments to the transfer of technology in the context of imperialism must be studied.

There are no scholarly studies of the technique and craft practices of the whole Middle East in the nineteenth and twentieth centuries. Indeed, the technical history of different regions of the modern Middle East is a little developed historical field. This article follows a thematic approach to analyse the permanent and changing features of technology in the Middle East.

Production Techniques

In the nineteenth and twentieth centuries, the imperialistic discourse idealized Western modernity that was to be brought to Middle Eastern societies. A very diverse range of craft industries shaped these societies and survives to the present day. However, some of the traditional craft techniques were transformed due to the opening of the economic market to the West. The importance of the changes undergone by craft techniques can be grasped, for example, through the histories of the glass and soap industries, the traditional pearl hunting in the Gulf of Persia, and the textile industry. Moreover, at the beginning of the twentieth century, the discovery of oil and its production by Western countries profoundly changed the economy and strategic interests of the Middle Eastern countries.

Craft Techniques and Traditional Manufacture

The production of glass dates back to 4000 BC. Glass is made from two main ingredients: silica (sand, sandstone) and soda ash or potash. Glass blowing dates back to Roman times. Previously, Syria and Egypt were known for their high-quality glass art. In the nineteenth and twentieth centuries, glass factories used some industrial methodes due to improvements in fusion techniques and chemistry. However, Damascus was still an important centre for the production of glass (Imam, Issa-Abyad, Métral, Naffakh, 1974). At the time, two types of glass workshops existed in Damascus: those without technical glass blowing tools and those who were 'semi-mechanized', where the work was done with the help of air compressors by three workers on an assembly line. Traditional glasswork buildings had a roof with wooden beams that was usually raised on pillars for ventilation, while more recent buildings have roofs made from non-flammable materials such as concrete or zinc (Imam, Issa-Abyad, Métral, Naffakh, 1974). Since the twentieth century, several Middle Eastern countries adopted the industrial production of glassware due to the low cost of energy and raw materials. For instance, in 2012, the French company Saverglass opened a factory in Ras al-Khaimah in the United Arab Emirates (Eglof, 2013 to produce luxury glass bottles.

Soap factories in Syria and Palestine, especially in the cities of Aleppo and Nablus, developed products using technological methods that local raw materials. For example, Nabulsi soap is made from olive oil, water, and soda ash extracted from the barilla plant of the Jordan river (Doumani, 1995). Aleppo soap adds the oil of the laurel berry to the recipe. Each Ottoman soap factory in Nablus comprised three levels. In the basement, the olive oil was stored in wells. On the ground floor, the soap was boiled in big vats. The boiled soap was thenspread out on the *mafrash*, a flat surface, where it dried thanks to cold air let in through large windows (Bontemps, 2009). The soap was cut into pieces on the first floor of Nabulsi soap factories, whereas the twenty-first-century soap factories of Aleppo do this on the ground floor.

The practice of pearl hunting in the Middle East dates back to the Neolithic period, and it was still pracised in the twentieth century (Piquet, 2013). The position of the region protects it from water currents, so that there are two areas suitable for pearl hunting: the Persian Gulf and the Red Sea. From 1850 to 1930, Bahrain Island was the first provider of natural pearls in the world. El-Kueit and El-Bidas were also two important pearl hunting areas at the time. Across the nineteenth century, almost 4000 ships and 30,000 individuals worked in pearl hunting (Cariño, Monteforte,

2006). In the Persian Gulf, the most popular type of oyster for pearl culture was *pinctada margaritifera erythraensis*.

Pearl hunting was less prevalent in the Dead Sea than in the Persian Gulf. Here, the pearl-producing oyster type was the *sadof* oyster. The fishing techniques in both areas were rather similar. In the Persian Gulf, tools and ships were funded by the *bunnias*, Hindu merchants who traded pearls in India. Divers, called *ghoas* or *sebs*, worked on ships led by a *nakhoda*, the captain of the crew. Divers wore a loincloth and finger protection made from leather known as *khubaat*. Almost forty times each day they dived and reached depths of up to ten metres. Each diver did so in an upright position with the help of a heavy stone that was tied to one foot with a rope, and a basket to collect the oysters was attached to the other foot as well as tied to the ship. Once the oysters had been collected, workers opened them and retrieved the precious pearls hidden in the flesh of the mollusks (Cariño, Monteforte, 2006). In the Middle East, pearl hunting techniques remained largely unchanged from Antiquity to the twentieth century. Pearling in the Middle East declined at the beginning of the twentieth century due to the popularity of oil production and the Japanese competition.

Carpets and Textile Manufacturing

Ottoman and Iranian carpets are the showpiece of Middle East craft techniques. The most ancient carpets known today were made in thirteenth-century Anatolia, whence the carpet-making skills spread across the Middle East and the world. The main materials used for oriental carpets were sheep's wool, and sometimes even silk and mohair (Quataert, 1986). During the nineteenth century, the demand for oriental carpets woven in the Ottoman and Persian Empires rose exponentially in Europe. This new market transformed the organization of carpet factories and the techniques in use.

In the Ottoman Empire, a large part of the commercially produced rugs was made in Western Anatolian villages. Oriental knotted carpets were no longer luxury products that only the upper classes in Europe could afford but accessible to a larger European clientele. The wool was cleaned, spun, and dyed by hand before being woven or knotted on traditional looms. At the end of the nineteenth century, steam-powered wool spinning factories were introduced in Western Anatolian villages (Quataert, 1986).

In Persia, carpets had been woven since the Middle Ages by nomadic peoples and villagers, but their export to Europe increased considerably during the second half of the nineteenth century (Seyf, 1990). Therefore, industrial-type carpet workshops were created in large cities such as Tabriz and Sultanabad at the end of the nineteenth century. Nevertheless, carpets were still largely woven by villagers and nomads in rural areas. Carpets were mainly produced by hand, which took a long time: one to two years for a high-quality carpet made by three or four workers weaving every day. The weavers used handlooms and spun the bulk of the yarn by hand. There were also workshops with ten to twenty looms. All weavers worked under the supervision of a master weaver, the *ustad*. The carpet pattern was traditionally known by the *ustad* and his weavers through chants, but in the nineteenth century weavers began to weave carpets from drawn patterns (Seyf, 1990). Additionally, in both Empires, the traditional dyes used were vegetal and animal, but due to the increasing European demand, weavers began to use synthetic dyes, which had been developed since 1856.

For centuries, Middle Eastern countries cultivated raw materials for textile production such as cotton and sheep's wool. Since the nineteenth century, the British government encouraged Egypt to focus its agricultural efforts on cotton production. The production of Egyptian cotton was mainly intended for export to European spinning factories (Marchal, 1956). In 1840, fabrics from Europe were the most prevalent on the Middle East market. Nevertheless, some traditional fabric manufacturing has remained in place, for example *ikat* textiles fabrics from Damascus and Aleppo (Chevallier, 1982). The dying technique was specific to *ikat* textiles: one part of the yarn is dyed before being woven, while the other retains its natural colour. However, the yarn was imported not from neighbouring countries, but from Manchester. Since the twentieth century, the Republic of Turkey gradually became a central country for the industrial production of textiles and clothing by adopting European and North American techniques.

Sericulture in the Middle East dates back to the medieval period. Silk production was particularly developed in Persia, Syria, Bursa (Hitzel, 2008), and Anatolia. Silk is made from the cocoon of the silkworm. The worm feeds on the leaves of the mulberry trees that grow on the mountain slopes of these regions. The cocoons are emptied at high temperature. Then, the fibres of the silkworms were spun into silk (Groffier, 1900). In the nineteenth century, the increase in the European demand for silk and the transformation of transport (e.g. railways) changed the silk culture in these regions. In the second half of the nineteenth century, the silk merchants from Lyon took an interest in Syrian silk. They were looking for good quality silk at a low price. Further, thanks to the development of steamboats, the exchange of cocoons and silk was safer and faster than before (Chevallier, 1982). The Syrian silkworm species disappeared between 1853 and 1873 due to diseases imported from Europe. From then on, a European species was gradually introduced.

Oil Production

Thanks to certain geological factors, Western Asia benefits from rich petroleum deposits in its soils. Mechanical drilling for oil was first developed in the United States in 1859 (Longrigg, 1961). Nineteenth-century developments in geology advanced research and petroleum exploration all over the world. The Sublime Porte knew of the importance of petroleum, as shown by the Armenian businessman Calouste Serkis Gulbenkian's report on the oil industry's potentialities which lead to the foundation of the Turkish Petroleum Company in 1912. The rise of the petroleum industry in the Middle East was connected to the major changes in shipping and production needs on a global scale. This is why European countries were interested in the exploitation of Middle Eastern oil resources. Experimental oil drilling took place at the beginning of the twentieth century. For example, William Knox d'Arcy (an Englishman) and A. L. Marriot (a Frenchman) were granted a concession by the Shah to extract petroleum from Persia at a time where the existence of these resources was merely speculative (Longrigg, 1961). However, in January 1908 the first oil deposits were discovered in vast quantities at Masjid-i-Sulaiman, in Persia. The British government decided to acquire a fifty-one per cent interest in the Anglo-Iranian Company. Little by little, European countries invested financially in Middle Eastern oil companies, and petroleum was also found in Iraq, Saudi Arabia, Kuwait, and in the Trucial States (which became the United Arab Emirates in 1971). The drilling techniques and equipment came from Europe in the early days of Middle Eastern petroleum exploitation. The drilling system used in the oil fields was rotary drilling, i.e. drilling the ground with a rotating head. Liquids, e.g. mud or water, are poured over the device that pierces the ground to cool it. Tubes and concrete are inserted into the borehole to maintain the well walls.

Urbanism and Communication in the Middle East

Urban planning and communication form both the framework for daily life and the means of exchange for the inhabitants of the Middle East. Many scientific works, especially in the fields of architecture and urbanism, have studied these areas. The Western countries tried to impose their views on city planning and exported to the Middle East new means of communication to reinforce their influence on Middle Eastern countries. Therefore, printing, photography, telegraphy, television, and the internet were developed further in Middle Eastern societies from the nineteenth century onwards.

Urbanism

As Chevallier explains, the Arab and Ottoman cities of the mid-nineteenth century shared common characteristics: they were organised around mosques, churches, the *khan* (i.e. a place for trade and hosting, Lamprakos, 2017), a *souk* (or *bazar* in Constantinople), courtyard houses, and fortresses enclosed by a wall (Chevallier, 1982). Due to the growing influence of European urban engineers during the nineteenth century, Middle Eastern cities gradually transformed to meet hygienic criteria and the demands of population growth, among other things. For example, during the first half of the nineteenth century, the khedive Muhammad Ali called on the expertise of European engineering to develop sanitary drainage plans in Cairo and to widen the streets. The Tanzimat reforms in the Ottoman Empire also tackled urban issues such as diseases and unsanitary conditions (Lafi, 2005).

Some cities in the Middle East were created to serve the needs of the population as well as to facilitate large-scale projects. This was the case for the city of Ismailia, located on the Isthmus of Suez, built as a support base for the Suez Canal construction site. The city's construction was launched in April 1862 and followed a utopian urban plan, of which the French engineers Bernard de Montaut, Jean-François Viller, and François-Philippe Voisin were in charge. All three were graduates of the École nationale des ponts et chaussées (French national school of civil engineering). They designed an open city, without walls, following orthogonal and diagonal lines. In the 1860s, its map was a landmark in the history of urban planning. Ultimately, a water channel encircled the city and formed a kind of enclosure around Ismailia (Montel, 1999). The city of Heliopolis, created on formerly plain ground ten kilometres northeast of Cairo, is another example of European urban planners' influence in Egypt. Connected to Cairo by two tramways lines and one electrified train line, the city was designed by Baron Edouard Empain and the Cairo Electric Railway and Heliopolis Oases Company. The city was built according to the principle of urban zoning: specific spaces were designed for each economic, social, and political function (Minnaert, Volait, 2003).

The imposition of French protectorates in the former Arab provinces of the Ottoman Empire in 1920 led the mandatory powers to consider the implementation of urban development plans, particularly in Beirut, Aleppo, and Damascus. In Beirut, the French government entrusted the 'Cabinet Danger', a French surveyors' office founded by Paul, Raymond and René Danger, with the responsibility of designing a hygiene plan for restructuring the city. The Danger urban plan was to subdivide the city according to its various intended purposes, including industries, housing, and the centre of political power (Ghorayeb, 1994).

The petroleum economy profoundly changed the landscape of urbanism in the Emirates in the second half of the twentieth century. The city of Dubai exemplifies these drastic architectural and urbanistic changes. In the 1950s and 1960s, Sheikh Rashid bin Saeed Al Maktoum decided to launch an urbanistic plan to transform the small village of Dubai into a city that would embrace the future economic and demographic growth of the Emirate (Katodrytis, Mitchell, 2015). The British architect John Harris drew the first masterplan for Dubai in 1959 and revised it in 1979 according to the city's needs. He planned a grid to organize a system of roads and infrastructures around the creek of Dubai. This plan encouraged the building of skyscrapers and led to the formation of very dense urban areas. In 1979, the Dubai World Trade Centre, designed by John Harris, was achieved, and it was the tallest building in Dubai for two decades.

Moreover, from the second half of the twentieth century onwards, several innovative architectural projects that combined traditional and new construction techniques were developed in the Middle East. As example, the King Abdul Aziz Historical Centre in Riyadh is an architectural complex that cost 182 million dollars and was inaugurated in January 1999. Designed by the Jordanian architect Rasem Badran and the Saudi Arabian architect Ali Shuaibi, it combines a palace, residences for the royal family, and administrative buildings. The architects used traditional building techniques from the Arabian Peninsula, such as sun-dried mud bricks and roofs made with tamarisk tree trunks covered with mud (Al-Asad, 2012).

Communication Techniques

This section will focus on development of the main communication tools of the nineteenth to twenty-first centuries, i.e. printing, photography, the telegraph, television, and finally the Internet in different parts of the Middle East. For the most part, these communication techniques do not come from the Middle East but the West. These techniques were adapted and suited to the needs of its local users.

The first Arabic type in a Middle Eastern country was created in Istanbul in Constantinople in 1728. It was only in 1798, during Napoleon's campaign, that the first Western printing house in the Middle East was introduced to Egypt. Egypt became the centre of book printing in the Middle East with the opening of the Bûlâq press in Cairo in 1822. The *Nahda* movement then accelerated the spread of book printing and the printing press in Middle Eastern countries during the nineteenth century (Sadgrove, 2004). At the same time, printing techniques in Europe changed substantially with the introduction of the steam-powered press and the creation of William Bullock's rotary machine press in the 1860s. These innovations gradually spread to the Middle East.

During the nineteenth century, various photographic processes were developed, and these new techniques aroused great interest in the Ottoman Empire. Newspaper articles reported on the daguerreotype, and books on photographic techniques were translated and circulated in the Empire (Beaugé, 1992). The daguerreotype is a photographic process that uses a silver-coated copper plate exposed to light. The silver plate with silver iodide produced an image. The exposure time was less than thirty minutes. During the first half of the nineteenth century, the photographic technique was introduced to the Ottoman and Persian Empires, particularly thanks to European travel photographerslike Fredéric Goupil Fesquet and Francis Frith. They mostly photographed biblical landscapes. Travelling photographers had to be resourceful in establishing dark rooms and using cameras in extreme weather conditions.

Some European photographers settled permanently in the Ottoman Empire to open photography studios. Among them was Félix Bonfils, a former bookbinder, who first opened a photography studio in Beirut, and then in Cairo and Alexandria (Woodward, 2003). Gradually, European photographers worked alongside local photographers such as Reif Efendi, who produced daguerreotypes, and the Abdallah brothers, who were official photographers of the Sultan of the Ottoman Empire in Constantinople (Beaugé, 1992).

The development of the electric telegraph around the world profoundly shortened communication times between one country and another. At the beginning of the nineteenth century, it took between five and eight months to send a letter from Great Britain to India (Headrick, 1988). Telegraph technology was considered a political tool that would strengthen imperial power and allow a more effective response to any problem or rebellion. The construction of telegraph lines that crossed the Middle East both undersea and overland was of the utmost strategic importance for the British government in order to communicate more efficiently with India. In the 1840s, the discovery of gutta-percha, natural latex from the Palaquium gutta tree, changed telegraph technology and allowed the development of submarine telegraph lines. Indeed, gutta-percha was a good isolator for the protection of copper cables. In 1848, Lieutenant Werner von Siemens, a Prussian artillery officer, invented a machine to coat cables with this natural latex. During the second half of the nineteenth century, the British government negotiated with the Ottoman Empire and the Egyptian government to obtain a concession to build telegraph lines that would join India to Great Britain. In 1856, a concession was granted by the Ottoman Empire and Egypt to Lionel and Francis Gisborne, the founders of the Red Sea and India Telegraph. They wanted to build a submarine telegraph line from Suez to Aden. This project was a failure due to technical issues: the cable was damaged because of improper copper wire insulation (Ahvenainen, 2011). This technical setback meant that the first effective electric telegraph lines in the Middle East were land lines. In 1858, a telegraph line connected Constantinople to Baghdad. The development of the telegraph in the Middle East raised the question of the sovereignty of the Ottoman and Persian Empires regarding the means of communication built on their lands. The Ottoman Empire used the British strategic interest in building telegraph lines on its lands as a means of developing communication in the Empire (Headrick, 1988).

In the second half of the nineteenth century, the expansion of television around the world also affected the countries of the Middle East. Hertzian and satellite television channels were created. Television has had an important impact on the political issues of Middle Eastern states (Guayybess, 2013). The pan-Arab channel Al Jazeera, which was launched in Qatar in 1996 (Gonzalez-Quijano, Guaaybess, 2009), is one example of the profusion of satellite channels targeting a multinational audience in the 1990s. Moreover, the internet spread in the Middle Eastern countries at the turn of the twenty-first century. Access to social networks such as Twitter or Facebook played a significant role in the events of the Arab Spring of 2010 and 2011.

Transport Technology

The nineteenth and twentieth centuries have undoubtedly seen a transport revolution (Bagwell, 1988). Sea, land, and air routes have been considerably expanded while transport times have decreased significantly. Above all, this transport revolution is a technological revolution with the development of steamboats, cars, and aircraft. Middle Eastern countries largely took part in the transport revolution. These transformations are related to the Western need to connect the different parts of the Middle East with each other to establish faster trade exchanges.

Land Transport

Camels and donkeys were among the oldest means of transport in most countries of the Middle East (Bulliet, 1975). During the nineteenth and twentieth centuries, the influence of European countries and trade led to the development of train and automobile transportation.

The railway was introduced in the Ottoman Empire in the second half of the nineteenth century for economic and strategic purposes. In 1856, the Ottoman Empire granted Europeans the first concession for building a railway line of 137 kilometres, from Izmir to Aydin in Anatolia. The same year, the British inaugurated a railway between Alexandria and Cairo (Özyüksel, 2014). Railway technology was then gradually transferred from Europe to the Middle Eastern states. For example, in 1869, 1338 kilometres of railways covered Egypt. This figure increased rapidly, to 4314 kilometres on the brink of the First World War (Issawi, 1982). Some parts of the Middle East developed the railway at a slower pace: Iraq only had 1320 kilometres of railways in 1914, whereas Iran did not have any railway network at all until 1928. The railways were developed to transport raw materials from the hinterland to the harbours.

However, the building of railways in the Middle East presented several technical challenges, such as the laying of the rails and the crossing of mountains and rivers.For example, the Hejaz railway, which ran from Damascus to Medina, was built from 1900 to 1908 to connect the furthestmost regions of the Ottoman Empire. This line also served to highlight Abdülhamid II's Pan-Islamist policy. The line was 1320 kilometres long and designed by engineers from Europe. According to Murat Özyüksel, in 1904, there were twelve Germans among the engineers working on the railway construction. Frenchmen, Austrians, Italians, and Belgians also took part in the Hedjaz railway project. This team of engineers was led by August Meissner, a German engineer. The wood for the 1,382,000 traverses was cut in the Ottoman forests, and one part of the iron was produced in the Ottoman foundry, while the other part — 10,000 irons bars — was imported from abroad. At first, the Naval Arsenal on the Golden Horn produced the rails, but the quality was so poor that it was decided to import manufactured rail from Belgium, Britain, and the USA. The construction of some sections of the railway faced some important technical issues due to the topography of the land crossed. In the section from Haifa to Dera, the Yarmuk river created several technical problems: eight tunnels, eighty-three bridges, and 246 aqueducts were built to enable the train to cross the plain (Özyüksel, 2014).

Maritime Transport

At the end of the eighteenth century, the Ottoman ships (125 tons) in the port of Alexandria were smaller than the European ships (164 tons) (Issawi, 1982). In the course of the nineteenth century, the invention of the steamship gradually changed maritime transport. As Daniel Headrick explains, four innovations which were developed further in the 1850s and the 1860s reduced the cost of the steamer: the screw propeller, the iron bull, the surface condenser, and the compound engine (Headrick, 1988). The advantages of iron boats over wooden boats lie in their lighter weight and their bigger capacity for transport. The construction of the Suez Canal was an essential part of the changes which occurred in shipping in the nineteenth and twentieth centuries. This 164-kilometre-long canal that pierced the Isthmus of Suez changed the distances of shipping routes. Inaugurated in 1869, the Canal reduced the distance between London and Bombay by forty-one per cent (11,619 kilometres through the Canal and 19,755 kilometres via the Cape route). In November 1854, Ferdinand de Lesseps founded the Compagnie universelle du canal maritime de Suez (Universal Company of the Maritime Canal of Suez). The construction work was monumental, with hundreds of workers digging the Canal by hand and loading wagons with displaced soil (Montel, 1999). The Suez Canal placed the Middle East at the heart of the world's shipping routes for a long time. When it opened, the Canal was six metres deep and twenty-two metres wide at the bottom. Soon it was too small for steamboats, and several projects to widen the canal were carried out. In 1924, the Suez Canal was forty-five metres wide at the bottom, seventy-three metres wide at the top, and 10.5 metres deep (Headrick, 1988).

Air Transport

Aviation experienced significant technical progress during the Second World War, allowing the development of air freight. The countries of the Middle East contributed to this development by building airport infrastructure including the Beirut airport, which opened in the 1950s with two asphalt runways (Butt, 2011). Middle Eastern countries also have national and private airlines such as Saudia in Saudi Arabia (which was created in 1945), Emirates in the United Arab Emirates (1985), and El Al in Israel (1948). Arab countries, particularly the United Arab Emirates, have emerged as key hubs for global air traffic. This is the case for Dubai's Al Maktoum International Airport, built in the middle of the desert from 2010 onwards, which has established itself as the third largest airport in the world in terms of passenger traffic.

Traditional craft production, transport, and communication techniques in Middle Eastern countries were developed over the course of several centuries. In the nineteenth century, the European influence accelerated both technological transformations and the industrial revolution in this region of the world. These transformations have led to a technological homogenization and to a greater connection between the Middle East and other countries of the world.

References

Abdulla, Rasha A., *The Internet in the Arab World: Egypt and Beyond* (New York, NY: Peter Lang, 2007).

Ahvenainen, Jorma, *The History of the Near Eastern Telegraphs before the First World War* (Helsinki: Finnish Academy of Science and Letters, 2011).

Al-Asad, Mohammad, *Contemporary Architecture and Urbanism in the Middle East* (Gainsville, FL: University Press of Florida, 2012).

Aslanapa, Oktay, *One Thousand Years of Turkish Carpets* (Istanbul: Eren, 1988).

Atasoy, Nurhan and Julian Raby, *Isnik: the Pottery of Ottoman Turkey* (London: Alexandria Press, 1989).

Bagwell, Philip, *The Transport Revolution* (London: Routledge, 1988).

Beaugé, Gilbert, 'Les Photographes turcs: jalons pour une histoire de la photographie en Turquie', *Cahier d'études sur la Méditerranée orientale et le monde turco-iranien*, 14 (1992), 113–32.

Beinin, Joel, *Workers and Peasants in the Modern Middle East* (Cambridge: Cambridge University Press, 2001).

Bontemps, Véronique, 'Naplouse, le savon et la ville. Patrimoine familial, travail ouvrier et mémoire au quotidien' (unpublished doctoral thesis, Université de Provence/Aix-Marseille-I, 2009).

Bourgey, André, *Industrialisation et changements sociaux dans l'Orient arabe* (Beyrouth and Lyon: CERMOC and Presses Universitaires de Lyon, 1982).

Bruant, Catherine, Sylviane Leprun, and Mercedes Volait, 'Figures de l'orientalisme en architecture', eds., *Revue des mondes musulmans et de la Méditerranée*, 73–74 (1996), 11–391.

Bulliet, Richard, *The Camel and the Wheel* (Cambridge, MA: Harvard University Press, 1975).

Butt, Gerald, *History in the Arab Skies: Aviation's Impact on the Middle East* (Nicosia: Rimal Publication, 2011).

Cariño, Micheline and Mario Monteforte, *Une histoire mondiale des perles et des nacres: pêche, culture, commerce* (Paris: L'Harmattan, 2006).

Chevallier, Dominique, *Villes et travail en Syrie, du* XIXe *au* XIXe *siècle* (Paris: Maisonneuve et Larose, 1982).

Cronin, Stephanie, ed., *The Making of Modern Iran: State and Society Under Riza Shah, 1921–1941* (London: Routledge, 2003).

Crouzet, Guillemette, 'Les Britanniques et l'invention du Moyen-Orient: essai sur des géographies plurielles', *Esprit*, 5 (2016), 31–46.

Deguilhem, Randi and Suraiya Faroqhi, eds, *Crafts and Craftsmen of the Middle East: Fashioning the Individual in the Muslim Mediterranean* (London: I. B. Tauris, 2005).

Djalili, Mohammad-Reza and Thierry Kellner, *Histoire de l'Iran contemporain* (Paris: La Découverte, 2017).

Donellan, Craig, *The Transport Revolution* (Cambridge: Independence Educational Publishers, 1996).

Doumani, Beshara, *Rediscovering Palestine: Merchants and Peasants in Jabal Nablus, 1700–1900* (Berkeley, CA: University of California Press, 1995).

Eglof, Emmanuel, 'Le Verrier Saverglass investit massivement au Moyen-Orient', *Le Figaro* (14 April 2013).

Elsheshtawy, Yasser, *Dubai: Behind an Urban Spectacle* (New York, NY: Routledge, 2010).

Faroqhi, Suraiya, *Artisans of Empire: Crafts and Craftspeople under the Ottomans* (London: I. B. Tauris, 2012).

Fontaine, Patrice, *Le Tapis persan ou le jardin de l'éternel printemps* (Paris: Éditions Recherche sur les civilisations, 1990).

Gharipour, Mohammad, *Contemporary Urban Landscapes of the Middle East* (New York, NY: Routledge, 2016).

Ghorayeb, Marlène, 'L'Urbanisme de la ville de Beyrouth sous le mandat français', *Revue du monde musulman et de la Méditerranée*, 73–74 (1994), 327–37.

Goodwin, Godfrey, *A History of Ottoman Architecture* (London: Thames and Hudson, 1971).

Gonzalez-Quijano, Yves, Guaaybess, Tourya (ed.), *Les Arabes parlent aux Arabes. La révolution de l'information dans le monde arabe* (Arles, Sindbad, Actes Sud, 2009).

Groffier, Valérien 'La Production de la soie dans le monde', *Annales de Géographie*, 44 (1900), 97–118.

Guayybess, Tourya, *Télévisions arabes sur orbite. Un système médiatique en mutation (1960–2004)* (Paris: CNRS Éditions, 2013).

Headrick, Daniel, *The Tentacles of Progress. Technology in the Age of Imperialism, 1850–1940* (New York, NY and Oxford: Oxford University Press, 1988).

Herbert, Gilbert and Silvina Sosnovsky, *Bauhaus on the Carmel and the Crossroads of Empire: Architecture and Planning in Haifa During the British Mandate* (Jerusalem: Yad Izhak Ben-Zvi, 2003).

Hitzel, Frédéric, 'Production et techniques de tissage de la soie à Bursa aux xve et xvie siècles', *Rives nord-méditerranéennes*, 29 (2008), 11–24

Howard, Roger, *The Oil Hunters: Exploration and Espionage in the Middle East, 1880–*1914 (New York, NY: Hambledon Continuum, 2008).

Hudson, Leila, *Transforming Damascus: Space and Modernity in an Islamic City* (London: Tauris Academic Studies, 2008).

Imam, Chafiq, Malak Issa-Abyad, Françoise Métral, Jean Métral, and Rabah Naffakh, 'L'artisanat du verre à Damas', *Bulletin d'études orientales*, 27 (1974), 141–81.

Issawi, Charles Philip, *An Economic History of the Middle East and North Africa* (New York, NY: Columbia University Press, 1982).

Kajja, Kamal, 'Al-Jazeera, phénomène ou leurre?', *Hérodote*, 133 (2009), 152–65.

Katodrytis, George and Kevin Mitchell, *UAE and the Gulf: Architecture and Urbanism* (New York, NY: John Wiley & Sons, Inc., 2015).

Lafi, Nora, ed., *Municipalités méditerranéennes. Les réformes urbaines ottomanes au miroir d'une histoire comparée. Europe méridionale, Maghreb, Moyen-Orient* (Berlin: K. Schwarz, 2005).

Lantz, François, *Chemins de fer et perception de l'espace dans les provinces arabes de l'Empire ottoman, 1890–1914* (Paris: L'Harmattan, 2005).

Lerner, Daniel, *The Passing of Traditional Society: Modernizing the Middle East* (New York, NY: Free Press, 1958).

Longrigg, Stephen Hemsley, *Oil in the Middle East: Its Discovery and Development* (New York, NY, Oxford: Oxford University Press, 1961).

Mantran, Robert, *Histoire de l'Empire ottoman* (Paris: Fayard, 2010).

Marchal, Jeanne, 'Le Coton en Égypte', *L'Information géographique*, 2 (1956), 56–60.

Maugeri, Leonardo, *The Age of Oil: the Mythology, History, and Future of the World's Most Controversial Resource* (Westport, CT: Praeger Publishers, 2006).

Minnaert, Jean-Baptiste and Mercedes Volait, 'Héliopolis au Caire', in *Villes rattachées, villes reconfigurées: XVIe–XVIe siècles*, ed. by Denise Turrel (Tours: Presses de l'Université François-Rabelais, 2003), pp. 335–65.

Mitchell, Timothy, *Rule of Experts: Egypt, Techno-Politics, Modernity* (Berkeley, CA: The University of California Press, 2002).

Montel, Nathalie, *Le Chantier du canal de Suez (1859–1869). Une histoire des pratiques techniques* (Paris: Presses de l'École nationale des Ponts et Chaussées, 1999).

Özyüksel, Murat, *The Hejaz Railway and the Ottoman Empire: Modernity, Industrialisation and Ottoman Decline* (London: I. B. Tauris, 2014).

Panini, Augusto, *Middle Eastern and Venetian Glass Beads: Eighth to Twentieth Centuries* (Milan and New York, NY: Rizzoli, 2007).

Piquet, Caroline, *Les Pays du Golfe de la perle à l'économie de la connaissance. Les nouvelles terres du libéralisme* (Paris: Armand Colin, 2013).

Quataert, Donald, 'Machine Breaking and the Changing Carpet Industry of Western Anatolia', *Journal of Social History*, 19.3 (1986), 473–89.

———, *Manufacturing and Technology Transfer in the Ottoman Empire, 1800–1914* (Istanbul: Isis Press, 1992).

Ragette, Friedrich, *Traditional Domestic Architecture of the Arab Region* (Stuttgart: Edition Axel Menges, 2006).

Reed, Stanley, *Oriental Rugs and Carpets* (New York, NY: Putnam, 1967).

Roper, Geoffrey, *The History of the Book in the Middle East* (Farnham: Ashgate, 2013).

Sadgrove, Philip, ed., *History of Printing and Publishing in the Languages and Countries of the Middle East* (Oxford: Oxford University Press, 2004).

Seyf, Ahmad, 'Carpet Manufactures of Iran in the Nineteenth Century', *Middle Eastern Studies*, 26.2 (1990), 204–13.

Shirazi, M. Reza, *Contemporary Architecture and Urbanism in Iran: Tradition, Modernity, and the Production of 'Space-in-Between'* (Cham: Springer International Publishing, 2018).

Volait, Mercedes, *Le Caire-Alexandrie, architectures européennes, 1850–1950* (Le Caire: IFAO/CEDEJ, 2011).

Woodward, Michelle, 'Between Orientalist Clichés and Images of Modernization. Photographic Practice in the Late Ottoman Era', *History of Photography*, 27.4 (2003), 363–74.

Yates, Chris, *Airport and Aviation Expansion: Africa and the Middle East* (Coulsdon: Jane's Information Group, 1998).

YAMINA BETTAHAR

The Integration Process of Science and Technology, between Dependence and the Desire to Innovate

The Case of Maghreb

Today, as crucial issues such as the preservation of 'natural resources', the fight against global warming, the energy transition, and the effects of uncontrolled globalization are being debated, it is useful to analyse the developmentalist policies implemented in the Maghreb in the 1970s (Assidon, 2002). Taking western countries as a reference, their aim was to speed up industrialization in order to provide a global and original response to underdevelopment; however, the results of these policies were mixed. The purpose of this chapter is to show that the rapid industrialization of these countries (i.e. Morocco, Algeria and Tunisia) is the most significant sign of contemporary transformations. Yet state-led reforms have done little to change social conditions, especially in the countryside. Currently these reforms are facing new challenges linked to globalization. My hypothesis is that the societies of these countries are living with policies adopted by generations that have turned out to be failures, and that these policies are now questioned by collective citizen mobilizations (notably the jasmine revolution in Tunisia or the *Hirak* movement in Algeria) while the youth propose innovative alternatives.

In order to understand the particular context of Maghreb countries without going back to the founding periods, I will begin the analysis in the 1960s, keeping in mind that local situations were rather different depending on whether a particular country was a protectorate or a colonial settlement. Tunisia and Morocco, both French protectorates (from 1881 and 1912 respectively), regained their independence in 1956. Algeria, which underwent a long period of colonization from 1830, became a French department in 1848. It did not gain its independence and sovereignty until 1962, after a particularly violent war of independence. These three configurations underline the diversity of colonial regimes and pathways towards independence. These diverse situations were shaped by different and unique contexts, including the particular system of government in place and the development policies that were favoured. As soon as political independence was obtained however, the leaders constructed a

Global History of Techniques, ed. by Guillaume Carnino, Liliane Hilaire-Pérez and Jérôme Lamy, Global Matters Culture Matérielles, histoire des techniques et circulation globale des savoirs, 9 (Turnhout: Brepols, 2024), pp. 83–98
BREPOLS PUBLISHERS 10.1484/M.TECHNE-EB.5.129802

narrative on the role of technology in society and as a vector of development. But beyond the common rhetoric, the contexts in which these discourses were deployed were distinctive rather than similar.

Prolegomena

An analysis of the place and role of science and technology in the Maghreb is only possible if we take into account the upheavals that affected Morocco, Algeria and Tunisia during the colonial period, regardless of the particular colonial regime in place. Countries in the Maghreb were part of a vast economic, social, cultural, and political web of relations between the dominant and the dominated (Turin, 1970).

The colonial period slowed down the development of science and technology in the countries concerned. Of course, for the Saint-Simonian utopian movement, which aimed to conquer the world through the mastery of technology, the colonies provided fertile ground for experimentation. In Algeria, the Saint-Simonians set out to design the technical conditions for a rational and productive colonization, and thereby laid the foundations for the colony's development. However, the vast development programme that was implemented and aimed at modernizing the country was almost exclusively at the service of the colonial project, access to equipment and training being primarily intended for the colonists rathe than the native population (Enfantin, 1843).

France's project of setting up colonial technical schools and its implementation in North Africa is now relatively well known thanks to a series of studies on the topic. It is clear that this educational system, especially the technical education initiated by the colonial authorities in the region, did not translate into the creation and development of an innovative technological system after independence, despite attempts to create specialized academic institutions, and despite the first establishment of scientific and technical courses intended to train students into qualified technical experts (e.g., in Tunis, the creation of a school of Arts and Crafts in 1857 or the Sadiki College in 1875; the colonial school of industry in Dellys, Algeria, in 1877; and the industrial and commercial school of Casablanca in 1917). Under the aegis of the French protectorates in Morocco and Tunisia and in the 'department' of Algiers, technical training was only perceived from a European perspective. In Algeria, the Industrial Institute, which was based on the model of the state-run schools of higher technical education in metropolitan France, was founded in 1924, and it was exclusively accessible to the children of Europeans, who later moved to France and entered the *grandes écoles* such as Polytechnique, Centrale, or les Mines. At the end of the nineteenth century the training of the first polytechnicians from the Maghreb was considered exceptional in a colonial system that remained essentially segregated, and particularly noteworthy here is the example of Cadi Chérif, the first Muslim Algerian polytechnician, who graduated from Polytechnique in 1887. Generally, the teaching of modern science and technology was not introduced in any significant way to local schools at this time, and the recruited pupils did not have sufficient knowledge to be able to follow the programmes in place. Indigenous students were admitted to the courses but merely tolerated in class, and their numbers were limited.

The colonial authorities were actually not proponents of the industrialization of the Maghreb because of the potential competition it could create for the French industry. Consequently, the technical schools responsible for training a workforce for the few local industries did not receive much support beyond maintaining a low level of education.

In fact, in the second half of the twentieth century, the French colonial authorities recruited mainly French engineers to contribute to the organization of the economic space of the three Maghreb countries as well as for harnessing natural resources. They introduced infrastructures and set up French state-owned companies in charge of exploiting resources (Bettahar, 2016). With the French executives the first technical instruments arrived. Algeria is particularly interesting, because here French engineers and technicians carried out the first astronomical, botanical, geological, mining, oil, and gas explorations, which provided an opportunity to import techniques that were adapted for the relevant research (e.g. drilling techniques in the Algerian Sahara).

It was only after the Second World War, and more particularly in the 1950s, that technical schools were transformed into engineering training schools. In Algeria, the Industrial Institute was reorganised by the decree of 13 December 1952, while a diploma for public works and building engineers was set up following the promulgation of the decree of 22 December 1952. By then, the idea of industrialization had gained ground. Technical education courses were extended to include technology, electronics, and electrical engineering.

In Tunisia, the system set up under the protectorate contributed to the early emergence of a highly qualified technical elite, some of whom continue to train students in the most prestigious *grandes écoles* in France. In Morocco, following the creation of the 'schools for the sons of notables' under the protectorate, Hubert Lyautey's legacy has continued for an astonishingly long time and continues to shape the production of contemporary scientific and technical elites, both francophone and francophile (Bettahar, 2016).

However, the history of science and technology is not being taught, and neither is the debt owed to the long histories of technical and technological developments, particularly in the Arab world (Rashed, 1997; Djebbar, 2001). If we consider the history of technology in Islam, it is clear that the complexities of this rich and vast field cannot be summarized in a few lines. However, as authoritative studies show with the help of various examples, medieval Islam was a prosperous and dynamic civilization thanks to the emergence and development of technologies adapted to local contexts (Nasr, 1968, 1979; Rashed, 1997; Djebbar, 2001). Muslim scholars, engineers, and technicians, both Arab and Persian, contributed to the development of modern technology in mechanical engineering, and civil engineering with irrigation and water supply systems (El Faïz, 2005). In his famous introduction to *Muqaddima*, the scholar Ibn Khaldun described in detail the cob technique (*tabya*), a process whereby a mixture of clay, water and natural fibres was used in construction, suggesting the existence of a Muslim technology. His work also showed that the

Arab-Muslim civilization revolved around circulation and commerce, as well as culture and science, and these exchanges were facilitated in particular by the use of paper, whose innovative manufacturing process was imported, as is well known today, from China (Jacomy, 1990).

It is difficult to provide a comprehensive overview of science and technology in the Maghreb since the publication of above-mentioned seminal works. The main obstacle is the fact that the existing data is incomplete and local contexts differ so much depneding on the socio-historical, economic, cultural, and political situations. Studies from the past few decades have focused on the public scientific policies and government reforms of scientific organization (Khelfaoui et al, 2006; Kleiche et al., 2008). They fail to provide the necessary socio-historical depth that allows for an analysis of the technological issues and challenges of the present and the future.

The era of decolonization in the Maghreb brought with it certain transformations and the desire to afford science and technology an important place within the framework of a proactive and emancipatory development.

The Age of Expectations: Developmentalist Policies

In the aftermath of their independence, Morocco, Algeria and Tunisia were at the center of a multitude of debates and multidisciplinary studies intended to investigate these new states, their growth, and the modalities of industrialization with models that could be applied to so-called Third World countries (Lacoste, 1959; Bettelheim, 1967; Amin, 1970; Bairoch, 1971). Regardless of their approach or method, most studies agreed that industrialization provided the best path towards development, and that a careful plan was necessary to achieve this goal. Local contexts also played an important role however.

In 1962, Morocco chose the 'liberal' economic option and attracted external capital by opening up access to its markets to foreign companies, with a system of loan guarantees, debt facilities, and regulatory exemptions. State interventionism did not vanish, but was mediated through directive planning, a customs and monetary policy. The adoption of a liberal-inspired market economy gave the State a significant role in the industrialization strategy and the so-called 'economic emergence' projects. Moreover, in addition to being an agricultural power, Morocco was the third largest producer and first exporter of phosphates in the world and therefore relied on its resources of phosphates to ensure a significant income in foreign currency. Finally, thanks to its dynastic continuity and a thousand-year-old national history, it possessed a rare stability within the Arab and African worlds (Vermeren, 2002). For this reason Morocco has benefited from continuous aid and support from Western countries and the European Union since the Cold War.

In Tunisia, the man who would become the country's first president in 1957, Habib Bourguiba, embodied the struggle for independence. From the day after independence onwards, the country opted for a non-capitalist path by advocating 'self-centered development'. From 1962 to 1964 and 1965 to 1968 two plans were drawn up, and with a third plan, drafted between 1969 and 1972, the country moved towards a gradual disengagement from the public sector. The leaders then attracted private capital in the industrial and tourism sectors.

In Algeria, following the internal struggles of the independence movements and as early as 1962, the National Liberation Front (FLN) became the party-state that was set to lead the authoritarian rebuilding of the country. Algeria, which had become a sovereign country, signed the historic agreements known as the 'Evian agreements' with France. The two sovereign countries jointly defined the guarantees and modalities of their future cooperation. In 1963, the new Algerian state showed a strong desire to take control of its own destiny. It defined the major strategic options for the implementation of industrialization supported by hydrocarbons (oil and gas) and their by-products. However, the country had been dependent on essentially French or international technical resources since the 1950s. The Algerian politicians whom independence brought to power planned the reconstruction and modernization of Algeria. Industrialization, which was the first priority of the State, defined all the economic and social transformations that followed. From an ideological standpoint, Algerian negotiators pronounced themselves in favour of a socialist option in the Tripoli programme as early as June 1962. The State made hydrocarbon exploitation the basis of a bold economic policy, founded upon the nationalization of established oil companies: the anticipated oil revenues were intended to finance the envisaged transformations — a challenge whose effects are being felt to this day.

When the first experiments, which were carried out between 1962 and 1965, failed, the Houari Boumediene government that resulted from the 1965 coup decided to reorient Algeria's development. The principles of development were preserved in the national charter that was adopted by referendum in 1976. They still fall under the socialist option, including its decolonizing and anti-imperialist corollaries. In practice, these stated choices needed to involve 'the recovery of natural wealth, the appropriation by the community of the major means of production, the balance of foreign trade, the financial independence of the state, the creation of a national market, as well as the mastery of technology' (Toumi, 1989: 68). Priority was given to heavy industry, and a choice was made to implement a 'self-centred and planned development' model that benefited the country itself. The development of industrial infrastructures were to be the engine of the economy, or, summed up in a pithy phrase, the 'industrializing industrialization'.

Focus on Algeria: Oil and the 'Industrializing Industrialization'

The Evian agreements defined the terms for the exploitation of Algerian oil and gas. France first pursued this exploitation through its flagship company, Sn Repal (Société nationale de recherche et d'exploitation de pétrole en Algérie), which discovered the deposits in 1956. Sn Repal gradually handed over control to the Algerian authorities, leading to the first nationalizations in 1967. It is within this framework that the country started the nationalization of refining and distribution activities with the creation in 1963 of a company (Sonatrach or Société nationale pour la recherche, la production, le transport, la transformation et la commercialisation des hydrocarbures, i.e. the national company for the research, production, transport, processing and marketing of hydrocarbons and their derivatives) which was placed at the head of the exploration (see the discovery of the site of El-Borma in the eastern region of Hassi-Messaoud for instance), drilling and distribution of petroleum and gas products. In 1980, the Algerian government created service companies and opened a petrochemical complex in the city of Skikda (Eastern Algeria), managed by Sonatrach.

The quadrupling of the price of oil between October 1973 and January 1974 had beneficial effects for the Algerian regime's new development policy. On a geopolitical level, the oil shock's symbolic value benefited Algeria and OPEC (Organization of Petroleum Exporting Countries) financially. The rise in oil prices was a consequence of the discovery of the interdependence between hydrocarbon producing countries whose strategic choices affect Western countries that are largely dependent on hydrocarbon. Now oil was weaponized in the field of international relations and disrupted the balance of power, a power previously held by the cartel of the 'seven sisters' that controlled the markets of oil supply and demand. These are the seven transnational companies of the Anglo-Persian Oil Company Cartel that dominated the world's oil industry from the the mid-1940s to the mid-1970s. Thus, the crisis caused by the oil 'stagflation' reinforced the role of the Algerian state, not only in its attempts to shape society, but also in its contributions to the organization of international relations according to its own rules.

To better understand the vicissitudes of the Algerian economy it is necessary to examine its industrial underpinings since the 1970s. Algerian socio-economic development was built around the concept of 'industrializing industries' that was introduced in the 1970s by the French economist Gérard Destanne de Bernis to the members of the Algerian government (Destanne de Bernis, 1971). This strategy originally served as a basis for the Soviet model. In its principles and applications, it aimed for an economic revolution based on industrialization, planning, and the collectivization of the countryside. With the implementation of soviet style plans, the model was to focus on industrial power and prioritize the development of heavy industries. When it was developed, this strategy was supposed to reduce the economic dependence of developing countries. The principles behind this meant relying on the production of those capital goods that were necessary to optimize growth. The greater the investment, the greater the future production of consumer goods, and eventually, thanks to a knock-on effect, this would lead to the economic modernization of the country.

Inspired by this model, Algeria adopted a voluntarist economic policy based on a centralizing state dirigisme which prioritized industrialization and the deployment of heavy industry. This was the case for steel, metal, metallurgical, mechanical, electrical, and electronic industries. These industries took the lion's share, with 62 companies employing an average of 2110 employees each, that is, employing 26%

of the industrial workforce and producing 36% of the industrial added value in 41 industrial branches (Bouyacoub, 2001).

In practice, it was a question of acquiring, through valuation on the international oil and gas market, the foreign currency that was likely to facilitate the import of the necessary means for setting up the industrial bases for global development. To do this, the country had to sell its gas and oil resources on the international market in order to be able to buy 'turnkey' or even 'market in hand' technical resources (Salem, Sanson-Hermitte, 1979). With this in mind, and in order to promote optimal conditions for growth, political leaders relied on the anticipated mobilization of large savings and on the production of the necessary capital goods. The greater the investment, the greater the future production of consumer goods.

By giving priority to industries located upstream in the production process, the Algerian government pursued several objectives. It gave itself the means to achieve effective economic independence by reducing external dependence while relying on external outlets. Thus, national resources were processed locally using the most advanced techniques, and the products obtained were used for the development of industries located downstream. Domestic trade increased. For example, the steel products of the Sns (National Steel Company) were used by Sonatrach for pipelines (El-Kenz, 1987). This was also the case in mechanical industries, where a part of the hydrocarbons was used as a national energy source without using European refineries. The petrochemical industries, in turn, supplied other industries, including basic chemicals for the establishment of other industries and fertilizers necessary for the modernization of agriculture.

The quest for technological independence thus led political and economic leaders to favour hydrocarbon exports, in conjunction to largescale injection of innovative technical knowledge (Germidis, 1976). To finance development programmes, the State used hydrocarbon revenues that increased substantially under the aegis of Sonatrach. Following the 1971 nationalization of the company and the increase of oil and gas revenues, several industrial hubs were established throughout Algeria.

The oil counter-shock of 1986, with the sharp fall in the price of oil, hit Algeria hard, since its oil revenues were the main source of foreign exchange earnings, but it also affected the neighbouring countries and caused an unprecedented general crisis in the 1990s — a social, political, and identity crisis which contributed to the impoverishment of the middle classes in particular. Since 2014, the oil market has become highly volatile and has greatly weakened the economies of producer countries. The gas market, too, has begun to decline. In general, fossil fuels have suffered from an overabundant supply from the United States, which has become the world's largest producer of hydrocarbons.

Within the Maghreb, the failure of the Algerian industrial model is the most obvious of all. It probably results from the ambitious objectives which were clearly expressed in the National Charter: 'The development of the country must create the necessary conditions for the construction and consolidation of a national economy, independent, integrated and self-centered, intensifying within it intersectoral relations and exchanges between branches' (National Charter, 1976). The stakes are high. The question of the necessary adaptation to a technical import system and of finding a common language that the techniques require need to be resolved. However, this has been a difficult question for Algerian rulers. The complexities are even greater when considering the consequences of the technical imports on the system of values and collective representations. The acceleration of imports and the influence of techniques is so rapid that it provokes increasingly ambiguous reactions. In his time, Georges Balandier stressed the need to take into account endogenous or induced character of technical development (Balandier, 1961). Transferring a complex of techniques and the associated knowledge from a system that designed it to a system to which it is foreign requires genuine translation. Moreover, these imports of techniques often have disruptive effects on the ground. For example, the manufacture of modern houses in areas where traditional construction techniques are still in use contributes to upsetting the social relations.

On a more basic level, Algeria has not been able to set up a high-tech industry either successfully or in a short period of time, partly due to a low level of literacy in its numerous population and a geopolitical context largely dominated by the Cold War. The recovery of national wealth turned into an economic war that the country could not win alone. The year 1986 saw the price of hydrocarbons quartered, and in the same year the demographic peak increased the socio-economic demand for housing, education, medical care, food, and transport. This conjunction of events was the trigger that caused the crisis of the 1990s. It put an end to the ambitious national project of economic and political independence.

After the euphoric enthusiasm of the developmentalist years and Algeria's prominent role in international and diplomatic contexts and, in times of crisis, in various international forums (i.e. within the movement of non-aligned countries, in OPEC as a member of the cartel of oil-producing countries — especially in 1973–74, after the oil shocks — and in the renegotiation of the international economic order), Algeria and its neighbouring Maghreb countries entered a period of instability that can be attributed in part to the near-failure of the economic models implemented in the 1960s and 1970s, but also to the uncertainties raised by globalization which has been experienced by local societies at all levels and has had harmful effects on the environment.

The early 1990s were marked by the increasing chasm between a growing population on the one hand and a shrinking economy on the other hand. Various studies have shown these phenomena to be aggravated by pressure from international organizations such as the IMF and the World Bank (Vandewalle, 1994; Guerraoui, Richet, 2009; Ben Romdhane, 2011). The indirect negative consequences of globalization have contributed to a new configuration of the relations between north and south and between the centre and the periphery. In fact, the suitability of national development strategies and those of private multinational firms, as well as the spread of capitalism in regions traditionally subject to the old international division of labour, have led to a thorough recomposition of the economic world map. Thus, 'nations that had hoped to be able to base their development on a better use of their natural resources without diversifying their export bases are being inexorably marginalized in international economic relations', with potentially disastrous domestic social and political consequences (Adda, 1998: 26). This is true for the three Maghreb countries, whose fate was sealed by the reconfiguration of the relations that started with their integration into globalization. In fact, for several decades now, competitive relations have been superimposed on the traditional relations that linked the countries of the north with the countries of the South. This recomposition of north-south relations likely contributes to the weakening of partner states. From this point of view, the innovative experiments that have been carried out for several years now, and particularly those done in Tunisia and Morocco within the framework of technopoles, deserve to be explored more fully.

An Educational and Scientific Policy at the Service of Development

Following the early student mobility in the aftermath of the Second World War, a voluntarist policy of sending students abroad was put in place by the Maghreb states within the framework of bilateral cooperation agreements with their main partner, France. Since there was a lack of managers for the number of technical jobs that needed to be filled, training abroad was extended to former Soviet bloc countries at the height of the Cold War, foremost among them the USSR. Despite strategies to diversify target partnerships in the Maghreb countries and the emergence of new zones of activity, France, the former colonial power, remained instrumental in accommodating students from the Maghreb (Bettahar, 2007, 2020a, 2020b). The French model for training the technical elite, recognized for the quality of its education, continued to be attractive to governments, and many members of this elite have been educated in engineering schools, both in the past (Polytechnique, Ponts & Chaussées, Centrale, Mines) and more recently. In addition, the linguistic proximity to France plays in the favour of future Moroccan, Algerian, or Tunisian elites, whose training in France has always been considered an educational and social asset.

But while training abroad has become an important aspect of Maghrebian students' education, significant reforms have been implemented in Morocco, Algeria and Tunisia to create and improve local higher education. From the early decades following independence onwards, a voluntarist institutional and centralized scientific policy was designed and implemented in the three countries. This was designed to establish the 'undisputed modernity' (Camau, 1989, 4) of state-led initiatives. Academic and scientific institutions were created or reorganised (for those inherited from the colonial period) to respond more effectively to the new social demands.

Since the 1990s, the governments of Morocco, Algeria and Tunisia have had to deal with problems created of the internationalization of higher education and of research systems in the context of a highly competitive global education market (following the Bologna process and the Lisbon strategy). They have therefore been resolutely committed to remodelling their scientific research systems, introducing major changes. The structural adjustment plans imposed by the IMF and the World Bank reveal a growing disjunction between the training profiles of higher education graduates on the one hand and their professional opportunities in industry and other fields on the other. The reforms had aimed to create collaborations between scientific training and research systems and the industrial sector in order to enable the latter to continually update its methods and techniques. This was the case in Morocco, where, with the support of the IMF, the WB, and UN organizations, a programme known as the 'Decade of Education' was implemented for the period between 1999 and 2009, while co-opted members of the Hassan II Academy of Science and Technology contributed to the considerations of national priorities regarding scientific policies and made proposals for research programmes, as well as arrangements for their funding, implementation and evaluation.

In recent decades, innovation in Morocco, Algeria and Tunisia has become a central issue (Rahmouni, 2014). However, innovation remains structurally determined by both economic, political and human factors external to companies' particular policies. Furthermore, these countries are faced with multiple barriers to innovation that cannot be addressed in detail within the scope of this article. It may suffice to cite here the difficult access to technological information of some remote areas, the shortage in qualified workers, and issues related to the appropriation of results and persistent asymmetrical access to information. A lack of interaction between the technical-scientific realm and production certainly also deserves a mention. In terms of innovation, it can be said, there is a great disparity between the activities of individual Maghreb countries (Filliatreau, 2010). However, all three countries explicitly consider innovation policy a national issue (Rahmouni, 2014; Amdaoud, 2016). Thus, a variety of programmes and policies are regularly implemented to foster innovation, both in the industrial sector and in higher education and scientific research. But in general, national innovation systems (NIS) and related schemes do not benefit from the communication networks responsible for their dissemination and diffusion. Also, the surveys and academic works that have begun to be published, often in the form of monographs and targeted sector studies, although very useful in the context of detailed case studies, are nonetheless fragmentary in nature and only concern targeted sectors without allowing a global view of the issue in question. In fact, what remains to be explored in much greater depth are the obstacles to innovation in the countries under consideration, and more generally in developing countries that, faced with western models of growth, wish to overcome their scientific and technological 'accumulated backwardness'. Indeed, little work has been done on the factors that prevent actitivites leading to innovation, known as 'barriers to innovation' (Bouguerra, 1993). These factors are of interest from a socio-anthropological point of view beyond mere economic issues — although these should not be downplayed.

Theoretical and Epistemic Issues

The socio-economic vicissitudes of the Maghreb countries should also be seen in the wider context of the Muslim World which has undergone a vast movement of reforms following their independence. These reforms have given rise to many controversies that have brought to the fore the question of reference models, mostly in a polemical manner. Within the framework of global history, these debates are based on the idea that modernity was born in Europe and that all other nations received this modernity from Europe, an idea that prevailed until the 1960s (König-Pralong, 2015). The debates on the superiority and inferiority of 'races' or religions, and the identification of a scientific 'backwardness' of countries in the Muslim world, raised the question of technical transfer and of the developing countries' ability to adopt and adapt these technologies. For Algeria in particular, was also the question of the transposition of ideas of social progress, based on the myth of science and technical progress bringing social progress to a society which for over a century had lived under colonial domination and segregation, resulting in unequal access to education and high levels of illiteracy (Turin, 1970). These debates highlight the complexities of the undertaking as well as the compromises that need to be made in such circumstances.

Viewed through the prism of 'catching up', policies were inspired by the socialist model (Vichnevski, 2000). Closer to our time, the emergence of developmentalist theories can be considered through the utilitarian philosophy of John Rawls (2009) or the Frankfurt School via one of its most fervent contemporary proponents, Jürgen Habermas (Habermas 1987, 1988). In its contemporary forms, developmentalism originally drew its references from the economic history of industrialized countries after the Second World War. Countries such as Germany, Japan, and the USSR explored the paths of its implementation by integrating three major paradigms: modernization, the idea that accelerated growth could only result from the expansion of industrial activities, and voluntarism, based at times on a declared political nationalism and always on the intervention in the allocation of resources of the State, called upon to correct the laws of the market that had previously distributed industry unequally throughout the world.

Turning the tables to put in place a voluntarist policy that is resolutely turned towards modernity; creating the conditions for the emergence of technical and social progress; and introducing the country into the international (capitalist) economy: this was the ambition of the Algerian government that was authoritarian but courted by Western leaders who were interested in procuring oil and gas supplies from Algeria. Making the best use of its strategic position and a favourable economic climate, the young state took advantage of the international rivalry to diversify its partners. It intended to establish its power by forming increasingly strong ties with Western countries (first and foremost France) while also choosing a socialist option that would similarly bring it closer to the countries of the Eastern bloc at the time (namely mostly the USSR).

It is interesting to recall that Algeria served as a vast laboratory for economic experimentation during its French occupation in the 1840s. The experimentation was conducted by one of the most ardent defenders of Saint-Simonism, Prosper Enfantin, within the framework of the 'Commission d'exploration scientifique de l'Algérie' (Enfantin, 1843). The reforms advocated by the Saint-Simonians for a Muslim country like Algeria occurred at the end of the nineteenth century, and thus at a time when a debate opposed the philosopher Ernest Renan and the intellectual Jamal a-Din al-Afghani, who was considered the founding father of Islamic modernism as well as the main actor of the Arab renaissance known as the Nahda and pan-Islamism (Renan, Al-Afghani, 2011; Ayduz, 2014). In the nineteenth century, the Nahda was a movement of modern Arab cultural 'renaissance' at once literary, political, scientific and religious. In a conference on the theme of 'Islam and Science', Renan criticized the states governed by Islam and their decadence, and he also defended the idea that Islam and modernity were incompatible (Balcou, 2015). Afghani, for his part, saw a political explanation, i.e. that of the despotism of the leaders of the time, against which he saw education as the means of bringing societies in the Muslim world into modernity. This position was highly topical, as the revolutions of the 'Arab Spring' have shown.

Maghreb countries have experimented with industrialization models with mixed results. Some analysts have blamed the protective and authoritarian interventionism of the welfare state that was introduced in these countries at the time of their independence. State intervention, which was omnipresent as well as bureaucratic and centralized from the economic sector through to education and scientific research, did not encourage the development of technological innovation. However, a question arises: would these countries have been able to develop a high-tech industry successfully and in such a short period of time, starting from a very different material base, with a large and poorly educated population and counting on the income from raw material?

Today, Maghreb countries are trying to increase their profile in an increasingly competitive international context. Morocco, faced with environmental problems, has embarked on an ecological transition, reinforced by the success of the COP22, which took place in Marrakech in November 2016, and by the participation of the Cherifian sovereign in the One Planet Summit in Paris in 2017. Ambitious projects in the field of renewable energies (some of which are supported by the French Development Agency) have been launched in recent decades (e.g. the Noor solar power plant in Ouarzazate and wind farms), and training institutes for renewable energy professions have been set up. This is also the case in Tunisia. Despite a tense economic, social, and political recent history, the country has relied on technopoles as real tools for development. The site of the El Ghazala Technopark, created in 1999 as the space for the development of high-tech activities, has become the country's technological showcase. Specialized in the field of ICT and microelectronics, the technopark has been able to take advantage of the availablility of engineers with degrees in computer science. The model of technopoles and incubators of innovative companies (Fayolle, Ben Salah, et al., 2010) has attracted other countries in the region such as Morocco (with the creation of CasaNearShore), Egypt, and, more recently, Mauritania and Madagascar (Huet, Viennois, et al., 2012). These research structures further complicate what is a truly global history of techniques governed by multiple temporalities and heterogeneous technical cultures.

References

Adda, Jacques, *La Mondialisation de l'économie. 2. Problèmes* (Paris: Éditions La Découverte, 1998).

Amdaoud, Mounir, 'La Construction d'un système national en Algérie: institutions et politique d'innovation', *Marché et organisations*, 2, 26 (2016), 49–91.

Amin, Samir, *Le Maghreb moderne* (Paris: Les Éditions de Minuit, 1970).

Assidon, Elsa, *Les Théories économiques du développement*, Series: Repères (Paris: La Découverte, 2002).

Ayduz, Salim, Ibrahim Kalin, and Caner Dagli, eds, *The Oxford Encyclopedia of Philosophy, Science, and Technology in Islam*, Series: Oxford Encyclopedias of Islamic Studies (Oxford: Oxford University Press, 2014).

Bairoch, Paul, *Le Tiers-monde dans l'impasse. Le démarrage économique du XVIIIe au XXe siècle* (Paris: Gallimard, 1971).

Balandier, Georges, ed., *Le Tiers-monde: sous-développement et développement* (Paris: PUF, 1961).

Balcou, Jean, *Ernest Renan. Une biographie* (Paris: Honoré Champion éditeur, 2015).

Ben Romdhane, Mahmoud, *Tunisie. État, économie et société. Ressources politiques, légitimation et régulations sociales* (Tunis: Sud Éditions, 2011).

Bettahar, Yamina, 'Institutionnalisation et professionnalisation de la science en Algérie: le cas des naturalistes coloniaux', in *Institutions and Societies for Teaching, Research and Popularisation, Proceedings of the xxth International Congress of History of Science*, ed. by Andrée Despy-Meyer (Turnhout: Brepols, De Diversis Artibus, 1997), pp. 207–27.

———, 'De l'ENSEP/ENSET à l'ENPO d'Oran. Recompositions et mutations d'une institution algérienne d'enseignement technique (1970–2012)', in *Le Siècle d'une grande école pour les sciences, les techniques, la société*, ed. by Florent Le Bot, Virginie Albe, *et al.* (Rennes: Presses Universitaires de Rennes, 2015), pp. 343–58.

———, 'La Contribution de l'École nationale supérieure d'électricité et de mécanique de Nancy (E.N.S.E.M.) à la formation des ingénieurs électriciens nord-africains (1900–1960)', in *Des ingénieurs pour un monde nouveau. Histoire des enseignements électrotechniques (Europe, Amériques), XIXe XXe siècle*, ed. by Marcela Efmertova and André Grelon (Brussels: P. I. E. Peter Lang, 2016), pp. 357–80.

———, ed., *Universités et grandes écoles. Circulations étudiantes et scientifiques d'hier à aujourd'hui* (Nancy: Pun-Edulor, 2020a).

———, 'L'Université d'Alger sous la Troisième République. Circulations académiques et scientifiques (1868–1950)', in *Repenser la 'mission civilisatrice'. L'éducation dans le monde colonial et postcolonial au XXe siècle*, ed. by Damiano Matasci, Miguel Bandeira Jerónimo, and Hugo Gonçalves Dores (Rennes: PUR, 2020b), pp. 69–103.

Bettelheim, Charles, *Planification et croissance accélérée* (Paris: F. Maspéro, 1967).

Bouguerra, Mohamed Larbi, *La Recherche contre le tiers-monde* (Paris: PUF, 1993).

Bouyacoub, Ahmed, 'La Politique industrielle: état des lieux et perspectives', in *Où va l'Algérie?*, ed. by Ahmed Mahiou and Jean-Robert Henry (Aix-en-Provence: Iremam / Paris: Karthala, 2001), pp. 185–205.

Camau, Michel, *La Tunisie*, Series: Que sais-je? (Paris: PUF, 1989).

Destanne de Bernis, Gérard, 'Les Industries industrialisantes et les options algériennes', *Revue Tiers Monde*, 47 (1971), 545–63.

Djebbar, Ahmed, *Une Histoire de la science arabe* (Paris: Points, Inédit Sciences, 2001).

El Faïz, Mohammed, *Les Maîtres de l'eau. Histoire de l'hydraulique arabe* (Arles: Actes Sud, 2005).

El-Kenz, Ali, *Le Complexe sidérurgique d'El Hadjar. Une expérience industrielle algérienne*, Series: Recherches sur les sociétés méditerranéennes (Paris: Éditions du CNRS, 1987).

Enfantin, Paul, *Colonisation de l'Algérie* (Paris: P. Bertrand, Librairie, 1843).

Fayolle, Alain, Amina Ben Salah, 'Efficacité des pépinières dans la création d'entreprise innovante: cas de la Tunisie', *Innovations*, 33 (2010), 157–79.

Filliatreau, Ghislaine, dir., *Indicateurs de sciences et de technologies. Rapport de l'Observatoire des sciences et des techniques, Edition 2010*, (Paris: Economica, 2011).

Germidis, Dimitri, *Le Maghreb, la France et l'enjeu technologique* (Paris: Éditions Cujas, 1976).

Guerraoui, Driss and Xavier Richet, eds, *Compétitivité et accumulation de compétences dans la mondialisation. Comparaisons internationales* (Paris: L'Harmattan, Questions contemporaines, 2009).

Habermas, Jürgen, *Théorie de l'agir communicationnel* (Paris: Fayard, 1987).

———, *L'Espace public. Archéologie de la publicité comme dimension constitutive de la société bourgeoise* (Paris: Payot, repr. 1988).

Huet, Jean-Michel, Isabelle Viennois, 'Les Technolopes, moteurs de développement', *L'Expansion Management Review*, 4.147 (2012), 112–19.

Jacomy, Bruno, *Une Histoire des techniques* (Paris: Points, Inédit Sciences, 1990).

Khaldûn, Ibn, *Le Livre des exemples. 1. Autobiographie. Muqaddima*, trans. by Cheddadi Abdesselam, Series: Bibliothèque de la Pléiade (Paris: Gallimard, 2002).

Khelfaoui, Hocine, ed., *L'Intégration de la science au développement. Expériences maghrébines* (Paris: Éditions Publisud, 2006).

Kleiche Dray, Mina and Roland Waast, eds, *Le Maroc scientifique* (Paris: Éditions Publisud, 2008).

König-Pralong, Catherine, 'Histoire médiévale de la raison philosophique moderne', *Annales, Histoire, Sciences sociales*, 70.3 (2015), 667–712.

Lacoste, Yves, *Les Pays sous-développés*, Series: Que sais-je? (Paris: PUF, 1959).

Nasr, Seyyed Hossein, *Sciences et savoir en Islam*, Series: La Bibliothèque de l'Islam (Paris: Sindbad, 1968).

Rahmouni, Mohieddine. ' Perception des obstacles aux activités d'innovation dans les entreprises tunisiennes ', *Revue d'économie du développement*, 22 (2014) 69–98.

Rashed, Roshdi, ed., *Histoire des sciences arabes*, vol. 3: *Technologie, alchimie et sciences de la vie* (Paris: Seuil, 1997).

Rawls, John, *La Théorie de la justice*, trans. by Catherine Audard (Paris: Points, Essais, 2009).

Renan, Ernest, and Jamal al-Din Al-Afghani, *Le Journal des débats, 1883*, 'L'histoire globale par les sources' (2011/10) http://blogs.histoireglobale.com/wp-content/uploads/2011/10 Renan-al-afghani.pdf [accessed 17 June 2021].

Salem, Mahmoud and Marie-Angèle Sanson-Hermitte, 'Les Contrats "clé en main" et les contrats produits en main', in *Technologie et rente de développement* (Paris: Librairies techniques, 1979).

Sampson, Anthony, *The Seven Sisters: The Great Oil Companies and the World They Shaped* (New York: Viking Press, 1975).

Speziale, Fabrizio, *Ernest Renan. L'Islam et la science, avec la réponse d'al-Afghânî* (L'Archange Minotaure, 2005), *Abstracta Iranica*, vol. 29 (2008), document 310, <http://journals.openedition.org/abstractairanica/30892> [accessed 21 May 2021].

Toumi, Mohsen, *La Tunisie de Bourguiba à Ben Ali* (Paris: PUF, 1989).

———, Le Maghreb (Paris: PUF, collection 'que sais-je?', 1992).

Turin, Yvonne, *Affrontements culturels dans l'Algérie coloniale. Écoles, médecines, religions, 1830–1880* (Paris: F. Maspéro, 1970).

Université d'Alger, *L'Impérialisme* (Alger: SNED, 1970).

———, *Rapports de dépendance au Maghreb* (Paris: CNRS, 1976).

Vandewalle, Dirk, 'Rupture avec le socialisme: libéralisation et privatisation économiques en Algérie', *NAQD*, special issue, *Réajustement structurel & systèmes politiques*, 7 (1994), 3–23.

Vermeren, Pierre, *Histoire du Maroc depuis l'indépendance* (Paris: La Découverte, Repères, 2002).

Vichnevski, Anatoli, *La Faucille et le rouble:la modernisation conservatrice en URSS* (Paris, Gallimard, 2000).

FRANCESCA BRAY, ALEKSANDRA KOBILJSKI

Technology and Modernization in East Asia

What Dragons Can Teach Us About Global History

China, Japan, South Korea, and Taiwan are now ultra-high-tech societies, acknowledged world leaders in such crucial and transformative sectors of contemporary technology as high-speed trains and electric vehicles, smart cities and smart phones, nano-electronics, AI and robotics, biotechnology and space programmes. East Asians today are confident that they belong to technologically advanced nations which are shaping the world's future. Indeed, when they visit Western countries today, many East Asians find them technologically backward. In the 1980s Japanese visitors were surprised by poor transport and telecommunications; today Chinese visitors find it quaint that people still rely on cash or plastic rather than apps or facial recognition for purchases, or that a short train journey from Los Angeles to San Diego, or from London to Birmingham, takes several hours.

Yet respect at home and abroad for East Asia's technological capacities has been hard won. Denigrated as technologically primitive in colonial times and stigmatized as producers of cheap and shoddy goods through the Cold War, until recently many East Asians still worried about catching up with the West, while Western experts were convinced that East Asian societies lacked the capacity for innovation: they could borrow, they could copy, but they could not invent (Cox, 2007).[1] By 1980, as Western fears of 'Japan as number one' took shape (Vogel, 1979), Japan was grudgingly admitted to the club of technically competent societies; the negative image was then transferred to China. These views were echoed in histories of technology that positioned the West as the modern world's uniquely creative centre, exporting science and technology to backward peripheries through commerce, colonial rule, transnational education, and development. Influential celebrations of European technological exceptionalism (Landes, 1969, 1999; Mokyr, 1992) were matched by critiques of Western technological imperialism which likewise highlight Europe as the fount of innovation (Adas, 1989; Headrick, 2010).

1 The notion that Japanese entrepreneurs were good at imitation can be traced back to the 1540s when it was first articulated by officials and generals of Chosŏn Korea (Carré 2019). In the Euro-American world the accusation that Japanese businesses had poor ethics and no scruples dates back to the late nineteenth century (Hunter 2016). Today these notions are regularly applied to China.

Global History of Techniques, ed. by Guillaume Carnino, Liliane Hilaire-Pérez and Jérôme Lamy, Global Matters Culture Matérielles, histoire des techniques et circulation globale des savoirs, 9 (Turnhout: Brepols, 2024), pp. 99–116
BREPOLS PUBLISHERS 10.1484/M.TECHNE-EB.5.129803

The orthodoxy of donor West and recipient Rest has been radically qualified in recent post-colonial histories of technology that decentre the West, for example by focussing on local appropriations of technology through the lens not of transfer but of user agency, co-production, and reconfiguration (Arnold, DeWald, 2012). The long-term rise of East Asia to technological prominence on the global scene offers some important additional elements to understanding the creativity of local technological actors, their aims, values, and the resources they mobilized. Moreover, having contributed so many vital technological innovations acknowledged to have transformed today's global technological cultures, from the Walkman concept of personal, portable music to the development of just-in-time networks of horizontally integrated production, the history of technology in East Asia encourages us to identify and integrate flows of influence from Rest to West into our histories. This is now a central element in global histories of technology in the early modern world (Finlay, 2010, on porcelain; Riello, 2013, on cotton; Schäfer, Riello, Molà, 2018), but it still seems to be assumed that once Europe began to industrialize the flows were all one way, radiating from the West. A key point is that unlike many other parts of the world East Asia did not fall under European colonial rule.[2] Technology imported directly or indirectly from the West certainly played an important role in modernizing the individual East Asian countries. But one can make a strong argument that local circulations of technology played a key role in forging East Asia as a region, first through the rise of Japan as colonial power, and post-1945 as an arena of strong mutual influence, dense cultural and technological exchange, and technocratic principles shared despite — or indeed because of — the region's political divisions and rivalries between Communist and 'free world' nations. Technology has played as important a role in making modern East Asia as it has for Europe, and these entangled histories are generating a growing, excellent literature upon which we draw for this chapter.

Our chapter begins with a historical outline emphasizing the role of technology in the making of modern East Asia. We then present four local cases that illuminate some specifics of East Asian histories while raising important general issues in the history of technology. The first case, on building the steel industry in Meiji (1868–1912) Japan, shows us the reworking in situ not of an imported technology, but of an essential raw material. The second, on agricultural development in Maoist China in the 1960s, looks at a radical socialist redefinition of what constitutes technological expertise. The third, on road construction in South Korea, shows how Cold War politics reconfigured a civil engineering project as a military campaign for national reunification. The fourth, on humanoid robots in contemporary Japan, shows cutting-edge AI engineering deployed as a 'retro-technology' to reproduce a conservative gender regime. We conclude with some reflections on what the East Asian experience contributes to innovative and critical global history.

---

2 China, which between 1842 and 1949 was obliged to concede a number of extra-territorial enclaves to the Western powers, has sometimes been categorized as a 'semi-colony', but retained its own sovereign government throughout.

A Brief History

When Britain, the United States, and other Western imperial powers began their military and economic intrusions into East Asia in the mid-nineteenth century, the region's technological backwardness — as perceived by the standards of the newly industrialized nations of the West — made it vulnerable. Westerners attributed such technological disparities to the superiority of their own civilization: lacking the curiosity, intellectual rigour, and technical creativity which had sparked Europe's scientific and industrial revolutions and the West's rise to global domination, other cultural traditions had not been capable of developing 'proper science' or its applications. In this racialized calculus machines served as 'the measure of men', ranking the world's peoples and justifying colonial rule as a means to bring progress to 'backward' societies (Adas, 1989).

Technology could enslave, or it could liberate. Imports of advanced technologies like railways and telegraphs allowed colonialists to extract resources and profits more efficiently and helped control the colonized. Notoriously, colonial masters resisted sharing technical expertise with subalterns: 'natives' were hands, not heads. Determined to avoid this fate, mid-nineteenth century politicians and reformers in East Asia identified science and technology as the key to national self-strengthening that would enable their countries to resist the incursions of Western powers. Anti-imperialism prompted the first phase of East Asian techno-nationalism, 'the belief that technology is a fundamental element that must be indigenized, diffused, and nurtured in order to make a nation rich and strong' (Samuels, 1994; Kelly, 2015: 50; Low, 2003).

The Chinese and Japanese governments began sustained campaigns to modernize their armies, navies, and communications systems, bringing in Western experts and equipment as a first step, but also sending students abroad to train as scientists, doctors, and engineers. Western-pattern universities and technical schools were founded, Western-style arsenals and factories constructed, railways and telegraphs imported and installed. Western governments, companies, banks, and missionary societies played an important role in providing capital and expertise although, unlike in colonial states, the governments of East Asia retained sovereign control.

Enfeebled by foreign predations and civil strife, China's progress towards technological modernization was patchy before 1949, when the country was unified as the People's Republic of China (Waley-Cohen, 1999; Elman, 2009; Bray, 2019). In Japan, the Meiji government, formed in response to the threat of Western ambitions, was strong enough to push through coordinated policies of militarization, industrialization, and infrastructural development that integrated the nation and brought the rapid rise of Japan both as a respected rival to the Western powers and as an imperial power with its own colonial territories (Morris-Suzuki, 1994). Acquiring Taiwan after defeating China in 1895, most of Manchuria after defeating Russia in 1905, and Korea through annexation in 1910, Japan deployed railways, mines, and dams, forestry projects and agricultural colonies as classic imperialist tools to knit together, exploit, and develop its territories, expand its control, and build international networks (Lee, 2013; Moore, 2013; McDonald, 2015).

Japan, like Germany, made a 'miraculous' recovery after its devastating defeat in 1945, reviving a techno-nationalist spirit to galvanize public commitment to reconstruction. First industries were reestablished. By the late 1950s a *denka būmu* (boom in domestic electrical appliances) had begun — as early as in the USA and well before many European nations (Low, 2003; Francks, Hunter, 2012; Oldenziel, Zachmann, 2009). Japan also found a ready export market through East and Southeast Asia for its relatively cheap appliances and its small-scale machinery, well-adapted to prevailing smallholder farming systems, and it now packaged large-scale infrastructural projects in the region not as colonial development but as aid (Moore, 2014; Choi, 2017). In terms of transformative innovation, by the 1970s Japan was gaining international respect as a pioneer of decentralized, horizontally-integrated 'just-in-time' production, co-ordinating many small units rather than seeking economies of scale through vertical integration; it was also recognized as a leader in electronics, both in excellence of production and in imaginative vision, as embodied for example in the Sony Walkman (Morris-Suzuki, 1994; Gay et al., 1997; Nakaoka; 2006; Francks, 2015, on the historical roots of this mode of organization). Spurred by Japan's success and building at least in part on Japanese colonial legacies of technical training and infrastructure, as well as on generous Cold War support from the United States, South Korea and Taiwan (Republic of China) followed similar trajectories. By the 1980s they were gaining fame as 'Asian Dragons' (Vogel, 1991). By the 1990s companies like Samsung were emerging as global leaders in electronics and digital devices, vying for market dominance and pace of innovation with Japanese giants like Sony.

Of the two communist states, North Korea and the People's Republic of China, it was North Korea that initially took the lead in successfully deploying technology to build the nation. Building on Japanese colonial infrastructure and with copious Soviet aid, the nation invested heavily in major industries, achieving 'world-beating growth rates in the 1950s and 1960s' and stirring fury in the breasts of South Korean leaders (Cumings 2008: 45). By the 1970s North Korea was purchasing advanced technology and turnkey plants from Japan and the West, about a decade before the People's Republic of China (PRC) adopted the same modernization strategy. Decline set in by the 1980s, however: communications technologies lagged, industries and agriculture suffered from insufficient energy supplies and the diversion of increasingly scarce resources into nuclear and military programmes (Cumings, 2008: 45–49; Armstrong, 2009).

The trajectory of the PRC was very different (Volti, 1982; Bray, 2019). In an initial period of classic Leninist-socialist technocracy, in close collaboration with the Soviet Union, Chinese policy was to build up heavy industry. During the Great Leap Forward of 1958 to 1961 and the Cultural Revolution of 1966 to 1976, Maoist policy of class struggle assigned expertise to the masses in both industry and agriculture (the military and space sectors were excepted). The Maoist period was hardly one of conventional technological progress, but it did lay an infrastructural base for take-off when technocracy was restored after Mao's death in 1976. Under the Four

Modernizations, introduced in 1978, national priority was assigned to building up agriculture, industry, technology, and defence. Thereafter, 'in little more than two decades [between 1980 and 2000], China made a leap of industrialization comparable to what took the U.S. a century' (Christopher Cullen, quoted in *China: The Dragon's Ascent*, 2000). In addition to bringing in technological equipment and expertise through joint ventures with Western, Japanese, and South Korean companies, a crucial factor in PRC technical development and integration into global circuits of technological exchange since the 1980s has been investment from small firms and multinationals of the supposedly enemy state, Taiwan—Foxconn being the pre-eminent example.

This sketch has emphasized the importance of intra-regional circulation of technology, expertise, and development styles or models in making the techno-nationalist states of the East Asia region, their rivalries, enmities, coalitions, and cohesion. In addition, all the East Asian nations have used and continue to use technology as a means to extend their global influence. Japan refurbishes its colonial links in South-east Asia through technical development aid for hydropower or agricultural projects (Moore, 2014). In the 1970s the PRC vied with Taiwan in exporting development projects to Africa (Lin, 2015; Siu, McGovern, 2017). Today the PRC 'One Belt One Road' (OBOR) project is building roads, ports, industries, and bases all through Eurasia along a generously defined Silk Road, in the process technically colonizing the continent through the diffusion of Chinese engineering standards (Cai, 2017: 11).

The Craft in Coking: Japan Builds a New Steel Industry

Steel was the miracle ingredient of Japan's spectacular post-1945 rise as a global industrial powerhouse. In the aftermath of a disastrous fifteen-year war, Japan poured energy into redeveloping its industrial capacity; by 1973 the country ranked as the world's largest steel exporter. Despite deep shifts in global steel markets, Japan still holds a steady second place in global steel exports (Yonekura, 1994, 2017).

Little attention has been paid to the shifts in technological culture that enabled Japan to expand its capacity to produce steel on a large scale and with fascinating degrees of specialization.[3] In the absence of readily available sources on links between the nineteenth-century worlds of artisanal and industrial steel production, most industrial history of steel offers a broad-brush genesis narrative grounded in the (eventually) successful of transfer of (mostly) German technology to Japan.

As such, industrial history of steel belongs to the genre of industrial technology history which has commonly focussed on the role of different government bodies in supporting Japan's industrialization and of entrepreneurs in shepherding the process. This bias is not limited to the steel industry and is for the most part an artefact of

3 Virtually no research to date addresses the question of what role (if any) traditions of artisanal steel production of high quality but limited scale played in Japan's rise as one of most important global actors of mass-produced boutique steels.

the kinds of primary sources marshalled: government documents and reports, and (auto)biographies of entrepreneurs or high-ranking bureaucrats.

By technology transfer, we mean purchasing equipment, hiring foreign experts, and sending Japanese students and technicians to train abroad. Here the assumption is that modern industrial technology is like a plug and play device: it can be plugged in anywhere — with some adjustments — and played for considerable profit. This line of argument, which suggests that it is by copying and imitating that Japan industrialized, resonates with contemporary anxieties of European and North American manufacturers about being stripped of their secrets by the supernatural abilityof their Japanese competitors to copy and adapt (Cox, 2007).

The work of Nakaoka Tetsurô remains a rare exception in the history of Japan's industrialization in showing a more nuanced understanding of the technological culture of modern Japan (Nakaoka, 2006). Building on Nakaoka's work, a small but growing body of research, based on the empirically informed understanding of what it meant to implement technical designs made elsewhere and keep them running constructively, challenges the notion of transfer.

In the grammar of modernity, steel was the alloy of strong nations, the stuff of railroads and battleships. The mastery of industrial-scale steel production was symbolically charged in Japan as elsewhere (Misa, 1998). However, one particularity of Japan's pursuit of industrial steel was the tension between the country's history of sophisticated medieval and early modern steel production, and the modernizing nation's relative handicap in appropriating the production protocols for steady, mass, standardised quality production required by the imperatives of modern state building, including warfare. At the heart of the handicap was the nature of the raw materials involved in blast-furnace steel making: iron ore and coke, as opposed to the iron sands and charcoal of Japan's familiar *tatara*-style iron and steel making. In particular, energy concerns loomed large in Japan because in the course of the nineteenth century, coke (a solid carbon fuel produced by processing coal) emerged as virtually indispensable in industrial-scale iron and steel production.

Like all other forms of fuel, coke is a product, not a raw material (Sougy, 2008; Jones, 2014). Manufacturing coke involves heating coals in a retort (coke oven) in the absence of air to expel moisture and volatile matter (impurities). In the 1880s, the standard coking technology was the so-called beehive method which consisted of feeding coals into a row of retorts for periods of between forty-eight and seventy-two hours, with no particular provision for recuperation of the fumes and effluvia of the manufacturing process. It turns out that the noxious emissions of the coking process could be turned into merchandise if they could be recovered: ammonia, coal tars, and benzene had significant market potential. Hence, the chemical industry was the key actor in introducing the so-called 'by-product coking' in which the toxic waste was repurposed as chemical components. The first commercially successful by-product coking plant was developed by Solvay & Cie in Belgium in the early 1890s.

Just when Solvay was venturing to set up by-product coking plants outside Belgium, on the other side of the planet Osaka was emerging as Japan's primary capital of industrialization, with a particular concentration of chemical factories (Abe,

2006).[4] Among them was a start-up chemical plant which set up a by-product coke manufacturing facility in late 1897, based on the purchase of coke ovens from Solvay & Cie.

As fine-grained studies of the supposed 'successes and failures' of Meiji industrialization reveal, buying equipment and even safely installing imported machinery was no guarantee that the technology would become operational (Kobiljski 2022). The choice of the costly Semet-Solvay ovens was a sign of ambition and chutzpah. Before the ovens could be made to work successfully so the enterprise could become profitable, engineers and technicians at the small Osaka start-up had to overcome a major obstacle inherent to the very idea of coking Japanese coals in Belgian ovens: Japanese coals had on average twice as many impurities as the coals which the ovens were designed to deal with (Four à coke Semet-Solvay: 63). The difference in quality of input assumed by the design of the coke ovens was a considerable challenge; no technical scenario for coking Japanese coals in those ovens existed in the technical literature of the day available in Japan (e.g. Fulton, 1895).

Though it is common to think of coal in the singular, not all coals are the same: their chemical composition can vary significantly. The quality of coal is not simply a question of a spectrum ranging from poor to excellent quality coals. Rather, it is a question of matching the coals to particular uses. Hence, in the nineteenth-century industrial context it is more appropriate to talk about coals rather than coal, and to keep in mind that there were no good or bad coals. Instead, there were coals more or less appropriate for a particular use.

Manufacturing metallurgical coke in Solvay ovens in Osaka using Hokkaido coals is manifestly not a story of purchasing, assembling, and adapting Belgian and British-produced equipment to fit the local conditions. Equally, it was not Japan's self-proclaimed poverty in mineral energy resources that made coke manufacturing such a challenge. Rather, it was a story of mismatch between the kinds of coals required by the specification of the ovens and the kinds of coals available in Japan. By the time the Osaka start-up entered the scene of coke manufacturing in 1898 to 1899, there was already a five-decades-long history of failure in solving the coke problem, most notably but not only at the Kamaishi Iron Works (Wittner, 2008). The coals prospected in China's northeast (Manchuria, soon to be colonized by Japan) were of roughly the same composition, therefore imperialism was not a cure for this particular problem.

Hence, in order to produce coke from Japanese coals in Belgian-made coke ovens, it was not the ovens that had to be adapted (as the classical technology transfer paradigm would suggest), but rather the input of raw materials had to be 'doctored' in a preparatory phase in order to give satisfactory results in the ovens that remained unchanged.

4 The industrial census data in 'Kôjo tsûran' (*Factories Overview*, 1904) indicate that in 1902, Osaka had twice as many chemical plants and employed twice as many workers in its chemical industry as Tokyo (Kobiljski 2022: 333).

The mix of anxiety and urgency typical of Meiji industrial ventures gave birth to a set of particular attitudes towards the lessons of past failures, standard production protocols, and the conventional order of operation (Kobiljski 2022). One of the first conventions Osaka engineers and technicians disregarded was the typical order of operation in preparing coals for coking and grinding the coal before washing. The second convention disregarded was that mandated by the historical patterns of coal sourcing: they opted to test a large sampling of coals from recently opened mines on the island of Hokkaido instead of working with long-established suppliers from the island of Kyūshu (Shimomura 1899). After two years of falling short of this goal, the Osaka start-up decided to disregard yet another convention and make coking a two-step process, instead of a single point operation. By 1903 the combination of different ways Osaka engineers and technicians worked around imported technology had resulted in a production protocol for a by-product coke. This involved not so much tinkering with the imported forms of the technology itself (whether artefacts or protocols), but rather a fundamental rethinking of all the basic suppositions hardwired into the technical design.

Expertise under Radical Socialism

Between 1949 and 1966 the communist government of the People's Republic of China pursued policies of agricultural reform and improvement (*nongshi gailiang*) as developed under the preceding Republic. Laboratory research and experimental stations pursued programmes of plant breeding and the development of suitable industrial inputs, notably machinery, fertilizers, and pesticides, and sought efficiency gains through increased scale of farming operations (Lavelle, 2015; Lee, 2010, 2015). Overall gains from this input-heavy style of modernization in the early PRC were unimpressive: progress was hampered by lack of capital resources, and by the mismatch between the scale economies of 'rationalized' agriculture and China's tiny family farms.[5] Yet this conventional techno-scientific model, in which farmers were recipients, not producers of new knowledge, was embraced by many PRC leaders, for it was globally endorsed as the way of the future not only in capitalist countries, but also in the Soviet Union, which at that point was a close ally providing considerable financial and scientific aid to the fledgling PRC.[6]

Mao Zedong and his close allies, however, vehemently opposed this consensus as a betrayal of all that China's peasant-proletariat revolution stood for. Modernizers were accused of 'taking the capitalist road' and 'considering technology as #1', when a truly revolutionary society should 'hew to the mass line' and 'place politics in command' (Saha, Schmalzer, 2016: 155). Mao proposed a radically different view of

---

5 Once landholdings were collectivized and managed as People's Communes, the mismatch of scale was largely overcome (Stavis, 1978).

6 Despite political antagonism, US and Soviet agronomic strategies were very similar and indeed drew upon each other for technological models (Fitzgerald, 1996).

science and social progress, in which theory was subordinate to practice, knowledge came through experience, and the working masses were the fount of authentic and effective knowledge.

In 1963, when China was still reeling in the aftermath of the disastrous Great Leap Forward, one group of political leaders saw the technocratic model of 'scientific farming' as the only way forward.[7] Experimental stations were encouraged to expand the scope of their extension work by increasing the number of 'model fields' (*yangbantian*) where new seeds and methods were tested and their benefits demonstrated to farmers (Saha, Schmalzer, 2016: 155). At the same moment two alternative, anti-technocratic experiments involving 'mass science' were taking shape.

The first involved what STS scholars call the co-production of knowledge. In 1964 'mass scientific experiment groups' (*qunzhongxing kexue shiyan xiaozu*) were introduced as a new and revolutionary means to promote scientific farming by bringing together 'expert research' and 'mass science'. These groups had a three-in-one (*sanjiehe*) composition of cadres (local officials), educated youth, and 'old peasants' (experienced but poor and often illiterate farmers):

> Cadres grasp all aspects of the situation, such that they can determine what kinds of experiment will serve production needs; old peasants are down-to-earth and unconcerned with profit and fame, understand the rhythms of production, and have a wealth of practical knowledge; and youth have technological knowledge and are accepting of new technologies, such that they 'dare to think and act'.[8]

This novel assemblage of competences was intended to produce science better suited to Chinese society's needs, revolutionizing expertise, and overturning class hierarchies. In a move that was truly iconoclastic within China's ingrained gender prejudices, not only old peasants but also rural women were now admitted into the magic circle of potential scientists.

Equally radical was the 'Dazhai model' (Zhao, Woudstra, 2007). Beginning in 1953, just after the village had been organised as a cooperative, the inhabitants of this arid, desperately poor, and environmentally degraded corner of Shanxi Province put together a ten-year plan to transform the landscape using the only resource they had: their own labour. The village headman, Chen Yonggui, led the campaign. Terraces were built, collapsed in the rains, were successfully redesigned and rebuilt, all without machines, cement, or visiting experts. Housing was improved, crops flourished in the painstakingly manured soil of the new terraces. Once the news spread, Dazhai was singled out by Chairman Mao as a model of self-reliance and proof of the capacity of the masses to produce viable science independently. Here the conventional knowledge flow was reversed: cadres came from all around China to 'learn from Dazhai'.

The revolutionary models of agricultural improvement were vehemently anti-technocratic, resisting or inverting 'elitist' expertise. Dazhai represented grassroots

---

7 Agricultural production collapsed during the Great Leap Forward, a Maoist project for self-sufficiency, when villages were urged or forced to produce their own iron, cement, and fertilizer (Wagner, 2011)..

8 Official report, quoted in Saha, Schmalzer, 2016: 156.

authority and a bottom-up flow of expertise in which party leaders learned from peasants, and a tiny village became a model for the nation. The three-in-one groups were equally radical, for they demanded active acceptance by elite members (educated youth, cadres) that uneducated peasants could work with them as equals. The flow of technical knowledge here was designed to eliminate class hierarchies and to serve the revolution. After the death of Mao in 1976, however, mass-science projects like Dazhai were quickly discredited, and China reverted to a model of technocratic governance, with professional scientists and engineers exerting a control over policy unequalled in other nations (Greenhalgh, 2008; Andreas, 2009).

The Road to Unification

In December 1970 the military dictator of South Korea, President Park Chung Hee, opened South Korea's first highway. Completed in a mere three years, the 430 kilometres of the Gyeongbu Highway brought Seoul, the capital, within a few hours' drive of the industrial city of Daegu and the nation's main port, Busan. The road was considered a triumph of what could be achieved by pure patriotic energy and determination, and a key step in the drive to develop and reunite the country (Jeon, 2010).

At the end of the Korean War in 1953 the land lay in ruins, its cities and countryside, factories and fields obliterated by bombing. After many centuries as a united country the nation was partitioned: the Democratic People's Republic of Korea (DPRK) in the North joined the communist bloc, while the Republic of Korea in the South allied with the Western powers.

Under occupation as a Japanese colony, between 1910 and 1945, the northern provinces of Korea, the site of rich mineral resources, were developed into an industrial zone, while the plains of the south were exploited principally as a settler colonialist rice bowl for the Japanese empire (Uchida, 2011; Lee, 2013). After the Korean War ended, with help from the USSR and China the DPRK was able to recover rapidly. Industrial production was restored to 1949 levels by 1959: the DPRK was an international success story of rapid industrial growth. For several years more, GNP and growth rates in the North surpassed the South (Kang, 2008: 142; map of reunified Korea as envisaged by Park in Jeon, 2010: 56, Figure 1). Park Chung Hee made it his mission to overtake the North; building up economic power, he believed, was the key to reunifying the country.

The Gyeongbu Highway was soon hailed as a symbol of the 'Miracle on the Han River', South Korea's post-war economic boom, and the symbolism has weathered well. In 2009 a national archive website repeated that the highway had served as 'a stepping stone for economic growth' (Jeon, 2010: 57). At the time, however, many viewed the highway as a misconceived vanity project. In a survey of Korean transport needs commissioned by the Korean government in 1966, the International Bank for Reconstruction and Development concluded that the money would be better spent on building up networks of smaller roads; it refused to finance the highway project.

Public opinion in Korea was also opposed: the project was extremely costly, and with so few motor vehicles most people believed it would serve no real purpose. Even for supporters, a serious practical obstacle was that the country lacked experienced engineers and construction companies. But Park was determined and forged ahead anyway. In a move evocative of the self-reliance policies of Maoist China and of North Korea, Park's government marshalled local resources to build the highway in record time, without foreign finance or expertise. Army engineers who had served in Vietnam but had never built a big road before were set to designing and building the highway, directing a workforce of civilian contractors and labourers, supplemented by troops, whose work came free. Everyone had to learn on the job, under extreme pressures of time, cost, and shortage of suitable equipment and materials.

'Make it first, fix it later' was the motto. The huge road literally cut corners to save time and money, built in straight lines where curves were advisable; the engineers cut back on necessary depth and width; concrete proved faulty; tunnels collapsed; asphalt was laid too thin. It is no coincidence that the devil-may-care cutting of corners (helpfully labelled 'planned deficiencies') echoed the bravado and dash of an offensive. Park galvanized the nation by framing the highway project as a military operation aimed at defeating the North. In addition to the army engineers, groups of soldiers (providing free labour) were assigned along with civilian work-gangs to work on construction. The frozen hills through which workers had to cut with shovels and picks were designated 'the enemy'. The campaign was victorious: the road was opened after just three years. The fact that it has been in perpetual reconstruction ever since has kept its patriotic symbolism alive and well through five decades (Jeon, 2010).

Japan's Androids: Reproducing Gender in Post-Human Terms

Since the 1970s Japan has been the robot capital of the world. The nation led the way in industrial robotic automation, closely followed by South Korea. In 2016 South Korea was global leader, with 478 units per 10,000 workers, Japan was second with 314 (Robertson, 2017: 17). Then came companion robots. The engaging Aibo dog was first released in 1999. Since then Japanese engineers have produced robots with humanoid features, skin textures, gestures, expressions, and reactions. Some look like glamorous young women, others like moon-men, others again are little more than a pair of sympathetic albeit alien eyes and a gentle voice mounted on a soft, lap-friendly cushion. In 2016 the Japanese Ministry of Economy, Trade, and Industry introduced the Robots for Everyone plan; the government is pouring money into robot development for teaching, substitution for human service staff, and above all for companionship and care, for old people and children (Frumer 2018: 157). The South Korean government is pursuing similar policies (Guevarra, 2015). In an era where the populations of all the advanced economies are greying rapidly, developments in Japan are being eagerly watched.

Here we analyze Japan's humanoid robots not as a technology opening new social horizons, but as a 'retrotech' that applies advanced technology in the service of traditionalism (Robertson, 2017: 23). These sophisticated AI devices constitute a technology of gender, the latest addition to a long line of domestic technologies presented in modern societies as offering personal emancipation to women while reproducing gender roles, and here pushed by a developmental state seeking a silver-bullet technological solution to a double-edged demographic crisis (Robertson, 2017: 17; Frumer, 2018). In the industrial era, new domestic technologies have typically been marketed as labour-saving devices that emancipate women by reducing the burdens of housework. But, as feminist historians have shown, expectations that domestic technology would emancipate women and rewrite gender roles were not met: around the world the fundamental responsibilities of married women to keep house, bear and raise children, and care for sick or old family members were not transformed but rather reinforced by more efficient technologies for preparing meals, doing laundry, or performing childcare (Cowan, 1983; Hayden, 2002).

While gadgets are quite effective at doing laundry, the intensely human tasks of care have proved extremely resistant to technological substitution. In the absence of public care provision or communal facilities, the only married women to escape are those who can afford to revert to the old practice of employing maids and nurses to perform these feminine duties in their place. In Californian middle-class homes the maids and carers are women, often undocumented, from Central America. In Hong Kong and Taiwan they are Filipinas, in the cities of the PRC they come from rural households. The Japanese government, however, allows only a tiny number of overseas nurses and carers (not maids) to enter the country each year. A general aversion to admitting immigrants to Japan is justified in the case of care professionals by the official argument, echoed loudly in the media, that it would be culturally very difficult for Japanese people to feel comfortable in intimate proximity with foreigners (Lan, 2016).

Almost sixty per cent of Japanese women under fifty work outside the home, but without home help to relieve the pressures of the double burden. The Japanese government is calling on more women to join the workforce to revitalize the economy (Prime Minister Abe's 'womenomics') (Robertson, 2017: 12), yet state provision of childcare facilities remains negligible. Increasing numbers of grandparents now live independently, hoping to reduce the burden on their children. This means they are unavailable for childcare, while often requiring help in their own homes. In consequence, marriage rates are falling, Japanese women are failing — or refusing — to have children, and those who have children find it difficult to continue working. Instead of responding to urgent public demand to expand daycare services, the response of the Abe government has been to seek a technological fix, working with AI companies to foster development of 'elder-care, child-care and household "maid" robots [to] lighten the unpaid workload of married women (or women contemplating marriage), making them more receptive to staying married (or getting married) and to having multiple pregnancies' (Robertson, 2007: 372; Robertson, 2017: 24).

Japanese politicians and AI engineers envisage domestic service robots as the ideal wife that a modern Japanese woman needs to combine work and family. By freeing women to bear and raise more children, the robots will by proxy take on the patriotic role of bearing children for the nation. The challenge of caring for fast-growing numbers of dependent elderly, a task almost universally assigned to women and largely devolved in most other rich nations to immigrants, is likewise hampered in Japan by strict immigration quotas and assertions that Japanese cannot relate comfortably to foreigners. Here, too, robots are seen as the ideal solution. There is a vast scientific and popular literature arguing that Japanese are culturally predisposed to relating to robots (see Robertson, 2017; Frumer, 2018 for a survey). In their policy pronouncements and publicity materials, the Japanese government and the AI industry imply that the Japanese public is already happily co-existing with robots and enjoying the interactions and services they offer. But although androids are routinely presented as being integrated into Japanese homes and other institutions, in fact they are still at the stage of experimental prototypes, and Japanese women are taking to social media to mobilize not for more robots, but for more daycare centres (Robertson, 2017: 30, 25).

In Conclusion

The Osaka coking story refutes deep-seated assumptions, dating from the late nineteenth century but alive and well today, that the Japanese were skilled at copying but lacked entrepreneurial spirit and innovative capacity. This dense and complex story of remaking not a machine but its energy source prompts us to rethink technology transfer as a useful category of historical analysis, while highlighting the need to think historically about apparently 'natural' materials like coal, copper, or silk (LeCain, 2017; Onaga, forthcoming).

The 'mass science' campaigns of Maoist China constituted a deliberately anti-technocratic campaign, intended to eliminate class-based hierarchies and revolutionize Chinese society by putting workers and peasants in control of knowledge creation. Critics claim that technocracy is particularly insidious because it effectively disguises political issues as neutral technical problems. Like the Japanese androids, the agricultural terraces of Dazhai draw our attention to the politics of technological design and the power structures of expertise.

In retrospect, the Gyeongbu Highway figures as a logical technical step in Korea's strategy of industrial growth, stimulating the development of motor transport, bringing main centres within easy reach, and lowering transport costs. The Highway did indeed provide infrastructure essential for rapid economic growth. But the fact that the construction project was enacted at the time as an urgent military campaign to defeat communism and reunite the country, together with the enduring material consequences of that urgency, remind us that symbolic goals and the intertwining of material and symbolic efficacy are integral dimensions of any history of technology.

In analyzing Japan's national policy of promoting humanoid robots in the domestic and care sectors as retrotech, we highlight one case among many where the dazzling technical sophistication of AI design and the weighty cognitive or ontological debates around its future serve to obscure the societal and personal politics engineered into these artefacts. The case of Japanese androids contributes new insights to the broader feminist literature on the gendering of technology and the contribution of supposedly emancipatory innovation to maintaining hierarchies.

We have chosen to present case studies located within specific national histories, but our general introduction, with its emphasis on shifting geographies and flows of knowledge and practice, West to East, East to East, East to West, and beyond, has made clear that East Asia is a case par excellence where focussing on individual nations distorts historical understanding of the past and the present. In the modern, connected world, there is no such thing as a 'Chinese typewriter' or an 'American missile' (Mullaney, 2016; Krige, 2019). Yet, while many historians ardently dispute the validity of regional histories, for the history of technology it does make sense to look at East Asia as a region: internal and external exchanges and circulation of knowledge were both formative, and the East Asia case refutes Western-centred history of technology by clearly illustrating that single-centred networks do not work as an explanation: global modernity has no untouched backwaters or independent states (Mullaney, 2017: 29–30).

References

Abe, Takeshi, *Kindai Ôsaka keizai shi [Economic History of Modern Osaka]* (Osaka: Osaka Daigaku Shuppankai, 2006).

Adas, Michael, *Machines as the Measure of Men: Science, Technology, and Ideologies of Western Dominance* (Ithaca, NY: Cornell University Press, 1989).

Andreas, Joel, *Rise of the Red Engineers: the Cultural Revolution and the Origins of China's New Class* (Stanford, CA: Stanford University Press, 2009).

Armstrong, Charles K., 'Kang Ho-je, Pukhan kwahak kisul hyŏngsŏngsa I [History of Science and Technology in North Korea I]', *East Asian Science, Technology and Society*, 3.4 (2009), 543–46.

Arnold, David, and Erich DeWald, 'Everyday Technology in South and Southeast Asia: An Introduction', *Modern Asian Studies*, 46.1 (2012), 1–17.

Bray, Francesca, 'Technology', in *The Making of the Human Sciences in China: Historical and Conceptual Foundations*, ed. by Howard Chiang (Leiden: Brill, 2019*), pp. 29–51.*

Cai, Peter, *Understanding China's Belt and Road Initiative* (Sydney: Lowy Institute, 2017) <http://hdl.handle.net/11540/6810> [accessed 22 November 2022]

Carré, Guillaume, *Avant la tempête: la Corée face à la menace japonaise 1530–1590* (Paris: Edition de Boccard, 2019).

China: *The Dragon's Ascent (Part 8: The Dragon Ascends: Creating China's Future)* (Films Media Group, 2000) http://www.films.com/id/ecTitleDetail.aspx?TitleID=6226 [accessed 22 November 2022]

Choi, Hyungsub, 'The Social Construction of Imported Technologies: Reflections on the Social History of Technology in Modern Korea', *Technology and Culture*, 58.4 (2017), 905–20.

Cowan, Ruth Schwartz, *More Work for Mother: the Ironies of Household Technology from the Open Hearth to the Microwave* (New York: Basic Books, 1983).

Cox, Rupert, ed., *The Culture of Copying in Japan: Critical and Historical Perspectives* (London & New York, NY: Routledge, 2007).

Cumings, Bruce, 'Historical Setting', in *North Korea: A Country Study*, ed. by Robert L. Worden, 5th edn (Washington, D.C.: Library of Congress, 2008), pp. 1–58.

Elman, Benjamin A., *On Their Own Terms: Science in China, 1550–1900* (Cambridge, MA: Harvard University Press, 2009).

Finlay, Robert, *The Pilgrim Art: Cultures of Porcelain in World History* (Berkeley, CA: University of California Press, 2010).

Fitzgerald, Deborah, 'Blinded by Technology: American Agriculture in the Soviet Union, 1928–1932', Agricultural History, 70.3 (1996), 459–86.

Francks, Penelope, 'Rice and the Path of Economic Development in Japan', in *Rice: Global Networks and New Histories*, ed. by Francesca Bray, Peter A. Coclanis, Edda L. Fields-Black, and Dagmar Schäfer (Cambridge and New York, NY: Cambridge University Press, 2015), pp. 318–34.

———, and Janet Hunter, *The Historical Consumer: Consumption and Everyday Life in Japan, 1850–2000* (New York, NY: Palgrave Macmillan, 2012).

Frumer, Yulia, 'Cognition and Emotions in Japanese Humanoid Robotics', *History and Technology*, 34.2 (2018), 157–83.

Fulton, John, *Coke: A Treatise on the Manufacture of Coke and the Saving of By-Products* (Scranton, Penn.: The Colliery Engineer Co., 1895).

Gay, Paul du, Stuart Hall, Linda Janes, Anders Koed Madsen, Hugh Mackay, and Keith Negus, *Doing Cultural Studies: The Story of the Sony Walkman* (Maidenhead and New York, NY: The Open University Press, 1997).

Goto, Y., *Kôjo tsûran kaidai [Comment on Factories Overview]*, 1986

Greenhalgh, Susan, *Just One Child: Science and Policy in Deng's China* (Berkeley, CA and London: University of California Press, 2008).

Guevarra, Anna Romina, 'Techno-Modeling Care: Racial Branding, Dis/Embodied Labor, and "Cybraceros" in South Korea', *Frontiers*, 36.3 (2015), 139–59, 195–96.

Hayden, Dolores, *Redesigning the American Dream: The Future of Housing, Work, and Family Life* (New York, NY and London: W.W. Norton, 2002).

Headrick, Daniel R., *Power over Peoples : Technology, Environments, and Western Imperialism, 1400 to the Present* (Princeton, N.J.: Princeton University Press, 2010).

Hunter, Janet, *'Deficient in Commercial Morality'? Japan in Global Debates on Business Ethics in the Late Nineteenth and Early Twentieth Centuries* (London: Palgrave Macmillan UK, 2016).

Jeon, Chihyung, 'A Road to Modernization and Unification: the Construction of the Gyeongbu Highway in South Korea', *Technology and Culture*, 51.1 (2010), 55–79.

Jones, Christopher F., *Routes of Power: Energy and Modern America* (Cambridge, MA: Harvard University Press, 2014).

Kang, David, *'The Economy'*, in *North Korea: A Country Study*, ed. by Robert L. Worden, 5th edn (Washington, D.C: Library of Congress, 2008), pp. 135–80.

Kelly, Dominic, 'Ideology, Society, and the Origins of Nuclear Power in Japan', *East Asian Science, Technology and Society*, 9.1 (2015), 47–64.

Kobiljski, Aleksandra, 'Energy Workarounds: Designing Coals for the Japanese Steel Industry, 1895–1911', *Technology and Culture*, 63.2 (2022), 326–48.

Krige, John, *How Knowledge Moves: Writing the Transnational History of Science and Technology* (Chicago, IL and London: University of Chicago Press, 2019).

Lan, Pei-Chia, 'Deferential Surrogates and Professional Others: Recruitment and Training of Migrant Care Workers in Taiwan and Japan', *Positions: Asia Critique*, 24.1 (2016), 253–79.

Landes, David S., *The Unbound Prometheus: Technological Change and Industrial Development in Western Europe from 1750 to the Present* (Cambridge: Cambridge University Press, 1969)

———, *The Wealth and Poverty of Nations: Why Some Are So Rich and Some So Poor* (New York, NY: W. W. Norton & Company, 1999)

Lavelle, Peter, 'Agricultural Improvement at China's First Agricultural Experiment Stations', in *New Perspectives on the History of Life Sciences and Agriculture*, ed. by Denise Phillips and Sharon Kingsland (Cham: Springer International Publishing, 2015), pp. 323–44.

LeCain, Timothy J., *The Matter of History: How Things Create the Past* (Cambridge and New York: Cambridge University Press, 2017).

Lee, Jung, 'Invention without Science: "Korean Edisons" and the Changing Understanding of Technology in Colonial Korea', *Technology and Culture*, 54.4 (2013), 782–814.

Lee, Seung-Joon, 'The Patriot's Scientific Diet: Nutrition Science and Dietary Reform Campaigns in China, 1910s–1950s', *Modern Asian Studies*, 49.6 (2015), 1808–39.

———, 'Taste in Numbers: Science and the Food Problem in Republican Guangzhou, 1927–1937', *Twentieth-Century China*, 35.2 (2010), 81–103.

Lin, James, 'Sowing Seeds and Knowledge: Agricultural Development in Taiwan and the World, 1925–1975', *East Asian Science, Technology and Society*, 9.2 (2015), 127–49.

Low, Morris, 'Displaying the Future: Techno-nationalism and the Rise of the Consumer in Postwar Japan', *History and Technology*, 19.3 (2003), 197–209.

McDonald, Kate, 'Asymmetrical Integration: Lessons from a Railway Empire', *Technology and Culture*, 56.1 (2015), 115–49.

Misa, Thomas J., *A Nation of Steel: The Making of Modern America, 1865–1925* (Baltimore, MD: Johns Hopkins University Press, 1998).

Mokyr, Joel, *The Lever of Riches: Technological Creativity and Economic Progress* (New York, NY and Oxford: Oxford University Press, 1992).

Moore, Aaron Stephen, *Constructing East Asia: Technology, Ideology, and Empire in Japan's Wartime Era, 1931–1945* (Stanford: Stanford University Press, 2013).

———, 'Japanese Development Consultancies and Postcolonial Power in Southeast Asia: The Case of Burma's Balu Chaung Hydropower Project', *East Asian Science, Technology and Society*, 8.3 (2014), 297–322.

Morris-Suzuki, Tessa, *The Technological Transformation of Japan : From the Seventeenth to the Twenty-First Century* (Cambridge: Cambridge University Press, 1994).

Mullaney, Thomas S., *The Chinese Typewriter: A History* (Studies of the Weatherhead East Asian Institute, Columbia University) (Cambridge, MA: The MIT Press, 2017).

———, 'Controlling the Kanjisphere: The Rise of the Sino-Japanese Typewriter and the Birth of CJK', *The Journal of Asian Studies*, 75.3 (2016), 725–53.

Nakaoka, Tetsurô, *Nihon kindai gijutsu no keisei: dentô to kindai no dainamizumu [The Formation of Modern Japanese Technology: A Dynamics of 'Tradition' and 'Modernity']* (Tokyo: Asahi Shinbunsha, 2006).

Oldenziel, Ruth, and Karin Zachmann, eds., *Cold War Kitchen: Americanization, Technology, and European Users* (Cambridge, MA: MIT Press, 2009)

Onaga, Lisa, *Cocoon Cultures: The Entangled History of Biology and Silk in Modern Japan* (Durham, NC: Duke University Press, forthcoming)

Riello, Giorgio, *Cotton: The Fabric That Made the Modern World* (Cambridge and New York: Cambridge University Press, 2013).

Robertson, Jennifer, 'Robo Sapiens Japanicus: Humanoid Robots and the Posthuman Family', *Critical Asian Studies*, 39.3 (2007), 369–98.

———, *Robo Sapiens Japanicus: Robots, Gender, Family, and the Japanese Nation* (Berkeley and London: University of California Press, 2017).

Saha, Madhumita, and Sigrid Schmalzer, 'Green-Revolution Epistemologies in China and India: Technocracy and Revolution in the Production of Scientific Knowledge and Peasant Identity', *BJHS Themes*, 1 (2016), 145–67.

Samuels, Richard J., *'Rich Nation, Strong Army': National Security and the Technological Transformation of Japan* (Ithaca, NY: Cornell University Press, 1994).

Schäfer, Dagmar, Giorgio Riello, and Luca Molà, eds., *Threads of Global Desire: Silk in the Pre-modern World* (Pasold Studies in Textile History) (Woodbridge, Suffolk, UK and Rochester, NY, USA: The Boydell Press, 2018).

Siu, Helen F., and Mike McGovern, 'China–Africa Encounters: Historical Legacies and Contemporary Realities', *Annual Review of Anthropology*, 46.1 (2017), 337–55.

Sougy, Nadège, *Les charbons de la Nièvre, la houillère de La Machine, ses produits et ses marchés (1838–1914)* (Grenoble: PUG, 2008).

Stavis, Benedict, *The Politics of Agricultural Mechanization in China* (Ithaca, NY: Cornell University Press, 1978).

Uchida, Jun, *Brokers of Empire: Japanese Settler Colonialism in Korea, 1876–1945* (Harvard East Asian Monographs 337) (Cambridge, MA and London.: Harvard University Asia Center, distributed by Harvard University Press, 2011).

Vogel, Ezra F., *The Four Little Dragons: The Spread of Industrialization in East Asia* (The Edwin O. Reischauer Lectures 1990) (Cambridge, MA and London: Harvard University Press, 1991).

———, *Japan as Number One* (Cambridge, MA: Harvard University Press, 1979).

Volti, Rudi, *Technology, Politics and Society in China* (Westview Special Studies on China and East Asia) (Boulder, CO.: Westview Press, 1982).

Wagner, Donald B., 'Background to the Great Leap Forward in Iron and Steel', 2011 <http://donwagner.dk/MS-English/MS-English.html> [accessed 29 November 2022]

Waley-Cohen, Joanna, *The Sextants of Beijing: Global Currents in Chinese History* (New York, NY: W. W. Norton & Company, 1999).

Wittner, David, *Technology and the Culture of Progress in Meiji Japan* (New York, NY: Routledge, 2008).

Yonekura, Seiichiro, *The Japanese Iron and Steel Industry 1850–1990: Continuity and Discontinuity* (New York, NY: St. Martin's Press, 1994).

———, 'Steel Exports Report: Japan', *Global Steel Trade Monitor*, March 2017, p. 1.

Zhao, Jijun, and Jan Woudstra, '"In Agriculture, Learn from Dazhai": Mao Zedong's Revolutionary Model Village and the Battle against Nature', *Landscape Research*, 32.2 (2007), 171–205.

DHRUV RAINA, OMPRASAD, NAYANI SARMA

Framing the History of Technology in India

From the Colonial Era to the Digital Age

In order to elaborate upon the historiography of technology in India, there is the need to delimit the universe of discourse. Since the volume itself is concerned with technology in the nineteenth and twentieth centuries, the field of discussion is well circumscribed. Furthermore, the objective of this historicization concerns a range of technological systems and regimes that developed in the wake of the Industrial and subsequent technological revolutions. Within economic history, there has been a long tradition of research that has engaged with the history of the Industrial Revolution and that of colonialism (Hobsbawm, 1999). Perhaps that would be a convenient starting point for this historiographical essay.

The modern history of science as different from the history of modern science of India was inaugurated by the British orientalists in the last decades of the eighteenth century, and was subsequently pursued by German and French scholars followed by Indians who jumped into the field in the last quarter of the nineteenth century (Colebrooke, 1817; Ray, 1902; Seal, 1915). There were natural differences between these traditions that extended to their historiographical moorings and objects of investigation. The modern history of technology dealing with the history of modern technology does not appear to have such distinguished origins. This does not mean that British or French orientalists were not interested in the precolonial technical sub-systems of the subcontinent (Deloche, 1980, 1983). Besides, recent studies on cotton and globalization during the first Industrial Revolution have extensively stressed that British and European growth largely relied on reinterpretation of the products and on the relocation of production in the West (Riello, Parthasarathi, 2009; Riello, 2013). East India Company officials followed by British colonial officials in the late eighteenth and first half of the nineteenth century were positively captivated by the diversity, competence, and technical heritage of the subcontinent (Baber, 1996). The travelogue of the French astronomer Le Gentil documents the speeds of the local caravels operating between the South China Sea and India's eastern coastline (Le Gentil, 1779–1781). In fact, this gaze was quite in contrast with the orientalist account of the sciences of India, at least the accounts as they began to appear in the second half of the nineteenth century.

These first-hand accounts by colonial officials of local technical practices cannot be considered history, though they fall at the intersection of ethnography and the

Global History of Techniques, ed. by Guillaume Carnino, Liliane Hilaire-Pérez and Jérôme Lamy, Global Matters Culture Matérielles, histoire des techniques et circulation globale des savoirs, 9 (Turnhout: Brepols, 2024), pp. 117–134
BREPOLS PUBLISHERS 10.1484/M.TECHNE-EB.5.129804

technical (Dharampal, 1971). In any case British orientalists were certainly not preoccupied with detailing the cultures of techniques in the region. But technology figured in significant ways in the literary and cultural imaginaries of the time. By the second half of the nineteenth century, with the gathering reaction of Victorian England to industrial culture, the arts and crafts movement surfaced as a requiem to industrial civilization (Triggs, 1902), and an Indian counterpart of this movement that possibly inspired Havell's work (Havell, 1907). In any case, the threat posed to traditional arts and crafts promoted a revival of scholarly interest and institutional pressure for its renewal through education (Raina, Habib, 1991).

The question concerning technology was possibly raised in attempts to understand the role of technology in India's colonization, and to understand the causes of India's economic and industrial backwardness and the West's political ascendancy (Naoroji, 1901; Dutt, 1908; for a recent look at the reversal, see Mishra, 2012). But it was only in the twentieth century, it could be claimed, that questions concerning technology and society in India began to be raised within disciplinary academic communities, though not by historians of technology proper. Studies on the history of technology sought to situate their place and role in the colonial economy and the processes of modernization and industrialization (Bagchi, 1972; Tyabji, 1995). Thus, writing about technology can be examined through three frames detailing this history and employing distinct historiographical and methodological protocols. These frames are also temporally distinct and overlap. An additional frame runs across these periods, a frame that offers a celebratory and triumphalist account of technology, imbued as it is with Baconian optimism. The publications appearing in the *Indian Journal of History of Science* on technology subscribe to the internal account of the unfolding of technological development wherein the logic of efficiency is the immanent logic of scientific development.

On the other hand, journals of history proper have more comprehensively covered the history of technology; in particular, two journals worthy of mention are the *Indian Economic and Social History Review* and *Studies in History*. In this case, as opposed to the previous one, the emphasis is on history proper rather than what concerns history of technology per se. Which leaves us to ask the question what the history of technology could be or is? Without being too normative and universalize the agenda of the history of technology, the minimalist objectives could be to study the nature and rate of technological change, and the identification of factors governing the rate of change. In the latter case, the history of technology becomes an interdisciplinary field intersecting with several of the social sciences and environmental science. These questions could be posed in a more generic sense for technologies everywhere, though at different times, or could as well be asked when localized to the specific Indian experience, as we do here. However, an overarching topos of technological change in the contemporary world is framed by the idea of globalization, which in turn manifests itself in Western 'diffusionism', and it is only with the emergence of the study of technological systems that scholars have begun to engage with an orchestra of 'mutually adjusting processes' (Miller, 2017).

Colonialism and Modernization

Fig. 1. Passenger train crossing the Dapoorie Viaduct (first railway bridge in India) from Bombay to Thane in 1855. https://scroll.in/article/669579/the-long-history-of-indian-railways-told-through-images

As far as India is concerned, the historiography of technology unfolds along three axes as suggested earlier. Thematically the first has to do with the anchoring of a new technological regime and technological systems in India in the late eighteenth and nineteenth centuries. The title of the two books by Daniel Headrick, namely *The Tools of Empire* and *Tentacles of Progress*, ground the coordinates of this frame, for modern technology in the hands of the colonial rulers became an instrument for the extension and consolidation of British rule in India. Headrick's first book, in fact, synthesizes and develops a perspective about technology and imperialism that was developing through the 1960s and 1970s, pointing out how changes in the motives and means of technological intervention helped colonial power consolidate their hold over the colonies in three phases of colonial expansion, namely penetration, conquest, and consolidation (Qaiser, 1982). These technological regimes he categorizes as tools of penetration (steamboat and quinine), tools of conquest (guns), and most importantly, consolidation (communication technologies). The work is of a comparative nature, discussing Africa and contrasts the process with what happened in India (Headrick, 1981).

At the same time, technology was also the agent for the transformation of a traditional society into a modern one. A great deal of this historical writing was premised on a kind of social determinism. By the time Headrick's books were published, and in the books themselves, the questions formulated became more nuanced and complicated, and so did the answers. Within this diffusionist model, while the East India Company and then colonial officials focussed on the consolidation of their power in India and expansion of their productive resources, the technical innovations introduced generated interest among the Indians. Detailed instances of the introduction of steam vessels and the railways, as well as the electric telegraph, illustrate the case (Sangwan, 1988) (Figure 1).

In a paper published several decades later, Headrick elaborated upon the dual role played by communication technologies in colonial India. The British colonial regime invested enormously in communication technologies and networks in the nineteenth century to bind their colonies into a cohesive empire. The expansion of communication technologies also played a decisive role in the emergence of nationalist sentiments among the natives by enabling a more regular flow and exchange of information across the colonies (Headrick, 2010). However, with the global turn in the history of technology, another perspective began to emerge in the first decade of the twenty-first century, which did not refute this historiography but examined the process of drawing new nations and communities into the web of communication. But more on that later.

In his second book, Headrick argued that the massive transfer of technology from the European imperial powers to their colonies in Asia and Africa led to underdevelopment rather than industrialization (Headrick, 1988). This was in agreement within the emergent historiography of economic history. The colonial regime focussed on the geographic transfer of technology instead of its cultural diffusion by limiting the access of colonies to technical education and discouraging non-European entrepreneurship. These were and became the focal themes of much research that followed and preceded the publication of Headrick's book. In a comparative perspective, the question of the transfer of technology and the industrialization of India, China, and Japan, Inkster's studies argued that the relative political sovereignty of China and Japan provided conditions that were far more salubrious for the adoption, diffusion, and adaptation of technology than did colonial India (Inkster, 1988).

Earlier studies were premised on a linear and reductionist relationship between technological invention and endowment, and on the other projects of colonial expansion and domination. Consequently, there was a brutal efficiency about the way in which technology changed the landscape of a 'traditional and backward society' into a modernizing one. As the inadequacies of technological determinism became evident, the faith in the inevitable unfolding of the logic of technological efficiency declined at least within the imaginaries of students of technology. Thus, while on the one hand the role of the railways established in British India is largely understood within a discourse of economic exploitation, on the other historians also recognize its significance in weaving together the Indian nation (Chatterjee, 2018; Kerr, 2006, 1995). But inasmuch as the colony was a laboratory and site of experimentation, the railways too served as such a space for both Indians and colonial officials. Removed from their communities, vegetarians travestied social taboos and experimented with other cuisines, while traditional social taboos against travelling began to erode. On the other hand, Europeans travelling the Indian railways were offered the spectacle of a 'nominal provincial Europe' on wheels. The social imaginary kept shifting over the next century. Commencing as a symbol of colonial oppression and exploitation, it was seized by Indians and blended into their dreams of nationhood (Chatterjee, 2018; Pillai, 2019). Moving away from the diffusionist turn, the newer studies on the railways have begun to emphasize the role of local knowledge(s) and human resources and understanding in the making of the railways in India (Mukhopadhyay, 2018).

The other side of the story, of course, is that this connectivity forged across the subcontinent by steam locomotive communication was rendered possible at an ecological price that entailed the denudation of large sub-Himalayan forest tracts — the timber was required for fuel and railway sleepers necessary for laying railway lines (Das, 2010). The nuance of the historian's argument was not merely an outcome of finer and detailed archival research, but the historian's craft was as much a product of the constantly shifting technological unconscious of the time.

We could argue that, beyond the scaffoldings offered by technological and social determinism, innovation studies have also preoccupied students of technology and society, but the two scaffoldings just mentioned have had such a captivating influence that a more contingently focussed evolutionary history has yet to emerge. One of the features of the historiography of this period is that, in the literature of the time, 'modern science and technology' is presented as a singular frame, with technology being considered an outcome of the application of scientific principles — in other words, modern science and technology were seen as components of a unitary system. On examination of nineteenth century writing in English, one reckons with different responses to the local knowledge(s) of the region, wherein the so-called sciences of India, when looked at through the epistemic frame of modern science, is looked at differently from technical practices. When these techniques are compared to industrial technology, does the perspective begin to diversify — the deficiencies of technical practices of the region in the light of industrially mass-produced goods are one frame through which technology is looked at. In the early decades of the twentieth century, in the Swadeshi movement and the rise of Gandhian thought, a critique of industrial civilization and the homogenization of the industrial product begins to be articulated (for the rise of economic thought in the articulation of nationalism, see Chandra, 2010; for the Gandhian critique, Gandhi, 2014).

From Decolonization to Postcolonial Hybridity

The frame of technological determinism was widely prevalent during the high tide of modernization theory in the social sciences. At this conjuncture of social theory, technology was seen as ideologically neutral and value-free: this, as others have argued, was concurrent with the Cold War ideology of science. The economic theories of Kondratieff and Rostow's input — output growth models dominated the unconscious of science and technology in the developing and decolonizing worlds of the 1950s and 1960s (Escobar, 2011: 269–89).

The paradigms of modernization and development theory were challenged by the failure of technology and development to deliver on their promise by the end of the 1960s and the early 1970s (Raina, 2003). The introduction of modern technological systems and the delayed industrialization of former colonial nations demonstrated the failure of technology transfer models to produce an economic miracle in these very societies and nations. This failure led to the recognition of the social and cultural life of technology (Elzinga, Jamison, 1986). This recognition was concurrent with the ascent in studies of the social construction of technology in the Northern Academy (Harding, 1998). We shall skip a discussion here of the sources of the challenge posed for modernization theory as well, and restrict our comments to the mention that these developments in social theory were deeply entangled. During the heyday of modernization theory the historian of Mughal India, Irfan Habib, had posed the Indian equivalent of the Needham question concerning technology and the non-emergence of capitalism and the industrial revolution in the region (Habib,

1969).[1] In the 1990s, the world systems theorist Andre Gunder Frank was to press for a reframing of the timeline of the Needham question given that sixty-three per cent of world markets were concentrated in India and China till the middle of the eighteenth century (Frank, 1997).

Globally, too, studies in 'technovation' had shifted focus from the inventor as technological hero to the social context of technological innovation as well as the failure of innovation (Buchanan, 1991; Braun, 1992). This movement corresponded in India with the beginning of the post-modern/post-colonial turn, as modern technological systems came to be seen as regimes for the domination of India by the West (Sachs, 1992; Escobar, 2011; within a modernist frame, an attempt to arrive at a balanced perspective is offered by Salomon, Lebeau, 1993). Modern science and technology had become an exploitative, patriarchal system for the economic exploitation of India and the destruction of its ecological and natural resources (Shiva, 1988). The *glasnost* in the social theory of science and technology spawned a new set of questions, concerns, and methods.

Thus, so-called 'Western' ideas of technology and culture and its Eurocentric assumptions were up for interrogation, and Claude Alvares in *Homo Faber* affirmed that Asian countries had their own cultures of technology that were disrupted and distorted by European colonialism (Alvares, 1979, 1992). Abandoning the frame of technological neutrality, the work was canonized as one of the clarion calls for resurrecting appropriate technologies that drew upon local technical knowledge that had been passed down over centuries. In other words, there arose the recognition that technological knowledge was passed on like genetic material or memes, and ways of engaging with them could be developed (Goonatilake, 1984, 1998; Reddy, 1978a). In other words, the history of technology was to provide political and value lessons for the future construction of technology and technological systems.

Naturally, in the process of decolonization the tenets of the diffusion of Western technology to India would have been replaced by a 'reverse diffusionism' that suffers from the same failings as diffusionism (Miller, 2017: 263). Miller's study on stationary steamboats and steam engines takes up the relocation of 'Western' technology in India, pointing out that constraints of costs, durability, and availability of materials and technical parts, and a number of other factors, resulted in the hybridization of steam engines and the concomitant enrichment of the repertoire of skills of Indians and Europeans in India (Miller, 2017: 270–71).

Departing from diffusionism and deficit models that sought to explain the tardy reception or the rejection of new technologies and technological systems into India in terms of local knowledge, learning, and cultural practices, the new studies animated this context, bringing it alive rather than approaching local and cultural contexts through the hermeneutics of impediments or suspicion (Sen, 1988). These studies drew upon more interactionist frameworks engaging with the co-constructedness of technological systems and regimes in which human and natural resources, cultural

---

1 Headrick too raises the question why the former colonies of the world failed to industrialize despite the massive technology transfers of the imperial era.

practices played a role in processes of domestication, appropriation of these systems as well as to processes of resistance and rejection (Bhattacharya, 1966). From the wider social context, three conceptually linked developments resonated within the history of technology whose outcome was the appearance of new genres and historiographical framings.

This was the second moment in the history of nineteenth- and twentieth-century technology. The referred-to 'crisis in development' and the expanding awareness of the negative externalities of industrial expansion and consumerism turned the focus to environmental degradation and the marginalization of indigenous communities in the developmental process (Raina, Chowdhury, Chowdhury, 1997). The concrete search for alternatives to renewable energy spurred on the movement for technological alternatives. In India, the idea of the social constructedness of technology — the rejection of what Hacking called the inevitability of the existing regime — gathered momentum in the search for alternatives (Reddy, 1978a, 1978b). Gandhi's *Hind Swaraj*, which was a concerted critique of late-nineteenth- and early-twentieth-century industrialism, was reread by neo-Gandhians who articulated a civilizational critique of 'Western technology'. Retrospectively seen, these framings were, in terms of East and West, a sort of post-colonial reconfiguration of an older binary of the colonizer and colonized (Nandy, 1980, 1983).

But by the late 1980s, this space created by neo-Gandhian thought opened up a window for thinking about sustainable futures. This belongs to the contemporary history of technology and requires a different starting point. In studying the multiple processes referred to earlier, the inadequacy of earlier framings of Indian responses to technology — be they anti-technology, traditionalist, Luddite — became evident and failed to capture essential elements of the process. As part of this interrogation, studies of India's industrial research system represented by the network of laboratories of the Council of Industrial and Scientific Research (CSIR) were also undertaken (Visvanathan, 1984). Shiv Visvanathan's doctoral study of the CSIR's National Physical Laboratory was one of the first laboratory studies on the subcontinent but made the important point of linking the sociology of science and technology to the sociology of development. At this new moment in the constructivist turn, the earlier critiques or rejection of technology and technological systems began to be seen as part of the response of the cultural assimilation of technology (Elzinga, Jamison, 1986). Technology became an important object of investigation in environmental history, and when the negative social and ecological impact of the green revolution began to show up, these critiques began to show up more substantially within the social studies of science and technology (Shiva, 1991; Yearley, 1988; Parayil, 1992). Thus, more celebratory and heroic accounts of technology made way for more analytical and critical accounts of technology. This created a space for the recovery of lost histories of technical practices and more substantial engagement with the nature of the relationship between science and technology and modern technological life forms (Alvares, 1979, 1992). If economic history, in part inspired by Marxism, had provided the intellectual impetus for understanding the nature of technology and technological change, now anthropology and the interdisciplinary science of ecology were doing so (Guha, Gadgil, 1993). Technology thus became a major exogenous factor in discussions on environmental history — though for many, analytically speaking, it was never entirely exogenous but endogenous to certain forms of life.

Part of the post-colonial turn was rendering visible those marginal and less than visible contributions and contributors erased by triumphalist and heroic narratives of technological history. By the 1980s, Roy Macleod and others had revealed that while the laboratory of science and technology was the colony, local knowledge and practices played a significant role in reconstituting the identity and place of modern technology (Russel, Macleod, 1979). Thus, the Thomason College of Engineering at Roorkee, now the elite Indian Institute of Technology, was established in the late nineteenth century in British India before an engineering college was established in England. This experience, down to the level of the creation of textbooks and syllabi, informed subsequent efforts in technical education in Europe (Mittal, 1986).

This was particularly manifest in the domain of technical education and education more generally. The earlier histories of technical education were anchored within a framework of the modernization of education that was diffusionist in its explanatory orientation, wherein the social milieu was either docile, receptive, or posed impediments, but did not actively shape the institutions or content of technical education. These explanatory forms dovetailed with a larger critique of the colonial project that evoked the economic motives of the colonial state both as enabling or disabling as the context required, as well as the demands of its industrial constituencies.

The postcolonial turn emphasized the active role of the social cultural milieu in shaping several of these colonial projects, resisting them when required, and appropriating them outside the precincts of state-led enterprises in order to establish colleges and universities through voluntarist efforts for nationalist ends (Raina, 2011). While the demand for agricultural education picked up momentum in the last decade of the nineteenth century, it was left to a German agronomist, Voelcker, to point out that it would be disastrous to dismiss local knowledge about agricultural practices, crops, and soils that had evolved over the centuries, was validated over historical time, and was dynamically changing. The report played a role in shaping the content of agricultural education in a limited way, for many of its recommendations were ignored due to the compulsions of the colonial state (Raina, 1997). Nevertheless, education was a domain wherein the demands for naturalization were the most pressing. European industrial houses and the incipient national bourgeoisie had quite distinct expectations from technical education and lobbied for their respective demands (Bhargava, 2018). The professionalization and expansion of the community of engineers was accelerated by the processes of industrialization and indigenization, and within decades of the freedom struggle and after, was shaped by the political project of nation building premised on scientific and technological self-reliance (Ramnath, 2017).

The rise of nationalism towards the last decades of the nineteenth century provided a new pivot around which the concerns of technology and nationalism were reflected in debates over engineering and mistrification — in other words, colonial economic interests had a role to play in regulating the rate of technology transfer (Habib, 1995). The First World War and the need to pump up Britain's war-torn economy led to the colonial government loosening its grip over the transfer of technology. Furthermore, by the end of the nineteenth century the character of the colonial state had begun to change — from a laissez-faire state of the nineteenth century to a state responsible to its citizens (Raina, Jain, 2004). This combination of factors significantly altered the circulation of some of the technologies from the West into India. Studies seemed to place greater emphasis on the central role of the colonial state and the role of state actors in the making of industrial and technology policy. By contrast, there were a variety of state actors and a number of non-state actors, such as colonial officials, officials employed by the rulers of indirectly ruled India, educationists, Indian entrepreneurs, landlords, and local rulers, who were involved in diverse projects of developing technological institutes (Ambirajan, 1995; Tyabji, 2011).

The general historiographical sentiment until the 1970s was that, directly administered, India was technologically more advanced than the rest, for they were better administered, and that the colonial state had invested more heavily in them. With the rise of regional studies and regional history in India in the 1970s, this broad historiographical trope was reversed in a qualified manner (Hettne, 1976). The history of technical education in the indirectly ruled states of Baroda, Cochin, Mysore, and Travancore unfurled along pathways quite at odds with directly administered India — something that continues to be reflected in educational systems of three of these four regions (Raina, 2012). In other words, these regions exhibited a trajectory of modernization that was, if not radically distinct, at least different from that of colonial modernity. Some of the national icons associated with the building of dams and the planning of cities were employed in these states — Chetput Pattabhiraman Ramaswamy Iyer and Mokshagundam Visvesvaraya, etc.

The history of technical education, then, between the last three decades of the nineteenth century and the first two decades of the twentieth century, revealed that programmes of technical education had to engage with local knowledge and the existing material culture of technical education. As a result, the establishment of institutions of technical education through voluntarist efforts by Indians reflected deeper engagements with the process of naturalization and domestication, and out of this engagement modern institutions of first technical and then technological education were assembled (see three chapters on technical education in Habib, Raina, 2004).

Secondly, the process of establishing technical and technological institutes that furnished the human resource capabilities for modernization was previously seen in diffusionist terms, or as one of cloning Western institutions. With the regional and postcolonial turns, the process of improvization, innovation, and acculturation were rendered visible. Furthermore, discussion on technical education among Indian educationists in the 1920s and 1930s suggests a concern that the curriculum for technical education was tailored for producing a cadre of professionals qualified to maintain government public works or fitting into existing systems of industrial production and not towards envisioning and developing new technical systems. These discussions were more or less concurrent with the rise of engineering science in the United States of America. In India, the institutionalization of engineering education on the MIT model manifested itself in the decade following the end of colonial rule (Bangaru, 2004). We shall return to the historiographical import of these developments.

Global Technoscience

By the early 1980s, India emerged as a player in the global world of technology having established credible nuclear and space research programme capabilities and installing its own version of the supercomputer. The development of its research facilities and with an ample human resource bank in advanced institutes of scientific and technological research distributed globally, it joined the genetic and information technology revolution and created its very own Silicon Valley in Bangalore (Saini, 2012). Two families of meta-narratives could be identified, both premised on a break with the linear theory of technological evolution and with linear historical narrative; and in its place we have a web of connected histories, intersecting timelines, and multiple rhythms of technological development. For example, Geijerstam's book entitled *Landscapes of Technology*, though dealing with the nineteenth century, examined the installation of two iron works in India in the second half of the nineteenth century by three Swedish engineers (Geijerstam, 2004). The author chose to adapt the ideas of socio-technical systems and social carriers of technology proposed by Thomas P. Hughes. In a comparative context the author unveiled an unknown history of the ironworks fabricated by these Swedish engineers and examined the differences in iron production between India and Sweden. The Swedish engineers in India benefitted from international networks of the iron-making trades, and knowledge moved fairly rapidly across continents despite what appears by contemporary standards to be fairly sluggish modes of communication. The socio-technical system that had been elaborated at the iron works was fairly brittle since the boundaries between the British management, the Swedish engineers, the British skilled workmen, and the Indian workmen were rendered vulnerable by the conflict-ridden nature of the relationship between the different groups of actors.

Secondly, the period was concurrent with the adoption of the Latourian concept of technoscience and the adapted hybrid concept of postcolonial technoscience (Latour, Woolgar, 1979). Taking a cue from postcolonial studies, Warwick Anderson proposed that the concept of postcolonial technoscience in science studies played the role of destabilizing received framings and concepts of the history of technology, bringing in more heterogeneous sources and puncturing universalist claims (Anderson, 2002). The insights provided scholars an opportunity to navigate the hitherto considered binaries of traditional techniques and modern technology. Furthermore, this required that the metropole and the colony were to be examined within the same analytical frame in order to overturn 'conventional accounts of Western technoscience'. This destabilization is concurrent with the 'decline of the nation state', which results in the recognition of a 'proliferation of hybrid identities' and the imperative adoption of interdisciplinary and multi-cited studies of technoscience.

In response to Anderson's essay, Itty Abraham argued that the importance placed on 'local indigenous knowledge' doesn't necessarily make the nation state go away but could on the contrary reproduce the nation or shall we add indigenism. Thus, a postcolonial technoscience which is just about contact zones of 'clashing knowledge' is incomplete without being placed in the political and institutional context, and as far as the Global South is concerned, there is an imperative need to investigate the actual practice of science and its material culture (Abraham, 2006b).

Fig. 2. This photograph shows India's first geo-stationary satellite Ariane Passenger Payload Experiment (APPLE), mounted on a bullock cart to provide a non-magnetic environment to figure out a communication problem. https://www.thebetterindia.com/204852/india-isro-historic-pictures-thumba-mangalyaan-vikram-sarabhai-kalam-pslv/.

At the same time, postcolonial theory of science and technology as a cognitive movement introduced notions of hybridity that rendered porous the boundaries between East and West, the modern and the so-called traditional (Figure 2). This porosity was premised on multidirectional knowledge flows, thereby not just highlighting the transformation of local knowledge(s) by the so-called global, but the reconstitution of the global by the local itself. The historiography of globalization provided a framework to look at multi-directional flows of knowledge and practices. Diffusionist models of technology transfer characterized the historiography of technology of an earlier era. Now the global Americanization of higher education was reflected in the founding of the Indian Institutes of Technology (IIT) (Sebaly, 1974; Bassett, 2009, 2016; Leslie, Kargon, 2006; Raina, 2017). The IITs themselves contributed to the human resource manpower in the information technology revolution and thereby in the internationalization of American education. This sense of achievement of the IITs was reflected in a series of commemorative volumes that were published on the completion of fifty years of the founding of the first generation IITs (Manchanda, 2008; Rajguru, Pant, 2003; Subbarao, 2008; Sukhatme, 2009).

The turn to the global history of technology opened up new spaces and problematics, as technology threw up new professions and communication processes. A relatively recent study coming in from the global history of communication explores the relationship between the history of technology and the history of journalism in colonial India. The introduction of telegraphy is reflected, in turn, in the development of the print media, more specifically the expansion of the reach of newspapers in the nineteenth and twentieth centuries. However, during this period, while the media played a role in drawing regions of peripheral visibility into the core, in India as elsewhere news was transmitted through a variety of communication technologies to a growing readership for very diverse political ends (Bonea, 2016).

Since the 1980s, the histories of space and the nuclear energy research programmes have begun to be written. But the historiography of these technologies and the associated technosciences for strategic reasons has been hedged within national histories of technological development. Thus, while the technologies, technological systems, and practices have been mobile and in circulation globally, the histories as written are splintered between the poles of international collaboration, indigenization, and competition (Narasimha, 1985).[2] In other words, these histories detail the complicated pathways of the indigenization of high-tech and innovation through mobilization of local skills, and wherever possible, through international collaboration. Within national contexts the strategic nature of these technologies dictated that the histories produced were internal and frequently authored by technologists, engineers, and scientists who were part of the associated organizations or had played a central role in the development of the programme. Scientists, given the nature of the technologies, too played a role in their development and in writing about them (Manoranjan, Radhakrishnan, 2012; Aravamudan, 2017; an exception within the genre is Raj, 2000; Sundaram, Krishnan, Iyengar, 1998). An extension of the space programme was the Satellite Instructional Site Experiment (SITE). Drawing upon different socio-technical imaginaries of global techno-science, from 'Atoms for Peace' in the 1950s, Indian and American scientific elites participated together in a satellite programme that legitimated the space programme (Siddiqi, 2020). The point is that, while ideologies of post-colonial development aligned along different axes of the Cold War world, utopian and pragmatic contingencies converged to inaugurate a space programme, the *raison d' être* ostensibly being to serve 'nation building and an egalitarian ethos' (Siddiqi, 2015: 45).

Even so, despite the fact that historians of technology proper have still to enter the domain, the flavour of international collaboration is difficult to miss, while on specific technologies there have been embargoes that have, in turn, compelled indigenization. But in the domain of space technology, India seemed to have benefitted from its relatively neutral stance during the Cold War. But the backdrop of the Cold War was the context of the politicization of nuclear and space technologies, reflected in the discourses on nuclear proliferation and disarmament. This provided a fertile ground for scholars from international politics to jump into the field and narrativize the political economy of technoscientific development in these areas (Perkovich, 1999, 2000; Abraham, 2006a). Robert Anderson's work offers us an ethnography of the network of Indian scientists and their colleagues in the West (Anderson, 2010).

---

2 While the history of science and technology was always concerned with the transmission and circulation of ideas, Narasimha's study of rockets in Mysore in the late eighteenth to early nineteenth century plots a sort of global history of rockets — with their beginnings as fire crackers in China, their subsequent deployment in warfare in Mysore by Tippu Sultan against the English East India Company, and then travelling on to Europe. Though one axis of the study addresses the concern about origins, the other is located within the history of incremental improvization and revolutionary change (Narasimha, 1985).

A pressing complaint of technologists who are also consumers of these histories is that the textual focus of these studies overrides the concerns with technical practices and material culture. If in the early period the focus was on capability building, import substitution, the celebration of the green revolution, attention is now being devoted to internationalization, sustainability, and eco-friendly development. The framework of globalization emphasized the circulation of technological knowledge and the mutual shaping of technological regimes and social systems.

In summary we might add that there is much that we have overlooked in sweeping the history of modern technology in India over the last two hundred years into three broad movements. For in doing so we have run the risk of generalization, when in fact each of these movements could be viewed as landscapes, each characterized by its diversity and nuance, and some of this persisting over time and landscapes. Secondly, much of what has been discussed is restricted to the introduction and stabilization of large technological systems. But the marginalization and/or the adaptation, modification, and modernization of pre-modern technologies in their encounter with modern technological regimes has not been sufficiently theorized in the historiography. These histories are as important but need to be framed in a more global perspective that discloses how elements of pre-modern techniques and practices may have been cannibalized and incorporated into modern technological systems.

References

Abraham, Itty, 'The Ambivalence of Nuclear Histories', *Osiris*, 21 (2006a), 49–65.

———, 'The Contradictory Spaces of Postcolonial Techno-science', *Economic and Political Weekly*, 41.3 (2006b), 210–17.

af Geijerstam, Jan, *Landscapes of Technology Transfer: Swedish Ironmakers in India, 1860–1864* (Skriftserie 42) (Stockholm: Jernkontorets Bergshistoriska, 2004).

Alvares, Claude, *Homo Faber: Technology and Culture in India, China and the West: 1500 to the Present Day* (The Hague: Martin Nijhof, 1979).

———, 'Science', in *The Development Dictionary: a Guide to Knowledge as Power*, ed. by Wolfgang Sachs (London: Zed Books, 1992), pp. 212–28.

Ambirajan, S., 'Science and Technology Education in South India', in *Technology and the Raj: Western Technology and Technical Transfers to India 1700–1947*, ed. by Roy Macleod and Deepak Kumar (New Delhi: Sage Publications, 1995), pp. 112–36.

Anderson, Robert S., *Nucleus and Nation: Scientists, International Networks and Power in India* (Chicago, IL: University of Chicago Press, 2010).

Anderson, Warwick, 'Introduction: Postcolonial Technoscience', *Social Studies of Science*, 32.5/6 (2002), 643–58.

Aravamudan, R. and Gita Aravamudan, *ISRO: A Personal History* (New Delhi: Harper Collins, 2017).

Baber, Zaheer, *The Science of Empire: Scientific Knowledge, Civilization and Colonial Rule in India* (New York, NY: State University of New York Press, 1996).

Bagchi, Amiya Kumar, *Private Investment in India 1900–1939* (Cambridge: Cambridge University Press, 1972).

Bangaru, Saroj, 'Debates in Technical Education: a Prelude to the Foundation of the Indian Institute of Technology (1930–1950)' (unpublished MPhil dissertation, Jawaharlal Nehru University, New Delhi, 2004).

Bassett, Ross, 'Aligning India in the Cold War Era: Indian Technical Elites, the Indian Institute of Technology at Kanpur, and Computing in India and the United States', *Technology and Culture*, 5.4 (2009), 783–810.

———, *The Technological Indian* (Cambridge, MA: Harvard University Press, 2016).

Bhargava, Prakrati, 'Modern Education and the Industrialization of Kanpur: Diverging Perspectives and Agendas, 1905–1946' (unpublished doctoral thesis, Jawaharlal Nehru University, New Delhi, 2018).

Bhattacharya, Sabyasachi, 'Cultural and Social Constraints on Technological Innovation: Some Indian Case Studies', *Indian Economic and Social History Review*, III.3 (1966), 240–67.

Bonea, Amelia, *The News of Empire: Telegraphy, Journalism and the Politics of Reporting in Colonial India, c. 1830–1900* (Oxford: Oxford University Press, 2016).

Braun, Hans-Joachim, 'Introduction to the Symposium on Failed Innovation', *Social Studies of Science*, 22 (1992), 213–30.

Buchanan, R. A., 'Theory and Narrative in the History of Technology', *Technology and Culture*, 3.2 (1991), 365–76.

Chandra, Bipan, *The Rise and Growth of Economic Nationalism in India: Economic Policies of Indian National Leadership 1880–1905* (New Delhi: Har-Anand Publishers, 2010).

Chatterjee, Arup K., *The Purveyors of Destiny: a Cultural Biography of the Indian Railways* (New Delhi: Bloomsbury, 2018).

Colebrooke, H. T., *Algebra with Arithmetic and Mensuration from the Sanscrit of Brahmagupta and Bhāscara* (London: John Murray, 1817).

Das, Pallavi, 'Colonialism and the Environment in India: Railways and Deforestation in 19th Century Punjab', *Journal of Asian and African Studies*, 46.1 (2010), 38–53.

Dharampal, Shri, *Indian Science and Technology in the Eighteenth Century: from Contemporary European Accounts* (New Delhi: Impex India, 1971).

Deloche, Jean, *La Circulation en Inde avant la révolution des transports* (Paris: Publications de l'École Française d'Extrême-Orient, 1980).

———, *Contribution a l'histoire de la voiture en Inde* (Paris: Publications de l'École Française d'Extrême-Orient, 1983).

Dionne, Russel and Roy Macleod, 'Science and Policy in British India, 1858–1914: Perspectives on Persisting Belief', in *Proceedings of the Sixth European Conference on Modern South Asian Studies* (Paris: CNRS, 1979), pp. 55–68.

Dutt, Romesh Chandra, *The Economic History of India in the Victorian Age*, 3rd ed. (London: Kegan Paul, 1908).

Elzinga, Aant and Andrew Jamison, 'The Other Side of the Coin: the Cultural Critique of Technology in India and Japan', in *Technological Developments in China, India and Japan*, ed. by E. Baark and A. Jamison (London: Macmillan, 1986), pp. 205–51.

Escobar, Arturo, 'Development and the Anthropology of Modernity', in *The Postcolonial Science and Technology Reader*, ed. by Sandra Harding (Durham, NC and London: Duke University Press, 2011), pp. 269–89.

Frank, Andre Gunder, *ReORIENT: Global Economy in the Asian Age* (Berkeley, CA: University of California Press, 1997).

Gandhi, M. K., *Hind Swaraj or Indian Home Rule* (Ahmedabad: Navajivan Publishing House, 2014).

Goonatilake, Susantha, *Aborted Discovery: Science and Creativity in the Third World* (London: Zed Books, 1984).

———, *Toward a Global Science: Mining Civilizational Knowledge* (New Delhi: Vistaar Publications, 1998).

Guha, Ramachandra and Madhav Gadgil, *This Fissured Land: an Ecological History of India* (New Delhi: Oxford University Press, 1993).

Habib, Irfan, 'Potentialities of Capitalist Development in the Economy of Mughal India', *The Journal of Economic History*, 29.1 (1969), 32–78.

Habib, Irfan S., 'Science, Technical Education and Industrialization: Contours of a *Bhadralok* Debate, 1895–1915', in *Technology and the Raj: Western Technology and Technical Transfers to India 1700–1947*, ed. by Roy Macleod and Deepak Kumar (New Delhi: Sage Publications, 1995), pp. 235–49.

——— and Dhruv Raina, *Domesticating Modern Science: a Social History of Science and Culture in Colonial India* (New Delhi: Tulika, 2004).

Harding, Sandra, *Is Science Multicultural? Postcolonialisms, Feminisms, and Epistemologies* (Bloomington, IN: Indiana University Press, 1998).

Havell, E. B., *Essays on Indian Art, Industry and Education* (Madras: G. A. Natesan & Co, 1907).

Headrick, Daniel, 'A Double-edged Sword: Communications and Imperial Control in British India', *Historical Social Research/Historische Sozialforschung*, 35 (2010), 51–65.

———, *The Tentacles of Progress: Technology Transfer in the Age of Imperialism, 1850–1940* (Oxford and New York, NY: Oxford University Press, 1988).

———, *The Tools of Empire: Technology and European Imperialism in the Nineteenth Century* (Oxford and New York, NY: Oxford University Press, 1981).

Hettne, Bjørn, *The Political Economy of Indirect Rule: Mysore 1881–1947* (London: Curzon Press, 1976).

Hobsbawm, Eric J., *Industry and Empire: from 1750 to the Present Day* (London: Penguin: 1999).

Inkster, Ian, 'Prometheus Bound: Technology and Industrialization in Japan, China and India Prior to 1914. A Political Economy Approach', *Annals of Science*, 45.4 (1988), 399–426.

Kerr, Ian J., *Building the Railways of the Raj 1850–1900* (Delhi: Oxford University Press, 1995).

———, *Engines of Change: the Railroads that Made India* (Moving Through History: Transportation and Society) (Westport: Praeger Publishers, 2006).

Latour, Bruno and Steve Woolgar, *Laboratory Life: the Social Construction of Scientific Facts* (Princeton, NJ: Princeton University Press, 1979).

Le Gentil, Guillaume, *Voyage dans le les mers de l'Inde (1760–1771), fait par ordre du roi, à l'occassion du passage du Vénus, sur le disque du soleil, le 6 juin 1761, & le 3 du même mois 1769* (Paris: Imprimerie Royale, 1779–1781).

Leslie, Stuart and Robert Kargon, 'Exporting MIT: Science, Technology and Nation Building in India and Iran', *Osiris*, 21.1 (2006), 110–30.

Manchanda, Rohit, *Monastery, Sanctuary, Laboratory: 50 years of IIT Bombay* (Mumbai: Trikkon, 2008).

Manoranjan, Rao and P. Radhakrishnan, *A Brief History of Rocketry in ISRO* (Universities Press, 2012).

Miller, David Phillip, 'The Relocation of Technology between East and West: Stationary Steam Engines and Steamboats in India in the Early Nineteenth Century' in *Science and Narratives of Nature: East and West*, ed. by Jobin M. Kanjarikat, Gordon McOuat, and Sundar Sarukkai (New York, NY and London: Routledge, 2017), pp. 261–302.

Mishra, Pankaj, *From the Ruins of Empire: the Revolt Against the West and the Remaking of Asia* (New York, NY: Allen Lane, 2012).

Mittal, K. V., *History of Thomason College of Engineering* (Roorkee: Roorkee University, 1986).

Mukhopadhyay, Aparajita, *Imperial Technology and Native Agency: a New History of Railways in Colonial India, 1850–1920* (London: Routledge, 2018).

Nandy, Ashis, *Alternative Sciences* (Delhi: Allied Publishers, 1980).

———, *The Intimate Enemy: Loss and Recovery of Self Under Colonialism* (New Delhi: Oxford University Press, 1983).

Narasimha, Roddam, *Rockets in Mysore and Britain, 1750–1850 A.D.* (Bangalore: Project Document DU 8503, National Aeronautical Laboratory, 1985).

Naoroji, Dadabhai, *Poverty and Un-British Rule in India* (London: Swan Sonnenschein & Co., 1901).

Parayil, Govindan, 'The Green Revolution in India: a Case Study of Technological Change', *Technology and Culture*, 33 (1992), 737–56.

Parthasarathi , Prasannan and Giorgio, Riello, *The Spinning World: a Global History of CottonTextiles, 1200–1850* (Oxford: Oxford University Press, 2009).

Perkovich, George, 'Bhabha's Quest for the Bomb', *Bulletin of the Atomic Scientists*, 56.3 (2000), 45–63.

———, *India's Nuclear Bomb: the Impact on Global Proliferation* (Berkeley, CA: University of California Press, 1999).

Pillai, Manu S., *The Courtesan, the Mahatma and the Italian Brahman* (Chennai: Context, 2019).

Qaisar, Ahsan Jan, *The Indian Response to European Technology and Culture* (New York, NY: Oxford University Press, 1982).

Raina, Dhruv, 'Engineering Science and the Indian Institutes of Technology: Reframing the Context of the Cold War and Science (1950–1970)', *Contemporary Education Dialogue*, 14.1 (2017), 1–22.

———, 'Institutions and Knowledge: Framing the Translation of Science in Colonial South Asia', *Asiatische Studien/Études Asiatiques*, 65.4 (2011), 945–67.

———, 'Scientism and Romanticism', in *Images and Contexts: the Historiography of Science and Modernity in India* (New Delhi: Oxford University Press, 2003), pp. 19–48.

———, 'Visvesvaraya as Engineer-sociologist and the Evolution of His Techno-Economic Vision', in *Dialogues Across Disciplines: from the NIAS Archives*, ed. by Roddam Narasimha and Dillip Ahuja (Bangalore: NIAS, 2012).

——— and Irfan Habib, 'Technical Institutes in Colonial India: Kala Bhavan, Baroda (1890–1900)', *Economic and Political Weekly*, 26.46 (1991), 2619–24.

——— and Ashok Jain, 'Big Science and the University in India', in *Science in the Twentieth Century*, ed. by John Krige and Dominique Pestre (London: Harwood Publishers, 2004), pp. 859–77.

Raina, Rajeswari S., 'Public Patronage and Political Neutrality in Agricultural Research: Lessons from British Experience', *The Economic and Political Weekly*, 32.39 (1997), 2473–85.

Raina, Vinod, Aditi Chowdhury, and Sumit Chowdhury, *The Dispossessed: Victims of Development in Asia* (Hong Kong: Arena Press, 1997).

Raj, Gopal, *Reaching for the Stars: the Evolution of India's Rocketry Programme* (New Delhi: Viking Press, 2000).

Rajguru, Suvarna and Ranjan Pant, *IIT: India's Intellectual Treasures* (Mumbai: Coffee table books. Indus Media, 2003).

Ramnath, Aparajith, *The Birth of an Indian Profession: Engineers, Industry and the State, 1900–1947* (New Delhi: Oxford University Press, 2017).

Ray, Praphulla Chandra, *A History of Hindu Chemistry* (Calcutta: Chuckervertty & Co and Kegan Paul, 1902).

Reddy, A. K. N., 'The Genetic Aspects of Western Technology', *Mazingira*, 5 (1978a).

———, 'Technologies appropriate to rural development', in *INSA Seminar on Contribution of Science to Rural Development* (New Delhi: 1978b).

Riello, Giorgio, *Cotton: The Fabric that Made the Modern World* (Cambridge: Cambridge University Press, 2013).

——— and Roy Tirthankar, eds, *How India Clothed the World: the World of South Asian Textiles, 1500–1850* (Leiden: Brill, 2009).

Sachs, Wolfgang, ed., *The Development Dictionary: a Guide to Knowledge as Power* (London: Zed Books, 1992).

Saini, Angela, *Geek Nation: How Indian Science is Taking Over the World* (London: Hodder and Stoughton, 2012).

Salomon, Jean-Jacques and André Lebeau, *Mirages of Development: Science and Technology for the Third World* (London: Lynne Rienner, 1993).

Sangwan, Satpal, 'Indian Response to European Science and Technology, 1757–1857', *The British Journal for the History of Science*, 21.2 (1988), 211–32.

Seal, B. N., *The Positive Sciences of the Ancient Hindus* (Delhi: reprint Motilal Banarsidad, 1915, reprinted 1985).

Sebaly, Kim Patrick, *The Assistance of Four Nations in the Establishment of the Indian Institutes of Technology, 1945–1970* (unpublished doctoral dissertation, University of Michigan, School of Education, 1973).

Sen, S. N., 'The Character of the Introduction of Western Science in India During the Eighteenth and Nineteenth Centuries', *Indian Journal of History of Science*, 1.2 (1988), 112–22.

Shiva, Vandana, *Staying Alive: Women, Ecology and Survival in India* (New Delhi: Kali For Women, 1988).

———, *The Violence of the Green Revolution* (London: Zed Books, 1991).

Siddiqi, Asif A., 'Making Space for the Nation: Satellite Television, Indian Scientific Elites, and the Cold War', *Comparative Studies of South Asia, Africa and the Midle East*, 35.1 (2015), 35–49.

———, 'Whose India? SITE and the Origins of Satellite Television in India', *History and Technology*, 36.3–4 (2020), 452–74.

Subbarao, E. C., *An Eye for Excellence: Fifty Years of IIT Kanpur* (New Delhi: Harper Collins, 2008).

Sukhatme, S. P., *Four Decades at IIT Bombay: Recollections and Reflections* (Bombay: 2009).

Sundaram, C. V., L. V.Krishnan and T. S. Iyengar, *Atomic Energy in India: 50 Years* (Government of India, Department of Atomic Energy, 1998).

Triggs, Oscar Lovell, *Chapters in the History of the Arts and Crafts Movement* (Chicago, IL: Bohemia Guild of the Industrial Art League, 1902.).

Tyabji, Nasir, *Colonialism, Chemical Technology and Industry in Southern India 1880–1937* (New Delhi: Oxford University Press, 1995).

———, 'Exemplar of Academia — Industry Interchange: the Department of Chemical Technology at Bombay University, 1915–1935', in *Science and Modern India: an Institutional History, c. 1784–1947*, ed. by Uma Das Gupt, Project of History of Science, Philosophy and Culture in Indian Civilization, 15(4) (New Delhi: Pearson Education India, 2011).

Visvanathan, Shiv, *Organizing for Science* (New Delhi: Oxford University Press, 1984).

Yearley, Steven, *Science, Technology, and Social Change* (London and Boston, MA: Unwin Hyman, 1988).

JOSHUA GRACE

African Technological Pasts Since 1800

By the mid-1950s, Vincent Njovu, known in Tanzania as 'the first driver of Tanganyika', had retired to a farm in southwestern Tanzania, where he participated in anti-colonial campaigns by helping fund a visit of that movement's figurehead, Julius Nyerere. In his memoir, *Dereva wa Kwanza Tanganyika* (the first driver of Tanganyika), Njovu describes his ascent from refugee, and later a houseboy in Dar es Salaam, to the operator of GT-1, an Albion lorry that received the British colony's first government license plate — thereby turning Njovu into the colony's 'first' African driver. Njovu described the car as the 'love of his heart', and given the car's pristine condition decades later, that love showed in the driver's care and skill. Though colonial officials doubted Africans' ability to appropriate machines like automobiles, Njovu was one of many operators who proved them wrong. After the Second World War, Njovu moved to the State House as the governor's driver. He then retired and joined anti-colonial efforts to see that his former passenger, the governor, permanently left East Africa. And thus, he found himself listening to Nyerere's plans for nationhood when one of his colleagues raised a question (Njovu, 1981: 40).

If maintaining nationhood required technological power in the twentieth century, as many believed, what would new nations who did not have that power do, the man asked? Nyerere took the question in stride by pointing out that plenty of British colonial officials did not know how to do the very things they associated with their racial and cultural identities. Nyerere gave the example of a recent match factory opened in the colony and used a British policeman standing guard at the event to make the point. Speaking in Kiswahili, and thus without the policeman's knowledge, Nyerere suggested it unlikely the *mzungu*, a term for a White person, knew how to make matches himself. Technologically speaking, the *mzungu* was the same as anyone in the crowd who could not do so. They were all technological equals in not knowing this process; their equality showed similarly in use — their expertise of matches started and ended with use; and they could all learn the process, too. As a philosophy of technology, empire associated culture and race with technological power, as numerous historians have detailed. And it tended to approach both elements — race and culture — as immutable and thus as relevant markers of the racial hierarchies undergirding colonial rule (Njovu, 1981; see also, Hecht 2012).

Global History of Techniques, ed. by Guillaume Carnino, Liliane Hilaire-Pérez and Jérôme Lamy, Global Matters Culture Matérielles, histoire des techniques et circulation globale des savoirs, 9 (Turnhout: Brepols, 2024), pp. 135–160
BREPOLS PUBLISHERS 10.1484/M.TECHNE-EB.5.129805

Nyerere's answer offered a different lesson to the crowd: whether for an individual or a nation, technological power was not confined to colonial identities or processes. Indeed, one can interpret the president as saying: if we've learned anything during colonial rule, being White or European did not make things work in the colony. We did! — and he could rightfully say so. Nyerere concluded that if an independent nation wanted to produce something, it could pursue those decisions by appropriating or making their own manufacturing process — decisions they could make as sovereign peoples. The point was not that an independent nation should pursue the science and technology associated with postwar development. Indeed, he generally cautioned against that belief over the next decade. Rather, he asserted that what Africans knew of their own colonial history undercut arguments of the period that linked Africa's perceived technological weakness to doubts about its readiness for independence.[1]

This answer must have resonated with Njovu because he had thought about this issue since he was a child. Made a refugee by the Maji Maji war — an uprising during German rule often analysed, as Njovu does, as a battle between indigenous and European weapons — the driver believed as a young boy, he says, that Africans could only defeat Europeans with the latter's weapons. But when he travelled the long distance from home to the coastal port of Dar es Salaam to attend secondary school, the government sent him with porters on vernacular routes due to the limited scope of motorized transport. Njovu marvels at porters' ability to read landscapes and calls them 'experts' of travel. Though he boards a train to complete the last part of his journey, the future driver does not offer the clear distinctions between indigenous and imperial tools found in the tools of empire approach to history. Both forms of mobility impressed him, and Africans could obviously do both. And thus, Njovu used driving to pursue social mobility for his family before contributing to the expulsion of colonial rule (Njovu, 1981; Grace, 2013).

These Tanzanian narrations of technological history fit broader patterns in Africa since the 1800s. Far from being overwhelmed by outside things and ideas, African societies had their own tools, knowledge, cultures of production and maintenance, and philosophies — already products of innovation and exchange within and beyond the geographic boundaries presently associated with, Africa (de Luna, 2017). Though few research projects centred technological history during the era of political decolonization, most schools of African historiography took for granted societies' ability to shape their own worlds by producing knowledge and things, even in constraining times. Emerging with and through independence movements between the late 1950s and early 1990s, widespread efforts to emphasize African agency in histories of any type emerged from new methodologies, such as oral history, that de-emphasized colonial documents as sources of African history — and by extension, writing as sine

1 See, for example, the introduction for a 1960 discussion which includes Julius Nyerere and Eleanor Roosevelt on 'Prospects of Mankind': Paul Noble, & National Film Network (Producers), & Davis, D. (Director). (1960). Africa-Revolution in Haste. [Video/DVD] WGBH Boston. Retrieved from https://video.alexanderstreet.com/watch/africa-revolution-in-haste.

qua non of knowledge complexity (Niane 1984; Kimambo, 1991). Scholars found plentiful evidence in interviews and in emerging fields of archaeology, but colonial ideas about Africa as a place of technological absence and need persist. For every *Black Panther*, a fictional tale based upon the fact of African technological power and creativity, there are many more books like Jared Diamond's *Guns, Germs and Steel*, a macro-level Euro- and North America-centric world history in which African societies feature for what they did not do: industrialize in ways that would have maintained the pre-industrial technological parity between Africa and Europe (Diamond, 1999; Akyeampong, Bates, Nunn, and Robinson, 2017: 13; as noted by Hecht, 2012 and Mavhunga, 2104, 2017).

This essay explores how and why the fact of African technological agency emerged as a critical component and contribution of African historiography since the 1950s. It shows that politicians, intellectuals, activists, and scholars have used the nexus of users and materiality to broaden which things, tools, people, places, institutions, and ontologies count not just in academic histories of technology, but in the broader regional and global movements African societies engaged, shaped, and in some cases, rejected. Its main contribution centres around the variety of historical alternatives users, thinkers, and politicians imagined and made material (in a hard material sense) since the early 1800s. At the broadest global level of this volume's scope, I centre alternatives to reiterate what Gabrielle Hecht, Chakanetsa Mavhunga, and Laura-Ann Twagira have already said: that dominant narratives about Africa's technological past — narratives often situated in a world-comparative context — have uncritically accepted colonial paradigms of history that placed the continent outside of meaningful geographies and temporalities of technological history (Hecht, 2012; Mavhunga 2014; Twagira, 2020). The point, here, is not only that world histories are incomplete without Africa, but that African histories of technology offer new ways to engage and imagine historical trends, on the African continent and beyond, over the past two centuries.

In turn, centring these alternatives provides an opportunity to reexamine what counts as a technology, who counts as user or expert, which ontologies — including associated definitions of materiality — figure as powerful and efficacious, and which sociotechnical and 'econotechnical' systems get linked to well-being or its opposite in historical scholarship (Masolo, 2017; Mavhunga, 2017a; Hughes, 1983: 29). The essay looks at these alternatives in four sections. The first sets a broader intellectual and historiographic context for exploring the relationship between technology, identity, and power in African history. The second and third sections explore more specific examples of technological identity and power between the early 1800s and 1950s while the fourth takes readers into the period of African independence.

African Pasts and the Technopolitics of Identity

Several years before Njovu heard Nyerere speak about technology and independence in colonial East Africa, Camara Laye's novel, *The Dark Child*, originally published in

1953 as *L'Enfant noir*, explored similar issues across the continent in Guinea through the maturation of its protagonist, Baba. Set in both rural and urban areas during the late colonial period, the novel revolves around the tension Baba feels between the opportunities for social mobility associated with European education and the pull of his father's workshop — a space where spirits and genies, snakes, griots, and healers play integral roles in knowledge creation and in one's ability to smith gold. Baba is so fascinated by the latter process that his mother reprimands him for his habit of listening to apprentices talk about their work at night instead of sleeping. Despite this interest, French colonial school — described as 'anguish' by the protagonist — ultimately leads Baba to Technical College in Conakry (80; see also, d'Avignon, 2022: 97–98). Far from being impressed by the content of this education, Baba complains to his uncle, 'I might well have never left my father's forge'. A restructuring of the school's education sees Baba change his opinion about the school and earn a proficiency certificate. The novel ends as Baba journeys to Paris by plane for more technical training against is his mother's wishes (Laye, 1954: 153).

What might appear as a clear colonial triumph conceals a much more complex story about technopolitical identities. Baba's journey from fascination with his father's workshop to technical certification in Conakry and his commitment to leave home for training in France owe almost exclusively to the opportunities for social mobility attached to colonial institutions and social hierarchies and not to technological power. To this point, Baba fears becoming a manual laborer whose work is defined by menial tasks and low pay and status. Baba roots this, on the one hand, in the superior education, pay, and status reserved for African clerical staff. On the other hand, Baba ascribes more substance and power to vernacular ways of knowing and built worlds. Whether at his own house — where Baba interacts with his father and apprentices — or in rice fields — where Baba watches his uncle's skill harvesting rice — vernacular forms of knowledge remain effective (and sometimes more effective than colonial ideas), relevant, and desirable to the protagonist (who, at times, regrets leaving the forge) (Laye, 1954). There is, then, no colonial technological triumph or even hollowing out of what existed before.

Laye's novel offers a productive departure point for exploring what Paul Edwards and Gabrielle Hecht call a 'technopolitics of identity'. They define this as 'the entanglement of technology with narratives of national and social identity'. In addition to having 'concrete political dimensions and material outcomes', technopolitical identities can be challenged, undone and reworked — in the case of this chapter, by interrogating colonial claims and by centring African spaces, identities, material worlds, and sources (Edwards, Hecht, 2010: 620). In Nyerere's case, the president discarded the colonial essentialism that linked technological power exclusively to European bodies and minds — a colonial technopolitical identity — while endorsing an alternative technopolitical identity rooted in political sovereignty and already existing forms of African technical excellence. Moreover, in Laye's case, readers engage technopolitical identities shaped by spirits, snakes, griots, and healers in ways that expands who and what counts in technological histories.

This essay applies a technopolitics of identity to broader regional and continental pasts. A field in the African history of technology has emerged only in the last decade or so, but African histories of knowledge, skill and expertise, and movement have populated studies since at least the 1960s. When the first generations of African historians turned their attention to pre-colonial kingdoms and trade routes, they provided an evidence-based answer to the mythic colonial histories that described Africa as a place without history. Technology and built worlds featured in this counterargument in oral histories and archaeological research, undercutting White supremacist claims that material achievement in any African society, whether iron smelting or the stunning stone-built worlds of Great Zimbabwe, had come from another, distant society some called, 'the lost white tribe' (Robinson, 2016). As scholars of iron work showed that multiple African societies had at least achieved parity in temperatures of smelting with other parts of the world, Djibril Tamsir Niane evoked the combination of metal production, salt trade, and mobility patterns to challenge prevailing assumptions about African societies' isolation from other continental or world regions — including the notion that African ideas, because they were weaker than external knowledge, were quickly abandoned after such exchange (Schmidt, 1997; Niane, 1984; Masolo, 2017; Mavhunga, 2017a).

At the same time, some historical scholarship struggled to place, and sometimes define, African cultures and ontologies of technology on their own terms. Writing during the heyday of modernization theory in the 1960s and 1970s, even scholars critical of such developmentalist approaches, including dependency theorists, tended to define technologies — and technologies of power, in particular — around Western tools, institutions, and assumptions of knowledge (Rodney, 1972) While this may seem problematic in hindsight, it is important to see this move within a historical context with two relevant features. First, discourses of 'technology' and its supposed absence in Africa had been weaponized against the continent since at least the 1880s, including justifications for European rule. Both the General Act of the Berlin Conference (Article 16) and the Brussels Conference Act of 1890 touted specifically European technologies and ideas as solutions to enslavement and as evidence of European claims to racial and culture superiority (Great Britain, 1890). This link was not lost on Aimé Césaire who, in 1955, flipped this original script by writing 'that between *colonization* and *civilization* there is an infinite difference' (emphasis original) (Césaire, 2001: 34). As one piece of evidence of this difference, Césaire described colonizing nations' willingness to exploit and destroy while also not building (or extending) the technological systems colonizers associated with 'civilization' to colonized populations. For Césaire, then, colonization's rupture came not from new industrial technologies but rather from a sociopolitical issue: a 'pseudo-humanism' found in Enlightenment discourse that departed from the various forms of civilizational interaction found in world history (Césaire, 2001: 37).

Césaire's distinction matters because world historical narratives of the period covered by this volume often assume a global scale technological divergence in which technological change in Europe and the United States, including tools of empire, become the driving forces of social, political, and economic change for the last two centuries. Typified by Daniel Headrick's *Tools of Empire*, this narrative does not necessarily suggest European industrial technologies were immediately effective; *Tools of Empire* sometimes shows the opposite and, in some cases, gestures to the precarious forms of technopolitical power railroads, guns, or steamboats created. Still, the book — and even more so Headrick's subsequent publication, *Tentacles of Progress* — associates technology in an imperial context almost exclusively with European institutions (schools and technical institutes), ideas, and things (Headrick, 1981, 1988). Consequently, explaining empire's occurrence rested upon technological processes that happened in Europe that were extended to colonial spaces through empire and which figure in world-scale causal narratives of change because they had not occurred in Africa. Moreover, this technological rupture became an anchoring point for a host of historical issues that extended beyond the period of European colonization. These include explanations for the uneven landscape of global economic development — including a belief in Africa's continued dependence on external things and ideas — the perceived failures of independent African states — which are often portrayed as unable to orchestrate large technological projects — as well as the continual centring of non-African places as the geographies of innovation and competence.

Second, the first historical scholarship on African technology emerged in the midst of debates across the continent about which tools and technological pasts should inform African futures after independence — including ongoing debates about which ideas, things, and networks could help sustain that sovereignty in a dynamic, often unfriendly, geopolitical environment (Hecht, 2011; Bloom, Miescher, Manuh, 2014). Most forms of modernization theory placed all useful knowledge and things outside the continent in a Cold War West and East. We see below that African leaders, intellectuals, communities, and some scholars rejected and reworked these ideas. Definitions of technology were never limited to the problematically narrow forms the editors address in the introduction. Yet the close relationship between technology and development since the postwar era in scholarship and international development policy gave modernization-theory-based definitions of technology explanatory power for present and historical trends on the continent (Mavhunga, 2017a). By the mid-1970s, about a decade and a half since the first wave of decolonization, technological absence (things nations had not yet been able to procure) and misuse (things African users and politicians were accused of using incorrectly) figured prominently in explanations for economic contraction across a wide ideological spectrum (Grace, 2019). Frantz Fanon's *The Wretched of the Earth*, published in English in 1963, described the national bourgeoisie, including bureaucrats, as a compradorial class which derived it status and power to purchase cars, suits, and other luxury items from links to former colonial powers (which tended to produce those items as well) (Fanon, 1963: 176).

As bureaucrats drove critiques from an academic left, the World Bank and International Monetary Fund seized upon a different aspect of African nations' technological cultures, also locating the problem in a supposedly bloated and inefficient state. The WB had endorsed state driven forms of development from the 1960s through the mid-1970s, even noting that because colonial rule had often left African nations with little useful infrastructure, a coordinating state needed to fill in to plug holes from the past. In addition to being short lived, that approach morphed into a technological criminalization of African states, often, without actual details of that misuse (which is not to say it did not occur but that the policy came before evidence). For the WB, the status of African states as technologically inappropriate actors need only be seen in markets distorted by state interventions and by planning from above. Structural Adjustment Policies sought to remedy this situation by limiting state intervention and planning — a return, at least in part, to the idea that African leaders did not possess the know-how to build their own societies and that external policies offered the remedy (Stein, 2008; Ha-Joon, 2022). This technological rupture acted differently than the imperial narrative discussed above. Instead of being driven by events that happened outside of Africa, accusations of state failure rooted incapacity to various places and institutions on the continent, ultimately ending a short-lived period defined by African-led and defined modernization.

Which brings us back to the broader technopolitical identities evident in Laye. The baseline his work establishes provides a different set of expectations about African technological histories. First, it shows that the representation of Africa as a void of things and knowledge is absurd. One need look no farther than a 1954 novel to know Africans have and tell much more complex histories of technology power, and identity — including those not tied to colonial histories. Second, by rooting itself in vernacular spaces, ideas, and concepts, it demonstrates that the successful representation of Africa as a technological void by European travel writers and administrators rested on narrow definitions of technology and knowledge — such as those covered in this volume's introduction — that othered African ways of knowing and doing despite evidence of African innovation and competence. Finally, and as a result, Laye's work underscores the possibility of using technology and built worlds to explore more complex ideas about power and things in African pasts. The next section uses secondary English sources to show the contours of a deep history of technology rooted in a variety of makers, innovators, users, material worlds, and ontologies of well-being.

African-centred Histories of Technology in the 1800s

A quick tour of East, West, Central, and Southern Africa shows just how many technological stories there are to tell during the 1800s. Such a tour also underscores the importance of rooting histories of technology in details of African use. In West and Central Africa, narratives about Africans and technology at the turn of the nineteenth-century used to suggest a clear relationship between imported guns and increased enslavement, or what some called the gun-slave cycle. Importation of guns — and horses — linked African societies to European imports in ways that fueled predation on the continent: power and protection required weapons which, in turn, African elites acquired through trading slaves with Europeans (Thornton, 1998). A classic case of technological determinism, the gun-slave cycle not only presented African societies as weak vis-à-vis global trade, but also gave the gun too much credit at the expense of African built worlds, technologies, and skills for both enslaving and resisting enslavement (Hall, 2022). Powerful polities such as Dohomey, Oyo, and Benin could make and trade for technologies of enslavement and move captives across long distances using various tools of mobility. Each also benefitted from close relationships between iron smelters, smiths, and royal power (Barnes, Ben-Amos, 1983; Herbert, 1994). African material worlds also tell powerful histories about communities who distanced or protected themselves from enslavement. In present-day Guinea-Bissau, Walter Hawthorne follows Balanta communities who moved to lowlands and erected intricate walls to protect themselves. Yorùbá-speaking Ọhọri communities who fled Dohomey's growth built a road in a forest with the permission of the forest's spirits, which subsequently protected against encroachment while allowing exchange (Hawthorne, 2003; Filippello, 2017).

A similar story about West and Central African knowledge applies to the Atlantic World, where Daniel Littlefield, Judith Carney, and Edda Fields-Black have shown the critical importance of African rice knowledge to New World risiculture (Littlefield, 1991; Carney, 2001; Fields-Black, 2014). Enslaved people in South Carolina likely taken from Kongo, John Thornton argues, brought military tactics — and had long familiarity with a variety of weapons before enslavement — that facilitated revolt, whereas Kevin Dawson uncovers the details of African swimming cultures in ship reclamations that required skilled divers (Thornton, 1998; Dawson, 2018). That African knowledge travelled and sustained (both in their replication in other environments and in sustaining communities of captives) matters, Littlefield noted in 1991, because African contributions to American history — and we could say the same for world history — were largely limited to culture, including music and art. Without condoning problematic divisions between soft and hard technical knowledge, Littlefield rightly observes that African agency too often entered the conversation only when it did not impinge upon, and thus challenge, the production of knowledge that shaped major regional narratives of economic and political development and its assumed Whiteness (Littlefield, 1991).

Picking up the story of enslavement, power, and materiality in the 1800s risks missing the underlying truths about the relationship between West and Central African material worlds and the globalizing economies of the Atlantic World, as Toby Green might point out. African mining and mining technology fueled regional, and later worldwide, trade both before and alongside European expansion. 'West African gold producers had provided the gold that financed Mediterranean economies,' and thus 'Portuguese voyages into the Atlantic were in many ways a response' to known African economic, political, and technological power (Green, 2019: 37). Green ultimately argues that this power entangled African societies with emergent forms of capitalism that were increasingly marked by violence, extraction, and political instability. The hard material world that predates and then shapes these entanglements is most important for the point at hand. In addition to guns, horses, and gold, Green demonstrates an inability to understand the emergence of modern economies without including cowrie shells (and other vernacular currencies), brass, copper, cloth, iron, vast infrastructures of movement, growing urban worlds, and new foods and forms of food production throughout these regions (see also, Kriger, 1999).

Moving to the Swahili Coast of East Africa, Abdul Sheriff demonstrates that monsoon-based boat movement helped create a *mare liberum*, an open ocean, constituting what he terms 'globalization with a difference' (Sheriff 2014: 11). With winds shifting every several months, monsoon mobilities created conditions for social tolerance and exchange that nevertheless recognized the stunning linguistic, religious, and ethnic diversity the winds brought together between China, the Arabian Peninsula, and East and Southern Africa. Or as he put it: the Swahili coast was a 'heterogeneous society in a constant process of homogenization' (Sheriff, 2014: 30). Historians of technology may be most interested in Sheriff's point that boats themselves were products of dispersed geographies — teak wood from India, mangrove poles from the Swahili Coast, and the Arabian Peninsula produced the vessels (Sheriff, 2010). We thus see a technology of mobility mirror and create social exchange. Such mobility confounds contemporary geographic notions about African agency and material things — bound, as they often are, to the notion that 'African' power stops at the coast (if it is recognized at all) or that societies on the continent are overwhelmed by outside forces. Indeed, if we step off the boats onto land, Prita Meier demonstrates that oceanic cultures of circulation both facilitated and were the object tof particular types of African material expression in stone and coral. Towers denoted political power and claims to local space (claims that were visible from the water) while in domestic spaces, colorful plates from across the Indian Ocean covered the walls of Swahili homes. Meier thus shows material forms of power composed of local resources and with items from thousands of miles away (Meier, 2016).

Both Sheriff and Meier contrast these sociotechnical exchanges in the Indian Ocean with assumptions about the inevitability of European empire and its technological forms. Sheriff draws upon Zheng He's journey to the western Indian Ocean. With ships three times larger than the Portuguese vessels that came only decades later, he undercuts might-makes-right stories of empire and technology that implicitly inform many nineteenth-century histories (in other words, Europe produced technological might; other places supposedly did not do the same). Zheng He could have subdued, or at least attempted to subdue, a number of littoral communities from China to the Red Sea. But exchange, not domination, drove these mobilities and the networks they created. Some forms of that exchange continued through British rule on the island as colonial subjects continued to use dhows to trade beyond the control of the colonial government (Gilbert, 2004; Sheriff, 2014). Meanwhile, Meier finds alternative expressions of commercial empire in Zanzibar's House of Wonders, constructed in the 1870s just before the formal onset of empire. Through an intricate analysis of Arabic script and mirrors adorning the building's walls, she asserts that the Omani sultan had no intention of capitulating to British power, but rather aimed to create an Islamic mercantile empire capable of competing with Europe. In both cases, built worlds centre alternative historical possibilities for East Africa during the 1800s (Sheriff, 2014; Meier, 2016).

One can go on at length about the material worlds evident in documentary, oral, and archaeological sources in Southern and Trans-Saharan Africa as well (as we see below). But the point remains the same: as Césaire asserted through distinctions between 'colonialism' and 'civilization', the 'rupture talk' that drove the former and which influenced scholarship during the nineteenth and twentieth-centuries flies in the face of cultures of technology that stretched across the continent and which influenced world history at multiple scales during the 1800s and 1900s (Césaire 2001: 34; Hecht, 2002: 719; Mavhunga, 2019: 10–23 and 371–20). Historian Chakanetsa Mavhunga writes that, on its own, recognizing African power and the intellectual histories that othered it is a big, and noteworthy, task. It requires 'reintellectualizing, de-thingifying, and rehumanizing' societies denigrated in much northern scholarship. It also, he points out, involves tackling problems of representation within the academy, where histories of technology have commonly been written by scholars from Europe and North America and published for those audiences (Mavhunga, 2019: 318).

Mavhunga thus argues that the point is not to simply add African technological histories to existing world historical narratives, but to write in ways that resonate with African societies and to 'open new analytical platforms'. Refusing to write about colonialism as a 'wholesale clearing out of ideas and practices', he situates histories in the language and analytical concepts of African societies: the fundamentals through which they created and sustained ways of knowing and doing (Mavhunga, 2019: 317). Hunting among Vashona in present day Zimbabwe, for example, involved spirits and animals as guides and trees as infrastructure. All acted as *nzira*, a Vashona word that fulfills dual purposes as 'ways': a material route for mobility and a way to act for a variety of forces, including humans, spirits, animals, trees. Writing about myriad *nzira* centres an ontology of knowledge and learning not hemmed in by disenchanted colonial and developmentalist discourses. And it worked in multiple ways. Migrant labourers avoided colonial networks through this gaze, while colonial regimes of conservation relied upon this knowledge and labour in attempts to enact government policies. Most importantly, though, *nzira* worked differently. In contrast to disenchanted colonial conservation strategies that linked African knowledge and action with poaching, *nzira* created and sustained interlocking, reciprocal webs that made such destructive practices nearly impossible (Mavhunga, 2014: 25).

It is one thing to say that African concepts and frameworks should matter for anyone researching African histories of technology and another to grant those centrality — thoroughly exploring their complexity and breadth through emic categories — in a field that tends to revert to Western norms and their lexicons (see Mavhunga, 2019: 10–23 and 371–20; d'Avignon, 2022). As Robyn d'Avignon puts it: 'While scholars have refuted assumptions of primitivism in African technological practice, we have only begun to interrogate how "technical" categories themselves may perpetuate these characteristics' (d'Avignon, 2022: 6; see also Mavhunga, 2019: 10–23 and 371–20). Mavhunga thus defines technology as ways of doing and knowing, a definition that does not constrain doing or knowing to Western institutions, disenchanted cosmologies, or ideas of actors and actants (Mavhunga, 2014). This move has become a core part of Africa's historiography of technology. Shadreck Chirikure questions whether precolonial sites of knowledge, including iron smelting and smithing and pottery, constitute laboratories — a term found throughout histories of science and technology. Chirikure answers, no, because histories of innovation do not require articulation with Western concepts or ontologies to count (Chirikure, 2017). Robyn d'Avignon tells a similar story by detailing the relationships between *jinne* (Maninke spirits), miners, and 'subterranean property claims' miners forged before gold mining. Noting the role of geology in histories of European empire and the general absence of a political geology for non-Western societies, d'Avignon traces the emergence of 'spirited geobodies' through human-spirit relationships in Greater Senegambia. Mining and access to resources went through *nyamadin* — a Maninka term for 'otherworldly energy' — whose location overlapped with 'cool and dark spaces' where gold sat. Access to these sites and claims upon resources required occult skill and knowledge of spiritual landscapes (d'Avignon, 2020: S22). None of this suggests unchanging African histories. Instead, it extends innovation and competence to African technical languages, ontologies, and ways of knowing (Masolo, 2017; Mavhunga, 2017b). As the next section shows, this applies to more recent time periods as well.

African Technological Work and Empire

Another part of Africa's technological past is rooted in more recent epochs and to technologies and systems with roots in Europe or North America. Yet it shares with longer-*durée* histories the fact of African societies' relationships and influence upon other parts of the world. And it assumes that technologies and ideas were not immutable mobiles (Latour, 1987) capable of travelling and acting upon African users as passive agents (Edwards and others, 2011). Instead, these technologies required existing social and technical structures both to work and find uptake among users. In the process of doing so, Africans not only shaped these tools and ideas to meet their own needs but also remade them in ways that diminish the weight of invention, design, and production in histories of technology — often stories with colonial or northern geographies — and thus alert us to alternative processes in world histories (Masolo, 2017; Edgerton, 2007; Mavhunga, 2017a; Hart, 2016). To excavate some of the key themes in this period, we begin with the period leading up to and under colonial rule — roughly, the 1860s to the 1950s — before exploring the technopolitics of independence since the late 1950s.

Though we have already tackled colonial rupture talk through the strength and scope of existing African cultures of technology, it is equally important to point out that European technological interventions were often weak or incomplete; that they were not always controlled or enacted by colonial states, officials, or related entities such as mission schools, and thus created unintended effects; and that when such projects were intimately connected to colonial projects, they often ended up relying upon African labour and ideas — new and old — to achieve basic function. Mavhunga observes that much about what was later termed 'colonial science' came from indigenous ways of working and living with the tsetse fly. More than a co-production of scientific knowledge during colonialism, long-held African knowledge and practices far exceeded officials' knowledge and reseach (Mavhunga, 2019). Laura Fair's history of cinema offers a different dimension to this process. She demonstrates that cinema culture in urban East Africa relied upon funding and enthusiasm from colonial subjects, responded to their tastes, and required their capital to create an urban public sphere under British protectorate rule in Zanzibar (Fair, 2018). Both examples situate Africans as producers (not just co-producers) of technical knowledge during colonial rule. Whether technical expertise on insects or the material construction of a public sphere of leisure, African knowledges and ideas were not just background noise that could fill in here and there for colonial lapses and oversights. Rather, Africans provided the content, impetus, and labour of knowledges often categorized as something colonial.

It is no coincidence that major post-war thinkers like Aimé Césaire and Walter Rodney linked empire and its technologies to 'pseudo-humanism' and 'underdevelopment', respectively. While Césaire observed that colonial governments rarely turned ideological promises of the civilizing mission into material things — even schools, roads, or hospitals that communities sometimes asked for — Rodney observes that the material instantiations of colonial technopolitics served metropolitan economic goals of resource extraction through railways and ports (including significant changes to ports and shipping to 'tie together colonies and metropologes') even as capital investments skipped huge parts of colonial society (Césaire, 2001; Rodney, 1972: 175). Scholars have sometimes pushed back against such structuralist descriptions of underdevelopment, noting that this framework grants historical agency largely to external forces (Cooper and Packard, 1997). But such work continues to resonate, especially technologically. Historians of Africa have already pointed out that colonial states were neither all-powerful nor all-knowing. Often operating on what Sara Berry terms 'hegemony on a shoe-string', governments invested minimally in built worlds or their maintenance, rarely materializing the technological claims of the civilizing mission even as they extracted resources, labour, and capital, as Césaire said decades earlier (Berry, 1992).

Colonialism thus produced a form of technopolitics from the late 1800s to the mid-1900s that was as notable for what it did not do, pursue, repair or maintain, or fund as for what it did. In South Africa, Keith Breckenridge illustrates that an emergent 'biometric state' aimed, but struggled, to tie extensive 'technologies of identification' to fingerprinting and vital registrations. Financial limitations, the state's distrust of the information provided by Africans, and limited reach into rural areas helped produce a bureaucratic system with 'no will to know' and 'limited curiosity' (Breckenridge, 2016: 8 and 15). Kenda Mutongi's history of buses in Nairobi, Kenya argues that African-centred motor mobilities did not emerge as a popular technology until after independence because such movement flew in the face of the settler colony's desire to control African bodies, especially within cities — a difficult project without motorized mobilities. Here, the danger of appropriation — that Africans could and would use vehicles in ways that challenged colonial hegemony — led to absence of a municipal technology (Mutongi, 2017). In some cases, colonial officials in East and West Africa chose automobiles as administrative tools in the 1920s because motor vehicles produced seasonal mobilities on cheap, 'temporary' roads constructed annually by tax default labourers. This limited capital investment made road and car desirable tools compared to railways and human porterage, the former requiring far more costly capital outlays and the latter higher labour costs (Freed, 2010; Grace, 2021).

Recent work has thus revisited some of the original takes on tools of empire to explore the nature of those tools from African experiences. Instead of seeking complete power or knowledge, as some imperial discourses claim, colonial regimes sought out and developed sociotechnical systems that maximized export profit (and production) and control over colonized populations at minimal costs to government and with almost no, or limited, capital investment in labourers or their societies (Rodney, 1972; Hart, 2016). New technologies — and one might say, colonial technological states — created sociotechnical systems that distributed the costs of empire onto African societies which subsequently made up for this absence through hard and forced labour, taxes, and through vernacular knowledge that filled in for these voids (Mavhunga, 2014). Even as colonial ideologies of development denigrated African knowledge and tools, their approach assumed African labour, knowledge, and built worlds would or could make things work when things fell apart and/or capital ran out — a common occurrence. Beyond fiscal concerns, this approach intersected with colonial concerns that reliably maintained technological systems could be appropriated for anti-colonial purposes, as Mhoze Chikowero (2014) and Marissa Moorman both demonstrate with radio (2019).

Moreover, pre-colonial (and in some cases, pre-1800) cultures of technology not only endured, but also provided the sociotechnical frameworks for the integration of outside technologies into African life and for African innovation within the colonial period (Krieger, 1999). Let's start with the former: the durability of existing cultures of technology and built worlds. Across the continent, African labourers and traders continued to use, and even expand, existing networks of mobility, including walking routes or waterways. For migrant labourers who worked for wages at European mines and plantations, walking on unmapped vernacular infrastructure provided literal material ways of voting with their feet through desertion — an act that both allowed them to seek higher wages and avoid tax authorities. Though most wage labourers were men, both women and men used these same networks to trade, creating niche economies that operated beyond both the colonial gaze and a government's power to intervene (Rockel, 2006; Mavhunga, 2014; Nkwi, 2014). In one of the more memorable examples of this infrastructure in East African history, Hehe Chief Mkwawa led German troops down to a dead-end path built for one purpose: ambush. At the same time, ontologies of power that required a broadened concept of infrastructure — as spirits, clouds, and rain — persisted. Rain and an authority's obligation to bring it, what Mavhunga calls a 'hydrological mobility', remained a critical component of African worlds (Mavhunga, 2014: 23; Sanders, 2008; Grace, 2021).

African societies also adopted new technologies — as they long had — often using them in unintended ways or remaking them in new vernacular spaces of innovation.

For example, histories of motor vehicles in West and East Africa demonstrate that existing material, social, and economic infrastructure allowed an almost immediate uptake of African vehicle use in ways that challenged not just colonial ideas about driving and the use of space, but also the technological control of economies. In Accra, drivers, farmers, and market women used trucks to create vernacular economies under colonial rule, and thus to pursue their own ideas of gendered adulthood (Hart, 2016). A similar phenomenon occurred in colonial Tanzania, where caravan routes from the early 1800s provided the material and social infrastructure for the expansion of car use. Though walking remained the most extensive form of mobility, African communities stitched motor, rail, and walking mobility into a multi-modal network of movement that drove economies colonial authorities did not control and aimed to squash through new 'carriage' restrictions. African users made these networks not just with existing knowledge of routes, but through vernacular institutions of driver and mechanic training which largely operated within new African spaces of innovation (Grace, 2021). The histories of farming and cooking technologies, bicycles, radio, and tellecommunications tell similar stories about Africans adopting tools with their own logics in mind and forming their own cultures of innovation to repair and remake each technology (Larkin, 2008; Twagira, 2020; Beck, 2009; Park, 2020).

Creativity within colonial rule does not preclude the destructive nature of European technopolitics. From roads and rails to large scale agriculture, pit latrines, weapons, passes, and mines, African communities associated a whole host of colonial technologies with coercion, slavery, torture, exploitation, and apartheid. Colonial labour and agricultural policies disrupted large-scale ecosystems, norms of food distribution, production, and exposed African labourers and their families to new forms of toxicity from working conditions and DDT alike. In Zambia, when motor vehicle transport slowly replaced cultures of food distribution that had long enabled communities to survive difficult times, famine ensued (Gewald, 2009). In South African gold mines and in uranium fields in Gabon and Madagascar, workers already exposed to a slew of occupational hazards were also exposed to radiation (Hecht, 2014). Mines, missions, and governments rarely built adequate sanitation infrastructure — falsely asserting that Africans would not or could not use it — but they disciplined African households for not building pit latrines according to specific government guidelines — measuring a pit's depth and width while collecting taxes, in some British cases — while turning excrement hauling into a denigrating and despised form of colonial domestic work (Hunt, 1999; Brownell, 2020). As Tasha Rijke-Epstein has shown for French-controlled Madagascar, colonial regimes of waste not only delegated maintenance onto colonial subjects, but also became a key material indication of subjects' often precarious place within colonial hierarchies (Rijke-Epstein, 2019).

One can nevertheless conclude that the colonial period involved various forms of African people and things working, including: previous practices that continued; forms of innovation that combined things old and new, indigenous and external; and systems built with the assumption that African labour would make up for a lack of hard material investments (Twagira, 2021). The importance of this work comes not only through the alternative histories of African innovation one can tell within the colonial period or the forms of continuity that stretch from pre- to post-colonial periods — a phenomenon which delinks stories of agency in world histories of technology from a heavy focus on (industrial) invention and manufacturing power. Moreover, it undercuts narratives that make European colonialism an inevitability through assumptions about technological power or weakness. With the latter based either upon European's perceived technological advances or Africa's perceived technological absence — or both — African technological work throughout colonial rule confounds those claims while pointing to sociotechnical alternatives for this period based upon the assumption of African power (Masolo, 2017).

Independence, Technology, and Development

The pursuit of political and economic independence through movements that sought to end colonial rule spanned about a half-century from the 1950s to the end of apartheid in South Africa in 1994. During this period, some of the same assumptions of colonial discourse found their way into ideologies of development that similarly positioned African societies as passive recipients incapable of creating or pursuing their own forms of sociotechnical well-being. As noted above, by the late-1970s, the perception that African leaders uncritically chased forms of technological development and consumption after achieving independence provided an extremely overgeneralized causal narrative for economic downtown and the broader limitations of African-led modernization schemes. There are, of course, valid concerns that the 'modernizing bureaucrats' who ran new nations constituted a disappointing governing class, including the more particular concerns that bureaucrats limited the technological options and imaginations available to new nations by following external models and privileging external tools and ideas (Cooper, 1997; Diouf, 1997; Mavhunga, 2014; Shivji, 1973).

There are also many stories about leaders, citizens, bureaucrats, and technologists who forged techopolitical identities irreducible to the more visible versions of modernization and development theory during the Cold War. For Kwame Nkrumah in Ghana, independence meant seeking scientific and technological sovereignty by 'joining the forward march of knowledge'. In pursuit of what Abena Osseo-Asare terms, 'scientific equity' and a historical example of afrofuturism 'linked to independence struggles', Ghanaian scientists sought equal access to nuclear technology while also condemning the Cold War's both-sided 'nuclear imperialism'. Despite limitations procuring the desired reactor, 'Nkrumah's approach provided a space for black African researchers to be at the helm of scientific work themselves' (Osseo-Asare, 2019: 1, 3, 171). Laura Ann Twagira shows that forging and maintain alternatives to well-being in Mali required looking beyond the state, external donors, and their priorities for production. In Mali, long seen as a potential regional breadbasket, Twagira centres the quotidian practices of women's food production — 'a grinding stone, a garden hoe, a fishing basket, or a cooking pot' — that not only sustained communities, but also made their lives more flavorful when mechanized food production schemes (repeatedly) failed during colonial and independent periods(Twagira, 2021:4).

Across the continent, the aforementioned Julius Nyerere warned against a close relationship between large technological systems and national development. In the nation's socialist blueprint, he distanced development from money, linking it instead to human-based values — work and sweat — and to tools most Tanzanians already had: hoes and axes, the emblems of the ruling party's flag. This technological philosophy became a core part of his definition of African socialism, or *ujamaa*. Because European 'orthodox socialism' was rooted in costly and exploitative forms of industrialization, Nyerere asserted that Tanzania need not undergo the sociotechnical changes equivalent to Europe's industrialization to pursue socialist forms of equality. He thus settled upon a technological hybrid that saw certain systems of Marx's phase of industrial capitalism — including electric systems, some mechanized agriculture, and transport — with a rooting in agricultural technologies and labour that would likely be placed in the German thinker's feudalist historical phase. To Nyerere, this offered a 'scientific socialism' that matched the historical conditions and tools of his nation and thus did not need to follow outside models. Nyerere would concede that both internal and external factors meant Tanzania did not arrive at an ideal place. But he nevertheless created a tradition of thinking about the relationship between tools and well-being differently that persists to the present (Nyerere, 1968; Brownell, 2020; Grace, 2021).

Even as new nations sought material alternatives to development, African leaders and citizens still linked the possibilities and constraints of political and economic sovereignty to their state's technological abilities. From laboratories, electric grids, cement-built infrastructure, and hoe factories to medical equipment, national airlines, state workshops, and technical institutes, African states felt pressure to deliver new technological worlds to their citizens in order to make the promises of anti-colonialism real (Osseo-Asare, 2019; Breckenridge, 2016; Green-Simms, 2017; Morton, 2019). As explored above, for many colonial subjects, imperial technopolitics were marked by what governments did not build or maintain and the delegation of that responsibility onto those with the least power. This did not mean that African societies expected immediate transformation; yet many citizens expected to see the fruits of indendendence through the construction and maintenance of schools, lighting, hospitals, and better transport systems. Doing so during the Cold War — which Peter Redfield observes, often 'ran hot' in Africa — meant these projects unfolded with real threats to their sovereignty as well (Redfield, 2011: 269). When Rhodesia's Unilateral Declaration of Independence halted Zambia's copper exports and oil imports, presidents Nyerere and Kaunda opened an oil pipeline from Tanzania's port at Dar es Salaam to Ndola in western Zambia as a technology of what the former called 'effective freedom' — a freedom manifest in built worlds instead of in political rhetoric (quoted in Grace, 2021: 185). Such literal nation building turned built worlds into dense sites of political commentary where citizens thanked governments for efforts and/or criticized their decisions (Morton, 2019).

In places like South Africa, where the fight against apartheid continued well into the 1990s, racialized apartheid policies were literally technopolitical in that they relied upon the construction and maintenance of infrastructure systems to enact and enforce policies of separate development. Antina von Schnitzler observes that much technopolitical effort under apartheid aimed to 'prevent the emergence of a (counter)public' through open urban spaces where political gatherings could take place. Resistance to apartheid thus also focussed upon technologies and systems. At times, this involved active destruction of apartheid's built worlds, including the transport systems that enforced geographic separation and inequality, or, as in Zimbabwe's fight for freedom, forging cultures of training and skill around revolutionary weaponry such as the AK-47 (von Schnitzler, 2016: 14; White, 2009; Mavhunga, 2011). But other forms of technopolitical claims-making emerged, too. Prepaid water meters in South Africa initially emerged in response to anti-apartheid rent boycotts and were later normalized after the end of apartheid in 1994 in response to 'mass nonpayment' and environmental concerns about saving water. Authorities imagined prepaid meters as tools of a more apolitical, streamlined, and rationalized governance than previous forms of racialized technopolitics. But users bypassed both the meters and their claims to neutrality, using them instead to directly engage the state over disappointments about the nature of technological citizenship in the wake of apartheid (von Schnitzler, 2016: 6).

As these cases show, the tight relationship between technology, development, and sovereignty — whether obtaining it from apartheid or maintaining it after political independence — figured prominently in debates about freedom, citizenship, and well-being. In some cases, such as Nigeria's oil economy, technological systems and the institutions that produced and maintained them — bureaucrats and multinational corporations — act as a bridge from a period of hope and optimism to critiques of African-initiated modernization efforts. Often presented as a sort of test case of African technological independence, Nigeria's declining oil fortunes brought attention to the destructive relationships between political elites, multinational corporations, and technocrats, or what Frederick Cooper calls the 'spigot economy' (Cooper, 2002: 170). Moreover, Nigeria's oil history aligned with other literature that expressed scepticism and critiques of interventionist states led by African bureaucrats pursuing large infrastructural projects, such as dams at high capital cost, and often through the destruction of rural livelihoods (Twagira, 2021). In this scholarship, the idea of technological independence emerges as a contradiction of sorts as the priorities of African modernization schemes and the structures of post-colonial states fail to produce the technological rupture independence seemingly offered (Cooper, 2002; Isaacman and Isaacman, 2013). Such studies raise questions about the affordances of the technological systems human societies have linked to development and well-being during the twentieth-century. As Prince Guma has recently argued, incomplete projects can also move attention beyond dichotomies of complete/incomplete and success/failure. 'Incompleteness' as an analytic, he suggests, helps one appreciate the infrastructural possibilities of breakdown, maintenance, and construction beyond the 'changing needs and demands of the technical experts and professionals' (Guma, 2022: 62).

While these critiques rightly ask scholars to imagine alternative forms of technological nation beyond top-down planning, they sometimes rest upon alternative visions of technological possibility and capacity that historical actors themselves did not see. Indeed, politicians and planners often raised the opposite concern: that colonialism had not produced infrastructures that facilitated the types of production, learning, mobility, and citizenship that nationhood required, as Césaire had pointed out above. Moreover, the Cold War context of the nation-building era proved to be an unfriendly environment to quickly make up for decades of colonial austerity. Noémi Tousignant calls this the 'problem of technical capacity', while Abena Osseo-Asare notes that such issues of capacity shaped, and ultimately limited, efforts in Ghana to create 'scientific and technological equity' through new national labs, research, and publication (Tousignant, 2018: 2; Osseo-Asare, 2019: 714). Scholars are still learning about the technological contours of the work of state — including the afterlives of colonial austerity — because much of the archival record has not yet been declassified. Yet work seems to suggest that most African nations received only a little more than a decade and a half to create technological sovereignty in the wake of empire before international finance organizations and other world development organizations linked economic contraction during this time period to technological misuse, especially among state elites (Cooper, 2002; Grace, 2021). In turn, this popularized the idea that Africa needs special, 'appropriate' technologies designed by foreign experts because African states and users cannot be trusted to design, pick, or use them appropriately. Marissa Mika has shown that this facilitates technological interventions based upon the idea that Africans cannot handle, and thus do not deserve, the best tools on the world market (Mika, 2020).

More recently, African history has emerged as a place to both rethink and unthink historical narratives of economic development, including the role of technology in them. Let me highlight two aspects of this move. The first has been to shift from a focus on states and leaders to a more expansive definition of users. Jennifer Hart's work on motor transport in Accra, Ghana, demonstrates that drivers forged economic and technological networks through which they pursued and sustained manhood in good and difficult times alike. In one of the book's main contributions, Hart pays close attention to the ways drivers created their own cultures of both skill and risk, thus taking readers beyond formulaic presentations of drivers as reckless technological agents unprepared for industrial modernity (as both colonial and post-colonial critiques asserted) (Hart, 2016). In a similar vein, Kenda Mutongi's history of bus transport in Nairobi, Kenya excavates the co-construction of urbanization and motorized mobilities. Formerly the centre of power in a settler colony that sought to control and reduce African spatial movement, urbanites in Nairobi combined bus mobilities and independence into a system of urban transport rooted in African capital and in the aesthetics of owners, operators, and passengers. As the city grew geographically, it often followed the *matatu* — an example of users planning and making the city from the street level (Mutongi, 2017). Across Kenya's southern border in Tanzania, a joint

Tanzania-China railway project similarly found uptake because the project resonated with rural livelihoods (Monson, 2011).

In Tanzania, Fair shows the power of cinema audiences and owners to create one of the continent's most vibrant movie-going cultures, including a socialist drive-in in Dar es Salaam. Though socialist authorities took a sceptical approach to the cinema due to its perceived capitalist associations, Fair uses oral histories and business archives to prove that movies were an iconic popular socialist technology of the independent period for men and women, young and old. Indeed, the popularity of blacksploitation, Hindi, and kung-fu movies decentres the West as a techno-cultural reference point on screen. As women (who could attend and fought for women's only shows) incorporated self-defense or new fashions into their repertoires, young men seized upon Bruce Lee's ethic of poverty to navigate difficult economic times with both dignity and pride (Fair, 2018). Recent work on cell phones and internet provides similar details about users' ability to shape technological changes they do not control. Internet cafés, for example, have emerged as sites for young people to forge identities amongst themselves and to mature into adults through access to local and transnational data. And though cell phones may seem clear evidence of global changes driving African cultures of technology, Toluwalogo Odumosu's research has shown that Nigerian engineers made unique mobile networks because 'they determined that a fully functional *Nigerian* network has to take into account real users and their particular practices' (emphasis original) (Burrell, 2012; Odumosu, 2017: 148).

In each of these cases, the nexus of technology and users opens up new avenues to consider continental and global histories of development and well-being since the late 1950s. Instead of a development narrative shaped almost exclusively by bureaucrats, politicians, or foreign consultants, development appears in this work through social, political, and cultural continuities with earlier vernacular concepts of well-being as well as a historical process influenced by users, their ideas, and ther innovations. That matters not just for telling social and cultural histories of development or for recognizing the way users can shape large technological processes, as other historiographies of technology have shown, but also for rethinking declensionist narratives of African economies after the mid-1970s. Critics of African states across the ideological spectrum looked at broken and breaking technologies as an indication that the period of African-initiated modernization had, or needed (in the case of SAPs), to come to end (Brownell, 2022). In this sense, broken technologies reflected broken politics. But users did not approach technopolitical processes this way. Instead, they repaired a variety of infrastructural systems — from electrical grids and mass transit to roads and computers. As Brian Larkin has shown, these repair cultures often came with sharp political critiques of the political and economic context that had made breakdown normative and which had delegated repair onto users, especially during the neoliberal period (Larkin, 2008). Still, the technological repair of systems often offered individuals and societies opportunities to remake political and economic systems, and it thus provides fruitful ground to rethink the relationship between technological function/dysfunction and political decline/repair.

If users and repair have helped scholars rethink the sources and possibilities of African development histories, others have pursued a more fundamental unthinking of development in contemporary histories of technology. Mavhunga's work has been central to this scholarship by demonstrating that scholars' ontological assumptions about technology, growth, or conservation are often drawn from theories of development itself and by scholars — and in fields — who live in places where development has long acted as a *telos* of history. Such scholarship thus risks normalizing, instead of questioning, developmentalist assumptions about technology, knowledge, and well-being — making them inevitable aspects of world history in ways that foreclose the possibilities of other forms of well-being. Mavhunga's work demonstrates that other ontologies of technology and well-being provide stark historical alternatives for narrating histories of possibility over the last two centuries (Mavhunga, 2014, 2019). Readers can also see that contrast in Julie Livingston's *Self-devouring Growth*. Livingston provides powerful snapshots of societies' commitments to economic growth as they manifest in industrialized meat markets, water shortages, and roads — all material symbols of a development that have increasingly become metrics of wealth and power in African societies — while recalling their distance from other historical modes of being and materiality in Southern Africa. Though cattle had long acted as symbols of wealth and rain (or the arrival of waters) as a legitimation of political power, ideologies of growth slowly reshuffled the material ontologies of power from sustainable to thoroughly destructive modalities (Livingston, 2019). For both authors, that project is not complete. But the point for world history is that this was not and need not be inevitable.

African technological histories since the 1800s fly in the face of what is most often expected from Africa in world history. But we must not let those expectations shape the stories about the continent's technological past. In addition to a pre-1800s baseline that both shaped and was shaped by regional and inter-continental engagements, the variety of African technological pasts make visible alternative approaches to world technological histories. While foremost among these is the inevitability of empire — based as it was in myths of African technological weakness — African history also provides opportunities to rethink who and what counts as an actor, actant, assemblage, or infrastructure in histories of knowledge and materiality. Such details are certainly critical to telling African-centred histories of technology, especially in situations in which spirits played important roles or in which repair, maintenance, or remaking offered opportunities for innovation — or in situations in which African knowledges travelled to Atlantic or Indian Oceans. Finally, these details reveal that African technologists had already forged alternative ideas of well-being and power before empire, and that they continued to do so as developmental ideas of well-being gained power in the twentieth century.

References

Austen, Ralph and Daniel Headrick, 'The Role of Technology in the African Past', *African Studies Review*, 26.3/4 (1983), 163–84.

Barnes, Sandra and Paula Ben-Amos, 'Benin, Oyo, and Dahomey: Warfare, Statebuilding, and the Sacralization of Iron in West African History', *Expedition*, 25.2 (1983), 5–21.

Beck, Kurt, 'The Art of Truck Modding on the Nile (Sudan): an Attempt to Trace Creativity', in *The Speed of Change: Motor Vehicles and People in Africa, 1890–2000*, ed. by Jan-Bart Gewald, Sabine Luning, and Klaas van Walraven (Leiden: Brill, 2009), pp. 151–74.

Berry, Sara, 'Hegemony on a Shoestring: Indirect Rule and Access to Agricultural Land', *Africa: Journal of the International African Institute*, 62.3 (1992), 327–55.

Bloom, Peter, Stephan Miescher, and Takyiwaa Manuh, eds, *Modernization as Spectacle in Africa* (Bloomington, IN: Indiana University Press, 2014).

Breckenridge, Keith, *Biometric State: the Global Politics of Identification and Surveillance in South Africa, 1850 to Present* (New York, NY: Cambridge University Press, 2016).

Brownell, Emily, *Gone to Ground: A History of Environment and Infrastructure in Dar es Salaam* (Pittsburgh: University of Pittsburgh Press, 2020).

Brownwell, Emily, 'Reterritorializing the Future: Writing Environmental Histories of the Oil Crisis from Tanzania', *Environmental History*, 27:4 (2022), 747–71.

Burrell, Jenna, *Invisible Users: Youth in the Internet Cafés of Urban Ghana* (Cambridge, MA: MIT Press, 2012).

Carney, Judith, *Black Rice: the African Origins of Rice Cultivation in the Americas* (Cambridge, MA: Harvard University Press, 2001).

Césaire, Aimé and Robin D.G. Kelley, *Discourse on Colonialism* (New York, NY: Monthly Review Press, 2000).

Chang H.J., *Kicking Away the Ladder: Development Strategy in Historical Perspective* (London: Anthem: 2002).

Chikowero, Mhoze, 'Is Propaganda Modernity?: Press and Radio for "Africans" in Zambia, Zimbabwe, and Malawi During World War II and Its Aftermath', in *Modernization as Spectacle in Africa*, ed. by Peter J. Bloom, Stephan F. Mieschere, and Takyiwaa Manuh (Bloomington, IN: Indiana University Press, 2014), pp. 112–35.

Chirikure, Shadreck, 'The Metalworker, the Potter, and the Pre-European African "Laboratory"', in *What Do Science, Technology, and Innovation from Africa Mean?*, ed. by Clapperton Mavhunga (Cambridge, MA: The MIT Press, 2017), pp. 63–77.

Cooper, Frederick, *Africa Since 1940: the Past of the Present* (New York, NY: Cambridge University Press, 2002).

———, 'Modernizing Bureaucrats, Backward Africans, and the Development Concept', in *International Development and the Social Sciences: Essays on the History and Politics of Knowledge*, ed. by Frederick Cooper and Randall Packard (Berkeley: University of California Press, 1997), pp. 64–92.

——— and Randall Packard, 'Introduction', in *International Development and the Social Sciences: Essay on the History and Politics of Knowledge*, ed. by Frederick Cooper and Randall Packard (Berkeley: University of California Press, 1997), pp. 1-41.

d'Avignon, Robyn, 'Spirited Geobodies: Producing Subterranean Property in Nineteenth-century Bambuk, West Africa', *Technology and Culture*, 61.2 (2020), 20–48.

Dawson, Kevin, *Undercurrents of Power: Aquatic Cultures in the African Diaspora* (Philadelphia, PA: University of Pennsylvania Press, 2018).

de Luna, Kathryn, *Collecting Food, Cultivating People: Subsistence and Society in Central Africa* (New Haven, CT: Yale University Press, 2017).

Diamond, Jared. *Guns, Germs, and Steel: The Fates of Human Societies* (New York: W. W. Norton & Company, 2005).

Diouf, Mamadou. 'Senegalese Development: From Mass Mobilization to Technocratic Elitism', in *International Development and the Social Sciences: Essays on the History and Politics of Knowledge*, ed. by Frederick Cooper and Randall Packard (Berkeley: University of California Press, 1997), pp. 291–319.

Edgerton, David, *The Shock of the Old: Technology and Global History Since 1900* (New York, NY: Oxford University Press, 2007).

Edward, Frank and Mikael Hård, 'Maintaining the Local Empire: the Public Works Department in Dar es Salaam, 1920–60', *The Journal of Transport History*, 41.1 (2020), 27–46.

Edwards, Paul N. and Gabrielle Hecht, 'History and the Technopolitics of Identity: the Case of Apartheid South Africa', *Journal of Southern African Studies*, 36.3 (2010), 619–39.

——— and others, 'AHR Conversation: Historical Perspectives on the Circulation of Information', *The American Historical Review* , 116:5 (2011), 1393–1435.

Fair, Laura, *Reel Pleasures: Cinema Audiences and Entrepreneurs in Twentieth-century Urban Tanzania* (Athens, OH: Ohio University Press, 2018).

Fields-Black, Edda, *Deep Roots: Rice Farmers in West Africa and the African Diaspora* (Bloomington, IN: Indiana University Press, 2014).

Fanon, Frantz, *The Wretched of the Earth* (New York, NY: Grove Press, 1963).

Filippello, Marcus, *The Nature of the Path Reading a West African Road* (Minneapolis, MN: University of Minnesota Press, 2017).

Freed, Libbie, 'Networks of (Colonial) Power: Roads in French Central Africa after World War I', *History and Technology*, 26.3 (2010), 203–23.

'General Act of the Conference of Berlin Concerning the Congo', *The American Journal of International Law*, 3.1 (1909): pp. 7–25.

Gewald, Jan-Bart, 'Peoples, Mines, and Cars: Towards a Revision of Zambian History, 1890–1930', in *The Speed of Change: Motor Vehicles and People in Africa, 1890–2000*, ed. by Jan-Bart Gewald, Sabine Luning, and Klaas van Walraven (Leiden: Brill, 2009), pp. 21–47.

Gilbert Erik, *Dhows & the Colonial Economy in Zanzibar 1860–1970* (Athens: Ohio University Press, 2004).

Grace, Joshua, *African Motors: Technology, Development, and the History of Development in Tanzania* (Durham, NC: Duke University Press, 2021).

———, 'Heroes of the Road: Race, Gender and the Politics of Mobility in Twentieth-century Tanzania', *Africa: Journal of the International African Institute*, 83.3 (2013), 403–25.

———, 'Mechanical Expression in a Broken World: Repair, Fun, and Everyday Life in Tanzanian Garages', in *Fun, Leisure, and Expressivity in Africa*, ed. by Oluwakemi Balogun, Lisa Gilman, Melissa Graboyes, and Habib Iddrisu (Athens, OH: Ohio University Press 2019), pp. 287–96.

Great Britain, Treaties, *General Act of the Brussels Conference Relative to the Slave Trade* (London: Her Majesty's Stationery Office, 1890).

Green-Simms, Lindsey, *Post-colonial Automobility: Car Culture in West Africa* (Minneapolis, MN: University of Minnesota Press, 2017).

Guma, Prince K, 'The Temporal Incompleteness of Infrastructure and the Urban', *Journal of Urban Technology*, 29.1 (2022): 59–67.

Hall, Rebecca, *The Hidden History of Women-Led Slave Revolts* (New York: Simon and Schuster: 2022).

Hart, Jennifer, *Ghana on the Go: African Mobility in the Age of Motor Transportation* (Bloomington, IN: Indiana University Press, 2016).

Hawthorne, Walter, *Planting Rice and Harvesting Slaves: Transformations Along the Guinea-Bissau Coast, 1400–1900* (Portsmouth, NH: Heinemann, 2003).

Headrick, Daniel, *The Tentacles of Progress: Technology Transfer in the Age of Imperialism* (New York, NY: Oxford University Press, 1988).

———, *Tools of Empire: Technology and European Imperialism in the Nineteenth Century* (New York, NY: University of Oxford Press, 1981).

Hecht, Gabrielle, *Being Nuclear: Africans and the Global Uranium Trade* (Cambridge, MA: MIT Press-Kindle Edition, 2012).

———, 'Introduction', in *Entangled Geographies: Empire and Technopolitics in the Global Cold War*, ed. by Gabrielle Hecht (Cambridge, MA: MIT Press, 2011), pp. 1–11.

———, 'Radioactive Excess: Modernization as Spectacle and Betrayal in Postcolonial Gabon', in *Modernization as Spectacle in Africa*, ed. by Peter J. Bloom, Stephan F. Mieschere, and Takyiwaa Manuh (Bloomington, IN: Indiana University Press, 2014), pp. 205–28.

———, 'Rupture-Talk in the Nuclear Age: Conjugating Colonial Power in Africa.' *Social Studies of Science*, 32.5–6 (2002): 691–727.

Herbert, Eugenia, *Iron, Gender, and Power: Rituals of Transformation in African Societies* (Bloomington, IN: Indiana University Press, 1994).

Hunt, Nancy Rose, *A Colonial Lexicon: Of Birth Ritual, Medicalization, and Mobility in the Congo* (Durham, NC: Duke University Press, 1999).

Isaacman, Allen F. and Barbara Isaacman, *Dams, Displacement and the Delusion of Development Cahora Bassa and Its Legacies in Mozambique, 1965–2007* (Athens, OH: Ohio University Press, 2013).

Kimambo, Isaria, *Penetration and Protest in Tanzania: the Impact of the World Economy on the Pare, 1860–1960* (London: James Currey, 1991).

Kriger, Colleen, *Pride of Men: Ironworking in 19th Century West Central Africa* (Portsmouth, NH: Heinemann, 1999).

Larkin, Brian, *Signal and Noise: Media, Infrastructure, and Urban Culture in Nigeria* (Durham, NC: Duke University Press, 2008).

Laye, Camara, *The Dark Child*, trans. by James Kirkup and Ernest Jones (New York, NY: Farrar, Straus, and Giroux, 1954).

Littlefield, Daniel, *Rice and Slaves: Ethnicity and the Slave Trade in Colonial South Carolina* (Urbana, IL: University of Illinois Press, 1991).

Livingston, Julie, *Improvising Medicine: an African Oncology Ward in an Emerging Cancer Epidemic* (Durham, NC: Duke University Press, 2012).

———, *Self-devouring Growth: a Planetary Parable as Told from South Africa* (Durham, NC: Duke University Press, 2019).

Masolo, D. A., 'The Place of Science and Technology in Our Lives: Making Sense of Possibilities', in *What do Science, Technology, and Innovation from Africa Mean?*, ed. by Clapperton Mavhunga (Cambridge, MA: The MIT Press, 2017), pp. 29–44.

Mavhunga, Clapperton Chakanetsa, 'A Plundering Tiger with Its Deadly Cubs? The USSR and China as Weapons in the Engineering of a "Zimbabwean Nation," 1945–2009,' in *Entangled Geographies: Empire and Technopolitics in the Global Cold War*, ed. by Gabrielle Hecht (Cambridge, MA: The MIT Press, 2011), pp. 23166.

———, 'Introduction: What do Science, Technology, and Innovation from Africa Mean?', in *What do Science, Technology, and Innovation from Africa Mean?*, ed. by Clapperton Mavhunga (Cambridge, MA: The MIT Press, 2017a), pp. 1–27.

———, 'The Language of Science, Technology, and Innovation: a *Chimurenga* Way of Seeing from *Dzimbahwe*', in *What do Science, Technology, and Innovation from Africa Mean?*, ed. by Clapperton Mavhunga (Cambridge, MA: The MIT Press, 2017b), pp. 45–62.

———, *The Mobile Workshop: Tsetse Fly and African Knowledge Production* (Cambridge, MA: MIT Press, 2018).

———, *Transient Workspaces: Technologies of Everyday Innovation in Zimbabwe* (Cambridge, MA: MIT Press, 2014).

Meier, Prita, *Swahili Port Cities: the Architecture of Elsewhere* (Bloomington, IN: Indiana University Press, 2016).

Mika, Marissa, 'The Half-life of Radiotherapy and Other Transferred Technologies', *Technology and Culture*, 61.2 (2020), S135–S157.

Monson, Jamie, *Africa's Freedom Railway: How a Chinese Development Project Change Lives and Livelihoods in Tanzania* (Bloomington, IN: Indiana University Press, 2009).

Moorman, Marissa, *Radio, State Power, and the Cold War in Angola, 1931–2002* (Athens: Ohio University Press, 2019).

Morton, David. *Age of Concrete: Housing and the Shape of Aspiration in the Capital City of Mozambique* (Athens: Ohio University, Press, 2019).

Mutongi, Kenda, *Matatu: a History of Popular Transportation in Nairobi* (Chicago, IL: University of Chicago Press, 2017).

Niane, Djabril Tamsir, 'Relationships and Exchanges Among the Different Regions', in *General History of Africa IV: Africa from the Twelfth to the Sixteenth Century*, ed. by D. T. Niane (London: Heinemann, 1984), pp. 614–34.

Njovu, Vincent, *Dereva wa Kwanza Tanganyika* (The First Driver of Tanganyika) (Dar es Salaam: Tanzania Publishing House, 1981).

Nkwi, Gam W. and Mirjam De Bruijn, 'Human Telephone Lines: Flag Post Mail Relay Runners in British Southern Cameroon (1916–1955) and the Establishment of a Modern Communications Network', *International Review of Social History*, 59.22 (2014), 211–35.

Nyerere, Julius, *Ujamaa: Essays on Socialism* (Dar es Salaam: Oxford University Press, 1968).

Odumosu, Toluwalogo, 'Making Mobiles African', in *What do Science, Technology, and Innovation from Africa Mean?*, ed. by Clapperton Mavhunga (Cambridge, MA: The MIT Press, 2017), pp. 137–50.

Osborn, Emily, 'Casting Aluminum Cooking Pots: Labour, Migration, and Artisan Production in West Africa's Informal Sector, 1945–2005', *African Identities*, 7.3 (2009), 373–86.

Osseo-Asare, Abena Dove, *Atomic Junction: Nuclear Power in Africa After Independence* (New York, NY: Cambridge University Press, 2019).

———, 'Scientific equity: experiments in laboratory education in Ghana', *Isis* 104.4 (2013): 713–41.

Park, Emma, '"Human ATMs": M-Pesa and the expropriation of affective work in Safaricom's Kenya', *Africa*, 90.5 (2020): 914–33.

Redfield, Peter. 'Cleaning Up the Cold War: Global Humanitarianism and the Infrastructure of Crisis Response', in *Entangled Geographies: Empire and Technopolitics in the Global Cold War*, ed. by Gabrielle Hecht (Cambridge, MA: The MIT Press, 2011), pp. 267–91.

Rijke-Epstein, Tasha, 'The Politics of Filth: Sanitation, Work, and Competing Moralities in Urban Madagascar, 1890s–1977', *The Journal of African History*, 60.2 (2019), 229–56.

Robinson, Michael, *The Lost White Tribe: Explorers, Scientists, and the Theory that Changed a Continent* (New York, NY: Oxford University Press, 2016).

Rockel, Steven, *Carriers of Culture: Labour on the Road in Nineteenth-century East Africa* (Portsmouth, NH: Heinemann, 2006).

Rodney, Walter, *How Europe Underdeveloped Africa* (Dar es Salaam: Tanzania Publishing House, 1972).

Roitman, Janet, *Fiscal Disobedience: an Anthropology of Economic Regulation in Central Africa* (Princeton, NJ: Princeton University Press, 2005).

Sanders, Todd, *Beyond Bodies: Rain-making and Sense-making in Tanzania* (Toronto: University of Toronto Press, 2008).

Schmidt, Peter, *Iron Technology in East Africa: Symbolism, Science and Technology* (Bloomington, IN: Indiana University Press, 1997).

Shivji, Issa, *The Silent Class Struggle* (Dar es Salaam: Tanzania Publishing House, 1973).

Sheriff, Abdul, *Dhow Cultures of the Indian Ocean: Cosmopolitanism, Commerce, and Islam* (New York, NY: Columbia University Press, 2010).

———, 'Globalisation with a Difference: an Overview', in *The Indian Ocean: Oceanic Connections and the Creations of Societies*, ed. by Abdul Sheriff and Engseng Ho (London: Hurst and Company, 2014), pp. 11–41.

Stein, Howard, *Beyond the World Bank Agenda: An Institutional Approach to Development* (Chicago: University of Chicago Press, 2008).

Storey, William, *Guns, Race, and Power in South Africa in Colonial South Africa* (New York, NY: Cambridge University Press, 2008).

Thornton, John, *Africa and Africans in the Making of the Atlantic World, 1400–1680* (New York, NY: Cambridge University Press, 1998).

Tousignant, Noémi, *Edges of Exposure: Toxicology and the Problem of Capacity in Postcolonial Senegal* (Durham, NC: Duke University Press, 2018).

Twagira, Laura Ann, *Embodied Engineering: Gendered Labor, Food Security, and Taste in Twentieth-Century Mali* (Athens: Ohio University Press, 2021).

———, 'Introduction: Africanizing the History of Technology', *Technology and Culture*, 6.2 (2020), S1–S19.

———, 'Machines That Cook or Women Who Cook? Lessons from Mali on Technology, Labor, and Women's Things', *Technology and Culture*, 61.2 (2020), S77–S103.

von Schnitzler, Antina, *Democracy's Infrastructure: Techno-Politics and Protest After Apartheid* (Princeton, NJ: Princeton University Press, 2016).

White, Luise, *Speaking with Vampires: Rumor and History in Colonial Africa* (Berkeley, CA: University of California Press, 2000).

———, '"Heading for the Gun": Skills and Sophistication in an African Guerrilla War', *Comparative Studies in Society and History*, 51.2 (2009), 236–59.

JAWAD DAHEUR

Central European Historical Writing on Technology, from 1945 to the Present Day

Institutional Contexts and Directions of Research

As a region of the European continent lying between the variously defined areas of Eastern and Western Europe, Central Europe is, rather than a physical entity, a concept of shared history. Though this common historical past goes far back in time to medieval and early modern imperial legacies, an important milestone for the region was the end of the First World War in 1918, when the three Central European Empires, i.e. the German, Austro-Hungarian, and Russian Empires, were dismantled and a number of independent states emerged in their place. After the Second World War ended in 1945, Europe was divided into a Western and a Soviet sphere of influence, separated by the Iron Curtain. Central Europe was subsumed as a satellite into the Socialist bloc and experienced strategies of communist modernization as well as a forceful integration into a large military and economic entity. During the Cold War, the term Central Europe was increasingly applied to the westernmost Warsaw Pact countries (Poland, Czechoslovakia, and Hungary) to specify them as communist states that were culturally tied to the West (Schöpflin, Wood, 1989). As pointed out by Maciej Górny, the idea of gathering the historiographies of these countries into a common regional narrative is far from obvious. The professional exchange of works and ideas within the region was, indeed, much less established than that between its particular parts and other historiographies, be they French, German, or Anglo-Saxon (Górny, 2018). Yet the comparison of Polish, Czech, Slovak, and Hungarian historical writing does make sense if we look at some secondary features. Some similarities also exist with the experiences of other communist countries in Central Eastern Europe, such as the German Democratic Republic, Romania, and Bulgaria. The aim of this chapter is to briefly outline the history and current state of research on the history of technology within the area by presenting and comparing the changes that have affected the discipline from 1945 to the present day. Though the history of technology has a long tradition in Central Europe, it was only after the Second World War that extensive resources were put into researching the field, along with the history of science. In the early post-war period, rebuilding Europe was a powerful motivation for the development of scientific research. In the Socialist Bloc, science and technology were given a predominant role in the shaping of the new society, in particular in order to assist the communist party and the state to improve

Global History of Techniques, ed. by Guillaume Carnino, Liliane Hilaire-Pérez and Jérôme Lamy, Global Matters Culture Matérielles, histoire des techniques et circulation globale des savoirs, 9 (Turnhout: Brepols, 2024), pp. 161–178
BREPOLS PUBLISHERS 10.1484/M.TECHNE-EB.5.129806

the material and cultural lives of the working people (Fleron, 1977). In this context, the breakthrough in the professionalization and support of the history of technology was associated with the emergence or recreation of the Academies of Sciences, which encompassed both the humanities and the natural sciences. From that time on, one of the features of historical writing on technology was the coexistence of a communist perspective with more or less nationalistic historical interpretations. The collapse of communist dictatorship throughout the Eastern Bloc in 1989 marked a clear turning point in determining the possibility of any return to Europe and the establishment of the liberal democracy and market economies. This chapter follows the chronology of changes that affected research and teaching in the field, with particular emphasis on the institutional contexts, thematic choices, and the realm of publication.

Outline of Institutional Developments in the Post-war Period

In the 1950s and 1960s, the research capacity and material conditions for the development of the history of technology were relatively favourable thanks to the new professional workplaces that emerged at that time. In Poland, a Committee for the History of Polish Science was established in 1952 at the Polish Academy of Sciences (Polska Akademia Nauk, PAN). A year later, it had five sections dealing with the history of social sciences, natural sciences, technology, medicine, urban planning, and architecture. Its organ, the *Kwartalnik Historii Nauki i Techniki Quarterly Journal for the History of Science and Technology*, was launched in 1956 (Pilatowicz, 2006). Another major step for the foundation of the Polish history of technology was the 1955 opening of a National Museum of Technology, after the previous one was destroyed during the war. Under the patronage of the Naczelna Organizacja Techniczna (Central Technical Organisation, NOT), the Museum was established in the Palace of Culture and Science in Warsaw (Figure 1). In 1963, the Committee was renamed Committee for the History of Science and Technology and attached to the Faculty of Social Sciences at the PAN. In 1965, during the eleventh International Congress of the History of Science in Warsaw and Cracow, discussions concerning international cooperation intensified. Three years later, Eugeniusz Olszewski (1911–1994), the leading figure in the Polish history of technology, became the first president of the International Committee for the History of Technology (ICOHTEC). A trained engineer, he had made outstanding efforts in consolidating the field in Poland, founding a chair of history of technology at the Warsaw University of Technology in 1956 and in his role as the vice-director of the Committee for the History of Science and Technology since 1958. In 1983, Olszewski, with others, founded the Polish Society for the History of Technology, hosted by the Museum of Technology in Warsaw (Jasiuk, 1985).

In Czechoslovakia, the starting point for the rise of research was the Commission for the History of Technology, which was established within the Czechoslovak-Soviet Institute in 1950. In 1953, it was integrated into the newly created Československá Akademie Věd/Česko-slovenská Akadémia Vied (Czechoslovak

Fig. 1. Exhibition hall in the National Museum of Technology in Warsaw (1964). Narodowe Archiwum Cyfrowe 3/51/0/10.4/409.

Academy of Sciences, ČSAV). A year later, the annual publication *Sborník pro dějiny přírodních věd a techniky* (*Proceedings for the History of Science and Technology*) was created. During the following years, attempts were made to unite all individuals active in the field. In 1959, a Commission for the History of Natural, Medical, and Technical Sciences was approved as a leading institution and attached to the sixth section of the ČSAV. History of technology was covered by several groups working on the history of metallurgy, mining, transport, cartography, and other fields (Jakubec, Efmertová, 2001). At the same time, a Department for the History of Natural Sciences and Technology was established at the ČSAV. In the 1970s, it drove the activity in the field to some extent (Folta, 1998). Another major actor was the Společnost pro dějiny věd a techniky (Society for the History of Science and Technology, SDVT), founded in 1965 to provide a platform for professionals and amateurs working in the area. After opening several local branches over the years, the SDVT organised many conferences and seminars. In 1968, the *Proceedings for the History of Science and Technology* were replaced by the quarterly *Dějiny vědy a techniky* (*History of Science and Technology*). Luboš Nový (1929–2017) became its chief editor. The development of research was also ensured by the National Technical Museum of Prague. Founded in 1908 as an association, it was nationalized in 1951 and became a major platform for science. Research activities relied in particular on its archives, which contained hundreds of technical documents and a large collection of photographs and posters. Since 1955 it has published a journal, the Sborník Národního technického muzea (*Proceedings of the National Technical Museum*) (Jílek, Majer, 1980).

Fig. 2. Ceremony for the 200th anniversary of the opening of the Mining Academy in Banská Štiavnica (1964). Virtuálna Banská Štiavnica Fotka č. 7987.

In the Slovak parts of the Republic, the development of the history of technology was based on specific institutions. At the Slovak Academy of Sciences (SAS), created in 1942 and re-established in 1953, research groups started to work on the history of technology, natural sciences, biology, and medicine. In 1960, they were united in the Oddelenie dejín vied a techniky (Department of the History of Science and Technology,ODVT) of the Academy's Historical Institute, placed under the leadership of Ján Tibenský (1923–2012), a well-established figure in the field in Slovakia. At the time, the main journal published in Slovak was *Z dejín vied a techniky na Slovensku* (*From the History of Science and Technology in Slovakia*). Up to 1989, research progressed thanks to the efforts of the ODVT, museums, and universities (Hudek, 2008). An important place for exchange was the Slovak Society for the History of Science and Technology, which was founded in 1966. In 1972, with five sections and a branch in Košice, it started to publish original monographs and translations of foreign works. During this period, there were institutions with which the ODVT worked closely, such as the Technical Museum in Košice, the Forestry and Woodworking Museum in Antol, and the Mining Museum in Banská Štiavnica (Figure 2).

In Hungary and in East Germany, institutionalization happened later than in Poland and Czechoslovakia. In 1948, the Műszaki és Természettudományi Egyesületek Szövetsége (Federation of Technical and Scientific Associations, MTESZ), was officially created to gather all the actors in the field. Within this framework, a gathering of individuals interested in the history of technology occurred on an informal basis (Németh, 1998). Although in the 1960s, some representatives of Hungarian history of technology were already widely acknowledged, such as László Makkai (1914–1989) and Walter Endrei (1921–2000), it was only in 1972 that the Magyar Tudományos Akadémia (Hungarian Academy of Sciences, MTA) formally established a research group for these topics within the Technical University of Budapest. In 1974, a Committee for the History of Science and Technology was set up by the Hungarian Academy of Sciences to organize the growing activity in the field (Szász, 2012). At the time, only one periodical was dedicated to the subject, i.e. the *Review of History of Technology*, which has been published since 1963 (Vámos, 2012). As in Poland and in Czechoslovakia, the Hungarian Museum of Science and Technology, which opened in 1973, supported the development of research as well as the protection of technical and industrial heritage. Under the direction of Ferenc Szabadváry (1923–2006), who is best known for his work in the history of chemistry, the Museum accumulated a collection of some 12,000 objects — manufactured products of various origins, precision machines, agricultural machines and tools, telecommunication equipment, photographs — and became a major place of research for scholars (Caswell, 1997).

In East Germany, the history of technology was slow to emerge as an established discipline. The territory of the Soviet occupation zone (which became the German Democratic Republic in 1949) did include several centres that had established research activity in the field recently, i.e. the Karl-Sudhoff-Institut in Leipzig, the Institute for History of Medicine and Natural Sciences in Berlin, and Ernst Haeckel's former institute in Jena, which had been turned into an Institute for the History of Zoology and Evolutionary Biology in 1945. But due to a lack of scholars and funds, only Leipzig remained active in the immediate post-war period (Meinel, 1998). At the turn of the 1950s, a professor at the Technical University of Dresden, Richard Woldt (1878–1952), made great efforts to establish the field in the academic landscape. Shortly before his death, he succeeded in forming an Institute for the History of Technology and Science. In 1952 it started its activities, later in cooperation with Chamber of Technology (Kammer der Technik, KdT), a communist party-affiliated organization founded 1946 in the Soviet occupation zone, and their joint task was to unite the engineers and scientists that were active in East Germany. Though it was well endowed with staff and budget, this institute did not meet the ideological requirements of the communist party and was eventually dissolved in 1963, after producing little scholarly work (Weber, 1995). In the meantime, in 1960, some institutional structures had been consolidated in Leipzig with the creation of the *Schriftenreihe für Geschichte der Naturwissenschaften, Technik und Medizin* (Publication Series for the History of Natural Sciences, Technology and Medicine), which would soon become akin to an official medium for East German historians of science and technology (Meinel, 1998). Nevertheless, in the 1960s Dresden continued to be important for the discipline, along with other research centres like the Freiberg Mining Academy (also a Technical University) and the Technical University of Karl-Marx-Stadt (Chemnitz-Zwickau) (Braun, 2000). Around that time, Rolf Sonnemann

(1928–2010), professor of economic history at the Technical University of Dresden, became a central player in the circles of both historians of science and historians of technology in East Germany. He contributed to the field's institutionalization by founding the Centre for the History of Technical Sciences in Dresden in 1978 (Buchheim, Sonnemann 1985; Hänseroth, Mauersberger, 1996).

Context and Directions of Historical Writing on Technology During the Socialist Period

In socialist Poland, Czechoslovakia, and Hungary, as well as in East Germany, the development of the history of technology much constrained, primarily by the regimes' ideological expectations. In the wake of Sovietization, the leading communist parties were determined to base all scientific production on Marxism and Leninism. Placed under direct governmental control, the newly founded Academies of Sciences were tasked with the centralization of research as well as its ideological control. Over the years, their roles became more complex, but the political context and the need for negotiation strategies with the party's demands continued to have an impact until 1989. Due to an often-tense political context, historians of technology were not spared by purges or marginalization, particularly in Czechoslovakia, where the period of normalization after the Prague Spring of 1968 was felt acutely (Górny, 2018). Some scientists had to leave, and the Department for the History of Natural Sciences and Technology was forced to accept new scholars and adjust its directions of research (Folta, 1998). Generally, however, historians of science and technology accommodated the marriage with Marxism well. Since research in the field was relatively new, they did not, in contrast to colleagues working on other areas of historiography, need to go through the process of dismantling 'legends' developed by previous 'bourgeois' scholarship (Hudek, 2008). Moreover, most scientists and engineers shared the empiricist and positivist epistemology promoted by Marxist historical materialism. In Hungary, historians of sciences did not initiate and were not interested in any radical criticism of this view, and most of them accepted it as a given (Palló, 2016). This was also the case in Poland and Czechoslovakia.

According to historical materialism, technological progress (combined with political revolution) was regarded as the driving force of history as shaping the relations of production, and hence the socio-political changes (Rapp, 1985). In Slovakia, Tibenský explored the history of technology according to this theoretical framework. In Poland, the leading figures in the field also complied with Marxism. In his lectures, Olszewski systematically referred to classical works like Engels' Dialektik der Natur (*Dialectic of Nature*, 1883) (Łotysz, 2009). In the first issue of the *Quarterly Journal for the History of Science and Technology*, he defined 'technology' and 'technical progress' based on Marx's Das Kapital (*Capital*, 1867–1883) and Lenin's writings (Olszewski, Pazdur, 1956). In East Germany, the history of technology formed part of a comprehensive concept of economic history and was placed under the firm control of Marxist historians like Jürgen Kuczynski (1904–1997) and his disciple

Wolfgang Jonas (1926–2010). It was mostly studied through the lens of the 'productive forces' and their role in history (Weber, 1995; Weber, Engelskirchen, 2000). For balance, however, it should be stressed that Marxist approaches were already being ridiculed in part by Central European researchers, especially in Poland and Hungary (Górny, 2018). Criticism then sharpened, in the wake of de-Stalinization. In 1957, for instance, the young Polish historian Bolesław Orłowski (b. 1934) highlighted the deliberate falsification of aviation history by Soviet scholars (Orłowski, 1957). In general, authorities overlooked political inadequacies as long as historians expressed loyalty to the regime and its ideology in their work. Consequently, many of them simply demonstrated their good will by quoting Marx and Lenin in their articles. Historians of technology, however, were generally less affected by such requirements than colleagues working on other branches of the historical sciences, primarily because their work connected with the natural sciences and engineering, fields which were reputed to be apolitical, positive, and factual. With time, the ideological influence of Marxism became milder, and authors were less likely to refer to it. This explains, for instance, why it was possible for many books that had been prepared in the 1980s to be published in the 1990s without significant adjustments (Hallon, 2008). In East Germany, leading researchers such as Rolf Sonnemann (1928–1910) and Gisela Buchheim (1931–1991), who founded the Centre for the History of Engineering Sciences in 1978, largely ignored the theoretical and methodological requirements of Marxism, which earned them constant criticism from their philosophical and economic colleagues (Weber, 1995).

Nevertheless, the party's ideological expectations had an impact at the least in the selection of topics for research. In accordance with historical materialism and class-struggle approaches, priority was given to research on living conditions and the technical environment of workers employed in the heavy industry. In Czechoslovakia, for instance, authorities particularly supported projects dealing with the history of extractive and industrial branches such as mining and metal processing (Hudek, 2008). In Poland, the history of coal mining was also a major field of interest for historians of technology, which was due to the importance of the sector in the socialist economy. Until the mid-1980s, the country was one of the world's top five coal producers. The coalfields of Upper Silesia drew workers and wealth, creating a conurbation of about three million people around Katowice. In the 1960s, the mining engineer Stanisław Kossuth (1893–1968) contributed to the history of coal mining in Upper Silesia underlining the local specificities in extraction methods (Kossuth, 1965, 1968). More generally, the history of mining increasingly became an object of study in the course of the numerous anniversary celebrations organised by the communist regime. Notable in this respect are the fortieth anniversary of the Geological Institute in Warsaw (1959), the twentieth anniversary of the Central Mining Institute in Katowice (1965), and the fiftieth anniversary of the Mining and Metallurgy Academy in Cracow (1969). Significant results were also generated in the context of the 150th anniversary of the Academic Mining School at Kielce, which had been founded by Stanisław Staszic. All these events led to the publication of monographs and collective studies (Pazdur, 1976). In the neighbouring German

Democratic Republic, the history of mining and metallurgy also played an important role in the development of the history of technology. The Freiberg Mining Academy (a Technical University) was a major research centre. In 1978, it hosted the ninth symposium of the ICOHTEC on the history of mining and metallurgy (Braun, 2000).

However, scientific production was also influenced by the everyday material conditions of research. Despite the institutionalization efforts of the 1950s and 1960s, only a few chairs in the history of technology were established in Poland, Czechoslovakia, Hungary, and East Germany. Further, support from the authorities did not always meet expectations. In Czechoslovakia, members of the SDVT constantly complained about inadequate facilities, limited paper supplies for their journal, and the lack of opportunity for publishing monographs (Winters, 1971). In East Germany, severe restrictions on publishing books and journals, as well as paper shortages and limited printing technologies, explain why the history of technology existed in part in the form of a 'grey literature' (Meinel, 1998: 92). When the journal *Z dejín vied a techniky na Slovensku* (From the History of Science and Technology in Slovakia) had to reduce its scope and began to be published irregularly due to its adverse financial situation at the beginning of the 1980s, the ODVT tried to address this gap by creating a new samizdat periodical (Hallon, 2008). The problem was not only financial, but also the fact that the history of technology was not a recognized subject at universities and engineering colleges. A recurrent grievance was the higher educational authorities' underestimation of the value of the discipline. The lack of trained personnel, in particular, was deplored (Olszewski, 1967; Tibenský, 1967). Around 1970, most of the work in the field was done by non-experts like natural scientists and engineers without historical education.

This scarcity of experts also explains why research efforts focussed on specific topics and periods. If we look at the articles published during the early years of the quarterly *History of Science and Technology*, for instance, it appears that about one-third of the articles on technology discussed materials and processes, especially metallurgy and mining, and about one-sixth civil engineering. Other topics such as transportation, agriculture, and military were barely or not at all represented. Chronologically, the majority of articles discussed the nineteenth and twentieth centuries, followed by the eighteenth century. There was little content on the Middle Ages and nothing on prehistory or antiquity (Winters, 1971). In East Germany, the dominant approach focussing on the history of productive forces in industrial periods left little room for the study of the pre-capitalist period, which was treated in a limited and stereotypical way (Hellige, 1978). In Poland, by contrast, most of the works published until around 1980 dealt with the early modern period, with topics including the history of geodesy, cartography, forestry, coal and salt mining, and the iron and textile industry, mostly before the end of the eighteenth century. Technological developments from the second half of the nineteenth century were far less explored (Nowak, 1982). In particular, there was almost no research on the technological history of socialist Poland (Olszewski, 1967).

Another problem for historians of Central European technology was the difficulty of going beyond regional or national studies and to address European or global evolutions. One reason for this was the relative isolation in which many were working. Though several international conferences took place in the region — such as the eleventh International Congress of History of Science in 1965 in Cracow or the ICOHTEC meetings of 1973 in Jablonna (Poland), in 1978 in Freiberg (East Germany), in 1982 in Smolenice (Czechoslovakia), and in 1986 in Dresden (East Germany) — scholars experienced great difficulty in establishing lasting connections with foreign colleagues. Behind the Iron Curtain, they often worked far apart from each other and from Western influences (Palló, 2016). In the 1950s already, contacts between historians of science and technology in East and West Germany became difficult, before coming to a standstill after the construction of the Berlin Wall in 1961. East German researchers were no longer allowed to receive Western periodicals or to submit papers to them (Meinel, 1998). That was also the case in post-1968 Czechoslovakia, where there was considerably less openness to international cooperation in general than in Poland and Hungary (Górny, 2018). While Central European science and technology journals published summaries in English, German, French, or Russian for most of their papers, local scholars only had limited access to foreign publications. As a result, it took a long time for a journal like *History of Science and Technology* to dispose of the 'provincialism' that it had proposed to overcome in its inaugural statement (Winters, 1971).

The Construction of National Histories of Technology: the Time of Synthesis (1970s to 1990s)

After 1970, plans to produce synthesis works on a national scale were on the rise. In Poland, preparatory work for an all-embracing national history of technology was initiated in the 1970s (Orłowski, 1973). Yet, fifteen years later, scholars were still working on partial synthesis on topics like aeronautics, machine construction, or metallurgy. For Eugeniusz Tomaszewski, the composition of a large synthesis was meant to be 'a priority for the next five year' (Tomaszewski, 1986: 172). Despite considerable effort, this task could not be accomplished before 1989, neither in Poland or Czechoslovakia, nor in Hungary. Initiatives towards synthetic publications partly stemmed from a will to celebrate the technological achievements and merits of each of these nations. In an article published in 1967, Tibenský, who had been previously working on the Slovak national revival in the nineteenth century, stressed the importance of science and technology in the history of the nation. In his view, although Slovakia was a small country, it still had 'contributed to significant moments, problems and personalities to the world history of science and technology' (Tibenský, 1967: 481). In the 1960s, similar views shaped Polish research. For instance, the discovery of the ancient Świętokrzyskie iron industry, which was active in the early few centuries of our era, had special significance from the point of view of a national narrative. Since it was the largest centre of this kind in 'barbaric' Europe — according to estimates, about 80,000 tons of iron were melted there — this research subject was made a priority, and its study consumed the lion's share of the Committee for the

History of Science and Technology's budget (Orłowski, 2003). In Czechoslovakia, historical research on mining developed because of the importance these sectors held in the country's industrial history. At a time when Slovaks made an effort to stress the greatness of their nation, topics such as the Mining Academy in Banská Štiavnica, which had been founded in 1763 by Empress Maria Theresa and was reputedly the world's first institution offering tertiary technical education, came to be of particular interest (Tibenský, 1967).

From the late 1960s onwards, publishing initiatives multiplied. In the year in which he was named Professor of Technology History at the Transport and Telecommunications Technical College in Győr, Horváth Árpád (1907–1990) issued a textbook on the general history of technology (Árpád, Pap, 1967). In Czechoslovakia, the first work to give an overview of the topic was published in 1974 under the direction of Luboš Nový. It covered technology used in the historical Czech and Slovak lands from prehistoric times to the end of the eighteenth century (Nový, 1974). In the following decade, a collective work covering the next two centuries of Czechoslovak history of technology was prepared with support from the Academy of Sciences. The first four volumes dealing with the period from 1800 to 1918 were issued (Jílek, 1983–1986). In the Slovak parts of the Republic, new publication projects were launched thanks to the commitment of Ivan Staněk (1915–1972), president of the Slovak Society for the History of Science and Technology and head of the SAS's editorial board (Tibenský, Hrochová, Mauerová, 1976). In 1979, Tibenský signed a book encompassing scholarship, invention, and institution building in a variety of fields, from the Renaissance to the present (Tibenský, 1979). Interestingly, Tibenský recognized the modest nature of Slovak accomplishments in science and technology in comparison with those of Western Europe. He also handled sensitive periods of the Slovak Republic during the Second World War and of the post-war Communist regime with some objectivity. In the 1980s, Tibenský finally managed to publish the first few volumes of the collective project dedicated to key personalities in Slovak science and technology. Although the idea was launched in the 1960s, it was only in 1986 that the first publication was completed, mostly because of the long time it took to find authors who were able to collect biographical data (Hudek, 2008). The first two volumes dealt with the period from the early Middle Ages to 1918 and included 214 portraits of scientists, engineers, and technicians (Tibenský, Pöss, 1986,1988). In Poland, a project of same nature was prepared with support from the NOT. The first volume, dealing with 134 figures from the Polish history of technology, was issued at the end of the decade (Skarżyński, Jakubowska, 1989). In East Germany, as mentioned above, the history of technology was to a large extent part of the 'history of productive forces'. In the second half of the 1980s, a group of historians published a three-volume work on that topic, covering the period from 1800 to 1945. Many pages were dedicated to technological aspects of mining, metallurgy, chemistry, and a wide range of other productive activities (Müller, 1985; Berthold, 1988; Lärmer, 1990).

The Reorganization of the Field After 1989

The events of the late 1980s and early 1990s that resulted in the end of communist rule in Central and Eastern Europe had an ambivalent effect on the discipline. There was, on the one hand, some continuity, particularly with regard to the main editorial projects which were pursued during the 1990s and the early 2000s. On the other hand, new problems emerged in the wake of the political and economic upheavals of the post-socialist period. In Czechoslovakia, after the communist system collapsed, the ČSAVs budget and personnel were curtailed. When the country was divided in 1992, the Czech and Slovak Academies became legally independent from each other, as they had been in their operations for many years. One became the Academy of Sciences of the Czech Republic (Akademie Věd České Republiky AV ČR), the other remained the SAV (Winters, 1994). All these changes much affected the working conditions for historians of technology. When the Academy was forced to reduce the number of its researchers, only a few were able to find new positions. While some found a new place of work in the Archive of the AV ČR, others, like Jaroslav Folta (1933–2011) were transferred to the National Technical Museum, an institution which took on an important coordinating role (Folta, 1998; Debru, 1999). In the former German Democratic Republic, budgetary problems in the context of the reunification also led to the lowering of institutional ambitions for the field. Of the three planned professorships in the history of technology in Dresden, for instance, only one was finally established. There was, however, no 'colonization from the West', and this one position remained in the hands of a member of the former Dresden School (Braun, 2000). In Poland, the 1990s marked a slow decay for the history of technology. Many of the previously constituted structures were dissolved, and the word 'technology' was even removed from the name of the Institute for the History of Science and Technology. In contrast to what happened in the Czech Republic and Slovakia, the National Museum of Technology in Warsaw had, even by the mid-2000s, not yet emerged as an institution capable of compensating for these shortfalls by developing far-reaching research activities (Pilatowicz, 2006).

After some difficult years, some publication projects initiated before 1989 were completed in the second half of the 1990s. Two volumes covering the Czechoslovak history of technology from 1918 to 1945 were published, and the book series dedicated to the pioneers of science and technology in Slovakia was supplemented with a third volume presenting 155 figures from the twentieth century (Jílek, 1995; Tibenský, Pöss, 1999). In Hungary, a study dealing with the technical aspects of industrialization in the country and abroad was published (Walter, Jeszenszky, 1993). A few years later, József Németh retraced the history of the MTESZ in a book issued for the Federation's fiftieth anniversary, which was, in fact, the occasion to gather extensive data about the last 150 years of Hungarian history of science and technology (Németh, 1998). In Poland, further efforts were made to populate the publication landscape more densely. In 1992, the Society for the History of Technology launched a new series entitled *Polish Engineers in the 19th and 20th Centuries* and also started publishing its annual publication in 1999.

One of the most important challenges of the 1990s remained to be the question of education, which had been a key issue for decades (Nový, Stasiewicz-Jasiukowa, 1981). After 1989, the senior pioneers of this research gradually left to retire. Since regular courses were scarce and not compulsory, a generational renewal was not guaranteed. Moreover, concerns about the low level of professionalization, which had been regularly expressed since the 1960s, were still present. In Poland, the history of technology remained mainly a topic of interest for individuals from other fields, mostly retired engineers (Orłowski, 2003). During the early1990s, Jaroslav Folta, who was in charge of the International Commission for Teaching History of Science and Technology, conducted a survey about the situation in the Czech and some Slovak academic institutions (Folta, 1996). In Poland, there was not much support available from universities to promote activities in the field (Debru, 1999). The situation was not any better in Hungary, and it was only in 2000 that a Department of Innovation Studies and History of Technology was established at the Budapest University of Technology and Economics, with Imre Hronszky (b. 1942) at its head (Hronszky, 2009). Though progress was made in the following years, leading to the creation of new courses in several Polish, Czech, Slovak, and Hungarian universities and engineering colleges, there is still no graduate programme specifically dedicated to history of science and technology in Central Europe today.

Newest Trends and Paradigms of the 2000s and 2010s

After the turn of the millennium, new themes and ways of writing the history of technology emerged. Topics such as the social context of technology, the position of women in science and technology, and the history of everyday life penetrated more deeply into the field, although some pioneering work can be traced back to the 1970s and 1980s (Hallon, 2008). For instance, Éva Vámos (1950–2015), who was the director of the Hungarian Museum for Science and Technology for many years, intensified her research on women in the history of science and technology. One of the other trends of the 2000s was also a growing interest in the technological developments of the twentieth century, especially during the communist period, which had been poorly handled previously. In 2003, three new volumes on the history of technology in the Czech Lands after 1945 were issued under the direction of Jaroslav Folta. Among the 180 authors, only four were professional historians, while the rest were eyewitnesses who had taken part in the development of their fields. The book recalled important issues of the socialist era, such as the expropriation of enterprises, the priority given to heavy industry, the complex economic relationships within the Eastern Bloc, and the limited cooperation with the West (Folta, 2003). Between 2010 and 2014, the National Technical Museum published four further volumes devoted to the history of science and technology in twentieth-century Czechoslovakia, of which three dealt with the socialist period (Janovský, Kleinová, Stříteský, 2010; Lorencová, Novotný, 2011; Lorencová, 2012; Kleinová, 2014). The work devoted to the years between 1970 and 1990, for instance, presented some of the evolutions that occurred between the 'normalization' — the restoration of firm party rule following the Warsaw Pact invasion of Czechoslovakia in 1968 — and the glasnost era of liberalization that began in the late 1980s, providing insights into important topics such as the rise of Czechoslovak computer technology, the contribution of Czechoslovak design to the car industry, and the evolution of domestic weapons production and technology during that period (Lorencová, 2012).

Works published after 2000 also illustrate a growing tendency to put national histories of technology into a larger narrative, taking into account historical developments on the European and global scale. These approaches were, of course, not completely new, but they took on particular significance in the context of what was often perceived as the 'return to Europe' of countries like Poland, Czechoslovakia, and Hungary. In 2004, the year in which the European Union grew with Central European countries of the Visegrád Group, a new book series entitled *Czech Technology in the Light of World Development* was launched under the direction of Jaroslav Folta with support from the NTM. In Poland, Bolesław Orłowski, after having published a near-ultimate compendium covering the history of technology in Poland (Orłowski, 2006), took up the challenge of offering a universal synthesis on the issue, which was probably the first attempt of this kind in the Central European historiography of technology. His book, published in 2010, takes a holistic perspective in the shape of an in-depth analysis of human civilization. It presents the influence that human needs, sizes of populations, shortages, and wars exerted on human creativity through the centuries and discusses the role of technological breakthroughs in developments such as globalization and democratization. Orłowski analyses the perception and variability of assessments of technical progress, as well as the ideological and nationalistic biases in its interpretation. The central chapters deal with various topics including energy (from the water wheel to atomic reactors), the 'improvement' of nature (through the use of materials, building engineering, and artificial lighting), and the struggle with time (by means of writing, calendars, and clocks) and space (via communication, observation devices, information transfer, and military technologies) (Orłowski, 2010).

Expanding the spectrum of geographical research contexts also led Central European technology historians to take an increasingly prominent part in building a truly European space of scholarship, open to approaches beyond national exclusivism and Western reductionism. One example is their contribution to research networks such as Tensions of Europe, which was launched in 1999 with the aim to study how technological processes connected and divided Europe, and thereby to develop a European history of technology. By 2004 the network included over two hundred scholars from twenty-one countries across the former Iron Curtain and organised its first international conference in Budapest. From then on, it has engaged in many research, education, and outreach activities, including the book series *Making Europe* that already counts six volumes. The European and even global narratives that emerged from these dynamics are now contributing to a better understanding of the technological interactions between various geographical regions of Europe and the rest of the world. For instance, by investigating the history of infrastructures, railroads, highways, energy systems, and communication networks, historians of technology offer new insights into the history of European integration, which was usually viewed as a political process. From this perspective, for example, the Cold War can be re-interpreted as 'a massive de-linking and relinking of transport, energy, and communication infrastructures in the very middle of Europe' (Misa, Schot, 2005: 9). More generally, Central and Eastern Europe appears as a research area that can contribute to a broad reflection on East–West relations in the history of technology, along with Western Europe and North America, Russia and Central Asia, and the Far East (Högselius, Dazhi, 2017). In this respect, historians of technology in Poland, Czechia, Slovakia, and Hungary will certainly be led to develop in-depth exchanges with researchers from other regions of Europe and the world, and reciprocally so, in the coming years.

References

Árpád, Horváth and János Pap, *Technikatörténet [History of Technology]* (Budapest: OMIKK, 1967).

Berthold, Rudolf, *Geschichte der Produktivkräfte in Deutschland von 1800 bis 1945*, vol. III: Produktivkräfte in Deutschland 1917/18 bis 1945 (Berlin: Akademi-Verlag, 1985).

Braun, Hans-Joachim, 'History of Technology in Germany: a Success Story?', *History of Technology*, 22 (2000), 255–64.

Buchheim, Gisela and Rolf Sonnemann, 'Zur Geschichte der Technikwissenschaften in der DDR', *NTM. Schriftenreihe für Geschichte der Naturwissenschaften, Technik und Medizin*, 22 (1985), 19–24.

Caswell, Lyman R., 'The Hungarian National Museum for Science and Technology', *Bulletin for the History of Chemistry*, 20 (1997), 46–49.

Debru, Claude, ed., *History of Science and Technology in Education and Training in Europe* (Luxemburg: Office for Official Publications of the European Communities, 1999).

Fleron, Frederic J., *Technology and Communist Culture. The Socio-cultural Impact of Technology Under Socialism* (New York, NY: Praeger, 1977).

Folta, Jaroslav, *Dějiny vědy a techniky: výuka na českých a slovenských vysokých školách [History of Science and Technology: Teaching at Czech and Slovak Universities]* (Prague: NTM, 1996).

———, 'O stanie badań, przygotowaniu badaczy oraz organizacji historii nauki i techniki w Czechach [On the State of Research, Education of Researchers and Organisation of the History of Science and Technology in Czechia]', *Kwartalnik Historii Nauki i Techniki*, 43.2 (1998), 113–24.

———, ed., *Studie o technice v českých zemích 1945–1992 [Studies on Technology in the Czech Lands 1945–1992]* (Prague: Encyklopedický Dům, 2003).

———, ed., *Česká technika na pozadí světového vývoje [Czech Technology in the Light of World Development]* (Prague: NTM, 2004).

Górny, Maciej, 'Historical Writing in Poland, Czechoslovakia, and Hungary', in *The Oxford History of Historical Writing*, vol 5, ed. by Axel Schneider and Daniel R. Woolf (Oxford: Oxford University Press, 2018), pp. 254–59.

Hänseroth, Thomas and Klaus Mauersberger, 'Das Dresdener Konzept zur Genese technikwissenschaftlicher Disziplinen — eine Bilanz', *Dresdener Beiträge zur Geschichte der Technikwissenschaften*, 24 (1996), 20–45.

Hallon, Ľudovít, 'Nové smery dejín techniky v slovenskej historiografii po roku 1989 [New Trends in the History of Technology in Slovak Historiography after 1989]', *Studia Bibliographica Posoniensia*, 3 (2008), 222–35.

Hellige, Hans Dieter, 'Grundlinien der Technikgeschichtsschreibung in der DDR', *Geschichte und Gesellschaft*, 4.2 (1978), 272–87.

Högselius, Per and Yao Dazhi, 'The Hidden Integration of Eurasia: East-West Relations in the History of Technology', *Acta Baltica Historiae et Philosophiae scientiarum*, 5.2 (2017), 71–99.

Hronszky, Imre, 'Research into and Teaching History of Technology and Methodology of History of Technology at the BME', *ICOHTEC Newsletter*, 52 (2009).

Hudek, Adam, 'Inštitucionalizácia výskumu v oblasti dejín vied a techniky na Slovensku od počiatkov po rok 1989 [Institutionalization of Research in the Field of History of Science and Technology in Slovakia from the Beginnings Until 1989]', *Studia Bibliographica Posoniensia*, 3 (2008), 211–20.

Jakubec, Ivan and Marcela Efmertová, 'Historiografie vědy a techniky v Čechách a možnosti mezinárodní spolupráce [The Historiography of Science and Technology in Czechia and the Possibilities of International Cooperation]', *Cahiers du CEFRES*, 21 (2001), 19–41.

Janovský, Igor, Jana Kleinová, and Stříteský Hynek, eds, *Věda a technika v Československu v letech 1945–1960 [Science and Technology in Czechoslovakia, 1945–1960]* (Prague: NTM, 2010).

Jasiuk, Jerzy, 'Pierwszy rok działalności Polskiego Towarzystwa Historii Techniki [First Year Activity of the Polish Society for the History of Technology]', *Kwartalnik Historii Nauki i Techniki*, 30.2 (1985), 427–29.

Jílek, František, ed., *Studie o technice v českých zemích 1800–1918 [Studies on Technology in the Czech Lands 1800–1918]*, 4 vols (Prague: Národní Technické Muzeum, 1983–1986).

———, ed., *Studie o technice v českých zemích 1918–1945 [Studies on Technology in the Czech Lands 1918–1945]* (Prague: Národní Technické Muzeum, 1995).

——— and Jiří Majer, *National Technical Museum Praha: Development, Collections, Studies* (Prague: Národní Technické Muzeum, 1980).

Kleinová, Jana, *Věda a technika v českých zemích mezi světovými válkami [Science and Technology in the Czech Lands between the World Wars]* (Prague: NTM, 2014).

Kossuth, Stanisław, *Górnictwo węglowe na Górnym Śląsku w połowie XIX wieku [Coal Mining in Upper Silesia in the Middle of the 19th Century]* (Katowice: Śląsk, 1965).

———, 'Zarys historyczny rozwoju metod eksploatacji w polskim górnictwie węglowym do 1914 roku [Historical Outline of the Development of Mining Methods in the Polish Coal Mining Industry Until 1914]', in *Monografia polskiego górnictwa. Systemy eksploatacji węgla*, ed. by M. Borecki (Katowice: Śląsk, 1968), pp. 46–67.

Lärmer, Karl, *Geschichte der Produktivkräfte in Deutschland von 1800 bis 1945*, vol I: Produktivkräfte in Deutschland 1800 bis 1870 (Berlin: Akademie-Verlag, 1990)

Lorencová, Ivana, ed., *Věda a technika v Československu od normalizace k transformaci (70.–90. léta 20. století) [Science and Technology in Czechoslovakia from Normalisation to Transformation (1970s–1990s)]* (Prague: NTM, 2012).

——— and Michal Novotný, eds, *Věda a technika v Československu v 60. letech 20. století [Science and Technology in Czechoslovakia in the 1960s]* (Prague: NTM, 2011).

Łotysz, Sławomir, 'Eugeniusz Olszewski: Engineer, Historian and Cofounder of the Committee', in *International Committee for the History of Technology 1968–2008*, ed. by Wolfhard Weber (Bochum: ICOHTEC, 2009), pp. 36–43.

Meinel, Christoph, 'German History of Science Journals and the German History of Science Community', in *Journals and History of Science*, ed. by Marco Beretta, Claudia Pogliano, and Pietro Redondi (Florence: Olschki, 1998), pp. 77–96.

Misa, Thomas J. and Johan Schot, 'Introduction: Inventing Europe. Technology and the Hidden Integration of Europe', *History and Technology*, 21.1 (2005), 1–19.

Müller, Hans-Heinrich, *Geschichte der Produktivkräfte in Deutschland von 1800 bis 1945*, vol II: Produktivkräfte in Deutschland 1870 bis 1917/18 (Berlin: Akademi-Verlag, 1985).

Németh, József, *Műszaki fejlődés, természettudományos műveltség: műszaki és természettudományi egyesületek szövetségének története, 1948–1998 [Technical Progress, Scholarship in Natural Sciences: the History of the Federation of Technical and Scientific Associations, 1948–1998]* (Budapest: MTESZ, 1998).

Nový, Luboš, ed., *Dějiny techniky v Československu (do konce 18. století) [History of Technology in Czechoslovakia (to the end of the 18th century)]* (Prague: Academia, 1974).

——— and Irena Stasiewicz-Jasiukowa, *Problems of Teaching the History of Science: Studies of Czechoslovak and Polish Historians of Science for the 16th International Congress on the History of Science* (Prague: Institute of Czechoslovak and General History, 1981).

Nowak, Tadeusz M., 'Stan badań nad historią techniki w Polsce [The State of Research on the History of Technology in Poland]', *Kwartalnik Historii Nauki i Techniki*, 27.1 (1982), 91–106.

Okey, Robin, 'Central Europe / Eastern Europe: Behind the Definitions', *Past & Present*, 137 (1992), 102–33.

Olszewski, Eugeniusz, 'Historia nauki i historia techniki w Polsce Ludowej [The History of Science and the History of Technology in the Polish People's Republic]', *Kwartalnik Historii Nauki i Techniki*, 12.4 (1967), 755–64.

——— and Jan Pazdur, 'Rola postępu technicznego w procesie historycznym [The Role of Technical Progress in the Historical Process]', *Kwartalnik Historii Nauki i Techniki*, 1.1 (1956), 47–70.

Orłowski, Bolesław, 'Historia techniki-fanaberia czy pozytywna konieczność? [The History of Technology — a Fad or a Positive Necessity?]', *Prace Komisji Historii Nauki Polskiej Akademii Umiejętności*, 5 (2003), 119–28.

———, *Historia techniki polskiej [History of Polish Technology]* (Radom: Instytut Technologii Eksploatacji, 2006).

———, *Powszechna historia techniki [Universal History of Technology]* (Warsaw: Mowia Wieki, 2010).

———, 'Przeciwko wypaczaniu historii lotnictwa [Against the Distortion of Aviation History]', *Kwartalnik Historii Nauki i Techniki*, 2.1 (1957), 168–70.

———, 'Wstępny projekt syntezy historii techniki polskiej [Introductory Project for a Synthesis of Polish History of Technology]', *Kwartalnik Historii Nauki i Techniki*, 18.1 (1973), 95–198.

Palló, Gábor, 'Blind Isolation: History of Science Behind the Iron Curtain', in *Science Studies During the Cold War and Beyond*, ed. by Elena Aronova and Simone Turchetti (New York, NY: Palgrave Macmillan, 2016), pp. 129–48.

Pazdub, Jan, 'Stan prac nad historią nauki i techniki górniczej [State of the Art in the History of Mining Science and Technology]', *Kwartalnik Historii Nauki i Techniki*, 21.4 (1976), 709–23.

Pilatowicz, Józef, 'Dzieje techniki na łamach *Kwartalnik Historii Nauki i Techniki* (1956–2005) [The History of Technology in *Kwartalnik Historii Nauki i Techniki* (1956–2005)]', *Kwartalnik Historii Nauki i Techniki*, 51.1 (2006), 165–93.

Rapp, Friedrich, 'Soviet-Marxist Philosophy of Technology', *Studies in Soviet Thought*, 29.2 (1985), 139–50.

Schöpflin, George and Nancy Wood, eds, *In Search of Central Europe* (Totowa, NJ: Barnes & Noble Books, 1989).

Skarżyński, Tadeusz and Anna Jakubowska, *Słownik biograficzny techników polskich [Biographical Dictionary of Polish Technicians]* (Warsaw: Sigma, 1989).

Szász, Zoltán, 'Történetírás és technikatörténet [Historiography and History of Technology]', in *A tudomány-, technika- és orvostörténet kutatásának irányai és intézményei az elmúlt 40 évben témában*, ed. by Éva Vámos and Lilly Vámosné Vigyázó (Budapest: MTESZ, 2012), pp. 19–23.

Tibenský, Ján, *Dejiny vedy a techniky na Slovensku [The History of Science and Technology in Slovakia]* (Martin: Osveta, 1979).

———, 'The History of Science and Technology in Slovakia', *Technology and Culture*, 8.4 (1967), 480–84.

———, Mária Hrochová, and Mária Mauerová, eds, *Biobibliografia prírodných, lekárskych a technických vied na Slovensku do roku 1850 [Biobibliography of Environmental, Medical and Technical Sciences in Slovakia until 1850]* (Martin: Matica slovenská, 1976).

——— and Odrej Pöss, eds, *Priekopníci vedy a techniky na Slovensku [Pioneers of Science and Technology in Slovakia]*, vol. I (Bratislava: Obzor, 1986).

——— and ———, eds, *Priekopníci vedy a techniky na Slovensku [Pioneers of Science and Technology in Slovakia]*, vol II (Bratislava: Obzor, 1988).

——— and ———, eds, *Priekopníci vedy a techniky na Slovensku [Pioneers of Science and Technology in Slovakia]*, vol. III(Bratislava: Obzor, 1999).

Tomaszewski, Eugeniusz, 'Kierunki rozwoju historii nauki i techniki w Polsce do roku 2000 [Directions for the Development of the History of Science and Technology in Poland until 2000]', *Kwartalnik Historii Nauki i Techniki*, 31.1 (1986), 169–82.

Vámos, Éva, 'Közel fél évszázados a Technikatörténeti Szemle [For Nearly Half a Century, the Technikatörténeti Szemle]', in *A tudomány-, technika- és orvostörténet kutatásának irányai és intézményei az elmúlt 40 évben témában*, ed. by Éva Vámos and Lilly Vámosné Vigyázó (Budapest: MTESZ, 2012), pp. 67–73.

Walter, Endrei and Sándor Jeszenszky, *Technikatörténet: 1760–1960. Az ipari forradalom határainkon túl és Magyarországon Szerk [History of Technology: 1760–1960. The Industrial Revolution Beyond our Borders and in Hungary]* (Budapest: ELTE, 1993).

Weber, Wolfhard, 'History of Technology in Germany after 1945: Institutions, Methods, Fields of Interest', *Icon*, 1 (1995), 148–71.

——— and Lutz Engelskirchen, *Streit um die Technikgeschichte in Deutschland 1945–1975* (Münster: Waxmann Verlag, 2000).

Winters, Stanley B., 'Dějiny věd a techniky [History of the Sciences and Technology]', *Technology and Culture*, 12.3 (1971), 487–90.

———, 'Science and Politics: the Rise and Fall of the Czechoslovak Academy of Sciences', *Bohemia. Zeitschrift für Geschichte und Kultur der böhmischen Länder*, 35 (1994), 268–99.

ERIK VAN DER VLEUTEN

Pluralizing 'Eurocentric' Technology Discourses 'Back Home'

Technology and Societal Challenges in Western Europe

If a global history of technology is to break with 'universalizing, diffusionist and Eurocentric models' and restore a plurality of conceptions of technique in diverse socioecological contexts, as this book's introduction desires, it might benefit from a discussion of what those 'models' were about. To that purpose this chapter revisits the history of 'Western' dominant discourses about technology in relation to society and nature, with a particular focus on engineering communities in Western Europe and its diaspora — notably North America and other former colonies.

As we shall see, successive engineering discourses have time and again emphasized the broader relationship between technology, society, and nature, and indeed often in quite universalistic terms. Since the birth of engineering as a civil discipline and profession in the decades around the turn of the nineteenth century, that discipline has been legitimated with reference to the transformative power of innovation to solve major economic, social, political, and — most recently — environmental challenges worldwide. Today such a discourse once more pervades the engineering community and beyond. '[P]eople face a host of global challenges that must be addressed through long-term and innovative education, research, and engineering solutions', state the presidents of the US National Academy of Engineering, the UK Royal Academy of Engineering, and the Chinese Academy of Engineering in a joint manifesto-like article in 2016 (Mote, Dowling, Zhou, 2016: 4). The three presidents explain how, over the past decade or so, the engineering community has translated today's key, global challenges, (as represented by e.g. the UN's Sustainable Development Goals and the EU's societal challenges programme) into the so-called 'Grand Challenges for Engineering', spurring the community to work on solutions so that 'human life as we know it can continue on this planet' (Mote, Dowling, Zhou, 2016: 4). This programme continues to make a great imprint on engineering education, research funders, and tech company mission statements worldwide.

That present-day discourse to save humanity and the planet through innovation may be regarded as sympathetic or self-interested depending on one's perspective. Either way it illustrates how universalistic discourses may monopolize problem and solution definitions, obscuring alternative problem experiences, definitions, and solutions of other social groups. This chapter will therefore not only revisit the history

Global History of Techniques, ed. by Guillaume Carnino, Liliane Hilaire-Pérez and Jérôme Lamy, Global Matters Culture Matérielles, histoire des techniques et circulation globale des savoirs, 9 (Turnhout: Brepols, 2024), pp. 179–194
BREPOLS PUBLISHERS 10.1484/M.TECHNE-EB.5.129807

of 'Western' dominant engineering discourses that supposedly became so influential throughout the world (indeed, so much so that it needs a global history turn to restore lost alternatives); it will also problematize these discourses' supposed universalism 'back home'. It does so by highlighting not one, but a number of potentially conflicting Western engineering discourses on technology and socioecological change. In addition, it discusses how various social groups appropriated those discourses, at times articulating their problems and solutions very differently and accordingly taking technical change, its societal implications, and its governance into widely diverging directions. In order to do so, this chapter must explicitly avoid *a priori* definitions of such concepts as 'technology', 'humanity', 'society', or 'grand challenges', and instead trace how diverse historical actors have filled these and related notions with meaning — and acted upon those meanings.

For want of space, the following sections focus on four pivotal 'Western' engineering discourses (each of which came in several varieties). These four discourses overlapped and co-existed in time and space. Still, they particularly resonated in different periods in (West) European history, and this chapter will discuss and unpack each discourse in the specific historical context that brought it to prominence. The second section discusses the societal promises of technology which, despite repeated technological conflicts and calamities, became widely shared and celebrated in the long nineteenth century. The third section addresses pessimistic discourses on technology's destructive 'unintended consequences' that rose to prominence in Europe's so-called thirty years' crisis between 1914 and 1945. The fourth and fifth sections discuss two sets of discourse on how to save technology's promises from its negative unintended consequences: technocratic innovation discourses thrived in the postwar decades, as did diverse (and mutually conflicting) brands of participatory innovation discourse that emerged in Europe's counterculture years but stretched deeply into the neoliberal age.

Before we continue, a brief note on geography is warranted. This chapter was tasked to discuss the history of technology within the mesoregion known (particularly during large parts of the twentieth century) as 'Western Europe'. It does so by discussing dominant engineering discourses on technology and societal change in that region, but in doing so, it draws primarily upon two decades of research in the pan-European history of technology association *Tensions of Europe*. That association birthed a transnational European history of technology conception that rejects *a priori* geographical delineations and instead spotlights connections and circulations across national, regional, and continental boundaries, notably including transatlantic and (post)colonial linkages (Misa, Schot, 2005; van der Vleuten, 2008). Accordingly, this chapter, though centring on engineering discourses found in what loosely (but never unproblematically) may be termed 'Western Europe', repeatedly traces the histories of those discourses beyond geographically or politically bounded notions of the subcontinent.

Promises and Appropriations

The first of these 'universalistic and Eurocentric' engineering discourses was born in the decades around the turn of the nineteenth century. That discourse underscored that even in the richest part of the world, in northwestern Europe, the great majority of people lived in poverty; hunger and malnutrition, poor clothing and housing, infectious diseases and low life expectancies ruled the day. However, in these decades a new professional breed of civil engineers — military engineering had much older roots — and other believers increasingly argued that modern technology of the kind pioneered in the British Industrial Revolution could and would change all that. The promise of progress through technology became a credo of the long nineteenth century, and by the end of that period the vast majority of Western European populations had indeed gained access to more and better food, housing, clothing, health, mobility, and energy than ever before.

Illustrating this discourse of promise, we here limit ourselves to two iconic examples from the first half of the nineteenth century. First, historians have credited the writings of Michel Chevalier in the early 1830s as a particularly pertinent and influential articulation of the ideology of social progress through technology. A recent graduate from Paris' prominent engineering schools, Chevalier combined the modernization thinking of young engineers of his generation with the hopes of the Saint Simonian 'religion of humanity' seeking peaceful roads to modernization and improving the human condition. As the Saint Simonian movement journal's editor, Chevalier gained a platform to articulate his views. Here he shared the problem analysis of his peers: Europe had been mired in violence and poverty for centuries, and political roads to modernization and improving the human condition had only triggered more violence — most recently in the July Revolution of 1830. His solution: divert military funds to building a transcontinental railway and steamship network, the British high tech of the time. This would liberate humans from political, economic, and natural constraints; unite peoples across nations, continents, class, and natural boundaries in common economic collaboration creating prosperity for all; and produce peace, for why attack those on whom one's prosperity relies. The message resonated widely. For example, young so-called 'state engineers' in the newborn Belgian nation cited Chevalier when initiating the world's first national railway network as 'an intimate link between future prosperity and the independence of the nation' (de Block, 2011: 89). And in the Netherlands, from which Belgium was separated, Chevalier was translated in the context of similar railway debates in the 1830s. Note that the translation's title did not follow the original *Système de la Méditerranée* [The Mediterranean System] but read *De ijzerbanen beschouwd als de voornaamste materiele middelen ter bevestiging van den vrede in Europa, en ter bestendiging van het geluk des menschdoms* [Railways, considered the most important means for peace in Europe and happiness for humanity]. The universalistic promise that technology creates prosperity and peaceful collaboration for 'humanity' subsequently accompanied connective technologies from telegraphy and telephony, electricity supply, motorways and aviation to the internet and social media (Högselius, Kaijser, van der Vleuten, 2015).

Some two decades later, the discourse of promise had taken firm root and was showcased in an even more iconic event — the Great Exhibition of the Works of Industry of All Nations of London in 1851. The first world expo put on display the promise of every thinkable breakthrough in technology and captured the imagination of its six million visitors and many more through reports of the event that followed. While Chevalier's version of technology's promise had focussed on prosperity and peace, the exhibition suggested much more concretely how a wealth of technologies would improve workers', public, and domestic lives. Newspaper editor Horace Greeley summarized the message in his *Art and Industry* (1853: 52):

> [i]n our discoveries in science, by our applications of those discoveries to practical art, by the enormous increase of mechanical power [...] we have [...] given to Society at large, to almost the meanest member of it, the enjoyments, the luxury, the elegance, which in former times were the privilege of kings and nobles [apitalization in the original].

Among the many visitors, foreign engineers and government officials took the promise back home to their respective countries, seeking to emulate the apparent success story.

This universal promise to 'humanity' and 'society' tended to obscure notions of social difference and inequality. It is therefore important to observe how diverse historical actor groups appropriated this discourse and translated it into specific innovation agendas fitting their own contexts and priorities. These appropriations partly overlapped and partly emphasized conflicting interests (van der Vleuten, Oldenziel, Davids, 2017).

For example, the discourse was adopted in very different political settings. The Belgian railway project already illustrated how national governments and state engineers throughout the subcontinent and beyond wed the promise of technology to political nationalism. National governments established 'state engineering' or 'public works' agencies (often modelled after the French *Corps des ponts et chausées*) executing or coordinating (through subsidy and concession schemes) the construction of national railway, waterway, roadway, and telegraphy networks, turning the nation into an economically competitive and politically governable entity. Thus emerged what has been called the 'infrastructure state' (Guldi, 2012). Urban governments likewise translated the promise to a municipal context: they set up municipal engineering departments or appointed city engineers to tackle such urban problems as health hazards and the unruliness of over-populated cities. In the famous words of Baron Georges-Eugène Haussmann (1809–1891) for the case of Paris, the point was to 'regularize the disordered city, to disclose its new order by means of pure, schematic layout [...] to give unity to and transform the operative whole' (as cited in Graham, Marvin, 2001: 55). That sentiment spread rapidly through international urban governance, engineering, and hygienist conferences. Though few city councils proceeded by means as radical as Haussmann's, many turned private roads into public spaces; replaced mazes of dead-end streets with thoroughfares with piped water supply, sewage, and possible tram rail, gas, and electricity infrastructure; and made ungovernable or rebelling quarters accessible to police enforcement (Hård, Misa, 2008).

By contrast, nascent experiments in transnational and global governance in the form of international engineering organizations and networks nominally adopted Chevalier's global ambitions of forging global telegraphy, railway, postal, geodetic, meteorological, and other connections to the benefit of 'humanity'; yet in practice these were often 'European' organizations where engineers and diplomats negotiated notions of global progress with mundane national interests. And in a fourth and extremely influential political appropriation, colonial administrations adopted the discourse of promise for governing British, French, Dutch, Portuguese, Spanish, Italian, Belgian, German, Danish, and Russian colonies. Railways, telegraphy, the postal system — the three great engines of social improvement as the Governor General of India called them — and more became core to Europe's 'civilizing mission', also involving the violent subjugation of colonial subjects at a time when Europeans governed more than eighty per cent of the globe (1914). Resistance groups therefore discovered an entirely different purpose of these technologies — as targets for sabotage and attacks (Kaiser, Schot, 2014; Högselius, Kaijser, van der Vleuten, 2015; Diogo, van Laak, 2016).

Businesspersons took the discourse of promise to a very different arena: whether or not they bought into political promises of social progress, technology certainly provided new business opportunities. The world's first public railway — the Liverpool–Manchester line of 1830 — had been initiated by corn and textile merchants seeking to decrease transportation times and costs. But when the railway company itself unexpectedly provided high returns on investment, investors and entrepreneurs discovered the railway company as a business opportunity in itself, igniting a railway boom that engulfed Britain and the world (British, and to a lesser degree French, Dutch and other shareholders, owned most of the world's technological infrastructure by 1914). This was a very different business model from Chevalier's idealistic diversion of military funds to infrastructure construction. Entrepreneurs and investors now closely followed science and technology breakthroughs in search of new opportunities, innovating traditional industries — such as the textile industry — and establishing entirely new ones such as the chemical and electrotechnical industries. The tech-firm ousted trade firms in the business landscape and did so in close interaction with the expansion and diversification of engineering from civil, mechanical, and construction engineering to e.g. chemical, electrotechnical, and industrial engineering. Large tech firms pioneered research labs and patent offices that further boosted their business fortunes.

Various user communities took technology's promise in still different directions, often seeking utility, entertainment, or empowerment for specific groups. For example, critics found that modern technology only delivered its promise to urban elites and middle classes. In response labour unionists, feminists, nutritionists, and others teamed up to take technology to the working class, and also to change hierarchical relationships between men and women and masters and servants. For example they developed communal social housing and facilities like shared kitchens, launderettes, stoves, and refrigerators. In a similar vein, farmer communities lamented how cities modernized and grew while the countryside became impoverished and depopulated. They searched for affordable power sources and machinery to reinvigorate rural economy and life; Poul la Cour's (1846–1908) wind-electric turbine and cooperatively owned village power stations to electrify the Danish countryside are a good example (Oldenziel, Hård, 2013; Hansen, 1985).

In sum, diverse and often antagonizing groups appropriated the universalist discourse of promise for very different contexts. Ironically, by taking the promise in so many different directions, they jointly built the 'universality' of the promise; as modern technology pervaded public and private life, many experienced great leaps in living conditions, personal health, income, education, and more, at least in Western Europe. These changes came with the invention, institutionalization, and reputation of the profession of engineering. There was disagreement about what 'engineering' was (science-based engineering taught at of the *École polytechnique* in Paris vs. shop floor training in the 'practical arts' in the UK, for instance), how it should relate to politics, and who could practice (e.g. should women be admitted?). But many agreed that 'the engineer is the king of our epoch', as the 1873 edition of the Larousse encyclopedia stated (as cited in Kohlrausch, Trischler, 2014: 65–66).

Broken Promises

The long nineteenth century saw its share of technology conflicts and failures, from early-nineteenth-century Luddism to the high-profile sinking of the Titanic at the eve of the First World War. Yet only in the decades following 1914 critique of technology became so prominent that it could challenge the dominant discourse of promise. Dystopian technology discourse came to stay, and never again could utopian technology discourse claim the stage alone.

The most visible instances of technology's broken promises during Europe's thirty-year crisis, undoubtedly, include two World Wars, the Holocaust, colonial violence, and the global economic crisis of the 1930s. Chevalier's influential notion that technological development and collaboration could forge peace was decisively compromised by the use of technology in war: the First World War was the first to be called 'an engineer's war' with reference to the introduction of machine guns, chemical warfare, tanks, bomber planes, electrocuting fences, and much more (Christie, 1922: 99). One might counter that those technologies were applied on comparatively modest scales, and in many respects the Great War was fought using traditional means. However, Chevalier's 'railroads for peace' were key to the unprecedented massacres in the trenches. Behind trench warfare were modern logistics: railways and modern telecommunications continued to feed tremendous amounts of soldiers, ammunition, and food into the Western front from both sides, like two giant conveyor belts. This tragedy had been long in the making. As a French senator, Chevalier had witnessed the military perversion of his life's work in the Franco-Prussian war of 1870. Prussia beat the French army within months using superior militaries logistics: railroads and telegraphy moved troops effectively to outnumber the French in every vital battle. Following that display of military might, all militaries on the subcontinent radically reformed military strategy around logistics. And as they formed alliances and anticipated each others' Military Travel Plans, a tightly coupled system of prescheduled and interlocking military actions emerged. It was triggered in 1914 and got unexpectedly stuck in trench warfare; the logistics machines, however, kept feeding in soldiers, producing some of the deadliest battles in human history (Bucholz, 1994).

Examples of technology's role in co-producing violent wars abound. Iconic is the story of chemist Fritz Haber, who received the 1918 Chemistry Nobel Prize for the nitrogen synthesis that binds atmospheric nitrogen to hydrocarbons — which was used in artificial fertilizers that helped feed starving world populations. But when German military command asked Haber to save the fatherland, the 'maker of bread from air' became 'Dr. Death', the inventor of chemical warfare. Haber was not alone in this endeavour; the German initiative was a response to French experiments with poison gas shelling, and soon all major militaries explored gas warfare — the American Chemical Society pledged the aid of its 15,000 members to the US chemical Warfare Service, for example. Similarly, the enthusiasm of theoretical physics discovering nuclear fission turned to the nightmare of nuclear warfare in the Second World War, which by then had already seen the technological horrors of carpet bombing, rockets, and of course the Holocaust's logistics. The notion of European 'civilization' became as tainted at home as its 'civilizing mission' was abroad — especially after missionaries and journalists exposed the torture, rape, and killing by the Belgian colonial army in Congo Free State, the British machine gunning of unarmed men, women, and children in Amritsar (India), the French bombing a peaceful demonstration in Vinh (Vietnam), the Italian extermination-through-bombing policies in Ethiopia, and much more. No longer could Europe's technology be unproblematically depicted as a great civilizer, and Europe itself as 'civilized' (Kohlrausch, Trischler, 2014; Diogo, van Laak, 2016).

Chevalier's notion of technology producing 'joint prosperity' was equally overturned, perhaps most significantly so during the Great Depression of the 1930s. Scientific management, the star invention of the young discipline of industrial engineering, and conveyor belts were supposed to benefit both capital and labour, increasing productivity as well as wages. But under financial stress, factory managers used these techniques to cut labour costs, leading to mass layoffs and unemployment. Besides, speeding up conveyor belts stressed out the remaining workers; Charlie Chaplin's movie *Modern Times* (1936), about a conveyor belt worker loss of sanity, represented an experience widely encountered in the US and Western Europe. However, workers were not the only ones to suffer. Technology's promise to business and entrepreneurship also backfired as the entrepreneurial world experienced mass bankruptcies. Besides, inventor-entrepreneurs in smaller companies were increasingly squeezed out by larger tech firms with their research labs and patent lawyer units — unless they were willing to take a role as suppliers to larger firms.

Technology's promise to liberate peoples from political, economic, and natural constraints, too, was turned upside-down. That broken promise was experienced by various user groups who had embraced modern technologies for liberation and empowerment. Electrical appliances came with the new danger of electrocution and fires, and in the case of gas lightning, railways, and above all automobiles, death by technology became a widespread phenomenon. Motorists were increasingly seen as joy riders, speed maniacs, and killers responsible for spiking traffic deaths by the non-motorist majority, and were increasingly regulated. So were bicyclists and pedestrians, who were assigned to specific segments of the streets and fined if they did not comply. The experience of freedom had taken a beating (Tenner 1997; van der Vleuten, Oldenziel, Davids, 2017). On a higher level of abstraction, philosophers, social critics, and engineers argued that technology itself now systematically threatened human individuality and freedom. For example, one of Germany's most prominent engineers, Walter Rathenau (1867–1922), felt that modern men and women were turning into mere 'cogs' in modern production and consumption systems. As early as in 1913 he had written about a mechanization of the world that produced a mechanization of the spirit. The notion of the enslavement of modern humans was further elaborated by critics such as José Ortega Y Gasset (1883–1955), Oswald Spengler (1880–1936), and the early Frankfurter Schule, while at the same time the enslavement of humans by robots and machines became a prominent theme in science fiction novels (Hughes, 2004).

The prestige of engineering took a similar beating. It is perhaps telling that in science fiction, scientists and engineers had traditionally figured as heroes, but now they increasingly figured as villains — together with businessmen, politicians, and criminals (Hirsch, 1958). Matters got still worse when their complicity in totalitarian regimes was widely exposed — even if that complicity was widely enforced by authorities that persecuted those who insisted that totalitarian engineering was often 'bad engineering' and that demanded allegiance of engineering organizations in return for their continued existence (note that only a small minority of Germany's 222,000 engineers joined the Nazi party (Kohlrausch, Trischler, 2014)). It was no longer uncontroversial to claim that engineers stood on the right side of history.

A New Hope: Technocracy

After the Second World War, several engineering discourses came to prominence that sought to rescue technology's benefits from its harmful potential. This chapter outlines the contours of two of these: the discourses of technocratic and participatory innovation.

Many today associate technocracy with a neglect of democratic accountability. However, in a postwar context, the technocratic innovation discourse centred on a different and very urgent political problem. Politicians of different stripes and commercial managers had only recently steered technology towards global war, worker exploitation, and a crash of the world economy. They had proven that technology was too dangerous a tool in their hands. In a postwar context of increasing nationalism and an emerging Cold War, this could lead to a Third World War — a nuclear war. Hence the call for a different breed of professionals to take charge. Engineers, architects, planners, and other expert groups were *not* trained to win battles of political ideology, moral righteousness, or profit-making. Instead, so it was argued, they were committed to scientific methods to define problems, analyse those problems, and optimize solutions.

Technocratic thought had deep roots into the nineteenth century (not least to Saint Simonianism and international organizations), and had gained a boost in the prewar technocracy movement, which had condemned political, financial, and criminal manipulation: tackling societal problems as engineering problems, 'there will be no place for Politics or Politicians, Finance or Financiers, Rackets or Racketeers' (Anon. 1937: 3). In the postwar decades that notion gained ever wider currency, and experts gained an unprecedented mandate from politicians and the public to make key technology decisions. This was not a *carte blanche*; the relations between experts, politics, and business were complex and remain an important historical research topic. But discursively, technocracy became a respected governance approach associated with successful postwar reconstruction, the avoidance of a Third World War, and unprecedented economic growth with vast welfare growth, almost zero unemployment, and more equal wealth distribution than ever before.

In the technocratic innovation discourse, three features were key to salvaging technology's benefits from its harmful potential (van der Vleuten, Oldenziel, Davids, 2017). First, technology should be de-politicized and de-commercialized. The notion that experts — not politicians or businessmen — should set innovation agendas translated into what became known as the 'linear model of innovation', inspired by prewar corporate research lab experiences and new science policies in the United States. The technological innovation trajectory was to start with 'basic' or 'fundamental' research, where experts engaged in 'undirected research' of 'fundamental problems'. Next came 'applied research', which transformed basic insights into usable products and processes. Finally, these would result in increased economic growth, health, and social welfare. Some historians have mistakenly criticized this model for not corresponding to the reality of innovation and social change, but that was not at all the point of the model; its point was to create leeway for experts to set innovation agendas — even though in practice, experts had to negotiate (and compromise) that leeway with all kinds of political and commercial interests (Balconi, Brusoni, Orsenigo, 2010). Thus followed the establishment of national research councils making taxpayer money available to fundamental research, presided over by experts, and new structures of fundamental research institutes such as the German Max Planck Institutes, the French Centre national de la recherché scientifique institutes, and similar institutes even under the Spanish and Portuguese dictatorships. The Organization for Economic Co-operation and Developmet OECD promoted the linear model to its Western members states, and international research institutes such as CERN and the European Space Research organization emerged.

In tech firms, similar models were celebrated to empower research lab experts in setting corporate innovation agendas. Central laboratories should e.g. receive direct funding (instead of funding from the company's business units to perform specific tasks) and hire academics to do unguided research within fields where the company was active (e.g. solid state physics or molecular spectra in the case of the Philips Electronics, where the postwar central lab director, theoretical physicist Hendrik Casimir (1909–2000), articulated a rather detailed linear model of innovation). The findings could then be translated into specific targets through 'target research' (e.g. finding a substance with given properties) and prototyping. Innovations coming out of such expert-run central labs were then handed to the business division's 'factory engineering' and 'application research' labs to adapt products for mass production and user demands.

A second element of the technocratic saving of technology from harmful consequences was to technify politics — defining, analysing, and solving societal problems not by lobbying, voting, arguing, fighting, or profit optimizing, but through scientific methodology. A particularly prominent scientific method in this context was the so-called systems approach (Hughes, Hughes, 2000; Lundin, Stenlås, Gribbe, 2010). Societal and business problems were extremely complex because of many interacting technical, social, economic, and environmental issues. Societal problems should therefore be modelled as systems with interacting elements which could subsequently be simulated. Manipulating selected elements could reveal — often counter-intuitive — system level responses and therefore help find policy options that led to beneficial rather than destructive results, and to system optimization. This methodology gained a decisive boost from Second World War British military experiences with Operations Research that simulated and optimized existing military systems; after the war experts — not least in the US, the UK, and Sweden — developed models for analysing and optimizing future systems (adding game theory, complex scenarios, and feedback loops in human–machine interactive systems in approaches such as systems analysis and system dynamics) for application to industrial problems (localization, investment, employment, or market decisions), urban problems, national economic, transport and energy system planning, and even world problems — consider the modelling of human–earth system interactions by Jay Forrester and the MIT computer lab that was behind the influential Club of Rome report *The Limits to Growth* (1968 to 1972).

Third and finally, the technocratic innovation discourse emphasized the great need for many and responsible engineers to take a leading role in industry and government. Throughout the Western world and beyond, the number of engineering schools multiplied. These engineers should be quality decision makers and future leaders, and engineering curricula were adapted accordingly. For one, theory ousted practice; engineering was presented more as a science than an art. For Gordon Brown, dean at the Massachusetts Institute of Technology MIT — the educational institute that many Western European engineers now looked to for inspiration — education trained 'the engineer's ability to relate seemingly unrelated events of nature, whether abstract or tangible, in quantitative ways, to make new and useful theories, materials, devices, complex systems, and especially systems in which men interact with machines' (as cited in van der Vleuten, Oldenziel, Davids, 2017: 125). Such interdisciplinary human–machine system competences required engineering science but also social sciences, life sciences, and humanities training. Another indication of the call for responsible engineers was a marked change in ethical codes of engineering institutions. No longer was loyalty to employers or making things work the chief virtue of engineers. Instead, not least inspired by work of the German engineering society with professional philosophers to process the war-time experience of Nazi collaboration, the chief value became loyalty to the dignity of human life and service to fellow humans (Mitcham, 2009).

The Participatory Fix

Around 1970, the technocratic consensus reversed almost completely and gave rise to a fourth prominent technology discourse — the discourse of participatory innovation. Critique of technocracy was an important ingredient of that discourse. Part of this critique came from social movements and counterculture activists rebelling against what they called the dominant Western technological world view. Environmental activist organizations such as Friends of the Earth (1969) and Greenpeace (1971) lamented that postwar technological systems had been optimized for exploiting nature; civil rights activists found that minority viewpoints had been ignored, and that experts acted beyond the democratic control of elected politicians. The peace movement stressed how the military–industrial–university complex had produced the nuclear arms race and perverse weaponry used in the Vietnam war horrors. Social critics revamped older critiques and argued that systems approaches prioritized rationality and order by stifling emotions, free expression, and communality; they felt that instead, humans, not the system, should come first. And further, incumbent players undermined technocratic thinking. For example, the US Department of Defense study Project Hindsight was pivotal in the invalidation of the linear model of innovation, finding that merely three per cent of the 710 key events leading to twenty crucial weapon systems had come from basic undirected research; 97 per cent came from applied research (Hughes, 2004; Wise, 1985).

These different sources of critique shared a notion that expert-run 'closed systems' needed opening up to other people, issues, values, and approaches. The linear model of innovation was in effect reversed: use and application should be the starting point for setting research and innovation agendas and priorities. And in order to make that happen, citizens, users, and other stakeholders or their representatives should participate directly in technological decision making and design. After all, the argument ran, it was they — not experts speaking on their behalf — who understood their problems and future needs best. And besides, those who had to live with the consequences of technology surely had a democratic right to also shape that technology.

As in the case of the technology discourses described earlier, we may recognize many different appropriations of the notion of participatory innovation (van der

Vleuten, Oldenziel, Davids, 2017). In national and local politics, for example, action groups organised protests to affect technology decisions about nuclear weapons and energy, motorways, airports, land reclamation projects, housing, and much more. Iconic or their success were the anti-nuclear protest marches drawing hundreds of thousands of protesters in many Western European countries in the late 1970s and early 1980s. Even before the 1986 Chernobyl disaster, nuclear energy policies were shelved, e.g. in Denmark and Austria, and several ongoing nuclear construction projects were cancelled, e.g. in Germany and Spain. While political protest often took a conflictual approach, a new breed of technology mediators sought to forge consensus amongst stakeholders. In their view, deliberation should produce a co-decision or co-design process. They developed the field of participatory technology assessment, which became influential e.g. in Denmark, the Netherlands, Germany, and Austria, and developed approaches such as citizen conferences to bring citizens voices to the technological decision making process, and roadmap or scenario workshops to initiate a conversation between corporate, environmental, and local stakeholders, forging a mutual understanding of each other's concerns and translating these into joint technological decision-making acceptable to all. And from the early 1980s, neoliberal thinkers also appropriated the participation discourse in still a different manner. That particularly influential appropriation opposed the left-wing notion of participation as a political right and posed that the welfare state made people passive onlookers waiting for the state to solve any problem they might experience. In the neoliberal vision of the 'participation society' consumer groups, patient organizations, companies, and other stakeholders would define and solve their problems themselves — and do so much more efficiently than the distant state apparatus could ever hope for. The neoliberal appropriation deeply politicized the notions of participation and participatory innovation, especially after cutting government expenses became an end in itself, and state functions were massively transferred to (often for-profit) private organizations.

Tech firms also appropriated the participatory innovation discourse in various ways. The linear model, as noted, was often reversed: management decided that henceforth business units would define innovation agendas based on marketing research and business opportunities, and hire central lab researchers to do specific research to that purpose. Innovation scholars recognized user innovation as a key resource for identifying future markets. Designers, for their part, appropriated the new discourse with notions such as user-centred design, including 'cooperative design' (coined by Scandinavian labour unions pioneering worker participation in e.g. factory automatization processes) and 'participatory design' (coined in the US in the context of introducing personal computing at the workplace). And action groups managed to press boardrooms to adopt Corporate Social Responsibility programmes. Finally, corporate research leaders opened up closed innovation systems; the notion of 'open innovation' suggested that multiple companies pool research and innovation resources and form innovation ecosystems.

Outside the commercial settings of tech companies, user communities, too, reclaimed agency and initiated a host of activities. The technocratic innovation narrative had induced user representatives to present themselves as 'experts' and as such join expert committees for developing social housing, for example. But in the participatory age, all kinds of user communities started, once again, to define and solve their own problems. Initiatives ranged from Danish schoolteachers and alternative energy groups (re)starting the grid-connected wind-electric turbine success story to later renewable energy communities; from Do It Yourself hardware for house improvement to the small house movement; from biketivists reclaiming automobile-congested urban streets to today's revival of cycling as a sustainable urban mobility mode; and from the 1970s suburban garage home computer builders to open-source software and app design by users.

Finally, the engineering community itself appropriated the participatory innovation discourse from the very start, when young engineers started to rebel and were at the forefront of all kinds of alternative technology movements. By the late 1960s older engineers also recognized that a humanizing technology turn was imminent and that engineering institutions also needed to 'open up' (Wisnioski, 2012). Initiatives ranged from establishing technology and society divisions at major engineering associations and Science Technology and Society programmes in engineering schools to increasing the enrolment of women in engineering education. In the technocratic age women's enrolment had been encouraged to increase the number of engineers, but in the context of so-called differential feminism this enrolment also had connotations of bringing 'female values' into engineering. The science shop movement of the 1970s promoted socially engaged students and professors to work on real-life practical problems posed by disadvantaged citizens, financially weak worker groups, or civil society groups; the idea spread in the 1980s across Western Europe and beyond the university world. While the European Commission in the early 2000s praised the initiatives for interdisciplinary cooperation on problems that mattered, in the Netherlands, where the movement had originated, the neoliberal turn meant that idealistic science shops disappeared or changed into 'knowledge valorization centres'. And with reference to the non-European world, the 'intermediate technology' or 'appropriate technology' movement sought to develop local solutions with and for local communities in the global South — solutions that preferably could be locally constructed, with local materials, and be locally maintained. Organizations such as Engineers Without Borders had similar agendas and were often student-led. These initiatives, too, increasingly felt the pressure on idealism of the neoliberal turn.

Today, technocratic and participatory innovation discourses often co-exist and mingle in novel ways. It is widely acknowledged that both had their strengths and weaknesses; both could be hijacked by political and commercial interests, for example. Current Grand Challenges for Engineering and 'save the planet through technology' discourses therefore call for the exploration and development of novel, nonlinear ways to govern technological decision making and design that would ideally combine the best of both worlds.

In the meantime, a global history of technique that seeks to escape 'universalistic' Western technology models and restore the plurality of technology meanings is certainly needed — especially at a time when global North-born concepts such as sustainable innovation, responsible innovation, and the Anthropocene once again seem to project global North priorities on global South situations with little regard for the plurality of ways of being, knowing, and relating that imply diverse problem definitions and solution search directions. Still, such a global history of technique also needs to consider that supposedly universalistic 'Western' technology discourses differed through time, have been appropriated very differently by diverse social groups, and remained thoroughly contested also 'back home' in Western Europe. Thus emerges a major challenge for global historians of technique: how to study the connected histories of technology and socio-ecological challenges in different places in the world in ways that simultaneously appreciate regional diversity and distinctiveness as well as transregional and transcontinental connections — material, institutional and discursive — that bring distant regional histories into mutual conversation (van der Vleuten 2019, 2020; de Hoop et al. 2022).

References

Anon., 'What is technocracy?', *The Technocrat*, 3.4 (1937), 3.

Balconi, Margherita, Stefano Brusoni, and Luigi Orsenigo, 'In Defence of the Linear Model: an Essay', *Research Policy*, 39.1 (2010), 1–13.

Bucholz, Arden, 'Armies, railroads, and information: the birth of industrial mass war', in: Jane Summerton, ed., *Changing large technical systems* (Boulder, Colo: Westview Press, 1994), 53–70.

de Block, Greet, *Engineering the Territory. Technology, Space and Society in 19th and 20th century Belgium* (unpublished doctoral dissertation, University of Leuven, 2011).

de Hoop, Evelien, Aarthi Sridhar, Claiton M. da Silva, Erik van der Vleuten, 'Historicising Entanglements: Science, Technology and Socio-Ecological Change in the Postcolonial Anthropocene', *Global Environment* 15.2 (2022), 195–08.

Christie, Alexander G., 'A Proposed Code of Ethics for all Engineers', *The Annals of the American Academy of Political and Social Science*, 101.1 (1922), 97–104.

Diogo, Maria Paula and Dirk van Laak, *Europeans Globalizing: Mapping, Exploiting, Exchanging. Making Europe: Technologies and Transformations 1850–2000*, vol. VI, (London and New York, NY: Palgrave, 2016).

Graham, Steve and Simon Marvin, *Splintering Urbanism: Networked Infrastructures, Technological Mobilities and the Urban Condition* (London and New York, NY: Routledge, 2001).

Greeley, Horace, *Art and Industry: as Represented in the Exhibition at the Crystal Palace* (New York, NY: Redfield, 1853).

Guldi, Jo, *Roads to Power: Britain Invents the Infrastructure State* (Cambridge, MA: Harvard University Press, 2012).

Hansen, Hans Christian, *Poul la Cour. Grundtvigianer, opfinder og folkeoplyser* (Askov: Askov Højskoles Forlag, 1985).

Hård, Mikael, and Thomas J. Misa, eds, *Urban Machinery: Inside Modern European Cities* (Cambridge, MA: MIT Press, 2008).

Hirsch, Walter, 'The Image of the Scientist in Science Fiction: a Content Analysis', *American Journal of Sociology*, 63.5 (1958), 506–12.

Högselius, Per, Arne Kaijser, and Erik van der Vleuten, *Europe's Infrastructure Transition: Economy, War, Nature. Making Europe: Technologies and Transformations 1850–2000*, vol. III, London and New York, NY: Palgrave, 2015).

Hughes, Thomas P., *Human-built World: How to Think about Technology and Culture* (Chicago, IL: University of Chicago Press, 2004).

Hughes, Agatha C. and Thomas P. Hughes, eds, *Systems, Experts, and Computers: the Systems Approach in Management and Engineering, World War II and After* (Cambridge, MA: MIT Press, 2000).

Kaiser, Wolfram and Johan W. Schot, *Writing the Rules for Europe: Experts, Cartels and International Organizations. Making Europe: Technologies and Transformations 1850–2000*, vol. IV, (London and New York, NY: Palgrave, 2014).

Kohlrausch, Martin and Helmuth Trischler, *Building Europe on Expertise. Innovators, Organizers, Networkers, Making Europe: Technologies and Transformations 1850–2000*, vol. II, (London and New York, NY: Palgrave, 2014).

Lundin, Per, Niklas Stenlås, and Johan Gribbe, eds, *Science for Welfare and Warfare. Technology and State Initiative in Cold War Sweden*? (Sagamore Beach, MA: Science History Publications/USA, 2010).

Mitcham, Carl, 'A Historico-ethical Perspective on Engineering Education: from Use and Convenience to Policy Engagement', *Engineering Studies*, 1.1 (2009), 35–53.

Misa, Thomas J. and Johan Schot, 'Inventing Europe: Technology and the Hidden Integration of Europe', *History and Technology*, 21.1 (2005), 1–19.

Mote, C. D., Ann Dowling, and Ji Zhou, 'The Power of an Idea: the International Impacts of the Grand Challenges for Engineering', *Engineering*, 2.1 (2016), 4–7.

Oldenziel, Ruth and Mikael Hård, *Consumers, Tinkerers, Rebels: the People who Shaped Europe. Making Europe: Technologies and Transformations 1850–2000*, vol. I, (Houndmills: Palgrave, 2013).

Tenner, Edward, *Why Things Bite Back: Technology and the Revenge of Unintended Consequences* (New York, NY: Vintage, 1997).

van der Vleuten, Erik, 'History and Technology in an Age of Grand Challenges: Raising Questions', *Technology and Culture*, 61.1 (2020), 260–71.

———, 'Technology, Societal Challenges, and Global Sustainability History', *ICON*, 24 (2019), 34–52.

———, 'Toward a Transnational History of Technology: Meanings, Promises, Pitfalls', *Technology and Culture*, 49.4 (2008), 974–94.

———, Ruth Oldenziel, and Mila Davids, *Engineering the Future, Understanding the Past. A Social History of Technology* (Amsterdam: Amsterdam University Press, 2017).

Wise, George, 'Science and Technology', *Osiris*, 2nd series, 1 (1985), 229–46.

Wisnioski, Matthew H., *Engineers for Change: Competing Visions of Technology in 1960s America* (Cambridge, MA: MIT Press, 2012).

ELENA KOCHETKOVA, JULIA OBERTREIS

The Achilles Heel of a Big Player

*Technology and Modernity in the Russian Empire, the Soviet Union and Russia (Nineteenth to Twenty-First Centuries)**

Julia Obertreis, a co-author of this chapter, passed away right before the volume was published. She was an excellent colleague and friend, and the field has lost a lot after she left us.

The history of technology of the Russian Empire and the Soviet Union — in its existing rudimentary form — has long been dominated by contradictory narratives: nation-centred success stories from the Russian and Soviet perspective on the one hand, and critical studies highlighting Russian and Soviet 'backwardness' and a lack of technological innovations on the other (Nolte, 1991; Hildermeier, 1987; a very critical and one-sided evaluation: Sutton, 1968–1973; less biased: Graham, 2013). This chapter addresses major developments in the history of technology in Russia and the Soviet Union as well as continuities in post-socialism, including the delicate but important subject of Russia's technological lag behind other countries, which, however, will not be approached from a typical Western-centric perspective. This task is not an easy one, since only little work has been done on an overview of the subject to date, although a number of specific studies exist.[1] The following questions will be given priority in this chapter: how were economic, social and political developments intertwined with technological trends and innovations? How have armed conflicts and wars affected technological developments? What role did transfers of technology, know-how, and experts to and from the Russian Empire and the Soviet Union play? Which technical innovations originated in Russia or the Soviet Union? The

* This chapter is dedicated to the memory of Larissa Zakharova, who provided inspiration for this chapter and passed away in 2019 before completing this piece. She was an outstanding historian of technology who significantly contributed to the history of communication and fashion in the Soviet Union.

1 Within the few existing surveys, the history of technology is often combined either with science (see e.g. Holloway, 2006), or with infrastructure and traffic (see e.g. Helmedach, 2009).

Global History of Techniques, ed. by Guillaume Carnino, Liliane Hilaire-Pérez and Jérôme Lamy, Global Matters Culture Matérielles, histoire des techniques et circulation globale des savoirs, 9 (Turnhout: Brepols, 2024), pp. 195–218
BREPOLS PUBLISHERS 10.1484/M.TECHNE-EB.5.129808

admiration of and enthusiasm for technology in some cases aquired the qualities of a cult. Which cults of technology emerged, and how were they expressed and promoted? What is the role of Imperial and Soviet industrial heritage in today's Russia? By addressing these questions, we will investigate the history of the Russian region in the *longue durée,* tracing its major transformations from the nineteenth until the twenty-first centuries. In relation to contemporary technologies and during different periods of its history, Russia repeatedly experienced a backwardness and weakness that was perceived and described by contemporaries from Russia/the Soviet Union itself. In different contexts, the deficiencies proved to be the Achilles heel of a (potential) big player in global history.

Technology in the Russian Empire

In the nineteenth century, one of the turning points in technological development, was, of course, the beginning of the industrialization of the vast Russian Empire. Due to the heavy demand for iron in railway construction and the military, the iron industry was one of the key industrial branches. Instead of the traditional charcoal for smelting, the state now urged the industry to use black coal. What has often been called 'catch-up industrialization' (*nachholende Industrialisierung*) — a term clearly expressing the perspective of Western Europe and the USA — reached a first climax in the 1890s and brought with it much innovation (Helmedach, 2009: 80).[2] However, most of the innovation was based on imported machinery and research done abroad. National industrial research remained relatively weak until the First World War (Holloway, 2006). While foreign investment opened up possibilities of innovation, the Russian government in general remained cautious about foreign (as well as Russian) stock companies and monitored their activities.[3]

More so than in most Western countries, in Russia industrialization was guided 'from the top'. This can be observed, for example, in the history of railway construction in Russia, which started in 1837 and experienced a boom in the 1890s. From 1890 to 1900 the distance covered by the railway in the Russian Empire increased from 29,057 to 51,808 kilometres and took the lead in Europe (Neutatz, 2013; Marks, 1991; Fadeev, 1994). Minister of finance Sergei Yulyevich Witte pushed the construction of the Trans-Siberian railway forward and generally promoted substantial state investment in railway construction as a means of furthering industrialization. The railway was of tremendous significance for Russia, and not only with regard to economic development. It 'shrunk' vast distances and, as in other countries, conveyed the impression of a general acceleration and an altered temporality. Therefore, it

2 Unless otherwise stated, all translations into English are the authors' own.

3 The sum total of foreign investments in the Russian Empire's industry (not taking into account investments in banks, municipal enterprises, and trade) equalled more than 1.3 billion rubles, which was about one third of all investments. Major lenders of capital were Great Britain, France, Belgium, and Germany, and the largest investments were in the mining, metallurgical, and metal working industries as well as in the chemical and electrical industries (Heller, 2005: 17).

Fig. 1. The Mosin rifle, a sample of 1891, 'Марка' no 1995, Michel no 2212.

was an expression of modernity (Cvetkovski, 2006; Sperling, 2011; Schenk, 2016). It also helped to establish new social orders, e.g. via the introduction of different classes on the train, and shaped towns and cities through the erection of railway station buildings, which were often elaborately built in the major cities and can be considered as the 'city gates of modernity' (Schenk, 2007). Moreover, in the eyes of the elites of the Empire, the railway brought with it the 'light' of civilization to different regions, among them Central Asia, which was conquered and annexed to the Empire in the course of the second half of the nineteenth century (Hofmeister, 2018). Alongside the grand railway construction, Witte promoted a protectionist policy to increase domestic industrial development, making it, in fact, more productive than it was before.

Despite reliance on foreign technologies, Russian inventors developed a few 'home-grown' innovations which became recognized internationally. After the Crimean War and in the context of the Great Reforms, the modernization of the military was a priority for the government (Benecke, 2006). In weaponry, the three-line rifle M1891 deserves special mention due to its transnational origin, its significance for the army in several wars, and its longevity and international distribution. This is an internal magazine-fed, bolt-action, five-shot military rifle known in Russia today as 'Mosin's rifle' (*vintovka Mosina*) and in the West as 'Mosin-Nagant' (Figure 1). It was developed from 1882, and the design that was eventually approved contained elements created and proposed by the Russian weapons engineer Sergei Ivanovich Mosin (1849–1902), the Belgian weapons engineer and manufacturer Henri-Léon Nagant (1833–1900), and two further Russians. The rifle was produced in vast quantities from the 1890s onwards and was used on a large scale for the first time in the Russo-Japanese War of 1904 to 1905. Although when compared with the Japanese standard rifle at the time, some of its shortcomings became obvious, no significant changes were made to the model due to inertia at the production factories. In the First World War, Mosin's rifle was again used as a standard issue weapon by the Russian army, and it continued to be used in the early Soviet period in the context of militarized mass-organizations. After the Second World War, it enjoyed a wide international distribution in the context of the Cold War, although it was no longer produced in the Soviet Union. Mosin's rifles and carbines were in use in wars in Korea and Vietnam as well as on both sides of the Soviet–Afghan War (Fedoseev, 2019).

Another Russian military-related invention of international significance emerged during the First World War, when all parties were confronted with gases used in chemical warfare. While several countries invented early gas masks independently from each other, the Russian chemist Nikolai Dmitrievich Zelinsky (1861–1953) and the technical expert Mikhail Ivanovich Kummant (life dates unknown) invented the first effective filtering gas mask with activated charcoal in 1915. This mask could absorb a wide range of poisons. This indicates the strong position of Russian chemistry in the international competition. Other Russians offered inventions at the same time as their colleagues in the West, e.g. parachutes, radios, and electric lamps, but most did not gain international recognition.

The Bolsheviks' Technological Drive

Beginning with the October Revolution (also known as the Russian and Bolshevik Revolution) in 1917 and the Civil War 1917–22, the Bolsheviks strove to realize their vision of a secular and centralized political system, which was thought of as a fairer alternative to Western, capitalist modernity (Calic, Neutatz, Obertreis, 2011). Science and technology were intended as the basis of an alternative modern state and were understood as integral elements in the realization of the grand enlightenment project of providing the population with social engineering while relying on reason alone. The 1920s saw a restructuring of the academic and research landscape by the Bolsheviks. The universities lost significantly in status, and the Academy of Sciences was granted the status of the 'highest scholarly institution' in the country. Further, more than seventy new research institutes were established, which, among other things, led to a narrow specialization of much of the research, especially in the natural sciences (Holloway, 2006: 553–4).

The Bolsheviks had a strong technocratic tendency and considered the introduction of new technologies crucial for upholding their rule successfully. The first nationwide general project of advancing technology was electrification. Propagated widely with Lenin's motto that the electrification of the whole country added to Soviet power equalled communism, the enormous undertaking was organised in 1920 via the GOĖLRO (*Gosudarstvennaia komissiia po ėlektrifikatsii Rossii*), the state commission for the electrification of Russia. The electrification campaign carried much symbolism, as electric light was associated with enlightenment, education, culture, and progress, which would jointly make up a new society. This was especially effective in rural areas, where the lightbulb was supposed to introduce 'culture' to the remotest peasant hut. In this (rural) context, electrification even had a sacred dimension. At the same time, it was a prerequisite for reconstruction after the First World War and the Civil War and for economic and industrial modernization. The programme foresaw the installation of thirty power stations within a period of ten to fifteen years. Ten of them were hydropower stations, among them the famous gigantic dam on the Dniepr at Zaporizhzhia (Ukraine), which became the largest hydropower station worldwide at the time. The revised electrification programme was fulfilled by 1933 (Haumann, 1974).

In the eyes of the Soviet leadership, new technologies were badly needed to overcome the disdained and much-cited 'backwardness' (*otstalost'*) that Soviet propaganda considered a remnant of the tsarist period. To turn Russia and the whole Soviet Union into a modern industrialized state that would differ fundamentally from the old agrarian state was one of the Bolsheviks' central concerns, and the quest for technology derived from this. Stalin expressed this clearly in a speech to the Central Committee in November 1928, at a time when the regime introduced the first five-year plan and with it central economic planning:

> We have assumed power in a country whose technical equipment is terribly backward [...]. At the same time we have around us a number of capitalist countries whose industrial technology is far more developed and up-to-date than that of our country. [...] In order to secure the final victory of socialism in our country, we must also overtake and outstrip these countries technically and economically. Either we will achieve this, or we will be forced to the wall (Stalin, 1928).

The (perceived) backwardness of the Soviet Union and its 'capitalist encirclement' served to legitimize the radical political change begun in 1928 to 1929 and the extraordinarily high speed of the industrialization that the regime embarked on at this time. With it came a 'large, well-funded, party-controlled R&D [research and development] effort' (Holloway, 2006: 558).

As science and technologies were at the centre of the Bolsheviks' attention, the role of engineers and technical workers became an important political issue. Many engineers were willing to cooperate with the new regime, not so much because of political sympathies (which also played a role), but rather because of their eagerness to finally realize the projects they had often worked on for some time before the Revolution and were not able to complete due to a lack of governmental approval or finance. Towards the end of the 1920s, the state's attitude towards the 'old' engineers, i.e. those who had received their training and entered a professional career before the Revolution, changed and they were now persecuted. The government's aim was to produce new, 'red' engineers. In an impressive campaign, representatives of the formerly lower strata of society, namely peasants and workers, were encouraged to receive technical training (Fitzpatrick, 1979). The engineering profession became diluted; in its place the entities of 'engineeral-technical workers' (ITR, *inzhenerno-tekhnicheskie rabotniki*) and 'specialists' were introduced and became fixed elements of socio-professional classification. The dark side of the great advancement that took place was the persecution of the 'bourgeois' engineers. Cruelly depicted in caricatures as vile snakes that ought to be crushed by proletarian boots, the prerevolutionary engineers became targets of campaigns against 'sabotage' and, more specifically, economic sabotage known as 'wrecking'. At the same time, their know-how was essential, and the regime made some effort towards making use of their expertise. As Susanne Schattenberg has shown, in 1930s novels, plays, and films model engineers were depicted as builders of the new, Soviet world, and they also often perceived themselves as such (Schattenberg, 2002).

Against the backdrop of a general educational boom in all Soviet-period decades since the (end of the) 1930s, more than one third of the 'specialists' who had received higher education had trained as an engineer. Technical training, which was often rather too specialized, opened doors to the positions of leaders in management, administration, and politics. Many of the leading personnel in the 1960s and 1970s had a background in technical engineering (Beyrau, 1993). The creation of 'red' engineers and their adoption of leading roles in society had become reality to some extent. The technocratic aspect of the Bolsheviks' worldview was thus embodied in millions of personal biographies.

Apart from electrification and the erection of large (hydro) power stations the production of steel was also a focus of the Soviet industrial modernization effort of the 1930s. The most well-known of the steel production sites is that of the then newly-founded city of Magnitogorsk (meaning Magnetic Mountain). Characteristic of these and other industrial construction sites is the impressive success stories and quick rise of production figures which Soviet propaganda displayed in numerous posters and all kinds of texts. These successes were achieved against the background of risky undertakings, such as the use of newly erected, not-yet finished blast furnaces, leading to a high rate of accidents and many human losses (Kotkin, 1995).

Aviation was another technological project highlighted by the Soviet regime. Before the October Revolution, there was already some enthusiasm for civil and military aviation in Russia as in other countries, and aeronautical clubs had sprung up. Much of the technological input came from Western countries, but Russia had the second largest airforce worldwide on the eve of the First World War. Thanks to the aircraft designer Igor Ivanovich Sikorsky (1889–1972) it also had the largest planes, like the first four-cylinder aircraft, the S-21 *Russky Vityaz* ('Russian warrior', translation S. Palmer); these proved that building multi-engine airplanes was possible and were redesigned during the war to serve as bombers (Palmer, 2006: 57–8). A specific Russian aviation culture had already emerged before the war, and Russian pilots had been praised for their 'Russian' qualities, which included being ready to sacrify their lives for the cause and collective priorities coming always before individual ones. Scott W. Palmer saw an 'iconic vision of the airplane' and a 'compensatory symbolism' at work, both before and after the October Revolution (Palmer, 2006: 283). In the Soviet context, the staging of international flights and the setting of new long-distance records, especially in the 1930s, conveyed the impression of technological competency and strength. Pilots like Valery Pavlovich Chkalov (1904–38) were praised as legendary warriors (*bogatyri*), 'falcons', or 'Stalin's eagles', serving as icons for building the pantheon of the Soviet nation. During the period of the early five-year plans, Soviet airplane production increased dramatically. Industrial espionage in the West contributed to modernization, and Soviet technicians improved designs from abroad (Palmer, 2006).

Technology for the Countryside and in Times of War

The Bolsheviks' modernization efforts referred not only to industry and cities but also to the countryside, where new technologies were supposed to be implemented as well. A good and relatively well-researched example is the introduction of tractors in the 1920s and 1930s. It signified the unification of industry with agriculture and was propagated by Vladimir Lenin as a way to attract peasants to the communist state system. In much of the coverage and propaganda on developments in agriculture, the large number of tractors being produced was highlighted as a sign of modernity and progress (Josephson, 2010; Miller, 1970). Several vast tractor factories emerged in Kharkiv (Ukraine), Chelyabinsk, and Stalingrad, while the Putilovets (Kirov) tractor factory in Leningrad had existed before the Revolution and was now being expanded. However, in the 1930s most tractors used in the Soviet Union were still manufactured in the USA. American technology played an important role in the production of tractors as well as in many other industrial branches in the Soviet era.[4] Against the background of the Great Depression, Western countries willingly accepted Soviet offers of exchanging personnel and know-how and provided interim funding in the form of credit. Apart from the late nineteenth and early twentieth centuries, there was, however, no import of capital, because foreign ownership of means of production in the new socialist state was an impossible concept (Lutz, 2011).

Quantitative success was most impressive in the non-Russian regions of the Soviet Union. While there had been only 'a handful' of tractors in Uzbekistan in 1920, their number rose to 24,200 by 1940 (Alimova, Golovanov, 2005: 228). In Central Asia, the introduction of tractors had the connotation for local peasants of getting away from their tiresome manual work and modernized not only agriculture, but also whole regions and their societies, which were perceived by Moscow as 'backward'. Different temporalities were ascribed by the Bolsheviks to the center, Moscow, which provided progress in the form of new technologies, and the non-Russian (and also to the Russian) peripheries which were supposed to be grateful for receiving them. A perception of imperial superiority attached to technological standards and innovations was in play, as had been the case before the Russian Revolution (Van der Straeten, Obertreis 2022). Tractors are also a good example of the difficulties surrounding the implementation of machinery. Many of the tractors delivered to Uzbekistan were not in use in the mid- and late 1930s. For most agricultural work,

4 Anthony Sutton maintains that from the 1930s until the end of the Second World War both the tractor and the automobile industries in the Soviet Union were 'completely dependent on foreign (primarily American) technology for design, construction, and equipment, as well as initial operation' (Sutton, 1968–1973: 194).

man- and cattle power were used instead. Machinery and equipment were not distributed according to local demand, there was a lack of technical know-how, and tractors and other machines (like combines, later harvesters) were not suited to the relatively small fields of Central Asia (Obertreis, 2017). The introduction of tractors also had a gender dimension: while it was certainly usually men who drove and maintained tractors, the 'traktoristka' (female tractor driver) became a symbol of dual progress, i.e. both the introduction of machinery to agriculture and the emancipation of women. Pasha (Praskovia Nikitichna) Angelina (1912–1959), was one of the first female tractor operators and became a well-known shock worker and Stakhanovite as well as one of the 'ordinary celebrities' of the Stalinist 1930s. Containing elements of a typical biography of Soviet celebrities of the 1930s, Angelina was elected deputy of the Supreme Soviet, wrote a book about her experience, and met Stalin in person (Fitzpatrick, 1999: 74–5).

The Great Terror of the years 1936 to 1938 had an impact on technological innovation within the Soviet Union. Many experts working in research institutes were imprisoned or even executed, and inexperienced individuals took their places. The Terror deeply affected the army as well, and research on radar systems and rockets came to a halt in the late 1930s and was only revived during the war. The German invasion of the Soviet Union in 1941 led to a relocation of many research institutes, which were moved to the Eastern parts of the country (Siberia, Central Asia). Scientists and technical experts not only fought in the war but also consciously worked for the victory of their homeland by improving weapons and production processes from the workplace (Holloway, 2006: 561–62). The Axis powers' fight-back in the following years was partly made possible by deliveries from the US (and to a much lesser degree, Great Britain and Canada) in the context of the Lend-Lease Act. The deliveries to the Soviet Union included planes, tanks, and military equipment like battlefield telephones. High-octane fuel that was needed for high-capacity planes was of special importance, and 'Jeeps and Studebaker trucks became the backbone of Red Army mobility' (Harrison, 1985: 29). Beyond weaponry, it was the delivery of secondary goods and foodstuffs that contributed to shortening the war and securing the allies' victory. Soviet soldiers humorously called the Lend-Lease tins with corned beef the 'second front'. The Western deliveries were also a source of technology import. However, the US held back its most innovative military technology (Overy, 2001: 276, 299–301, 316; Havlat, 2017).

At the same time, Soviet research and development generated military technology that was characterized by its longevity, efficiency, and comparably easy use. Multiple rocket launchers, known internationally as 'Katyusha', were designed by a Soviet research institute specialized in rocket technology, the 'reactive scientific research institute' (RNII, *Reaktivnyi nauchno-issledovatel'skii institut*). Katyusha was in use from the beginning of the German invasion onwards and was known and feared in the German *Wehrmacht* as 'Stalin's organ' (*Stalinorgel*) because of its appearance and the whistling sounds it made. It proved highly effective in firing at broad targets. During the war, the Soviet Union produced more than 10,000 launchers and more than twelve million ballistic missiles. The immense firing power of the reactive artillery contributed significantly to the Soviet victory (Stache, 1987: 78). One of the multiple rocket launcher's advantages was that it could be placed on all different kinds of motor vehicles, and it was installed on American trucks when they were delivered to the Soviet Union. In this way, Soviet technology and American deliveries were combined in a successful way. Katyusha-like multiple rocket launchers were exported from the Soviet Union to a number of countries in the postwar period for decades and used internationally, similarly to Mosin's rifle, in many of the major wars of the twentieth century.

Soviet Technologies and the Cold War

The Second World War stimulated the development of many technologies, particularly in chemistry and physics, which later had a great impact on different aspects of social life. The Cold War with its ideological fronts created urgency and stimulated technological innovation even further (Noble, 2011). It was an accelerator for technological development internationally, inspiring different countries to invest, transfer, and spy on the latest achievements. Cold War technologies both posed a nuclear risk and held societal promise, as they almost simultaneously inspired the invention of military weapons and everyday consumer products. The rivalry with the Western world led the Soviet leaders to rely even more on technological development and further kindled their desire to make the USSR a technological empire. The Soviet government emphasized the need for rapid technological, scientific, and industrial transformations, but this was not a linear path towards success.

After the Second World War under Stalin's leadership the government continued to invest heavily in the military sector (Popov, 2001; Lazarev, 2002; Gregory, 2003). The annexed Finnish, Baltic, German, and Japanese territories enabled the Soviet government to acquire technologies that were needed for the military and the industrial sector (Kochetkova, Pokidko, 2019, 270). The new state borders included a few dozen newly acquired sawmilling and pulp and paper making enterprises, which were important for the manufacture of gun powder and other products of value to the military. German engineers and scientists were forcefully moved to the USSR after the war, and their expertise played a significant role in the development of Soviet military technology (Oleynikov, 2000; Siddiqi, 2009). Intelligence was another important source for technical information and the development of the Soviet Union's technologies. Even during the Second World War, the Soviets knew much about American research on the atomic bomb project. The first Soviet nuclear weapon was completed in 1949 with a successful testing of the atomic bomb, a few years after America's atomic bombings of Japan. With this 'superweapon' the Soviet Union became a superpower in the Cold War, contributing to the spread of fears and risks. Alongside the military project, Soviet engineers developed large-scale research on the 'peaceful' use of atoms in the production of electricity and for other applications in science, medicine, agriculture, and industry. Like Western countries which developed atomic energy, Soviet science, technology, and propaganda promoted the image of

Fig. 2. Soviet Agat-9 computer, the last mass-produced model.

the peaceful atom, and the Soviet Union heavily relied on the use of nuclear power in energy production (Brown, 2013). The Soviet state was the first to launch a nuclear power station in Obninsk in 1954.

Nikita Khrushchev's decade of leadership from 1955 to 1964 was a time of great technological achievements in both military-related contexts and the consumer-oriented industry. The successes in Soviet space engineering — the first launch of Sputnik in 1957 and human flight to outer space in 1961 — were a blow to American technological efforts and significantly increased trust in science and technology in both Soviet society and abroad. These successes spurred a cult around cosmic flight and cosmonauts in the Soviet Union and other socialist-governed countries. Yuri Gagarin and other cosmonauts, including the female cosmonaut Valentina Tereshkova, were depicted as a new generation of heroes and made it possible for the regime to generate enthusiasm about technological developments. Gagarin's image in particular worked as a link between different groups of the Soviet population and a powerful role model. Mastering technology and entering the heavenly spheres combined into a very attractive heroic model which integrated elements of the pre-Soviet and early Soviet aviation cults. Moreover, Gagarin became a kind of peace ambassador, travelled to many countries, and represented a positive image of the Soviet Union (Gestwa, 2009; Andrews, Siddiqi, 2011; Maurer et al., 2011).

In addition to the successes in the military and space spheres, the Soviet Union had some great achievements in computing: for example, in 1950, Sergey Lebedev introduced the first electronic computer to Europe (Gerovitch, 2002). In the 1950s, the physicians and experts of quantum electronics Aleksandr Prokhorov and Nikolai Basov were the first to develop masers and lasers, for which they, together with the American Charles Townes, received a Nobel Prize in 1964 (Gross, Herrmann, 2007) (Figure 2).

The institution mainly responsible for the development of technology was the State Committee for Science and Technology (*Gosudarstvennyi komitet po nauke i tekhnike*, GKNT), established in 1948 as the "state committee on introducing advanced technologies into the people's economy (*Gosudarstvennyi komitet po vnedreniiu peredovoi tekhniki v narodnoe khoziaistvo, Gostekhnika*), which changed its names many times, gathered research on new technologies, and was responsible for experimentation on innovations at research institutions and industrial enterprises (Autio-Sarasmo, 2011, 2016). The government attempted to intensify the innovation processes, considering them the premise for economic development (*Holloway, 2006*). However, while Soviet science occupied leading positions globally, in particular in physics and mathematics, technologically the USSR lagged behind (Holloway, 2006: 572). Except for the military, physics, the hydropower industry, and space, the Soviet Union did not manage to take the lead in the technological competition and the scientific-technological advancement that unfolded globally. By the late 1960s it even lost the lead in many technological fields. In particular, as early as in 1955 and despite Lebedev's early computer, top political leaders knew the country would not manage to compete with the USA in terms of production and its computers would be unable to compete on the international market (Reynolds, 2010).

Nevertheless, technological achievements reassured Soviet leaders that it was possible to build 'the material basic of communism' quickly and to hasten the move from socialism to communism. In 1961 Khrushchev declared the country's ability to establish communism in just a couple of decades. In his concept of communism, the rise of living standards was implied, and alongside ideological aspects, the material well-being of the population played an important role. Khrushchev's liberalization and intention to make everyday life more comfortable shifted the government's priorities to the consumer sphere to a certain extent. The state leader himself now stated that it was particularly necessary to supply citizens with beautiful and high-quality clothes. As Larissa Zakharova shows in her study of Soviet fashion design, the Soviet textile industry easily combined its own functional fashion with Western styles (Zakharova, 2019). Mass production had changed the patterns of consumption, focussing on consumer needs and rendering both kitchens and guns important to the Cold War rivalry (Oldenziel, 2005). Technological modernization was to serve not only the increase of production but also the creation of modern technologies, ranging from milk packages and clothes to refrigerators and modern kitchens. The Soviet leadership aimed to build a modern society to respond to the rapidly developing Western consumerism, and here technology played a crucial role. The famous Kitchen Debate between Nikita Khrushchev and Richard Nixon at the 1959 opening of the American National Exhibition in Moscow showed that home and consumer goods were a top priority in the Cold War competition.

Khrushchev himself believed in the power of chemistry, but this was also a belief of many engineers and industrial scientists. Chemistry was considered an applied tool for the improvement of food packaging and conservation, among other things. Communism was declared to be achieved by the combination of Soviet power and chemicalization while under Lenin it was electric power rather than chemistry that had been seen as the key. By the mid-1950s the regime had mostly completed the electrification of the main housing infrastructures of the country, especially in large cities, and now the development of chemistry was the new priority. In the 1950s and 1960s, living standards had increased compared to the previous decade (Reid, 2005). Shortages remained a key problem of Soviet economic supply and everyday consumption though. People used things longer than they were supposed to and stored things they had no immediate use for in their homes. In the shortage economy, the materiality of life depended much on the shadow economy and on individual inventiveness and material exchanges between people from the USSR and the West, during tourist visits or on other occasions (Orlov, Popov, 2016; Golubev, 2020).

Repairs and do-it-yourself (DIY) practices were widespread in (post-)Soviet society and provided important technological fixes in everyday life. Therefore, the Soviet society has been called a 'repair society' (*obshchestvo remonta*) (Gerasimova, Chuikina, 2004: 71). DIY contained elements of creative work, and the official propaganda on Soviet science and technology found an expression in them. This becomes most obvious in the example of the *liubiteli*, the amateurs, some of whom were organised in "circles" (*kruzhki*). They created high-tech radios in the inter-war period and photographic cameras from the 1950s onwards, and boosted the technical development of these products (Lovell, 2011). DIY pertained to many more products, and analysing them contributes to our understanding of social spheres and spaces. When a military engineer created fairy lights for Christmas for his daughter using lights originally intended for a space rocket at his workplace, or when a citizen of Leningrad sewed an entire outfit for going out to dance from a linen bedsheet bought at a shop on the Nevsky Prospect, the borders between professional activities and hobbies, and between materials, knowledge, and skills in the workplace, shops, and homes were blurred (Vasilyeva 2012, 2019). DIY did not just rely on individual skills and knowledge but was also propagated in books and journals that taught individuals how to create a lampshade, a thermos flask, or a shower, and how to repair furniture or a toilet. DIY and repair practices were indispensable for a functioning Soviet economy that was itself in constant need of 'repair'. This was a typical feature of societies and economies in socialist-governed countries, but comparative studies are yet to be written.

Soviet Global Technological Connections

As we know from recent research, the Cold War was not reduced to military tensions and competition, but it also entailed contacts, encounters, and the cooperation of various actors, ranging from officials to engineers, scientists, scholars, and artists (Autio-Sarasmo, Humpreys, Miklóssy, 2010; Autio-Sarasmo, Miklóssy, 2011; Gorsuch, Koenker, 2013; Sanchez-Sibony, 2014; Mikkonen, Koivunen, 2015). Basically, the Soviet Union never stopped to import Western technologies, which it considered as the means of building its own technological base, despite the export controls and economic embargo introduced by the USA in 1949 (Cain, 2007). Following a certain degree of liberalization of the regime, Khrushchev's leadership spoke openly about the need for competition with the West on the one hand, and for learning about the highest achievements in Western technology on the other. In this sense, it simultaneously learned from the West and challenged it. In both aspects, progressive technology was the priority: it was crucial for building up communism and for fulfilling the yearly plans and for 'beating the West'. At the same time, as it had done before the Second World War, the Soviet state relied heavily on importing technologies in different forms: as patents, within industrial and trade fairs, and in various forms of scientific-technical cooperation, including visits to foreign enterprises and research institutes made by Soviet officials, engineers, and scientists. The USSR signed agreements regarding scientific-technical and economic-scientific-technical cooperation with many countries. These included a Soviet-Finnish agreement on scientific-technical cooperation signed in 1955, an agreement on economic-scientific-technological cooperation with Austria in 1968, a similar agreement with France in 1966, and a protocol on scientific cooperation and exchange between the Soviet Academy of Sciences and the Academy of Finland in 1971, among others. In 1972 alone, Soviet institutions cooperated with 420 foreign companies (Bykov, 1974). In the mid-1970s, the Soviets closely cooperated with the United States on space exploration, which they perceived to be the implementation of the politics of the so-called peaceful co-existence, the official doctrine declared in 1956. In 1975, the USSR and the USA carried out the Apollo-Soyuz Project, which was the first joint US–Soviet space flight. This served not only the purposes of scientific experimentation but also political purposes, to show that both sides pursued a policy of détente. At the same time, the Soviet leadership aimed to develop close connections with other socialist countries, in particular cooperating within the Comecon or CMEA (Council for Mutual Economic Assistance, *Sovet Ekonomicheskoi Vzaimopomoshchi*) (Müller, Jajeśniak-Quast, 2017). Among these international cooperations were large-scale, prestigious infrastructural and technological projects, such as the Druzhba pipeline and the transnational electricity grid 'Mir' which incorporated the USSR, Mongolia, GDR, Bulgaria, Poland, Romania, Hungary, and Czechoslovakia.

The Soviet leaders used peaceful co-existence and technological aid as the ideological framework for extending the Soviet impact on recently decolonized countries of the 'Third World'. The Soviet Union appeared 'to many people to offer a vision of modernity that was more attractive than Western capitalism, especially in the 1920s and 1930s' (Holloway, 2006: 577). Soviet influences were embedded into various technological projects, ranging from the construction of hospitals and schools to inviting students to the USSR and providing bank loans. The extent of Soviet technological aid was significant and included the reconstruction of the Aswan Dam in Egypt, hydropower electric technologies in Ghana and Angola, geological investigations, a cement factory in Mali, support for state farming in Ethiopia, and other projects (Westard, 2007; Mavhunga, 2011; Iandolo, 2014; Kochetkova et al., 2017). In 1959 Khrushchev proclaimed with enthusiasm that the whole world would be socialist, without any state borders (Jersild, 2014). However, the Soviet government aimed to retain the dominance of the Soviet Union and of the Russian language in particular. In 1960, the Peoples' Friendship University (*Universitet Druzhby Narodov*) also called Patrice Lulumba University was established in Moscow and attracted many students from former colonies. A special case was China, due to its size and the geopolitical situation. In the 1950s, before the Sino-Soviet split, the Soviet Union invested heavily in relations with this People's Republic. Thousands of Chinese individuals moved to the Soviet Union to receive training in industry and agriculture. Around 8,000 to 10,000 Soviet advisors were sent to China. They furthered Chinese developments and were depicted in some of the Chinese propaganda posters as superior white engineers, contrasted with the tanned Chinese workers, thus expressing a clear (civilizational) hierarchy that was characteristic of the perception of the Soviet–Chinese relationship at the time (Matten, 2019).

Transfers of knowledge and technologies also occurred within the Soviet Union itself. One example is the modernization of agriculture and irrigation in Central Asia. In the 'Hungry Steppe', a sizeable irrigation project was launched in the second half of the 1950s, for which a special trust was created, and a generous amount of machines and know-how was provided by the political and economic centre. This did not mean that everything was delivered to the periphery ready to use. Experiments in 'rationalization' measures were usually welcome by the authorities. For example, irrigation pans made of concrete were produced on the spot in the Hungry Steppe in order to speed up the erection of irrigation systems. The 'hungrysteppers' (*golodnosteptsy*) were connected to Uzbekistan's capital Tashkent and to Moscow via personal networks that secured central support (Obertreis, 2017). Indigenous managers were involved in this kind of modernization projects but it shouldn't be forgotten that important political decisions on economy and infrastructures were taken in Moscow, not in Tashkent or Dushanbe or Tiflis, the republics' capitals.

From the Demise of the USSR to the Present

Although later Soviet leaders continued to provide other socialist states with sizeable money loans to secure their political sympathies, by the 1980s, 'technology and information had become the Soviet Achilles heel' (Reynolds, 2010: 378). Not only did the Soviet economy not produce revolutionary innovation, it even failed to provide its citizens with products and technologies that were already available in other countries. In the mid-1980s, the total number of households with phones in the USSR was only one-sixth of that in the United States, despite its larger population (Reynolds, 2010: 391). To receive a home phone connection or purchase a car, Soviet consumers needed to wait for many years. In general, the Soviet Union revealed its inability to translate many inventions into technological innovations with just a few exceptions, including the Kalashnikov rifle that was invented in 1947 (Graham, 2013). Despite significant achievements in computer and laser development, the Soviets did not commercialize these. The Hungarian economist Janos Kornai has spelled out how Soviet adaptations of some foreign innovations lagged behind. Thus, cellophane was invented in France in 1917, adopted by the USA in 1924, and only in 1936 by the

USSR; and polystyrene was introduced in Germany in 1930, transferred to the USA in 1933, and only came to the USSR in 1955 (Kornai, 2012).

Scholars provide different explanations for the question why the Soviet economic system did not result in the industrial dissemination of most domestic inventions. Among these explanations are underdeveloped democratic institutions and the problem of human rights, barriers between research and production, and the overcentralization of the economy (Bailes, 1978; Berliner, 1988; Graham, 1993, 2013). The industry suffered from a lack of scientific research despite many attempts by the state to increase the involvement of science in production. Industrial enterprises established scientific-technical societies and were encouraged by the state to cooperate more closely with research institutions. In practice, however, this was complicated and often resulted in mutual ignorance. Industrial enterprises were under much pressure to fulfill the five- (or seven-) year plans and were not motivated to undertake risky experiments (Sokolov, 2009).

Environmental consequences of Soviet industrialization and the use of natural resources contribute to our understanding of Soviet technology (Weiner, 1999; Gestwa, 2003; Josephson et al., 2013). Environmental concerns and a new understanding of the use of resources emerged in the Soviet Union from as early as the 1970s. Various kinds of pollution and the importance of developing contamination prevention facilities became a political issue — even if environmentalism 'Soviet style' differred greatly from the Western environmental movements of the period (Obertreis, 2018; Mazanik, 2019). Technological deficiencies and priorities of production led to the overexploitation of natural resources such as forests, minerals, and water, among others. For instance, in the mid-1950s, the Soviets harvested 302 million cubic metres of wood, while the USA harvested 320 million cubic metres. At the same time, the USA produced four times more plywood, six times more paper, and twenty times more cardboard than the Soviet Union. Harvesters left lots of the Soviet wood behind in forests as waste, increasing the problem of deforestation. At the same time, many enterprises attempted to recycle waste into pulp, paper, and construction materials. Some Soviet industries successfully recycled used paper, glass, and other waste, including attempts to use wood and sawmilling waste instead of timber in industrial operations, but this did not create a stable infrastructure for recycling and production from waste resources for the entire country (Kochetkova, 2018).

The development of nuclear energy, mining, pulp and paper production and other types of production carried serious environmental risks. Even though the Chernobyl catastrophe of 1986 is notorious for its ecological consequences, smaller yet frequent accidents took place in Soviet industrial enterprises even earlier (Brown, 2013: 6). Soviet investigations conducted by research institutions showed dangerous levels of pollution in water and air as early as in the late 1940s. In particular, the research led by the All-Union Institute of Paper and Pulp [*Vsesoiuznyi institut tselluilozno-bumazhnoi promyshlennosti*] proved the extremely dangerous nature of the pulp making sector, the third most important in the country based on the number of employees, and the significant damage it caused to rivers and lakes. The construction of a cellulose and paper making plant at Lake Baikal and another on River Selenga intensified the search for new technological instruments for water treatment facilities in the 1960s, in particular, the so-called biological method used by some companies. The establishment of the Baikal and Selenga enterprises provoked environmental protests that were part of the slowly developing protests against pollution and nature destruction (Weiner, 1988, 1999; Brain, 2011; Breyfogle, 2015).

In the end, the Soviet technological system as a whole was not competitive in the context of rapid technological transformation. The Soviet regime kept highly important technologies secret and did not invest into developing their wider use, especially for those which were initially seen as important for the military industry. By the very end of the Soviet era, the shortage of essential goods and the unfolding economic and political crises became top-priority questions for Soviet society, while technologies and innovations were put on the back burner.

After the demise of the socialist regime in 1991, Russia went through difficult times: many scientists and professionals emigrated, while dozens of industrial and research enterprises went bankrupt. At the same time, the 1990s were fruitful in terms of Russian scientists' and technicians' international contacts and cooperations: many projects led by and involving Russian specialists received foreign funding. The strongest areas of research in Russia remained chemistry, nuclear power, metallurgic manufacturing, and outer space engineering, which was, to a large extent, due to vast previous Soviet investments in these fields. In 2002, President Vladimir Putin signed the development plan for science and technology, which ran until 2010 and stated that the transition of the existing economy to an innovative economy was a priority, even though the government spent much less money on science than some other countries did. As in the Soviet period, the new Russian leadership aimed to overcome the technological gap with the West. In 2009, President Dmitry Medvedev initiated the establishment of Skolkovo, a technopolis near Moscow which was inspired by the Silicon Valley. The project was heavily funded by state and private investors for developing research centers of excellence in informational technologies in particular and building transport and social infrastructure in Russia. The priorities at Skolkovo were biomedical, energetic, IT, space and nuclear technologies, which were seen in Russia as the main springboard for an innovation economy (Crescenzi, Jaax, 2017; Rowe, 2014).

However, Russia has remained an economy dependent on raw natural resources — a path it embarked on in the 1960s — and the number of innovative technologies produced was too small: today, the share of Russian knowledge-intensive technologies on the global market is less than one per cent. Some connect this to the lack of democratic institutions which hinder individual initiatives and workable business infrastructures (Inglehart, Welzel, 2005). Nowadays, democracy as a desired goal is not a word of the Russian political language anymore but it became a word to describe a 'hostile' Western world.

In post-socialism, Soviet industrial heritage had various outcomes: some companies were restructured, like the Svetogorsky pulp and paper making plant on the border with Finland, while others were bankrupted and left abandoned, like the textile factory *Krasnoe Znamia* in St. Petersburg. Yet others were transformed into cultural objects and creative spaces in the form of the famous '*Vinzavod*', a centre for contemporary arts in Moscow established at the site of the former Moscow Bavaria brewery. Industrial heritage produces connections between different layers of time, building the materiality of the past into the present urban and social landscapes.

To sum up, technology played an important role in policy making in the Russian Empire, the Soviet Union, and in post-socialist Russia, and exemplified attempts to build an economically stronger society. Stimulated by the concept that Russia was lagging behind the West, the tsarist regime invested in railway construction and other prioritized technologies. From the nineteenth century onwards, some military innovations became widespread internationally. After 1917, for many decades and in various forms, the drive for achieving strong scientific results characterized the Soviet rule. Despite the declared Soviet opposition to the West, technological modernization was heavily supported by transfers from Western countries. The United States and Germany in particular provided a source of know-how, technological designs, industrial plants, and also engineers. Overall, industrialization significantly increased the role of technology in Soviet society and led to big successes in constructing separate industrial branches. At the same time, despite the regime's efforts to support nature conservation and reforestation programmes, and despite the fact that environmentalism grew stronger in the last Soviet decades, the level of industrial pollution remained high.

Technology played a crucial role in forming the Soviet society, which to a large extent was produced by material things. The material world of Soviet people was shaped by economic shortages and possibilities created by the shadow economy. Do-it-yourself and amateur inventions contributed to this world, representing a specific Soviet type of creative practices.

From the perspective of the Soviet leadership, the country struggled continuously with its backwardness in comparison to the world's technologically most advanced states. We can detect this struggle against backwardness and for technological advance in various periods of the Soviet era: it motivated post-war economic recovery, was at the core of the attempt to leap from socialism to communism, and it also drove modernization during perestroika. The Soviet leaders saw technological success as an important tool in the competition between East and West and in increasing the prosperity of the Soviet people. Transferring and adapting Western knowledge and technologies was part and parcel of Soviet technological politics and development, and implied that the Soviet Union had many connections with other countries and international organizations. During the Cold War, particularly from the late 1950s onwards, the Soviet leadership aimed to extend its ideology and influence and to show the attractiveness of the Soviet path to modernity to other regions, in particular the 'developing countries' in the Middle East, Asia, and Africa. Soviet technological impact in these countries, and also in the 'peripheries' of the Soviet Union, was immense.

After the demise of socialism, the Russian science and technology sector was debilitated in the face of an economic crisis and emigration, but in the early 2000s, technological innovation became a priority of Russian politics. Even so, much of

Russian technological development is resting on the oil and gas economy and the Soviet industrial heritage which, however, is gradually vanishing due to changing economic structures. In the face of Russia's full scale military action against Ukraine which unfolds as this chapter is being finalised, questions about the role of the military and military technology and Russian/Soviet/Russian imperial domination (attempts) come up again. The history of Russia's military should not be seen as a niche topic for fans of militaria and weapons but deserves a full historical investigation — in all its ramifications. As today's Russian leaders are concentrating their attention on the military and financial support for non-military technologies is waning, the shape of Russia's (non-)technological future becomes visible.

References

Alimova, Dilorom and Aleksandr Golovanov, 'Uzbekistan', in *History of Civilizations of Central Asia*, ed. by Adle Chahryar, Madhavan K. Palat, and Anara Tabyshalieva (Paris: UNESCO Publishing, 2005), pp. 225–46.

Andrews, James T. and Asif A. Siddiqi, *Into the Cosmos: Space Exploration and Soviet Culture* (Pittsburgh, PA: University of Pittsburgh Press, 2011).

Autio-Sarasmo, Sari, 'Knowledge Through the Iron Curtain: Soviet Scientific-technical Cooperation with Finland and West Germany', in *Reassessing Cold War Europe*, ed. by Sari Autio-Sarasmo and Katalin Miklóssy (London and New York, NY: Routledge, 2011), pp. 66–82.

———, 'Technological Modernization in the Soviet Union and Post-Soviet Russia: Practices and Continuities', *Europe-Asia Studies*, 68.1 (2016), 79–96.

———, Brendan Humpreys, and Katalin Miklóssy, eds, *Winter Kept Us Warm: Cold War Interactions Reconsidered* (Helsinki: Alteksanteri Institute, 2010).

——— and Katalin Miklóssy, *Reassessing Cold War Europe* (London and New York, NY: Routledge, 2011).

Bailes, Kendall, *Technology and Society under Lenin and Stalin: Origins of the Soviet Technical Intelligentsia, 1917–1941* (Princeton, NJ: Princeton University Press, 1978).

Benecke, Werner, *Militär, Reform und Gesellschaft im Zarenreich. Die Wehrpflicht in Russland 1874–1914* (Paderborn: Schöningh, 2006).

Berliner, Joseph, *Soviet Industry from Stalin to Gorbachev: Essays on Management and Innovation* (Aldershot: Edward Elgar, 1988).

Beyrau, Dietrich, *Intelligenz und Dissenz: die russischen Bildungsschichten in der Sowjetunion 1917–1985* (Göttingen: Vandenhoeck und Ruprecht, 1993).

Breyfogle, Nicolas, 'At the Watershed: 1958 and the Beginnings of Lake Baikal Environmentalism', *Slavonic and East European Review*, 93 (2015), 147–80.

Brain, Stephen, *Song of the Forest. Russian Forestry and Stalinist Environmentalism, 1905–1953*, (Pittsburgh, PA: University of Pittsburgh Press, 2011).

Brown, Kate, *Plutopia: Nuclear Families, Atomic Cities, and the Great Soviet and American Plutonium Disasters* (Oxford: Oxford University Press, 2013).

Bykov, Aleksandr, 'Nauchno-tekhnicheskoe sotrudnichestvo sotsialisticheskikh stran', vol. 17, 1974 <https://www.booksite.ru/fulltext/1/001/008/080/451.htm> [accessed 21 February 2019].

Cain, Frank, *Economic Statecraft During the Cold War: European Responses to the Soviet Union Trade Embargo* (London: Routledge, 2007).

Calic, Marie-Janine, Dietmar Neutatz, and Julia Obertreis, 'The Crisis of Socialist Modernity — The Soviet Union and Yugoslavia in the 1970s', in *The Crisis of Socialist Modernity. The Soviet Union and Yugoslavia in the 1970s* (Göttingen: Vandenhoeck & Ruprecht, 2011), pp. 7–27.

Crescenzi, Ricardo and Alexander Jaax, 'Innovation in Russia: the Territorial Dimension', *Economic Geography*, 93 (2017), 66–88.

Cvetkovski, Roland, *Modernisierung durch Beschleunigung: Raum und Mobilität im Zarenreich*, (Frankfurt am Main: Campus Verlag, 2006).

Fadeev, Gennadii, *Istoriia zheleznodorozhnogo transporta Rossii*, vol. I, 1836–1917 (St Petersburg: Peterburg. gos. un-t putey soobshch, 1994).

Fedoseev, Semion, Dal'nost, bystrota i metkost', *Vokrug Sveta*, 4.2763 (2004) <http://www.vokrugsveta.ru/vs/article/339/> [accessed 13 December 2019].

Fitzpatrick, Sheila, *Everyday Stalinism: Ordinary Life in Extraordinary Times. Soviet Russia in the 1930s* (Oxford: Oxford University Press, 1999).

———, 'Stalin and the Making of a New Elite, 1928–1939', *Slavic Review*, 38.3 (1979), 377–402.

Gerasimova, Ekaterina and Sofya Chuikina, 'Obshchestvo Remonta', *Neprikosnovennyi zapas*, 34.2 (2004), 70–77.

Gerovitch, Slava, *From Newspeak to Cyberspeak: a History of Soviet Cybernetics* (Cambridge, MA: MIT Press, 2002).

Gestwa, Klaus, 'Kolumbus des Kosmos. Der Kult um Jurij Gagarin', *Osteuropa*, 59 (2009), 121–52.

———, 'Ökologischer Notstand und sozialer Protest. Der umwelthistorische Blick auf die Reformunfähigkeit und den Zerfall der Sowjetunion', *Archiv für Sozialgeschichte*, 43 (2003), 349–84.

Golubev, Alexey, *Things of Life: Materiality in Late Soviet Russia* (Ithaca, NY: Cornell University Press, 2020).

Gorsuch, Anne E. and Diane P. Koenker, eds, *The Socialist Sixties: Crossing Borders in the Second World* (Bloomington and Indianapolis, IN: Indiana University Press, 2013).

Graham, Loren, *Lonely Ideas: Can Russia Compete?* (Boston, MA: The MIT Press, 2013).

———, *The Ghost of the Executed Engineer: Technology and the Fall of the Soviet Union* (Harvard, MA: Harvard University Press, 1993).

Gregory, Paul, 'Soviet Defense Puzzles: Archives, Strategy and Underfulfilment', *Europe-Asia Studies*, 55.6 (2003), 23–37.

Gross, Andreas J. and Thomas R. Herrmann, 'History of Lasers', *World Journal of Urology*, 25.3 (2007), 217–20.

Harrison, Mark, *Soviet Planning in Peace and War, 1938–1945* (Cambridge: Cambridge University Press, 1985).

Haumann, Heiko, *Beginn der Planwirtschaft: Elektrifizierung, Wirtschaftsplanung und gesellschaftliche Entwicklung Sowjetrußlands 1917–1921* (Düsseldorf: Universitätsverlag, 1974).

Havlat, Denis, 'Western Aid for the Soviet Union During World War II: Part I', *Journal of Slavic Military Studies*, 30.2 (2017), 290–320.

———, 'Western Aid for the Soviet Union During World War II: Part II', *Journal of Slavic Military Studies*, 30.4 (2017), 561–601.

Heller, Klaus, 'Geschichte des modernen Unternehmertums in Russland bis 1917', *Digitales Handbuch zur Geschichte und Kultur Russlands und Osteuropas* vol. 17 (2005), ed. by Abteilung für Geschichte Ost- und Südosteuropas des Historicums der LMU München, Institut für Slavische Philologie der LMU München, pp. 16–17. <https://epub.ub.uni-muenchen.de/760/1/Heller_Unternehmertum.pdf> [accessed 20 December 2019].

Helmedach, Andreas, 'Technik, Verkehr, Umwelt', in *Studienhandbuch Östliches Europa*, vol. II, *Geschichte des Russischen Reiches und der Sowjetunion*, ed. by Thomas M. Bohn and Dietmar Neutatz (Köln: Böhlau, 2009), pp. 79–88.

Hildermeier, Manfred, 'Das Privileg der Rückständigkeit. Anmerkungen zum Wandel einer Interpretationsfigur der neueren russischen Geschichte', *Historische Zeitschrift*, 244. 1 (1987), 557–604.

Hofmeister, Ulrich, *Die Bürde des Weißen Zaren. Russische Vorstellungen einer imperialen Zivilisierungsmission in Zentralasien* (Stuttgart: Franz Steiner Verlag, 2018).

Holloway, David, 'Science, Technology, and Modernity', in *The Cambridge History of Russia*, vol. III, *The Twentieth Century*, ed. by Ronald Suny (Cambridge: Cambridge University Press, 2006).

Iandolo, Alessandro, 'Imbalance of Power: the Soviet Union and the Congo Crisis, 1960–61', *The Journal of Cold War Studies*, 16.1 (2014), 32–55.

Inglehart, Ronald and Christian Welzel, *Modernization, Cultural Change and Democracy: the Human Development Sequence* (Cambridge: Cambridge University Press, 2005).

Jersild, Austin, *The Sino-Soviet Alliance: an International History* (Chapel Hill, NC: The University of North Carolina Press, 2014).

Josephson, Paul R., et al., eds, *An Environmental History of Russia* (Cambridge and New York, NY: Cambridge University Press, 2013).

———, *Would Trotsky Wear a Bluetooth? Technological Utopianism under Socialism, 1917–1989* (Baltimore, MD: The Johns Hopkins University Press, 2010).

Kalinovsky, Artemy M., *Laboratory of Socialist Development. Cold War Politics and Decolonization in Soviet Tajikistan* (Ithaca, NY and London: Cornell University Press, 2018).

Kochetkova, Elena, 'Industry and Forests: Alternative Raw Materials in the Soviet Forestry Industry from the Mid-1950s to the 1960s', *Environment and History*, 24.3 (2018), 323–47.

———, David Damtar, Lilia Boliachevets, Polina Slyusarchuk, and Julia Lajus, *Soviet Technological Projects and Technological Aid in Africa and Cuba, 1960s 1980s*, in Basic Research Programme. Series HUM 'Humanities', 143 (2017) <https://papers.ssrn.com/sol3/papers.cfm?abstract_id=2921129> [accessed 21 January 2023].

——— and Pavel Pokidko. 'Soviet industrial production and waste dispersal: a case study of pulp and paper plants on the Karelian Isthmus, 1940s–1980s', Scandinavian Economic History Review, 67.3 (2019), 282–69.

Kornai, Janos, *Razmyshleniia o kapitalizme* (Moscow: Izdatel`stvo Instituta Gaidara, 2012).

Kotkin, Stephen, *Magnetic Mountain. Stalinism as a Civilization* (Berkeley, Calif.: University of California Press, 1995).

Lazarev, Valery and Paul R. Gregory, 'The Wheels of a Command Economy: Allocating Soviet Vehicles', *Economic History Review*, 55.2 (2002), 324–48.

Lovell, Stephen, 'How Russia Learned to Listen. Radio and the Making of Soviet Culture', *Kritika: Explorations in Russian and Eurasian History*, 12.3 (2011), 591–615.

Lutz, Martin, *Siemens im Sowjetgeschäft. Eine Institutionengeschichte der deutsch-sowjetischen Beziehungen 1917–1933* (Stuttgart: Franz Steiner Verlag, 2011).

Marks, Steven, *Road to Power. The Trans-Siberian Railroad and the Colonization of Asian Russia, 1850 1917* (London: I.B. Tauris, 1991).

Matten, Marc A., 'Turning Away from the Big Brother: China's Search for Alternative Sources of Knowledge During the Sino-Soviet Split', in *Moving Knowledge-The Soviet Union and China in the Twentieth Century, Comparativ. Zeitschrift für Globalgeschichte und vergleichende Gesellschaftsforschung*, ed. by Marc A. Matten and Julia Obertreis, 29.1 (2019), 64–90.

Maurer, Eva, Rüthers, Monica, Richers, Julia and Carmen Scheide, eds, Soviet Space Culture. Cosmic Enthusiasm in Socialist Societies (London and New York, NY: Palgrave Macmillan, 2011).

Mavhunga, Clapperton, 'A Plundering Tiger with its Deadly Cubs? The USSR and China as Weapons in the Engineering of a "Zimbabwean Nation" 1945–2009', in *Entangled Geographies. Empire and Technopolitics in the Global Cold War*, ed. by Gabrielle Hecht (London and Cambridge, MA: The MIT Press, 2011), pp. 231–66.

Mazanik, Anna, 'Environmental Change and the Soviet Media before 1986. Dissident and Officially Sanctioned Voices', in: *Climate Change Discourse in Russia Past and Present*, ed. by Marianna Poberezhskaya and Theresa Ashe (London and New York: Routledge, 2019), pp. 32–49.

Mikkonen, Simo and Pia Koivunen, eds, *Beyond the Divide: Entangled Histories of Cold War Europe* (New York, NY and Oxford: Berghahn, 2015).

Miller, Robert F., *One Hundred Thousand Tractors: The MTS and the Development of Controls in Soviet Agriculture* (Cambridge, MA: Harvard University Press, 1970).

Müller, Uwe and Dagmara Jajeśniak-Quast, eds, *Comecon Revisited. Integration in the Eastern Bloc and Entanglements with the Global Economy* (Leipzig: Leipziger Universitätsverlag, 2017).

Neutatz, Dietmar, *Träume und Alpträume: eine Geschichte Russlands im 20. Jahrhundert* (Munich: C.H. Beck, 2013).

Noble, David, *Forces of Production: a Social History of Industrial Automation* (New Brunswick, NJ and London: Transaction Publishers, 2011).

Nolte, Hans-Heinrich, 'Tradition des Rückstands. Ein halbes Jahrtausend "Russland und der Westen"', *Vierteljahresschrift für Sozial- und Wirtschaftsgeschichte*, 78 (1991), 344–64.

Obertreis, Julia, *Imperial Desert Dreams. Cotton Growing and Irrigation in Central Asia, 1860–1991* (Göttingen: Vandenhoeck & Ruprecht, 2017).

———, 'Soviet Irrigation Policies under Fire. Ecological Critique in Central Asia, 1970s–1991', in *Eurasian Environments. Nature and Ecology in Imperial Russia and Soviet History*, ed. by Nicholas B. Breyfogle (Pittsburgh, PA: University of Pittsburgh Press, 2018), pp. 113–29.

Oldenziel, Ruth, Adri Albert de la Bruhèze, and Onno de Wit, 'Europe's Mediation Junction: Technology and Consumer Society in the 20th century', *History and Technology*, 21.1 (2005), 107–39.

Oleynikov, Pavel V., 'German Scientists in the Soviet Atomic Project', *The Nonproliferation Review*, 7.2 (2000), 1–30.

Orlov, Igor and Alexey Popov, 'Imported Goods and the Spirit of Consumerism. Consumption in Soviet Tourism Abroad', *The Soviet and Post-Soviet Review*, 42.2 (2016), 182–218.

Overy, Richard, *Die Wurzeln des Sieges. Warum die Alliierten den Zweiten Weltkrieg gewannen*, trans. by Jürgen Charnitzky (Stuttgart and Munich: Deutsche Verlags-Anstalt, 2001).

Palmer, Scott W., *Dictatorship of the Air. Aviation Culture and the Fate of Modern Russia* (Cambridge: Cambridge University Press, 2006).

Popov, Vasilii, 'Stalin i sovetskaia ekonomika v poslevoennye gody', *Otechestvennaia istoriia*, 3 (2001), 61–67.

Reid, Susan, 'The Khrushchev Kitchen: Domesticating the Scientific-technological Revolution', *Journal of Contemporary History*, 40.2, (2005), 289–316.

Reynolds, David, 'Science, Technology, and the Cold War', in *The Cambridge History of the Cold War*, ed. by Melvyn Leffler and Odd Arne Westad, vol. III, *Endings* (Cambridge: Cambridge University Press, 2010), pp. 378–99.

Richers, Julia, 'Himmelssturm, Raumfahrt und "kosmische" Symbolik in der visuellen Kultur der Sowjetunion', in *Die Spur des Sputnik. Kulturhistorische Expeditionen ins kosmische Zeitalter*, ed. by Igor J. Polianski (Frankfurt am Main: Campus, 2009), pp. 181–209.

Rowe, Elna Wilson, 'The Future, the Foreign and the Public–Private Divide: Socio-political Discourses around Skolkovo', *Journal of Eurasian Studies*, 5 (2014), 39–47.

Sanchez-Sibony, Oskar, *Red Globalization: the Political Economy of the Soviet Cold War from Stalin to Khrushchev* (Cambridge: Cambridge University Press, 2014).

Schattenberg, Susanne, *Stalins Ingenieure. Lebenswelten zwischen Technik und Terror in den 1930er Jahren* (Munich: Oldenbourg Wissenschaftsverlag, 2002).

Schenk, Frithjof Benjamin, 'Bahnhöfe: Stadttore der Moderne', in *Sankt Petersburg. Schauplätze einer Stadtgeschichte*, ed. by Karl Schlögel, Frithjof Benjamin Schenk, and Markus Ackeret (Frankfurt/Main: Campus Verlag, 2007), pp. 141–57.

———, *Poezd v sovremennost'. Mobil'nost' i sotsial'noe prostranstvo Rossii v vek zheleznykh dorog* (Moscow: Historia Rossica, 2016).

Schwartz, Matthias, 'Bote des Weltalls, Ikone des Fortschritts. Heroische und postheroische Figurationen des ersten Kosmonauten Jurij Gagarin', in *Zaal Andronikashvili* (Berlin: u.a. (Hg.), Kulturheros. Genealogien. Konstellationen. Praktiken, 2017), pp. 334–65.

Siddiqi, Asif, 'Germans in Russia: Cold War, Technology Transfer, and National Identity', *Osiris*, 23.1 (2009), 120–43.

Sperling, Walter, *Der Aufbruch der Provinz: die Eisenbahn und die Neuordnung der Räume im Zarenreich* (Frankfurt am Main: Campus Verlag, 2011).

Sokolov, Andei, 'Rezhimnost' na sovetskikh predpriiatiiakh', in *Rezhimnye liudi v SSSR*, ed. by Tamara Kondrat`eva and Andrei Sokolov (Moscow: ROSSPEN, 2009), pp. 99–127.

Stache, Peter, *Sowjetische Raketen im Dienste von Wissenschaft und Verteidigung* (Berlin: Militärverlag, 1987).

Stalin, J. V., Ob industrializatsii strany i o pravom uklone v VKP(b): Rech' na plenume TsK VKP(b) 190go noiabria 1928 g. ['On the Industrialisation of the Country and the Right Deviation in the C.P.S.U.(B.)'] (speech, plenum of the C.P.S.U.(B.), 19 November 1928) <https://www.marxists.org/russkij/stalin/t11/t11_26.htm> [accessed 21 January 2023].

Sutton, Antony C., *Western Technology and Soviet Economic Development*, 3 vols (Stanford, CA: Hoover Institution Publications, 1968–1973).

Vasilyeva, Zinaida, 'Do-it-yourself Practices and Technical Knowledge in Late Soviet and Post-Soviet Russia', *Tsantsa*, 17 (2012), 28–32.

———, 'From Skills to Selves: Recycling 'Soviet DIY' in Post-Soviet Russia' (unpublished doctoral dissertation, University of Neuchâtel, 2019).

Weiner, Douglas, *A Little Corner of Freedom* (Berkeley, CA: University of California Press, 1999).

———, *Models of Nature: Ecology, Conservation, and Cultural Revolution in Soviet Russia* (Bloomington, IN: Indiana University Press, 1988).

Westard, Odd Arne, *The Global Cold War: Third World Interventions and the Making of Our Times* (Cambridge: Cambridge University Press, 2007).

Zakharova, Larissa, 'Dior in Moscow: a Taste for Luxury in Soviet Fashion under Khrushchev', in *Pleasures in Socialism: Leisure and Luxury in the Eastern Bloc*, ed. by Susan Reid and David Crowley (Evanston, IL: Northwestern University Press, 2010), pp. 95–120.

Zaloga, Steve and James Grandsen, *Soviet Tanks and Combat Vehicles of World War Two* (London: Arms and Armour Press, 1984).

JÉRÔME BAUDRY

Technology in North America

Nation Building, Industrialization, and Technoscience

While it is increasingly common to measure national leadership and the power of countries in terms of their technological dominance, perhaps in no other nation than the United States of America has technology become so entangled with the country's self-image and ideological fabric (Kelley, 1990). Indeed, it is often said that the United States has enjoyed a particular 'romance' with technology; however, this love affair has never been smooth and quiet (Cowan, Hersch, 2018). From early-nineteenth-century fears that the machine would conflict with the pastoral ideals of the nation (Marx, 1964) or its belief in republican values (Kasson, 1973) to the race for technological supremacy during the Cold War and to late-twentieth-century anxiety about Japanese and Chinese economic miracles, technology has been a constant preoccupation of political and intellectual debate in the United States. More importantly perhaps, beyond the recurring polemics between politicians, engineers, and economists — which took place also in Canada (Francis, 2009) — in no other country has the life of ordinary citizens been so thoroughly and consciously technological: from Mississippi steamboats to the car culture of the 1950s (Flink, 1975) and to the personal computer revolution, Americans have more often than not embraced enthusiastically, even emotionally (Nye, 1994), the promises of technology, in ways unseen elsewhere. Technology might be said to have been so crucial in the history of the United States precisely because only there did the politics of technology move beyond ideology to become a lived experience.

Such a development was not by all means written from the start. During the modern period, from the time of the American Revolution (1765–1783) to the beginning of the twenty-first century, the country changed spectacularly, shifting from an economy based predominantly on agricultural and other extractive industries to, by the end of the nineteenth century, an economy centred on the export of manufactured goods, and after World War II to a highly complex knowledge economy. Along with this transformation, and perhaps not unrelated to it, the country expanded geographically in a tremendous way as the frontier gradually moved west, uniting the continent from the Atlantic to the Pacific ocean, far from the initial thirteen states of the American Republic. Technology played a crucial role in this process, providing the networks and nodes that would unify the nation. Paradoxically, even after the westward expansion had come to an end, technology proved even more momentous,

Global History of Techniques, ed. by Guillaume Carnino, Liliane Hilaire-Pérez and Jérôme Lamy, Global Matters Culture Matérielles, histoire des techniques et circulation globale des savoirs, 9 (Turnhout: Brepols, 2024), pp. 219–234
BREPOLS PUBLISHERS 10.1484/M.TECHNE-EB.5.129809

whether by enabling American twentieth-century expansionism and interventionism, or by serving as a metaphorical frontier that promised to be endless.

In this inseparably geographical, technological, and political process, the United States benefitted not only from the technological culture of the European colonizers and immigrants, but also from that of the original inhabitants of the land, the Native Americans, whose knowledge of maize culture proved invaluable to the first settlers, and from that the slaves, who brought their knowledge of rice culture from Africa to the southern plantations. What has been called the 'transfer' (Stapleton, 1987) or 'transit of technology' (Hindle, 1964) from Europe to North America rarely amounted to the pure transfer and adoption of already existing technologies, even in the early modern period. European technological culture, once severed from its local infrastructure and displaced in the new material and social conditions of North America, developed in unprecedented ways, from the American axe to the Kentucky rifle (Baudry, 2016). Yet it is perhaps only at the beginning of the modern period, especially in the decades following the American Revolution, that technology in North America started to take on a distinct character. In this chapter, among a dazzling array of possible directions, I have chosen to focus on three narratives that appear to me essential to understand the development and specificity of American technology: technologies of the frontier, industrialization, and technoscience. Although the chapter deals overwhelmingly with the United States, whenever it is possible, references are made to the history of technology in Canada (Jarrell, Ball, 1980; Jarrell, Hull, 1991; Jones-Imhotep, Adcock, 2018).

Nation Building and Technologies of the Frontier

In the late eighteenth century, the American Revolution had brought together thirteen colonies which varied considerably in legal, economic, and cultural character. Unification was first political, through the adoption of a common Constitution in 1787 and 1788, but it was also technological. For many, notably the Secretary of the Treasury Albert Gallatin in his 1807 *Report on the Subject of Public Roads and Canals*, connecting the different parts of the nation through improved transportation systems appeared crucial, especially since the country's future development seemed to hinge on westward expansion. Although the British government had made some attempts during the colonial period to build roads in order to consolidate the empire, at the turn of the nineteenth century roads in the United States were in a rather dismal state. The political situation of the new nation, with its weak federal government, left transportation matters in the hands of the individual states, and coordination between them rarely materialized. Following English precedent, it was mainly left to companies to build and maintain private improved roads, called 'turnpikes', which allowed travellers and shippers to use them in exchange for a toll. In the United States, the Philadelphia and Lancaster Turnpike Company was the first such firm to operate in 1794, but due to its financial success it was quickly joined by many other corporations, which totalled several thousands in the 1810s (Pursell, 2007). The only federal exception was the Cumberland, later called the National Road, whose construction began in 1808 and which soon connected the Potomac and Ohio rivers, providing the main transport path to the West for thousands of settlers. Roads and highways largely remained a local responsibility until the First World War.

However, by the time the National Road reached Columbus, Ohio, in 1833, many thought that canals would provide a better method of transportation. Native Americans and colonists alike had depended on rivers for quick transportation, but several fall lines, including one running north to south with rapids and waterfalls, were serious obstacles to transportation. Once again, in the wake of the eighteenth-century canal craze in England, many canal projects had been envisaged before the Revolution, but only in the 1780s and 1790s did some of them materialize. The Erie canal, which was forty feet wide and 363 miles long when it opened in 1825, was a much-celebrated construction work and a major engineering achievement for the young nation (Bernstein, 2005). Most importantly, it was a key breeding ground for an entire generation of American engineers, whose training in the field at Erie made up for the paucity of institutions for engineering education at the time (Calhoun, 1960). On top of being a testament to the importance of the 'culture of improvement' in America (Sheriff, 1996), the Erie canal demonstrated how technology could rearrange political and economic power, establishing New York City as the American metropolis as well as the North's economic ascendance over the South (McGreevy, 2009). In the second half of the twentieth century, megaprojects such as the St Lawrence Seaway even exceeded the scale of the nation, unifying the North American continent beyond the borders of the United States and Canada (Macfarlane, 2014).

Steamboats, which soon replaced mules and horses as the principal motive power on American canals, offer an interesting viewpoint on the mythology of invention in the history of American technology. The quest to develop a viable steam-propelled boat attracted individuals who later became prominent icons in the American pantheon of inventors, such as John Fitch (1743–1798), Robert Fulton (1765–1815), and Oliver Evans (1755–1819), but the idea of steam propulsion for boats was not novel at all, being nearly as old as the steam engine itself. Of course, from a technical point of view, the proper proportions of engine to boat and the best way to apply the power still had to be worked out; however, only the right combination of money, mechanical talent, and political support — all things which Robert Fulton managed to gather in 1807 for his triumphant demonstration on the Hudson River — could lead to successful 'invention' (Figure 1). By 1830, the standard western steamboat engine, operating on high pressure and without condensing steam, had been established. Contrary to its eastern counterpart, which was mainly used for passengers, the iconic western (and southern) steamboat of Mark Twain fame was used for freight (Gudmestad, 2011). The number of such boats in circulation rapidly increased, leading to two important unintended effects. First, the high demand for and construction of engines for boats had a spill-over effect, strengthening the engine industry also for stationary use by mills and factories. For example, the city of Cincinnati, which saw nearly ten steamboats call each day, became a centre not only of steamboat building and repair, but also of general machine work and of machine

Fig. 1. *Steamboat Travel on the Hudson River* (1811–*c.* 1813), drawing by Pavel Petrovich Svinin, Collection of the Metropolitan Museum of Art, New York.

tool manufacture. Second, the problem of boiler explosion, which had led to the loss of 520 steamboats by 1850, led the federal government to fund scientific research, in the form of a grant given to the Franklin Institute of Philadelphia in 1830, and to regulate the industry beyond the state level (Paskoff, 2007). As we shall see later in the chapter, this prefigured what eventually became a defining feature of modern American technology in the late nineteenth century and especially in the twentieth century — federal investment in and regulation of technoscience.

However, in the end, the railroad was the technology that proved to be the most consequential for uniting the nation, geographically but also culturally, with a strong imaginary of the railroad developing both in the United States (Douglas, 1992) and in Canada (den Otter, 1997). The first locomotives and locomotive engines came from England, and British engineers were instrumental in transplanting the technology to America (Carlson, 1960), where the straight lines and quarried-rock foundation stones of English railroads had to be adapted to the rugged terrain and the harsher winters of the country. By 1840, the United States had become a solid manufacturer of locomotive engines, even exporting them to Russia, Austria, and several German states. The railroad shaped the new nation, connecting cities and farmlands in the East and opening up the West for conquest settlement (White, 2012). Cities would prosper if they were connected to the railroad, such as Chicago (Cronon, 1991), and the superiority of the Union during the Civil War was linked to its dense and efficient railroad network (Thomas, 2011). Because of the scale of operations, by the second half of the nineteenth century railroad companies became the first modern corporations (Chandler, 1977), creating huge technological bureaucracies that left their uniformizing imprint on much of modern life (Trachtenberg, 1982) — notably on time itself, which was standardised nationally in 1883 and organised along time zones in order to accommodate the scheduling and communication needs of the new rail empires.

Neither improved roads nor canals, steamboats, railroads, or even telegraphs and cars are distinctly American technologies, since they were imported from Europe, especially from Great Britain, along with the men, technological knowledge, and know-how required to operate them. Yet only in the particular North American context were these technologies imbued with so much significance. Far from being mere 'transportation technologies', these technologies of the frontier supported the westward expansion of the United States and helped to build a unified nation. In doing so, they took on cultural and symbolic meanings — the 'culture of improvement', the triumph of man over nature, and America's civilizing mission through technology (Adas, 2006) — that still play a prominent role today in how Americans value and feel about technology.

Industrialization in America

In the same way that the 'technologies of the frontier' just discussed were rarely of American origin, early industrialization in the United States depended on the 'transit of technology'. The circulation of technologies and technological men was particularly strong in the textile (Jeremy, 1981; Tucker, 1984) and steam engine (Pursell, 1969) industries, as well as in civil engineering (Stapleton, 1981). Of course, no 'transfer' went without adaptation, change, and further invention; indeed, once transplanted to North America, these technologies soon endowed the 'American Industrial Revolution' (Hindle, Lubar, 1986) with distinctive features. For example, because the population was limited in size and the westward expansion of the frontier provided ample cheap land for new immigrants, the industrial labour force was small and wages were high; consequently there was a marked incentive for inventors, engineers, and manufacturers to develop technologies and management techniques that saved labour. Several iconic inventions in the early nineteenth century, such as Oliver Evans's 'bucket elevator' and 'hopper boy' (Ferguson, 1980), and Eli Whitney's 'cotton gin', were attempts to mechanize and automatize labour-intensive operations in flour milling and cotton textile production. As small and large machine shops opened in many towns all over the settled parts of the country, a strong machine tool industry developed, accelerating the pace of industrialization.

Although Whitney's plan and attempts to achieve parts interchangeability in the manufacturing of muskets never came to fruition, later in the nineteenth century his initial idea became a fixture of American industrialization (Smith, 1981). The goal of the 'American system of manufactures', as it was called by the English commissioners who visited the New York Industrial Exhibition in 1853, was to achieve the greatest uniformity of product, relying on special-purpose machines with jigs and gauges and an unskilled workforce to produce similar parts of accurate dimensions. One example of such specialized machines was Thomas Blanchard's gun stock lathe, which could be used to make identical copies out of a single pattern (Cooper, 1991). Also called 'armory practice', the American system of manufactures depended in no small part on the military, which was willing to invest financially and technologically in an enterprise that largely exceeded civilian needs at the time — setting up a pattern which proved decisive for US technology, as we shall see in the third part of the chapter. Initially developed at the federal arsenals of Springfield and Harpers Ferry (Smith, 1977), the practice and its corresponding technologies soon spilled over to other industries, such as the machine tool industry (Rosenberg, 1962) and the consumer goods industry (Hoke, 1990), producing sewing machines, clocks, watches, typewriters, bicycles, and eventually cars.

More than any other product, it was actually the bicycle that forged the link between nineteenth-century armoury practice and twentieth-century mass production (Hounshell, 1984). The two manufacturing techniques of sheet metal stamping and electric resistance welding were initially developed in the bicycle industry but were soon carried over into automobile manufacture, notably by Henry Ford. Ford and the Ford Company are often credited with having 'invented' the techniques of mass production, although several iconic Fordist manufacturing techniques, such as the assembly lines (the moving lines of subassemblies and final chassis units), were already employed in other industries, for instance in Chicago's meatpacking industry (Nye, 2013). 'Fordism' nonetheless spread rapidly to other countries, epitomizing the modern and distinctly American way of industrial production (Schlombs, 2019). Recent historiography has shown that this picture of American industrialization as characterized essentially by mass production is perhaps too one-sided, insisting, on the contrary, on the fact that diversity, specialty goods, and flexibility in production (batch production) remained important in several sectors, such as in the textile, woodworking, and metalworking trades (Scranton, 1997). Instead of a unanimous embrace of mass production, and later automation (Noble, 1984), 'two cultures of American industry' coexisted side by side (Scranton, 1995).

The culture of mass production in America encompassed more than new technologies and manufacturing machines, as the values of mechanization, interchangeability, and standardization were thought to apply not only to machines but also to men. If capital was evolving towards more productivity and efficiency, there was another factor of production, labour, that needed to be reformed (Haber, 1964). Testifying before Congress in 1912, Frederick Winslow Taylor framed the movement for Scientific Management (later called 'Taylorism') as a way to discipline labour and end class struggle: productivity gains would lead to such great 'surplus' and abundance that the social question would become meaningless (Hays, 1959). A mechanical engineer by training, Taylor purported to apply the 'scientific' (and therefore supposedly neutral) method of observation and measurement to the work process; armed with stopwatches and, later, motion-picture cameras, engineers recorded the traditional methods of work, broke them down into uniform units, and reconstituted them into new work processes that management would enforce (Kakar, 1970; Kanigel, 1997). From the flowcharts of Henry L. Gantt to the basic movements ('therbligs') of Frank B. and Lilian M., Taylor's disciples created a host of concepts, tools, and techniques that helped them to further their rationalizing credo (Nelson, 1992), although the practical endeavours of scientific management did not go without friction (Aiken, 1960). In the end, the Taylorist pursuit of efficiency was far from purely ideational, as it had to be inscribed materially in the very architecture of work (Biggs, 1996) through the 'visible hand' of engineers, managers, and regulators (Osman, 2018).

By 1913, on the eve of the First World War, the United States had become the world's leading industrial nation. As for the 'technologies of the frontier' discussed in the first part of this chapter, industrialization in North America — including Canada (Hull, 2018) — depended in no small part on technologies that originated in Europe; however, it was quite distinctly driven by an ethos of rationalization that put a premium on organization, coordination, and regulation — that is, on ways of managing technology (and labour) rather than on specific technologies. At the turn of the twentieth century, displacing the earlier myth of the lone and genius inventor (Hughes, 1989), the cult of the engineer helped to support a technocratic vision of democracy that assigned to 'science', with its alleged universality and neutrality, the task to remake the social world in place of politics, always uncertain and inefficient (Layton, 1971; Akin, 1977; Millard, 1988). Although such a rosy picture was rapidly disproved by the Great Depression (Bix, 2000), the belief in the superior authority of technological science, rooted partly in the American path to industrialization, remains, until today, a distinct characteristic of America's relation to technology.

Towards Technoscience

In the advent of the 'scientific' engineer at the turn of the twentieth century and the correlative downfall of the lone inventor, other forces were at play than just the collective belief in the values of the Progressive Era. The increasing complexity of existing technologies and the growing power of scientific theory over natural phenomena meant that science, technology, and industry had become mutually dependent — a fact that several scholars have studied under the name of 'technoscience'. Although such a development was not specific to North America — it also gained traction in Europe starting in the late nineteenth century —, due to several factors the advancement of technology in the United States unfolded quite uniquely in the novel form of technoscience. One such factor was the corporatization of technological research (Noble, 1977). By the mid-1870s, even Thomas Edison, the quintessential American 'lone' inventor, had felt the need to recruit a research team, hiring not only technicians and engineers but also scientists who had earned their PhDs from top colleges in the country and abroad. One of the first and most consequential industrial research laboratories was the one founded at General Electric in 1900 (Wise, 1985). With important work on vacuum tubes, X-rays, and the improvement of incandescent lamps, the laboratory quickly grew from a staff of eight in 1908 to nearly 300 in 1917. Beyond electricity and the lighting industry, telegraphy (Israel, 1992), telephony (Reich, 1985), and chemistry (Hounshell, Smith, 1988) were fields in which industrial research blossomed in the early twentieth century. When the United States entered into the First World War, more than 350 industrial laboratories had been established in the country.

Twentieth-century wars — first the Great War, and later even more so the Second World War — had far-reaching consequences on technological development, driving American technology more firmly in the direction of technoscience (Smith, 1985). Before the First World War, there had been only limited forays into the management of science and technology by the federal government (Dupree, 1957), with the creation of several conservation agencies in the late nineteenth century, including the Forest Service, or the attempt to subsidize the agricultural and engineering experiment stations (Kevles, 1971) set up throughout the country. The Great War greatly expanded the federal government's involvement and eventually shaped the future of many technologies, for example aircraft (Vander Meulen, 1991), which benefitted from the creation of the National Advisory Committee on Aeronautics in 1915 (Roland, 1985) and a few years later of the Langley Aeronautical Laboratory (Hansen, 1987), one of the first national laboratories in the United States (Westwick, 2003). In Canada, national research also took off during the First World War with the creation of the National Research Council (Eggleston, 1978). The Second World War accelerated this trend (Zimmerman, 1986), especially in the United States with the establishment of the Office of Scientific Research and Development in 1941 (Pursell, 1979). The unprecedented amount of federal funding the government was ready to invest in military and military-related research benefitted a small number of top science and engineering institutions, notably the Massachusetts Institute of Technology, as well as a host of industrial contractors, such as AT&T, General Electric, or RCA. This close association between the federal government (especially the army), industry, and elite engineering schools survived the end of the war and thrived during the Cold War under the name of the 'military-industrial-academic' complex (Leslie, 1993). By the mid-1960s, one third of all research and development costs in the United States was assumed by the federal government. In the end, the governmentalization of (science and) technology also extended to supposedly independent expert bodies, such as the National Academy of Sciences (Boffrey, 1975), leading to concerns and controversies about the supposed neutrality and universality that the figure of the scientific engineer was thought to embody.

American technology, especially in the form of massive technoscientific projects, grew tremendously under the era of the military-industrial-academic complex. One such project was of course the Manhattan Project, which was entrusted to the Army Corps of Engineers and brought together, in utmost secrecy, military engineers, scientists, and industrial firms during the Second World War in order to produce the first atomic bomb (Rhodes, 1987). Although the news that two nuclear weapons had been dropped on Hiroshima and Nagasaki was generally received with dismay and horror, pessimism soon gave way to a new wave of technological enthusiasm when a new 'Atomic age' was declared, with its promise of free energy and other new civilian technologies (Boyer, 1985; Bothwell, 1988; Smith, 1991). The same assurance that technologies developed during war — either hot or Cold — would naturally lead to non-military spin-offs and consequently to increased abundance for the American people undergirded most of the post-war big technoscientific endeavours. Wartime innovations in jet propulsion and rockets ultimately led to the development and

Fig. 2. *Buzz Aldrin with Apollo 11 Lunar Module on the Moon* (1969), photograph by Neil Amstrong, Collection of the Metropolitan Museum of Art, New York.

launch of satellites and to the Apollo programme, which succeeded in 1969 in landing humans on the Moon (Koppes, 1982; McDougall, 1985) (Figure 2). Another wartime and then Cold War technology, the computer, which was initially developed at MIT and at the University of Pennsylvania to help compute data and solve problems in such areas as ballistics, eventually found its way into the homes of American (and global) consumers as it moved from the mainframe to the personal computer (Edwards, 1996; Remond, Smith, 2000). However, success stories such as that of the computer tended to elude the fact that the opportunity cost to civilian technology was high. A number of critics pointed to the fact that Cold War military technologies developed thanks to colossal subsidies from the federal government hurt significant parts of the country's economy (Markusen, Campbell, Dietrick, 1991) and stifled real innovation, with 'a government contract', according to President Dwight D. Eisenhower in his 1961 farewell address, 'becom[ing] a substitute for intellectual curiosity' (Pursell, 2007: 279).

The American military-industrial-academic complex proved particularly powerful in shaping twentieth-century technology in the form of 'technoscience', through the technique of 'managed research' (Cowan, Hersch, 2018: 258). The corporatization of technological research, the investment — both financial and regulatory — of the federal government in technological matters, and finally the global wars of the century conspired to give rise to unprecedented technological achievements, from the atomic bomb to the computer. Here again, it is in what historian Thomas P. Hughes has called 'large technical systems', rather than in discrete technological innovations, that a distinctly American way of doing technology seemed to emerge. More precisely, what perhaps constituted America's most important technological achievement lay in the managerial and technological innovations, such as 'systems engineering' and 'operations research', that allowed the management of highly complex projects (Hughes, 1998). The military was instrumental in these developments, and the belief in the power and transferability of those managerial procedures was so strong that Vice President Hubert H. Humphrey simply declared in 1967 that 'techniques that are going to put a man on the moon are [...] exactly the techniques that we are going to need to clean up our cities' (Humphrey, 1968: 19).

Although America's interaction with technology over the last two centuries has been successful in many ways — bringing about the geographical and cultural unification of the country through 'technologies of the frontier', transforming the economy from predominantly extractive to highly industrialized and supporting the nation's geopolitical dominance through gigantic technoscientific projects — what has been frequently seen as a 'romance' started to turn sour in the 1960s. In 1962, Rachel Carson told the American public that the pesticide DDT, which had been praised for eradicating malaria in several parts of the world, was in fact leading to ecological devastation, contaminating food chains and killing songbirds, thus creating a 'silent spring' (Carson, 1962). A few years later, the attorney and activist Ralph Nader revealed that American automobiles were 'unsafe at any speed' (Nader, 1965), detailing how Detroit automakers such as General Motors knowingly produced unreliable cars. In 1970, social critic Lewis Mumford radically changed his mind about technology and progress, declaring that the combined forces of capitalism and militarism created technological systems or 'megamachines' that 'suppress[ed] many of the most essential characteristics of organisms, personalities and human communities' (Mumford, 1970: 37). Finally, analysing the 1979 nuclear accident at the Three Mile Island power plant, sociologist Charles Perrow concluded that such an event was not an exceptional case but simply a 'normal accident', because highly complex technological systems were inherently 'risky enterprises [with] catastrophic potential' (Perrow, 1984).

Such critical and intellectual reflections on the dangers of high technology did, of course, not arise unexpectedly. The rise of the environmental movement, along with the energy crisis of the mid-1970s, had suggested that technology was perhaps not the relevant answer to all kinds of social problems. Meanwhile, the Vietnam War quagmire had shown how America's destructive and sophisticated technology, from napalm to carpet bombing and Agent Orange, could be out of place and out of scale, eventually failing to achieve the geopolitical goals of the nation. The 1986 *Challenger* disaster, during which the eponymous space shuttle broke apart after seventy-three seconds in flight, and even more so the 2003 *Columbia* disaster, when the shuttle disintegrated during atmospheric entry, confirmed the fragility of highly complex technological systems and led to a major slowdown in American space programmes. Finally, the era of 'reaganomics' in the 1980s and the end of the Cold War, with the collapse of the USSR in 1991 and the alleged victory of the United States, dramatically altered the structure of the military-industrial-academic complex when the federal government curbed funding and substantially withdrew from direct investment in technoscientific matters.

With such critical developments, it is nonetheless unclear if an entirely new political economy of technology has emerged in the United States, replacing the earlier regime of technoscience based on the close alliance between the federal government, industry, and academia. The crisis of American industrialism in the form of Fordism and Taylorism, under pressure from the 'lean production' model of Japanese manufacturers in the 1970s and 1980s, has certainly led to the adoption of more flexible forms of production and technological management. At the same time, the federal government and Congress have encouraged the commodification of technological research, notably with the 1990 Bayh-Dole Act, which has allowed federally funded academic institutions and projects to engage in profitable patenting, spawning a new culture of start-up companies and innovation parks. If these transformations are a testament to the enduring American capacity for designing new ways of managing technological innovation, it must be said that, despite their much-celebrated hype, their success in generating new and important technologies remains to be evaluated. Today, in the early twenty-first century, anxieties — whether about economic competition with China, military and technological dominance, or the environmental and climate crisis — seem to have taken over optimism as the main mode of American engagement with technology.

References

Adas, Michael, *Dominance by Design: Technological Imperatives and America's Civilizing Mission* (Cambridge, MA: Harvard University Press, 2006).

Aiken, Hugh G. J., *Scientific Management in Action: Taylorism at Watertown Arsenal, 1908–1915* (Princeton, NJ: Princeton University Press, 1960).

Akin, William E., *Technocracy and the American Dream: the Technocrat Movement, 1900–1941* (Berkeley, CA: University of California Press, 1977).

Baudry, Jérôme, 'Rencontres et politique de la technique en Amérique du Nord', in *Histoire des techniques. Mondes, sociétés, cultures (XVIe–XVIIIe* siècles), ed. by Guillaume Carnino, Liliane Hilaire-Pérez, and Aleksandra Kobiljski (Paris: Presses Universitaires de France, 2016), pp. 231–43.

Bernstein, Peter L., *Wedding of the Waters: the Erie Canal and the Making of a Great Nation* (New York, NY: W. W. Norton & Co., 2005).

Biggs, Lindy, *The Rational Factory: Architecture, Technology, and Work in America's Age of Mass Production* (Baltimore, MD: Johns Hopkins University Press, 1996).

Bix, Amy Sue, *Inventing Ourselves Out of Jobs: America's Debate over Technological Unemployment, 1929–1981* (Baltimore, MD: Johns Hopkins University Press, 2000).

Boffrey, Philip M., *The Brain Bank of America: an Inquiry into the Politics of Science* (New York, NY: McGraw-Hill, 1975).

Bothwell, Robert, *Nucleus: the History of Atomic Energy Canada Limited* (Toronto: University of Toronto Press, 1988).

Boyer, Paul, *By the Bomb's Early Light: American Thought and Culture at the Dawn of the Atomic Age* (New York, NY: Pantheon Books, 1985).

Calhoun, Daniel C., *The American Civil Engineer: Origins and Conflict* (Cambridge, MA: Harvard University Press, 1960).

Carlson, Robert E., 'British Railroads and Engineers and the Beginnings of the American Railroad Development', *Business History Review*, 34.2 (1960), 137–49.

Carson, Rachel, *Silent Spring* (Boston, MA: Houghton Mifflin, 1962).

Chandler, Alfred D., *The Visible Hand: the Managerial Revolution in American Business* (Cambridge, MA: Harvard University Press, 1977).

Cooper, Carolyn C. *Shaping Invention: Thomas Blanchard's Machinery and Patent Management in Nineteenth-century America* (New York, NY: Columbia University Press, 1991).

Cowan, Ruth Schwartz and Matthew Hersch, *A Social History of American Technology* (New York, NY: Oxford University Press, 2018).

Cronon, William, *Nature's Metropolis: Chicago and the Great West* (New York, NY: W. W. Norton & Co., 1991).

den Otter, Andy Albert, *The Philosophy of Railways: the Transcontinental Railway Idea in British North America* (Toronto: University of Toronto Press, 1997).

Douglas, George H., *All Aboard! The Railroad in American Life* (New York, NY: Paragon House, 1992).

Dupree, Hunter A., *Science in the Federal Government: a History of Policies and Activities to 1940* (Cambridge, MA: Harvard University Press, 1957).

Edwards, Paul, *The Closed World: Computers and the Politics of Discourse in Cold War America* (Cambridge, MA: MIT Press, 1996).

Eggleston, Wilfrid, *National Research in Canada: the NRC, 1916–1966* (Toronto: Clarke, Irwin, 1978).

Ferguson, Eugene S., *Oliver Evans: Inventive Genius of the American Industrial Revolution* (Greenville, DE: Hagley Museum, 1980).

Flink, James J., *The Car Culture* (Cambridge, MA: MIT Press, 1975).

Francis, R. Douglas, *The Technological Imperative in Canada: an Intellectual History* (Vancouver: UBC Press, 2009).

Gudmestad, Robert H., *Steamboats and the Rise of the Cotton Kingdom* (Baton Rouge, LA: Louisiana State University Press, 2011).

Haber, Samuel, *Efficiency and Uplift: Scientific Management in the Progressive Era* (Chicago, IL: University of Chicago Press, 1964).

Hansen, James R., *Engineer in Charge: a History of the Langley Aeronautical Laboratory, 1915–1958* (Washington, DC: United States Government Printing Office, 1987).

Hays, Samuel P., *Conservatism and the Gospel of Efficiency: the Progressive Conservation Movement, 1890–1920* (New York, NY: Atheneum, 1959).

Hindle, Brooke and Steven Lubar, *Engines of Change: the American Industrial Revolution, 1790–1860* (Washington, DC: Smithsonian Institution Press, 1986).

Hoke, Donald R., *Ingenious Yankees: the Rise of the American System of Manufactures in the Private Sector* (New York, NY: Columbia University Press, 1990).

Hounshell, David A., *From the American System to Mass Production, 1800–1932: the Development of Manufacturing Technology in the United States* (Baltimore, MD: Johns Hopkins University Press, 1984).

——— and John Kenly Smith, *Science and Corporate Strategy: DuPont R&D, 1902–1980* (New York, NY: Cambridge University Press, 1988).

Hughes, Thomas P., *American Genesis: A Century of Invention and Technological Enthusiasm, 1870–1970* (Chicago, IL: University of Chicago Press, 1989).

———, *Rescuing Prometheus* (New York, NY: Pantheon Books, 1998).

Hull, James, 'The Second Industrial Revolution in Canadian History', in *Made Modern: Science and Technology in Canadian History*, ed. by Edward Jones-Imhotep and Tina Adcock (Vancouver: University of British Columbia Press, 2018), pp. 125–42.

Humphrey, Hubert, 'HHH: on the Space Program', *Aerospace Technology*, 20 May 1968.

Israel, Paul, *From Machine Shop to Industrial Laboratory: Telegraphy and the Changing Context of American Invention, 1830–1920* (Baltimore, MD: Johns Hopkins University Press, 1992).

Jarrell, Richard A. and Norman R. Ball, eds, *Science, Technology, and Canadian History* (Waterloo, ON: Wilfrid Laurier University Press, 1980).

——— and James Hull, eds, *Science, Technology, and Medicine in Canada's Past: Selections from Scientia Canadensis* (Thornhill, ON: Scientia Press, 1991).

Jeremy, David J., *Transatlantic Industrial Revolution: the Diffusion of Textile Technologies Between Britain and America, 1790–1830s* (Cambridge, MA: MIT Press, 1981).

Jones-Imhotep, Edward and Tina Adcock, eds, *Made Modern: Science and Technology in Canadian History* (Vancouver: University of British Columbia Press, 2018).

Kakar, Sudhir, *Frederick Taylor: a Study in Personality and Innovation* (Cambridge, MA: MIT Press, 1970).

Kanigel, Robert, *The One Best Way: Frederick Winslow Taylor and the Enigma of Efficiency* (New York, NY: Viking, 1997).

Kasson, John F., *Civilizing the Machine: Technology and Republican Values in America, 1776–1900* (New York, NY: Grossman Publishers, 1973).

Kelley, Robert, *The Shaping of the American Past*, vol. I, 5th ed. (Englewood Cliffs, NJ: Prentice-Hall, 1990).

Kevles, Daniel J., 'Federal Legislation for Engineering Stations: the Episode of World War I', *Technology and Culture*, 12.2 (1971), 182–89.

Koppes, Clayton R., *JPL and the American Space Program: a History of the Jet Propulsion Laboratory* (Yale, CT: Yale University Press, 1982).

Layton, Edwin T., *The Revolt of the Engineers: Social Responsibility and the American Engineering Profession* (Cleveland, OH: Press of Case Western Reserve University, 1971).

Leslie, Stuart W., *The Cold War and American Science: the Military-Industrial-Academic Complex at MIT and Stanford* (New York, NY: Columbia University Press, 1993).

Macfarlane, Daniel, *Negotiating a River: Canada, the US, and the Creation of the St Lawrence Seeway* (Vancouver: University of British Columbia Press, 2014).

Markusen, Ann, Peter Hall, Scott Campbell, and Sabina Dietrick, *The Rise of the Gun Belt: the Military Remapping of Industrial America* (Oxford: Oxford University Press, 1991).

Marx, Leo, *The Machine in the Garden: Technology and the Pastoral Ideal in America* (New York, NY: Oxford University Press, 1964).

McDougall, Walter A., *The Heavens and the Earth: a Political History of the Space Age* (New York, NY: Basic Books, 1985).

McGreevy, Patrick, *Stairway to Empire: Lockport, the Erie Canal, and the Shaping of America* (Albany, NY: State University of New York Press, 2009).

Millard, J. Rodney, *The Master Spirit of the Age: Canadian Engineers and the Politics of Professionalism* (Toronto: University of Toronto Press, 1988).

Mumford, Lewis, *The Pentagon of Power* (New York, NY: Harcourt Brace, 1970).

Nader, Ralph, *Unsafe at Any Speed: the Designed-in Dangers of the American Automobile* (New York, NY: Grossman Publishers, 1965).

Nelson, Daniel, ed., *A Mental Revolution: Scientific Management since Taylor* (Columbus, OH: Ohio State University Press, 1992).

Noble, David F., *America by Design: Science, Technology, and the Rise of Corporate Capitalism* (New York, NY: Knopf, 1977).

———, *Forces of Production: a Social History of Industrial Automation* (New York, NY: Knopf, 1984).

Nye, David E., *America's Assembly Line* (Cambridge, MA: MIT Press, 2013).

———, *American Technological Sublime* (Cambridge, MA: MIT Press, 1994).

Osman, Michael, *Modernism's Visible Hand. Architecture and Regulation in America* (Minneapolis, MN: University of Minnesota Press, 2018).

Paskoff, Paul F., *Troubled Waters: Steamboat Disasters, River Improvements, and American Public Policy, 1821–1860* (Baton Rouge, LA: Louisiana State University Press, 2007).

Perrow, Charles, *Normal Accidents: Living with High-Risk Technologies* (New York, NY: Basic Books, 1984).

Pursell, Carroll W., 'Thomas Diiges and William Pearce: an Example of the Transit of Technology', *The William and Mary Quarterly*, 21.4 (1964), 551–60.

———, *Early Stationary Steam Engines in America: a Study in the Migration of a Technology* (Washington, DC: Smithsonian Institution Press, 1969).

———, 'Science Agencies in World War II: the OSRD and Its Challengers', in *The Sciences in the American Context: New Perspectives*, ed. by Nathan Reingold (Washington, DC: Smithsonian Institution Press, 1979), pp. 359–78.

———, *The Machine in America: a Social History of Technology*, 2nd ed. (Baltimore, MD: Johns Hopkins University Press, 2007).

Reich, Leonard S., *The Making of American Industrial Research: Science and Business at General Electric and Bell, 1876–1926* (Cambridge, MA: Cambridge University Press, 1985).

Remond, Kent C. and Thomas M. Smith, *From Whirlwind to MITRE: the R&D Story of the SAGE Air Defense Computer* (Cambridge, MA: MIT Press, 2000).

Rhodes, Richard, *The Making of the Atomic Bomb* (New York, NY: Simon & Schuster, 1987).

Roland, Alex, *Model Research: the National Advisory Committee for Aeronautics, 1915–1958* (Washington, DC: Scientific and Technical Information Branch, National Aeronautics and Space Administration, 1985).

Rosenberg, Nathan, 'Technological Change in the Machine Tool Industry, 1840–1910', *Journal of Economic History*, 23.4 (1962), 414–43.

Schlombs, Corinna, *Productivity Machines: German Appropriations of American Technology from Mass Production to Computer Automation* (Cambridge, MA: MIT Press, 2019).

Scranton, Philip, *Endless Novelty: Specialty Production and American Industrialization, 1865–1925* (Princeton, NJ: Princeton University Press, 1997).

———, 'The Politics of Production: Technology, Markets, and the Two Cultures of American Industry', *Science in Context*, 8.2 (1995), 369–95.

Sheriff, Carol, *The Artificial River: the Erie Canal and the Paradox of Progress, 1817–1862* (New York, NY: Hill and Wang, 1996).

Smith, Merritt Roe, 'Eli Whitney and the American System of Manufacturing', in *Technology in America: a History of Individuals and Ideas*, ed. by Carroll W. Pursell (Cambridge, MA: MIT Press, 1981), pp. 45–61.

———, *Harpers Ferry Armory and the New Technology: the Challenge of Change* (Ithaca, NY: Cornell University Press, 1977).

———, ed, *Military Enterprise and Technological Change: Perspectives on the American Experience* (Cambridge, MA: MIT Press, 1985).

Smith, Michael L., 'Advertising the Atom', in *Government and Environmental Politics: Essays on Historical Developments since World War Two*, ed. by Michael J. Lacey (Baltimore, MD: Johns Hopkins University Press, 1991), pp. 233–62.

Stapleton, Darwin H., 'Benjamin Henry Latrobe and the Transfer of Technology', in *Technology in America: a History of Individuals and Ideas*, ed. by Carroll W. Pursell (Cambridge, MA: MIT Press, 1981), pp. 34–44.

———, *The Transfer of Early Industrial Technologies to America* (Philadelphia, PA: American Philosophical Society, 1987).

Thomas, William G., *The Iron Way: Railroads, the Civil War, and the Making of Modern America* (Yale, CT: Yale University Press, 2011).

Trachtenberg, Alan, *The Incorporation of America: Culture and Society in the Gilded Age* (New York, NY: Hill and Wand, 1982).

Tucker, Barbara M., *Samuel Slater and the Origins of the American Textile Industry, 1790–1960* (Ithaca, NY: Cornell University Press, 1984).

Vander Meulen, Jacob A., *The Politics of Aircraft: Building an American Military Industry* (Lawrence, KA: University Press of Kansas, 1991).

Westwick, Peter J., *The National Labs: Science in an American System, 1947–1974* (Cambridge, MA: Harvard University Press, 2003).

White, Richard, *Railroaded: the Transcontinentals and the Making of Modern America* (New York, NY: W. W. Norton & Co., 2012).

Wise, George, *Willis R. Whitney, General Electric, and the Origins of U.S. Industrial Research* (New York, NY: Columbia University Press, 1985).

Zimmerman, David, 'The Organization of Science for War: the Management of Canadian Radar Development, 1939–1945', *Scientia Canadensis: Canadian Journal of the History of Science, Technology, and Medicine*, 10.2 (1986), 93–108.

DAVID PRETEL

Latin America in the Global History of Technology*

Until recently a marginal field of study, the history of technology in Latin America has lately shown a remarkable vitality in both volume and scope. The growing interest in this region's technological past is evident from the number of historiographical discussions written during the last decade or so (Portuondo, 2009; Saldaña, 2013; Castro, 2020; Pretel, Inkster, Wendt, 2019). Such essays, some of them from the disciplinary position of science and technology studies (Medina, da Costa Marques, Holmes, 2014; Kreimer, Vessuri, 2018; Vaccarezza, 2011; Ragas, 2020), have provided critical overviews of the shifting perspectives, themes, and methodologies in recent scholarship from the colonial period to the present day. To some extent, recent writings have taken on the task of shattering diffusionist assumptions and offering a corrective to earlier 'dependency' accounts. That shift in focus has helped to highlight Latin America's role as a laboratory of technological modernity while making visible the ingenuity of local actors and the importance of place. Yet, at the same time, most scholars have taken too much attention away from the economic consequences of this region's uneven technological development and the external factors behind it.

Rather than providing yet another general survey of this broad topic, what follows is a discussion of some of the main features of the prevailing historical narrative of the past two decades along with some of the limits of this narrative. As it is not possible within the scope of this chapter to cover all the publications on the topic, nor all of the economic sectors or technological developments in nineteenth- and twentieth-century Latin America, the second part of this essay concentrates on the studies that have framed this region's complicated technological past within its transnational, international, and global contexts. This thematic selectivity, regrettable in some ways, may help highlight the technological links and networks spanning Latin America itself as well as Latin America and the rest of the world. The final part of this essay argues for the need to Latin-Americanize the global history of technology.

* I would like to thank Justin Castro and Manuel Contreras for feedback on an early draft of this essay.

Global History of Techniques, ed. by Guillaume Carnino, Liliane Hilaire-Pérez and Jérôme Lamy, Global Matters Culture Matérielles, histoire des techniques et circulation globale des savoirs, 9 (Turnhout: Brepols, 2024), pp. 235–252
BREPOLS PUBLISHERS 10.1484/M.TECHNE-EB.5.129810

The New Narrative and its Limits

The history of technology as it pertains to Latin American countries is a growing field, particularly the English-language scholarship. Admittedly, it is an undefined area of research in which many of the contributions are written by scholars other than historians. Disciplinary boundaries and a plurality of research interests explain why we still lack a book-length synthesis of Latin America's technological past. Such an endeavour would require a challenging interdisciplinary approach that puts into dialogue a fragmented body of literature where each subfield (e.g. social studies of science and technology, economic history, history of science and technology, environmental history, labour history) has developed in relative isolation. It would need to consider not only literature written by scholars outside Latin America, but also Latin American research agendas and the dispersed works written in Spanish and Portuguese. Constrained by scholars' preference for nation-centred narratives, Latin American histories of technology addressing intra-regional disparities and inter-connections remain a minority. Moreover, comparative analyses that approach the history of technology in Latin America by contrasting it with other regions are, for the most part, still lacking.

The broad-ranging writings of social scientists dominated the conversations on the history of technological change in the region until at least the late 1980s. Such scholarship portrayed Latin America as dependent on industrial technology and ultimately a long-term victim of foreign hegemony in the region (Suárez, Yoguel, 2020). Sidestepping such a rigid structuralist argument, more recent socio-cultural narratives have been conceived largely in reaction to many of those earlier studies that emphasized the region's lack of technological autonomy. By bringing to the fore localized examples of Latin American ingenuity, experimentation, and research, recent historiography on Latin American science and technology highlights the agency of a wide range of actors — experts and laypeople, locals and foreigners — who adapted and used machines and tools, including small-scale artefacts and consumer goods. Current historiography also contrasts with earlier studies in that it is mostly concerned with very detailed cases rather than broader understandings of technological development at the national or regional levels (Castro, 2020; Pretel, Inkster, Wendt, 2019).

The lines of research have multiplied. Increasingly, scholars are concerned not only with the study of disruptive high-tech innovations and large technological systems, but also with the study of everyday artefacts, practical know-how, and indigenous epistemologies. From this perspective, a variety of ordinary places, sites, and institutions — from jungles to households — are presented as spaces of technological innovation, adaptation, and coproduction of knowledge. Here well-crafted empirical cases — usually microhistories and qualitative analysis — are the necessary starting points. By stressing that the technology in use in Latin America has not necessarily been passively imported from abroad, new work has successfully challenged both the centre–periphery dichotomy and state-driven modernization models (Medina, da Costa Marques, Holmes, 2014). Alternatively, historians are providing a critical socio-cultural exploration of the processes of adaptation, reproduction, and transformation of imported technologies to specific Latin American settings. This approach has ultimately resulted in studies of so-called hybrid technologies, creole expertise, and localized practices in a myriad of sectors, including hydraulics, mining, and agriculture.

Historians and historically minded scholars have for some time shown a renewed interest in the politics of technologies in Latin America, particularly participation of experts in state control, diplomacy, and development plans (Buckley, 2017; Medina, 2011). This tendency is apparent in studies of the role of communication and information technologies — telegraph, radio, cybernetics, television — where engineers and scientists have played a part in advancing political movements and national projects. This wave of scholarship is revisiting historiographical debates, offering a different entry point into — and new meaning for — the political issues that have traditionally dominated Latin American historiography, such as, for example, military conflicts and social revolutions. A telling example is the role of engineers and other workers — in areas such as radio broadcasting, irrigation projects, and railway infrastructures — in the progression of the Mexican Revolution and its post-revolutionary aftermath (Castro, 2016; Guajardo, 1999; Wolfe, 2017; Hill, 2017).

There has concurrently been a call to disentangle the relationship between technological culture and identity, revealing the shifting meanings that technologies and expertise have assumed across the region (Lucena, 2007; Tinajero, Freeman, 2013; Dalton, 2018). Take, for instance, the case of the Volkswagen Beetle produced, from the 1960s, in Mexico and Brazil — countries where this car became not only a commercial success but also a national icon (Wolfe, 2010; Rieger, 2013). Another emerging theme is the role of technological systems such as radios, waterworks, and roads in the making of borderland cultures, particularly technological encounters along the Mexico–United States divide (Robles, 2019; Alvarez, 2019). However, historians have been slow to explore the Americanization of Latin America's technological culture after the Second World War as manifested in, for example, technical education and engineering styles. Indeed, regional comparisons in engineering histories have yet to be drawn. Some historians, and even engineers, have written on the rise of the engineer profession in different national contexts of the region (Contreras, 1994 and 1995; Castro, Garza, 2021). Fewer have situated the history of engineering in Latin America in comparative or global terms.

In another vein, some historians have started to examine the advent of technological modernity in Latin America, showing the cultural impact — and unanticipated problems — of new artefacts and infrastructures such as electric power, mobility machines, and transit systems (Montaño, 2022; Gallo, 2006). A case in point is Willie Hiatt's exploration of the arrival of aeroplanes in the Peruvian Andes, which serves the author to interrogate the broader cultural and political transformations brought by global modernity (Hiatt, 2016). Also salient in the emerging historiography is the study of pre-Columbian technological systems and other technologies — including materials, plants, and animals — that, with minor changes or through creolization, have remained in continuous use for over five hundred years (Alexander, 2019;

Candiani, 2014). By making clear that in Latin America, as elsewhere, old and new technologies coexist, these works serve as an excellent corrective to the traditional grand narrative of linear technological progress and unidirectional knowledge diffusion.

The field abounds with studies examining the environmental consequences of large engineering projects, public works, and extractive industries, from dams to mining exploitations (Wolfe, 2017; Gómez, 2020). In this regard, the enviro-technological approach to Latin American history proves immensely fruitful. By seeing Latin America's natural environment as more than just a source of commodities, these studies emphasize the imbrication of nature and technology, and therefore challenge conventional economic, political, and social histories. The profound ecological impact of the invasive technologies delineated in some of these works may well be extended to other issues, such as the penetration of rainforests and the industrialization of urban ecosystems. However, environmental histories cannot ignore that extractive technologies and infrastructures are necessary evils for resource economies and as a means for transport and mobility.

The study of consumer goods and users' groups is another fairly new development. Some pioneering works have examined the mass consumption of agricultural tools, telephones, and household gadgets designed elsewhere and the novel recombination of these artefacts with local technologies. In these studies, everyday local actors — some of them without power or at the fringes of society — and not foreign experts, are the leading players behind technological transformations. A good example is the examination of the widespread use of machetes among popular classes in rural Colombia during the second half of the nineteenth century (Otero-Cleves, 2017). This study shed new light on the impact of localized advertising strategies and customization in an era of accelerated globalization. The active role of rural communities in building alternative telephone networks during the twentieth century also makes for a fascinating case study (Berth, 2019). What is still lacking is further examination of Latin America's cultures of repair, recycling, and maintenance (Velho, Ureta, 2019).

While the dominant social-cultural direction in the field introduces plurality through closer attention to everyday practices, human decisions, and the experience of technology, it gives only passing acknowledgement to the role of Latin American technologies and technological activity in broader processes of global economic change. The focus has shifted from technological innovation as the domain of foreign societies to technological innovation as an inherent attribute of Latin America societies (Pretel, Inkster, Wendt, 2019). With few exceptions, much of the new body of literature has failed to engage with pressing questions of economic inequality and poverty. Nor does it address the links between technological policies and Latin American paths of development. Still, a few scholars have managed to complement localized cases with a broader analytical examination of pressing issues. Notable exceptions are Edward Beatty's *Technology and the Search for Progress in Modern Mexico* (2015) and Eve Buckley's *Technocrats and the Politics of Drought and Development in Twentieth-century Brazil* (2017).

For their part, most economic and business historians have moved away from broad theoretical models revolving around Latin America's technological dependency. Since the 1990s, studies have concentrated on the patterns of technological trade (Tafunell, 2009), manufacturing companies (Gómez-Galvarriato, 2013), and national innovation systems (Cimoli, 2000). The spread of transportation infrastructure and its effects on mobility and market integration is also a central topic in the economic literature. In this regard, economic and business historians have long been interested in exploring the early history of Latin America's railways — as well as their eventual privatization and decline (Coatsworth, 1981; Miller, 1986; Summerhill, 2003; Kuntz, 2015). These studies have made clear that the initial construction of the major railway lines in countries such as Mexico, Cuba, and Argentina was largely dependent on materials, capital, and engineers from abroad. The majority of such works on railways, however, tends to offer only passing attention to technological issues — avoiding the 'black box' of knowledge generation, adoption, and use — with few exceptions (Guajardo, 2010). Although such histories of economic life are inconsistent with former historical-structuralist views, they have not put to rest the ghost of technological dependence. If anything, some economic historians have made clear that institutional and policy factors at the national level, rather than systemic world constraints, must account for Latin America's technological failures and successes (Beatty, 2015).

After many years of overlooking patent records, lately much effort has gone into the aggregation of patenting data for nineteenth- and twentieth-century Latin America (Beatty, Pineda, Sáiz, 2017). Broadly speaking, the central focus of these overarching studies has been a detailed medium-term (sometimes comparative) analysis of inventing activity according to economic sector, type of applicants, and geographical location — including the relative importance of patents granted to foreign applicants. In such studies, property rights are presented as essential drivers of technological innovation — and patent systems as modernizing institutions — without questioning the ultimate benefits for the countries of the region. Meanwhile, the contested political and economic integration of Latin American countries into the international patent framework since the late nineteenth century has received relatively little attention and deserves further treatment (Pretel, 2022).

Technology and Globalization in Latin America

Latin America is far from a cohesive region. It is a vast area with a multiplicity of nations, cultures, and languages as well as a diversity of historical landscapes. Still, it is meaningful to see Latin America as a geohistorical region that shares a common identity and historical trajectory, particularly a pervasive colonial legacy and similar patterns of nation-state development (Moya, 2011). Latin America's pivotal role in the process of globalization has been among the most distinctive features of its history since colonial times, though its centrality has been underrepresented in the growing body of global histories of the nineteenth and twentieth centuries (Brown, 2015; Langer, 2004).

Latin America's place in the global history of technology is intriguing and uncertain. The region not only has a distinct technological culture, but it is characterized by an uneven distribution of technologies. As Maria Portuondo observes, the history of technology in Latin America is ambiguous, as in many ways it 'offsets triumphs with failures, independence with dependence, and progress with decay' (Portuondo, 2009, p. 500). Moreover, the idea that this region has lacked innovative activity and an independent technological history is still commonplace. Eduardo Galeano's words 'the goddess technology does not speak Spanish' are still ingrained in the narratives of some historians (Galeano, 1997).

The international dimension of the history of technology is not just important in itself, but can illuminate broader issues of Latin America's past and present. On the one hand, connections with and dependencies on the wider world during postcolonial times explain many of this region's patterns of technological transformation. Certainly, Latin America's technological relationships with the United States and Europe were asymmetrical during the nineteenth and twentieth centuries. On the other hand, a salient characteristic of Latin America's postcolonial history has been the region's quest for technological autonomy, either in the form of nationalist technology policies or through transnational collaborations among experts.

After the mid-1820s, during the aftermath of Latin America's political independence, technological development became a central policy feature of the early Latin American republics (Saldaña, 2006). The end of colonialism in the region brought significant advances in the legal framework that, in turn, fostered engineering education, the professionalization of scientists and technicians, and the development of public works. After a short period of weakened connections with the wider world, Latin American nations resumed their substantial reliance on imported technology. Although ostensibly local or national projects, the expansion of infrastructures (telegraphs, bridges, railways, roads, water supply) mobilized global technologies, materials, and engineers. In other words, during the mid-nineteenth century, the political freedom of Latin American nations did not mark a radical break with earlier dynamics of technological transformation. In times of seemingly dramatic political changes, technological histories of mid-nineteenth-century Latin America have lent themselves to narratives of continuity — an issue that deserves further exploration.

The region's intensified integration into the world economy occurred during the second half of the nineteenth century. The quest for industrial progress drew Latin American republics to import manufacturing machinery, as is clear from industrial histories of — and patent and trade data for — Mexico, Brazil, and Argentina, the largest economies in the region (Suzigan, 2000; Beatty, 2015 Pineda, 2009; Rochi, 2005). However, nowhere were the effects of technological transformation more visible than in commodity production. International integration often meant exportation of equipment, production methods, and machinery traded for agricultural and mining products. Export commodities dominated both processing machinery and infrastructure that opened up natural resources to foreign interests.

The increasing production of cash crops and mineral resources between the 1880s and the Great Depression spurred infrastructural development in the region. New infrastructure connecting commodity frontiers with modern harbours opened up peripheral regions to foreign capitalists with their eyes set on rich natural resources. The expansion of railways plugged Latin American commodity frontiers into global trade networks. For instance, in the late nineteenth century, the production of bananas brought the construction of railways further into large plantations in the hinterlands of Costa Rica, Panama, Honduras, Colombia, and Ecuador. Another case in point is the introduction of refrigerated ships from the 1870s that allowed for the long-distance trade of highly perishable products such as meat and bananas. Furthermore, the critical innovation of mechanical refrigeration explains the rise of Argentina as the world leader in the export of beef and mutton. At the same time, railways were national projects that facilitated the internal transportation of passengers and had ambivalent effects in terms of nation-state building and economic integration (Lewis, 2008; Zanetti, García, 1998).

Alongside the spread of infrastructure technologies for commodity production came foreign expertise. Beginning in the mid-nineteenth century, a large number of expatriate technical experts built professional careers in Latin America and the Caribbean. During the years of export-led growth, Latin American nations had open economies, highly integrated into international circuits of production and trade that provided transnational career opportunities for scientific and technological professionals. Engineers, botanists, and agronomists, among others, were crucial in tropical agriculture and large-scale mining. These networks of expertise working in the region were transnational, but their activities were distinctively Latin American, as their primary task was to adapt new techniques and methods to the local environment and economic needs (McCook, 2002; Curry-Machado, 2011).

The case of Porfirian Mexico is especially telling. During the period from 1876 to 1911, after decades of technological 'backwardness', this country experienced a substantial transformation in its material conditions (Beatty, 2015). A new legal framework incentivized foreign investment in manufacturing as well as in public infrastructures such as communication and transport technologies. However, with certain noteworthy exceptions, technological transformations remained inferior to those of industrialized states, and the introduction of foreign technologies was characterized by limited assimilation of technological capabilities.

The mechanization of Mexican production — such as, for example, in textiles manufacturing — gave rise to a new dualism between economic sectors embracing industrial technologies and others continuing to use traditional technologies. Porfirian Mexico was also characterized by expanding material influence of the United States. By the 1911 Revolution, Mexico was the first destination overall for US foreign investment. American capital owned over 70% of Mexican mines and foundries, and American firms had built over 80% of Mexico's railway lines. Notwithstanding the Mexican Revolution, the presence and involvement of American industrialists and engineers in Mexico would remain significant throughout the 1920s and beyond.

Fig. 1. Unloading Cane at a Modern Cuban Mill (*c.* 1900–1906). Detroit Publishing Co, P. *Unloading Cane at a Modern Mill.* Cuba. Courtesy of the Library of Congress (Accessed 14/02/2022) https://www.loc.gov/item/2016798979/

Patterns of technological change and transfer in the colonies of the Spanish Caribbean were even more ambiguous during the nineteenth century. On the one hand, Puerto Rico and Cuba struggled to attain technological freedom (Elena, Ordóñez, 2000). By 1850, sugar production in Cuba and Puerto Rico had become an industrial enterprise that required extensive processing machinery purchased from the United States, France, and Britain as well as transnational machinists and engineers (Curry-Machado, 2011). On the other hand, there were significant innovations on the domestic front, with slave plantations becoming sites of technological experimentation that triggered the development of hybrid sugar and railway technologies adapted to these islands' environmental conditions (Rood, 2016) (Figure 1).

In the years following independence from Spain, the Spanish-speaking Caribbean accelerated the mechanization of the mass production and chemical control of sugar production (Dye, 1998). However, this industry continued to be dependent on the practical expert knowledge of invisible skilled labour (Singerman, 2017). The Cuban Republic, born in 1902, absorbed over two thirds of all US exports of sugar manufacturing machinery and emerged as a major destination for US materials, equipment, and durable goods. US investment in Cuban copper and iron mining likewise picked up during the first decade of the twentieth century, along with the further expansion of railways and other large civil engineering projects on the island.

The technological imprint of US engineering in Latin America is also well illustrated by the case of the construction of the transoceanic Panama Canal connecting the Atlantic and the Pacific. This gigantic natural infrastructure — comprised of a set of dams, gates, locks, and control stations — was designed and supervised by

Fig. 2. Workers building the Panama Canal (1903). Harris & Ewing, photographer. Panama Canal Panama Canal, 1913. Courtesy of the Library of Congress (Accessed 14/02/2022), https://www.loc.gov/item/2016864862/

American engineers but built by thousands of invisible and poorly-trained working people from the Isthmus of Panama and other parts of the Caribbean, many of whom died by accident or from disease (Greene, 2010; Carse, 2014; Maurer, Yu, 2011). This canal system, notwithstanding its critical role in global shipping and commerce, has been managed and maintained by local experts and workers ever since its inauguration in 1914. The Panama Canal can rightly be celebrated as a historical landmark in the global history of engineering, even in the history of economic globalization; however, its construction and administration also reflect the central role of local knowledge and labour — as well as environmental constraints — in the rise of the new global economic order at the turn of the twentieth century (Figure 2).

Even as some Latin American nations collaborated with the United States on massive engineering projects, Brazil, Argentina, and Mexico introduced regulations and institutions that aimed to break their technological dependence on the world economy. During the Great Depression and its aftermath, Latin American nations adopted developmental policies that prompted a renewed interest in technology. From the late 1930s, Brazil, Argentina, and Mexico introduced regulations and institutions with the aim of, among other things, ending their technological dependence on the world economy. These included the nationalization of natural resources as well as incentives for the transfer of science-based technologies to Latin American nations (Adler, 1987; Bastos, Cooper, 1995). The developmental policies increased autonomy and domestic technical capabilities in a context of "import substitution industrialization (ISI)" (Katz, 2000). However, even when ISI policies incorporated advanced technology in some strategic sectors, this generally happened without any sustained investment in technological training, experimentation, or practical research (Hirschman, 1968). What happened in the region during the 'developmentalist' era continued to reflect the specific material realities of Latin America and the way that Latin American countries had integrated into the international economy in the late nineteenth century.

Regional technological and scientific collaboration strengthened during the late 1930s and the post-war period — although admittedly it started before. The creation and consolidation of new engineers' networks and professional associations throughout Latin America correlated with a higher level of regional expertise and an increasing professionalization in strategic sectors. This genuine maturity of Latin American engineering reflected a regional reaction to foreign intrusions concomitant with the nationalization of subsoil resources and the development of civil engineering, mining, and irrigation projects (Contreras, 1994, 1995; Buckley, 2017; Castro, Garza, 2021). Examples included the establishment of the South American Union of Engineering Associations in 1938 and the South American Petroleum Institute in 1941. The economic and diplomatic involvement of regional communities of engineers varied from sector to sector. In the oil industry, for instance, the consolidation of state oil companies in Mexico, Brazil, Bolivia, and Argentina favoured professionalization and institutionalization among regional petroleum engineers and made them more inclined to act as technical diplomats (Zuleta, 2018).

During the post-war decades, one of the main concerns among public actors — including policymakers, technologists, economists, and state elites — was Latin America's pervasive technological dependence (Kreimer, Vessuri, 2018). In the writings of 'dependency' theorists (e.g. Andre Gunder Frank, Fernando Enrique Cardoso, Celso Furtado), Latin American history ultimately became global history. Instead of enclosing the region's history, these authors argued that international economic inequalities explained much of Latin America's past and present. Latin America was presented as a dependent economy that imported manufactured goods and exported raw materials. From this perspective, popular in the 1960s, manufacturing and technological flows went primarily in one direction: from industrialized states to Latin America. While the influential dependency theory arguably oversimplified Latin America's complex reality and unique history — and delivered few empirical results — it does provide useful analytical tools and central concepts with which to understand this region's asymmetrical technological relations.

The transnational dimension of techno-scientific collaboration during Latin America's Cold War is intriguing. Often framed as development programmes, foreign technical assistance in the region mobilized transnational materials, knowledge, and heterogenous expert communities. Scientists, engineers, and bureaucrats — some working for bilateral or multilateral agencies — became active geopolitical actors that shaped Latin America's technological policy. Organizations such as the Rockefeller Foundation, among others, contributed to foreign hegemonies in Latin America and reinforced the region's asymmetrical relationships with global powers like the United States and the Soviet Union (Cueto, 1994). However, transnational experts working in Latin America were not mere puppets enacting rival political agendas but active agents working on the ground under specific technological and scientific conditions. The practical implementation of projects in energy production, cybernetics, or biomedical research, just to name a few, was contingent on local priorities and constraints. Take, for example, the rise of nuclear science in Latin America in the 1950s and 1960s. In Brazil, Mexico, and Argentina, the development of nuclear technology did not just derive from the region's dependent relationship with foreign powers but reflected certain local imperatives and regional competition for hegemony in atomic energy (Hurtado de Mendoza, 2005). Similarly, the expansion in the manufacturing of fertilizers during Latin America's Green Revolution was enhanced by the intraregional transfer of technologies of nitrogen production and the collaboration of agronomists, especially between Mexico, Colombia, and Peru (Cushman, 2013: 282–89).

The early adoption of computers and digital technologies in Latin America again reflects a tension between technological autonomy and dependence. Latin American states, universities, armed forces, and communication firms made notable efforts in appropriating knowledge on computer hardware and software (Langer, 1989; Costa Marques, 2015 Gerstenberger, 2019). These efforts included new regulation, research projects, professional training, and transitional collaborations with multinational computer manufacturers such as IBM. From the 1990s, as in other world regions, the widespread use of new information and communication technologies steadily expanded. This transition increased the efficiency of business administration and services throughout the region. It also transformed politics and social relations, even in the educational and military spheres. However, internet penetration rates have been lower than in the developed world, resulting in an increasing inequality in information access and education.

Latin-Americanizing the Global Narrative

In 2009, the historian Lissa Roberts put out a timely call for the history of science and technology to be integrated into global history by situating our investigations 'in the contexts of globally extensive circulation and locally intensive theatres of exchanges' (Roberts, 2009, p. 24) . In this regard, fresh scholarly work has made clear that Latin Americans not only have a long history of technological ingenuity but have been active in the adaptation of knowledge and artefacts coming from elsewhere. However, historians still have an incomplete understanding of the place of Latin America in the global history of technology. Texts considering Latin American histories of technology within a global framework have been slow to emerge, with local and national approaches remaining the most common. Neither Latin America as a region nor its individual nations are considered central in the majority of global histories of technology for the post-1800 period, although this situation is rapidly changing.

There are many benefits to contextualizing Latin America's technological and scientific past within the global scene, as colonial historians have already demonstrated (Cañizares-Esguerra, 2017). Like other recent publications, this volume confirms that global history can shed light on technological differences and interrelationships among different regions of the world. Yet this approach comes with certain risks. Indeed, another valuable lesson to be drawn from recent works is that a more global perspective on Latin America's technological past entails moving beyond one-way exchanges and looking at reciprocal influences as well as to south-south connections. Of course, globalization is not synonymous with global history. Infrastructures and artefacts can divide societies and have deglobalizing effects. Donald Trump's project of constructing a divisive wall at the Mexican-US border is a telling example of this. It is, in short, inaccurate to see the global history of technology as the history of the internationalization of artefacts and knowledge — or, worse, as merely a history of global technological diffusion.

Drawing attention to the global entanglements of Latin American history of technology does not preclude a study of localized histories, rather the opposite. The most promising works are those that investigate the global by looking at the local. For example, Gabriela Soto-Laveaga's *Jungle Laboratories* (2009) is largely about the transnational co-production of a global commodity — the contraceptive pill. Yet it addresses not only transnational connections but also the central contribution of rural Mexicans through their expert knowledge of the extraction of a wild yam called barbasco. By tracing the role of local practices in the construction of globalized pharmaceutical knowledge and steroid manufacturing industries, Soto-Laveaga places Latin America squarely at the centre of the global history of science, technology and medicine, and equally importantly asserts the agency of subaltern actors. She also challenges the 'image of the laboratory as a pristine space that excluded "dirty and uncivilized" locals' (Soto-Laveaga, 2009: 11) and, therefore, raises methodological questions about which sources can be used to study less visible or even marginal people.

Just recently, scholars have begun to explore the notion of 'itineraries' to problematize Latin America's post-war history of technology. Gisela Mateos and Edna Suárez (2019) propose an alternative approach to the study of technological circulation and movement during the twentieth century. Concentrating on the 'materiality of travel' of radioisotope techniques, these Mexican historians show how the universalization and standardization of nuclear knowledge during the Cold War happened in the context of 'asymmetrical, nonreciprocal exchanges between countries' (Mateos, Suárez, 2019, p. 347). But at the same time, in tracing the material itineraries of precise resources, standards, instruments, and professionals throughout Latin America, Mateos and Suárez offer a critical approach to foreign technical assistance, one that privileges domestic agency and global movements in unexpected places. Similarly, Andra Chastain and Timothy Lorek, in a recent edited volume (2020), also embrace the concept of the itinerary and offer fascinating examples on the transnational co-production of technology and politics in Latin America, thereby calling into question the distinction between international and domestic expertise.

As may be clear by now, recent historiographical surveys, as well as ongoing research, are proof that the history of technology in Latin America is an important topic. The past decade has seen the emergence of a Latin American history of technology that is more plural and addresses topics that are not limited to conventional analysis of technological trade and transfer. The emerging body of scholarship conveys that the region and its people have had a technological life of their own, marked by the struggle for autonomy and the co-production of artefacts and practices. In other words, Latin America has both developed a distinctive technological culture and been a theatre of technological modernity. Further, scholars have shown that the history of technology in Latin America is important not only because of what it tells us about material transformations within the region itself, but also because it offers us a different perspective on traditional debates and historical processes more generally, particularly the history of globalization and transnational connections.

References

Adler, Emanuel, *The Power of Ideology: the Quest for Technological Autonomy in Argentina and Brazil* (Berkeley, CA: University of California Press, 1987).

Alexander, Rani T., *Technology and Tradition in Mesoamerica After the Spanish Invasion* (Albuquerque, NM: University of New Mexico Press, 2019).

Alvarez, C. J., *Border Land, Border Water: a History of Construction on the US–Mexico Divide* (Austin, TX: University of Texas Press, 2019).

Bastos, Maria Ines and Charles Cooper, *The Politics of Technology in Latin America* (London: Routledge, 1995).

Beatty, Edward, *Technology and the Search for Progress in Modern Mexico* (Oakland, CA: University of California Press, 2015).

———, Yovanna Pineda, and Patricio Sáiz, 'Technology in Latin America's Past and Present: New Evidence from the Patent Records', *Latin American Research Review*, 52.1 (2017), 138–49.

Berth, Christiane, Bringing Communication to the Countryside: Rural Telephony in Latin America, 1900–1985', *History of Technology*, 34 (2019), 65–88.

Buckley, Eve, *Technocrats and the Politics of Drought and Development in Twentieth-century Brazil* (Chapel Hill, NC: University of North Carolina Press, 2017).

Candiani, Vera, *Dreaming of Dry Land: Environmental Transformation in Colonial Mexico City* (Stanford, CA: Stanford University Press, 2014).

Cañizares-Esguerra, Jorge, 'On Ignored Global Scientific Revolutions', *Journal of Early Modern History*, 21.5 (2017), 420–32.

Castro, J. Justin, 'History of Technology in Nineteenth and Twentieth Century Latin America', *History Compass* 18, no.3 (2020), 1–11.

———, *Radio in Revolution: Wireless Technology and State Power in Mexico, 1897–1938* (Lincoln, NE: University of Nebraska Press, 2016).

——— and James Garza, *Technocratic Visions: Engineering, Technology, and Society in Mexico, 1876–1946* (Pittsburgh, PA: University of Pittsburgh Press, 2021).

Carse, Ashley, *Beyond the Big Ditch. Politics, Ecology, and Infrastructure at the Panama Canal* (Cambridge, MA: MIT Press, 2014).

Chastain, Andra B. and Timothy W. Lorek, *Itineraries of Expertise: Science, Technology, and the Environment in Latin America's Long Cold War* (Pittsburgh, PA: University of Pittsburgh Press, 2020).

Cimoli, Mario, ed., *Developing Innovation Systems: Mexico in a Global Context* (London and New York, NY: Continuum, 2000).

Coatsworth, John, *Growth Against Development: the Economic Impact of Railroads in Mexico* (DeKalb, IL: Northern Illinois University Press, 1981).

Contreras, Manuel E., 'Ingeniería y Estado en Bolivia durante la primera mitad del siglo XX', in *Saberes andinos: ciencia y tecnología en Bolivia, Ecuador y Perú*, ed. by Marcos Cueto (Lima: Instituto de Estudios Peruanos, 1995), pp. 127–58.

———, *Tecnología moderna en los Andes: minería e ingeniería en Bolivia en el siglo XX* (La Paz: Biblioteca Minera Boliviana, 1994).

Costa Marques, Ivan da, 'History of Computing in Latin America', *IEEE Annals of the History of Computing* 37, no. 4 (2015), 10–12.

Cueto, Marcos, ed., *Missionaries of Science: the Rockefeller Foundation and Latin America* (Bloomington, IN: Indiana University Press, 1994).

Curry-Machado, Jonathan, *Cuban Sugar Industry: Transnational Networks and Engineering Migrants in Mid-nineteenth Century Cuba* (New York, NY: Palgrave Macmillan, 2011).

Cushman, Gregory T., *Guano and the Opening of the Pacific World: a Global Ecological History* (Cambridge: Cambridge University Press, 2013).

Dalton, David S., *Mestizo Modernity: Race, Technology, and the Body in Postrevolutionary Mexico* (Gainesville, FL: University of Florida Press, 2018).

Dye, Alan, *Cuban Sugar in the Age of Mass Production* (Stanford, CA: Stanford University Press, 1998).

Elena, Alberto and Javier Ordóñez, 'Science, Technology, and the Spanish Colonial Experience in the Nineteenth Century', *Osiris*, 15.1 (2000), 70–82.

Galeano, Eduardo, *Open Veins of Latin America: Five Centuries of the Pillage of a Continent* (New York, NY: Monthly Review Press, 1997).

Gallo, Ruben, *Mexican Modernity: the Avant-Garde and the Technological Revolution* (Boston, MA: The MIT Press, 2006).

Gerstenberger, Debora, 'Challenging Martial Masculinity: the Intrusion of Digital Computers Into the Argentinian Armed Forces in the 1960s', *History of Technology*, 34 (2019), 165–86.

Gómez, Rocio, *Silver Veins, Dusty Lungs: Mining, Water, and Public Health in Zacatecas, 1835–1946* (Lincoln, NE: University of Nebraska Press, 2020).

Gómez-Galvarriato, Aurora, *Industry and Revolution: Social and Economic Change in the Orizaba Valley, Mexico* (Cambridge, MA: Harvard University Press, 2013).

Greene, Julie, *The Canal Builders: Making America's Empire at the Panama Canal* (New York, NY: Penguin, 2010).

Guajardo, Guillermo, 'Tecnología y campesinos en la Revolución mexicana', *Mexican Studies/Estudios Mexicanos*, 15,2 (1999), 291–322.

———, *Trabajo y tecnología en los ferrocarriles de México: una visión histórica, 1850–1950* (Mexico: Consejo Nacional para la Cultura y las Artes, 2010).

Hiatt, Willie, *The Rarified Air of the Modern: Airplanes and Technological Modernity in the Andes* (New York, NY: Oxford University Press, 2016).

Hill, Jonathan, 'Circuits of State: Water, Electricity, and Power in Chihuahua, 1905–1936', *Radical History Review*, 127 (2017), 13–38.

Hirschman, Albert O., 'The Political Economy of Import-substituting Industrialization in Latin America', *The Quarterly Journal of Economics*, 82.1 (1968), 1–32.

Hurtado de Mendoza, Diego, 'Autonomy, Even Regional Hegemony: Argentina and the "Hard Way" Toward the First Research Reactor (1945–1958)', *Science in Context*, 18.2 (2005), 285–308.

Katz, Jorge, 'The Dynamics of Technological Learning During the Import-substitution Period and Recent Structural Changes in the Industrial Sector of Argentina, Brazil, and Mexico', in *Technology, Learning, and Innovation: Experiences of Newly Industrializing Economies*, ed. by Linsu Kim and Richard R. Nelson (Cambridge: Cambridge University Press, 2000), pp. 307–34.

Kreimer, Pablo and Hebe Vessuri, 'Latin American Science, Technology, and Society: a Historical and Reflexive Approach', *Tapuya: Latin American Science, Technology and Society*, 1.1 (2018), 17–37.

Kuntz, Sandra, ed., *Historia mínima de la expansión ferroviaria en América Latina* (Mexico: El Colegio de Mexico, 2015).

Langer, Erick D., 'Generations of Scientists and Engineers: Origins of the Computer Industry in Brazil', *Latin American Research Review*, 24.2 (1989), 95–111.

———, 'Placing Latin America in World History', *Hispanic American Historical Review*, 21 (2004), 393–98.

Lewis, Colin M., 'Britain, the Argentine and Informal Empire: Rethinking the Role of Railway Companies', in *Informal Empire in Latin America: Culture, Commerce and Capital*, ed. by Matthew Brown (Oxford: Wiley-Blackwell, 2008), pp. 99–123.

Lucena, Juan C., 'De Criollos a Mexicanos: Engineers' Identity and the Construction of Mexico', *History and Technology*, 23.3 (2007), 275–88.

Mateos, Gisela and Edna Suárez, 'Technical Assistance in Movement: Nuclear Knowledge Crosses Latin American Borders', in *How Knowledge Moves: Writing the Transnational History of Science and Technology*, ed. by John Krige (Chicago, IL: Chicago University Press, 2019), pp. 346–67.

Maurer, Noel and Carlos Yu, *The Big Ditch: How America Took, Built, Ran, and Ultimately Gave Away the Panama Canal* (Princeton, NJ: Princeton University Press, 2011).

McCook, Stuart G., *States of Nature: Science, Agriculture, and Environment in the Spanish Caribbean, 1760–1940* (Austin, TX: University of Texas Press, 2002).

Medina, Eden, *Cybernetic Revolutionaries: Technology and Politics in Allende's Chile* (Cambridge, MA: MIT University Press, 2011).

———, Ivan da Costa Marques, and Christina Holmes, eds, *Beyond Imported Magic. Essays on Science, Technology, and Society in Latin America* (Cambridge, MA: The MIT Press, 2014).

Miller, Rory, 'Transferring Techniques: Railway Building and Management on the West Coast of South America', Working paper no.7, University of Liverpool, Institute of Latin American Studies (1986), pp. 1–66.

Montaño, Diana J., *Electrifying Mexico* (Austin, TX: University of Texas Press, 2022).

Moya, José C., 'Introduction: Latin America — The Limitation and Meaning of a Historical Category', in *The Oxford Handbook of Latin American History*, ed. by José C. Moya (New York, NY: Oxford University Press, 2011), pp. 1–24.

Otero-Cleves, Ana María, 'Foreign Machetes and Cheap Cotton Cloth: Popular Consumers and Imported Commodities in Nineteenth-century Colombia', *Hispanic American Historical Review*, 97.3 (2017), 423–56.

Pineda,Yovanna, *Industrial Development in a Frontier Economy: the Industrialization of Argentina, 1890–1930* (Stanford, CA: Stanford University Press, 2009).

Pretel, David, Ian Inkster, and Helge Wendt, 'Technology in Latin American History: Perspectives, Scales and Comparisons', *History of Technology*, 34 (2019), 1–22.

———, 'Latin America and the Emergence of an International Patent System", *Artefact. Techniques, histoire et sciences humaines*, no. 16 (2022), pp. 269–87.

Portuondo, María, 'Constructing a Narrative: The History of Science and Technology in Latin America', *History Compass* (2009), pp. 500-522.

Ragas, José, 'Science and Technology in Modern Latin America', (Oxford: Oxford Bibliographies, 2020).

Rieger, Bernhard, *The People's Car: a Global History of the Volkswagen Beatle* (Cambridge, MA: Harvard University Press, 2013).

Roberts, Lissa R., 'Situating Science in Global History: Local Exchanges and Networks of Circulation', *Itinerario*, 33.1, (2009), 9–30.

Robles, Sonia, *Mexican Waves: Radio Broadcasting Along Mexico's Northern Border, 1930–1950* (Tucson, AZ: University of Arizona Press, 2019).

Rochi, Fernando, *Chimneys in the Desert: Industrialization in Argentina During the Export Boom Years, 1870–1930* (Stanford, CA: Stanford University Press, 2005).

Rood, Dan, *The Re-invention of Atlantic Slavery: Technology, Labor, Race, and Capitalism in the Greater Caribbean* (Oxford: Oxford University Press, 2016).

Saldaña, Juan José, 'La historiografía de la tecnología en América Latina: contribución al estudio de su historia intelectual', *Quipu. Revista Latinoamericana de Historia de las Ciencias y la Tecnología*, 15.1 (2013), 7–26.

———, 'Science and Freedom: Science and Technology as Policy of the New American States', in *Science in Latin America: a History*, ed. by Juan José Saldaña (Austin, TX: University of Texas Press, 2006), pp. 151–62.

Singerman, David, 'The Limits of Chemical Control in the Caribbean Sugar Factory', *Radical History Review*, 127 (2017), 39–61.

Soto-Laveaga, Gabriela, *Jungle Laboratories: Mexican Peasants, National Projects, and the Making of The Pill* (Durham, NC: Duke University Press, 2009).

Suárez, Diana and Gabriel Yoguel, 'Latin American Development and the Role of Technology: an Introduction', *Economics of Innovation and New Technology*, 29.7 (2020), 661–69.

Summerhill, William R., *Order Against Progress: Government, Foreign Investment, and Railroads in Brazil, 1854–1913* (Stanford, CA: Stanford University Press, 2003).

Suzigan, Wilson, *Industria brasileira. Origem e desenvolvimento* (São Paulo: Editora Hucitec/ Editora da Unicamp, 2000).

Tafunell, Xavier, 'Capital Formation in Machinery in Latin America, 1890–1930', *The Journal of Economic History*, 69.4 (2009), 928–50.

Tinajero, Araceli and Brian J. Freeman, *Technology and Culture in Twentieth-century Mexico* (Tuscaloosa, AL: University of Alabama Press, 2013).

Vaccarezza, Leonardo Silvio, 'Ciencia, tecnología y sociedad: el estado de la cuestión en América Latina', *Ciência e Tecnologia Social*, 1.1 (2011), 42–64.

Velho, Raquel and Sebastián Ureta, 'Frail Modernities: Latin American Infrastructures Between Repair and Ruination', *Tapuya: Latin American Science, Technology and Society*, 2 (2019), 428–41.

Wolfe, Joel, *Autos and Progress: the Brazilian Search for Modernity* (Oxford: Oxford University Press, 2010).

Wolfe, Mikael, *Watering the Revolution: Environmental and Technological History of Agrarian Reform in Mexico* (Stanford, CA: Stanford University Press, 2017).

Zanetti, Oscar and Alejandro García, *Sugar and Railroads: a Cuban History, 1837–1959* (Chapel Hill, NC and London: University of North Carolina Press, 1998).

Zuleta, Cecilia, 'Engineers' Diplomacy: the South American Petroleum Institute, 1941–1950s', in *Technology and Globalisation. Networks of Experts in World History*, ed. by David Pretel and Lino Camprubí (London: Palgrave Macmillan, 2018), pp. 341–70.

Part 2

Artefacts, Processes, Sectors

GUILLAUME CARNINO

Introduction

While the other sections of this book are based on the complex interactions between different issues, the following section focuses on individual fields to reveal their inner complexities. Contrary to the common mythology of invention (in which a supposedly genius act sparks off all further developments), the techniques presented here are essentially characterized by their implementation and scale, so that they sometimes constitute a field in their own right, with its own rules and technical capital — akin to Bourdieu's cultural capital (Bourdieu, 1980). The studies grouped together in this section highlight two important points.

On the one hand, structure and context causally take precedence over invention: although invention may be a necessary condition, it is never sufficient. In other words, an artefact, process, or technical sector is never a given, it only becomes culturally significant (to the point of being the focus of whole chapters in a book!) in the process of its deployment. It should be noted that a technological sector may not cohere with the entire socio-economic world, but that it always requires a certain compatibility with it. When a technical system is born, it often needs to adapt to the existing system in order to be used within it.

On the other hand, once established as a sector or even a field, artefacts tend to transform the world to which they were previously adapted. These transformations are also at the heart of this section. In other words, once an artefact has been incorporated into the existing system to respond to pre-existing needs and trends, it becomes a constitutive force in itself for the new world that gave birth to it, to the point of "ex-apting" the universe (rather than ad-apting to it) so that it can be shaped according to its own dynamics.

In biology, adaptation — the term used to describe how function is at the origin of the organ creation, for example an elephant's ears, which developed to allow for temperature regulation — and exaptation — the term used to describe how an organ on the other hand can creat function, for example, the feathers that some dinosaurs developed for thermal homeostasis, which later made flight possible — occur cyclically and sporadically (Gould, Vrba, 1982). Similarly, technical evolution does not follow a pre-established pace or direction, but occurs as technical transformations emerge from socio-economic needs, and these transformations in turn generate new needs, etc. In short, this is neither about explaining everything through social determinism or technical determinism, but rather about understanding that these are interdependent, although they rarely occur simultaneously. Thomas P. Hughes (1988) has insisted a new technology generally has to conform to pre-established political, economic, and socio-cultural customs in order to develop further, whereas the opposite is often the case once a technological sector has been established; its own requirements then become those of a part of society, something spokespersons for the sector are there to ensure. One can, therefore, think of the technical evolution of the past two centuries (and perhaps even beyond) as being sometimes more technologically-motivated and sometimes more socially-motivated (in a broad sense), with the constant change obscuring which factor is more in play, even if sometimes one factor becomes structurally more dominant.

Naturally, the case studies gathered here cannot be reduced to this pattern, but it should nevertheless be noted that their richness shows that — beyond the great discourses on determinism (both social and technical) — multiple causality is, as always in history, *de rigueur*, and that a complex technical set is always both a part and a product of our world today. Importantly, readers of this second part of the present volume will be struck by the growth over time (if only in sheer scale) of the devices and processes described. Today's world is clearly the result of exponential technological growth (on which economic growth has also been rested) and of an increasingly complex socio-technical reality, neither of which showing any sign of slowing gown, despite the environmental disasters for which they are largely responsible. The fact that the pace of change accelerates ever faster even as ecological concerns have appeared evidences our societies are materially constrained by the technological complexity of the material processes that underpin our daily lives.

The order of the following chapters is not significant, beyond the simple fact that we have tried to make the transitions from one to the other as smooth as possible, so that if read from beginning to end the section reads as smoothly as possible, despite the different writing styles.

The first chapter, written by Jonas van der Straeten and Heike Weber, queries the global question of technical temporalities by suggesting contemporary technological phenomena are characterized by heterochrony.

The second chapter, written by Kevin Troch, is devoted to mining, whose ongoing evolution has been crucial from an economic and material point of view. Mining is at the heart of manufacturing and appears to alter not only ecosystems (due to pollution), but also landscapes and cities, with reconfigurations that can be either positive or cataclysmic.

Heavy industries, a sector that consumed mining products and a major player in globalization, are at the center of Florence Hachez-Leroy's chapter. This booming sector has seen infrastructures grow dramatically in both economic and technical terms for nearly two centuries.

In their chapter, Anaël Marrec and Mahdi Khelfaoui show that energy consumption has been increasing since the nineteenth century and directly informs all contemporary production, to the point of becoming deeply entwined with political aims, social organizations, and economic activities.

Jérôme Lamy and Arnaud Saint-Martin argue that it is fruitful to combine the fields of aviation and space exploration in an unprecedented history of aerospace, where actors, schemes, and systems of funding are intertwined to bring about a new world that is organised by and for economic and military power.

Shifting the focus to mobility on land rather than space, Arnaud Passalacqua emphasizes two factors: on the one hand, the powerful inertia of transportation systems, leading to any change taking at least a decade to be implemented and on the other hand, the process of standardization across different countries, as a result from the global interconnection of transportation systems. Specifically, the author shows that socio-technical standardization grows with the wider deployment of transportation systems.

In their chapter on maritime technology, Géraldine Barron and Julia Lajus maintain that a technical unit (such as a sailing ship) is also co-dependent on and mutually-constitutive of other systems (social, political, and economic). Thus for instance the development of steam boilers and then diesel engines threatened previously-established social links characteristic of older technologies. In the same way, the container (a minor technical innovation) and electricity (viewed here as a technical and informational instrument on a ship) deeply changed the world of fishing and maritime freight, leading to today's globalization.

Water as an energy source was the preferred instrument of industrial production from the Middle Ages to the nineteenth century, but Raphaël Morera and Elisabeth Mortier demonstrate that over the last two centuries, its status has gradually changed. Water has gradually become a product in its own right around which environmental, agricultural, and subsistence issues continue to grow, as contemporary pollution levels make it difficult to be optimistic about the future.

Like water, food has been a particularly difficult area of technological change since the nineteenth century. Ludovic Laloux identifies a series of major transformations (including temperature control for food preservation and consumers' changes in taste and behaviour that led to the development of sweet products, for example, as well as a proliferation of additives), with organic agriculture as the most recent of these developments.

Fabien Knittel dispels any preconceived notions about the progress of agriculture by showing that the choice of a technique often depends on multiple factors. Efficiency is only one of these, political and cultural factors being often actually more influential. This leads the author to suggest that the Green Revolution, which upset previous socio-technical choices, was also an instrument of political and economic domination.

In a radically culturalist perspective, Vincent Kuitenbrouwer looks at the history of communication techniques to show that, from the telegraph to the telephone, the radio or television and the Internet, all these global systems always also reflect specific political and cultural choices.

The text chapter attempts to look beyond the event-driven histories of computers and networks to offer an interpretative framework that considers socio-technical, political, and economic factors, in order to show that it is not invention but its deployment that is important, and that the nature and extent of digital development directly depends on the world it creates and in which it takes place.

By interweaving banking and technology histories, Sabine Effosse and Laure Quennouëlle-Corre reveal how modern technologies, from credit cards to account management on smartphones, via the use of paper money, are the real agents of globalization. Their focus on FinTechs, which combine digital technology and banking, allows us to see the extent to which the world economy is grounded in technology.

The diversity of historiographical approaches to civil engineering is amply illustrated in the chapter written by Dominique Barjot, who questions the idea of "progress" and shows how national traditions (French and English in particular) have influenced the evolution of construction practices and systems.

Finally, Audrey Millet recalls the importance of the textile industry in the industrialization process from the nineteenth century to the present day (when high-tech innovations influence the production of fabrics) while pointing out the dangerous links between clothing, exploitation through subcontracting on a global scale, and world-wide pollution.

References

Bourdieu, Pierre, *Questions de sociologie* (Paris: Minuit, 1980).

Gould, Stephen Jay and Elizabeth S. Vrba, 'Exaptation — a Missing Term in the Science of Form', *Paleobiology*, 8 (1982), 4–15.

Hughes, Thomas P. and Renate Mayntz, eds, *The Development of Large Technical Systems* (Frankfurt am Main/Boulder, CO: Campus Verlag/Westview Press, 1988).

JONAS VAN DER STRAETEN, HEIKE WEBER

Technology and its Temporalities, a Global Perspective*

During the lockdown phases of the global COVID-19 pandemic we have become eyewitnesses to the changing spatio-temporal constitution of society, technology, and the everyday. Around the globe, hospitals emerged in a matter of weeks, textile and car factories shifted production to hygienic or medical equipment, and mask production was accelerated, while other economic fields or traffic came to a standstill once a region was struck by the coronavirus. Global just-in-time production chains broke down, planes were kept in parking position, and consumers began stockpiling food as supermarket shelves needed weeks to be refilled with toilet paper or other staples. Working conditions for telecommuting — which had not been a viable alternative previously — were rapidly implemented, and while essential (German: 'systemrelevant') workers in medical care, administration, or medical industry faced an acceleration of work, others experienced an unsettling deceleration of life. For months, the extremes of having to pause or speed up, of hope in a future sustainable post-growth economy and narratives of crash, decline, and crisis dominated public discourse on a global scale.

The COVID-19 pandemic distorted the complex ways in which myriads of concurrent cycles and rhythms of technological systems and artefacts are interwoven across different temporal and spatial scales. It rendered visible those temporal dimensions of technology that are otherwise hidden, overlooked, or taken for granted. At the same time, it raised the question of long-term impact. Only time will tell if the pandemic leaves any lasting impact on societies, economies, technologies, and the environment. Regardless of the virus, conditions once defined as constant or cyclical, such as the North's permafrost (that currently proves to be less permanent than assumed) or the Earth's biogeochemical 'cycles' (that have been superimposed by human-made material flows), are undergoing changes. Young climate activists such as

* This paper has profited from the research project 'A Global History of Technology, 1850–2000' (funded by the European Research Council) at the Technical University of Darmstadt as well as the international workshop 'Technology's Temporal Regimes: Past, Present and Futures', organised by Heike Weber, Armin Grunwald, Annette Leßmöllmann, and Marcus Popplow at the Institute of Technology Futures (ITZ) (KIT, 29–30 November 2018). We would like to thank both institutions for their support.

Global History of Techniques, ed. by Guillaume Carnino, Liliane Hilaire-Pérez and Jérôme Lamy, Global Matters Culture Matérielles, histoire des techniques et circulation globale des savoirs, 9 (Turnhout: Brepols, 2024), pp. 261–280
BREPOLS PUBLISHERS 10.1484/M.TECHNE-EB.5.129811

Greta Thunberg, for instance, feel their future lives affected by past and present coal power plants, car engines, planes, and the throw-away society's consumption patterns. The so-called 'Earth Overshoot Day' attempts to quantify such temporalities; the date at which humanity has consumed more resources than the Earth can renew during an entire year moved from late July and early August during the 2010s to 22 August in 2020, but was only due to the economic cutbacks caused by the COVID-19 pandemic (Earth Overshoot Day, 2022).

Our current experience of temporal shifts around the globe challenges the straightforward common narrative of modern technology as an instrument for measuring and controlling time and of its role in accelerating modern life. In this essay, we therefore discuss existing frameworks of interpretation that may provide a guide to this complex set of time dimensions and perceptions. To grasp the intangible idea of time and relate it to the historical study of technology on a global scale, we propose the conceptual triad of temporalities, heterochrony, and chronography.

First, we will make a case for studying technology via 'temporalities' rather than simply 'time'. This shift, we argue, offers a useful heuristic for overcoming the ontological separation of technology and time as two separate entities (Champion, 2019: 247). In analogy to materiality or spatiality, temporality is a fundamental condition of technology that cannot be understood in isolation from its other qualities. Different technologies have their own rhythms, timelines, and temporal dependencies, some of which are defined by humans (e.g. the concept of product life-cycles), while others evade human perception and control, as becomes visible in failed attempts to synchronize traffic, manage the decay of infrastructures, or deal with technology's long-term legacies. Simply put, the notion of technology's temporality calls upon historians to situate time in technology, rather than just situating technology in time.

Second, the temporalities of technology do not readily lend themselves to existing ideas of universal timelines, periods, caesuras, and transitions. Every ensemble of artefacts and practices that we find in a specific geographical and historical setting tells its own story about a technology's temporalities. We encounter site-specific configurations of technologies with different genealogies and in different stages of use — including those that exist only in the form of defunct or out-of-use legacies or have entirely merged into other technologies. We find that each of these ensembles is characterized by its own interplay of various rhythms, speeds, and dynamics of change. Therefore the temporalities of a specific technology are never universal but relational, conditional, and site-specific: the same technology can be 'new' in one setting, and 'old' in another. It can accelerate certain processes (e.g. traffic) in one setting and decelerate it in another. In what follows, we use the term 'heterochrony' for technology to describe the heterogeneity of its respective temporalities in different settings.

Third, we propose the 'chronographical' perspective as a conceptual approach for a historiography of technology that considers the heterochrony of technology more seriously by following its signatures in the landscapes of our technological world.

The chapter is structured as follows: we begin by summarizing current trends in the debate on time and temporality in the humanities and lay out how an emerging body of literature has broadened our understanding of technology's visible and hidden temporalities. In the second section, we turn to the heterochrony of technology, the specific timelines of technology and the ways they are entangled. The third section then introduces the chronographical perspective as a conceptual approach that can help us make sense of the site-specific, heterochronous ensembles of practices and artefacts that we can observe across the world. The fourth section turns to the challenge of the global. How can we relate our study to technology's heterochrony with the study of connections and relations between different spaces or places as it is typically done in global history? Finally, the fifth section focusses on those temporalities which extend into the far future and the role of historians of technology in debates on the global injustices related to technology's 'afterlife' and lasting legacies.

The Pluri-temporalities of a Global World: a Review of Recent Literature

The question of time has been a recurrent theme in the study of history. In the course of the twentieth century, scholars have developed multiple pathways to approach it. To name but a few, Fernand Braudel introduced the tripartite structure of *geohistoire, longue durée,* and *histoire événementielle* as distinct temporal dimensions of history (Braudel, 1949). E. P. Thompson's seminal work on the effect of timekeeping on the work discipline of industrial workers stands out within a scholarly tradition claiming that mechanized and standardised time technologies came to structure all aspects of human life in modern societies (Thompson, 1967). Reinhart Koselleck and François Hartog, among others, synthesized the research on time into basic reflections on the meaning of time and time periods in historiographical writing (Koselleck, Tribe, 2004; Hartog, 2012; Landes, 1983). A change of time cultures, they argued, is constitutive to modernity as a historical period. Recently this idea has come under scrutiny, as the humanities — after the spatial turn of the late twentieth century — are, once again, reassessing the dimension of time (Champion, 2019). No matter whether this renewed interest in time qualifies for a 'temporal turn' (Gange, 2019) or not, it has substantially added to the conceptual toolkit for approaching temporality.

When we ask about the specific relationship between time and technology, we distinguish two general directions: a first cluster of studies has asked how societies experienced, perceived, or modified time alongside the appropriation of certain technologies; a more recent, second strand of research has turned to the question of non-human actors and their 'temporal agency'.

The history of time and time perception has long been an established field of research. A large body of literature has argued that modern technologies such as mechanical clocks, trains, and the telegraph were key to processes such as the standardization of time or the compression of space and time (Schivelbusch, 2014; Wenzlhuemer, 2015; Harvey, 2011). In a similar vein, recent works have examined how mobile phones, the internet, and social media have brought about fluid time, flexible coordination practices, and feelings of being 'always available' and 'perpetually present' (Wajcman, 2015; Castells, 2009; Weber, 2008). In view of the ongoing move to the digital in all spheres of daily life, some authors have raised far-reaching concerns about the impact of this trend on historiography and historical consciousness in general. Gumbrecht, for instance, argues that the digital age is characterized by simultaneities. The sequential experience of a temporal 'before' and 'after', he claims, is thus dissolving into what he terms a 'broad present' (Gumbrecht, Born, 2010). Advocates of digital history, by contrast, proclaim that digital tools can help historians overcome the chronological arrangement of history (Tanaka, 2019).

When it comes to authoritative works on time in late modernity, a Western bias still dominates, and with it, the metaphor of acceleration. For the sociologist Hartmut Rosa, late modernity is characterized by a dynamic acceleration of social change and life (Rosa, Trejo-Mathys, 2013). Environmental historians have identified a 'great acceleration' from the 1950s onwards, when resource extraction and use grew exponentially (McNeill, Engelke, 2016). At the same time, scholars in mobility studies emphasize that every form of mobility can also generate conditions of immobility, disruption, standstill, or deceleration (Adey et al., 2014). Synchronizing and scheduling the different components of traffic and transport systems usually requires periods of waiting, queuing, and transit (Kellermann, 2020). Lastly, the research on technologies of acceleration has been complemented by studies on technologies that are directed at the deceleration of certain processes. Modern conservation and cryotechnologies, for example, have been specifically developed to slow down 'natural' time processes such as decay (Friedrich, Höhne, 2017; Radin, 2017; Radin, Kowal, 2017).

Studies on non-Western regions have emphasized the heterogenous appropriations of 'modern' time. They show how different time regimes co-existed, conflicted with, or influenced each other in (semi-)colonial settings such as Egypt, Lebanon, or the Indian subcontinent (Barak, 2013; Ogle, 2015; Wishnitzer, 2015; Nanni, 2012). The export of modern technologies of time, such as the telephone or the train schedule, did not just move in one direction. Quite to the contrary, these apparently hegemonic technologies were resisted, reappropriated, and sometimes even turned against the agenda of modernization they were supposed to promote. For example, the various 'countertempos' that On Barak identified for colonial Egypt were not just alternatives to empty mechanical time (Barak 2013: 5). They emerged out of a creative engagement with imported time technologies while being technological and modern themselves (Barak, 2013: 83). The 'temporalities of waiting' have recently been taken up by the Africa studies community as a heuristic for understanding the conflictual temporalities of a continent which is torn between the hope for a better future and the multifold everyday experiences of waiting (Stasik, Hänsch, Mains, 2020). The authors provide a stark reminder of the inverse relationship between the distribution of waiting time and the distribution of power. After all, controlling other people's time is an important mode of exercising power (Göttlich, 2015). Such studies on 'temporal globality' (Gange, 2019: 301) put into question the triumphant narrative of the global dissemination of Western standard time. Allegra Fryxell therefore calls for studies of the multiplicity of time or the 'pluritemporalities' of the modern world in all their complexity, while avoiding the 'value-laden terms of historical periodization' (Fryxell, 2019: 297).

A second, largely separate perspective supported by scholars of the environmental humanities, the history of technology, STS, and anthropology, brings the temporalities of materials, of the 'made', of the environment, or of the entire Earth into play. Infrastructures, for instance, have been shown not only to produce time but also to be products of time (Engels, 2020). They comprise differently aged elements and are in need of synchronization or of maintenance or repair over time (Schabacher, 2021; Anand, Gupta, Appel, 2018; Joniak-Lüthi, 2019). Maintenance and repair have belatedly moved into focus. Steve Jackson has argued, for example, that history of technology should apply a 'broken world thinking' which takes 'erosion, breakdown, and decay, rather than novelty, growth, and progress' as a starting point (Jackson, 2014: 221). Technologies are subject to wear and tear, but also to abandonment, demolition, dismantling,disposal, and ruination — temporal dimensions that Heike Weber captures with the term 'unmaking' (Weber, 2014). The question of how technology of the past has been unmade opens up a vast field of research, but current studies are almost exclusively limited to the issue of waste, most notably hazardous waste and its problematic long-term legacies (Müller, 2019).

In a similar vein, Krebs and Weber have recently argued for a temporal perspective on technology which analyses 'how societies value and treat a technology in respect to time and how societies not only"make" and "use" technology but also "unmake" it' (Weber, Krebs, 2021: 15). Such time relations have multiplied as modern technologies have increasingly become subject to temporal reasoning: geologists calculate resource availabilities, engineers measure durabilities, economists conceptualize product life-cycles, technology assessment and technology foresight anticipate technological futures, consumer technologies are said to have 'lifespans' and to become 'obsolescent', to name but a few examples (Weber, 2019).

Environmental history, in turn, has scrutinized the lasting legacies of technologies (e.g. Jarrige, Le Roux, 2017). Within social ecology, LTSER (long-term socio-ecological research) asks about past and future long-term legacies of human interventions into natural systems (Winiwarter, Schmid, Haberl, Singh, 2016). The extremely long timelines of radioactive decay or the persistence of toxic chemicals have resulted in temporal chasms. Winiwarter and others classify abandoned technological sites according to the longevity of their respective legacy and the prospective burden of maintenance requirements which later generations will have to shoulder (Winiwarter, Schmid, 2020; Winiwarter, Schmid, Haberl, Singh, 2016); 'wicked legacies' is their term for those legacies which are more or less in need of care into perpituity. The sociologist Barbara Adam has identified a fundamental temporal disjuncture between nature and environment on the one hand, and between technology, science, industry, and labour on the other while analysing their respective 'timescapes' (Adam, 1998: 19). These timescapes include aspects of timing, tempo, duration, moments of succession or simultaneity, and other temporal characteristics like seasonal rhythmicity or latency periods. According to Adam, 'industrial time' and the unique temporalities of environmental phenomena generate temporal clashes, just as the long-term effectiveness of environmental hazards defies the short-term thinking of politics and policies. (Adam, 1998: 11).

Finally, the debates on 'Deep History' and the Anthropocene have contrasted (human) historical time concepts with geological times (Trischler, 2016; Ellis, 2018; Bonneuil, Fressoz, 2016). The 'technofossil' idea even merges geological and human time: human-made debris such as plastic waste is conceived of as 'index fossils', i.e. as material indicators forcaesura in the planet's history (Westermann, 2020; Zalasiewicz et al., 2014). Gabriele Hecht has recently argued the need for altogether new narratives and analytic modes to capture the complex 'interscalar' connections of time and space at stake. In her study of a mine in Mounana (Gabon) which had been extracting uranium ore for France's nuclear plants since the late 1950s, different scales of 'deep time' and 'human time' intersect in conflicting or unexpected ways. Hecht makes these intersections and entanglements visible by conceiving of the uranium-bearing rocks as an 'interscalar vehicle' between spaces and times. When, for instance, French engineers spotted traces of self-sustaining fission dating back to the uranium's distant past, the mine turned into a site of international scientific endeavour, since these traces promised to provide valuable lessons learned for the treatment of nuclear waste. At the same time, long-term health hazards among workers or inadequate remediation activities on the work site were ignored (Hecht, 2018). Hecht is among the few scholars who have critically argued against the global holisms of the Anthropocene debate, such as its reference to 'humanity' or 'planetarity', that are blind for regional and social inequalities and history's situatedness in place and time.

The Heterochrony of Technology

The renewed research interest in the global pluri-temporalities has confused many of the timelines that have long been taken for granted in the historical study of technology. We now turn to the question of how thinking through temporalities can help in crafting narratives of technological change that are more global in their geographical scope and more critical of teleological and triumphalist narratives of modernity and its related time concepts. To address these questions, we need to move beyond the exhaustive research on time perception, decentre the study of the temporalities of technology from anthropocentric analytical categories, and turn to the more 'hidden' timelines behind technological artefacts and practices.

In doing so, we can take up some of the few voices in the history of technology that have made a case for the study of technological longevity and continuity vis-à-vis innovation, novelty, and change. With a focus on technical infrastructures, Svante Lindqvist already argued in 1993 that what he termed the 'technological landscape' is shaped by technologies that are in the later stages of their life-cycles. While Lindqvist's concept of a technological landscape was a rather inanimate backdrop to human activities, however, David Edgerton later directed the scholarly gaze to the world of 'technology-in-use' and made a forceful argument about the significance of allegedly old technologies at different moments in history and in different places (Edgerton, 2008). Edgerton also added a global dimension to his argument by pointing out that outdated technologies from the West often reappear in the 'poor world' in a hybridized form as 'creole technologies' (Edgerton, 2008: 43).

These observations challenge ideas of technological change as a successive process, in which existing technologies are rarely simply substituted or abandoned in favour of newer ones. Animal traction by horses, for example, formed the backbone of urban mobility in European metropoles at the turn of the twentieth century. During the Second World War, more horses were used than in any other war previously and proved to be more important than the German V-2 rocket (Edgerton, 2008). Many cities rely on century-old sewer systems, and even the recent digitization of Germany's telecommunication system worked with copper cables from the era of analogue telephony (Henrich-Franke, 2014).

A glance at the history of different 'energy landscapes' (Jones, 2014; Hasenöhrl, Meyer, 2020) around the globe shows the inadequacy of thinking through universal caesurae for 'humanity', for example the beginning of a 'fossile age' in parallel to the industrial revolution. In the Global South, for instance, 'traditional' energy sources, in particular biomass, dominate (Singh et al., 2012), even if non-western settings such as the Ottoman Empire were early testing grounds for adopting coal (Barak, 2020). And while Great Britain had entered the so-called 'fossil age' by 1800, coal remained of minor importance for energy consumption in South European countries until the late nineteenth century (Zumbrägel, 2020).

To make sense of these heterogenous and interwoven timelines of technology, we argue that popular metaphors such as the binary notions of 'old' and 'new' technologies or the coinage of a 'simultaneity of the non-simultaneous' ('Gleichzeitigkeit des Ungleichzeitigen') (Schmieder, 2017; Landwehr, 2012) are too reductionist and essentialist. What is more, in the study of technology it is still the history of Western industrialization that serves as a benchmark for defining technological anachronisms. The three-stone fire that a Maasai herder in East Africa might use to cook, along with a mobile phone for communication, raises more eyebrows than an inhabitant of New York who uses age-old infrastructures such as the metro, streets, or canalization and relies on walking as a main means of locomotion.

As a more neutral and comprehensive concept we propose to employ the term 'heterochrony' (literally 'different timing') usefully for the study of technology. A term with origins in evolutionary developmental biology, heterochrony generally describes any difference in the timing or duration of a developmental process. Applied to the study of technology, heterochrony implies that the temporalities of a technology are not stable but change over time and across different settings. We can derive three major theories from the notion of 'heterochrony' for the historical study of the temporality of technology. First, historical timelines of a technology are never universal but can be only understood in relation to and in their embeddedness in a specific geographical setting. Second, in order to develop truly global and more balanced narratives of technological change, we therefore need to study the temporalities of technology in multiple, especially non-Western settings and find ways to relate them to each other. Third, in doing so, we need to take the full 'life-cycle' of a technology into account, especially the processes of reuse, 'unmaking', repair, decommissioning, and decay that have received little attention so far (Weber, 2014). But how can such an endeavour be put into practice? There is a need for conceptual tools that factor in both the embeddedness of these timelines in social and material environments on the one hand, and the way they are influenced by the connections, flows, and encounters studied by global history on the other.

Temporalizing the Technological Landscape

For this purpose, the concept of the technological landscape may provide a useful conceptual point of departure. Going beyond Lindqvist's narrow definition, Francesca Bray understands the technological landscape as 'the repertoire and distribution of skilled material practices and technical artefacts that a society draws upon to function' (Bray, 2016: 8). If we follow Tim Ingold's seminal essay on the temporality of the landscape, the landscape must be understood as being perpetually in the making, as a process rather than a given physical space or result of its deliberate construction (Ingold, 1993). It takes and changes its form by incorporating a complex mesh of human practices, natural cycles, and material transformations, all of which come with their distinct temporal characteristics. To understand the role of technology in shaping the landscape, we need to uncover those temporal characteristics embedded in or ascribed to the material practices, existing knowledge, and technical artefacts — as discussed earlier — and relate them to each other and to other temporalities of the landscape (environmental, political, economic) (van der Straeten, 2021).

We thus suggest differentiating between the chronology and the chronography of that landscape. A chronological account is aimed at identifying significant events, periods, or transitions within the landscape and situate them on a linear, universal timeline. For a long time, this timeline was defined through a teleological progressivist and anthropocentric understanding of time that found its most consequential form in modernization theory. Accordingly, history of technology was an innovation-centric chronology of the technological landscape, or, as we argue in more detail below, of Western landscapes of technology. While the chronology perspective helps us to bring order into the changes of the technological landscape by positioning them in a regular system of dated time intervals, it offers few conceptual entry points to study the heterochrony of technology, beyond the trivial notion of the 'simultaneity of the non-simultaneous'.

By contrast, a chronographical perspective on the technological landscape (from the Greek *gráphein*, to write, to draw) is based on the notion that this landscape constitutes an enduring record of the processes of making, using, and 'unmaking' of technology and their respective temporalities (Bensaude-Vincent 2018; Ingold 1993). A chronographical approach, as proposed by Bernadette Bensaude-Vincent (2019) will therefore aim for a broader study of the technological landscape as an archive or memory of these processes. It invites us to probe deeper into the visible and hidden temporalities of different technologies and their signatures in the landscape. Such signatures are found in shared material practices that emerge from the manifold individual engagements with technology in the institutions that perpetuate social configurations in dealing with a specific technology, material conditions reflecting processes of creation and decay, or recurring cycles and routines — to name just a few. The distinction between chronology and chronography might be regarded as homologous to that of time and temporality, as 'a blurring of what might seem determined, an entangling of time with action, a refusal of subject-object divisions' (Champion, 2019: 247).

Needless to say, the distinction between chronology and chronography is an ideal-typical one. Ultimately, a chronographical perspective will also have to be translated into one, or a set of, chronological accounts if it is not to be ahistoric. Research therefore cannot ever entirely commit to the one or the other. A few examples given below, however, illustrate that the prosed distinction can inspire research designs that are better suited to uncover the heterochrony of technology in different settings, including the global.

The Global Scale: a Chronographic Approach

While the chronographical approach highlights the embeddedness and entanglement of the heterochronous timelines of technology within specific environments, it does not address the connections and relations between these different spaces or places. Ingold's concept of landscape as such avoids the question of geographical scale, or rather it denies its significance (Ingold, 1993: 155). With a different motivation, global historians have attempted to move beyond predetermined geographical units — for example, the state, the city, or the empire — as a framework for writing history. Rather than using these entities as a point of departure, Angelika Epple argues for global history as a 'relational history of entities. In other words, [...] a history that shows how entities change continuously through their involvement in relations' (Epple, 2018: 395). What does this the mean for the study of technology and its temporalities on the global scale? First, a relational history of technology is not a history of relations, a concept underlying conventional histories on the diffusion or transfer of technology. Second, instead of predefined geographical units, a global history of of technology should follow the practices and processes that shape the technological landscape across different scales. Such a perspective invariably shows that the temporalities of a technology in different locations are entangled with each other or even mutually constitutive.

Global historians have provided instructive case studies in this regard. Roland Wenzlhuemer, for example, has shown how the construction of telegraph lines decoupled communication from transport and thereby altered the temporal relations within and between distinct localities (Wenzlhuemer, 2015). The telegraph ranks high on a list of technologies which global historians have studied for their 'globalization effects' (Osterhammel, 2014: 715). A focus on these technologies, however, narrows the view to a specific subset of temporal relations, as it tends to prioritize connection over embeddedness, socio-cultural aspects of time over the hidden temporalities of technology, the question of 'what is genuinely novel about a certain technology' (Wenzlhuemer, 2020: 76) over the question of technological longevity and continuity, and convergence over divergence. While these technologies often cause short-term disruptions of temporal relations, these disruptions seem to be hardly more than bumps on a historical road on which different landscapes and their respective temporalities converge into a global pattern of a great acceleration (Bayly, 2009: 451–87).

How can we then write narratives that seriously discuss the heterochrony of technology on a global level? If the path leads through relational histories of technological landscapes, an important first step is to increase the pluralism of individual studies on the temporality of technology in different places around the globe.

Having discussed the growing literature on the temporality of technology in the so-called Global North, we can now turn to the technological landscapes of the Global South, which have received much less scholarly attention. In doing to, we can ask how useful the chronographical (in contrast to chronological) perspective is for transcending a derivative perspective on technology in the non-Western world. The chronographical perspective, we argue, can help differentiate our understanding of technology's timelines on the global scale — and particularly in non-Western settings — in three ways.

The chronographical perspective can, first, balance established narratives of global technological change by uncovering the persistence of indigenous technologies vis-à-vis allegedly more advanced imported technologies. The global history of sugar, for example, has long been presented as a textbook case of globalization, mobility, and growing uniformity of a consumer product, and the mechanization and rationalization of its production (Mintz, 1985). In a study on India, one of the world's largest producers and consumers of sugar, Mikael Hard tells a strikingly different story of continuity in sugar-making (Hård, 2023). Despite attempts by the British colonial administration to mechanize sugar production, itinerant sugar boilers ('khandsaris') in Northern India continued to rely on established, pre-industrial, methods of making a variety of indigenous sugar products that were sold on local markets rather than exported (Amin, 1984). Again, the temporalities of the technological landscape in rural India, the cycles of sugarcane cultivation, the timelines of capital accumulation, and the constancy of cooking practices and taste offer convincing explanations why sugar producers refrained from introducing capital-intensive technologies and continued to rely on customary ways of making sugar. In their widely acclaimed book *Poor Economics*, Duflo and Banerjee have provided convincing economic explanations why the temporalities of accumulating capital and scaling up production prevent small businesses in the so-called poor world from abandoning established local technologies (Banerjee, Duflo, 2012: 205–34). Therefore, persistence and continuity are to be understood not as an absence of change, but as an — often quite effective —

mechanism of keeping a technology in sync with other temporalities embedded in the landscape.

There are other examples for the persistence of indigenous technologies in the socialist world. While the global export of socialist architecture has received much scholarly attention, the widespread practice of the self-help construction of houses throughout the USSR has not (Smith, 2010). In their study on the transformation of Samarkand's urban landscape in the former Soviet Republic of Uzbekistan, Van der Straeten and Petrova show that private house builders relied on traditional vernacular building practices using adobe bricks, even after the start of the mass housing campaign there around 1960 (van der Straeten, Petrova, 2021). While industrial construction materials such as fibre cement found their way into old Islamic neighbourhoods, adobe remained the most important construction material. In Korea, the traditional *ondol* floor heating system came under pressure from proponents of the Western-style radiator heating system, but made a comeback in the 1970s in a version that relied on industrially manufactured materials such as briquettes, cement, and plastic pipes (Lee, 2011). Rather than exoticizing non-simultaneity from a Western perspective, these examples point out the need to probe deeper into the hybrid material configurations that the heterochrony of technology generates in different landscapes.

Second, the chronographical perspective can replace singular timelines of an anonymous 'technology' with a differentiated study on the temporalities of the concrete material practices and artefacts associated with it. The history of the Bedford TJ truck serves as an example here. Its global chronology can be summarized as follows: originally designed in the UK in the 1950s, the truck's production there ended in 1987 when the Bedford company closed after acknowledging the technologically 'advanced' competition from other manufacturers. Hindustani motors in India then took over the production of the truck, which continued to enjoy great success overseas. In his ethnography on truck modding in Sudan, Kurt Beck's provides a fascinating glimpse into the technological landscape in which the imported Bedford trucks — originally designed for on-road use — are disassembled and entirely rebuilt into off-road vehicles for long-distance transport that runon rough roads through the desert (Beck, 2009). A chronography of that landscape may uncover the temporalities of road transport in Sudan; of makeshift, second-hand, and brand new car parts; of the lives and careers of mechanics and drivers that are often related by kinship; and of the way that practices of driving, repairing, maintaining, rebuilding, training, and learning become interwoven in an almost inextricable way (Beck, 2009).

While a global chronology of the Bedford TJ truck relegates Sudan to the status of a final destination for an out-of-date Western truck model, a chronography of the local technological landscape places the truck as a technical artefact at the junction of multiple entangled timelines. It also calls for critical reflection of both the potential and the limits of Edgerton's concept of a creole technology as one that is 'derived, but different from, the originating case' (Edgerton, 2008: 43). The concept can prove useful to study the global chronology of certain technologies. From the chronographical perspective, however, it attaches too much significance to a technology's place of origin and too little to everything else that makes it work within the technological landscape — for example by synchronizing it with its other temporalities. Rather than transcending the derivative perspective on technology in the 'poor world' (Edgerton, 2008: 43), an aganda that Edgerton promotes along with others, like the African historian of technology Clapperton Mavhunga (Mavhunga et al., 2017), the concept of creole technology might therefore rather hinder its realization.

Third, the chronographical perspective calls us to critically revisit the nexus between novelty, power, and the movement of technology on a global scale. Following in the footsteps of Daniel Headrick's book *The Tools of Empire* (Headrick, 1981), a plethora of studies have described how Western powers have capitalized on the uneven availability and affordability of 'new', 'modern', or 'advanced' technologies across the globe for military conquest, political control, and economic gain. In the last two decades, this narrative has come under scrutiny in various in many ways (for an overview, see van der Straeten, Hasenöhrl, 2016). The chronographical perspective adds an important aspect to this criticism by showing that it is not only 'new' technology that sets things, people, and capital in motion and shifts the balance of power. An illustrative example is the emergence of an extensive global network of the second-hand trade, for example for cars (Beuving, 2004) or clothing (Brooks, 2019; Fontaine, 2008). This trade capitalizes on the diverging modalities of technological obsolescence within different technological landscapes. In addition to consumer taste or state regulation, it is its embeddedness in a wider technological ensemble that decides when a technology becomes obsolete.

As we follow global networks of reuse, recycling, and disposal, we also encounter the new forms of exploitation, injustice, and violence. In parallel to the industrial countries' 1970s environmental turn, certain regions in the Global South became the 'ghost acres' of mass production and consumption, as Simone Müller has coined the waste disposal sites beyond national jurisdiction (Müller, 2019), just as sugar from the West Indies and cotton from the Americas and India once fed the British industrialization. In a form of 'garbage imperialism' — characterized by uneven power relations, informal networks and commodity flows, obscure subcontracting, migrant work of an unorganised poor, grey or black markets, and, all too often, illegal activities — waste of production and consumption ends up in regions and times far away from the spaces and moments of the production and use of goods.

'Unmaking' ships in the now legendary shipbreaking yards of Bangladesh and India, for example, results in local long-term health hazards and environmental destruction, e.g. when the ships' asbestos is reused as building material, oil is drained into the sea, or waste is dumped. But Ayushi Dhawan's work on the Alang Ship Breaking Yard in Gujarat (India) demonstrates that the global reuse, reappropriation, and dismantling of technologies involves further conflictual temporalities (Dhawan, 2021). In the months of dismantling, the discarded ships provide short-term jobs as well as lasting materials for subsequent reuse, including scrap that moves on to local steelworks or clocks, cutlery or fire extinguishers, which are sold on second-hand markets. Dhawan thus describes the ships' 'unmaking' as 'toxic and life-giving at the same time' (Dhawan, 2021, 285).

The Heterochrony of the Future: a Global Challenge

We conclude our essay by with some general reflections on those temporalities which extend into the far future, for example through their legacies and the long-term consequences left behind by established technologies or by outdated, abandoned, or discarded technologies. Modern technology, from nuclear plants to plastic bags, is deeply intertwined with both the past and the future. In contemporary consumer societies, disposable plastic bags and PET bottles represent ephemeral items, often used just once, but their materiality comes with long-term implications. The production, consumption, and disposal of such short-lived items connect a deep past — the hundreds of thousands of years through which fossil resources have accumulated — with an unknown future confronted with plastic's global environmental consequences.

An unknowable time dimension per se, the future is highly contested (Jasanoff, Kim, 2015): futures are desired, imagined, or forecast, and the resulting "sociotechnical imaginaries" are powerful tools to mobilize resources and actors. But what is the role of historians of technology in the debates on temporalities reaching into the future? For one thing, in the course of the twentieth century, modern optimism on progress and a 'technological fix' to social and environmental problems gave way to a more nuanced, multifaceted interpretation of technology's development, as more and more past technologies have turned into problems in their own right. Mass motorization and waste disposal, for instance, generated unforeseen environmental consequences on increasingly broader temporal and spatial scales. For another thing, studying the heterochrony of the past, we have good reasons to believe that the future will be also characterized by an heterogenous handling of technologies and their potential legacies.

Unintended long-term consequences of technologies are nothing new, but in the course of the twentieth century they have assumed unprecedented geographical and temporal dimensions. Some of them are effective in stealthy or invisible ways, and such 'slow violence' as Robert Nixon has termed it, often affects the poor or disempowered regions with social and ecological damage (Nixon, 2011). In late-twentieth-century engineering, remediation and so-called 'after-care' projects have become genuine measures to remedy after-effects from previous technological activities — from coal mining to contaminated sites of previous industrial production. Engineering is increasingly engaged in 'repairing' effects of past technological decisions — from car catalysts to current phosphorus recovery in municipal wastewater — and thereby often reinvents or reappropriates former traditional practices such as recycling, constructing with wood and loam, or producing energy from biomass. But the most dramatic and lasting 'side-effects' of modern technology — such as climate change, nuclear fallouts, the dispersion of plastic particles, or the massive alteration of the flows of carbon dioxide, nitrogen, or phosphorus, which have set off the Anthropocene debate — are less easy to remedy. Not only do they represent challenges on a global scale, but also challenges which will extend into a 'deep' future: they will interfere with future generations to be yet born.

The 'technofossils' of the past — the accumulated 'debris' and 'remains' of 'humanity' left behind predominantly by industrialized societies — are said to number thirty trillion tons of material within 170,000 different 'synthetic mineral' material categories (Ellis, 2018: 148). But what does this mean in a more nuanced heterochronic perspective? And how can we tackle the need to 'exnovate' technologies on the global scale? Past examples include materials such as DDT, chlorofluorocarbons (CFCs) (Armitage, 2019), asbestos, or the phasing out of nuclear power generation, all of which demonstrate the heterochrony not only of innovation, but also of exnovation — as shown, India is willing to integrate asbestos into its economy (Dhawan, 2021), and many nations aspire to build their first nuclear power plant.

The time factors involved in modern technologies, their production, use, reuse, and their 'unmaking', are not merely an academic problem. Long-term, unknown, or unknowable environmental effects stand in conflict with short-term economic and political decisions. Moreover, any transition towards a more sustainable technological landscape needs to overcome the inertia and path dependencies of large-scale production and mass consumer practices that have normalized over time, but also with diverging prospects on future technological developments. On a global scale, the climate crisis illustrates the global and generational injustices caused by these conflicting temporalities. While an increasing number of governments, companies, and individuals argue for the need to rapidly decarbonize economies and to 'transform' societies to make them sustainable, these narratives of crisis and a 'great transformation' inevitably reflect a Western-centric perspective, since there is no global consensus on the adequate grammatical tense to describe industrial modernity. Even among the OECD countries, only a handful of governments have adopted the future perfect tense to describe the fate of some of some key technologies such as coal power plants. Governments of many non-Western countries, tellingly labelled with the present participles of 'developing countries' and 'emerging economies', have an unwavering belief in the economic promises of carbon-intensive industrialization trajectories. Among the most visible displays of the impossibility to find a common language are the annual United Nations Climate Change Conferences. Tasked with finding a global consensus on climate change, these global forums regularly fail to reconcile the various temporalities of a global problem which plays out locally in myriad different ways.

References

Adam, Barbara, *Timescapes of Modernity: the Environment and Invisible Hazards* (London and New York, NY: Routledge, 1998).

Adey, Peter, David Bissell, Kevin Hannam, Peter Merriman, and Mimi Sheller, eds, *The Routledge Handbook of Mobilities* (London and New York, NY: Routledge Taylor & Francis Group, 2014).

Amin, Shahid, *Sugarcane and Sugar in Gorakhpur: an Inquiry into Peasant Production for Capitalist Enterprise in Colonial India* (Oxford and Delhi: Oxford University Press, 1984).

Anand, Nikhil, Akhil Gupta, and Hannah Appel, eds, *The Promise of Infrastructure* (Durham, NC and London: Duke University Press, 2018).

Armitage, Kevin C., 'The Locked Door: Thomas Midgley Jr, Chlorofluorocarbons, and the Unintended Consequences of Technology', in *A Field on Fithethe Future of Environmental History*, ed. by Mark D. Hersey and Ted Steinberg (Tuscaloosa, AL: University of Alabama Press, 2019), pp. 209–35.

Banerjee, Abhijit V. and Esther Duflo, *Poor Economics: a Radical Rethinking of the Way to Fight Global Poverty* (New York, NY: PublicAffairs, 2012).

Barak, On, *On Time: Technology and Temporality in Modern Egypt* (Berkeley, CA: University of California Press, 2013).

———, *Powering Empire: How Coal Made the Middle East and Sparked Global Carbonization* (Oakland, CA: California: University of California Press, 2020).

Bayly, Christopher Alan, *The Birth of the Modern World: 1780–1914, Global Connections and Comparisons* (Malden, MA: Blackwell, 2009).

Beck, Kurt, 'The Art of Truck Modding on the Nile (Sudan): an Attempt to Trace Creativity', in *The Speed of Change: Motor Vehicles and People in Africa, 1890–2000*, ed. by Jan-Bart Gewald, Sabine Luning, and Klaas van Walraven (Leiden and Boston, MA: Brill, 2009), pp. 149–74.

Bensaude-Vincent, Bernadette, 'Of Times and Things. Technology and Durability', in *French Philosophy of Technology*, vol. XXiX, ed. by Loeve and Nijssen (Cham: Springer International Publishing, 2018), pp. 279–98.

——— "Clashing Temporalities." Presentation at Darmstadt University of Technology, December 13, 2019.

Beuving, J. Joost, 'Cotonou's Klondike: African Traders and Second-hand Car Markets in Bénin', *Journal of Modern African Studies.*, 42.4 (2004), 511–37.

Bonneuil, Christophe and Jean-Baptiste Fressoz, *L'Événement anth ropocène: la terre, l'histoire et nou* (Paris: Éditions Points, 2016).

Braudel, Ferdinand, *La Méditerranée et le monde méditerranéen à l'époque de Philippe I I* (Paris: Armand Colin, 1949).

Bray, Francesca, 'Flows and Matrices, Landscapes and Cultures', *Icon*, 22 (2016), 8–19.

Brooks, Andrew, *Clothing Pthety: the Hidden World of Fast Fashion and Second-hand Clothes* (London: Zed Books, 2019).

Castells, Manuel, *Mobile Communication and Society: a Global Perspective; a Project of the Annenberg Research Network on International Communication* (Cambridge, MA and London: MIT Press, 2009).

Champion, Matthew S., 'The History of Temporalities: an Introduction', *Past & Present*, 243.1 (2019), 247–54.

Dhawan, Ayushi, 'The Persistence of SS France: Her Unmaking at the Alang Shipbreaking Yard in India', in *Histories of Technology's Persistence: Repair, Reuse and Disposal*, ed. by Stefan Krebs and Heike Weber (Bielefeld: transcript, 2021), 263–86.

Earth Overshoot Day, "Earth Overshoot Day 2022 Home - #MoveTheDate." Accessed November 29, 2022. https://www.overshootday.org/.

Edgerton, David, *The Shock of the Old: Technology and Global History Since 1900* (London: Profile Books, 2008).

Ellis, Erle C., *Anthropocene: a Very Short Introduction* (Oxford and New York, NY: Oxford University Press, 2018).

Engels, Jens Ivo, 'Infrastrukturen als Produkte und Produzenten von Zeit', *NTM* Journal of the History of Science, Technology and Medicine, 28.1 (2020), 69–90.

Epple, Angelika, 'Calling for a Practice Turn in Global History: Practices as Drivers of Globalization/s', *History and Theory*, 57.3 (2018), 390–407.

Fontaine, Laurence, ed., *Alternative Exchanges: Second-hand Circulations from the Sixteenth Century to the Present* (New York, NY and Oxford: Berghahn Books, 2008).

Friedrich, Alexander and Stefan Höhne, 'Regimes of Freshness: Biopolitics in the Age of Cryogenic Culture', *Medicine Anthropology Theory*, 3.3 (2017), 112–54.

Fryxell, Allegra, 'Time and the Modern: Current Trends in the History of Modern Temporalities.' *Past & Present* 243.1 (2019): 285–98. https://doi.org/10.1093/pastj/gtz012.

Gange, David, 'Time, Space and Islands: Why Geographers Drive the Temporal Agenda', *Past & Present*, 243.1 (2019), 299–312.

Göttlich, Andreas, 'To Wait and Let Wait', *Schutzian Research*, 7 (2015), 47–64.

Gumbrecht, Hans Ulrich and Frank Born, *Unsere breite Gegenwart* (Berlin: Suhrkamp, 2010).

Hård, Mikael, *Microhistories of Technology: Making the World*. [S.l.]: Palgrave Macmillan, 2023.

Hartog, François, *Régimes d'historicité: présentisme et expérience du temps* (Paris: Points, 2012).

Harvey, David, *The Condition of Postmodernity: an Enquiry into the Origins of Cultural Change* (Malden, MA: Blackwell, 2011).

Hasenöhrl, Ute and Jan-Henrik Meyer, 'The Energy Challenge in Historical Perspective', *Technology and Culture*, 61.1 (2020), 295–306.

Haumann, Sebastian, Martin Knoll, and Detlev Mares, eds, *Concepts of Urban-environmental History* (Bielefeld: Transcript, 2020).

Headrick, Daniel R., *The Tools of Empire: Technology and European Imperialism in the 19th Century* (New York, NY: Oxford University Press, 1981).

Hecht, Gabrielle, 'Interscalar Vehicles for an African Anthropocene: On Waste, Temporality, and Violence', *Cultural Anthropology*, 33.1 (2018), 109–41.

Henrich-Franke, Christian, 'Alter Draht'—'neue Kommunikation': die Umnutzung des doppeldrahtigen Kupferkabels in der Entwicklung der digitalen Telekommunikation', in *Umnutzung: Alte Sachen neue Zwecke*, ed. by Stephan Habscheid (Göttingen: V & R Unipress, 2014), pp. 97–112.

Ingold, Tim, 'The Temporality of the Landscape', *World Archaeology*, 25.2 (1993), 152–74.

Jackson, Steven J., 'Rethinking Repair', in *Media Technologies: Essays on Communication, Materiality, and Society*, ed. by Kirsten A. Foot, Pablo J. Boczkowski, and Tarleton Gillespie (Cambridge, MA: The MIT Press, 2014), pp. 221–39.

Jarrige, François and Thomas Le Roux, *La Contaminat ioft monde: une histoire des pollutions à l'âge indus triel* (Paris: Éditions du Seuil, 2017).

Jasanoff, Sheila and Sang-Hyun Kim, *Dreamscapes of Modernity: Sociotechnical Imaginaries and the Fabrication of Power* (Chicago, IL: University of Chicago Press, 2015).

Jones, Christopher F., *Routes of Power: Energy and Modern America* (Cambridge, MA: Harvard University Press, 2014).

Joniak-Lüthi, Agnieszka, "Introduction: Infrastructure as an Asynchronic Timescape." *Roadsides* 1 (2019), 3–10. https://doi.org/10.26034/roadsides-20190012.

Kellermann, Robin, 'Waiting (For Departure)', in *Handbook of Urban Mobilities*, ed. by Ole B. Jensen, Claus Lassen, Vincent Kaufmann, Malene Freudendal-Pedersen, and Ida Sofie Gøtzsche Lange (Abingdon, VA and New York, NY: Routledge, 2020).

Koselleck, Reinhart and Keith Tribe, *Futures Past: On the Semantics of Historical Time* (New York, NY: Columbia University Press, 2004).Landes, David Saul, *Revolution in Time: Clocks and the Making of the Modern World* (Cambridge, MA: Belknap Press of Harvard University Press, 1983).

Landwehr, Achim, 'Von der "Gleichzeitigkeit des Ungleichzeitigen"', *Historische Zeitschrift*, 295.1 (2012), 1–34.

Lee, Keun-sung, *Hangung Hyeondae Apateu Ondorui Hyeongseong [Modern Korea Apartment Ondol Formation: Focused on Korea National Housing Corporation Apartment]* (Seoul: Seoul National University, 2011).

Mavhunga, Clapperton Chakanetsa, Geri Augusto, Shadreck Chirikure, Chux Daniels, Ron Eglash, Ellen Foster, Garrick E. Louis, D. A. Masolo, Neda Nazemi, and Toluwalogo Odumosu, *What Do Science, Technology, and Innovation Mean from Africa?* (Cambridge, MA: The MIT Press, 2017).

McNeill, John Robert and Peter Engelke, *The Great Acceleration: an Environmental History of the Anthropocene Since 1945* (Cambridge, MA: Harvard University Press, 2016).

Mintz, Sidney Wilfred, *Sweetnessand Power: the Place of Sugar in Modern History* (New York, NY: Viking, 1985).

Müller, Simone M., 'Hiddentheernalities: the Globalization of Hazardous Waste', *Business History Review.*, 93.1 (2019), 51–74.

Nanni, Giordano, *The Colonisation of Time: Ritual, Routine and Resistance in the British Empire* (Manchester: Manchester University Press, 2012).

Nixon, Robert, *Slow Violence and the Environmentalism of the Poor* (Cambridge, MA: Harvard University Press, 2011).

Ogle, Vanessa, *The Global Transformation of Time: 1870–1950* (Cambridge, MA: Harvard University Press, 2015).

Osterhammel, Jürgen, *The Transformation of the World: a Global History of the Nineteenth Century* (Princeton, NJ: Princeton University Press, 2014).

Radin, Joanna, *Life on Ice: a History of New Uses for Cold Blood* (Chicago, IL and London: The University of Chicago Press, 2017).

——— and Emma Kowal, eds, *Cryopolitics: Frozen Life in a Melting World* (Cambridge, MA: The MIT Press, 2017).

Rosa, Hartmut and Jonathan Trejo-Mathys, *Social Acceleration: a New Theory of Modernity* (New York, NY: Columbia University Press, 2013).

Schabacher, Gabriele, 'Time and Technology: The Temporalities of Care', in *Hardwired Temporalities. Infrastructures, and the Patterning of Time*, ed. by Axel Volmar and Kyle Stine (Amsterdam: Amsterdam University Press, 2021), 55–76.

Schivelbusch, Wolfgang, 'The Railway Journey: The Industrialization of Time and Space in the Nineteenth Century (Oxford: Blackwell, 1979).

Schmieder, Falko, 'Gleichzeitigkeit des Ungleichzeitigen', *Zeitschrift für kritische Sozialtheorie und Philosophie*, 4.1–2, 2017, 325–63.

Singh, Simron Jit, Fridolin Krausmann, Simone Gingrich, Helmut Haberl, Karl-Heinz Erb, Peter Lanz, Joan Martinez-Alier, and Leah Temper, 'India's Biophysical Economy, 1961–2008. Sustainability in a National and Global Context', *Ecological Economics: the Journal of the International Society for Ecological Economics*, 76.100 (2012), 60–69.

Smith, Mark B., *Propthe of Communists: the Urban Housing Program from Stalin to Khrushchev* (DeKalb, IL: Northern Illinois University Press, 2010).

Stasik, Michael, Valerie Hänsch, and Daniel Mains, 'Temporalities of Waiting in Africa', *Critical African Studies*, 12.1 (2020), 1–9.

Tanaka, Stefan, *History Without Chronology* (Lever Press, 2019).

Thompson, Edward Palmer, 'Time, Work-discipline, and Industrial Capitalism', *Past & Present*, 38 (1967), 56–97.

Trischler, Helmuth, 'Das Anthropozän. Eine Herausforderung für die Geschichte der Wissenschaften, Technik unft helt', *NTM* Journal of the History of Science, Technology and Medicine, 24.3 (2016), 309–35.

van der Straeten, Jonas, "The Rhythms Behind Change. Historiography and the Temporality of Non-Western Technological Landscapes." *Technikgeschichte* 88. 2 (2021), 191–96. https://doi.org/.10.5771/0040–117X–2021–2-191

——— and Ute Hasenöhrl, 'Connecting the Empire: neue Forschungsperspektiven auf das Verhältnis von (Post)Kolonialismus, Infrastrukturen und Umwelt, *NTM* Journal of the History of Science, Technology and Medicine, 24.4 (2016), 355–91.

——— and Mariya Petrova, 'Mud Bricks in a Concrete State: Building, Maintaining and Improving One's Own House in Soviet Samarkand', in *Histories of Technology's Persistence: Repair, Reuse and Disposal*, ed. by Stefan Krebs and Heike Weber (Bielefeld: transcript, 2021).

Wajcman, Judy, *Pressed for Time: the Acceleration of Life in Digital Capitalism* (Chicago, IL and London: The University of Chicago Press, 2015).

Weber, Heike, *Das Versprechen mobiler Freiheit: zur Kultur- und Technikgeschichte von Kofferradio, Walkman und Handy* (Bielefeld: transcript Verlag, 2008).

———, '"Entschaffen": Reste und das Ausrangieren, Zerlegen und Beseitigen des Gemachten', *Technikgeschichte*, 81.1 (2014), 1–32.

———, 'Zeitschichten des Technischen: Zum Momentum, „Alter(n)"und Verschwinden von Technik,' in *Provokationen der Technikgeschichte. Zum Reflexionsdruck historischer Forschung*, ed. by Martina Heßler, and Heike Weber (Paderborn: Schöningh 2019), pp. 107–50.

Wenzlhuemer, Roland, *Connecting the Nineteenth-century World: the Telegraph and Globalization* (Cambridge: Cambridge University Press, 2015).

———, *Doing Global History: an Introduction in Six Concepts* (London: Bloomsbury Academic, 2020).

Westermann, Andrea, 'A Technofossil of the Anthropocene: Sliding Up and Down Temporal Scales With Plastic', in *Power and Time: Temporalities in Conflict and the Making of History*, ed. by Dan Edelstein, Stefanos Geroulanos, and Natasha Wheatley (Chicago, IL and London: The University of Chicago Press, 2020), pp. 122–44.

Winiwarter, Verena and Martin Schmid, 'Socio-natural Sites', in *Concepts of Urban-environmental History*, ed. by Sebastian Haumann, Martin Knoll, and Detlev Mares (Bielefeld: Transcript, 2020), pp. 33–50.

———, ———, Helmut Haberl and Simron J. Singh, 'Why Legacies Matter: Merits of a Long-term Perspective', in *Social Ecology: Society–nature Relations Across Time and Space*, ed. by Helmut Haberl, Marina Fischer-Kowalski, Fridolin Krausmann, and Verena Winiwarter (Cham: Springer, 2016), pp. 149–68.

Wishnitzer, Avner, *Reading Clocks, Alla Turca: Time and Society in the Late Ottoman Empire* (Chicago, IL: University of Chicago Press, 2015).

Zalasiewicz, Jan, Mark Williams, Colin N. Waters, Anthony D. Barnosky, and Peter Haff, 'The Technofossil Record of Humans', *The Anthropocene Review*, 1.1 (2014), 34–43.

Zumbrägel, Christian, 'Urban Energy Consumption, Mobility and Environmental Legacies', in *Concepts of Urban-environmental History*, ed. by Sebastian Haumann, Martin Knoll, and Detlev Mares (Bielefeld: Transcript, 2020), pp. 167–89.

KEVIN TROCH

Prospecting, Extracting, and Post-mining Techniques in the Anthropocene Era

Mines, and by extension the extractive industries, constitute a technical environment. Living and working in underground mines is impossible without technical systems such as shaft linings, ventilation, and dewatering.

Given that society today increasingly depends on mining resources, the history of extractive techniques and their economic, social, and environmental implications is both relevant and important to our future. The nineteenth century was pivotal in the history of our reliance on underground resources. It was then that Western Europe, followed by the rest of the world, ceased being an economy based on organic energy to become an economy based on mineral resources, and especially coal. Coal accounted for less than 3% of world energy consumption in 1800 and increased to 56% in 1913. The technical choices made during the nineteenth century, such as the development of steam engines, railways, and electricity, generated a growing demand for mineral and fossil raw materials and caused the mining system to develop exponentially. This progress and these technical trends are largely at the heart our seemingly unquenchable thirst for mineral and fossil fuels (Gray, 2017).

Prior to the mid-eighteenth century, mining was relatively limited throughout the world. The European production rose sharply when Great Britain and then France developed coal mines. At the beginning of the nineteenth century, coal production amounted to a few tens of millions of tons before increasing exponentially from the 1840s onwards, first in Europe, then in North America, and finally in the rest of the world. For example, the United States of America produced about 176,000 tons of coal in 1810, 8.5 million tons in 1850, 40 million tons in 1870, 270 million tons in 1900, and 680 million tons in 1918, and have produced more than 1 billion tons of coal each year since the 1990s. Coal mining in the German Ruhr region rose from 2 million tonnes in 1850 to 22 million tonnes in 1880 and 60 million tonnes in 1900 before peaking at 130 million tonnes in 1940 (Shulman, 2015). Coal is even more important for the development of the state in the case of China, where coal mining proliferated from the 1950s onwards, rising from 500 million tonnes produced in around 1960 to 3.5 billion tonnes in 2012 (Bardi, 2014).

The increased availability of coal made it possible to produce iron in greater quantities and thus reduce its cost, as well as, importantly, to power steam engines

Global History of Techniques, ed. by Guillaume Carnino, Liliane Hilaire-Pérez and Jérôme Lamy, Global Matters Culture Matérielles, histoire des techniques et circulation globale des savoirs, 9 (Turnhout: Brepols, 2024), pp. 281–296
BREPOLS ✠ PUBLISHERS 10.1484/M.TECHNE-EB.5.129812

(Gray, 2017). World iron ore extraction reached 500 million tonnes in the 1950s and expanded from 1.5 billion tonnes in the 2000s to 2.5 billion tonnes in the 2010s. New minerals were also extracted from the second half of the nineteenth century onwards thanks to the appearance of new technical developments. This was the case, among others, for bauxite, from which aluminium is produced. The Hérouelt-Hall process for manufacturing aluminium by electrolysis was introduced in 1887, and since then bauxite production has steadily increased, from 18 tons in 1887 via 450 tons in 1913, 886 tons in 1920, over 4,000 tons in 1930, then over 10,000 tons in the 1940s, 60,000 tons in the 1970s, to finally more than 100,000 tons per year since the 1990s. Between 75 and 120 billion tons of mineral materials have been extracted annually since the beginning of the twenty-first century. In volume this equals about 18,000 times the volume of rock and soil displaced by the 1883 eruption of Krakatoa in Indonesia or twice the volume of Mount Fuji in Japan (Bardi, 2014). The United States of America and China alone extract and move several billion tons of minerals each year. This growing need for mineral resources is driving the evolution of the mining system as a whole. While a full discussion of the history of mining techniques of the nineteenth to twenty-first centuries is largely beyond the scope of this chapter, excellent reference works on the subject are listed in the References at the end.

This chapter approaches the history of mining techniques chronologically around three themes, following the processes of mining operations: first, prospecting, i.e. the search for mineral deposits of sufficient quality and quantity to be economically profitable; then mining; and finally processing, that is, the fate of mining territories after extraction has ceased. These three stages are closely linked. They do not form a clear sequence, but rather intersect with each other. Since mining work is nomadic by nature and time sensitive, operators do not wait until the deposit is completely exhausted before carrying out prospecting, and post-mining problems often arise during the period of activity of a site. Finally, it should be noted that this chapter gives priority to large European mining complexes while also taking medium and small-scale operations into account, as well as extractive industries located outside Europe.

Prospecting

The term 'prospecting' is used to describe all actions related to the search for one or more mineral deposits and, once a discovery has been made, to determining whether or not their exploitation might be profitable. Prospecting on mines saves both time and money, as it determines the nature of deposits and the directions of the veins. Prospecting also contributes to the development of knowledge about the geological structure of explored areas and helps to produce geological maps that are useful for the extractive industries.

Since the nineteenth century and until today, prospectors have been using the sounding technique. The invasive technique of borehole mining involves drilling vertically into the ground to a depth where a significant sample for the identification of the nature of the subsoil can be removed. The drillers achieve this with the help of a series of connected metal rods. At the beginning of the nineteenth century, the use of the drill hole was widespread. The drill bit is a metal rod with a hardened metal cone at its end, which is encrusted with diamond dust to enable it to drill through hard rock. A motor creates rotation in and pressure on the rods, and the combination of the two allows the drill bit to sink into the ground. The drill cores are then brought to the surface and analysed. This technical innovation is relatively low in cost as the drill bit can be reused for prospecting on other sites. Another type of drilling used in the nineteenth century was percussion drilling, in which crushed rock and mud are brought to the surface by injecting water into the well (Debrabant-Malolepszy, Dumont, Six, 2013).

In the course of the twentieth century, other prospecting techniques were developed, such as electromagnetic and seismic methods. They take advantage of the physical properties of rocks and minerals to detect the structure of the subsoil indirectly. Seismic prospecting uses explosive charges, usually dynamite, whose explosion generates waves that propagate through the subsurface. When they reach the boundary between two layers of rock, some of the waves are reflected back to the surface, where they can be recorded by geophones. The analysis of these waves provides information about the structure of the terrain. The seismic method also uses trucks that generate low amplitude seismic waves by vibration. The development of space technologies and the increased number of satellites in the second half of the twentieth century, particularly since the 1970s, have made it possible for mining companies to carry out large-scale prospecting based on satellite data and to identify potentially exploitable geological structures. Finally, the development of directional drilling from the 1940s onwards made it possible to drill holes with interlocking rods that can be oriented in different directions. This technique is mainly used in oil operations and those that use hydraulic fracturing.

Exploitation

Exploitation is the primary phase in the life of an extractive industry operation. Mineral materials are extracted by various processes in order to be marketed. To reach them, operators may use either of two types of methods, either digging underground galleries and bringing the minerals to the surface, or excavating the surface of the ground in a so called 'open pit' mine, which is essentially a hole dug in the ground. These techniques are combined to ensure that the subsoil is controlled for mining. Here it is important to remember that the mining system in use up to the 1830s was relatively similar to that of the Early Modern period, and that the new techniques did not automatically replace the old ones. Several technical generations often coexisted within the same mining territory, and sometimes within the same mining operation (Veyron, 1999).

Digging

The first operation consists of digging a well. In European mines, this was traditionally done with the method of pumping water that is in the shaft during the drilling. A perfectly watertight casing was placed in position to stabilize the shaft and thus formed, in fact, the lining of the well walls. This was a kind of watertight tube that moved vertically through the waterlogged rock to reach the exploitable mineral veins. It had to be strong and sufficiently watertight to prevent water from seeping into the shaft and thus from flooding the galleries. At the beginning of the nineteenth century, the casing was made of oak wood, but in the course of the nineteenth century, cast iron became more widespread, which made it possible to widen the cross-section of the shaft and thereby increase the yield. In mines without water the walls were often covered with clay bricks. In the 1930s, concrete, which was solid and easy to manoeuvre, was increasingly used to ensure the stability and watertightness of deep shafts (Demey, 2016).

Another technique without dewatering was developed by the French engineer Jacques Triger in 1839. The process used compressed air, mainly in wet mines located in aquifer areas where the pumping of mine water was insufficient to keep the shafts dry. Compressed air was sent into the shaft with an injector in order to keep the water at the bottom. Workers entered the well through an airlock and then descended in a pressurized bell to dig the well deeper in order to reach the coal seams. The cuttings were then evacuated through the same airlock. Given the pressure, the process was dangerous for the workers, especially since the impact of such a pressurized environment on the body was unknown at the time. Numerous cases of decompression sickness occurred when this process became widespread throughout the extractive industries and the construction sector (Rainhorn, 2014).

Other systems were introduced in the second half of the nineteenth century, when the mining industry increased production exponentially. The Kindt and Chaudron system used in the collieries of northern France from 1867 onwards combined two technical innovations: the shaft was dug like a borehole, and a metal casing made of cast iron rings was gradually lowered into it, reducing the cost and time required to create a shaft. Other sinking techniques involved the solidification of the soil. For example, the Poetsch process used in Belgian Limburg between 1900 and 1910 made it possible for deep wells to be sunk into wet and unstable terrain by digging a series of boreholes in a circle around the well and placing tubes filled with coolant inside them. These tubes froze the ground and thereby made it possible to sink the well. The mines also used ground cementing as early as the end of the nineteenth century. Here slow-setting cement was injected into the ground to harden it by sealing cracks in the rock. These processes made it possible to exploit deposits that had previously been considered inaccessible (Troch, 2018).

An underground mining site is divided into floors or levels. On each level a hook is installed in the shaft, and it is from these points that the main drifts are dug to reach the deposit. These galleries are essential, because they ensure ventilation and enable the miners to circulate. To open the drifts, companies mainly use explosives today, but previously — and even in small-scale artisanal mines today — digging was done manually, with shovels and picks. For the method using explosives holes are drilled into the rock, explosives are inserted and set off, and then the debris is cleared and the roof of the gallery shored up to prevent collapse. This was dangerous and often fatal for those in charge of lighting the fuses.

In the nineteenth century things changed thanks to two technical innovations. The first appeared in England in 1831 with the development of the safety fuse for the ignition of explosives, making their use less hazardous. The second, major innovation appeared between 1860 and 1875 with the invention and marketing of dynamite. Gun powder was soon replaced by dynamite and other security explosives. Firing became even less dangerous for workers with the adoption of electric explosives. While previously blasting had to be carried out at close range, technical innovations in the second half of the nineteenth century allowed workers to operate from a safe distance (Centre historique minier de Lewarde, 2013).

The support is put in place as the galleries are being dug, using all the techniques that ensure the durability of the galleries. Supporting prevents rock falls and absorbs the pressure of the upper ground in underground mines. Until the end of the Second World War, the support structure was made from wood, mainly fir wood. In addition to the fact that fir wood is cheap, easily transportable, and can be carved into any shape needed, it also has the advantage that it cracks before breaking, so that miners are warned before an impending collapse of galleries. Pine species are also resistant to the heat and humidity that are characteristic of the mining environment. Wood was largely abandoned in the 1950s as material for support. In the 1920s and 1930s metal retaining structures appeared, which consist of height-adjustable struts and, unlike wooden retaining structures, can be salvaged. In the 1960s, props were gradually replaced by the walking support, which consists of four hydraulic jacks that form a pile. It moves along with the progress of the work and is better able to resist the pressures of the ground above.

Digging into the earth leads to a permanent struggle with water. It seeps into the galleries; it floods quarries and open-pit mines; and if a miner or a machine pierces an underground water reservoir, the mine threatens to flood. The first steam pumps appeared as early as 1710 in British collieries and then on the European continent. Gullies were dug in the galleries to carry the water to an overground reservoir. The pump, which was installed on the surface, brought the water up out of the galleries. In the course of the nineteenth century, these gullies were gradually replaced by pipes. The improvement of centrifugal pumps, their reduction in size, and then the use of electric pumps made it possible to install dewatering installations at the bottom of the mines.

One consequence of the dewatering of mine water is that it disturbs the hydrogeological balance. For example, in the middle of the nineteenth century, the Compagnie des mines d'Anzin encountered the Torrent, a vast underground saltwater lake left over from a prehistoric sea. Between 1847 and 1867, 800,000 cubic metres of water annually were extracted by three pits equipped with pumps to manage the Torrent.

This continuous pumping modified the local hydrogeological balance and disturbed the geological equilibrium, causing damage to the surface (Troch, 2018).

Finally, the galleries require lighting and ventilation. Atmospheric conditions are generally poor in underground mines, and the deeper the mine, the more natural ventilation is insufficient to effectively carry air from the surface to the bottom. Therefore, the air quality must be measured constantly so that the ventilation of the drifts can be adjusted. In fact, ventilation includes all the technical means used to establish and maintain appropriate conditions of temperature, humidity, and air quality for the miners' work. Until the 1850s, ventilation was provided by a double shaft system and by fireplaces installed at the bottom of one of these shafts to create air flow: as warm air pushed the stale air of the mine to the surface, fresh air was sucked down the other shaft and into the galleries. During the second half of the nineteenth century, centrifugal ventilators became widely used, including the high-speed ventilator invented in 1855 by Théophile Guibal (a professor at the Mons School of Mines) which was used to ventilate deep shafts. This vacuum-creating device allowed for fresh air to be sucked down to the bottom of the mine. This ventilator allowed mine operators to exploit deposits at a depth of more than 1,000 metres as well as to make the galleries larger. The air quality was monitored to avoid the harmful gases that are present underground, such as carbon monoxide and methane. Measuring equipment was developed in the course of the twentieth century, and after the 1950s monitoring stations were installed at the surface. They monitor atmospheric conditions using sensors placed at the bottom of pits (Cooper-Richet, 2011).

Lighting underground working sites was a challenge for mine operators prior to the widespread use of electric lighting. In mines, open-flame lamps provided effective lighting but posed a danger to miners. An explosion with fire could occur if the flame came in contact with a flammable gas. In 1815, the British chemist Humphry Davy created the first safety lamp. Surrounded by a fine wire mesh, it prevented the flame from coming in contact with underground gases, but its illuminating power was very low. This lamp was further optimized throughout the nineteenth century. The wire gauze was replaced with a glass cylinder, and from the 1880s onwards, miners' lamps were powered by petroleum oil, which surpassed previously used fuels in lighting power and longetivity. At the same time, in the 1860s and 1870s, electric lamps began to appear when tungsten wires replaced carbon in light bulbs. They were powered by lead accumulators, which made them heavy and bulky. The first electric lamps for miners were marketed by Thomas Edison between 1913 and 1916. They worked with a battery attached to the miner's belt, and these batteries were both rechargeable and easy to use. Henceforth, the lighting followed the miner's movements. This type of lamp was first used in the United States, then in Great Britain, and finally, in the rest of the world after the Second World War.

Extracting

Extraction is the main activity of a mine. It involves different techniques and methods that differ from one operation to another, depending on the structure of the deposit and the nature of the extracted materials.

The extraction is done from the shaft. The use of steam engines and then electric motors in the mines allowed the mines to be concentrated around one or two shafts, which no longer had to be filled in and dug out as the deposit was exploited by the machines operating the pumps and elevators. A network of galleries was then dug around a central shaft. This network sometimes sprawled further once diesel or electric locomotives were introduced in the twentieth century, followed by motor vehicles.

In the nineteenth and twentieth centuries, different types of workings or operations were used in underground mines. They were adapted to the thickness, inclination, and hardness of the veins. The working sites could be located above the main galleries, with the mined ore descending by gravity, or horizontally. In this type of mining site, material is removed by shovelling and rolling, or by mechanical means, e.g. on belt or armoured conveyors, or with the oscillating chute from the 1930s. The wagons in which the ores and waste rock were transported were first pushed by men, and then by horses, before being automated in the 1870s with the introduction of mechanical dragging. The wagons were then hooked to endless floating chains pulled by a winch set in motion by a steam engine or electric motor. Finally, depending on the nature of the overhanging land and the concentration of buildings on the surface, mines could either be exploited entirely and then be backfilled, or they can be exploited using the chamber and pillar technique, where pillars of rock and ore are left in the galleries to support their roofs (Demey, 2016).

For a long time, extraction was exclusively manual. The miners used picks, flat tools, hammers, and chisels. The material was transported from the mining site to the shaft using barrels and baskets, and then in wagons that were pushed on railway tracks. The barrels were lifted by horse-drawn carousels. These structures were called 'barritel', or wheel machines, and were in use until the 1840s. At the end of the nineteenth century, the introduction of compressed air as an energy source allowed the mechanization of extraction. In the 1860s, cutters were introduced in mines with thick and regular veins. These were motorized drilling machines that made it possible to extract large quantities of ore in a short time. In mines where these machines could not be used, compressed air-powered jackhammers were introduced. They were generated on the surface by piston compressors driven by steam or electric motors, and then transported to the bottom through steel or rubber pipes. After the Second World War, the use of cutters became widespread throughout the world's mining industries (Cooper-Richet, 2011).

The introduction of use of electricity from 1880 to 1900 and its widespread use in the 1930s radically changed the appearance of mines. With this source of energy, the cumbersome, noisy, and dangerous steam engines vanished, and their wheel, piston, and belt systems were replaced by alternating current turbines. As a result, the workspaces within the surface buildings were reorganised. For example, the Koepe system of extraction pulleys, which had first been developed in 1877 and were later adapted to electric motors, profoundly changed the appearance of the headframe overhanging the shafts, and even made it disappear. The wooden carousels were replaced by steel roller frames that were tens of metres high. The headframes built after the Second World War were made of concrete, which was more economical and stronger than steel. The headframes had been used to move miners and ores in barrels up and down the shaft until the 1850s, followed by elevator shafts that were pulled by wire ropes. Pumps and ventilators vanished from the surface and were placed in the galleries as early as the 1890s. Their electric motors were kept in explosion-proof housing to avoid contact with explosive gases.

The mechanization of extraction in the twentieth century has had significant consequences in the social, occupational health, and environmental arenas, with some variation depending on the products exploited. While the mining machines rendered the work underground less arduous and dangerous, they also devalued the miner's profession. The use of jackhammers, cutters, and other automated devices no longer required any specific skills. The speed of felling and sorting replaced the miners' skills. Moreover, from the inter-war period onwards (i.e. once the use of jackhammers and cutters became widespread), the use of low-skilled and inexpensive labour became widespread in European mines (Rainhorn, 2014). In other parts of the world, such as the silver mines in Zacatecas, Mexico, the miners opposed the introduction of European and North American mining techniques as they feared that these apparatuses would downgrade their skills and their status (Gomez, 2020). By contrast, the declining imperial regime of Qing China tried to adopt these Western techniques by employing European and American engineers and using mining machines and techniques (Wu Shellen, 2015). However, the machines wreaked havoc with the miners' health. Dust permeated the galleries and affected the miners' lungs. Respiratory diseases like silicosis swiftly spread among coal miners as the mechanization of work progressed. These instruments also emitted a continuous loud noise and caused deafness and nervous fatigue in the workers. In asbestos or uranium mines, dust was also a danger to miners and could cause cancer (Hecht, 2012).

Mechanization also caused environmental problems. The increase of the rate and volume of extraction also increased the disturbance of geological balances and disrupted the landscape. Consequently, land subsidence happened on an unprecedented scale, causing significant damage to the surface, including flooding, the appearance of subsidence lakes, the destabilization of buildings, cracks in the ground, and induced earthquakes. The accumulation of excavated material after sorting operations caused vast slag heaps, creating hills where there were none before, as in Belgian Hainaut and northern France (Troch, 2018). In metal mines, for example in copper or gold mining, these slag heaps may contain heavy metals and harmful minerals such as arsenic and mercury. Bauxite mine dumps consist of harmful red sludge containing, among other things, lead, cadmium, and mercury, which can seep into drinking water. Mine waste from uranium mining is radioactive and poses environmental and health challenges.

Techniques in Open-pit Mining

Developments in mining technology, including the use of mechanical excavators and dynamite in the late nineteenth and early twentieth centuries, enabled the development of open-pit mining in many parts of the world.

As Timothy LeCain and Brian Leech explain in their respective monographs on copper mining in Butte, Montana, mining companies took advantage of the surge in demand and price for copper in the late nineteenth century — the period in which the world was electrified — to exploit very low-grade copper deposits; this was done on the order of 2% per ton of earth around 1900 (LeCain, 2009; Leech, 2018). These low-grade deposits also became profitable through the use of industrial explosives in combination with excavators as well as the increased rate of ore extraction and processing. In this process the underground mines were abandoned, since there, the costs of drilling and maintaining drifts hampered production, and mining copper was only done with open-pit methods. These mines were large craters from which the soil had been removed and which were then treated with harmful chemicals to separate the copper from the waste rock. The soil contaminated by the copper processing process were stored in piles and became toxic piles of waste rock. Open-pit mining was soon applied to coal, gold, diamond, and gemstone mines throughout the rest of the world. The introduction and improvement of bucket excavators and large transport trucks as early as the 1920s made open-pit mining even more profitable, so that it became the norm in remote and sparsely populated areas (Debrabant-Malolepszy, Dumont, Six, 2013).

Dynamite and improved industrial explosives made it possible to exploit deposits that had previously been inaccessible. Since the 1950s, the mountaintop removal technique has been used to develop coal seams in mountainous regions. This is the case, among others, in the Appalachians, where mountains have gradually been destroyed to allow coal mining.

Processing

Useful minerals are rarely pure and are not always of consistent quality. For example, it is very rare to find native copper or silver; the metal is often combined with other substances within a mineral. Manufacturers must therefore treat the ore before marketing it.

During the twentieth century, the development of coal washing plants or coal preparation plants contributed to the increase in world coal production. Here the fuel was separated from the waste material via a complex system of vibrating screens and basins of dense liquor, usually magnetite. The washing of ores is a complex and expensive operation, which fulfils industrialists' requirements for the purest possible ores. Dense media are used to separate industrial and domestic coal from the slurry-slurry being defined as all sludge from a washhouse, regardless of its concentration, fineness, clean coal content, or origin. The slurry is dispersed across settling basins, resulting in significant pollution. Polluted water can spread to the land around the ponds; the slag heaps made up of residue from the sorting and washing plants are toxic; and the water that runs along these slag heaps is laden with pollutants which can infiltrate the water table and flow into streams and drinking water reservoirs (Troch, 2018).

In metal mines with low-grade ore deposits — this is somewhere in the region of less than 1% ore per ton of rock these days — or those with ore of poor quality, concentrators are used to process enormous amounts of soil and rock to recover useful minerals. These mines also use settling ponds, sometimes vast ones, to separate the ore from undesired substances. This technique causes major environmental problems, because the water in these ponds is full of harmful particles such as arsenic or lead (LeCain, 2009).

Another process that is problematic from an environmental point of view is leaching. This technique dates back to the Middle Ages and involves treating an ore in a series of chemical reactions meant to separate useful ore from its gangue and then separate it from worthless materials. It is mainly used in metal mines including copper, nickel, and uranium mines. In this technique, the gangue is crushed and stored in piles on land surrounded by dikes. A dissolving solution, often cyanide, is spread over the pile and then percolated into it. As it percolates, the solution agglomerates the useful ore, which is then recovered at the bottom of the pile for processing in enrichment units to extract the useful material. Another leaching technique called 'in situ leaching', has been applied on a large scale since the 1960s. The process is similar to heap leaching except the solution is distilled directly into the underground deposit through a wellbore. Once percolation is complete, the leach product is pumped through a second wellbore. While this technique avoids the creation of large contaminated dumps and settling ponds, it presents risks of groundwater pollution.

The processing of uranium poses another problem, which has been discussed, among others, by Gabrielle Hecht in her study of uranium mining in Africa. Once the ore is crushed, a series of chemical treatments transforms the raw ore into a form of uranium oxide called 'yellowcake'. The tailings produced in this process (e.g. chemical products) were often turned into sludges, which could then be stored in sealed dump lakes, piled up in the open air, or discharged directly into rivers, which turns them into serious hazards to both health and the environment (Hecht, 2012).

Small-Scale and Artisanal Mining

The world of the extractive industries is very diverse, including not only large mining complexes but also small-scale mines. The operators of the latter extract materials like gemstones, precious metals, decorative or collector's minerals, or minerals intended for industrial use such as cobalt and coltan.

While these mines have been in existence for a long time, they flourished throughout the world particularly from the second half of the nineteenth century onwards. The best-known examples are the California gold rushes of 1849, those in Australia of 1851, and those in the Witwatersrand, of South Africa in 1886; but there were also rushes for other types of deposits such as diamonds and silver. Mines were opened according to circumstances; for example, small mines were opened in Europe during the Second World War to produce the minerals needed by the metallurgical industries, such as manganese or fluorite, due to the shortages caused by the war. In these small artisanal mines, compared to industrial mines, the techniques were simple. Shafts were dug with shovels and excavators and explosives, the minerals were extracted with motorized machines or with pickaxes, hammers, and chisels. Ores were frequently treated with toxic materials, which could cause localized health and environmental problems (Lahiri-Dutt, 2018).

In mines for collectors' minerals, precise techniques are used, and the rate of extraction is slow. Miners use hammers, sledgehammers, and chisels as well as, sometimes, grinders and other tools that can slice through the matrix without damaging the crystals. This type of mining may take place in old industrial mines such as the Sweet Home silver mine in Colorado, which opened in 1873 and is now being mined for its rhodochrosites for the collectors' market, or the fluorite mines in Durham County and Cumbria in northern England, where industrial mining in the eighteenth and nineteenth centuries was replaced by mining for collectors. In some areas, these mines provide significant additional income for local people, such as the vanadinite mines of Mibladen in Morocco.

Legacy: Post-mining Techniques

Post-mining is a term generally used to refer to the phase following the partial or complete cessation of extractive activities in a given territory.

Compared to the eighteenth century and even the early nineteenth century, the environmental footprint of the extractive industries is changing. In addition to the expansion of surface buildings and industrial railway networks, environmental impacts are increasing. In the early twentieth century, mine subsidence reached the order of several metres. Man-made water surfaces expanded, as did subsidence lakes and marshes, and settling ponds for the shunting and washing plants. The slag heaps grew in weight, volume, and height, and some of them rose to 30 or 40 metres in height at the beginning of the twentieth century.

Between 1890 and 1910, European mining engineers and coal mines tried to limit the impacts of subsidence on the surface using various techniques, from backfilling to the installation of water lifting pumps that would need to operate for several centuries to stabilize the water level in subsided areas threatened by flooding. These rehabilitation techniques were in the tradition of 'technological fixes', that is, technical solutions characteristic of the late nineteenth and twentieth centuries. These techniques are intended to mitigate the effects of mines on the surface, but ultimately, they increase the exploited areas and add to the environmental damage (Troch, 2018).

Take, for example, the hydraulic backfilling that was popular in coal mines in the first half of the twentieth century (Figure 1). Theoretically, 90% of the volume of coal mined can be backfilled. The method first appeared in the United States, in the anthracite mines of Pennsylvania, before establishing itself in Europe, specifically in

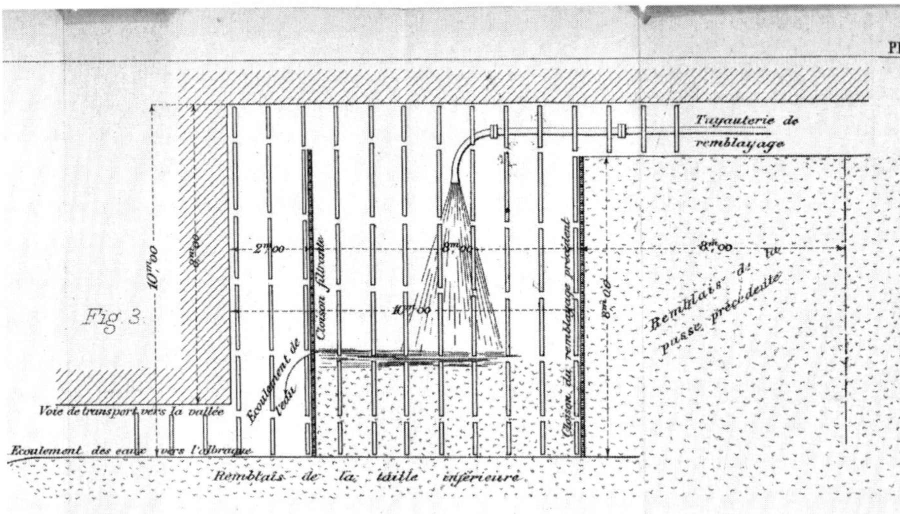

Fig. 1. Diagram illustrating the hydraulic backfill process. Baijot, Achille and Adolphe Demeure, 'Notes recueillies dans quelques mines du bassin Rhénan-Westphalien, dans le Nord et dans le Pas-de-Calais sur le remblayage hydraulique', *Revue universelle des mines, de la métallurgie, des travaux publics, des sciences et des arts appliqués à l'industrie*, 4th series, 26 (1909), ill. 4.

Prussian and Austrian Silesia in the 1890s. In 1900, hydraulic backfilling was used systematically in the Myslowitz mines in Germany. The practice spread rapidly across the Nord-Pas-de-Calais basin in the years 1900 to 1905. In Belgium, the introduction of hydraulic backfilling was more complex because of the nature of the deposits.

In hydraulic backfilling, the voids are filled by sending the waste material, mixed with water, directly into the galleries through a system of underground steel pipes that are supplied by electric pumps and lined with ceramics to limit wear. These backfills are mostly composed of washing plant shales, boiler slag, crushed pit stones, burned and unburned waste rock, crushed chalk, pure quartz sand, or clayey sands. The embankments solidify together with water to form a kind of compact sludge. The water is filtered through a dam that holds the fill in place. The use of machinery reduces the duration of the operation.

Hydraulic backfilling is also financially advantageous for the coal mines because it requires fewer workers than manual backfilling. It introduces machinery into a branch of mining that had not previously used machines, and also makes it possible to increase production capacity for a given area and to extend the working sites in depth and width. With hydraulic backfilling it is possible to carry out work underneath houses and railways, and to exploit parts of the deposit that would otherwise have been abandoned because they provided roof support and mitigated subsidence. Hydraulic backfilling, supposed to ensure the relative stability of the surface, primarily enables the collieries to maximize coal extraction.

The History of Extractive Techniques in the Anthropocene Era

The extractive techniques of the late twentieth and early twenty-first centuries also deserve historians' attentions. While the principle of extraction has not fundamentally changed, the volume of mining resources extracted during this period and the rate at which this is done are unprecedented in the history of mankind. Never before has so much mineral material been exploited in the world, and never before have we needed so many minerals and metals to maintain a certain standard of life (Bardi, 2014). In the twenty-first century, the scope of extraction is also profoundly different from that of the nineteenth and twentieth centuries (Black, 2018). If quarries are still relatively numerous on the European continent, mining is relatively limited, as it is confined to a few remaining coal mines and metal or mineral product mines. Mining moved from Europe to the Americas at the beginning of the nineteenth century, then to Africa at the end of the nineteenth century, and most recently to Oceania and the Far East since the second half of the twentieth century. China is the only nation that extracts the vast majority of the mineral resources needed to support contemporary lifestyles, including rare earths as well as aluminium, copper, iron, lead, tungsten, molybdenum, and vanadium. Far from being a reality of the past, the mining sector is thriving and is key to characterizing the new geological era of the Anthropocene, an era in which human activities have such an impact on the planet that they will leave their mark on the geological layers.

For example, prospecting and extraction methods are now adapted by the extractive industries to computer technology and automation, and extraction machines and transport trucks can now be controlled remotely. Processing facilities are being upgraded to the point where deposits containing less than 0.01% ore can be profitably exploited. The resulting, enormous mines are visible on satellite images, creating gaping holes in the Earth's surface. The demand for minerals is such that techniques for seabed mining are being investigated. The first deep-sea mines are expected to be operational at the end of the 2020s. Their environmental impact is yet unknown. Mining techniques are ultimately part of what Paul Josephson calls 'brute force technologies', that is, techniques of the domination and conquest of nature (Josephson, 2002). They are even at the heart of it. Their history therefore deserves our full attention.

The history of extractive techniques is complex because mining covers different realities. Yet the techniques used by the extractive industry from the nineteenth century to the present day have one thing in common, namely the extraction and processing of an ever-increasing volume of mineral resources. As we have seen, prospecting and mining techniques evolved rapidly from the 1840s onwards, gradually replacing the previous processes and techniques. The massive use of coal by industrial societies that furthered the production of cheap iron and steel led to the creation of new demands for mineral materials. The widespread electrification of the mining sites and the automation of task from the beginning of the twentieth century onwards allowed mining to expand. It also allowed work to be around the clock. The development of industrial explosives in the nineteenth century, and above all the introduction of dynamite, opened up new mining perspectives by opening up the possibility of open-cast mining and mountaintop removal. Ore processing techniques developed since the end of the nineteenth century have allowed the exploitation of deposits with increasingly lower ore grades, of the order of less than 1% for rare and precious minerals (LeCain, 2009). All of this aids the exploitation of almost all underground resources. Opposite these large mining complexes, there is also a whole network of small-scale and artisanal mines in which more limited techniques are employed (Lahiri-Dutt, 2018).

The unprecedented movement of mineral materials brought about by the development of the mining sector in the nineteenth and twentieth centuries has created many environmental problems that are difficult for mining operators to manage. Despite increasingly sophisticated backfilling techniques, mine subsidence irretrievably affects the surface of mining areas. Heavy metal slag heaps and settling ponds pose a danger to the surrounding population. The gaping holes created by open-pit mines are only rarely filled and rehabilitated. In some mining areas, for example northern France and Belgium, water pumping systems have had to be set up with the prospect of several centuries worth of pumping to avoid certain subsided areas being flooded (Aprile et al., 2015). The forthcoming opening of a new extractive frontier with seabed mining points to new economic and ecological challenges for the twenty-first century.

References

Aprile, Sylvie, Mathieu de Oliveira, Béatrice Touchelay, and Karl-Michael Hoin, eds, *Les Houillères entre l'état, le marché et la société. Les territoires de la résilience, XVIIIe–XXIe siècles* (Villeneuve d'Ascq: Presses Universitaires du Septentrion, 2015).

Bardi, Ugo, *Extracted. How the Quest for Mineral Wealth is Plundering the Planet. The Past, Present, and Future of Global Mineral Depletion. A Report to the Club of Rome* (White River Junction, VT: Chelsea Green, 2014).

Black, Megan, *The Global Interior. Mineral Frontiers and American Power* (Cambridge, MA: Harvard University Press, 2018).

Centre historique minier de Lewarde, *Des machines et des hommes. Émergence et mise en œuvre des innovations techniques dans les mines de charbon* (Lewarde: Centre historique minier de Lewarde, 2013).

Cooper-Richet, Diana, *Le Peuple de la nuit. Mines et mineurs en France (XIXe–XXIe siècle)* (Paris: Perrin, 2011).

Debrabant-Malolepszy, Virginie, Gérard Dumont, and Audrey Six, *Des machines et des hommes. Les techniques d'exploitation dans les mines du Nord-Pas-de-Calais* (Lewarde: Centre historique minier de Lewarde, 2013).

Demey, Thierry, *Sur les traces du diamant noir. Histoire du bassin minier franco-belge* (Brussels: Guide Badeaux, 2016).

Gomez, Rocio, *Silver Veins, Dusty Lungs. Mining, Water, and Public Health in Zacatecas, 1835–1946* (Lincoln, NE: University of Nebraska Press, 2020).

Gray, Steve, *Steam Power and Sea Power: Coal, the Royal Navy, and the British Empire, c. 1870–1914* (Cambridge: Cambridge University Press, 2017).

Hecht, Gabrielle, *Being Nuclear. Africans and the Global Uranium Trade* (Cambridge, MA: The MIT Press, 2012).

Josephson, Paul R., *Industrialized Nature. Brute Force Technology and the Transformation of the Natural World* (Washington, DC: Island Press, Shearwater Books, 2002).

Lahiri-Dutt, Kuntala, ed., *Between the Plough and the Pick: Informal, Artisanal and Small-scale Mining in the Contemporary World* (Canberra: The Australian National University Press, 2018).

Leech, Brian, *The City that Ate Itself. Butte, Montana, and Its Expanding Berkeley Pit* (Reno, NV: University of Nevada Press, 2018).

LeCain, Timothy J., *Mass Destruction. The Men and Giant Mines that Wired America and Scarred the Planet* (New Brunswick, NJ: Rutgers University Press, 2009).

Rainhorn, Judith, ed., *Santé et travail à la mine, XIXe–XXIe siècle* (Villeneuve d'Ascq : Presses Universitaires du Septentrion, 2014).

Shulman, Peter, *Coal and Empire. The Birth of Energy Security in Industrial America* (Baltimore, MD: The Johns Hopkins University Press, 2015).

Troch, Kevin, 'Ne pas grever l'avenir au bénéfice du présent. Une histoire environnementale de l'extraction du charbon en Belgique et en France de la fin du 18e siècle à l'entre deux-guerres: un développement non soutenable. L'exemple du Couchant de Mons et du Valenciennois' (unpublished doctoral thesis, University of Lille and University of Namur, 2018).

Veyron, Thierry, *Essai sur l'évolution technique des houillères françaises et belges: 1800–1880* (Paris: L'Harmattan, 1999).

Wu Shellen, Xiao, *Empires of Coal: Fueling China's Entry into the Modern World Order, 1860–1920* (Stanford, CA: Stanford University Press, 2015).

FLORENCE HACHEZ-LEROY

Heavy Industries

From Steel and Molecules to the Cloud

The expression phrase 'heavy industry' appeared following the acceleration of industrial development from the nineteenth century onwards, in contrast tow what was categorised as 'light' industry. The distinction was based on the levels of investment necessary for the construction and operation of the industries concerned. In this construct, the sectors associated with the production and transformation of raw materials, such as mining, metallurgy, electrometallurgy, papermaking, and first-stage chemical processing industries were thought of as 'heavy industries'. As their techniques evolved, certain other industries came to qualify for this classification as a 'heavy' industry, for example, the IT industry in the twenty-first century. While this category offers certain analytical advandages for economists and financial experts, it is difficult from the point of view of the history of technology to give a coherent account of a term which gathers industries of a highly heterogeneous nature.

Historians have paid particular attention to the iron and steel industries. After all, the uses of ferrous metals are both ancient and multifarious. The history of papermaking is another sector with a long history and a rich literature. In both these cases, the geographical territories covered by historical studies are vast, in keeping with the diversity of the techniques employed and the relative ease of their implementation. Unlike these two sectors, non-ferrous metals have elicited fewer studies to date. The chemical industry is another special case. The exceptional variety of production techniques and products, especially since the 1950s, and the broad range of spheres (from the food and textile industries to the cosmetic and health industrie) and contetxts (both civilian and military) where its output is being put to use — has given rise to much research, even if important areas remain unexplored.

Since it is such an extensive field, I have decided to provide an overview of it rather than a aim for a comprehensive study although I will explore some points in more detail. The technical history of the heavy industries has often been presented through the discoveries that marked their development (Kranzberg, Pursell, 1967; Daumas, 1978; Gille, 1978; Caron, 1985). However, the prism of innovation is not adequate for a correct account of the history of technology. Uses and practices on the receiving end of these industries must also play a key role. And the question of the environmental impact of these industries, which is sometimes catastrophic, must also be taken into account. After briefly reviewing the classic historiography on

Global History of Techniques, ed. by Guillaume Carnino, Liliane Hilaire-Pérez and Jérôme Lamy, Global Matters Culture Matérielles, histoire des techniques et circulation globale des savoirs, 9 (Turnhout: Brepols, 2024), pp. 297–314
BREPOLS PUBLISHERS 10.1484/M.TECHNE-EB.5.129813

the question, I will examine the recent approaches to heavy industries that address uses and the environment and conclude by looking at industrial archaeology, which raises questions that help shed light on the history of heavy industries and their technologies.

Heavy Industries, Inventors, and Innovations

Often the history of the production of iron, cast iron, and steel is centred on innovations — Beaumont Neilson's hot blast, Bessemer's steel process, Siemens-Martin's open-hearth process, Paul Héroult's electric furnace — which are seen as many historic milestones in the the chronicle of iron and steel production techniques (Edgerton, 1998). Spurred on by the development of the metallographic sciences in the early nineteenth century, the quality and quantity of production increased especially in the United Kingdom (Smith, 1965). A series of improvements were introduced to production processes, such as the use of the hot blast in the furnace or the recovery of gases at its top in order to reintroduce them to the furnace.

Iron and steel making are considered key among the profound technical changes that took place at the end of the eighteenth century. This sector was also at the heart of the industrial crisis that started in the 1970s. Historical research has been carried out at different levels — local, national, regional, and comparative — alongside monographic studies on particular companies or sites (Gille, 1964; Kahlow, 1991). The international and global approaches have highlighted the early development of the sector and how it has been shaped by a series of exchanges at the technical, financial, and human levels (Moreau, 2005; Barthel et al., 2019). This sector was of particular interest to historians during the second half of the twentieth century. A first wave of studies in the 1950s and 1960s was established in particular through the publication of the periodical *Revue d'Histoire de la sidérurgie*, which concentrated on the modern era. A second wave of literature was spawned by the crisis of the late 1970s in Europe and the United States. The 1990s then produced more studies on the sector as a whole or on particular companies of the time, this time with an approach in line with that of the *Business History* journal (Kipping et al., 2001; Kipping, 1996, 1997; Lindner, Pestre, 1996). Finally, the beginning of the twenty-first century introduced global approaches covering the early modern and modern periods (Evans, Rydén, 2005; Dillmann et al., 2011).

The rolling technique, which is very important in this sector, has been studied particularly at a series of national scales. An overview of the continuous use of hot casting between 1920 and 2000, which was published in 2012, showed how the technique, once transferred from the United States to Europe, was then improved at a local level by European plant suppliers, and how finally the plant designs spread globally, and particularly in Asia (Aylen, Ranieri, 2012). If metallurgical science was dominated by European research until WWII, US companies caught up in the inter-war period to become key competitors in the field. The CorTen (short for 'corrosion resistance and tensile strength') steel developed by the US Steel Corporation in 1933 exemplifies this expertise. This weathering steel, which is lighter in weight and had been designed for rail applications before being also used in the construction the US Steel Tower in Pittsburgh in 1971, has been taken up enthusiastically in architecture and contemporary art around the world since the end of the twentieth century.

The historiography of non-ferrous metals is comparatively sparse with some metals having received more interest than others (Garçon, 1998; Turnbull, 2008; Bencivengo, 2014). The most prominent amongst the non-ferrous metals in relation to its presence in the earth's crust as well as its uses is aluminium, which has received particular attention since 1986 for the centenary of the Heroult/ Hall process (Peterson and Miller, 1986). It is worth noting here that aluminium production uses mining activities for extracting bauxite and chemistry for obtaining alumina, and that it requires considerable electrical energy for electrolysis, thus necessitating the construction of dams and hydro-electric or thermal power plants. In France, a first series of studies on this topic was published by geographers and industrial historians from the 1950s onwards (Gignoux, 1955; Veyret-Verner, 1956; Chabert, 1973; Morsel, 1973). The history of electrolysis and of the industrial developments of aluminium on an international scale was the object of a substantial publication in 1964, written by engineers at Pechiney (Barrand, Gadeau, 1964).

During the same period, some work was done in the United States on the history of the Alcoa company and the inventor of the eponymous process (Carr, 1952; Cowles, 1958; Edwards, 1955). In 1986, the centenary of the invention of the Hall/ Héroult process prompted new research and publications, initially in the United States (Nunn, 1986; Smith, 1988; Graham, Pruitt, 1990) and then in France, thanks to the creation of a dedicated institute for history financed by the aluminium sector (Grinberg, 2017). This was the context in which former Pechiney engineers, with the help of academic historians, published a technical history of aluminium production in France (Morel, 1991; Plateau, 2004).

Aluminium production techniques have remained traditional in so far as the process is still essentially the same as it was in 1886 although improvements have naturally been made to the process with the introduction of new building materials, new electrolytic cells, and new ways in which materials are introduced into the latter. France has been in the lead in this area (Carbonell, 2012). The characteristically monopolistic structure of its industry has encouraged studies in the form of case studies (Grinberg, Hachez-Leroy, 1997; Hachez-Leroy, 1999; Lesclous, 1999; Le Roux, 1998). Since 2000, the geographical areas covered in the historiography have expanded to cover northern Europe, Latin America, Japan, and Africa, althought in these cases the approaches are often linked to economic history (Romainville, 2009; Fridenson, Hachez-Leroy, 2017, Awono Zinga, 2017; Miwa, 2017; Rougier, 2017).

The building of steam engines required by the industrial development of the second half of the nineteenth century led to considerable progress in the realm of mechanical engineering (Burstall, 1965). Internal combustion engines and petrol-driven engines, steam locomotives and railway rolling stock, machine tools of ever greater efficiency and ever larger dimensions (like the steam hammer developed in about 1840 by François Bourdon), hoisting devices like port cranes (powered from about 1850 onwards), as well as other transformative inventions suggest how specific machines grew to be large in scale. New apparatus was developed, like the overhead crane used for lifting heavy items towards the end of the 1890s, as was new equipment for railway building, like the steam-powered excavator (first patented in 1859) or devices for tunnelling, such as tunnel boring machines (created in 1845 for the Frejus railway tunnel). After the Second World War, similar developments took place when the creation of networks of pipelines involved the invention of combined machines that could dig trenches and lay pipes. Gradually, machines in engineering workshops were driven by hydraulic systems before being powered by electricity, which afforded greater power and improved the smoothness and regularity of operations. Rolling mills that flattened slabs of metal were originally operated by teams of workers who placed the slabs successively on either side of the rolls, but in the 1940s these were replaced by hot and cold rolling mills comprising a series of stands that allowed thick and voluminous slabs of metal to be flattened more rapidly. The processing of metal sheeting, which had first appeared in 1909 for the production of rails, was further improved in the 1950s, particularly for aluminium, with the invention of a process allowing for the continuous casting of liquid metal, which produced thin sheeting at a much faster rate. This development was aided by aeronautical inventions, such as the development of the Concorde at the end of the 1960s, and by the growth of the food packaging industry for mass markets. At the end of the nineteenth century, electrical engineering also emerged as an industrial sector in its own right, and the earliest institutions for electrotechnical training appeared in Great Britain and the United States. From the 1950s onwards, the entire field was transformed by the development of punch card technology and then microelectronics, facilitating the automation of work processes.

It is difficult to attempt writing an overview of the history of the chemical industy post-World War II, whereas such summaries are readily available for the nineteenth century and the period up to 1945 (Haber, 1958, 1971; Ihde, 1964; Liebenau, 1984; Brock, 1993; Sakudo, 2011; Langlinay, 2017). Dominated by German firms prior to 1914, the sector subsequently became more competitive, and many industrial research laboratories were gradually set up in other countries. The American industry, which originally relied on German raw materials and German patents, became independent during the First World War and developed its own products (Hounshell, Smith, 1988; Ndiaye, 2001). Case studies of different companies and of their departments of research and development provide some idea of the complexity of their research and industrial processes (Cayez, 1988; Bertrams, 2007; Bertrams et al., 2013). Historical research into different countries' chemical industries has underlined the East and West connections, as illustrated by the case of Hou Tê-pang, who received a grant to study chemical engineering at the Massachusetts Institute of Technology and obtained his doctorate at the University of Columbia in 1921 (Reardon-Anderson, 1986). Upon his return to China, he contributed to the development of the Chinese chemical industry, but still maintained close contact with the community of American chemical researchers. In 1922, the Golden Sea Research Institute of Chemical Industry, the first laboratory of its kind in China, was founded. The large number of biographies of chemists that has been published underlines the importance attached by past as well as present-day historians to identifying historical inventors (Dumas, 1965; Defries, 1970; Viel, 2006). Such research has the merit of underlining the uninterrupted circulation of knowledge between scientists that has taken place since the eighteenth century. More recent work has moved on from this potentially rather hagiographic approach to a cultural appreciation of the profession as a whole (Grelon, 1991; Fell, 2000; Knight, Kragh, 1998). A handful of publications have provided a global vision of the sector since 1945 (Bensaude-Vincent, Stengers, 1993; Travis et al., 1998; Bensaude-Vincent, 1998; Galambos et al., 2006).

Innovations in the chemical sector have undoubtedly received most attention because their possible applications reach across all human activities: health, food, transport, agriculture, etc. In the first quarter of the twentieth century, the production of the main intermediary chemicals, such as ammonia or nitric acid, was brought to industrial levels at the same time as synthetic fertilizers were developed. In the second half of the twentieth century, oil refinery developed into the production of different derivatives resulting in the manufacture of plastics and pharmaceuticals and caused a major break. Gradually, the distinction between general chemistry, inorganic chemistry, and organic chemistry (based on coal and petroleum) gave way to a classification by concepts and methods, i.e. molecular and biological chemistry, materials chemistry, and interfacial chemistry. Within molecular and biological chemistry, fine chemistry stands out: it concerns complex products with high added value (dyes, cosmetics, flavourings, pharmaceuticals, and phytosanitary products). Materials chemistry includes part of the former organic mineral chemistries. The field is still concerned with metals and alloys, but also with polymers, e.g. in innovations such as polymer alloys and biodegradable materials. Interfacial chemistry harnesses chemical reactions happening at the interface of two phases of matter — liquid-solid, solid-solid, solid-gas. Here, heterogeneous catalysis is crucial to many industrial processes: refining, petroleum chemistry, the synthesis of polymers, the production of fertilizers, etc. (Ihde, 1964; Bensaude-Vincent, Stengers, 1993; Lowe, 2016).

Two further sectors are worth considering here, that is, papermaking and communications technologies. The first witnessed an important technical change towards the middle of the nineteenth century with the substitution of wood for textiles as its raw material and the important influence of electrical energy. Due to the relative simplicity of papermaking paper mills were modestly sized and numerous at first, before larger production plants became necessary (André, 1991, 1996, 2013; Amsalem, 1983; Angus, 1976). The market in paper pulp has seen a specialization of specific regions worldwide, e.g. as tree growing or pulp production regions, and for the paper manufacture itself.

The second abovementioned field, techniques of communication, have started a whole new branch of heavy industry with the development of computer technology and, more specifically, the internet (Abbate, 1999; Le Crosnier, 2008). The transport and storage of data now requires impressive infrastructures, and enormous capital investment is mobilized today that is comparable to that invested in the transportation of electricity. Transatlantic cables, satellites and underwater data-centres — all these aspects of the industry are still evolving. The end of the twentieth century was marked by the expansion of digital techniques and digital objects, which have led to major changes in contemporary society (Carnino, Biagini, 2010). Algorithms themselves have become technical objects which historians of technology must now take into consideration, and for which novel issues of conservation are being raised (Di Cosmo, 2017; Abiteboul, Dowek, 2017; Di Cosmo, Hachez-Leroy, 2018).

Forms of Organization and Modes of Production

By their very nature, heavy industries are capital-intensive companies, which must take measures to secure their supplies of raw materials and develop their markets. From the late nineteenth century onwards, certain arrangements were made and agreements reached between producers at both national and international levels. In France these are called *cartels*, in Germany *Konzerne*, in the United States 'trusts', and in Japan *zaibatsu*. Depending on the sector, these organizations have turned out to be both solid and lasting. Their dominant positions have been vigorously contested in some countries, e.g. in the United States, where the Sherman antitrust act was passed in 1890. In Europe, it was only after the 1957 Treaty of Rome that the countries of the newly formed European Economic Community could come to an agreement with regard to fighting trust organizations. Lawyers were particularly interested in these arrangements both during the inter-war period and in the years prior to the Treaty of Rome. There was debate as to the positive or negative effects of such organizations for other firms and consumers. Historians became interested in these forms of organization particularly thanks to the initiative of Akira Kudo and Terushi Hara at the eighteenth Fuji Conference at Tokyo (Kudo, Hara, 1992). Research showed that the arrangements that were the most solid were those that were formalized in oligopolistic sectors such as was the case for aluminium or cellulose, and that such organizations here could remain reliably stable for several decades. Many were based in Switzerland. At the turn of the twenty-first century the European Union established significant tools to combat this type of trusts, which had resurfaced not only in industrial sectors but also in the service industries and banking (Schröter, 2013; Hachez-Leroy, 2015).

As far as their internal organization is concerned, prior to the First Word War companies moved towards vertical, upstream integration and then switched to downstream integration after the Second World War, before giving way to externalization and sub-contracting from the 1980s onwards, which was done under the influence of the growing financialization of capital. Alongside a concentration on 'core business', tasks that are seen as peripheral, such as transport services, logistics, or accountancy, are increasingly outsourced. Between 1984 and 2003, the proportion of companies in France that sub-contract work grew from 60% to 87%. Certain sectors, such as aeronautics or the motor car industry, have sub-contracted for a long time (Baudry, 2003). The construction of homogeneous political and economic spaces such as the European Union, which eliminates internal customs barriers, has had a considerable impact on the organization of industrial production, particularly where heavy industries are concerned. Even prior to the Second World War, European companies could often be distinguished from their American counterparts by the fact that they were still family-owned and family-run, like Krupp, Siemens, Vickers, and Schneider. In contrast to the United States, since the nineteenth century, European countries also had had a strong tradition of state intervention and programmes of nationalization. German railways were nationalized in 1880. This interventionist tradition declined overall after 1945, but the rates of decline varied from one country to another. The construction of the European Union was accompanied by measures which had a strong influence on the iron and steel industry, the arms industry, and on aircraft construction. Many large-scale European companies now share out their supply and production activities between different countries according to financial considerations such as the cost of labour or rates of taxation. The availability of raw materials or energy is no longer the main issue (Cassis, 2003). This is by no means limited to Europe. The liberalization and globalization of the economy which began in the 1970s was also based on commercial exchanges between businesses, representing between a third and a half of all the commercial exchanges in the world. The different stages in the manufacturing of goods, just as in the provision of services, have now been shared out between companies' different sites (Michalet, 2007). Regulations and world-wide governance can no longer operate independently from multinational companies, and in particular those involved in heavy industries, including the internet. At the turn of the twenty-first century, the relationships between the different actors — states, companies, and both governmental and non-governmental international organizations — became considerably more complex with regard to questions of law and power.

Uses and the Environment

The history of uses and users represents a field that has not to date been investigated properly for the industrial products of heavy industry, be they finished or semi-manufactured (Scranton, 1997; Oudshoorn, Pinch, 2003; Chatriot, Chessel, Hilton, 2004; Chauveau, 2006; Oldenziel, Hård, 2013). For metals, this includes the history of alloys and the history of metallography. Research on steel has mainly been concerned with imposing structures such as bridges or buildings, or occasionally smaller objects such as food tins (Bruegel, 2002; Mioche, Woronoff, 2006). For aluminium, there are no detailed studies allowing us to understand the main stages and conditions of the development of transformation techniques such as wire drawing or stamping, or the processes used for giving aluminium a special surface or colour. Generally, the long-term and international histories of ferrous and non-ferrous metallic alloys remain sparse (Cobb, 2010; Chemello et al., 2019). Approaching the question from the perspective of the market has offered a more global vision on technological developments, encompassing the technical, cultural, and social aspects of the materials studied (Friedel, 1983; Meikle, 1995; Hachez-Leroy, 1999; Schatzberg, 2003).

The development of new products or new substances also involves taking measure of the industrial risks involved (Boullet, 2006; Varaschin, 2007; Chambless, 1985) and transparency with users and customers, even at the risk of alienating certain customers (Østby, Zachmann, 2011; Oreskes, Conway, 2010). This is the reason why technological history cannot be separated from environmental history (Smith, 1997; Boyd, 2001; Fressoz, 2012). Heavy industries have had the most significant environmental impact. Pollution, occupational diseases, and the food safety are questions that are at the heart of heavy industries (Lefaux, 1968; Needleman, 1991; Homburg et al., 1998; Massard-Guilbaud, 2010; Buzzi et al., 2006; Hachez-Leroy, 2013; Exley, 2013). The applications of chemistry are immensely varied, but studying specific controversies allows us to work our way back from the product to its manufacture and design (Schuck, 1987; Rainhorn, 2012; Pessis, Sezin, Bonneuil, 2013; Snow, 2017; Blanc, 2007). This is the case, for example, for bisphenol A in babies' feeding bottles and additives in foodstuffs, such as nitrites (Meikle, 1997; Vogel, 2009; Hachez-Leroy, 2019). These new perspectives balance out the often rather over optimistic vision of technological history that tends to dominate the existing literature (Sheller, 2014).

The Contributions of Industrial Archaeology

In the 1950s, industrial archaeology began to open up new research perspectives and new ways of understanding the history of technology (Buchanan, 1972; Daumas, 1980). Heavy industries are particularly relevant here because of their scale, complexity, and environmental impact, if we think of the mining and metallurgical region of Essen in the Ruhr area or the copper producing region of Chile for instance (Preite, 2011; Woronoff, 2015). The high visibility of such sites has given rise to some stimulating thought about the notion of technical and industrial landscapes as part of the broader notion of the cultural landscape such as that developed by UNESCO (Preite, 2005; Michon, 2008). Nominations for UNESCO world heritage status need to be made in more and more detailed forms, and this has provided opportunities to reconsider the history of technology in a more global approach, comparing different geographical spaces, sometimes without the West acting as the sole reference. For the definition of criteria for outstanding universal value, technical history is essential, as is the question of its effects and consequences. Moreover, how should one take into consideration environmental pollution or industrial waste products (Preite, 2018; Beazley, 2007)? The inclusion of the nuclear testing site on Bikini Atoll to the UNESCO world heritage register (2010), like the addition of the German and Czech zones where uranium was mined and transformed (2019), are stimulating developments in the history of technology and its representations (Albrecht, 2017). The perceptions of industry and technology and their place in society can also provide new approaches to the history of technology (Woronoff et al., 2002; Pierrot, 2010). The combination of archival sources with field observations can add to our knowledge about technology and the ways in which we understand it (Fluck, 2008; Pillet, 2009; André, 2017). Additionally, the issue of preserving material traces offers many opportunities for developing our understanding of technical history. What should be kept, and how? The conversion of industrial buildings for different uses is important and raises many questions (Douet, 2012). How do we decide which sites should be redeveloped, which should be preserved in their original state? What should be kept in a former industrial site in order for its industrial past to remain legible? How can changes to the buildings' fabric be done in a way that is reversible? There are also important debates regarding the formation of museum collections, whether objects should be put in collections with other objects of their kind, or whether they should be kept in the context in which they were found (Ballé, Cuenca, Thoulouze, 2010). Does a steam engine tell the same story when it enters a museum collection as when it was in use? And how 'authentic' is a machine tool when we know that the very essence of industrial operation is to maintain, repair, and improve machines throughout their life in use (André, 2005; Buffa, 2015)? Looking at objects themselves can also be a fertile approach (Amouretti et al., 1984). This is the case, for example, for paper manufacturing. The analysis of watermarks provides information about the history of the paper's production and the techniques and people involved in the industry (Coural, 2011; Wenger E, 2022). Oral history and intangible cultural heritage are two vital aspects for the contemporary history of technology (Ruess, 2016).

Finally, today, digital tools newly inform our investigations of technical objects and their study, whether this is primarily to serve the interpretation of the objects for the general public, through 3D models or virtual reality, or used for recording the object, for example, in databases (Laroche, 2009; Benoit et al., 2009; Saou-Dufrene, Barbier, 2014). The vast dimensions of heavy industry sites often requires digital techniques for their interpretation, e.g. to see how these sites functioned, and for their preservation in memory when their physical conservation is deemed impossible. Digital tools can play a key role in preserving, and conveying the complexity of technological systems (Fisher et al., 2018).

Finally, it is important here to underline the fact that heavy industries are all part of a global economy, particularly since they now benefit from the unprecedented development of inexpensive maritime transportation (Daudin, 2003). Nine billion tonnes of merchandise were transported by sea in 2015, compared to one billion in 1960 and 4 billion in 1990 (United Nations Conference on Trade and Development, 2018). Most raw materials and agricultural produce are now carried in containers, and the size of container ships has grown from a capacity of 1700 TEU in the 1960s to 11,000 TEU since 2006. (TEU stands for 'twenty-foot equivalent unit' and refers to the length of a container.) Tankers could carry 5000 tonnes of crude oil in the 1950s, but by the 1970s, thanks to Ultra Large Crude Carriers, they reached a capacity of 500,000 tonnes. Cargo ships, carriers of liquefied natural gas, ore carriers, oil tankers, phosphate carriers, container ships and other bulk carriers — these ships are now specialized and have multiplied to the point of representing the principal transportation mode for raw materials today (80% in 2012). Oil is still the leading product, followed by mineral ores (Levinson, 2006; Solly, 2007; Rodrigue, 2017).

References

Abbate, Janet, *Inventing the Internet* (Cambridge, MA: The MIT Press, 1999).

Abiteboul, Serge and Gilles Dowek, *Le Temps des algorithmes* (Paris: Éditions le Pommier, 2017).

Albrecht, Helmuth, 'The Heritage of Uranium Mining in the German-Czech Ore Mountains', in, *Patrimoine industriel*, Special issue, *Entreprises et histoire*, 87, ed. by Florence Hachez-Leroy, (2017), pp. 88–106.

Amouretti, Marie-Claire, Georges Comet, Claude Ney, and Jean-Louis Paillet, 'À propos du pressoir à huile: de l'archéologie industrielle à l'histoire', *Mélanges de l'École française de Rome, Antiquité*, 96.1 (1984), 379–421.

Amsalem, Michael, *Technology Choice in Developing Country: the Textile and Pulp and Paper Industries* (Cambridge, MA: MIT Press, 1983).

André, Louis, *Aristide Bergès, une vie d'innovateur, de la papeterie à la houille blanche* (Grenoble: Presses Universitaires de Grenoble, 2013).

———, 'Les forges de Paimpont retrouvées', *Patrimoine industriel*, 70, (2017), 44–45.

———, 'L'invention de la machine à papier et les débuts du papier continu en France', in *Technique et papier peint*, ed. by Gérard Schmidt and Valérie Perain (Mulhouse: Société Industrielle de Mulhouse, 1991), 87–99.

———, 'La machine à papier PM1 de Bischofszell (Suisse)', *L'Archéologie industrielle en France*, 47 (2005), 61–65.

———, *Machines à papier. Innovation et transformations de l'industrie papetière en France, 1798–1860* (Paris: EHESS, 1996).

Angus, John, *Papermaking Pioneers: A History of New Zealand Paper Mills Limited and its Predecessors*, (Mataura, NZ: New Zealand Paper Mills, 1976).

Aylen, Jonathan and Ruggero Ranieri, eds, *Ribbon of Fire: How Europe Adopted and Developed US Strip Mill Technology (1920–2000)* (Bologna: Pendragon, 2012).

Awono Zinga, Guy Grégoire, 'L'introduction de l'aluminium dans la vie quotidienne au Cameroun', in *L'Aluminium, matière à création, XIXe-XXe siècles*, ed by Patrick Fridenson, and Florence Hachez-Leroy (Tours: PUFR, 2017), pp. 135–52.

Ballé, Catherine, Catherine Cuenca, and Daniel Thoulouze, eds, *Le Patrimoine scientifique et technique: un projet contemporain* (Paris: La Documentation française, 2010).

Barrand, Pierre and Robert Gadeau, eds, *L'Aluminium par les ingénieurs du groupe Pechiney*, 2 vols. (Paris: Eyrolles, 1964).

Barthel, Charles, Ivan Kharaba et Philippe Mioche, eds, *Les Mutations de la sidérurgie mondiale du XXe siècle à nos jours* (Bern: Peter Lang B., 2014).

Baudry, Bernard, ed., *Économie de la firme* (Paris: La Découverte, 2003).

Beazley, Olwen, 'A Paradox of Peace: the Hiroshima Peace Memorial (Genbaku Dome) as World Heritage', in *A Fearsome Heritage: Diverse Legacies of the Cold War*, ed. by John Schofield and Wayne Cocro (Walnut Creek, CA: Le Coast Press, 2007), 33–50.

Bencivengo, Yann, *Nickel: la naissance de l'industrie calédonienne* (Tours: PUFR, 2014).

Benoit, Serge, Shadia Kilouchi, Alain Michel, and Stéphane Pouyllau, 'Usines 3D. La simulation pour questionner les sources et les vestiges de l'histoire industrielle', in *Virtual Retrospect 2009*, ed. by Robert Vergnieux and Caroline Delevoie (Pessa: Ausonius, 2009), pp. 31–40.

Bensaude-Vincent, Bernadette and Isabelle Stengers, *Histoire de la chimie* (Paris: La Découverte, 1993).

———, *Éloge du mixte. Matériaux nouveaux et philosophie ancienne* (Paris: Hachette Littérature, 1998).

Bertrams, Kenneth, 'Converting Academic Expertise into Industrial Innovation: University-based Research at Solvay and Gevaert, 1900–1970', *Enterprise & Society*, 8.4 (2007), 807–41.

———, Nicolas Coupain, and Ernst Homburg, *Solvay: History of a Multinational Family Firm* (Cambridge: Cambridge University Press, 2013).

Blanc, Paul B., *How Everyday Products Make People Sick: Toxins at Home and in the Workplace* (Berkeley, CA: University of California Press, 2007).

Boullet, Daniel, *Entreprises et environnement en France de 1960 à 1990. Les chemins d'une prise de conscience* (Geneva: Librairie Droz, 2006).

Boyd, William, 'The Forest is the Future? Industrial Forestry and the Southern Pulp and Paper Complex', in *The Second Wave: Southern Industrialization from the 1940s to the 1970s*, ed. by Phil Scranton (Athens, GA: University of Georgia Press, 2001), pp. 16–218.

Brock, William H., *The Norton History of Chemistry* (New York, NY: W. W. Norton, 1993).

Bruegel, Martin, 'How the French Learned to Eat Canned Food, 1809–1930s', in *Food Nations: Selling Taste in Consumer Societies*, ed. by Warren Belasco and Philip Scranton (New York, NY: Routledge, 2002), pp. 113–30 .

Buchanan, Robert Angus, *Industrial Archaeology in Britain* (Harmondsworth: Penguin, 1972).

Buffa, Géraud, 'Taper sur la machine?', *Patrimoine industriel*, 66/67 (2015), 164–77.

Burstall, Aubrey F., *A History of Mechanical Engineering* (Cambridge, MA: MIT Press, 1965).

Buzzi, Stéphane, Pierre-André Rosental, and Jean-Claude Devinck, *La Santé au travail, 1880–2006* (Paris: La Découverte, 2006).

Carbonell, Mauve , *Le blanc et le noir. Cinquante ans de recherches sur la production d'aluminium au LRF* (Aix-en-Provence: REF2.C Éditions, 2012).

Carnino, Guillaume and Cédric Biagini, *Les Luddites en France: résistance à l'industrialisation et à l'informatisation* (Montreuil: L'échappée, 2010).

Caron, François, *Le résistible déclin des sociétés industrielles* (Paris: Librairie Perrin, 1985).

Carr, Charles C., *Alcoa: An American Enterprise* (New York, NY: Rinehart & Co, 1952).

Cassis, Youssef, ed., *L'Entreprise européenne,*Special issue, *Entreprises et histoire*, 33 (2003).

Cayez, Pierre, *Rhône-Poulenc, 1895–1975* (Paris: Armand Colin, Masson, 1988).

Chabert, Louis, 'L'Aluminium en Maurienne', *Revue de géographie alpine*, 61 (1973), 31–62.

Chambless, William, *Fifty Years of Research and Service. Haskell Laboratory for Toxicology and Industrial Medicine* (Wilmington, DE: Chambless William Print, 1985).

Chatriot, Alain, Marie-Emmanuelle Chessel, and Matthew Hilton, eds, *Au nom du consommateur. Consommation et politique en Europe et aux États-Unis au XXe siècle* (Paris: La Découverte, 2004).

Chauveau, Sophie, ed., *Consommer en masse*, Special issue, *Vingtième Siècle*, 91 (2006).

Chemello, Claudia, Malcolm Collum, Paul Mardikian , eds, *Aluminum: History, Technology, and Conservation: Proceedings from the 2014 International Conference* (Washington, DC: Smithsonian Scholarly Press, 2019).

Cobb, Harold M., *The History of Stainless Steel* (Materials Park, OH: ASM International, 2010).

Coural, Natalie, ed., *Le Papier à l'œuvre* (Paris: Hazan, Musée du Louvre, 2011).

Cowles, Alfred, *The True Story of Aluminum* (Chicago, IL: Regnery, 1958).

Daudin, Guillaume, 'La logistique de la mondialisation', *Revue de l'OFCE*, 87.4 (2003), 409–35.

Daumas, Maurice, *L'Archéologie industrielle en France* (Paris: Robert Laffont, 1980).

———, *Histoire générale des techniques* (Paris: Presses universitaires de France, 1978).

Defries, Amelia, *Pioneers of Science: Seven Pictures of Struggle and Victory* (Freeport, NY: Books for Libraries Press, 1970).

Di Cosmo, Roberto, 'Software heritage: pourquoi et comment construire la grande bibliothèque du code source', *Entreprises et histoire*, 87.2 (2017), 69–75.

———, Florence Hachez-Leroy, Pierre Paradinas, eds, *L'informatique*, Special issue, *Patrimoine industriel*, 73, (2018).

Dillmann, Philippe, Liliane Hilaire-Pérez, and Catherine Verna, eds, *L'Acier en Europe avant Bessemer* (Toulouse: CNRS, 2011).

Douet, James, ed., *Industrial Heritage Re-tooled: the TICCIH Guide to Industrial Heritage Conservation* (Camborne: TICCIH, 2012).

Dumas, Jean-Louis, 'Liebig et son empreinte sur l'agronomie moderne', *Revue d'histoire des sciences et de leurs applications*, 18.1 (1965), 73–108.

Edgerton, David, 'De l'innovation aux usages. Dix thèses éclectiques sur l'histoire des techniques', *Annales. Histoire, Sciences Sociales*, 4/5 (1998), 815–37.

Edwards, Junius, *The Immortal Woodshed: the Story of the Inventor who Brought Aluminum to America* (New York, NY: Dodd, Mead, 1955).

Evans, Chris and Göran Rydén, eds, *The Industrial Revolution in Iron. The Impact of British Coal Technology in Nineteenth-Century Europe* (London: Routledge, 2005).

Exley, Chris, 'Human exposure to aluminium', *Environmental Science*, 15 (2013), 1807–16.

Fell, Ulrike, *Disziplin, Profession und Nation. Die Ideologie der Chemie in Frankreich von Zweiten Kaiserreich bis in die Zwischenkriegszeit* (Leipzig: Leipziger Universitätsverlag, 2000).

Fisher, Tom , '3D Representations of Heritage — Understanding the Leavers Lace Machine through Documents, Ethnography and Animation', in *Patrimoine industriel, industrial heritage in the 21st century, new challenges*, ed. by Florence Hachez-Leroy, 1 (2018), pp. 316–24.

Fluck, Pierre, 'L'Archéologie industrielle, en sous-sol et en élévation', *Historiens & géographes*, 401 (2008), 205–11.

Fressoz, Jean-Baptiste, *L'Apocalypse joyeuse. Une histoire du risque technologique* (Paris: Seuil, 2012).

Fridenson, Patrick and Florence Hachez-Leroy, eds, *L'Aluminium, matière à création, XIXe–XXIe siècles* (Tours: PUFR, 2017).

Friedel, Robert, *Pioneer Plastic: the Making and Selling of Celluloid* (Madison, WI: University of Wisconsin Press, 1983).

Galambos, Louis, Takashi Hikino, and Vera Zamagni, eds, *The Global Chemical Industry in the Age of the Petrochemical Revolution* (Cambridge: Cambridge University Press, 2006).

Garçon, Anne-Françoise, *Mine et métal: 1780–1880: les non-ferreux et l'industrialisation* (Rennes: PUR, 1998).

Gignoux, Claude-Joseph, *Histoire d'une entreprise française* (Paris: Hachette, 1955).

Gille, Bertrand, *Histoire des techniques* (Paris: Gallimard, 1978).

———, *La Sidérurgie française au XIXe siècle* (Geneva: Droz, 1964).

Graham, Margaret B. W. and Bettye H. Pruitt, *R & D for Industry: a Century of Technical Innovation at Alcoa* (Cambridge: Cambridge University Press, 1990).

Grelon, André, ed., *La Chimie, ses industries et ses hommes, Culture technique*, 23 (1991).

Grinberg, Ivan and Florence Hachez-Leroy, eds, *Industrialisation et société en Europe occidentale de la fin du XIXe siècle à nos jours. L'âge de l'aluminium* (Paris: Armand Colin, 1997).

———, 'Histoire, patrimoine et entreprise: réflexion sur trente ans d'expérience de l'Institut pour l'histoire de l'aluminium', *Entreprises et histoire*, 89.4 (2017), 9–20.

Haber, Ludwig F., *The Chemical Industry During the Nineteenth Century* (Oxford: Clarendon Press, 1958).

———, *The Chemical Industry, 1900–1930: International Growth and Technological Change* (Oxford: Clarendon Press, 1971).

Hachez-Leroy, Florence, *L'Aluminium français. L'invention d'un marché, 1911–1983* (Paris: CNRS, 1999).

———, 'Aluminium in Health and Food: a Gradual Global Approach', *European Review of History*, 20.2 (2013), 217–36.

———, 'Cartels et ententes: les vieux démons persistants de l'économie', in *France-Japon, regards croisés. Mélanges en mémoire de Terushi Hara*, ed. by Dominique Barjot and Patrick Fridenson (Paris: PUPS, 2015), pp. 117–37.

———, *Menaces sur l'alimentation. Emballages, colorants et autres contaminants alimentaires, XIXe–XXIe* siècles (Tours: PUFR/PUR, 2019).

Homburg, Ernst, Anthony S. Travis, and Harm Schröter, eds, *The Chemical Industry in Europe, 1850–1914: Industrial Growth, Pollution and Professionalization* (Dordrecht: Kluwer, 1998).

Hounshell, David A. and John Kelly Smith Jr., *Science and Corporate Strategy: Du Pont R&D, 1902–1980* (New York, NY: Cambridge University Press, 1988).

Ihde, Aaron J., *The Development of Modern Chemistry* (New York, NY: Dover Publications, 1964).

Kahlow, Andreas, 'Materials in 19th Century Germany', *History and Technology, Materials in 19th–20th Century Europe*, 3/4 (1991), 255–66.

Kipping, Matthias, Ruggiero Ranieri, and Joost Dankers, 'The Emergence of New Competitor Nations in the European Steel Industry: Italy and the Netherlands, 1945–65', *Business History*, 43.1 (2001), 69–96.

———, 'How Unique is East Asian Development? Comparing Steel Producers and Users in East Asia and Western Europe', *Asia Pacific Business Review*, 4.1 (1997), 1–23.

———, 'Inter-Firm Relations and Industrial Policy: the French and German Steel Producers and Users in the Twentieth Century', *Business History*, 38.1 (1996), 1–25.

Knight, David and Helge Kragh, eds, *The Making of the Chemist. The Social History of Chemistry in Europe, 1789–1914* (Cambridge: Cambridge University Press, 1998).

Kranzberg, Melvin and Carroll W. Pursell, *Technology in Western Civilization* (New York, NY: Oxford University Press, 1967).

Kudo, Akira and Terushi Hara, *International Cartels in Business History. Proceedings of the 18th Fuji Conference Tokyo* (Tokyo: Univ. of Tokyo Press, 1992).

Langlinay, Erik, 'L'industrie chimique française et ses mutations, 1900–1931' (unpublished doctoral thesis, EHESS, 2017).

Laroche, Florent, 'Une nouvelle forme de capitalisation des connaissances grâce à l'archéologie industrielle avancée', *Documents pour l'histoire des techniques*, 18 (2009), 51–60.

Le Crosnier, Hervé, 'Internet, une industrie lourde', *Le Monde diplomatique*, 653 (2008), 19B–19.

Lefaux, René, *Practical Toxicology of Plastics* (Cleveland, OH: CRC Press, 1968).

Le Roux, Muriel, *L'Entreprise et la recherche. Un siècle de recherche industrielle à Pechiney* (Paris: Rive Droite, 1998).

Lesclous, René, *Histoire des sites producteurs d'aluminium: les choix stratégiques de Pechiney, 1892–1992* (Paris: Presses de l'École des Mines, 1999).

Levinson, Marc, *The Box: How the Shipping Container Made the World Smaller and the World Economy Bigger* (Princeton, NJ: Princeton University Press, 2006).

Liebenau, Jonathan, 'Industrial R&D in Pharmaceutical Firms in the Early Twentieth Century', *Business History*, 26.3 (1984), 335–40.

Lindner, Stephan H. and Dominique Pestre, *Innover dans la régression. Régions et industries menacées de déclin* (Paris: CRHST/Cité des Sciences et de l'industrie, 1996).

Lowe, Derek B., *The Chemistry Book: from Gunpowder to Graphene, 250 Milestones in the History of Chemistry* (New York, NY: Sterling, 2016).

Massard-Guilbaud, Geneviève, *Histoire de la pollution industrielle, France, 1789–1914* (Paris: EHESS, 2010).

Meikle, Jeffrey L., *American Plastic: a Cultural History* (New Brunswick, NJ: Rutgers University Press, 1995).

———, 'Material Doubts: the Consequences of Plastic', *Environmental History*, 2.3 (1997), 278–300.

Michalet, Charles-Albert, *Mondialisation, la grande rupture* (Paris: La Découverte, 2009).

Michon, Perrine, 'L'Opération de régénération des Docklands: entre patrimonialisation et invention d'un nouveau paysage urbain', *Revue Géographique de l'Est*, 48.1–2 (2008).

Mioche, Philippe and Denis Woronoff, eds, *L'Acier en France: produits et marchés, de la fin du XVIIIe à nos jours* (Dijon: Éditions universitaires de Dijon, 2006).

Miwa, Hajime, 'The Rise and Decline of Smelters and the Success of Aluminium Fabrication in Japan', *in L'Aluminium, matière à création, XIXe–XXIe siècles*, ed by Patrick Fridenson, and Florence Hachez-Leroy (Tours: PUFR, 2017), pp. 105–14.

Moreau, Alain, *L'Acier en mouvement. Transformation des marchés mondiaux 1874–2004* (Paris: Presses de l'École des mines, 2005).

Morel, Paul, *Histoire technique de la production d'aluminium* (Grenoble: PUG, 1991).

Morsel, Henri, *La Société d'électrochimie, d'électrométallurgie et des aciéries d'Ugine, 1889–1922* (Paris: Imprimerie La Ruche, 1973).

Ndiaye, Pap, *Du nylon et des bombes: DuPont de Nemours, le marché et l'état américain* (Paris: Belin, 2001).

Needleman, Herbert L., ed., *Human Lead Exposure* (Boca Raton, FL: CRC Press, 1991).

Nunn, Emily, *The Centenary of Hall's Discovery* (Oberlin, OH: Oberlin College, Office of Communications, 1986).

Oldenziel, Ruth and Mikael Hård, *Consumers, Tinkerers, Rebels. The People Who Shaped Europe* (Basingstoke: Palgrave Macmillan, 2013).

Oreskes, Naomi and Erik M. Conway, *Merchants of Doubt: How a Handful of Scientists Obscured the Truth on Issues from Tobacco Smoke to Global Warming* (New York, NY: Bloomsbury Press, 2010).

Østby, Per and Karin Zachmann, 'Food, Technology, and Trust, an Introduction', *History and Technology: An International Journal*, 27.1 (2011), 1–10.

Oudshoorn, Nelly and Trevor Pinch, eds, *How Users Matter. The Co-construction of Users and Technology* (Cambridge, MA: The MIT Press, 2003).

Peterson Warren S., et Ronald E. Miller (eds), *Hall–Héroult Centennial: First Century of Aluminum Process Technology 1886–1986.* (Warrendale: Metallurgical Society, 1986)

Pessis, Céline, Topçu Sezin, and Christophe Bonneuil, eds, *Une autre histoire des Trente Glorieuses. Modernisation, contestations et pollutions dans la France d'après-guerre* (Paris: La Découverte, 2013).

Pierrot, Nicolas, 'Les Images de l'industrie en France: peintures, dessins, estampes, 1760–1870' (unpublished doctoral thesis, Université Paris 1, 2010).

Pillet, Frédéric, *Saint-Quentin: histoire et patrimoine industriels* (Fontainebleau: Vif-Argent, 2009).

Plateau, Jean, 'Ce que l'aluminium doit à Henri Sainte-Claire Deville. Éclaircissements sur des travaux controversés', *Cahiers d'histoire de l'aluminium*, 32/33 (2004), 29–50.

Preite, Massimo, 'Conserving and Managing the Mining Landscape: the Amiata Mines (Italy) as an Essential Component of the European Mercury Heritage', in *Patrimoine industriel, industrial heritage in the 21st century, new challenges*, ed. by Florence Hachez-Leroy, 1 (2018), 49–55.

———, 'Géographie, histoire et représentation du paysage industriel', *Industrial Patrimony*, 13 (2005).

———, 'Piombino's Iron and Steel Heritage as a Resource for Urban Development', *Industrial Patrimony*, 26 (2011).

Rainhorn, Judith, 'The Fight Against Lead Poisoning in the Paint Industry: a Comparison Between the French and American Experiences, 1900–1940', in *At Work in the World: Proceedings of the Fourth Conference on the History of Occupational and Environmental Health*, ed. by Paul Blanc and Brian Dolan (San Francisco, CA: University of California Press, 2012), pp. 169–72.

Reardon-Anderson, James, 'Chemical Industry in China, 1860–1949', *Osiris*, 2 (1986), 177–224.

Rodrigue, Jean-Paul, *The Geography of Transport Systems* (New York, NY: Routledge, 2017).

Romainville, Michel, 'Les Routes africaines de l'aluminium', *Techniques & Culture*, 51 (2009), 74–97.

Rougier, Marcelo, 'The Challenges of the Aluminium Industry in Argentina (1930–2010)', in *L'Aluminium, matière à création, XIXe–XXe siècles*, ed by Patrick Fridenson, and Florence Hachez-Leroy (Tours: PUFR, 2017), pp. 95–104.

Ruess, Céline, ed., *Les Témoignages dans les musées industriels. Entre mémoire et patrimoine* (Liège: Maison de la métallurgie et de l'industrie de Liège, 2016).

Sakudo, Jun, *Les Entreprises de la chimie en France de 1860 à 1932* (Brussels: Peter Lang, 2011).

Saou-Dufrene, Bernadette and Benjamin Barbier, eds, *Heritage and Digital Humanities: How Should Training Practices Evolve?* (Berlin: LIT Verlag, 2014).

Schatzberg, Eric, 'Symbolic Culture and Technological Change: the Cultural History of Aluminum as an Industrial Material', *Enterprise & Society*, 4.2 (2003), 226–71.

Schröter, Harm G., 'Cartels Revisited. An Overview on Fresh Questions, New Methods, and Surprising Results', *Revue économique*, 64.6 (2013), 989–1010.

Schuck, Peter H., *Agent Orange on Trial: Mass Toxic Disasters in the Courts* (Cambridge, MA: Harvard University Press, 1987).

Scranton, Philip, *Endless Novelty. Specialty Production and American Industrialization, 1865–1925* (Princeton, NJ: Princeton University Press, 1997).

Sheller, Mimi, *Aluminum Dreams. The Making of Light Modernity* (Cambridge, MA: The MIT Press, 2014).

Smith, Cyril Stanley, ed., *The Sorby Centennial Symposium and the History of Metallurgy* (New York, NY: Gordon and Breach, 19765).

Smith, George David, *From Monopoly to Competition: the Transformations of Alcoa, 1888–1986* (Cambridge: Cambridge University Press, 1988).

Smith, Maureen, *The US Paper Industry and Sustainable Production: an Argument for Restructuring* (Cambridge, MA: MIT Press, 1997).

Snow, Bradley D., *Living with Lead: an Environmental History of Idaho's Coeur D'Alenes, 1885–2011* (Pittsburgh, PA: University of Pittsburgh Press, 2017).

Solly, Raymond, *Tanker: the History and Development of Crude Oil Tankers* (London: Chatham, 2007).

Travis, Anthony S., Harm Schröter, and Ernst Homburg, eds, *Determinants in the Evolution of the European Chemical Industry, 1900–1939: New Technologies, Political Frameworks, Markets and Companies* (Dordrecht: Kluwer, 1998).

Turnbull, Les, *The History of Lead Mining in the North East of England* (Hexham: ERGO Press, 2008).

United Nations Conference on Trade and Development, *Review of Maritime Transport 2018* (New York, United Nations, October, 3rd, 2018).

Varaschin, Denis, ed., *Risques et prise de risque dans les sociétés industrielles* (Brussels: P. I. E. Peter Lang, 2007).

Veyret-Verner, Germaine, 'Une industrie en pleine expansion: l'aluminium', *Revue de géographie alpine*, 44.2 (1956), 311–42.

Viel, Claude, *Henri Moissan, 1852–1907: pharmacien, premier français prix Nobel de chimie* (Paris: Pharmathèmes édition-communication, 2006).

Vogel, Sarah A., 'The Politics of Plastics: the Making and Unmaking of Bisphenol A "Safety"', *American Journal of Public Health*, 99, supplement 3 (2009), 559–66.

Wenger, Emanuel, *The Memory of Paper*, https://www.memoryofpaper.eu, access 23/04/2022

Woronoff, Denis, 'Le patrimoine sidérurgique français', *Patrimoine industriel*, 66/67 (2015), 36–51.

———, eds, *Les Images de l'industrie de 1850 à nos jours* (Paris: Comité pour l'histoire économique et financière de la France, 2002).

ANAËL MARREC, MAHDI KHELFAOUI

Energy in the History of Technology

Energy is central to the history of nineteenth- and twentieth-century technologies. The use of energy was at the heart of the industrial processes that shaped the nascent industrial society, from the extraction, transport, and processing of natural resources to the manufacture and consumption of goods and services. The improvement and development of new energy technologies embodied in 'converters', *i.e.* machines that transform a primary energy source into a form of energy that can be used for specific purposes (Debeir, Deléage, Hémery, 2013), have since become increasingly important in different economic, social, and political contexts. The development of new extraction methods, the extension of transport and distribution networks, and the multiplication of possible end uses made energy a core and ubiquitous feature of the landscapes, lifestyles and imaginaries of the "thermo-industrial civilization" (Gras, 2003). The notion of global energy systems reflects the transnational dimension of energy technologies in the modern era: physical infrastructures, commercial enterprises, social and political institutions, climate and environmental issues have taken on a global dimension that justifies the relevance of a global approach to the history of energy and its technologies (Podobnik, 2006; Massard-Guilbaud, 2018). Systemic approaches have also made it possible to consider energy from a global perspective and highlighted the inextricable relationship between energy sources and the organization of human societies (Sieferle, 2001; Debeir, Deléage, Hémery, 2013).

Despite its widely acknowledged importance, energy has only recently been considered a proper object of historical investigation, as evidenced by the fact that the first academic journal dedicated to the history of energy, *The Journal of Energy History*, was only created in 2018 (Massard-Guilbaud, 2018). Major historians of technology have identified the crucial role of energy technologies in world history, particularly focusing on the evolution of energy converters, on the development of energy networks, and on the successive innovations that shaped contemporary energy systems (Mumford, 2016; Daumas, 1962; Gille, 1978). Revolving around the notion of 'Industrial Revolution', these narratives provided detailed accounts of the technical improvements that led to increasingly powerful and efficient converters. But the history of energy has not been written within the context of the history of technology alone. Social and economic historians have also considered energy technologies as a major factor in the history of civilization. Following Karl Marx, who associated

Global History of Techniques, ed. by Guillaume Carnino, Liliane Hilaire-Pérez and Jérôme Lamy, Global Matters Culture Matérielles, histoire des techniques et circulation globale des savoirs, 9 (Turnhout: Brepols, 2024), pp. 315–336

BREPOLS PUBLISHERS 10.1484/M.TECHNE-EB.5.129814

the hand mill with feudal societies and the steam mill with industrial capitalist ones, several authors distinguished periods of human history according to their different modes of energy production (Cipolla, 1961; Smil, 1994; White, 2007). Technology-centered narratives of energy history have also been put into perspective by analyses that gave more room to social, political, and cultural factors. Energy technologies have thus been recognized as the basis of socio-technical systems that structured power relations (Hughes, 1983; Hecht, 1998; Mitchell, 2011; Malm, 2016), while the evolution of large energy systems has been interpreted as a succession of momentums set by social and technological forces (Hughes, 1987). The imaginary of power and productivism rooted in nineteenth century's thermodynamics have also been recognised as an integral part of contemporary energy systems (Gras 2003; New Daggett, 2019; Marrec, 2018), as has the culture that has accompanied new modes of consumption (Beltran & Carré, 2016 ; Mathis, 2019 ; Mathis, 2021).

Energy Sources and Uses

Many historical works have studied energy in a segmented way, approaching it via different sources, technologies, or economic sectors (Massard-Guilbaud, 2018). Most widely consumed energy sources and carriers, such as coal, oil, gas, water, and uranium, have been analysed from these perspectives. To understand the evolution of energy technologies and systems on a global scale, a possible first approach is the analysis of global energy production and consumption by type of resource (Etemad et al., 1991; Podobnik, 2006; Kander, Malamina, Warde, 2013). Based on the compilation of worldwide data, this type of analysis reveals trends that reflect global changes in energy production and consumption. First, energy consumption has grown exponentially since the nineteenth century. Moreover, the emergence of new energy consumption patterns has been largely cumulative: new energy uses have rarely completely erased older uses, and energy sources have added to each other, changing their relative share in the overall balance, while not necessarily decreasing in absolute terms.

The quantitative accounts of energy history indicate that organic power provided by animals, humans, and wood were the most important sources of energy in the global energy balance at the beginning of the nineteenth century. Other so-called 'traditional' technologies have long been used alongside innovative technologies in industrialized societies, and they were improved, even if they were not, or no longer, associated with the idea of modernity. They sometimes played a key role in industrialization processes, as was the case, for example, for hydraulic power in the French and Italian Alps or in New England (Benoit, 2020; Parinello, 2018; Steinberg, 1994), or for wooden coal and wind power in different French regions (Benoit, 2020, Terrier, 2020). Coal overtook wood from 1880 onwards before losing importance in the 1920s to the benefit of oil and then natural gas. Oil was of vital importance at the turn of the twentieth century and became the major source of energy in the world's energy balance in the 1960s. Natural gas followed a similar trend but remained less important in quantitative terms. Atomic energy has been supplying electricity grids since the 1950s. Despite early promises of providing an energy 'too cheap to meter' (*i.e.* produced at virtually no cost), nuclear energy continues to be globally less important than fossil fuels for reasons related to cost, safety, and social oppositions.

Energy statistics are useful for a historic overview of production and consumption trends. However, they tend to erase the diversity of energy sources over time and across geographic areas and obscure the large variety of energy uses across cultures and socioeconomic contexts (Fressoz, 2013). Indeed, global energy production and consumption trends have been influenced by several factors: technological (innovations), environmental (geographical distribution and availability of energy resources, landscape mutations and soil, air and water pollutions), geopolitical (wars, international competition, energy policies), economic (world prices and markets), social (social demand and oppositions, labour movements), and cultural (such as the accepted definitions of what is technological modernity).

The Development of Energy Systems and Evolution of Primary Converters

The notion of energy system refers to the interdependence between different socio-technical infrastructures that transform or convert a primary source of energy (coal, petroleum, biomass, wind, etc.) for a final use (electricity, heating, cooling, etc.) (Debeir, Deléage, Hémery, 2013). The development of modern energy systems has been based on profound changes in the shape, characteristics, configuration, and use of energy converters. The following main features can be identified in these evolutions: the increase in power outputs; the increasing distance between primary sources and final users; and the stability and flexibility of the delivered power. Power, transport capacity, and regularity have thus become the main performance criteria of modern energy converters. These trends are reflected in the extension of conversion chains and transport infrastructures, the increasing number of secondary converters, the limitation of converters to single energetic functions, and the shift from converters directly connected to end users to converters connected to energy networks and grids. While in the early nineteenth century, most energy uses were local and direct, energy converters have undergone important reconfigurations since then: some were adapted to new uses, others were abandoned, while new converters, like solar and nuclear energy systems, also appeared.

Fossil Fuel and Electric Systems

The coal system was the first to develop in the nineteenth century. This system existed through relationships between technologies of mining, railway transport, and iron and steel production, with the steam engine as a central element in this synergy (Gille, 1978). Coal systems combined power, consistency, and transport capability, and became the reference model of modern converters (Gras, 2003). The coal system spread in England from mines to textile factories in the early decades of the nineteenth century, and then to European and North American industries. It extended its influence to colonial empires, thanks to railways and steam navigation networks, and allowed the transportation of raw materials to metropolitan areas which, in turn, exported manufactured goods (Barak, 2020). In factories and in cities, gas networks appeared for industrial, urban, and domestic lighting at the beginning of the nineteenth century. They multiplied in the first three decades of the nineteenth century in Europe and in the United States. They became a structuring element of cities, connecting end users to large coal gasification plants and gasometers through underground networks (Williot, 2003), and contributed to transforming cities into a vast network of services (Lopez, 2019). Denser than coal, oil allowed the use of internal combustion engines that were lighter than the steam engine. Extracted from the ground by drilling and transported by sea, oil also allowed lower production costs (Mitchell, 2011). Navy ships became increasingly powered by oil, and gasoline-powered automobiles appeared at the end of the nineteenth century. Fostered during each world war, oil-fuelled cars spread around the world and became a symbol of modernity, shaping cities, suburbs, and countrysides with dense road and highway networks.

At the end of the nineteenth century, electricity networks made industrial and urban energy systems more complex (Hughes, 1983). They not only made it possible to transmit power, but also provided greater flexibility than the steam engine. Electricity grids multiplied in cities and spread progressively to rural areas, being interconnected through power plants, transformers, and high-voltage lines (Cohn, 2017). At first competing with gas for lighting, electric grids supplied industrial machines in factories from the end of the nineteenth century. Then they supplied private homes where household appliances appeared in the inter-war period before experiencing a dramatic increase in the post-war period (Beltran, Carré, 2016). The early supply of electric grids by coal-fired and hydropower plants was supplemented in the 1960s by natural gas and nuclear plants. Finally, in the 1990s, several factors contributed to the integration of renewable energies into electricity networks: the institutionalization of environmental issues, the liberalization of electricity markets (Évrard, 2013a, 2013b), as well as the densification of electric networks. More recently smart grid projects carried out by several governments and energy companies have aimed at increasing the number of supply points from the smallest scale (cities) to the largest (continents), thus blurring the boundaries between producers and consumers, and optimizing energy flows through storage, regulation, and control of consumption (Picon, 2015).

Human and Animal Energy

Historians have pointed out the persistent and central role of bodies in global energy production. In the nineteenth and twentieth centuries, physical labour was still central to economies worldwide. Many tools and machines for the transport of products were man-powered, whether in domestic or agricultural environments, in factories, mines, and construction sites (Nye, 1998; Kander, Malamina, Warde

2013; Jarrige, 2016, 2019). With mechanization, new forms of industrial production divided human labour according to gender and race (New Daggett, 2019). While fossil-fuelled machines contributed to the process of abolishing slavery (Nikoforuk, 2012), forced labour changed in form, yet persisted throughout the nineteenth and twentieth centuries on plantations, in the war industry, and for major transport infrastructures such as roads, rails, and canals (Langewiesche, 2004; Flory, 2017; Tiquet, 2019).

Regarding animal power, horses became more important than ever in the nineteenth century's European and Northern American countryside and cities (Deluermoz, Jarrige, 2017; Nye, 1998). Horse husbandry and traction technology underwent many improvements (MacShane, Tarr, 2007), and urban spaces became organised around horse-drawn trams, cars, and postal systems (Roche, 2008). Despite the increasing use of 'inanimate' energy converters, muscle powered devices remained important for a variety of uses in the twentieth century, especially in agriculture. For example, the number of draught horses and mules continued to increase in the United States until 1918 and was only surpassed by tractors in 1945 (Olmstead, Rhode, 2001). In European Mediterranean countries, draught animals represented a significant portion of energy use until the 1960s, while human power remains important until today (Kander, Malamina, Warde, 2013). For Inuit populations in the Baffin region, sled dogs were the main means of transportation until tribes settled and replaced sleds with snowmobiles in the 1950s (Tester, 2010). The bicycle is an example of a man-driven engine that was invented in the nineteenth century and became an important mode of transport in cities in the late twentieth century (Gaboriau, 1991).

Wood and Biomass Converters

Plant-fuelled converters, especially wood-fuelled converters, are the oldest and most widespread in the world. Despite the rise of coal, wood was used for industrial purposes until the late nineteenth century in several European countries (Kander, Malamina, Warde, 2013; Benoit, 2006) and in the United States (Nye, 1998). Wood has been used in domestic settings worldwide until today, heating homes in rural areas and being central for domestic heating requirements in countries including Sweden and Brazil (Santos, 2018). Its consumption has undergone a revival since the two oil shocks in industrialised countries, with the appearance of pellets and the multiplication of wood-burning stoves and community heating networks (Kander Malamina, Warde 2013). In the inter-war period and during the Second World War, wood was also used in some oil-deprived countries, such as France, Sweden, and Brazil, to generate mechanical power for automobiles through gasifiers (Brannstrom, 2005; Ekerholm, 2014; Molles, 2018). "Biomass", *i.e.* organic materials, were also the object of new conversion processes in the twentieth century. Biogas production from wood, animal and plant products developed first in European North American cities, and from the 1940s onwards in India and some South-East Asian countries where numerous farms were equipped with small-scale anaerobic digesters (Van Brakel,

1980). In the 1970s, the oil crises stimulated innovation in the field of biomass, which essentially consisted of adapting old techniques (gasification, methanisation, combustion) to larger scales and to the production of fuels (alcohol, gas) and electricity. Energy recovery processes have also been used to produce agrofuels, which have been widely used in Brazil (Santos, 2013). However, the renewable and local aspects of these resources have been highly questionable (Évrard, Poinsot, 2013).

Wind, Solar and Hydraulic Converters

Wind converters were of great importance in the nineteenth and twentieth centuries. At the beginning of the nineteenth century, sailing was the predominant form of transport in the European and North American navies. The replacement of sailing boats with steam engines in the naval forces and in the merchant navy took several decades, and multiple hybrid forms appeared, as sailing continued to be a more convenient way of propelling ships for a long time (Marnot, 2016; Barron, 2016). In the nineteenth century, 'Dutch' windmills were converted to steam mills or abandoned (Rivals, 1987). However, they still played a central role in some European regions, such as Lille in France where they were used for vegetable oil production until 1880 (Terrier, 2020), and in Denmark where they produced electricity for small town networks until the mid-twentieth century (Bruyerre, 2020). Pumping wind turbines spread in the Western American countryside from the 1870s onwards; they were used for domestic and agricultural needs, and numbered in the hundreds of thousands at the end of the nineteenth century (Righter, 1996). In French rural areas and former French colonies, several forms of small wind turbines were used until the 1930s before declining due to the extension of electricity networks (Marrec, 2018). Powerful wind generators have been connected to electricity grids since the 1970s. Solar converters first appeared in the form of thermal engines using fluids at low vaporization temperatures. Some prototypes were developed in France and in the USA between 1860 and 1900 (Kryza, 2003; Jarrige, 2010). The American Sun Power Company set up a pumping facility for irrigation in Egypt in the 1910s. At the turn of the century, solar water heaters were in widespread use in California. Between 1940 and 1970, solar and wind converters found outlets in non-electrified areas, and were part of international cooperation programmes (Teissier, 2015; Bouvier, 2016). In some African countries the use of solar motors spread between the late 1960s and 1980s (Girardier, 1995). Solar water heaters, photovoltaic panels, and wind turbines spread in communities of various sizes, from individual homes to entire neighborhoods. From the 1970s, they became part of counter-cultural and environmental movements searching for autonomy from large technological systems (Lopez, 2014).

At the beginning of the nineteenth century, river converters and some tidal converters enabled rural and urban communities to pump water, grind and press grains, and operate bellows and saws (Rivals, 1987; Silveira, 2019). In regions with abundant water currents and little coal availability, hydropower was a crucial driving force of industrialization during the first half of the nineteenth century, and it has remained an important source of primary energy with the development of electricity networks (Nye, 1998). New damming techniques, combined with the invention of turbines, allowed the building of increasingly powerful dams (Benoit, 2006). Some river mills and a few tidal power mills however continued to operate in European rural areas until the 1960s (Charlier, Ménanteau, 1997). But, in general, the hydromecanical use of water tended to decrease from the end of the nineteenth century, when electrical networks made it possible to transport energy over long distances (Parinello, 2018). Since then, most of new hydropower plants have been connected to the electricity grid. Major hydroelectric projects, such as the Niagara Falls dam, the James Bay dam, or the Three Gorges dam, are among the most important engineering infrastructures built in the past century.

Old uses of renewable energies have persisted in time and new uses have appeared. The study of these so-called marginal energy systems has allowed historians to uncover their previsouly unrecognised contributions to industrialization processes, and to discover unsuspected energy consumption modes (Jarrige, Vrignon, 2020). Thus, the recent rise of renewable energies in national energy programmes does not mean that they don't have a long history. Also, considering their 'alternative' character today must not obscure the fact that they were and still are dominant energy sources in various socio-cultural contexts. Finally, while renewable energies are often viewed as techno-political drivers of emancipation (Lopez, 2019), they could just as much reinforce power relationships and serve capitalist accumulation (Marrec, Teissier, 2020).

Environmental Impacts of Energy Systems

Energy systems have significantly shaped landscapes and environments. In rural and urban areas, visible and invisible infrastructures have grown to cover vast areas of the planet: coal and uranium mines, oil wells, dams, power plants, electric networks and pipelines, wind farms, nuclear waste treatment and storage sites, gas urban networks, etc. These infrastructures have created persistent health hazards and environmental risks (global warming, atmospheric pollutions, nuclear hazards, deforestation, etc.). The ways in which energy systems have affected the environments and lifestyles of the general population and workers have received much attention from environmental historians, who aimed to move 'beyond the nature and culture binaries' (Evenden, 2018: 700) and merge 'physical energy and social power' (Andrews, 2008) in single narratives. Indeed, as highlighted by envirotechnical approaches, environmental impacts of energy infrastructures are closely intertwined with their political and social features (Pritchard, 2011 by Parinello, 2018). Histories of rivers and hydroelectric damming among many examples (White, 1996; Bodon, 2003; Pritchard, 2011), and of coal and uranium mines and cities (for example : Rosental, Devinck, 2007; Rainhorn, 2014; Keeling, 2010; Hecht, 2014), show how social groups (workers and neighboring inhabitants) were especially affected by the activities linked to energy systems. Environmental works have also studied the destruction of tropical forests by the domestic and industrial uses of wood (Pádua, 2002; Dean, 1996), while others have analysed the historical roots of global warming from the rise of fossil-fuel-based economies in the nineteenth century (Ruddiman, 2005; Bonneuil, Fressoz, 2016; Malm, 2016; Cavert, 2016).

Social Drivers of Energy History

The history of energy is one of physical forces and infrastructures which are deeply embedded in social and cultural practices. Making sense of how different energy converters and production sources shaped society and how, in turn, society shaped energy infrastructures and uses calls for an analysis of the social drivers of energy history. For instance, how individuals and groups contributed to defining the uses of electricity has been well documented in studies on the cultural impact of the introduction of electric light in urban and rural areas (Schivelbusch, 1988). On the other hand, the growth of energy consumption, which greatly contributed to the rise of consumerism in capitalist societies, allowed the emergence of particular cultural practices and lifestyles, such as the so-called American way of life (Huber, 2013), the individual use of cars, the proliferation of household appliances and, more recently, of electronic devices (Bouvier, 2018). More generally, the history of energy has been greatly enriched by approaches that have tried to explain the development of energy infrastructures and the diffusion of energy devices through exploring social, environmental, economic and political factors.

Energy Companies and Their Markets

For the last two centuries, the history of energy systems has been deeply intertwined with the history of energy companies. The first coal companies were created in the mid-seventeenth century (Hatcher, 1993), but major corporations only appeared a century and a half later, accompanying the economic boom of the Industrial Revolution (Church, Hall, Kanefsky, 1986; Geiger, 1971). In the mid- and late nineteenth century, British and American coal companies forged close ties with iron, railroad, and steamboat companies, and became prime examples of industrial and corporate capitalism. The first gas lighting companies appeared at the beginning of the nineteenth century and ushered in the electrification of urban centres. Oil companies were created later in the same century but were active in more remote areas, near extraction sites. Rising to the status of states within a State, as energy production became a highly strategic and capital-intensive industry, public and private energy corporations had not only an economic role, but also a geopolitical one. Unsurprisingly, due to their size and influence, multinational oil companies have attracted most of the attention of economic and business historians, followed by the large public electric utilities that managed national monopolies after the end of the Second World War.

Relying on different physical transport networks, electricity and oil production formed different types of enterprises and markets: oil was rapidly dominated by multinational corporations and regulated by international regimes and markets, while electricity came to be dominated by regional or national companies and regulated at the state level by market monopolies. The oil industry has been historically dominated by a group of British and American oil companies, the so-called Seven Sisters (Sampson, 1975), which were organised in a cartel and whose activities were closely related to the imperial domination of Western powers over the Middle East and Latin America (Galpern, 2009; Cote, 2016). The international oil market was neither 'free' nor competitive, since an oligopoly of multinational corporations determined prices and shares, kept the technical expertise, and controlled most of the refining infrastructures and oil tanker fleets (Debeir, Deléage, Hémery, 2013). Until the 1960s, the Seven Sisters controlled more than 85% of crude oil production outside the United States, the Soviet Union, Canada, and China (Cutler, 2009). The geopolitical developments of the Cold War, the creation of the Organization of the Petrolium Exporting Companies (OPEC) and the nationalization of hydrocarbons eventually led to introducing a more diverse group of actors, including national governments and public oil companies, which sought to challenge the dominance of multinational corporations (Beltran, 2010; Cantoni, 2017).

The history of the electricity sector shows that electricity markets were created and evolved way differently than oil markets. From the end of the nineteenth century onwards, electric component manufacturers and international financial holding companies played a crucial role in the global electrification of the world (Neufeld, 2007). In particular, they played a paramount role in the creation of public utilities which benefitted from municipal or regional monopolies and managed distribution networks. On the eve of the First World War and during the inter-war years, manufacturers, supported by the influx of financial capital from banks, controlled electric utilities across countries and continents, in Western Europe, Russia, South America, colonial Africa and Asia (Hertner, Nelles, 2007; Armstrong, Nelles, 1986; Myllyntaus, 1997; Barjot et al., 2002). From the inter-war period onwards, governments became more involved in the strategic planning of electricity production at the national level, which lead to more stringent regulations by the public authorities (Armstrong, Nelles, 1986), as well as the gradual replacement of multinational foreign investments with public ownership through nationalization. This global phenomenon, labelled as the 'domestication process' of the electricity industry (Hausman, Hertner, Wilkins, 2008), accelerated after the Second World War and eventually led to the dominance of nation states over the electricity industry. From the 1980s, when electricity markets became increasingly subjected to neoliberal deregulation (Hirsh, 1999) and while grids were transnationally integrated (Lagendijk, 2008), what appeared to be a new trend at the time was, in fact, only the resurgence of a 'past pattern of private sector international involvements in public utilities' (Neufeld, 2007: 176).

The Rise and Fall of the State: from Energy Nationalism to the Liberalization of Markets

From the beginning of the twentieth century, the issue of energy became increasingly entangled with the evolution of international relations. Energy markets, such as the European coal market, were already influenced by national economic rivalries before the First World War (Fremding, 1995). International rivalries also existed in the oil trade since the beginning of the twentieth century (Frank, 2009), but took an unprecedented importance during the Cold War (Perovic, 2017). International tensions around energy issues were of course not limited to the Western and Eastern blocks, as illustrated by the case of OPEC countries. In countries under colonial rule, energy technologies were mainly used to extract and transport raw resources to metropolitan areas. The processes of decolonization reconfigured power relationships into new forms, which resulted in the development of major resource extraction and energy projects led by companies of the former ruling country that maintained ties of political, economic, and technological dependency with the former colony (Hecht, 2012; Bouvier, 2016).

In the beginning of the 1970s, energy independence and national independence became near-synonyms. In Western countries, the first oil crisis in 1973 constituted a political turning point that triggered the creation and implementation of national long-term energy policies (Venn, 2016). These policies covered issues such as the secure access to domestic and foreign supplies, the long-term planning of supply and the forecasting of demand, energy costs, and energy conservation. In Germany, for instance, the oil crisis transformed energy policy 'from relatively limited interventions into a comprehensive policy aimed at securing low-cost energy supply, while respecting environmental protection' (Renn, 2016: 227). In the United States, the federal government created a series of policies and regulatory bodies, such as the Energy Policy and Conservation Act in 1975, the Department of Energy in 1977, the Public Utility Regulatory Policies Act in 1978, and the Energy Security Act in 1980 (Parra, 2004). The French government launched the *Plan Messmer* in 1974, which accelerated its nuclear programme and projected the installation of 50 GW of nuclear electricity by 1980. In many European countries as well as in the USA, renewable sources of energy, such as wind, solar, and geothermal energy, benefitted from unprecedented government funding after the oil crisis.

State planning of energy production was correlated with state control over technological choices. Thus, energy nationalism also materialized in what has been called 'technological nationalism', understood as the combination of 'the idea of technological progress with the sentiments and goals of nationalism' (Adria, 2009: 45). In the course of the twentieth century, and especially after the Second World War, great technological achievements were used by governments to promote national sovereignty. Historian David Nye (1994) has coined the notion of 'technological sublime' to illustrate how the construction of emblematic energy infrastructures, such as power plants and large hydroelectric dams during the New Deal period in the United States contributed to the process of nation building. The development of nuclear energy in Japan was also strongly encouraged through technological nationalism (Low, 2003), as was the case in France (Hecht, 1998) and the Soviet Union (Josephson, 1995).

Between 1945 and 1975, a period known in French history as the *Trente Glorieuses*, national energy companies and electric utilities were driven by an overwhelming emphasis on technical excellence (Callon, 1979; Faucher, 1999). Benefitting from commercial monopolies as well as an easy access to public financial resources, they often opted for original technological solutions and riskier projects than those that the market would have required. Nuclear energy technologies greatly benefitted from this particular context (Bothwell, 1988; Finon, 1989). However, in the 1980s energy markets became more influenced by neoliberal policies. Several electric utilities around the world were partially or fully privatized, state monopolies were broken, and public markets were open to private competition (Winksel, 2002). The management of large technological public companies, including energy utilities, no longer sought technological autonomy as a primary goal but rather commercial performance (Muller, 1989; Khelfaoui, 2020).

Energy Workers and Social Movements

Historians have long been interested in the forms of labour and social organization that developed in different energy sectors. Existing scholarship on the history of energy workers has concentrated on blue collar workers, such as coal miners and, to a lesser degree, oil field workers. These studies have explored such themes as the organization of work, life around the workplace, and labour conflicts (Benson, 1980; Shifflett, 1995; Andrews, 2008). Among the historical studies that have tried to analyse the relationship between the 'materiality of energy' (Jones, 2016: 2) and the political and social dimension of labour, Timothy Mitchell's *Carbon Democracy* (2011) is one of the most influential. Mitchell's main thesis is that the different physical production and circulation networks of coal and oil in the late nineteenth and early twentieth centuries led to different forms of labour and political organization of workers, as well as different levels of opposition to public authorities and private company owners.

Historians have also analysed broader forms of social opposition to different energy projects and infrastructures, for instance by investigating how indigenous populations were directly affected by hydroelectric or large oil projects and infrastructures, and how they tried to oppose them. Many studies have also been devoted to contemporary environmental movements, especially anti-nuclear movements which mostly emerged in the 1970s (Babin, 1984; Wellock, 1988; Topçu, 2007). Their actions often transcended the boundaries of a single country (Le Renard, 2018) and were not limited to opposing government energy policies, since they also promoted alternative models of energy production and consumption (Évrard, 2013b). Other studies have also brought to light strategies used by governments and energy industrials to control critical discourses and alternative sources of knowledge generated by these environmental movements (Barthe, Callon, Lascoumes, 2001; Topçu, 2013).

Histories of white collar energy workers, from individual inventors to the technocratic elites formed by organised groups of engineers, scientists, and policy experts, have focussed on topics such as technological designs and innovation, technological visions, and the contested role of technical expertise in the shaping of energy politics (Hughes, 1989; Carlson, 1991; Feldenkirchen, 1994; Hecht, 1996). These studies thus shifted the historian's attention from the mine and the production plant to the office and the laboratory. Energy professionals competed over the creation and control of the legitimate forms of knowledge that shaped the evolution of different industries. For instance, Frehner (2011) showed how pioneering US oil prospectors, who were self-taught individuals with practical skills, came to be gradually replaced by scientifically trained geologists and geophysicists in the early decades of the twentieth century.

The idea that technological designs embody political visions and moral values was explored by Gabrielle Hecht through the notions of 'technopolitics' and 'technopolitical regimes' (Hecht, 1998). Engineers' professional identities are not only shaped by a homogeneous technological culture, but also by political contexts and ideological discourses that differ from one country to another, a point that has been particularly highlighted as far as the nuclear industry goes (Johnston, 2012; Schmid, 2015; Hecht, 1998; Jasanoff, Kim, 2009). In the wind energy sector, different engineering visions resulted in different technological designs, as shown by the contrasted development of turbine technologies in Denmark, Germany, and the United States. These designs varied not only according to national technological styles, but depending on the degree of confidence, ambition, and cultures (craft-based vs engineering based) displayed by the engineering communities of each country (Heymann, 1998). In the United States, promoters of solar and wind technologies associated their designs with environmental values, a decentralized vision of the organization of society, and a simpler way of life, which they contrasted with the more centralized and less democratic model of society they attributed to the use of nuclear energy and fossil fuels (Laird, 2003).

Conclusion

As highlighted by Nye (2010), just how embedded energy is in our daily practices may only become apparent when a sudden and unexpected loss of energy occurs. The omnipresence of energy and its ubiquitous character makes it a multidimensional historical object, one that can be analysed from multiple standpoints. In addition to the history of technology, economic, social, and environmental history approaches have added new contributions to the understanding of the role played by various actors in the upheavals caused by the development of energy systems. The profound changes caused by the development of energy systems include the dismantling and reconfiguration of local and global economies, the transformations of international relations, the loss of traditional cultures and the emergence of new social practices, the formation of new territories, the creation of major ecological imbalances, etc.

In addition to more traditional analytical categories (social, economic, technical, environmental, political), new notions are currently furthering the historiographical renewal in historical studies of energy. As it has become the subject of an important political consensus, the notion of energy transition has been subjected to historical research (Smil, 2010; Fouquet, 2010). While providing a useful global perspective on the evolution of energy systems, it has also been criticized for hiding the fact that the history of energy transitions is one of energy additions and enduring interdependencies rather than radical shifts (Fressoz, 2013). Nationally centered narratives of energy history are now being complemented by transnational approaches that seek to analyse the geopolitical influences exerted by the circulation of energy fluxes between countries and territories (Högselius, 2013; Perovic, 2017; Boon, 2018).

The history of energy has also been dominated by Western-centered narratives, while the study of 'peripheral' regions has mostly been limited to that of technological transfers from industrialized to developing countries (Myllyntaus, 1991; Ushida, 1991) or the exploitation of energy resources by colonial powers and multinational corporations. However, a more thorough understanding of the actual global energy system and its possible future transformations requires a more detailed analysis of what has existed outside industrialised countries. Thus, traditional uses of energy, which are usually absent from global energy balances, should also be considered in the global picture of energy history.

Finally, the history of energy is currently renewed from the perspective of gender. The gendered dimension of energy uses has long been limited to the historical study of electric household appliances (Cowan, 1985), but has been recently the subject of new investigations questioning the broader agency of women in energy shifts (Harrison Moore, Sandwell, 2021; Mathis et al., 2021). Finally, one must not forget that history is by far not the only discipline of the social sciences and humanities that has developed an interest in energy issues. For some years now, researchers have been calling for approaches that combine theoretical concepts and methodologies from the disciplines of history, anthropology, sociology, political science, and economics. Such multidisciplinary approaches will certainly help push further the study of such a multifaceted and complex object as energy (Sovacool, 2014; Hirsh, Jones, 2014).

References

Adria, Marco, *Technology and Nationalism* (Montreal and Kingston: McGill-Queen's University Press, 2009).

Andrews, Thomas, *Killing for Coal: America's Deadliest Labor War* (Cambridge, MA: Harvard University Press, 2008).

Armstrong, Christopher and Henry Vivian Nelles, *Monopoly's Moment: The Organization and Regulation of Canadian Utilities, 1830–1930* (Philadelphia, PA: Temple University Press, 1986).

Babin, Ronald, *L'Option nucléaire. Développement et contestation de l'énergie nucléaire au Canada et au Québec* (Montréal: Boréal Express, 1984).

Barak, On, *Powering Empire. How Coal Made the Middle East and Sparked Global Carbonization* (Oakland: University of California Press, 2020).

Barjot, Dominique, Arnaud Berthonnet, Daniel Lefeuvre and Sophie Cœuré, *L'Électrification outre-mer de la fin du* XIXe *siècle aux premières décolonisations* (Paris: Publications de la Société française d'histoire d'outre-mer et Fondation EDF, 2002).

Barron, Géraldine, 'Dévoiler la vapeur. La construction navale au XIXe siècle', in Bouvier, Yves and Léonard Laborie, *L'Europe en transitions: énergie, mobilité, communication,* XVIIIe–XXIe *siècles* (Paris: Nouveau Monde Éditions, 2016), pp. 119–40.

Barthe, Yannick, Michel Callon, and Pierre Lascoumes, *Agir dans un monde incertain. Essai sur la démocratie technique* (Paris: Seuil, 2001).

Beltran, Alain, *A Comparative History of National Oil Companies* (Brussels: Peter Lang, 2010).

———, 'Introduction: L'énergie dans l'histoire, l'histoire de l'énergie', *Journal of Energy History*, 1 (2018), <http://energyhistory.eu/node/84> [Accessed 2022/03/18].

——— and Patrice Carré, *La Vie électrique. Histoire et imaginaire (*XVIIIe–XXIe *siècles)* (Paris: Belin, 2016).

Benoit, Serge, *D'eau et de feu: forges et énergie hydraulique. XVIIIe-XXe siècle. Une histoire singulière de l'industrialisation française* (Rennes: Presses Universitaires de Rennes, 2020).

Benson, John, *British Coal-miners in the Nineteenth Century: A Social History* (New York: Holmes & Meier, 1980).

Bodon, Virginie, *La Modernité au village: Tignes, Savines, Ubaye. La submersion de communes rurales au nom de l'intérêt général, 1920–1970* (Grenoble: Presses universitaires de Grenoble, 2003).

Bonneuil, Christophe and Jean-Baptiste Fressoz, *The Shock of the Anthropocene: The Earth, History and Us* (London, Brooklyn: Verso Books, 2016).

Boon, Marten, *Multinational Business and Transnational Regions. A Transnational Business History of Energy Transition in the Rhine Region, 1945–1973* (London: Taylor and Francis, 2018).

Bothwell, Robert, *Nucleus. A History of Atomic Energy of Canada Limited* (Toronto: University of Toronto Press, 1988).

Bouvier, Yves, 'Consommateurs d'énergies, un concept-frontière pour l'histoire de l'énergie', *Journal of Energy History*, 1 (2018), <energyhistory.eu/en/node/86 [Accessed 2022/03/18]>.

———, 'Renouveler l'Europe de l'énergie depuis les années 1950. Une perspective européenne sans politique commune', in *L'Europe en transitions: énergie, mobilité, communication, XVIIIe–XXIe siècles* (Paris: Nouveau Monde Éditions, 2016), pp. 227–70.

———, 'Les transitions énergétiques dans l'histoire, entre succession des techniques et sédimentation des enjeux', in *Les défis énergétiques du XXIe siècle. Transition, concurrence et efficacité au prisme des sciences humaines*, ed. by Yves Bouvier (Brussels: P. I. E. Peter Lang, 2012), pp. 23–36.

Brannstrom, Christian, 'Was Brazilian Industrialisation Fuelled by Wood? Evaluating the Wood Hypothesis, 1900–1960', *Environment and History*, 11.4 (2005), 395–430.

Bruyerre, Philippe, *La Puissance du vent, Des moulins à vent aux éoliennes modernes,* (Toulouse: Presses universitaires du Midi, Histoire et Techniques, 2020).

Callon, Michel, 'L'État face à l'innovation technique: le cas du véhicule électrique', *Revue française de science politique*, 29 (1979), 426–47.

Cantoni, Roberto, *Oil Exploration, Diplomacy, and Security in the Early Cold War* (London: Routledge, 2017).

Carlson, Bernard, *Innovation as a Social Process: Elihu Thomson and the Rise of General Electric, 1870–1900* (Cambridge: Cambridge University Press, 1991).

Cavert, William, *The Smoke of London: Energy and Environment in the Early Modern City* (Cambridge: Cambridge University Press, 2016).

Charlier, Roger H. and Loïc Ménanteau, 'The Saga of Tide Mills', *Renewable and Sustainable Energy Reviews*, 1.3 (1997), 171–207.

Church, Roy, Alan Hall and John Kanefsky, *The History of the British Coal Industry. Volume 3: 1830–1913, Victorian Pre-Eminence* (Oxford: Oxford University Press, 1986).

Cipolla, Carlo M., 'Sources d'énergie et histoire de l'humanité', *Annales, sociétés, civilisation*, 16.3 (1961), 521–34.

Cohn, Julia A., *The Grid. Biography of an American Technology* (Cambridge, MA: The MIT Press, 2017).

Cote, Stephen, *Oil and Nation: A History of Bolivia's Petroleum Sector* (Morgantown, WV: West Virginia University Press, 2016).

Cowan, Ruth Schwartz, *More Work for Mother: The Ironies Of Household Technology* (New York: Basic Books, 1985).

Cutler, Cleveland, *Concise Encyclopedia of the History of Energy* (San Diego: Elsevier Science & Technology Books, 2009).

Daumas, Maurice, ed., *Histoire générale des techniques* (Paris: Presses universitaires de France, 1962).

Dean, Warren, *A ferro e fogo. A história e a devastação da Mata Atlântica Brasileira* (São Paulo: Companhia das Letras, 1996).

Debeir, Jean-Claude, Jean-Paul Deléage and Daniel Hémery, *Une Histoire de l'énergie. Les servitudes de la puissance* (Paris: Flammarion, 2013).

Deluermoz, Quentin and François Jarrige, 'Introduction. Écrire l'histoire avec les animaux', *Revue d'histoire du XIXe siècle*, 54 (2017), 15–19.

Ekerholm, Helena, 'Cultural Meanings of Wood Gas as Automobile Fuel in Sweden, 1930–1945', in *Past and Present Energy Societies. How Energy Connects Politics, Technologies and Cultures*, ed. by Nina Eckerholm Möllers and Karin Zachmann (Bielefeld: Transcipt Verlag, 2014), pp. 223–48.

Etemad, Bouda, Jean Luciani, Paul Bairoch and Jean-Claude Toutain, *World Energy Production 1800–1985* (Geneva: Librairie Droz, 1991).

Evenden, Matthew, 'Beyond the Organic Machine? New Approaches in River Historiography', *Environmental History*, 23 (2018), 698–720.

Évrard, Aurélien, *Contre vents et marées. Politiques des énergies renouvelables en Europe* (Paris: Presses de Sciences Po, 2013a).

———, 'Les énergies renouvelables et l'électricité. À propos d'un conflit entre un secteur et une alternative de politique publique', *Écologie & Politique*, 49 (2013b), 67–80.

Feldenkirchen, Wilfried, *Werner von Siemens: Inventor and International Entrepreneur* (Columbus, OH: Ohio State University Press, 1994).

Ferdinand, Malcolm, *Une Écologie décoloniale. Penser le monde depuis le monde carribéen* (Paris: Seuil, 2019).

Finon, Dominique, *L'Échec des surgénérateurs. Autopsie d'un grand programme* (Grenoble: Presses Universitaires de Grenoble, 1989).

Fouquet, Roger, 'The Slow Search for Solutions: Lessons from Historical Energy Transitions by Sector and Service', *Energy Policy*, 38 (2010), 6586–96.

Frank, Alison, 'The Petroleum War of 1910: Standard Oil, Austria, and the Limits of the Multinational Corporation', *American Historical Review*, 114 (2009), 16–41.

Frehner, Brian, *Finding Oil: The Nature of Petroleum Geology, 1859–1920* (Lincoln: University of Nebraska Press, 2011).

Fremding, Rainer, *Anglo-German Rivalry on Coal Markets in France, the Netherlands and Germany, 1850–1913* (Research Memorandum: University of Groningen, 1995).

Fressoz, Jean-Baptiste, 'Pour une histoire désorientée de l'énergie', *Entropia, revue d'étude théorique et politique de la décroissance*, 15 (2013), 173–87.

Gaboriau, Philippe, 'Les trois âges du vélo en France', *Vingtième Siècle*, 29 (1991), 17–34.

Galpern, Steven, *Money, Oil, and Empire in the Middle East. Sterling and Postwar Imperialism, 1944–1971* (Cambridge: Cambridge University Presss, 2009).

Geiger, Reed, *The Anzin Coal Company 1800–1833: Big Business in the Early Stages of the French Industrial Revolution* (Newark, NJ: University of Delaware Press, 1971).

Gille, Bertrand, ed., *Histoire des techniques: Technique et civilisations. Technique et sciences* (Paris: Gallimard, 1978).

Girardier, Jean-Pierre, 'Les Pompes solaires thermodynamiques et leur histoire', in *L'Énergie solaire en France*, ed. by Alexandre Herléa (Paris: Éditions du CTHS, 1995), pp. 127–43.

Gras, Alain, *Fragilité de la puissance, se libérer de l'emprise technologique* (Paris: Fayard, 2003).

Hertner, Peter and Henry V. Nelles, 'Contrasting Styles of Foreign Investment. A Comparison of the Entrepreneurship, Technology and Finance of German and Canadian Enterprises in Barcelona Electrification' *Revue économique*, 58.1 (2007), 191–214.

Harrison Moore, Abigaïl and Ruth W. Sandwell (eds.), *In a New Light: Histories of Women and Energy* (Montreal, Kingston: McGill-Queen's University Press, 2021).

Hatcher, John, *The History of the British Coal Industry,* vol. 1, *Before 1700: Towards the Age of Coal* (London: Clarendon Press, 1993).

Hausman, William, Peter Hertner and Mira Wilkins, *Global Electrification: Multinational Enterprise and International Finance in the History of Light and Power, 1878–2007* (Cambridge: Cambridge University Press, 2008).

Hecht, Gabrielle, *Being Nuclear. Africans and the Global Uranium Trade* (Cambridge, MA: The MIT Press, 2012).

———, *The Radiance of France: Nuclear Power and National Identity after World War II.* (Cambridge, MA: The MIT Press, 1998).

———, 'Rebels and Pioneers: Technocratic Ideologies and Social Identities in the French Nuclear Workplace, 1955–69', *Social Studies of Science,* 26 (1996), 483–530.

Heynmann, Matthias, 'The Shaping of Wind Technology Styles in Germany, Denmark, and the United States, 1940–1990', *Technology and Culture,* 39.4 (1998), 641–70.

Hirsh, Richard, *Power Loss: The Origins of Deregulation and Restructuring in the American Electricity Utility System* (Cambridge, MA: The MIT Press, 1999).

——— and Christopher Jones, 'History's Contributions to Energy Research and Policy', *Energy Resesearch & Social Science,* 1 (2014), 106–11.

Högselius, Per, *Red Gas: Russia and the Origins of European Energy Dependence* (London: Palgrave Macmillan, 2013).

Huber, Matthew, *Lifeblood: Oil, Freedom, and the Forces of Capital* (Minneapolis, MN: University of Minnesota Press, 2013).

Hughes, Thomas, *American Genesis: A Century of Invention and Technological Enthusiasm, 1870–1970* (London: Viking, 1989).

———, 'The Evolution of Large Technological Systems', in *The Social Construction of Technological Systems,* ed. by Wiebe Bijker, Thomas Hughes, and Trevor J. Pinch (Cambridge, MA: The MIT Press, 1987), pp. 51–82.

———, *Networks of Power. Electrification in Western Society, 1880–1930* (Baltimore, MD: Johns Hopkins University Press, 1983).

James, Frank, 'Mad Dogs and (Mostly) Englishmen: Colonial Relations, Commodities, and the Fate of Inuit Sled Dogs', *Inuit Studies, Études/Inuit/Studies,* 34.2 (2010), p. 129–47.

Jarrige, François, 'Actionner les pétrins. Les trajectoires énergétiques de la boulangerie française (1780–1930)', in *Sous le soleil. Systèmes et transitions énergétiques du Moyen Âge à nos jours,* ed. by Geneviève Massard-Guilbaud and Charles-François Mathis (Paris: Publications de la Sorbonne, 2019), pp. 311–28.

———, 'Métiers de chiens. Le moteur quadrupède ou l'intolérable travail canin, 1850–1950', *Jef Klak-Critique sociale & expériences littéraires,* 3 (2016), 132–42.

———, '"Mettre le soleil en bouteille": les appareils de Mouchot et l'imaginaire solaire au début de la Troisième République', *Romantisme,* 150 (2010), 85–96.

——— and Vrignon, Alexis (eds.), *Face à la puissance. Une histoire des énergies alternatives et renouvelables à l'âge industriel* (Paris: La Découverte, 2020).

Jasanoff, Sheila and Sang-Huyn Kim, 'Containing the Atom: Sociotechnical Imaginaries and Nuclear Power in the United States and South Korea', *Minerva,* 47 (2009), 119–46.

Johnston, Sean, *The Neutron's Children: Nuclear Engineers and the Shaping of Identity* (Oxford: Oxford University Press, 2012).

Jones, Christopher, 'Petromyopia: Oil and the Energy Humanities', *Humanities*, 5 (2016), 1–10.

Josephson, Paul, ''Projects of the Century' in Soviet History: Large-Scale Technologies from Lenin to Gorbachev', *Technology and Culture*, 36 (1995), 519–59.

Kaijser, Arne, 'Driving on Wood: the Swedish Transition to Wood gas During World War Two', *Technology and Culture*, 37/4 (2022), 468–86.

Kander, Astrid, Paolo Malanima and Paul Warde, *Power to the People: Energy in Europe Over the Last Five Centuries* (Princeton, NJ: Princeton University Press, 2013).

Keeling, Arn, '"Born in an Atomic Test Tube": Landscapes of Cyclonic Development at Uranium City, Saskatchewan', *The Canadian Geographer*, 54 (2010), 228–52.

Khelfaoui, Mahdi, 'De la logique d'arsenal à la logique de marché: Les transformations institutionnelles d'Énergie atomique du Canada limitée, 1975–1994', *Journal of Canadian Studies/Revue d'études canadiennes*, 54.1 (2020) 108–36.

Krasnodębski, Marcin, 'Upscaling Forest Waste: The French Quest for Fuel Autarky after World War I', *Technology and Culture*, 62.1 (2021) 107–27.

Kryza, Franz, *The Power of Light. The Epic Story of Man's Quest to Harness the Sun* (New York: McGraw-Hill, 2003).

Lagendijk, Vincent, *Electrifying Europe: The Power of Europe in the Construction of Electricity Networks* (Amsterdam: Aksant, 2008).

Laird, Frank, 'Constructing the Future: Advocating Energy Technologies in the Cold War', *Technology and Culture*, 44 (2003), 27–49.

Langewiesche, Katrin, 'Les limites du reboisement au Burkina Faso et au Bénin. Du travail forcé à l'approche participative', *Politique africaine*, 96 (2004), 196–211.

Le Renard, Claire, 'The Superphénix Fast Breeder Nuclear Reactor: Cross-border Cooperation and Controversies', *Journal for the History of Environment and Society*, 3 (2018), 107–44.

Lopez, Fanny, *Le Rêve d'une déconnexion. De la maison autonome à la cité auto-énergétique* (Paris: La Villette, 2014).

———, *L'ordre électrique. Infrastructures énergétiques et territoires* (Genève: Métis presses, 2019).

Macshane, Clay and Joel A.Tarr, *The Horse in the City, Living Machines in the Nineteenth Century* (Baltimore: Johns Hopkins University Press, 2007).

Malm, Andreas, *Fossil Capital: the Rise of Steam Power and the Roots of Global Warming* (London, New York: Verso Books, 2016).

Marnot, Bruno, 'À voile et à vapeur. Les ports maritimes en transition au XIXe siècle', in *L'Europe en transitions: énergie, mobilité, communication, XVIIIe–XXIe siècles*, ed. by Yves Bouvier and Léonard Laborie (Paris: Nouveau Monde Éditions, 2016), pp. 141–68.

Marrec, Anaël, 'Histoire des énergies renouvelables en France, 1880–1990' (unpublished doctoral thesis, University of Nantes, 2018).

——— and Teissier, Pierre, "Les énergies alternatives aux prises avec les politiques conservatrices", ed. by François Jarrige and Alexis Vrignon, *Face à la puissance, une histoire des énergies alternatives et renouvelables à l'âge industriel* (Paris: La Découverte, 2020), pp. 238–48.

Massard-Guilbaud, Geneviève, 'From the History of Sources and Sectors to the History of Systems and Transitions: How the History of Energy has Been Written in France and Beyond', *Journal of Energy History/Revue d'histoire de l'énergie*, 1 (2018), http://energyhistory.eu/node/88 [Accessed 2022/03/17].

Mathis, Charles-François, Virgili, Fabrice and Williot, Jean-Pierre, 'Foyers domestiques, genre et énergies : enjeux et perspectives', *Journal of Energy History/Revue d'histoire de l'énergie*, 6 (2021), http://energyhistory.eu/node/279 [Accessed 2023/07/10].

———, ——— and ———, 'Foyers domestiques, genre et énergies : enjeux et perspectives', *Journal of Energy History/Revue d'histoire de l'énergie*, 6 (2021), http://energyhistory.eu/node/279 [Accessed 2022/02/17].

———, ——— and ———, 'Foyers domestiques, genre et énergies : enjeux et perspectives', *Journal of Energy History/Revue d'histoire de l'énergie*, 6 (2021), http://energyhistory.eu/node/279 [Accessed 2022/02/17].

———, *La Civilisation du charbon* (Paris : Vendémiaire, 2021).

Mitchell, Timothy, *Carbon Democracy: Political Power in the Age of Oil* (New York, NY: Verso Books, 2011).

Molles, Camille, 'Imposer ou promouvoir. Les destins croisés de l'alcool-carburant et des gazogènes', in *L'Europe en transitions: énergie, mobilité, communication, xviiie–xxie siècles*, ed. by Yves Bouvier and Léonard Laborie (Paris: Nouveau Monde édition, 2018), pp. 73–82.

Muller, Pierre, 'La transformation des modes d'action de l'État à travers l'histoire du programme Airbus', *Politiques et Management Public*, 7.1 (1989), 247–72.

Mumford, Lewis, *Technique et civilisation* (Paris: Parenthèse, 2016).

Myllyntaus, Timo, 'Electrical Imperialism or Multinational Cooperation? The Role of Big Business in Supplying Light and Power to St Petersburg before 1917', *Business and Economic History*, 26 (1997), 540–49.

———, *Electrifying Finland* (London: Macmillan, 1991).

Myrna, Santiago, *The Ecology of Oil: Environment, Labor, and the Mexican Revolution, 1900–1938* (Cambridge: Cambridge University Press, 2009).

Neufeld, John, 'Multinational Enterprise and International Finance in the History of Light and Power, 1880s–1914', *Revue économique*, 58 (2007), 175–90.

Nikiforuk, Andrew, *The Energy of Slaves: Oil and the New Servitude* (Vancouver: Greystone Books, 2012).

Nye, David, *American Technological Sublime* (Cambridge, MA: The MIT Press, 1994).

———, *Consuming Power. A Social History of American Energies* (Cambridge, MA: The MIT Press, 1998).

———, *When the Lights Went Out: A History of Blackouts in America* (Cambridge, MA: MIT Press, 2010).

Olmstead, Alan M. and Paul W. Rhode, 'Reshaping the Landscape: The Impact and Diffusion of the Tractor', *American Journal of Economic History*, 61.3 (2001), 663–98.

Pádua, José Augusto, *Um sopro de destruição. Pensamento político e crítica ambiental no Brasil escravista (1786–1888)* (Rio de Janeiro: Jorge Zahar, 2002).

Parinello, Giacomo, 'Systems of Power: A Spatial Envirotechnical Approach to Water Power and Industrialization in the Po Valley of Italy, ca.1880–1970', *Technology and Culture*, 59.3 (2018), 652–88.

Parra, Francisco, *Oil Politics: A Modern History of Petroleum* (London: I.B. Tauris, 2004).

Perovic, Jeronim, *Cold War Energy: A Transnational History of Soviet Oil and Gas* (London: Macmillan, 2017).

Picon, Antoine, *Smart Cities: A Spatialised Intelligence* (Chichester: Wiley, 2015).

Podobnik, Bruce, *Global Energy Shifts. Fostering Sustainability in a Turbulent Age* (Philadelphia, PA: Temple University Press, 2006).

Pritchard, Sara B., *Confluence. The Nature of Technology and the Remaking of the Rhône* (Cambridge, MA: Harvard University Press, 2011).

Renn, Otwin, 'Coal, Nuclear and Renewable Energy Policies in Germany: From the 1950s to the Energiewende', *Energy Policy*, 99 (2016), 224–32.

Righter, Robert W., *Wind Energy in America. A History* (Norman: University of Oklahoma Press, 1996).

Rivals, Claude, *Le Moulin et le meunier dans la société traditionnelle française* (Ivry: Serg; Paris: Nerger-Levrault, 1987).

Roche, Daniel, *La Culture équestre de l'occident, XVIe–XIXe siècle. L'ombre du cheval*, vol. 1, *Le Cheval moteur. Essai sur l'utilité équestre* (Paris: Fayard, 2008).

Rosental, Paul-André and Jean-Claude Devinck, 'Statistique et mort industrielle. La fabrication du nombre de victimes de la silicose dans les houillères en France de 1946 à nos jours', *Vingtième Siècle. Revue d'histoire*, 95 (2007), 75–91.

Ruddiman, William F., *Plows, Plagues, and Petroleum: How Humans Took Control of Climate* (Princeton, NJ: Princeton University Press, 2005).

Sampson, Anthony, *The Seven Sisters: The Great Oil Companies & the World they Shaped* (New York, NY: Viking Press, 1975).

Santos, Gildas M., 'Energy in Brazil: a Historical Overview', *Journal of Energy History/Revue d'Histoire de l'Énergie*, 1 (2018), energyhistory.eu/en/node/56 [Accessed 2022/03/17].

Schivelbusch, Wolfgang, *Disenchanted Night: The Industrialization of Light in the Nineteenth Century* (Berg, 1988).

Schmid, Sonja, *Producing Power: The Pre-Chernobyl History of the Soviet Nuclear Industry* (Cambridge, MA: The MIT Press, 2015).

Shafiee, Katayoun, *Machineries of Oil: An Infrastructural History of BP in Iran* (Cambridge, MA: The MIT Press, 2018).

Shifflett, Crandall, *Coal Towns: Life, Work, and Culture in Company Towns of Southern Appalachia, 1880–1960* (Knoxville, TN: The University of Tennessee Press, 1995).

Sieferle, Rolf Peter, *The Subterranean Forest: Energy Systems and the Industrial Revolution* (Cambridge: The White Horse Press, 2001).

Silveira, Ana Claudia, 'Les moulins à marée du Portugal (XIIIe–XVIIIe siècle): une ressource énergétique pour les expéditions maritimes', in *Sous le soleil. Systèmes et transitions énergétiques du Moyen Âge à nos jours*, ed. by Charles-François Mathis and Geneviève Massard-Guilbaud (Paris: Publications de la Sorbonne, 2019), pp. 63–78.

Smil, Vaclav, *Energy in World History* (Boulder, CO: Westview Press, 1994).

———, *Energy Transitions: History, Requirements, Prospect* (Santa Barbara, CA: Praeger, 2010).

Sovacool, Benjamin, 'What are we Doing Here? Analyzing Fifteen Years of Energy Scholarship and Proposing a Social Science Research Agenda', *Energy Research & Social Science*, 1 (2014), 1–29.

Steinberg, Theodore, *Nature Incorporated, Industrialization and the Waters of New England* (Amherst: University of Massachussets Press, 1994).

Tester, Frank James, 'Can the Sled Dog Sleep? Postcolonialism, Cultural Transformation and the Consumption of Inuit Culture', *Journal of Marxism and Interdisciplinary Inquiry*, 3 (2010), 7–19.

Teissier, Pierre, 'Fours et maisons solaires de Mont-Louis-Odeillo. Interstices, intersciences et internationalismes de la recherche contemporaine', in *Dans le champ solaire. Cartographie d'un objet scientifique*, ed. by Guy Boistel and Stéphane Le Gars (Paris: Hermann, 2015), pp. 181–219.

Terrier, Didier, 'Moulins à vent, moulins à huile', in *Face à la puissance. Une histoire des énergies alternatives et renouvelables à l'âge industriel (XIXᵉ–XXᵉ siècles)*, ed. by François Jarrige and Alexis Vrignon (Paris: La Découverte, 2020), pp. 57–70.

Tiquet, Romain, *Travail forcé et mobilisation de la main-d'œuvre au Sénégal. Années 1920–1960* (Rennes: Presses universitaires de Rennes, 2019).

Topçu, Sezin, *La France nucléaire. L'art de gouverner une technologie contestée* (Paris: Seuil, 2013).

———, 'Les Physiciens dans le mouvement antinucléaire: entre science, expertise et politique', *Cahiers d'histoire*, 102 (2007), 89–108.

Ushida, Hoshimi, 'The Transfer of Electrical Technologies from the US and Europe to Japan, 1869–1914', in *International Technology Transfer: Europe, Japan and the USA, 1700–1914*, ed. by David Jeremy (Aldershot, Brookfield: Edward Elgar, 1991), pp. 219–41.

Van Brakel, Jaas, *The Ignis Fatuus of Biogas. Small-Scale Anaerobic Digesters ("Biogas Plants"): A Critical Review of the Pre-1970 Literature* (Delft: Delft University Press, 1980).

Venn, Fiona, *The Oil Crisis* (London: Routledge, 2016).

Wellock, Thomas, *Critical Masses: Opposition to Nuclear Power in California, 1958–1978* (Madison: The University of Wisconsin Press, 1998).

White, Leslie A., *The Evolution of Culture. The Development of Civilization to the Fall of Rome* (Walnut Creek, CA.: Left Coast Press, 2007).

White, Richard, *The Organic Machine: The Remaking of the Columbia River* (New York, NY: Hill and Wang, 1996).

Williot, Jean-Pierre, 'La Respiration des gazomètres ou la consommation de gaz manufacturé à Paris de 1860 à 1960', *Flux*, 51 (2003), 83–88.

Winskel, Mark, 'Autonomy's End: Nuclear Power and the Privatization of the British Electricity Supply Industry', *Social Studies of Science*, 32 (2002), 439–67.

JÉRÔME LAMY, ARNAUD SAINT-MARTIN

Aeronautics and Aerospace

Globalization and Technical Junctions

Two major features characterize the historical, sociological, and anthropological studies on aeronautics and space technologies. On the one hand, they tend to analyse aeronautics and aeropace separately as if the two fields were distinct, whereas they have much in common with many techniques moving from one to the other. On the other hand, the prevailing self-referential approach (see Lemonnier, 2004) tends to analyse all social, political, and cultural phenomena as resulting solely from technological innovation. This approach may be useful in that it reveals some details of how these technical systems operate, but it fails to see the whole picture. In this article, we aim to avoid these two pitfals by highlighting both the interconnections between aeronautics and spatial technologies and the way they connect to the economic, political, cultural, and scientific programmes that have been instrumental to their development. This connection is neither natural nor obvious, but rather the result of a socio-historical construct whereby technologies are conceived of as systems.

The very distinction between airspace and outer space derives from a cognitive and social construct which at times has contributed to blurring the differences between the fields of aeronautics and space technology. Theodore von Kármán proposed to locate the eponymous physical border of the Earth's atmosphere at 100 km (about 62 miles) above the Earth's mean sea level. Although this 'Kármán line' was accepted by aeronautics authorities in the 1950s, particularly by the International Aeronautics Federation, it has neither been universally adopted — the United States determined it to be at 50 miles — nor considered undeniable (Jonathan C. McDowell (2018) for instance placed it at a distance of 80 km), and the issue still causes taxonomical quarrels.

Propulsion is the other connection and technical issue linking the two domains that comes to mind. Mastering the right conditions and tools for moving vehicles and machines of varying size and weight whether in the air or outer space has always proved to be a technical challenge, ever since the invention of aircraft and the beginning of space travel. The knowledge and techniques of the early 'Aerospatiale' came from the military; this was the case, for instance, for applied mathematics, ballistics, mechanics, and projection as well as research and development on engines, aircraft, and rocket design. The different types of technologies involved in various propulsion systems naturally prompted decades of specific research (on turbo engines, turbo jets,

Global History of Techniques, ed. by Guillaume Carnino, Liliane Hilaire-Pérez and Jérôme Lamy, Global Matters Culture Matérielles, histoire des techniques et circulation globale des savoirs, 9 (Turnhout: Brepols, 2024), pp. 337–350
BREPOLS PUBLISHERS
10.1484/M.TECHNE-EB.5.129815

liquid-propellant rocket engines, etc.); while the physical mastery of flight conditions was also based on twin technologies whether it be in order to transport human beings in aircraft cabins or in spaceships, or to carry cargo in a plane or in the warheads of missile launchers. This was particularly true at the height of the Cold War, when intercontinental ballistic missiles from which the 'Space Race' conventional launchers were derived delivered thermonuclear warheads.

This chapter considers the links that were established in this way as well the most significant differences between these technological fields. These connections are determined by economic, political, and cultural factors and vary according to place and time. In order to make the reconsideration of this long history of air and space transport more manageable, we have chosen to focus on five specific areas central to both aeronautics and space technology, which will allow us to study their connections: (1) the development of an ever-changing industrial sector, (2) early militarization, (3) the prevalence of political decisions in the applications of technical systems, (4) specific technological trajectories, and (5) a notable tendency to create spectacular projects, a feature of an age marked by hubris and what some have called the 'technological sublime' (Nye, 1994). An investigation of the intersection of aeronautics and aerospace does not really fit into mainstream historiography. The landmarks of a chronological history of clearly defined technological sectors are naturally not included in the following analysis. Similarly, rather than referring to the pantheon of the leading figures of the hero histories of aeronautics and space, which often lend themselves to hagiographic or patriotic narratives (Chadeau, 1987), we have chosen to focus on those processes that eventually led some areas of activities to develop into stable technical systems, industrial organizations, and public and private administrative organizations. These are connected due to their joint expertise and procedural norms, and due to their similar if not identical values as we will examine below.

Defence Spending

Along with the notion of 'militarization', a common trope of geopolitical literature is the idea of the dependence of aeronautics on state structures turned 'space powers'. From the experimental, pioneering stage before 1914 to the concentration of the sector into a few major firms at international level today, aeronautics represents a very specific economic activity. Politics has been prominent within it, and from a very early stage. After the First World War, British aviation clearly was more advanced than other countries. The leaders of the Royal Air Force took advantage of their colonial conquests to establish 'a relatively cheap means of imperial control in sparsely populated new "mandates" in the Middle East' (Edgerton, 2013: 33); scientists and engineers as well as postal communications moved around a global empire. Even Western countries with smaller colonial territories relied on the aeroplane. The 1930s Italian fascist regime, for example, saw aviation as part of the 'constitution of material networks that enabled colonial control through transport and speed' (Caprotti, 2011: 398).

In warfare, aviation was an indispensable part of the military strategy of defence and attack. In some countries, the extent of the aeronautics sector was even demarcated by governmental decisions. That was the case in the 1930s in the USSR, where the army implemented 'the Douhet doctrine of strategic offense, which emphasized the role of heavy bombers in modern warfare' (Bailes 1976: 79). Nevertheless, this techno-political orientation was first and foremost intended to assert the 'legitimacy' of Stalin's power, rather than serve an actual military strategy.

The war effort completely transformed aviation and its manufacture. Vast orders were placed to produce planes, and manufacturers wanted to be able to fulfil them. The history of the aeronautics industry is the story of an increasing prominence of major competitive groups, initially at the national, and then at the international level. In this respect, the company that was to become Lockheed Martin is a good example of the establishment of a 'military-industrial complex' from the 1950s onwards, a complex which required a great adaptability to market downturns and steady political support (Hartung, 2010). For Boeing, its 'dual civil-military' business was a factor of stability. This duality has enabled these different sectors to adapt constantly (Depeyre, 2013: 62). In France, Dassault's focus on military aeronautics activities was based on a strategy of 'diversification' particularly in 'the command of military weaponry systems' (Carlier, 2013: 30, 32).

Because engines are used for military purposes, their design and development as well as their use and maintenance are classified as 'top secret'. The mystery in which they are shrouded has contributed to turning some planes whose technological set up is only partially known into quasi myths. Aerial or spatial reconnaissance programmes, e.g. the CORONA programme of satellites deployed under the CIA authorities between 1959 and 1972 (Day, 1998), were not disclosed until decades after they were conducted. The technology used in the field of aerial reconnaissance was re-used in space. The military strategy at work during the Cold War was based on cross-geopolitical power surveillance and contributed to the connection of air and space technologies and to the globaliation of surveillance systems.

Countries' preparation for a generalized conflict over the course of thirty years, from the 1950s to the 1980s standardised and at times merged the air and space sectors. As early as 1959, American forces carried out an unsuccessful missile strike attempt against the satellite *Explorer 6* from a B47 bomber (Bille, Lishock, 2004: 169). In 1977, the American army plans to use F15 fighter planes to destroy Soviet satellites (Gottfried, Lebow, 1985: 151). By contrast, the Strategic Defence Initiative proposed by the Reagan administration was geared towards driving the Soviet Union into an arms space race rather than merging air and space technologies.

Industrial Sectors and Golden Opportunities

Recent literature on the early aeronautics industry has uncovered the significant part played by small-sized organizations and and how public funding was not necessarily always the key to success. From the late nineteenth to early twentieth century, the 'first stable aeroplanes' were actually manufactured by small and medium size companies (Olivier, 2017). In the United States, the Wright brothers worked on a small scale and initially worked on both design and manufacture. Public funding (which could, at times, be significant, e.g. in the case of Samuel Pierpont Langley) did not necessarily grant success — far from it. The main reason was that early aeronautics was not organised according to the principles of contemporary industrial capitalism, but rather had developed within a sports culture based on exploits and feats. It only developed into an actual industrial and commercial sector requiring specific technological developments during the interwar period (Bilstein, 1969: 411), although this development was uneven (Heitmann, 1988: 327).

The concentration of production into an extremely small number of companies such as Boeing or Airbus able to oversee the large-scale manufacture of reliable civil and military aircraft was supported by the implementation of a network of factories located on different continents (Zuliani, Jalabert, 2005). And yet, this network was not systematically internationalized. Europe illustrates the limits of partner states' pro-active policies towards industrial integration. Attempts to set up a European economy of military aviation (which materialized in the Eurofighter project) were hindered by 'existing national options' (Hoeffler, Mérand, 2015: 76).

Beyond asserting a nation's power and being a symbol of imperial and military domination, aviation and aeronautics have become a political instrument of economic conquest since 1945. Because the sector is highly concentrated, governments are commonly directly involved in capital-intensive technological transfers or trade alliances. In the 1970s, the French President Georges Pompidou lobbied diplomatically to support the partnership between SNECMA (a French aircraft engine manufacturer which was later renamed Safran Aircraft Engines) and General Electric, which was formed to build the CFM-56 engine and to create the French-American CFM International joint venture. Beyond business interests, aeronautics is strategically important in international relations and protects the independence of those countries that manage to be self-reliant in the field. International agreements are therefore overseen, and even controlled, by governments who consider them necessary to the preservation of their power, and also to safeguard national industries against the risk of disintegration.

Astronautics, by contrast, started around the 1950s, following the early experiments with ballistic missiles. In the United States as in the USSR, the sector was organised under the structure of the state authorities, who were highly invested in it due to nuclear escalation and the Space Race. The armies' and NASA's purchasing and sub-contracting policies greatly influenced the setup of a 'private' industry, for which aeronautics was fertile ground and which helped introduce the notion of an 'aerospace sector' and supported its implementation (Bromberg, 1999). In Europe, national agencies have also been instrumental in organizing the industrial sector of astronautics. The European Space Agency, which was established in 1975 after several failed cooperative efforts, sought to develop the Ariane programme by combining an ambitious launch policy for the European space community with a relatively strong French control over the future of the companies tasked with building and operating the launcher (Krige, Russo, Sebesta, 2000: 455–90).

These technical systems rely on a number of critical infrastructures and dedicated organizations. On the ground, airports are entirely dedicated to ensuring the continuity of air transport operations, and their sites include runways, terminals, control towers, maintenance facilities, and garages, as well as customs and security systems (more stringent post-11 September 2001), air passenger traffic control, and cargo management. They also have the necessary connections from this self-contained world to the surrounding environment, e.g. other transport systems, and are integrated into the landscape, the local or regional economy, and into the geopolitics of countries. By contrast, access to military airports and the spaceports of governmental agencies or armies is not within everyone's reach. This is not surprising, as these comprise strategic facilities, destruction technologies, or launch sites for missiles and rockets. These sites are nevertheless embedded into specific sites which in turn they transform, like Kennedy space centre in Florida or the Guyanese spaceport in French Guiana. The latter's location in Kourou is emblematic of the relationship between launch facilities and imperialism (French Guiana became a French colony in the sixteenth century): the launch site extends, in a highly technical form, the French colonial project of controlling a distant territory that is administered according to the principles of the metropole (Redfield, 2000, 2002). In addition to these infrastructures, which open up airspace and outer space, an entire regulatory environment has developed gradually, regulating and constraining movements. The air and space regulations implemented by governmental or non-governmental bodies to ensure viable traffic on airways and space orbits. Therefore, these aerospace industries and all their necessary additional practicalities are interdependent, changing the face of the world in the course of the twentieth century. They have now become vital parts of industrial and post-industrial societies. Intercontinental travel has become common for some parts of the population, just like the flows of information and the interconnected 'global village' that had been imagined in the 1950s have materialized in part via satellite telecommunications and 'cultures in orbit' (Parks, 2005).

Aerospace Strategies

As mentioned previously, aeronautics technology has a strong military component. The global conflicts of the twentieth century advanced the development of aircraft and embedded weapons. As early as the 1910s, 'military commissions' in France supported the country's small and medium sized companies in the sector, while the army had four suppliers — Blériot, Farman, Voisin, REP — which were expected to produce around forty aeroplanes each. In 1911 the number had gone up to five —

Blériot, Farman, Nieuport, Deperdussin and Breguet — each with 57 orders (Olivier, 2017: 216). Although public military orders consistently underpin the system of aeronautics capitalism, relations between the army and aeronautics were far from smooth in the early twentieth century. In Prussia, during the First World War, the 'nationalization of industry was never even a remote possibility because [...] the army firmly believed in competition among private enterprises' (Morrow, 1977: 51). Similarly, in the United States, the army was strongly averse to aeronautics innovations. At the beginning of the twentieth century, Major General George Owen Squier tried to convince military staff that the recent innovations in aeronautics were worth considering; above all, he wished to adopt the model set up by the British Advisory Committee for Aeronautics (BACA) (Gross, 1990: 287), which had been carrying out top-level research in aeronautics in Britain since 1909. After the National Advisory Committee for Aeronautics (NACA) finally emerged in the United States in 1915, Squier continued to try to convince the army to create 'a giant American aircraft production program' (Gross, 1990: 292), but 'was alarmed by the army's highly fragmented aeronautical engineering efforts' (Gross, 1990: 296). It is therefore clear that creating military structures dedicated to aeronautics research was not an obvious path, even during the First World War. And yet, the First World War dramatically changed the range of long-term possibilities opened by aeronautics to military action. In wartime Britain, 'British aviation was relatively backward compared with that of France' (Fearon, 1985: 23). The 'crucial change was that, by 1918, flight was an intrinsic part of national defence, as large aircraft could fly to Berlin and return, the possibility of carrying fare-paying passengers over long distances was now more than a visionary gleam — it was a reality' (Fearon, 1985: 24).

Therefore, war served as a demonstration to Western governments that aeronautics could both contribute to convincingly building a strong geopolitical power while harnessing the emerging market of long-distance transport. The post-1945 period saw the continuation of this competitive logic marked as it was by countries trying to develop nuclear weapons as the ultimate deterrent. That is why the United States, the USSR, and France developed what Claude Carlier and Sylvain Champonnois have called 'the nuclear triad', i.e., 'strategic bombers, ground-based ballistic missiles, [and] nuclear-powered ballistic missile submarines (SSBN)' (Carlier, Champonnois, 2014: 361). Aviation was now part of a scheme which ruled geostrategic relations during the cold war. Over the course of six decades, aeronautics had thus moved from an exploratory sector to a vital instrument of political prestige and power. As outlined above, this gradual inclusion of aeronautics in public military programmes was complemented by the development of a dedicated industrial sector which was dependent on public procurement and defence spending.

Trajectories of Technological Invention

Aeronautics has undergone significant technical changes over the course of more than a century. These changes concern both specific parts of aircrafts (e.g. retractable landing gear and wings) and propulsion itself which has always acted as driving force of technological development. Historians in the field of techniques have highlighted that the aeronautical sector is uniquely defined by the way innovations emerge and integrate into existing processes. What follows focusses on some iconic examples resulting from two types of modelling.

Walter G. Vincenti proposed a general theory of aeronautical engineering including its learning processes and knowledge, and the way they relate to production (Vincenti, 1990). Notably, he explained the invention of the retractable landing gear as the result of a 'variation-selection' process. For that purpose, he examined the work of John K. Northrop, 'one of the most creative and influential airplane designers in the quarter-century following the mid-1920s' (Vincenti, 1994: 2). Northrop's Alpha aircraft had 'fixed landing gear' (Vincenti, 1994: 6). The idea of a retractable landing gear was mentioned in a report to the 'technical advisory of United Aircraft' (Vincenti, 1994: 7). But trials carried out at the California Institute of Technology convinced Northrop 'to adopt the pants-type gear in place of a retractable design' (Vincenti, 1994: 7–8). The subsequent versions of Northrop's aircraft (particularly XFT-1) led to a multitude of tests and experiments. Vincenti emphasises that Northrop did realize that a retractable landing gear was more operational, but that his choice of a 'fixed gear looks to be a temporary deviation from an engineering mainstream' (Vincenti, 1994: 14). The retractable gear did not result from an initial technical difficulty but was part of a 'variation-selection model' characterised by two successive 'phases', one 'hidden', the other 'overt' (Vincenti, 1990: 22). Consequently, engineers created 'a variety of variants'; of the overt variants, among others, the fixed landing gear and the retractable gear were retained as feasible (Vincenti, 1990: 23). After using the fixed gear, the daily control of difficulties actually made it possible to implement the retractable gear which had remained on the list of the engineers' solutions (Vincenti, 1990: 24).

The other technological model, which was proposed by F. G. Geels, relies on an analysis of co-evolutions and 'multi-level dynamics in transitions'. The iconic illustration of this model is the shift from propeller to turbojet between the 1930s and the 1970s (Geels, 2006). The way innovations develop can be traced at three levels: niche innovations that 'act as "incubation room" for radical novelties', the 'meso-level' innovations which are 'formed by socio-technical regimes', and those on the 'macro-level [...] formed by the socio-technical landscape' which are linked 'to aspects of the exogenous environment' (Geels, 2006: 1004). The breakthrough therefore moves from a constrained niche innovation to a class of learning processes for engineers before becoming a 'new technology' (Geels, 2006: 1004). Geels noted that in the inter-war period, passengers 'complained about noise, vibration, turbulence and airsickness', which led airline companies to demand 'larger aircraft that could fly higher to reduce turbulence and airsickness. They also wanted greater speed and longer range' (Geels, 2006: 1007). However, in this regard, propellers imposed limitations, as they could not reach high speed levels. Both engineers and scientists worked on 'the turboprop, using the rotational power of a gas turbine to spin a shaft that powered a propeller', but aeronautics specialists did not believe in this particular innovation. Turbojets then 'emerged independently from the work of Franck Whittle in Britain, Hans von Ohain and Herbert Wagner in Germany', who were 'relative outsiders to the aviation regime' (Geels, 2006: 1008). In that case, the scope of the technological domain was restricted due to the aeronautical community's lack of involvement. The war 'created a window of opportunity for the jet engine pioneers' (Geels, 2006: 1008); then a potential market for the gradual appropriation of the technological breakthrough opened up. Turbojets became part of the aeronautics technical apparatus after the Second World War, when they were adopted by civil aviation (Geels, 2006: 1010). 'Expanding markets created space for further stretching of existing aircraft designs, leading to the Super Constellation (1950) and DC-7 (1953)' (Geels, 2006: 1011). The turbojet was one step that led aeronautics innovation to develop gradually and spread commercially to the world.

The constraints of outer space travel triggered aerospace research. Technological research and dangerous experiments sending flights to the brink of outer space have now become part of the legend and of popular culture imagery with films such as *The Right Stuff* (1983). The US Air Force's X-15, a rocket-powered aircraft, revealed the great skills of pilots as well as the dangers they ran during the flights (Evans, 2013). Concepts for futuristic spaceplanes form part of the history of utopias about 'colonising' space, but realism prevailed for a long time — until the careful introduction of space shuttles happened and circum-terrestrial space flights became routine in the post-*Apollo* period. The design and manufacture of this extraordinary and supposedly reusable vehicle turned out to be extremely complex. The 1986 Challenger accident marked a tragic turning point in the growth of the sector. Spaceplane projects were not dropped entirely, but restricted to suborbital 'tourism' under the Karman line, as was the case, for example, for *SpaceShipOne* which won the 2004 XPRIZE — a distant echo of the Orteig Prize won by Charles Lindbergh in 1927.

These two instances of technological transformation clearly highlight the specific nature of aeronautics, i.e. the abundance of engineering proposals that is a necessary in the preparation for the following selecting stage, with the upscaling of innovation depending on the capacity of engineers (in the case of turbojets, military engineers) to work on the breakthrough and prepare its roll-out (Figure 1).

Hubris and Extravagance

Finally, aeronautics and space technologies, just like so many other technologies, are characterized by excess. Here, hubris does not consist of mobilizing huge scientific, technical, financial, and military resources or 'big science' alone. Air and space experiments also concern the physical risks taken by pilots and passengers. Mikael Hård and Andrew Jamison describe the Wright brothers' early experiments at the very beginning of the twentieth century: 'With the help of wind tunnels, and expert advice from scientists in Washington about aerodynamic principles, the Wright brothers were the ones who won the race to achieve manned flight. It was one of the oldest forms of hubris' (Hård, Jamison, 2005: 97).

Fig. 1. STS-70 Rollout NASA on The Commons.

Aeronautics and space hubris led to extreme experiments in which human bodies were confronted with unfamiliar and potentially dangerous physical conditions. From aviation pioneers to the first astronauts putting their lives at risk in space capsules that were being tested, experimental practices in air flight and space have always had a similar share in excess.

Examples of gigantism in aircraft design are plentiful, from the Hughes H-4 Hercules to Airbus A380, and of course Stratolaunch, the world's biggest airplane designed by Scaled Composites. Flying these giants required extremely powerful engines that swallowed much fuel. There again, the design of this type of aircraft took advantage of the techniques mobilized by the army during the Cold War. The engineers' creativity was boundless. The technological competition between the two superpowers pushed the experimentation with nuclear energy, e.g. in the Aircraft Nuclear Propulsion Project and its incredibly long flights (James, 2000) or the 'nuclear ramjets for unmanned bombers or cruise missiles' in the Pluto project (Hacker, 1995: 85). These technologies were certainly promising, but their implementation could be risky. Despite its positive results, the Pentagon put an end to the programme in 1964 (Hacker, 1995: 89), just like Eisenhower had stopped the ANP in 1956 (James, 2000: 173). For these programmes, engineers were invited to push their imagination to the very limits of the possible in engineering. The Lockheed division of Advanced Development Projects, also nicknamed 'Skunk Works', was known for its bold projects of legendary aircrafts, notably the U-2 tested on the famous Area 51, or, more recently, the F-22 Raptor. Whether it be designing stealth aircrafts, RPAVs (remotely piloted aerial vehicles), or ICBMs (Inter-Continental Ballistic Missiles), or working on projects of military satellites, engineers have summoned all their inventiveness and the spirit of competition to create the most lethal weapons.

From the time when V2 plans were brought back to America to the colossal Saturn-5, the Defence and NASA space programme engineers achieved the construction of an extraordinarily complex technical system. Soviet engineers were just as proactive and developed the N-1 launcher *Herkules*, which, although it was abandoned in 1974 after a number of failures, manifested unprecedented technological research work. Research in design grew to be extremely sophisticated. Countless spectacular and futuristic projects were designed by engineers. For example, projects relating to space stations triggered the imagination of developers of passenger travel via spacecraft from the very beginnings of the space age, and an 'astroculture' emerged (Neufeld, 2018: 137). The early visions were always extremely bold and contributed to making 'spaceflight' tangible; at the time, this was as much a matter of 'conquering' space as a matter of travel. In the 1970s, after the the achievement of the lunar landing of *Apollo* in 1969 which captured imaginations worldwide, there was much speculation about the next steps. Settlement projects varied from visions of escapes aboard vast space settlements, akin to the O'Neill 'cylinders' in *The High Frontier* (1976), to settlement on Mars, which is still a source of inspiration for thousands of space enthusiasts in the United States. The fantasy of humans who, when extracted from their natural environment, finally enjoy a fulfilled life in the cosmos is an ancient dream that is now pursued by those known as the 'New Space' entrepreneurs; from Jeff Bezos to Elon Musk, 'visioneers' combine the hubris of high-tech engineering with utopian visions, to dream up and engineer the bright new worlds of a forthcoming future. (McCray, 2013). In the meantime, prior to the advent of these 'new frontiers', both entrepreneurs and space agencies have been trying to keep up the pace and develop oversized super heavy-lift launch vehicles like the *Space Launch System* implemented by the NASA or the *Super Heavy/Starship* launcher designed by SpaceX to transport the first 'settlers' to Mars.

The five main sectors characterizing aeronautics (industrial involvement, the politicization of technology, close links with the military sector, the specificities of technical evolutions, and hubris) apply to space as well. Moreover, these five distinctive features closely intersect: the military-industrial complex is a major provider of large-scale projects, as demonstrated in the case of ANP and the Pluto project. Similarly, some innovations would not have emerged without the involvement of military engineers or without the creation of a civilian market that is strongly regulated and backed up by governments. This interrelation reached its highest levels in the space sector immediately after the Second World War.

This long history is full of glorious stories, failed ambitions, worldwide celebrations, ruined projects, models, and rusted metal sheets. The growing environmental concerns of the younger generations in the twenty-first century puts the future of these technologies into question. Although aircraft engines are more fuel efficient than they used to be, and although the experiments on solar planes help move away from the use of fossil fuels to propel planes, the climatic impact of air transport is clearly a subject of concern for those organizations in charge of expanding it, especially as air traffic keeps growing worldwide, foreshadowing congested airways and hub airports. The filling of circumterrestrial space with satellites also raises problems of management and pollution. Before the 2000s, this was a matter of indifference (space was 'open'), but now the increasing volume of space debris is a source of worry for the space community (Saint-Martin, 2016) — the prospect of collisional cascading (the Kessler syndrome) is no longer a sci-fi scenario, and the governance of orbital systems firmly needs to be adjusted. Apart from this, the militarization of space has become a strategic issue for world leaders.

In spite, or rather because of these obstacles, imagination and anticipation have reappeared in the last few years. The futuristic fantasies of space exploration, which had faded from the 1990s onwards, are back in the news, although the outcome of this 'new Space Race' is both uncertain and fragile. For the most enthusiastic proponents of this renewed space race, it will make it possible to escape from spaceship Earth in the hope that mankind will be able to survive in the farthest reaches of the solar system in Gerard O'Neill-like orbital settlements. In such scenari, strategic and technological hubris continues to fuel investments in aeronautics and space. Therefore, the globalization of these two sectors relies on geopolitical configurations of confrontation (similar to what underpinned the Cold War or colonization), but also on the industrial strategies of major corporate groups (since both space and aeronautics require significant resources), as well as on clearly stated state strategies (since air and earth surveillance are a prerequisite of territorial sovereignty), and on specific technological trajectories. The global pattern which emerges at the junction of these different determinants shows space and aeronautics to be more than ever the two sides of the same technological coin.

References

Bailes, Kendall E., 'Technology and Legitimacy. Soviet Aviation and Stalinism in the 1930s', *Technology and Culture*, 17.1 (1976), 55–81.

Bille, Matt and Erika Lishock, *The First Space Race. Launching the World's First Satellites* (College Station, TX: Texas A&M University Press, 2004).

Bilstein, Roger E., 'Technology and Commerce. Aviation and the Conduct of American Business, 1918–29', *Technology and Culture*, 10.3 (1969), 393–411.

Bromberg, Joan L., *NASA and the Space Industry* (Baltimore, MD: Johns Hopkins University Press, 1999).

Burigana, David, 'L'Accord SNECMA/General Electric et les origines de CFM International. Succès 'Global' franco… ou euro-américain? Les dessous diplomatiques d'une affaire techno-industrielle', *Histoire, économie & société*, (2010), 85–101.

Carlier, Claude, 'Dassault Aviation: de la renaissance à une diversification cohérente', *Entreprise et histoire*, 73 (2013), 30–43.

——— and Sylvain Champonnois, 'L'Adaptation de l'armée de l'air aux stratégies et aux armes nouvelles (1945–1958)', in *Histoire de l'armée de l'air et des forces aériennes du XVIIIe siècle à nos jours*, ed. by Jean-Marc Olivier (Toulouse: Privat, 2014), pp. 309–63.

Caprotti, Federico, 'Visuality, Hybridity, and Colonialism: Imagining Ethiopia Through Colonial Aviation, 1935–1940', *Annals of the Association of American Geographers*, 101.2 (2011), 380–403.

Chadeau, Emmanuel, *De Blériot à Dassault, l'industrie aéronautique en France (1900–1950)* (Paris: Fayard, 1987).

Day, Dwayne A., John M. Logsdon, and Brian Latell, eds, *Eye in the Sky: the Story of the Corona Spy Satellites* (Washington, DC: Smithsonian Books, 1998).

Depeyre, Colette, 'Boeing Boeing: la dualité civil-militaire source d'un rebond stratégique dans l'ère post-guerre froide', *Entreprise et histoire*, 73 (2013), 58–73.

Edgerton, David E., *England and the Aeroplane. Militarism, Modernity and Machines* (London: Penguin Books, 2013).

Evans, Michelle, *The X–15 Rocket Plane: Flying the First Wings into Space* (Lincoln, NE: University of Nebraska Press, 2013).

Fearon, Peter, 'The Growth of Aviation in Britain', *Journal of Contemporary History*, 20.1 (1985), 21–40.

Geels, Frank W., 'Co-evolutionnary and Multi-level Dynamics in Transitions: the Transformation of Aviation Systems and the Shift from Propeller to Turbojet (1930–1970)', *Technovation*, 26 (2006), 999–1016.

Gottfried, Kurt, and Richard Ned Lebow, 'Anti-Satellite Weapons: Weighing the Risks', *Daedalus*, 114.2 (1985), 147–70.

Gross, Charles J., 'George Owen Squier and the Origins of American Military Aviation', *The Journal of Military History*, 54.3 (1990), 281–306.

Hacker, Barton C., 'Whoever Heard of Nuclear Ramjets? Project Pluto, 1957–1964', *Icon*, 1 (1995), 85–98.

Hård, Mikael and Andrew Jamison, *Hubris and Hybrids: a Cultural History of Technology and Science* (New York, NY: Routledge, 2005).

Hartung, William D., *Prophets of War: Lockheed Martin and the Making of the Military-Industrial Complex* (New York, NY: Nation Books, 2010).

Heitmann, John A., 'Getting Places in a Hurry: the Development of Aviation in Long-era South Louisiana', *Louisiana History: The Journal of the Louisiana Historical Association*, 29.4 (1988), 325–42.

Hoeffler, Catherine and Frédéric Mérand, 'Avions de combat. Pourquoi n'y a-t-il pas d'européanisation', *Politique européenne*, 48 (2015), 52–80.

James, Carolyn C., 'The Politics of Extravagance. The Aircraft Nuclear Propulson Project', *Naval War College Review*, 53.2 (2000), 158–90.

Krige, John, Arturo Russo, and Lorenza Sebesta, *A History of the European Space Agency 1958–1987*, vol. 2, *The story of ESA, 1973 to 1987* (Noordwijk: ESA Publications Division, 2000).

Lemonnier, Pierre, 'Y a-t-il un chamane dans le cockpit? Sur quelques travaux d'histoire et de sociologie de l'aéronautique', *Technique & Culture*, 42 (2004).

McCray, Patrick. *The Visioneers: How a Group of Elite Scientists Pursued Space Colonies, Nanotechnologies, and a Limitless Future* (Princeton, NJ: Princeton University Press, 2013).

McDowell, Jonathan C., 'The Edge of Space: Revisiting the Karman Line', *Acta Astronautica*, 151 (2018), 668–77.

Morrow, John H. Jr, 'Industrial Mobilization in World War I: the Prussian Army and the Aircraft Industry', *The Journal of Economic History*, 37.1 (1977), 36–51.

Neufeld, Michael, *Spaceflight* (Cambridge, MA: The MIT Press, 2018).

Nye, David, *American Technological Sublime* (Cambridge, MA: The MIT Press, 1996).

Olivier, Jean-Marc, 'Produire les premiers avions fiables, une affaire de PME (fin du XIXe siècle–1914)', *Marché et organisations*, 30 (2017), 207–19.

O'Neill, Gerard K., *The High Frontier. Human Colonies in Space* (New York: William Morrow, 1976).

Parks, Lisa, *Cultures in Orbit: Satellites and the Televisual* (Durham, NC: Duke University Press, 2005).

Redfield, Peter, 'The Half-Life of Empire in Outer Space', *Social Studies of Science*, 32.5–6 (2002), 791–825.

———, *Space in the Tropics. From Convicts to Rockets in French Guiana* (Berkeley, CA: University of California Press, 2000).

Saint-Martin, Arnaud, 'Du big sky à l'espace pollué: l'effet boomerang des débris spatiaux', *Mouvements*, 87 (2015), 36–47.

Vincenti, Walter G., 'The Retractable Airplane Landing Gear and the Northop 'Anomaly': Variation-Selection and the Shaping of Technology', *Technology and Culture*, 35.1 (1994), 1–33.

———, *What Engineers Know and How They Know It* (Baltimore, MD: The Johns Hopkins University Press, 1990).

Zuliani, Jean-Marc and Guy Jalabert, 'L'Industrie aéronautique européenne: organisation industrielle et fonctionnement en réseaux', *L'Espace géographique*, 34 (2005), 117–44.

ARNAUD PASSALACQUA

Mobility Techniques

Between Globalization and Local Solutions

It may seem somewhat counterintuitive to want to return to the technical dimensions of the history of mobility given the way the field has tended to evolve away from this approach. The most recent developments in the field of what is now known as 'mobility studies' indeed have diverged further and further away from the technical dimensions of the systems that are being studied. The recent interest in uses rather than production, in visual culture, gender or in the political approaches to mobility studies tend to ignore questions of technique. Admittedly for a long time, the more cultural and political approaches to mobility, now at the center of the literature, had been totally absent from it. The work done in the last fifteen years to some extent redresses this absence. It may also be understood as a result of scholars wanting to differentiate themselves from the large, vocal community of old train, plane or car enthusiasts.

While the technical aspects of mobility may currently receive limited attention, issues related to connections and the circulation of ideas, of models, and systems is currently at the heart of the history of mobility. The creation of the journals *Mobilities* (2006) and *Transfers* (2011), which have come to play an influential part in mobility studies, is one indication of this, as is the recent publication of works centering on transport (Mom, 2015) sometimes with a global perspective (Albert de la Bruhèze et al., 2016). This transnational approach appeared relatively recently in a field that has long emphasized national or even local mechanisms and frameworks at work in urban transport, an approach aligned with the very organization of archives.

The global perspective of transport specialists has been reinforced by the shift towards a more social, cultural and political approach to the history of mobility (Divall, Revill, 2005), but the historiography of the past should not be reduced to a focus on the national. Because companies themselves have often positioned their activities within international markets, some studies in economic history adopted this focus early on (McKay, 1976). However, it is true that the framework provided by archives may have encouraged researchers to concentrate on local or national spaces and write case studies that may be localized in their scope but still constitute crucial contributions to the field (Caron, 1997–2017).

This article will attempt a survey of the global history of mobility techniques based on several hypotheses. On the one hand, it will attend to the technical dimension of

Global History of Techniques, ed. by Guillaume Carnino, Liliane Hilaire-Pérez and Jérôme Lamy, Global Matters Culture Matérielles, histoire des techniques et circulation globale des savoirs, 9 (Turnhout: Brepols, 2024), pp. 351–362
BREPOLS PUBLISHERS 10.1484/M.TECHNE-EB.5.129816

mobile systems, both in relation to the very material dimension of these techniques and to their links to other human sciences, a reminder that techniques of mobility should also be considered in relation to the recent turn in mobility studies. On the other hand, I will demonstrate that generally, this history of mobility can only be understood as a history of transfers and circulation, even at the most local level. The purpose of this article is to discuss the dynamics of homogenization at work in this field because it is a history of techniques which themselves circulate. My objective is to demonstrate that these circulations are more powerful than one might think, insofar as they involve objects that are part of everyday life and rooted in specific environments. But they are also weaker than expected, as the ways of moving and the representations of the different systems still greatly vary across the world.

After a theoretical survey of the history of mobility as a history of techniques, I will investigate land and passenger transport systems in the two dimensions of their use, i.e. the global and the local one, although some systems belong to both. Freight transport, which to a large extent belongs to a different historiography, can only be touched upon here. Similarly, I have chosen to focus on transfers, while other aspects of mobile systems, such as their environmental impacts and safety issues won't be studied in detail. Neither will the global question of their effectiveness, although today this question leads many to question the place mobility infrastructures have come to occupy in our societies, in view of the opposition and dependencies they have created as a result of globalization.

The History of Mobility as a History of Techniques

Today's links between digital communication networks and mobility systems — which is not without precedent, as illustrated by the links between telegraphs and railways — might almost cause us to forget the materiality of mobility. However, even in a self-driving car, booked and paid for by smartphone, guided by GPS, electric and safe, a human being must at some point be involved, if only as passenger for it to be a means of transportation. This cooperation between the infrastructure and the person moving, which is intrinsic to the very definition of transportation, makes the human body a central element of the equation. Mobility necessarily involves a level of materiality.

Therefore, any mobility system must take into account the characteristics of the human body, in particular its dimensions and what it can withstand, e.g. in terms of acceleration, noise, or temperature, as well as its physiological needs (food, etc.). The person involved, being endowed with reason, also has human concerns, representations, and desires.

The history of mobility is therefore always shaped, if only indirectly, by the material reality of the human beings moving around. In keeping with the idea that technique can be defined as the mediation between human beings and their environment, mobility systems are therefore perfect examples of technical systems.

These systems evolve according to the context of each era and the different geographical areas in which they are located. Following the pre-industrial period, a time when the energy sources involved in transportation were either human (e.g. in walking), animal (e.g. in horse-drawn vehicles), or elemental (wind power and river currents), the use of fossil or even nuclear fuels have unleashed humans' ability to move around by removing many obstacles. Speeds as well as distances covered have increased significantly, transport costs have reduced. The services are now reliably available around the clock even despite geographical and meteorological challenges.

These developments seem to point to a widespread speeding up of the pace of lifestyles in industrial societies (Rosa, 2013) which is probably only illusory, since the distribution of human activities and settlements across the globe, people's travel needs and relationship to time have also changed. Indeed, some have even argued there has been an overall stability of travel times (Zahavi, 1974) or that other definitions of speed should be taken into account such as generalized speed (Illich, 1974).

While top speeds have increased, slow speeds have persisted, and the use of pre-industrial systems has even intensified. Thus societies have gone from uniform slowness to differentiated speed (Ollivro, 2000). The evolution of the technical ecosystem of mobility does not lead to the replacement of one technique by another but rather to their superimposition. Since few of them have disappeared entirely, they overlap leading to overlapping identities as well — in the late twentieth century, pedestrians were often described as motorists who had managed to park their cars. In the face of competing innovations existing techniques usually evolve by finding a space of relevance — even a niche — that allows them to continue: for example, horse-powered mobility turning into horse racing at the beginning of the twentieth century, cycling becoming an iconic sports in the media in the second half of the century, or, most recently, scooters, which stopped being associated with childhood to be used by many adults to move around the city. Meanwhile walking, the simplest of mobility systems, remains central to the daily routine of most people, including in cities that promote new modes of mobility such as Paris, where it still accounts for half of all journeys made.

This coexistence of old and new technologies (Edgerton, 2006) is reinforced by the fact that it is extremely difficult for a technology to become integrated into people's modes of travel, which partly explains the resilience of the old technologies, once they are established. The history of mobility therefore probably comprises more techniques that have never seen the light of day than those that have disappeared. To succeed, a system must, among other things, be reliable, correspond to a demand, be cost-effective, safe, practical, and comfortable, and fit into the existing overall landscape. All these criteria are difficult to meet, depending as they do on governance, geography, existing networks, paths of dependence, social representations, and other factors.

This calls into question the idea that the history of mobility can be told as a story of progress, with increasing speeds being reached or a growing number of people traveling. Are individuals more mobile today than they were in the past? This question is probably not that relevant. It would also be difficult to describe a society that is more advanced than another in terms of mobility. Unprecedented speeds often also mean unprecedented dependence on resources and expertise. Would nineteenth-century European colonialism have taken place without steam-powered vessels or telecommunications cables?

Indeed, it is local contexts and situations that lead to particular configurations of how people of different countries and cities travel. Considering mobility systems in terms of Marcel Mauss's famous formula, which defines a technique as an action that is 'effective and traditional', is one way of questioning them (Mauss, 1973). In the manner of François Sigaut (Sigaut, 2003), we can try to describe a mobility system that did not meet the three characteristics described by Marcel Mauss. What would such a system be if not an action? It would be a spontaneous, natural displacement, such as that of the stars or, on a smaller scale, those caused by natural phenomena on earth (e.g. storms or currents). This mobility would be a sort of unvoluntary, uncontrolled drifting. What would a non-traditional travel system be like? That is not rooted in social practices, or in geographical contexts, the physical frameworks that make its existence possible. What would be a system that no one knew how to use? Arguably systems are not universal but used by specific social groups. Last, what would an ineffective travel system look like, that is a travel system that produced no effect? It would be a stationary system, a system which has broken down or cannot be implemented. This system would only be a draft, a plan, or built as a prototype, but not implemented or used. However, it is not only in terms of a person's physical movement between two points that the effectiveness of the system must be judged. There are other criteria at stake such as cost, pollution, energy consumption, etc., while a system of travel may also result from decisions that are not only related to offering a new means of going from A to B but reflect a political stance, or stem from support for a specific industry. In this sense, effectiveness depends on all actors considered, which are not simply the people traveling.

Several major mobility systems have developed on a global scale, creating links from city to city, from country to country, and from continent to continent. The origins are ancient — one just needs to think of the Roman Empire operating via the Mediterranean maritime trade and its famous land routes. The industrial period introduced new, competing, and complementary networks, first the railways and then air transport, while navigation and road travel also continued to develop. The scale of these systems has often made them archetypes of large technical systems (Hughes, Mayntz, 1988). The movement of equipment, energy, capital, and, of course, passengers and goods they allowed has made these systems central to the functioning of industrial societies. Railway systems are a particularly famous example since they transformed many existing frameworks, such as the relationships between production and consumer markets, notions of time, and hierarchies between different cities. With the steam engine as both the result of and the key to the industrialization process, the transformative impact of railways has gone far beyond that of facilitating transport.

However, this system, which was adopted in many countries, is not unform. Different choices were made for the rail gauge and, despite the introduction of national standardization movements in the nineteenth century, they are still quite different from one country to another today. The same applies to loading gauges, signalling, traffic directions, and other technical aspects that are more or less easy to change, so that with the advent of high speed trains some countries, such as Japan and Spain, have had to develop systems that are incompatible with their historical infrastructure. Land systems, in contrast to air systems, continue to be subject to local parameters issues that make them potentially less uniform.

However, land systems are involved with certain forms of globalization, and the driving forces behind these deserve to be considered. One might think that political conditions must be met in order for such movements to take place, but ultimately they turn out to be quite flexible, as illustrated by the case of industrial cooperation across the Iron Curtain, as in the case of Fiat (Fava, 2013). The ability of mobility objects to transfer to different contexts is very strong here, even if meanings obviously change in the process. Therefore, this force is based on dynamics that are much more powerful than mere political conditions, namely on the simultaneous emergence of similar needs, in particular because of the nineteenth-century shift in Western industrial and urban society that took place in the 1820s and 1830s, as both railways and urban transport were being developed.

These conditions, however, are not enough to explain the existence of movement. Economic actors are also at work. Ever since the invention of the railway, transnational actors have contributed to developing systems far away from their usual place of activity. They were driven by the rise of capitalism whereby Western capital was invested in railway projects in, for example, Russia and the Ottoman Empire, as well as by the simultaneous emergence of colonialism, which underpinned the creation of railway infrastructures (Hurd, Kerr, 2012) and later cultural representations around the car or driving (Ponsavady, 2018).

We can also question the role of transport systems themselves in their own expansion. Do networks not follow a rationale of growth which might be constrained by economic and political conditions, but in which generally speaking growth is the norm? European high-speed rail projects, in contrast to the closures of old lines that had been ongoing for a long time, can be seen as an expression of this dynamic of network expansion, as can the creation of road or rail infrastructures whose monumental, even spectacular, architecture turns them into political statements about friendship or reconciliation between countries, as for example the Øresund Bridge (Löfgren, 2007) or the bridges of friendship between Thailand and Laos or Russia and North Korea. By contrast, war is also a vector for the expansion of networks and the expansion of mobile systems, as illustrated for example by the reuse of aircrafts and road vehicles after the First World War.

Mobility systems are, therefore, the source of international cooperation movements, some of which are the oldest international governance mechanisms, including the Central Commission for the Navigation of the Rhine, which was founded in 1815. Other bodies that have regulated transport networks include the Universal Postal Union (1874), the International Union of Railways (1922), and the International

Fig. 1. Between technical transfer, advertising promotion, and globalization stakes: Citroën-Kégresse caterpillar cars of the Citroën's Central Asia Mission, known as the Yellow Cruise, in circulation in Asia (1931) (*Le Monde colonial illustré* (1931) © Paris, Bibliothèque nationale de France).

Civil Aviation Organization (1947). As in the related field of communications, this is where technical diplomacy operates (Laborie, 2010).

But beyond their governance, networks can also be examined via straightforwardly political projects at the international level. This is the Saint Simonian idea that networks are bearers of both peace and economic prosperity (Chevalier, 1832). European projects are among the best examples for this, whether for roads (Schipper, 2008) or rails (Anastasiadou, 2011), but a project such as the Interstate Highway programme in the United States (Rose, 1990) also comes to mind. It helped forge links between the federal states and, in this sense, was underpinned by a project about improved mobility while it also fulfilled a political agenda.

Conversely, globalization is also key to mobility. It offers larger markets to companies that can then reduce unit costs, particularly in the car industry (Figure 1). This is risky nevertheless, and the history of mobility is also the history of many failures or semi-successes in exports, such as that of the Shinkansen (Hood, 2007).

But it is not only objects that circulate, but also the users of these transport systems, who can become agents of transfer. From this transnational perspective, tourism plays an important role. It can transform a country and influence the development of its internal mobility, as in Switzerland (Tissot, 2000). Some areas are shaped by the influx of foreign visitors, as illustrated by the Pan-American Highway (Booth,

2014). The infrastructure then engages with the landscape to construct an image that circulates and ends up being iconic.

Thus, even the automobile system, one of the very icons of the globalization of transport, takes on various forms depending on the country involved, particularly because it develops in relation to the landscape and local people's social practices. Road systems show vast ideological differences as well, as illustrated by the history of highways, which in the contexts of fascist regimes were developed for a small elite, whereas in the United States they had a much more democratic agenda (Moraglio, 2017). And the same car can have vastly different meanings according to local contexts. One of the most famous car models worldwide, the Beetle, was designed in Nazi Germany but became an icon of Californian hippie communities in the 1960s before becoming the symbol of resilient mobility in the face of economic crises in Latin America (Rieger, 2013). Even today, at a time when major mergers are arranged between car manufacturers and when common development platforms for cars exist, there are still significant differences between local car markets guided by specific tastes, habits, and constraints whether legislative, geographical or related to available resources (Doulet, 2018).

Local Networks: Transfers and Differences

Unlike telecommunication techniques, which seem to cancel distances, physical mobility continues to be characterized by its materiality. This means scale is particularly important. Mobility at the local level is very different from that at the global level. This differentiation was already clear in pre-industrial times, when local mobility was based on walking and animal traction, and global mobility was mainly carried out by horses and ships. Industrialization made this difference more prominent by increasing the diversity of solutions for urban mobility and by greatly extending the size of urban areas, and therefore the distances covered, sometimes even colliding the local and the global scales.

Sometimes systems that were designed for the global arena have been overtaken by the local, as illustrated by the creation of stops and stations in the suburbs of large urban areas, along railway routes initially designed for interurban connections, in the nineteenth century. Conversely, although less commonly, some systems designed for local use take on a global dimension; there are, for example, some cases in border areas, where local urban transport has taken on an international dimension because it creates a cross-border connection. This is the case with the recent Strasbourg-Kehl tramway that operates between France and Germany.

The car clearly occupies a peculiar position in relation to the issue of scale: at an early stage it developed as a means of covering long distances while offering an efficient local service, in the city and then in the countryside, thanks to the seamless interweaving of local and more global road systems. For the most part, cars have not really developed differently depending on the scale at which they are used. The object is often the same for all, also because it is part of a system of representations where roadworthiness is valued, even when driven around the city. Its effects are closely linked to its multiscalar positioning, since the car tends to both expand space locally and compress it globally.

Behind these geographical scales are also different-size infrastructures, different numbers of potential users and various investment capacities, which explains why systems generally operate at a certain scale. The plane is the best example here, as its ground infrastructures are so enormous, contrary to the impression of lightness generally communicated by airline advertisements featuring an image of an aircraft flying in the sky. Airports, which are global network nodes, are also macroscopic objects with complex services on a local scale (Roseau, 2012).

Since industrialization involved a shift towards urbanization, most commonly the local scale has also become the urban scale. However, it is in cities that the greatest diversity of mobility techniques has become available. Often characterized by specific local constraints (uneven territory, canals, etc.), this ecosystem daily serves a large population that tends to naturalize the solutions in place. This can be linked to the know-how required to use local networks, which can only be acquired through daily practice. External perspectives also tend to define these transport solutions as characteristic and many become symbols of the city they serve, even if they are not very original. Now London has become associated with double-decker buses, who remembers that they used to be the norm in several major European cities? Through photography tourists also play a role in this process whereby local technical solutions are turned into heritage, as with the now iconic London Routemaster bus.

Local mobility also sees actual exceptions, whether in the form of systems that were never subsequently deployed (such as the Wupperthal suspended metro or the magnetic levitation train at Shanghai airport) or systems that have survived their almost total dismantlement, such as the Santa Justa lift in Lisbon or the Hong Kong double-decker trams. Other cities, from Rome to Los Angeles, are known for their lack of successful solutions to their transportation problems. Localism here is pushed to idiosyncratic extremes.

More generally, urban systems are, in fact, made up of a small number of solutions, although their forms and even their functions may differ from one city to another. They are very closely related, and what is most striking about them is not their diversity but the shared difficulties encountered by many innovations when establishing their place in the urban space. Shared taxis illustrate this idea perfectly. Although their precise use can vary, they are found in contexts as different as Ukraine, Argentina, or Nigeria, while some even operated in the streets of London and Paris in the 1930s (Dienel, Vahrenkamp, 2018).

Despite differences in topography and population density, and despite so many socio-cultural discrepancies, cities actually use a relatively small number of systems to allow their inhabitants to move around. This explains why links have existed between cities in different countries and despite the seeminlgy diverse local contexts. The exchange of capital, expertise, and rolling stock between cities began with the development of industrial era transport systems, especially because of the international positions of the major nineteenth-century capitalist groups. This was accelerated by the emergence of markets held by international engineering firms — the famous Moscow metro was built with the help of British experts, and previously, the first London omnibus company had been founded in Paris with French capital in 1856.

The case of the bus rapid transit (BRT), which has seldom been studied, illustrates these traffic flows well. The omnibus is generally considered to have been invented in Nantes in 1826, while people who deployed this technique have been identified in Paris (1828), London (1829), and New York (1830). A diffusionist pattern therefore seems to have been at work. Nevertheless, the Nantes innovation was immediately developed further into an omnibus/railway hybrid with the introduction of the tramway in New York in 1832. The tram then struggled to be implemented outside the United States (Clarsen et al., 2009). At the beginning of the twentieth century, when the omnibus became motorized, it became commonplace in cities across world, and in 1974 it was reinvented in a Brazilian conurbation, Curitiba, where it was used as a surface metro system with stations whose architectural design was inspired by the Beaubourg centre in Paris. Known as 'BRT', this system then spread, from Colombia to California, reaching Nantes in 2006, which became the first French city to adopt this solution (under the name of 'Busway') fully. This example shows how much the history of urban transport is rooted in technical transfers.

City tramway systems, although also subject to transfers — from the mid-twentieth-century adoption of PCC vehicles (Demoro, Kashin, 1986) by many networks around the world to the successful and lucrative export of the French tramway in the 2000s (Gardon, 2018) — have sometimes failed to be transferred. In its historic, nineteenth-century form, the tramway met with resistance, and was confined to the United States during the 1830s and 1840s, before gradually gaining a foothold in Europe from the 1850s onwards. There was some reluctance, or even resistance to tramways in Europe however, with interest groups successfully lobbying to ensure that trams would not be implemented in the most prestigious areas of cities such as in London and Paris. Paradoxically, the tramway world also offers the best examples of transfer, because it can be standardised to a degree that avoids having to engage with designs for specific local conditions as is the case for subways. The industrial interest groups which have backed the development of tramways since the nineteenth century have managed to harness the social and political dynamics in favour of its deployment.

More recently, bike sharing systems illustrate the circulation of a mobility system that barely needs adapted to the local context while the dockless electric scooters have developed complete independence. The new mobility systems partake in the competition between cities as well as the standardization of tourist destinations, where cities no longer require to be distinctive. Not only have cities lost part of their local character under the influence of the world's leading highstreet brands, but urban mobility itself has become standardised, as illustrated by tour guide companies such as Fat Tire Tours Bikes, the Big Bus Company, and others. Of English or Latin origin, the terms 'tramway' and 'bus', which came into use in the nineteenth century, may have anticipated this trend by their adoption in these or similar forms in a very large number of languages. It is therefore not only the object itself, but also the idea, or even language and terminology, that is involved in transfer dynamics.

It should be noted however that contacts and exchanges between cities do not necessarily lead to homogenization, as illustrated by Paris and London whose transport systems are both highly developed but very different (Lopez-Galviz, 2019). These differences owe both to the specificities of the local context and to the different political choices made in the two countries. This is especially true when the users of the transport systems are not its customers as when public bodies have become stakeholders, an involvement which in the nineteenth century met with resistance.

Thus, while mobility techniques undoubtedly circulate, there remains a key question about whether these global circulations lead to standardization. Here, as in the political arena, globalization does not mean homogenization. Many systems exist only in certain parts of the world. For example, passenger trains have become extremely marginal in the United States due to the rise of car and air travel, while they are still part of mobility practices in Europe or Asia, even in countries with comparable rates of car ownership. Tuk-tuks and rickshaws were strongly embedded in their local contexts before some forms were used differently, including in the West, for tourists or by delivery services. Other practices seem more limited to specific regions. This is the case for leisure aviation, which is highly developed in the United States, or the use of helicopters for private urban travel, which is reserved for certain urban areas, for example in Brazil.

Therefore, if networks on a global scale, from air navigation to container carrier navigation, require a level of standardization, differences persist on more local scales, from railways to road signs. And although on the ground the systems feel altogether different, interestingly at this level too transnational actors and exchange logics can also be found.

References

Albert de la Bruhèze, Adri, Martin Emanuel, Ruth Oldenziel, and Frank Veraart, eds, *Cycling Cities. The European Experience. Hundred Years of Policy and Practice* (Eindhoven: Foundation for the History of Technology, 2016).

Anastasiadou, Irene, *Constructing Iron Europe. Transnationalism and Railways in the Interbellum* (Amsterdam: Amsterdam University Press, 2011).

Booth, Rodrigo, 'Turismo, panamericanismo e ingeniería civil. La construcción del camino escénico entre Viña del mar y Concón 1917–1931', *Historia Santiago*, 47.2 (2014), 277–311.

Caron, François, *Histoire des chemins de fer en France 1837–1997* (Paris: Fayard, 1997–2017).

Chevalier, Michel, *Système de la Méditerranée* (Paris: Le Globe, 1832).

Clarsen, Georgine and Gijs Mom, Peter Norton and Gordon Pirie, eds, *Mobility in History: Themes in Transport* (Neuchâtel: Alphil, 2010).

Demoro, Harre and Seymour Kashin, *An American Original: the PCC Car* (Glendale: Interurban Press, 1986).

Dienel, Hans-Liudger and Richard Vahrenkamp, 'For a Social History of Shared Taxi Services: Some Notes', *The Journal of Transport History*, 39.1 (2018), 7–11.

Divall, Colin and George Revill, 'Cultures of Transport: Representation, Practice and Technology', *The Journal of Transport History*, 26.1 (2005), 99–111.

Doulet, Jean-François, *Atlas de l'automobile: globalisation et nouveaux horizons* (Paris: Autrement, 2018).

Edgerton, David, *The Shock of the Old. Technology and Global History Since 1900* (London: Profile Books, 2006).

Fava, Valentina, 'La Fiat e la Autovaz di Togliatti. Alla ricerca del fordismo perduto', *Historically. History Laboratory*, 9 (2013), 239–67.

Gardon, Sébastien, ed., *Quarante ans de tramway en France* (Paris: Libel, 2018).

Hood, Christopher P., 'Bullets and Trains: Exporting Japan's Shinkansen to China and Taiwan', *The Asia-Pacific Journal*, 5.3 (2007), 1–7

Hughes, Thomas P. and Renate Mayntz, eds, *The Development of Large Technical Systems* (Frankfurt/Main: Campus, 1988).

Hurd, John and Ian J. Kerr, *India's Railway History. A Research Handbook* (Leiden: Brill, 2012).

Illich, Ivan, *Energy and Equity* (NewYork, NY: Harper & Row, 1974).

Laborie, Léonard, *L'Europe mise en réseaux. La France et la coopération internationale dans les postes et les télécommunications: années 1850-années 1950* (Brussels: PIE Peter Lang, 2010).

Löfgren, Orvar, 'Island Magic and the Making of a Transnational Region', *Geographical Review*, 97.2 (2007), 244–59.

Lopez-Galviz, Carlos, *Cities, Railways, Modernities: London, Paris, and the Nineteenth Century* (London: Routledge, 2019).

Mauss, Marcel, 'Techniques of the body', *Economy and Society*, 2 (1973), 70–88.

McKay, John P., *Tramways and Trolleys. The Rise of Urban Mass Transport in Europe* (Princeton, NJ: Princeton University Press, 1976).

Mom, Gijs, *Atlantic Automobilism. Emergence and Persistence of the Car, 1895–1940* (New York, NY: Berghahn, 2015).

Moraglio, Massimo, *Driving Modernity. Technology, Experts, Politics, and Fascist Motorways, 1922–1943* (New York, NY: Berghahn, 2017).
Ollivro, Jean, *L'Homme à toutes vitesses. De la lenteur homogène à la rapidité différenciée* (Rennes: Presses universitaires de Rennes, 2000).
Ponsavady, Stéphanie, *Cultural and Literary Representations of the Automobile in French Indochina* (Basingstoke: Palgrave Macmillan, 2018).
Rieger, Bernhard, *The People's Car. A Global History of the Volkswagen Beetle* (Cambridge, MA: Harvard University Press, 2013).
Rosa, Hartmut, *Social Acceleration: A New Theory of Modernity* (New York, NY: Columbia University Press, 2013).
Rose, Mark H., *Interstate: Express Highway Politics 1939–1989* (Knoxville, TN: University of Tennessee Press, 1990).
Roseau, Nathalie, 'Airports as Urban Narratives, Toward a Cultural History of Global Infrastructures', *Transfers*, 2.1 (2012), 32–54.
Sigaut, François, 'La formule de Mauss', *Techniques & Culture*, 40 (2003), 153–68.
Schipper, Franck, *Driving Europe. Building Europe on Roads in the Twentieth Century* (Amsterdam: Amsterdam University Press, 2008).
Tissot, Laurent, *Naissance d'une industrie touristique. Les Anglais et la Suisse au XIXe siècle* (Lausanne: Payot, 2000).
Zahavi, Yacov, *Travel Time Budgets and Mobility in Urban Areas* (Washington, DC: U.S. Department of Transportation, 1974).

GÉRALDINE BARRON, JULIA LAJUS

Sailing and Fishing

When Technical Innovation Reshaped the Seas

In 1842, the French naval officer Pâris wondered how the history of mankind could be fully understood or written if historians kept turning a blind eye to the history of ships, which are the tools which made it possible to travel, discover, and conquer new lands as well as trade and fight. He called for a global history of seafaring in which the technical point of view would be complementary to economic, social, political, and environmental approaches. Not being a historian himself, he strived to keep a record of crafts and ships from all over the world and across all periods in order to create an archive of technical knowledge (Barron, 2019). Ethnographers quickly became interested in navigation but it would take more than a century for historians to finally take an interest in navigation from a technical point of view, and even longer to move beyond the large sailing ships to the 'maritime revolution' of the nineteenth century (Masson, Battesti, Service historique de la Marine, 1987). Some national histories of ships and shipbuilding do exist, especially for Britain and the USA (Slaven, 2013; Smith, 2018; Reid, 2020). There is also a series of books on 'warships in the age of sailing' that includes the history of British, French, Dutch, and Russian ships (Bender, 2014; Winfield, 2014; Winfield, Roberts 2017; Sozaev, Tredrea 2010), but this is descriptive rather than analytical. Nevertheless, the technological history of navigation remains a secondary topic in maritime history, which has recently developed new trends of research in line with global history (O'Hara, 2009; Fusaro, 2010), focusing on seas and oceans as links within a connected world. New fields such as trans-oceanic or Atlantic history and new concepts like 'maritimity' and 'maritimization' have been the focus of fruitful research and numerous publications. The recent volume titled *Oceanic Histories* integrates oceans into world history and envisages world history across oceans and maritime spaces. The authors invite other historians to 'undercut the imperial, military and national inclinations of much maritime historiography' (Armitage, Bashford, Sivadundaram, 2017: 12). A technological approach would serve this purpose too, as maritime historians who focus on economic and environmental aspects still focus too rarely on the techniques that underpin oceanic connections, especially in recent times.

Emphasizing the fact that a ship is an ideal vantage point from which to consider global connections and disconnections, quite a few historians actually look beyond its commercial, geopolitical, or symbolic meaning to examine the development of its

Global History of Techniques, ed. by Guillaume Carnino, Liliane Hilaire-Pérez and Jérôme Lamy, Global Matters Culture Matérielles, histoire des techniques et circulation globale des savoirs, 9 (Turnhout: Brepols, 2024), pp. 363–378
BREPOLS PUBLISHERS 10.1484/M.TECHNE-EB.5.129817

technical characteristics. The technical history of the ship has long been a subject attended to by engineers and naval officers. Admiral Serre (1884) wrote that, in order to understand and recreate ancient sailing techniques, archeologists and historians must become sailors, or sailors must become historians.[1] Would this statement still be accurate with regard to modern ships? It is not uncommon even today to come across books by naval architects, sailors, or engineers who are also historians (Ferreiro, 2010, 2020). Like the history of maritime technology in general, the history of fisheries technology has been written mostly by fisheries specialists (Stansby, 1963; Sahrhage, Lundbeck, 1992) who need to understand the changes in vessels and gear for fisheries management (Misund, Kolding, Fréon, 2002). While the technological history of seafaring obviously requires a good understanding of technical issues, it also needs a wide and interdisciplinary approach that includes the history of economics, of industry, and of geopolitics, as well as the social history of labour. There is also a strong interest in the technological side of fisheries from economic and social maritime historians (Robinson, 1996, 2000; Holm, Starkey, 1999; Wright, 2000; Kolle et al., 2017), some of whom have an interest in technology (Sverrisson, 2002; Heidbrink, 2011). Over the last decade, technology has also gained importance within marine environmental history for the purpose of explaining the effect of fisheries on ecosystems and fishers' communities (Bolster, 2012; Finley, 2017; McKenzie, 2018) as well as explaining changes in fishing capacity in relation to risks of overfishing (Engelhard, 2016).

This chapter focuses on innovations in open sea ship design and fishing techniques that have both profoundly changed seafarers' practices and had major impacts in many fields while, it must be remembered, traditional techniques continue to be used extensively in many areas. The technical history of navigation is one of a permanent hybridization and adaptation to environments and purposes such as exploration, war, trade, sports, and the exploitation of sea resources. It paves the way for a technological history of seafaring.

The Modern Ship

The nineteenth century saw a major shift in open-sea navigation: within a few decades, several breakthroughs laid the ground for the development of the modern ship and brought about profound changes in many areas. However, in the general perception the disruptive innovations of the time tend to eclipse the long series of incremental improvements that unfolded throughout the period and make it difficult to summarize the technical history of ocean exploitation of the last two centuries. The ship itself is a complex technical system (Gille, 1978) which has a number of essential qualities: it has to float, be balanced and able to be steered into a given direction, to carry its crew as well as the food and drink for the same for the entire period of being

1 'Il faut donc, pour arriver à la restitution des bâtiments de guerre de l'Antiquité, que les archéologues deviennent marins, ce qui paraît difficile, ou que les marins acquièrent l'intelligence des documents écrits et figurés réunis par les savants'.

at sea, and it also needs to carry cargo, be it passengers, goods, or military apparatus. External energies like wind, waves, and currents have a direct effect on vessels. When wind blows on the sails, propulsion is rather straightforward, even though navigation in large ships has become a complex system to enable vessels to gain speed and exactitude in direction. This became even more complex when the system had to produce its own energy and convert it into the motion needed to rotate a thruster that caused the vessel to move forward or backward, at a given speed, and in a given direction.

The first disruptive innovation was the introduction of an engine into the hull that did not depend on external energy, but rather on fuel that was brought on board to propel the ship. As early as the eighteenth century, many scientists and engineers in different countries designed boats powered by steam to sail upstream on rivers, but the technology only became efficient and safe enough to navigate the open sea in the early decades of the following century (Griffiths, 2000). Steam, generated by boilers burning coal and heating water, pushed a piston back and forth inside a cylinder; this movement was converted into a rotational force that set a propeller into motion. The earliest steamboats were propelled by paddlewheels adapted to still waters like rivers, lakes, or harbours; but they were too bulky and fragile to operate in rough waters. Sails remained indispensable at sea, where the paddlewheels remained in place as a secondary propeller, but this combination turned out not to be a viable solution in that the ships did not function optimally with either one of the propelling systems, let alone their simultaneous use.

An alternative to the paddlewheels appeared in the middle of the nineteenth century: the screw — for which many countries claim paternity — proved to be the true maritime propeller and was adopted once and for all on ocean-going vessels (Graham, 1958; Lambert, 1999). The screw propeller was less fragile than the paddlewheels and gave more stability to the vessel thanks to its position under water and in line with the longitudinal axis of the hull. Meanwhile, the shape, size, and number of screw propellers have evolved, as well as the materials they are made of, and the blades can be fixed or rotating, but the main principles remain the same today. In bow thrusters, the propellers are positioned on the side of the ship for propulsion at right angle from its longitudinal axis. Bow thrusters improve the maneuverability, quicken the docking, and reduce the use of tugs which is a very convenient addition to many kinds of vessels which require easy docking capacities. The pods deployed on certain types of ships in the final decade of the twentieth century are steerable and can be disconnected from the rudder.

In the ship's complex technical system, the evolution of the engine has coincided with profound structural changes. Switching to steam power required ship design to evolve, because weight distribution changed greatly and lowered the centre of gravity, which made ships more susceptible to rolling. The rostrum inspired by ancient galleys and tried on ironclads as an offensive weapon in the 1860s was a draft of bulbous bows which would eventually become one of the many solutions to the problems of balance and hydrodynamics (Ferreiro, 2011).

Several issues inherent to engine-driven ships emerged in the nineteenth century and are still of concern today. First of all, a vessel is designed to carry a cargo, be it goods, people, or artillery, whether stored below or on deck. The rigging had not impaired the loading capacity, but boilers and engines did: the freight on steamships was now competing for space with the propelling system and its fuel. Therefore, ship design and outfitting needed to strike a delicate balance between self sufficiency and efficiency. The greater the fuel consumption, whether for reasons of engine efficiency, speed, or the distance to be covered, the greater the loading capacities, and consequently, the ship's dimensions. The size of a wooden hull is mechanically limited by a length-to-width ratio of 4 or 5. Long ships are subject to torsion phenomena, especially when loaded with heavy machinery in their centre; moreover, stability and resistance to wave impacts are essential for a ship. The way out of this conundrum was the iron hull, which could reach a ratio of 6 and weigh less than 38% of the ship's total weight, whereas the wooden hulls represented half the tonnage. After 1870, innovation in the metalworking industry reduced this ratio to a third, but the growth in size encountered a new technical limit: a ship of 90,000 tons would only be sturdy enough to support itself, but not its cargo. After 1930 this limit fell again when riveting was replaced by welding and new materials like aluminum were introduced (Robert, 2008).

The propeller and its shaft were less bulky than the paddlewheels and left more space for the cargo, but what really changed the game with the screw-propelled vessel was the concurrent evolution of the engine system. At the end of the nineteenth century, masts and sails gradually disappeared from steam ships because of the increased efficiency of the engine system as a whole. Improvements in the metalworking industry and research in thermodynamics made it possible to create and harness high-pressures, paving the way for more efficient engines and boilers: ships gained speed and became more cost effective, and at last they were able to carry less coal than freight as well as cover longer distances without having to refuel. The transatlantic Blue Riband competition recorded the highest speeds in passenger service and thus can serve indirectly as a measure of regular innovation and technical improvement regarding speed. Speed was then, and still is today, an economic issue, and its cost has always been meticulously monitored by shipping companies in their financial profitability calculation (Hyde, 1975; Berneron-Couvenhes, 2007; Meyer, Stahlbock, Voß, 2012; Levinson, 2016). However, issues like soaring fuel prices or growing ecological concerns can render speed a secondary concern or even bring back technologies that do not require additional energy like the sails that shipping companies presently try to reintroduce on some cargo ships.

Regarding boilers and engines, the early steamships proved that land-based techniques seldom transfer easily to the marine environment and its specific constraints not least the mechanical strain due to constant rolling, pitching, and tossing which cause violent thumping of parts, vibrations, and torsion of the hull, but also the chemical alterations caused by salt water and damp. In the early nineteenth century most of the sailors' and ship designers' knowledge was largely empirical, as scientists had not yet solved questions of fluid mechanics and heat principles. Innovation in ocean-going vessels has always been a story of trial and error, and of minor or major adaptations, but in the nineteenth century, the technical innovation and science changed at such a fast pace that, by the time a ship was launched, her technology was often already obsolete, especially in terms of engines and efficiency. Innovations focused on the power of the machine in order to gain speed and autonomy. First the use of high pressures, then the introduction of double- and triple-expansion Compound systems, of condensers and superheaters, and, in the twentieth century, the substitution of coal by oil.

The introduction of the diesel engine on the eve of First World War enabled ships to get rid of boilers and their drawbacks in terms of space and manpower. When the diesel engine was introduced on ships, it competed with steam turbines which were used, for example, on the British ironclad *Dreadnought* in 1906. Steam turbines were smaller and lighter than traditional engines and allowed direct drive of the propeller shafts, which was a feature that saved significant amounts of space. While they were lighter, needed less maintenance, and produced less vibration, they nevertheless consumed more fuel, which presented a disadvantage in times of rising energy costs. Both diesel engines and steam turbines could be used as primary engines for a new type of propulsion by electric transmission. The US Navy pioneered turbo-electric propulsion in the 1910s, but steam turbines had to compete with the cheaper diesel engines. They were used on many American cargo ships and tankers that were built during the Second World War, but afterwards mainly survived on high-speed ships and some military vessels until the 1980s, when they were replaced by gas turbines. Apart from these exceptions, diesel engine gradually replaced steam engines between the two World Wars. Nuclear power started being used in marine and submarine propulsion after the Second World War. Nuclear propulsion required neither the stacking of fuel nor regular refuelling; it provided extraordinary autonomy, if at a cost that was prohibitive for the merchant navy. Nuclear propulsion therefore only survived on military ships and submarines, and on Arctic icebreakers which needed a high degree of autonomy. Today the trend is to eliminate the use of oil and replace it with alternative energies like liquefied natural gas or electricity provided by batteries instead.

Diesel-electric propulsion appeared in 1912 and offered many advantages over steam engines: it adapted to the most efficient pace of the screw, produced less vibration, could easily change speed, was more reliable, and no longer depended on the propeller shaft for their positioning inside the hull, which saved much space. The electric transmission also gave flexibility and efficiency in the conduct of ships. In the 1970s, the advent of power electronics represented a new game-changing innovation, as it enabled all types of vessels to adopt diesel-electric propulsion and transmission as well as manage all on-board electric equipment; this improved fuel efficiency and enabled the development of all-electric ships (AES) (Skjong et al., 2015). Submarines, for their part, almost always relied on electricity for purposes of propulsion and sound reduction, key to their military use and stealth tactics. The technology reached a first degree of maturity with the *Gymnote* in 1888, but successive generations of submarines retained one major drawback: they could not dive for long because they needed to supply the engines with air and allow the crew to breathe. It was not until the end of the Second World War that research focused on air-independent propulsion (AIP), which provided increased diving autonomy.

The impact of electricity should be specially considered beyond the propulsion system, because it led to a series of apparently minor changes which had a substantial impact in the long run. Electricity was first used for lighting. Lamps were used inside ships from the 1860s onwards, but its uses expanded and became more efficient a couple of decades later, when incandescent bulbs made it possible and convenient to replace oil or gas with electric lamps while also reducing the fire risk. Electric lighting certainly provided visual comfort and improved the life of crew and passengers. Its usefulness went well beyond mere comfort. Although navigation lights were universally adopted in 1849 to prevent ships from colliding, it was the invention of electric searchlights that truly revolutionized the conditions of navigation. After 1870, electric arc searchlights were first used on warships to detect torpedoes, and they soon proved to be sufficiently powerful for nocturnal operations like docking, loading, and the unloading of cargo, thus expanding potential working hours. In 1885, the company managing the Suez Canal authorized nighttime navigation for ships equipped with projectors, rendering obsolete the project of doubling the capacity of the canal to absorb the increase in traffic and thus proving the economic value of lighting in reducing crossing times. Electric lighting also improved the communication between ships: the Aldis lamps enabled the long-range and nightly transmission of information using Morse code, which survived in the era of wireless radiotelegraphy from the very beginning of the twentieth century onwards. Electric energy powered many other appliances like ventilators, rudder control systems, call bells, and order transmitters such as the Chadburn, which was invented in 1865 and initially mechanical. It was also used for navigation by powering gyroscopic compasses and electricity-powered logs. Finally, it improved the life of sailors, helping to reduce time spent on chores, but also the size of crews, for example when dishwashers or vegetable peelers were introduced in the US Navy in 1907 (Robineau, 2019). The production and management of electrical power was always an issue, which improved slightly with the introduction of power electronics in the 1970s.

As the size and weight of ships and metal apparatus increased, sailors, who could no longer rely on their muscular strength alone to operate the ship, were assisted and later replaced by mechanically powered deck machinery such as capstans, winches or cranes. Cargo ships were gradually equipped with their own cranes to load and unload the goods stored in the holds. Until the 1960s, the trend was to build merchant vessels with specific features adapted to the bulk cargo, be it bananas or ores, oil, cars, or refrigerated goods; others would carry breakbulk cargo composed of various products and containers of multiple shapes and sizes. Some ships were also designed for their destinations, depending on the port infrastructures available in Africa, Asia, Europe, or America. By contrast, others were built without such specifications and intended for the tramp trade. In 1956 the American engineer Malcolm McLean considered two alternatives for the time-consuming task of loading and unloading cargo manually: either driving loaded trailers aboard ro/ro (roll-on/roll-off) ships or refitting tankers with new spar decks that would accommodate trailers detached from their running gear and hoisted aboard by shoreside cranes. *Ideal X*, the first converted tanker, could carry fifty-eight trailers. McLean eventually converted cargo ships like the *Gateway City* into all-container ships equipped with onboard cranes, where boxes could be stacked one on top of another (Cudahy, 2006). Several technical considerations underpinned this pragmatic idea: it was necessary to strengthen the structure of the containers so that they would not collapse under the weight of those above them, and to join the piles together to avoid damage at sea. This was made possible by corner castings linked together with twist locks, and these patents were released in 1963, so that they could become an international standard. 1968 ISO recommendations reinforced standardization in defining sizes for the boxes; tonnage was no longer the only relevant category, as container cargo could now be measured in TEU (the 'twenty-foot equivalent unit'), a unit based on the dimensions of boxes twenty feet long and eight feet wide. Today's more commonly used forty-foot containers equal two TEUs. The adoption of container boxes prompted an evolution and major standardization in the design of cargo ships. The biggest container ships still have a length/width ratio of 6 but reach 232,618 gross tons. The storage is no longer limited to the hold, since up to 24,000 TEU containers can pile up to twenty levels and render the hold panels obsolete, imposing new building constraints to ensure the rigidity of ships that are 400 metres long and sixty metres wide. By comparison, the 1957 *Gateway City* was 140 metres long, twenty metres wide and could hold 226 boxes. Gigantism does not only affect container ships: ultra large crude carriers, also known as supertankers, are today's largest freight carriers, with lengths exceeding 450 metres and tonnages of 250,000. Those giant ships, like any other vessel, have to float, be steerable and able to manoeuvre in an unstable and sometimes rough environment, but also be able to navigate key passages between oceans, like the Panama and Suez Canals. Giant ships also need to land and dock to load or unload. They are part of an intermodal system whose nexus is the port. As a whole, the chain of transport had to adapt to the evolution of cargo ships, but containers also had an impact on the industry by giving rise to supply chain management and just-in-time manufacturing (Levinson, 2016); their dependence on the smooth operation of transport was highlighted in 2021 when the *Ever Given* ran aground and blocked the Suez Canal. The gigantism of ships has redefined the geography of ports and continues to do so today, as it requires deeper waters, longer quays, and adapted infrastructures. The evolution of the ship is thus strongly intertwined with that of its environment and in particular with quays, dockyards, warehouses, and storage areas, cranes, rail and road networks, fuel, water, and food supply infrastructures, it is also linked to the automation of labour. It is also linked to sea routes which themselves impact on the length of voyages, a question with high economic and geopolitical stakes. Steam navigation has opened the way for long-anticipated inter-seas canals such as the Suez or Panama Canals: they were key when speed and clock-ruled navigation, combined with refuelling issues, economic demand, and a new stage in globalization demanded that sea routes be redefined. They have responded to the increase in tonnages with widening works, but now impose structural limits on ship architects, and these constraints have led to the establishment of maximum dimensions for ships in relation to certain passages: Panamax, Suezmax, and even Malaccamax for the twenty-five-metre-deep Malacca Strait. Oversized ships are required to use alternate sea routes. Manpower and its evolution in size and skill should also be considered: mechanization reduced the size of crews drastically, from a thousand men on board an ironclad in the second half of the nineteenth century to a dozen in supertankers or huge container ships. Sailors on board have gradually been replaced by machines for power work and by computers for navigation monitoring and cargo management.

The evolution of shipbuilding techniques is not as straightforward as it might appear from this quick overview of two centuries of technical developments, and it is vital to take both economic and geopolitical contexts into consideration as well. The *Great Eastern* offers a good example of the joint use of new and established techniques in the attempt to solve several problems at the same time, i.e. those of size, speed, cost-effectiveness, and fuel autonomy. This liner was launched in 1858 (Figure 1), and designed by Isambard Kingdom Brunel to carry 4,000 passengers between Europe and Asia or Australia without refuelling. It was a real behemoth, a hyperbole of a ship, 211 metres long, weighing over 20,000 tons, with two motor engines — one for the screw propeller and one for the paddlewheels — five funnels and five masts, and thus the biggest ship of the century. But as Andrew Lambert put it, she was 'a magnificent technical solution to a problem that did not exist' (Lambert, 2011): there was no market for such a large cargo, coal mining in Australia reduced the need for coal storage, and more efficient engines made newer and smaller ships faster. The *Great Eastern* proved to be an economic failure but remained a symbol of the race for innovation and the all dominant drive to link continents with maximum cargo and at maximum speed that led to gigantism. Rather tellingly, the ship ended up being converted into a cable-laying vessel for the first transatlantic telegraphic lines.

Another striking example would be the two commercial failures that Malcolm McLean experienced with consecutive generations of container ships. In 1972 his steam-powered SL–7 were the largest and fastest container ships of their time, but they used large quantities of oil, and due to the oil crisis, the cost of the oil necessary to ensure its speed became unsustainable. A few year later, McLean signed a contract with Daewoo to deliver twelve larger but diesel-powered container ships with slow-speed performance in order to recoup his investiments, but these ships proved too slow to remain competitive in a growing shipping market (Cudahy, 2006). Innovations and technical choices proved to be highly dependent on market conditions.

Fishing

It is well-known that modern fishing technology has grown out of the technologies used in artisanal fisheries. Industrial fisheries technologies that began to develop from the last quarter of the nineteenth century onwards, following the introduction of steam power, co-existed with artisanal technlologies and continue to do so in some

Fig. 1. *The Great Eastern*. Robert Charles Dudley, 1892. The Atlantic Telegraph Cable Fleet Assembled at Berehaven, 1866. New York, The Metropolitan Museum of Art.

regions, so that there is a continuum between the most primitive and most advanced versions of fishing equipment (Misund, Kolding, Fréon, 2002). This chapter, however, focuses on a different history of industrial fisheries technologies that manifest obvious transformations.

Major changes in fisheries technologies roughly fall into the following categories: vessels and propulsion; fishing gear; navigational aids and fish detection; and fish processing, storage, and distribution (Heidbrink, 2011: 58). Vessels and gear are of particular interest here. The main types of fishing gear that were retained and became more efficient with the industrialization of fisheries are different varieties of nets and, to a lesser extent, longlines. Of the nets, the trawl was the first to be transformed. The trawl is a bag-shaped net, and its efficiency is determined by the width of its mouth. The beam-trawl, which was the most prominent gear of the 1890s, was usually propped open by a large wooden beam. The net consisted of four sections, the back, the belly, and two wings, and was dragged behind the boat along the bottom of the sea (Robinson, 1996: 14–15, 36) (Figure 2). The more efficient otter trawl was invented in the 1860s, and here large rectangular boards are used to keep the mouth of the trawl net open. The 1870s are known as the heyday of the sailing trawler in England, when trawlers were especially numerous in the North Sea. Trawls were carried by smacks that were considerably improved in the mid-nineteenth century by being changed from single-masted cutters into two-masted ketches which were able to carry more gear, and the beams of the trawls increased from 38 to 48 feet (from approx. 11.6 metres to 14.6 metres) (Robinson, 1996: 66–68).

Fig. 2. Beam-Trawl. Collins, J. W., 1889. 'Beam-Trawl Fishery of Great Britain, with Notes on Beam-Trawling in Other European Countries, etc.', *Bulletin of the United States Fish Commission*, 7, 1887, Washington, DC, Government Printing Office.

The most important development in marine technology during the nineteenth century was the emergence of the steamship. The first specially-built steam power trawler began to operate from Hull in 1881, and within a little more than ten years, such trawlers were common in the North Sea (Robinson, 1996: 91). These early engine-powered trawlers often continued to require sails, because the engines were inefficient, and coal or wood fuels were too expensive. Improvement arrived in the late nineteenth century with the design of otter board trawls that offered better control of nets as well as better fuel efficiency. At the end of the 1870s, steam capstans emerged which released fishers from hauling trawl by hand, and rope warps were replaced with warps made from steel that permitted trawls to work in deeper water; also, more expensive iron smacks began to replace their wooden predecessors, which had also grown in size (Robinson 1996: 68). In many countries, e.g. Norway, Canada, the USA, and Northern Russia, fishermen using hooks and lines opposed the use of otter trawls (Finstad, Lajus, 2011; Wright, 2000; McKenzie, 2015), but only few succeeded to postpone their introduction or significantly reduce their numbers.

Evidently, only large companies could afford the new trawling technology, and small inshore fisheries changed to new power gradually during the last decade of the nineteenth and first quarter of the twentieth centuries, first adopting the hot bulb engine and later the internal combustion engine. Vessel design changed from open boats to decked vessels and from wood to steel, but fishing was done over the side rather than over the stern. The First World War marked a major turning point in the development of fisheries technology due to the loss of a considerable number of fishing vessels, so that the rebuilding of vessels *en masse* after the War led to significant modernization (Heidbrink, 2008). In Russia, for instance, the first trawling fleet was a redesign of twelve British mine sweeping trawlers, which were converted into fishing trawlers (Finstad and Lajus 2011: 230).

The next major changes occurred after the Second World War, when shipbuilding developed and internal combustion engines, such as the diesel engine, had become cheaper. The fundamental change of the 1950s was connected with the mass introduction of stern ramp factory freezer trawlers. While in Western Europe both types of trawlers co-existed for many years, the stern trawlers were rapidly developed in the

Soviet Union and other socialist countries that went into long distance fisheries in the 1950s. Key changes that increased capacity were related to the introduction of synthetic fibre as a material for nets, which consequently were stronger, so that the production of much larger nets was possible.

By the mid-1970s, the revolutionary introduction of mid-water trawl became possible due to the development of echo sounding. Echo sound technology was first transferred from the navy to fisheries immediately after the war, but only the new technology of the 1970s made it possible to find not only demersal fish near the bottom of the sea, but also pelagic fish. The catching of pelagic fish had been a problem for a long time: in the pre-industrial period they were caught with beach seines (large nets) operated by hand from the shore and on small boats. Also, drift nets were (and still are) in use, many of which are several kilometres long. Significant innovation for catching pelagic fish came with the introduction of purse seines that capture fish shoals by surrounding them with a vast net. Although purse seines are known to have been used by sardine fishers since in the nineteenth century at least (Iitaka, 1971), their industrial implementation was also closely connected with the introduction of echo sounding that located the fish shoals to be harvested. Purse seiners' typical equipment usually consists of a power block or triple roller (triplex) and storage equipment for hauling and stowing the net on board, while on larger vessels, the fish is pumped onto the ship.

Since the 1950s the relative cost of fuel has dropped markedly, which has facilitated the development of long-distance offshore fisheries. Hull designs changed to fit the better, more powerful engines that proved economical to operate once the fuel was relatively cheap. The fuller hull profiles had the advantage of providing more storage and working space (Gartside, Kirkegaard, 2009). The introduction for the oceanic long-distance fisheries of a factory ship modeled after what had happened in whaling took the facilities for processing and freezing caught fish directly on board. Soviet fisheries, as well as some German and British companies, advanced the development of this vessel and its freezing equipment, while other nations were reluctant to introduce this technology (Heidbrink, 2011). Often a factory ship was used as a mother ship that collected and processed fish from smaller vessels. That was of particular advantage to the nations without direct access to productive fishing grounds, although this technology evidently led to overfishing and international conflicts over access to fish stocks.

Throughout the last two centuries, technical progress at sea has been tremendous, and this chapter has given only a glimpse into its many dimensions. The development of regulations alongside the intensification of traffic or of positioning and communication systems that gradually connected the ship to its environment as well as to the shore also deserve examination, as do the questions of sea routes and national and international maritime domains. A technical history of shipping and fishing needs to be global, as it closely connects technical issues with broader layers of historiography. Even though ships and their apparatus have long been considered insignificant, and the objects subsumed by their functions, they do appear relevant to global history and world oceanic histories from technical as well as economic, environmental, and social points of view. Some authors stress the historical importance of ships as 'historical arenas' (Dusinberre, Wenzlhuemer, 2016) where one can observe mobility and immobility, transit and exchange, and the opening up and closing in of the world. While historically, fishing vessels and gear were developed to increase catch per unit, the effort involved and environmental concerns became only an issue when pressure was exerted by local communities or national and international agencies; in the present century, strict regulations for the prevention of overfishing and by-catch have become very important for fishing technologies. While this article has emphasized the major innovations which permitted the expansion of traffic and catches as well as the standardization of techniques and a worldwide network of shipyards, ports, and routes, the global vision must not conceal the richness of local solutions that have been born from various influences and are often more successfully adapted to ecological and environmental issues.

References

Armitage, David, Alison Bashford, and Sujit Sivasundaram, *Oceanic Histories* (Cambridge: Cambridge University Press, 2017).

Barron, Géraldine, *Edmond Pâris et l'art naval: des pirogues aux cuirassés* (Toulouse: Presses universitaires du Midi, 2019).

Bender, James, *Dutch Warships in the Age of Sail 1600–1714. Design, Construction, Careers and Fates* (Barnsley: Seaforth Publishing, 2014).

Berneron-Couvenhes, Marie-Françoise, *Les messageries maritimes. L'essor d'une grande compagnie de navigation française, 1851–1894* (Paris: PUPS, 2007).

Bolster, W. Jeffrey, *The Mortal Sea: Fishing the Atlantic in the Age of Sail* (Cambridge, MA and London: The Belknap Press of Harvard University Press, 2012).

Cudahy, Brian J. *Box Boats: How Container Ships Changed the World* (New York, NY: Fordham University Press, 2006).

Dusinberre, Martin and Roland Wenzlhuemer, 'Editorial. Being in Transit: Ships and Global Incompatibilities', *Journal of Global History*, 11.2 (2016), 155–62.

Engelhard, Georg H., 'On the Need to Study Fishing Power Change: Challenges and Perspectives', in *Perspectives on Oceans Past. A Handbook of Marine Environmental History*, ed. by Kathleen Schwerdtner Máñez and Bo Poulsen (Dordrecht: Springer Netherlands, 2016), pp. 89–101

Ferreiro, Larrie D., *Bridging the Seas: The Rise of Naval Architecture in the Industrial Age, 1800–2000* (Cambridge, MA: The MIT Press, 2020).

———, *Ships and Science: The Birth of Naval Architecture in the Scientific Revolution, 1600–1800* (Cambridge, MA: The MIT Press, 2010).

———, 'The Social History of the Bulbous Bow', *Technology and Culture*, 52.2 (2011), 335–59.

Finley, Carmel. *All the Boats on the Ocean: How Government Subsidies Led to Global Overfishing* (Chicago, IL: University of Chicago Press, 2017).

Finstad, Bjorn-Petter and Julia Lajus, 'The Fisheries in Norwegian and Russian Waters, 1850–2010', in *A History of the North Atlantic Fisheries. From the 1850s to the Early Twentieth-First Century* (Bremen: Hauschild, 2011), pp. 226–37.

Fusaro, Maria, 'Maritime History as Global History? The Methodological Challenges and a Future Research Agenda', *Research in Maritime History*, 43 (2010), 267–82.

Gartside, D. F. and I. R. Kirkegaard, 'A History of Fishing', in *Encyclopedia of Life Support Systems (EOLSS)*, vol. 2, *The Role of Food, Agriculture, Forestry and Fisheries in Human Nutrition* (Paris: UNESCO, 2009), pp. 70–104.

Gille, Bertrand, ed., *Histoire des techniques: technique et civilisations, technique et sciences* (Paris: Gallimard, 1978).

Graham, G. S., 'The Transition from Paddle-Wheel to Screw Propeller', *The Mariner's Mirror*, 44.1 (1958), 35–48.

Griffiths, Denis, *Steam at Sea: Two Centuries of Steam-Powered Ships* (London: Conway Maritime Press, 2000).

Heidbrink, Ingo, 'From Sail to Factory Freeze: Patterns of Technological Change', in *A History of the North Atlantic Fisheries*, ed. by David J. Starkey and Ingo Heidbrink (Bremen: Hauschild, 2011), pp. 58–71.

———, 'Marketing and Technological Changes in the German Deep-Sea Fishing Industry: A Response to the Crisis during the Interwar Period', in *Crisis and Transition: Maritime Sectors in the North Sea Region 1790–1940*, ed. by Lars U. Scholl and David Williams (Bremen: Hauschild, 2008), pp. 102–11.

Holm, Poul and David J. Starkey, eds, *Technological Change in the North Atlantic Fisheries*, Studia Atlantica, 3 (Esbjerg: Fiskeri- og Søfartsmuseet, 1999).

Hyde, Francis E., *Cunard and the North Atlantic 1840–1973: A History of Shipping and Financial Management* (London and Basingstoke: The Macmillan Press, 1975).

Iitaka, Yunosuke, 'Purse Seine Design and Construction in Relation to the Fish Behaviour and Fishing Conditions', in *Modern Fishing Gear of the World*, ed. by Food and Agriculture Organization (London: Fishing News, 1971), pp. 253–56.

Kolle, Nils, Alf Ragnar Nielssen, Atle Døssland, and Pål Christensen, *Fish, Coast and Communities: A History of Norway* (Bergen: Fagbokforlaget, 2017).

Lambert, Andrew, 'John Scott Russell. Ships, Science and Scandal in the Age of Transition', *The International Journal for the History of Engineering & Technology*, 81.1 (2011), 60–78

———, 'Responding to the Nineteenth Century: The Royal Navy and the Introduction of the Screw Propeller', *History of technology*, 21 (1999), 1–28.

Levinson, Marc, *The Box: How the Shipping Container Made the World Smaller and the World Economy Bigger*, 2nd ed. (Princeton, NJ and Oxford: Princeton University Press, 2016).

Masson, Philippe, Michèle Battesti and Service historique de la Marine, *La Révolution maritime du XIXe siècle* (Paris: Lavauzelle, 1987).

McKenzie, Matthew, *Breaking the Banks: Representations and Realities in New England Fisheries, 1866–1966* (Amherst, MA: University of Massachusetts Press, 2018).

———, 'Modernity and Marine Environmental History: Cultural Influences on the Mechanization of Georges Bank Fisheries, 1905–1912', in *Fluid Frontiers: New Currents in Marine Environmental History*, ed. by John R. Gillis and Franziska Torma (Cambridge: The White-horse Press, 2015), pp. 105–16.

Meyer, Jasper, Robert Stahlbock and Stefan Voß, 'Slow Steaming in Container Shipping', in *2012 45th Hawaii International Conference on System Sciences*, (IEEE, 2012), pp. 1306–14.

Misund, Ole Arve, Jeppe Kolding and Pierre Fréon, 'Fish Capture Devices in Industrial and Artisanal Fisheries and Their Influence on Management', in *Handbook on Fish Biology and Fisheries*, vol. 2, ed. by Helen Ashton (Oxford: Blackwell Publishing, 2002), pp. 13–36.

O'Hara, Glen, '"The Sea Is Swinging Into View": Modern British Maritime History in a Globalised World', *The English Historical Review*, CXXIV.510 (2009), 1109–34.

Reid, Phillip, *The Merchant Ship in the British Atlantic, 1600–1800, Continuity and Innovation in a Key Technology*, Technology and Change in History, 18 (Leiden and Boston: Brill, 2020), pp. 222–40.

Robert, Pascal, 'Histoire des techniques électriques employées sur les navires civils au XIXe et au début du XXe siècle' (unpublished doctoral thesis, Université de Nantes, 2008).

Robineau, Didier, 'L'Introduction de l'électricité dans la marine militaire 1880–1935' (unpublished doctoral thesis, Université de Nantes, 2019).

Robinson, Robb, 'Steam Power and Distant-water Trawling', in *England's Sea Fisheries: the Commercial Sea Fisheries of England and Wales since 1300*, ed. by David J Starkey; Chris Reid, and Neil Ashcroft (London: Chatham, 2000), pp. 206–16.

———, *Trawling: The Rise and Fall of the British Trawl Fishery* (Exeter: University of Exeter Press, 1996).
Sahrhage, Dietrich and Johannes Lundbeck, *A History of Fishing* (Berlin and Heidelberg: Springer Verlag, 1992).
Serre, Paul, 'La trière athénienne', *Mémoires présentés par divers savans à l'Académie royale des sciences de l'Institut de France, et imprimés par son ordre: sciences mathématiques et physiques*, 28.4 (1884), 1–38.
Skjong, Espen, Egil Rødskar, Marta Molinas, Tor Arne Johansen, and Joseph Cunningham, 'The Marine Vessel's Electrical Power System: From its Birth to Present Day', *Proceedings of the IEEE*, 103.12 (2015), 2410–24.
Slaven, Anthony, *British Shipbuilding 1500–2010: A History* (Lancaster: Carnegie Publishing, 2013).
Smith, Crosbie, *Coal, Steam and Ships: Engineering, Enterprise and Empire on the Nineteenth-Century Seas* (Cambridge: Cambridge University Press, 2018).
Sozaev, Eduard and John Tredrea, *Russian Warships in the Age of Sail 1696–1860: Design, Construction, Careers and Fates* (Barnsley: Seaforth Publishing, 2010).
Stansby, M. E., *Industrial Fishery Technology: A Survey of Methods for Domestic Harvesting, Preservation, and Processing of Fish Used for Food and for Industrial Products* (New York, NY: Reinhold Publishing Corporation, 1963).
Sverrisson, Árni, 'Small Boats and Large Ships: Social Continuity and Technical Change in the Icelandic Fisheries, 1800–1960', *Technology and Culture*, 43.2 (2002), 227–53.
Winfield, Rif, *British Warships in the Age of Sail, 1817–1863: Design, Construction, Careers and Fates* (Barnsley: Seaforth Publishing, 2014).
——— and Stephen S. Roberts, *French Warships in the Age of Sail, 1626–1786: Design, Construction, Careers and Fates* (Barnsley: Seaforth Publishing, 2017).
Wright, Miriam Carol, *A Fishery for Modern Times: the State and the Industrialization of the Newfoundland Fishery, 1934–1968* (Oxford: Oxford University Press, 2000).

RAPHAËL MORERA, ÉLISABETH MORTIER

Hydraulic Technology in the Nineteenth and Twentieth Centuries

Water Harnessed by Industry

Water-use technologies draw on several fields of expertise. An expert of hydraulic work can successively wear the hat of an engineer, mechanic, institutional decision-maker, doctor, and water-user. According to the geographer Pierre Gourou, hydraulics is about management: it concerns the relationships between societies and their territories, and between technical systems and institutions (Gourou, 1940). Hydraulic technologies and infrastructure open up limited trajectories for both physical systems and the values and functioning of societies.

These two aspects of water-use technologies have marked the historiography of the field. At the end of the nineteenth century, the history of hydraulics was inseparable from the history of civilization and progress. Studies on hydraulics were incorporated into much broader fields, such as geography, and sought to distinguish societies by showing how their achievements expressed their ways of being in the world (Sorre, 1943). Since the 1970s that trend has given way to a more technical approach to hydraulics, with the aim not of drawing distinctions between groups of people, but rather shedding light on the role of hydraulics in the processes of economic development and in technical innovation and change (Hunter, 1979). In France, Serge Benoit's research has significantly contributed to this shift, as have several studies on dams and their role in economic growth (Benoît, 2020). This historiographical current frequently includes research on hydraulics policies. Since Karl Marx and Karl Wittfogel, the history of hydraulic works has taken on a highly political aspect (Wittfogel, 1957). Despite being historically invalidated, an association between political forms and the size of infrastructure has sparked countless discussions combining political science and the history of technology (Briant, 2002). This association has recently been extended through research on the environmental issues related to water management. As a result, since the 1980s hydraulics historiography has attended to a much greater extent to environmental history through studies examining water-related conflicts in the context of pressures on water resources (Biggs, 2012; Worster, 1985).

This change in focus responds to contemporary issues and brings together various currents of the history of hydraulics. However, it also overlooks the processes through which water has become a central component of industrial economies. While water

Global History of Techniques, ed. by Guillaume Carnino, Liliane Hilaire-Pérez and Jérôme Lamy, Global Matters Culture Matérielles, histoire des techniques et circulation globale des savoirs, 9 (Turnhout: Brepols, 2024), pp. 379–390
BREPOLS PUBLISHERS 10.1484/M.TECHNE-EB.5.129818

was extensively used in pre-industrial techniques, the function of hydraulics in technical systems has profoundly changed since the nineteenth century. Indeed, whereas in the eighteenth century water was, above all, a technical means, a component of the production process, industrialization caused it to become a product itself. Water played a decisive role in the industrialization process before, in turn, being seized by that dynamic of commodification. In this regard, differences in pollution levels in various parts of the world are particularly revealing. In places where water is treated before being returned to its natural cycle in the environment, rivers are much less polluted than in areas where it is not treated. Let us now look at how water-use techniques have been incorporated into industrial production since the nineteenth century, first initiating and then completely transformed by it.

Age-old Hydraulics and the Industrial Revolution

Ancient China and the Near East stand out for their lastingcapacity to harness the power of water for production purposes. In China, water mills were used for industrial purposes, in particular for providing energy for metalworking (e.g. blacksmiths' bellows and hammers) (Viollet, 2005). In Europe, beginning in the first century AD, water mills powered by rivers and streams were used to grind wheat (Brun, Fiches, 2007). Wooden Roman wheels measured between 1.5 and 3 metres in height and between 25 and 70 centimetres in width (Viollet, 2005). Developed in Africa in the fourth century, the horizontal wheel, driven by water falling from a great height, was another means of generating hydropower. In the Roman Empire, aqueducts powered waterwheels that provided the required energy for flour mills and sawmills. In the Western world, during the Middle Ages and in the Early Modern era, hydromechanical power was used in various hydraulic structures such as dams, canals, and aqueducts to supply sufficient water to turn vertical or horizontal wheels. Hydraulics played a decisive role in the growth of different trades and industries, in particular mining (Shaw, 1984; Gribbon, 1947).

In the eighteenth century, the uses of waterpower were diversified, allowing for the increased mechanization of manufacturing. Hydraulic wheels that were connected to looms for weaving cotton and wool were developed in the eighteenth and nineteenth centuries. After the invention of the spinning jenny in 1767, the English inventor Richard Arkwright (1732–1792) created the water frame in 1769: the first spinning frame for the production of cotton thread powered by a waterwheel (Jumeau, 2001). In 1772 the Englishman John Whitehurst invented the hydraulic ram, which could raise water using the power of flowing water. Motor-driven pumps to draw water from wells and watercourses appeared in the nineteenth century and greatly increased the efficiency of water collection. Steam pumps, which were powered by coal, then diesel, and finally electricity, worked by means of a set of pistons and valves. In 1779 Samuel Crompton combined the spinning jenny and the water frame to make the spinning mule, a machine that harnessed hydraulic power to produce strong yarn (Builder, 1939). At the time, vertical wheels were being improved through the use of iron for their blades (Viollet, 2005).

In the 1830s the Frenchman Benoît Fourneyron invented the water turbine. Its principle was simple: falling water moved the turbine, with the water flowing from the inside to the sides. In 1832 Fourneyron's first industrial turbine, which had a diameter of 2.2 metres and a power of 35 kilowatts, was used to work the bellows of a blast furnace in the Jura department of France. Fourneyron's turbine was adopted across Europe and the United States (Cardwell, 1995) and inspired variant models. In the 1850s a turbine invented by the British engineer James B. Francis improved the process by making the water supply more efficient, with the flow moving from the outside to the inside (Viollet, 2005).

These inventions enabled hydraulics to play a leading role in the Industrial Revolution (Hunter, 1979). In the United States, during the early years of its independence, economic and industrial development depened greatly on the ability to control water. The Lowell mills, built at the confluence of the Concord and Merrimack rivers in Massachusetts, made use of hydraulic techniques that were cutting-edge in terms of both installations and mechanics (Gross, 1993). For France, Serge Benoît's research has highlighted the importance of water for economic growth during the first two-thirds of the nineteenth century (Benoît, 2020). Indeed, all of continental Europe made significant use of hydraulic resources during that period. The British path for development, centred on coal, can be seen as both a model and an exception. In this vein, recent research has shown the extent to which the transition from water to coal and steam was primarily motivated by managerial considerations rather than by the inefficiency of hydraulic technology (Malm, 2016).

Industry and the Changing Uses of Water

After contributing significantly to the nineteenth-century industrial boom, hydraulic technologies were then transformed by industrial discoveries and innovations. From the 1850s onwards, a revolution in materials opened up new prospects for developers and investors. Concrete, developed in the 1840s, began to be used in civil architecture in 1852. The growth of the concrete industry soon enabled the construction of monumental hydraulic structures and paved the way for the rise of the hydroelectric industry. This development was amplified by advances in the steel industry. Increased production capacities and the improved reliability of procedures made it possible to design complex hydraulic networks that could withstand intense pressure. Finally, from a less monumental but nonetheless fundamental angle, the development of rubber, plastics, and other synthetic materials allowed for the creation of radically new water conveyance systems.

The history of hydroelectricity illustrates how a technical system was reconfigured and opened up as a result of innovations in materials. From the 1880s onwards, hydroelectric power was further developed, particularly in mountainous regions. The various turbines invented in the course of the century began to be connected to generators with belts, with the result that hydroelectricity gradually replaced hydromechanics. The small cities of Europe and North America were the first to benefit from hydroelectricity to bring electric light to their inhabitants. In 1881 a hydroelectric plant harnessed the power of Niagara Falls (Viollet, 2005). Throughout the twentieth century, hydroelectric power spread in the Western world. Dams, artificial lakes, and canals were built to create flow rates and height differences that were sufficient for operating hydroelectric plants. The power of water could then turn a turbine connected to an alternator to produce electricity.

During the inter-war period, hydroelectricity became a tool for modernization and for recovering from the economic crisis of 1929. Its development dramatically changed territories and helped strengthen the political leadership bolster the political clout of major nations. The histories of the United States and the young URSS overlap in surprising ways. The Tennessee Valley Authority, established within the context of the New Deal to revive the American economy, essentially consisted in the industrialization of a southern state through the transformation of its hydrography and the construction of several dams to regulate the course of the Tennessee River (Callahan, 1980; Culvahouse, 2007). Reinforced concrete thus became a material means of implementing policies on farming, navigation routes, and industrialization. The USSR and Ukrain followed a closely related rationale in the reconfiguration of the course of the Don River was reconfigured to support a policy of industrialization through electrification (Rassweiler, 1988).

The technical trajectory that arose between the end of the nineteenth century and the Second World War was strengthened in the second half of the twentieth century. Large-scale hydraulic works imposed themselves as symbols of and the condition for economic development and political affirmation. The Aswan Dam, for example, embodies the independence of Egypt. The monumental size of the Three Gorges Dam expresses China's quest for power (Pomeranz, 2015). Despite the environmental damage it causes and its eviction of people from their homes, hydroelectricity still plays a significant role in economic policies. According to the International Energy Agency, hydroelectricity accounts for 16.6 per cent of electricity production in the world today (International Energy Agency). The share of water in energy production, however, should be more accurately evaluated. Indeed, water is indispensable in some industries, particularly in the nuclear energy sector since the 1970s. Water is needed to cool down the reactors. In France, for example, the electric utility company EDF extracts water for its nuclear power plants from groundwater tables and the sea.

Changes that occurred in materials development also had a major impact on small-scale hydraulic works. This effect was particularly noticeable in the area of agriculture. The combination of industrial metallurgy and the technical mastery of rubber and then plastic made it possible to design much more efficient and complex irrigation systems at the end of the nineteenth century. Furthermore, mechanical pumps could convey pressurized water over large distances and regularly send it up into water towers, which were erected in an increasing number of areas.

Supplying Water of High Quality

In the nineteenth century, various traditional techniques continued to be used to raise water from wells. They primarily harnessed the motive force of humans, animals, and water. Water-lifting wheels had been used since Antiquity, in particular the overshot wheel, which was a wheel with trough-like compartments that could draw water for domestic and agricultural uses (Viollet, 2005). In Egypt, and more broadly in the Arab and Persian world, the use of the *sâqia*, a water-lifting wheel, was very common weel intothe nineteenth century. It was a wooden vertical wheel to which buckets were attached to draw water from a brick well. The wheel turned using the traction force of an animal such as a bovine or camel, which was sometimes attached to a wooden axle or a horizontal wheel to facilitate the motion (Laferrière, Ménassa, 1974). The *shadoof*, also widespread in the Arab world and Eastern Europe, is a leverage device for lifting water in a bucket through human strength. It has existed since the second millennium BC and is still used today (Viollet, 2005). In the context of irrigation, the *noria*, a hydraulic wheel with large buckets, draws water from rivers and streams for use in irrigation. The *noria* moves by means of the hydraulic power of rivers and streams (Viollet, 2005).

Outside the Western world, in the nineteenth century, stored rainwater and runoff constituted two of the main sources of domestic water consumption. Rainwater harvesting cisterns dating from Antiquity exist in Jerusalem, for example, and other types of reservoirs can be found in rural and urban areas throughout the world (Lemire, 2010). Examples include the cisterns used by the Bedouin in the Negev Desert well into the twentieth century (Evanari, Shanan, Tadmor, 1982). Another very sophisticated method of water conveyance has been used in the Arab and Persian world since Antiquity: the *qanât*. This is a series of dug channels and vertical shafts for collecting water that has seeped underground. These hydraulic constructions were used in Iran, for example, well into the twentieth century (Goblot, 1963). Globally, water storage benefitted from innovations made in the area of materials in the second half of the twentieth century. Concrete, metal, and plastics, in particular high-density polyethylene, became widespread around the world (Maison, Mansotte, Margueron, 2010).

Water supply sources for cities and rural areas in nineteenth-century Western Europe are well known. In the 1820s, for example, water from the Thames and its tributary, the New River, provided the inhabitants of London with water for domestic use (Mille, 1854; Ward, 2003). Public fountains, standpipes, pumps — essential urban monuments at the time — and waterways going through cities made up the main drinking-water resources in urban areas in the West. Private wells, whose water was sometimes of questionable quality, constituted the norm for a long time. Taking advantage of technical expertise, cities began to bring in water from farther and farther away. In some cases, particularly in arid environments, fossil water was used, or yet-untapped springs were sought. Water diviners' age-old techniques prevailed for a long time and even received renewed interest at the end of the nineteenth century, undergoing major developments in British protectorates in the Near East

(Bensaude-Vincent, 2002). The most significant changes, however, arose from developments in hydrogeology. In the nineteenth century, the techniques developed to locate groundwater advanced at the same pace as the structuring progress in hydrogeology. As a result, mere understanding of the formation of groundwater was surpassed by the ability to analyse and quantify the behaviour of water tables (Castany, 1991). In Dijon in 1856, the hydrogeologist Henri Darcy conceptualized the Darcy Law, thus setting out the foundations of underground hydrodynamics. Other hydrogeologists, such as the Frenchman Jules Dupuit in the 1860s and the German Adolph Thiem at the beginning of the twentieth century, made considerable advances in hydrogeology and methods for collecting water. Exploratory drilling and bore-hole drilling became ways of reaching the piezometric level of groundwater tables thanks to improved geophysical and geological knowledge.[1]

The ancient technique of drilling underwent remarkable developments in the nineteenth century (Castany, 1991; Lyonnet, 2008). Percussion drilling, commonly used in the nineteenth century, was inexpensive and did not require any drilling fluid such as mud or water. Petroleum extraction in Europe and the United States began in the 1850s and led to the introduction of a new technique, rotary drilling, which was highly valued in the hydraulics sector at the beginning of the twentieth century (Mylander, 1953). This method consists of lowering a rotating rod towards the ground from a derrick and destroying the rock through its weight and rotation. This method requires the use of drilling fluids — water, mud, or air — to cool down the rotating rod, to transport the drill cuttings back to the surface and to slow down the surge of artesian water from the sources reached (Lyonnet, 2008). Unlike the first method, this technique allows for deep drilling by controlling the upward flow of artesian water. A third method appeared in the nineteenth century, specially developed for drilling into hard ground: down-the-hole (DTH) drilling. The rotary drill bit is driven by percussion using high-pressure compressed air. These three techniques are suitable for drilling and capturing groundwater in different types of soils.

From a sanitary point of view, access to drinking water was a major health issue in nineteenth-century cities (Frioux, 2013). However, progress in hydrology was accompanied by a certain degree of mistrust regarding groundwater, i.e. the fear that this water might be contaminated with harmful substances that had seeped down from the surface. In the nineteenth century, purification and disinfection processes sought to eliminate all immediate risks of illness. The Germans popularized water ozonation in Europe. This complex and costly technique very quickly had to compete with bleach, which was added to water to purify it (chlorination) but gave it a peculiar taste. Finally, during the First World War, Philippe Buneau-Varilla developed a new method in Verdun that used smaller doses of chlorine than in chlorination (100 microgram versus 1 milligram per litre), which was referred to in French as *verdunization*. In the 1920s and 1930s, this method spread throughout European cities and to the French colonies despite its occasionally disappointing results. In

1 The piezometric level of a groundwater table is its highest point.

the twentieth century, chlorine prevailed as the most commonly used disinfectant after water treatment and filtering for human consumption. Indeed, chlorination is indispensable for the distribution of water through urban pipe networks. Today, more than three quarters of the drinking water supplied through distribution networks in the world is treated with chlorine (Centre d'information sur l'eau, 2019). Water purification and disinfection in the twentieth century was not only a response to an urgent health imperative, but also a means of providing water with a balanced level of minerals to prevent long-term chronic illness. Since the end of the Second World War, various Western countries and large cities around the world have developed water quality standards (Barraqué, 2004).

The removal and subsequent treatment of wastewater before its release into the natural cycle of water in the environment are a corollary to purification. In 1850s London, every street was equipped with a sewerage system. Most houses were linked to it through an earthenware pipe into which all wastewater flowed (Mille, 1854). This water was then released into a tributary of the Thames. A lack of precautions and water treatment caused cholera epidemics, for example in the Broad Street district in 1854 (Chapelle, 2005). When epidemiology was still in its infancy, the British doctor John Snow attributed the spread of disease to water. For most of the nineteenth century, Parisian cesspits were used to collect faecal matter and wastewater. Capable of holding between 20 and 30 cubic metres, the cesspits were located at the bottom of cellars and connected to the upper storeys of houses by soil pipes (Jacquemet, 1979). Due to the considerable handling costs, these watertight barrels were often not emptied and replaced by Parisian septic tank maintenance workers until after overflowing into the cellar, a situation that caused foul odours and posed health risks (Jacquemet, 1979). Transported by horse and cart, the contents were emptied into roads near Paris and then turned into fertilizer. During the Second French Empire, clay sewer pipes were gradually replaced with pipes made from cast iron to ensure a better seal. Ducts were enlarged to facilitate their cleaning. On average they were 2.5 metres high and 1.3 metres wide (Frioux, 2013). Buildings were connected to the sewerage system via a *tinette-filtrante* ('filtering tub') whose function was to prevent black water, defined as waste water containing faeces and urines, from entering the pipes (Chatzis, 2014).

It was not until the 1850s that sewage treatment measures were introduced on a city-wide scale. In Paris, this project — a vast underground collection system — was designed by Eugène Belgrand. At the end of the nineteenth century, the mains drainage system, which collected wastewater and rainwater in the same pipes, was adopted in large cities in Europe and the United States (Chatzis, 2014). This system was extended to urban areas in the twentieth century. However, it is still not present in rural areas of sub-Saharan Africa and developing countries (United Nations Organization, 2018).

The use of wastewater for irrigating fields, a practice dating from the Middle Ages, continued into the nineteenth century (Herment, 2017). In the following century, new sewage treatment methods were developed, i.e. septic drain fields and the activated sludge process (Frioux, 2013). These methods were then combined in the twentieth century to create the model of water treatment plants. In the case of septic drain fields, wastewater is filtered through a thick layer of sand. The activated sludge technique, which was developed in Manchester, involves injecting air into the wastewater to accelerate the decomposition of organic waste. The procedures used in water treatment plants are well established (Deutsh, Vuillerme, 2003). The water is first pre-treated by screening, de-oiling, and degreasing. Next, the water is decanted and treated with activated sludge, which is a bacterial treatment. Lastly, the water is clarified through a final decantation before being returned to watercourses. In the twenty-first century another, notably costly, step is sometimes used: phosphorus treatment after bacterial treatment.

Despite appearances, and despite the dominant technological discourse, water continues to occupy an important place in contemporary technical systems. Its uses have considerably evolved: from a tool, water has developed into a product without ever ceasing to be a driving force. Its abundance and fluidity automatically put it at the centre of production processes. The dramatic effects of global warming on water have raised serious questions about the future of industrial and technological systems. Changes to hydrological regimes and the rise in runoff temperatures thus represent major challenges for the nuclear industry. Other signs attest to the importance of water — water continues to attract investment and highly imaginative technological projects. Desalination, or desalinization, uses two methods to remove the salt from sea water or brackish water: thermal-based desalination through distillation, or membrane-based desalination through reverse osmosis (Cotruvo, Fawell, Voutchkov, 2010; El-Sheikh, 1988). Mekorot, the national water company of Israel, has developed a water supply network called the National Water Carrier that uses five desalination plants to produce enough desalinated water for up to seventy-five per cent of domestic water consumption (Mekorot). Since the beginning of the twenty-first century, disinfection using ultra-violet rays has been developed, but it is still a costly process and does not exclude the use of chlorine.

Given the scarcity of water today, hopes are that technology can generate it. Indeed, more than ever, water is the subject of demiurgic dreams. During the 2008 Olympic Games in Beijing, for example, experiments to tackle pollution included cloud seeding, whereby silver iodide and dry ice were shot into clouds to induce rain (Niquet-Cabestan, 2008). Such efforts, however, above all reveal the spread and renewal of a technological discourse that refuses to acknowledge environmental reality.

References

Barraqué, Bernard, 'Normes de qualité de l'eau: histoire et prospective', *15e Journées scientifiques de l'environnement. Usages de l'eau: synergies et conflits* (2004), pp. 1–14. <https://hal.archives-ouvertes.fr/hal-00686867> [accessed 25 April 2019].

Benoit, Serge, *D'eau et de feu: forges et énergie hydraulique (XVIIIe–XXe siècles). Une histoire singulière de l'industrialisation française* (Rennes: Presses universitaires de Rennes, 2020).

Bensaude-Vincent, Bernadette, 'Des rayons contre raison? L'essor de la radiesthésie dans les années trente', in *Des savants face à l'occulte*, ed. by Bensaude-Vincent, Bernadette et Blondel, Christine (Paris: La Découverte, 2002).

Biggs, David Andrew, *Quagmire: Nation-building and Nature in the Mekong Delta* (Seattle, WA: University of Washington Press, 2012).

Briant, Pierre, ed., 'Politique et contrôle de l'eau dans le Moyen-Orient ancien', *Annales: Histoire, Sciences Sociales*, 3 (2002), pp. 515–63.

Brun, Jean-Pierre and Jean-Louis Fiches, *Force hydraulique et machines à eau dans l'Antiquité romaine* (Naples: Centre Jean Bérard, 2007).

Builder, Albert, *L'Eau potable … où trouver l'eau, comment la capter, la stocker, l'épurer, la distribuer dans l'habitation, l'atelier, la ferme* (Paris, Liège: Ch. Béranger, 1939).

Callahan, North, *TVA: Bridge over Troubled Water* (South Brunswick, NJ: A. S. Barnes, 1980).

Cardwell, Donald, *Wheels, Clocks, and Rockets: a History of Technology* (New York, NJ: W. W. Norton and Company, 1995).

Castany, Gilbert, 'Origine et évolution des concepts des eaux souterraines', *in Travaux du Comité français d'histoire de la géologie*, 5 (Paris: Comité français d'histoire de la géologie, 1991), pp. 1–7.

Centre d'information sur l'eau ('water information centre'): <https://www.cieau.com> [accessed 7 March 2019].

Chapelle, Franck, *Wellsprings: a Natural History of Bottled Spring Waters* (New Brunswick, NJ: Rutgers University Press, 2005).

Chatzis, Konstantinos, 'Alimenter en eau et assainir les immeubles parisiens, 1850–1930: la généralisation du "système Belgrand"', *Flux*, 97–98 (2014), 30–36.

Cotruvo, Joseph, Voutchkov, Nikolay, Fawell, John, Payment, Pierre, Cunliffe, David, Lattemann, Sabine, eds, *Desalination Technology: Health and Environmental Impacts* (Boca Raton, FL: CRC Press, 2010).

Culvahouse, Tim, *The Tennessee Valley Authority. Design and Persuasion* (New York, NY: Princeton Architectural Press, 2007).

Deutsh, Jean-Claude and Martine Vullierme, 'L'Évolution des techniques', *Flux*, 52–53 (2003), 17–26.

El-Sheikh, M. I., 'L'agressivité du climat et le problème de l'eau potable au Kuwait (Arabie)', *Annales de Géographie*, 540 (1988), 150–70.

Evanari, Michael, Leslie Shanan, and Naphtali Tadmor, *The Negev: the Challenge of a Desert* (Cambridge, MA: Harvard University Press, 1982).

Frioux, Stéphane. *Les Batailles de l'hygiène: villes et environnement de Pasteur aux Trente Glorieuses* (Paris: Presses Universitaires de France, 2013).

Goblot, Henri, 'Dans l'ancien Iran, les techniques de l'eau et la grande histoire', *Annales. Economies, sociétés, civilisations*, 18.3 (1963), 499–520.

Gourou, Pierre, *L'Utilisation du sol en Indochine française* (Paris: Centre d'études de politique étrangère, 1940).

Gribbon, H. D., *The History of Water Power in Ulster* (New York, NY: A. M. Kelley, 1969).

Gross, Laurent F., *The Course of Industrial Decline: the Boot Cotton Mills of Lowell Massachusetts, 1835–1955* (Baltimore, MD: John Hopkins University Press, 1993).

Herment, Laurent, 'Vidanges et fertilisants: le cas de la poudrette parisienne au milieu du dix-neuvième siècle', *Journal for the History of Environment and Society*, 2 (2017), 95–126.

Hunter, Louis C., *A History of Industrial Power in the United States*, vol. I, *Water Power in the Century of Steam Engine* (Charlottesville, VA: University of Virginia Press, 1979).

International Energy Agency, <https://www.iea.org> [accessed 9 March 2019].

Jacquemet, Gérard. 'Urbanisme parisien: la bataille du tout-à-l'égout à la fin du XIXe siècle', *Revue d'Histoire Moderne et Contemporaine*, 26.4 (1979), 505–48.

Jumeau, Alain, *L'Angleterre victorienne: documents de civilisation britannique du XIXe siècle* (Paris: PUF, 2001).

Laferrière, Pierre and Laïla Ménassa, *La Sâqia: technique et vocabulaire de la roue à eau égyptienne* (Cairo: IFAO, 1974).

Lemire, Vincent, *La Soif de Jérusalem: essai d'hydrohistoire (1840–1948)* (Paris: Publication de la Sorbonne, 2010).

Lyonnet, Philippe, 'Les principales techniques de forage', in *Le Forage d'eau au XXIe siècle*, ed. by Pierre Pouchon (Paris: AHSP, 2008), pp. 5–8.

Maison, Dominique, François Mansotte, and Thomas Margueron. 'L'alimentation en eau potable en Guyane: problématique et solutions appropriées', *Santé Publique*, 2 (2010), 181–92.

Malm, Andreas, *Fossil Capital: the Rise of Steam Power and the Roots of Global Warming* (London: Verso, 2016).

Maurel, Alain, *Le Dessalement de l'eau de mer et des eaux saumâtres et autres procédés non conventionnels d'approvisionnement en eau douce* (Paris: Technique et Documentation, 2001).

Mekorot, 'The New National Water Carrier', Mekorot <https://www.emsmekorotprojects.com/project/the-new-national-water-carrier/> [accessed 6 March 2019].

Mille, Adolphe Auguste, *Rapport sur le mode d'assainissement des villes en Angleterre et en Écosse, présenté à M. le préfet de la Seine par M. Mille* (Paris: Vinchon, 1854).

Mylander, Harvey A., 'Oil Field Techniques for Water Well Drilling', *Journal — American Water Works Association*, 45.7 (1953), 764–72.

Niquet-Cabestan, Valérie, 'La Chine face au défi des Jeux Olympiques', *Politique étrangère*, 73.2 (2008), 253–66.

Pomeranz, Kenneth, 'Les eaux de l'Himalaya: barrages géants et risques environnementaux en Asie contemporaine', *Revue d'histoire moderne & contemporaine*, 62 .1 (2015), 7–47.

Rassweiler, Anne Dickason, *The Generation of Power: the History of Dneprostroi* (Oxford: Oxford University Press, 1988).

Shaw, John, *Water Power in Scotland, 1550–1870* (Edinburgh: J. Donald, 1984).

Sorre, Maximilien, *Les Fondements biologiques de la géographie humaine: essai d'une écologie de l'homme* (Paris: Armand Colin, 1943).

United Nations Organization, 'Il faut accroître les investissements pour une couverture universelle des services d'assainissement (OMS)', (UNO, 1st October 2018) <https://news.un.org/fr/story/2018/10/1025472> [accessed 5 March 2019].

Viollet, Pierre-Louis, *Histoire de l'énergie hydraulique: moulins, pompes, roues et turbines de l'Antiquité au xxe siècle* (Paris: Presses de l'école nationale des Ponts et chaussées, 2005).

Ward, Robert, *London's New River* (London: Historical Publications, 2003).

Wittfogel, Karl A, *Oriental Despotism: a Comparative Study of Total Power* (New Haven, CT: Yale University Press, 1957).

Worster, Donald, *Rivers of Empire: Water, Aridity and the Growth of American West* (New York, NY: Pantheon, 1985).

LUDOVIC LALOUX

Food Technology

Food Preservation, Packaging, and Transport

From the ancient amphorae used to carry wines across the Mediterranean to the tureens of Nérac foie gras sent from the Aquitaine Basin to decorate the table of the sixteenth-century Danish king Christian II, both processed and unprocessed foods have long been sent across the world (Pitte, 2010; Meyzie, 2010). For centuries, desiccation techniques, fermentation, and also the use of fat, alcohol, salt, and sugar have made it possible to preserve food stuffs for consumption at a later date, whether in preparation for periods of dearth, or to sustain populations both at the end of winter, when supplies were exhausted, and in early spring, when the first agricultural products sometimes proved insufficient (Flandrin, Montanari, 1996). However, during famines, wars, or other troubled times, stored food products may not be enough to supply the population. The search for new processes to meet the food needs of a growing population, through the development of optimized food conservation, packaging, and transport, has therefore long been essential. At the turn of the nineteenth century, Nicolas Appert's mastery of the use of heat to preserve food opened promising new perspectives. In addition to the continued use of natural preservatives (sugar, salt, fat, and acid) as well as freeze drying, the nineteenth century laid the foundations for food preservation techniques based on hot or cold temperatures. This, in turn, led to a shift in lifestyles and new opportunities to avoid scarcity provided that, from producer to consumer, the appropriate technical preservation equipment existed and basic hygiene measures were implemented. Finally, in the twentieth century, new innovations such as the development of artificial preservatives, the improvement of vacuum technology that extended the shelf life of products, or the arrival of off-season products, led to questions as to whether these modernizations were, indeed, improvements.

Temperature Control for the Preservation of Foodstuffs

In terms of food preservation, the great novelty of the nineteenth century was temperature control, both for heat and for cold. This then led to the development of several processes for the preservation of milk in particular.

Global History of Techniques, ed. by Guillaume Carnino, Liliane Hilaire-Pérez and Jérôme Lamy, Global Matters Culture Matérielles, histoire des techniques et circulation globale des savoirs, 9 (Turnhout: Brepols, 2024), pp. 391–402
BREPOLS PUBLISHERS 10.1484/M.TECHNE-EB.5.129819

Heat and Food Preservation

Born in Champagne, France, Nicolas Appert first became a confectioner and then a distiller before carrying out his first experiments on food preservation using heat for sterilization in 1795. Appert took particular care to protect the preserved food from the surrounding air in order to allow its subsequent consumption. This preservation technique did not alter the food's qualities. For long term storage, he used hermetically sealed containers, in particular champagne bottles, which were the strongest glass containers at the time. The process invented by Appert destroyed all microbes and bacteria. In 1802, he founded the world's first food cannery in Massy, near Paris where he employed about 50 people to produce tinned fruit and vegetables. Subsequently, the use of tin cans manufactured in a factory in England provided sturdier and optimized packaging. From 1806 onwards, Appert supplied the French Imperial Navy with tinned goods. Published in 1810, his book *The Book of All Households or the Art of Preserving All Animal and Plant Substances for Several Years* is based on his invention of the sealed metal box which protected food from the surrounding air after being sterilized by heat. Henceforth the term 'appertization' was used to describe this method, which allowed for the preservation of food in glass jars as well as in tin and aluminum cans.

Around 1820, sardines in oil were the first seafood to be tinned, and they are considered the first foodstuff that was produced industrially. Nobbed and cooked, sardines were packed with vegetable oil, most commonly olive oil from Mediterranean countries. At the time this was an innovative product with soft meat and a taste that was significantly different from the fresh sardines that were consumed in fried form by low-income European populations. For several decades, canned sardines remained a rare and luxurious product that was only eaten by wealthy people around the world and by members of public bodies, such as the navy. However, using both heat for cooking and oil for preservation, technological progress in manufacturing made it possible to increase the volume produced and reduce prices by the beginning of the twentieth century. From then on, the consumption of tinned sardines in oil became more and more popular. Tinned tuna, mackerel, and anchovies followed suit. Seafood could now be preserved with techniques than salt and be easily consumed on all land all year round.

Nicolas Appert's research anticipated that of Louis Pasteur in the second half of the nineteenth century. Pasteur's eponymous method, pasteurization, heated food to a temperature of just under one hundred degrees Celsius, which was followed by sudden cooling to destroy pathogenic microorganisms and the enzymes responsible for autolysis. It preserves the consistency, appearance, and taste of food and is applied mainly to milk, fruit juices, and tinned vegetables (Freedman, 2007). The process also preserves the food's proteins, vitamins, minerals, and flavour to such a degree that pasteurized products qualify as 'fresh' products. However, pasteurization fails to remove all microorganisms from the food and does not eliminate fungal spores. Non-sterile, pasteurized foods therefore contain small amounts of germs. Another weakness of pasteurization is that the shelf life of pasteurized products does not exceed a few days, and that they must be kept in a cold environment (of no more than 4°C) in order to limit the proliferation of germs that remain present in the product.

Refrigeration and Freezing for Food Preservation

Cold has been used for food preservation since ancient times. However, in the nineteenth century, the production of artificial cold was introduced. In 1834, the US citizen Jacob Perkins filed a patent for steam compression, not intended for cooling food directly but for producing ice to keep food cool. Later, in 1858, Charles Tellier built a machine to produce cold using liquefied ammonia gas. Subsequently, he installed a mechanical compression machine using liquefied gas at the Menier chocolate factor in Noisiel, near Paris.

In 1876, under Tellier's leadership, *Le Frigorifique*, the first ship with refrigerated holds, left Rouen for Buenos Aires with 30 tonnes of meat. This was organised to demonstrate the use of cold for the transport of perishable foods. After a journey of 12,000 kilometers across the Atlantic Ocean, which was achieved in 105 days, the meat reached its destination in a well-preserved state. The low temperature of the ship's holds, kept between 0 and minus 2 degrees Celsius, caused artificial drying, with a dehydration of the meat by 25%, which, in turn, prevented the development of microorganisms. In 1877, the Frenchman Ferdinand Carré transported frozen meat at minus thirty degrees Celsius from Argentina to France, again by boat. In February 1880, the cargo ship *Strathleven* brought the first load of Australian meat to London: 134 tons of frozen mutton and beef that had survived a 24,000 kilometers journey of nine weeks. These experiences truly mark the beginning of the cold chain. Argentinians, Australians, and New Zealanders quickly understood the importance of these initiatives for exporting meat to the European continent.

Produced in the United States, the first electric refrigerators began to be installed in homes in 1913. However, they only increased slowly in number, and their actual dissemination began in Western countries in the 1950s. Refrigerators allow food to be stored for several days at a temperature of around 4 to 5 degrees Celsius. From the 1960s onwards, freezers were also installed in homes in the West, primarily for farmers who filled them mainly with the meat and vegetables they produced. Thanks to a temperature of minus 18 degrees Celsius, these devices allow food to be preserved for several months. However, it was important to be able to bring down the temperatures of foods to freezing point fast since clow cooling causes the formation of large crystals that alter the taste of the products. According to Danish cooling expert Ottesen, the good preservation of frozen food requires it to be cooled rapidly at a low temperature. Then the ice crystals that form remain very small in size, thus keeping the structure of the food intact. Inspired by the traditional techniques of the Inuit of Labrador, who performed quick freezing by exposing the products of their hunting and fishing activities to the polar wind at around minus 40 degrees Celsius, quick freezing was developed for the industry. With these temperatures applied in the food industry, the process creates an excellent conservation of food products and their transport with adapted means of transportation.

Milk Preservation

While Louis Pasteur was interested in the preservation of milk, its first pasteurization was done by the German chemist Franz von Soxhlet in 1886. Until the beginning of the twentieth century, cow's milk, a fragile product, was more accessible to rural inhabitants than to urban dwellers. Hygiene problems made it difficult to provide regular supplies of milk to cities. However, in 1933, the development of the method of preserving milk by heating it for two seconds at an ultra-high temperature, between 138 and 158 degrees Celsius, opened up new commercial opportunities. Once cooled, the milk was poured into an aseptic container and would keep for a long period of time as long as it did not come into contact with air. However, it was not until 1951 in Lund, Sweden, that Ruben Rausing designed tetrahedral cardboard packaging to easily store and transport liquids. The following year, this system found its first industrial application: the Lundaortens Mejeriförening dairy used it to package milk-based products. As early as 1953, polyethylene was introduced as a plastic coating for cardboard to ensure that it could hold liquids. A very thin layer of aluminum was added to protect the milk from oxygen and light, and the continuous filling method offered maximum hygiene.

In the 1960s, this type of packaging became truly sterile thanks to the fact that the machine itself was sterilized. The process ensures that the sterile product is packaged well. Hermetically sealed in opaque packaging, the product can be stored for six months at room temperature. Brickshaped packaging also made it easier to store and created the possibility for a wider dissemination of the product. First developed in Sweden, this type of packaging revolutionized consumption patterns due to its practicality, both for the transport and the storage of the product and for its use by consumers. Within a few years, via the development of specific machines, this new packaging not only spread across Western Europe, but also to countries such as the Soviet Union (in 1961) and the United States (in 1962). Most countries around the world adopted it, not only for milk, but also for fruit juices and various other products such as tomato concentrate.

However, until the twenty-first century, Americans preferred milk that was simply pasteurized (at 62 to 88 degrees Celsius for 15 to 20 seconds), which had a barely altered taste, unlike the quick ultra-high temperature treatment, which gives milk a slight caramelized taste by burning some of its natural sugars (lactose) (Valenze, 2012). At the turn of the twenty-first century, the development of microfiltration offered an alternative that combined the authentic taste of fresh milk with its nutritional qualities and long-term storage. Indeed, on one hand, milk can be filtered through ceramic filters whose pore diameters in the microfiltration membrane are only 0.5 micrometre 'effective' (effective meaning here that their actual diameter is often a slightly larger) stopping all bacteria. On the other hand, this milk can be stored for up to six months when heated up to only 96 degrees Celsius for six seconds, as this neutralizes the action of the endogenous enzymes which would otherwise spoil the milk. This new process allows milk to retain its natural qualities so that it is impossible for consumers to differentiate between fresh pasteurized milk and microfiltered long life milk.

Lifestyle Changes

Changes in eating habits began in the nineteenth century. A decrease in the consumption of bread and other centuries-old basic products was accompanied by a larger diversification in the food consumed (Klein, Watson, 2016). The development of the agri-food industry created both a more varied offer of products and new means of preservation, while improved packaging encouraged the distribution of food over longer distances.

Meat Extract and Freeze drying

In his *Treaty on Organic Chemistry*, published in Paris in 1840, the German chemist Justus Liebig (1803–1873) laid the scientific foundations for the mineral nutrition of plants. Liebig is particularly well known today for his experiments in chemistry applied to food. While cooking pieces of meat in water, he realized that the taste and smell of the meat transferred to the broth, while the rest of the meat lost its flavour and became hard. He concluded that the major nutrients were now found in the broth rather than in the cooked meat.

In Liebig's opinion, the obtained broth was a particularly beneficial product for health. Slowly cooked in water, minced meat yielded a very tasty broth that Liebig concentrated by using a water bath to achieve a soft and brown texture. Discussed in his 1844 *Chemische Briefe*, the meat extract thus produced, which was easily soluble in water, made it possible to reconstitute the broth. It was then easy to add a few small pieces of conserved vegetables, finely chopped and dried to ensure their preservation. Convinced of the usefulness of his work on meat, Liebig even had the idea of treating a young girl suffering from cholera by administering meat treated with a diluted solution of hydrochloric acid to facilitate digestion and prevent vomiting. The girl survived this disease, which was often fatal. In this way, Liebig's research on food also entered the medical field.

Prior to Liebig's scientific work, other ready-to-use soups had existed, however without similar nutritional qualities. In Europe, the severe agricultural problems of 1845 to 1847 and the famines that followed underlined the importance of having products with a long shelf life. Georg Christian Giebert, a railway engineer based in Brazil, read the *Chemische Briefe* and proposed industrial production to Liebig in order to produce more meat extract in a meat processing plant that he had purchased in Uruguay. This was the beginning of the industrialization of a process that had used rather traditional methods until then. Subsequently, production increased and more factories were established that could turn 100 kilograms of fresh meat into 4 kilograms of meat extract. Further scientific research led to the production of soup bars made from meat extracts, vegetable powders, and condiments. At the end of the nineteenth century, four firms competed for the market in ready-made soups: Liebig (Germany), Knorr (Germany), Maggi (Switzerland), and Amour (United States). Subsequently it was possible to have sufficient amounts of food in winter and early spring, when fresh vegetables are in short supply, thanks to reconstituted soups and meat extracts that could be stored without any problems.

The Democratization of Sugar Consumption

In sugar consumption one commonly distinguishes between the pursuit of a sweet taste on the one hand, and the need to preserve food on the other. In nature, sugar presents itself in the forms of honey, sugar cane, as well as in the form of birch and maple syrups whilst it is present in carrots and other vegetables. In the eighteenth century, under the sponsorship of the kings of Prussia, Marggraf and Achard showed that beets could also yield sugar (Laloux, 2020). As a consequence sugar cane and beet were in competition with each other in the nineteenth century, with the aim to produce more and more sugar to meet the growing demand. Thus sugar, which had been a luxury product at the beginning of the century, gradually became more accessible and its price was reduced. Foods associated with sugar such as coffee, tea, and chocolate, and also lemonades and ice cream, stimulated consumption, albeit sometimes without the consumer's knowledge; one iconic example is Coca-Cola, a drink developed in 1886 by John Pemberton, a pharmacist in Atlanta in the United States, one-tenth of which is sugar.

Further, once Van Houten in Amsterdam had invented the solubilization of cocoa to produce chocolate powder, chocolate makers multiplied innovations to encourage customers to consume their sugar-rich products. Lindt developed fondant chocolate with the conching process; Nestlé became best known for its milk powder; Poulain introduced vanilla paste; Banania developed its chocolate powder made from banana and barley malt extract; Neuhaus invented filled chocolate; and Mars contributed to the innovations with its milk chocolate bar that was filled with confectionery and caramel (Shapiro, 2015). At beginning of the twentieth century, due to the growing demand for chocolate on the world market, cocoa production increased significantly. Cocoa plantations are now becoming more numerous in Africa and Asia, exceeding the Latin American plantations in size.

Once a taste for sweet foods has been established, the appeal persists even when the price of sugar increases, because the flavours and foods preferred by one's palate change little over time. Thus, while initially a luxury product, sugar became a staple. At the beginning of the nineteenth century, its use was limited to one or two kilograms per person per annum in northwestern Europe, and by 1900, yearly sugar consumption reached 45 kilograms in Great Britain, 30 kilograms in the United States, 24 kilograms in Switzerland, and 23 kilograms in Denmark, but only 14 kilograms in Germany, 7 kilograms in Russia, and 3 kilograms in Italy. Eating habits such as the high consumption of jams, marmalades, and sweet sauces among the British explain their high level of sugar consumption. Apart from periods of war, sugar consumption in industrialized countries increased steadily until the 1970s and 1980s, reaching an average of about 35 kilograms per year per person, before falling back to around 25 kilograms at the beginning of the twenty-first century. Due to changes in dietary trends, the addition of sugar to products has been widely criticized. This explains the decrease in consumption. More precisely, it is the addition of sucrose, commonly referred to as sugar which has been the target of criticism. But the agri-food sector incorporates other sugars, such as glucose, into industrial fruit juices, and fructose and glucose syrups into apple compotes that are presented as low-sugar foods. Actually, the reduction in sugar in these apple compotes concerns only sucrose; paradoxically, the total sugars, including glucose and fructose, may be higher. Despite all criticism, sugar has become a truly common consumer product.

Changes in Food Packaging

The transport of food products requires care, both in terms of packaging and storage, in order to preserve them as well as possible. The quality of the packaging is essential, otherwise the products will be damaged (Hachez-Leroy, 2019). Until the twentieth century, the longstanding use of natural materials such as clay, leather, wood, cork, as well as fibres (hemp, flax, straw, and wicker) and processed products such as metals, paper, cardboard, and glass, continued being used. In the second half of the twentieth century however, the increasing use of a new product, plastic, changed the game. As a result, most of the traditional protective materials disappeared, mainly because of their weight or the difficulty they entailed for handling, but also because materials needed not to react chemically with the products that were being transported. Indeed, any toxicity or modification of the physico-chemical and organoleptic characteristics of food products must be avoided.

However, cardboard, metal, wood, paper, and glass continued to be used because of adaptations made to these products. Thus, particularly for beverages such as beer, fruit juice, or soft drinks, metal is widely used, but now in much lighter form, taking away the weight issue. Thanks to some countries' continued communication of instructions for the recycling of beverage bottles, glass continues to be used. Paper in the form of cooking paper is also still being used as it is both waterproof and resistant to high temperatures. In this form, it is particularly suitable for packaging greasy products such as pastries, butter, margarine, and even pre-cooked poultry that can be reheated in either a traditional or microwave oven. In fact, the most significant developments probably concern cardboard. It is manufactured from natural and renewable materials, based on starch, straw (preferably from wheat), or cellulose fibre. Now made with double or even triple flutes, it has become more resistant and lightweight. Cardboard is suitable for carrying fruit and vegetables, as long as it is treated to resist humidity. It competes with light wood, which is still being used for crates. At the beginning of the twenty-first century, the development of micro-flute cardboard packaging facilitated packaging, creating packaging appreciated for its elegance, particularly for products such as luxury chocolates.

Plastic materials are widely used, either in rigid form for crates, or in flexible form for packaging. Since the beginning of the twenty-first century, the proliferation of plastic bags and packaging has led governments to inroduce legislation to reduce or even eliminate their use. The introduction of bioplastics, which are sometimes presented as an answer to the problem, is not the solution. Bioplastics are six to seven times more expensive than older plastics. In addition, their lack of strength means that they can only be used once. Today, plastic in its various forms, paper and cardboard are the main materials used for packaging food products.

New Improvements?

Food additives are very useful for the preservation and transport of food. However, they have numerous drawbacks, which explains the importance of trying to limit heir use. Various emerging research and processes show alternatives.

Food Additives: An Essential Role?

Food additives are grouped into several families: texturizing agents, anti-caking agents, antioxidants, colourings, preservatives, artificial sweeteners, and flavour enhancers. Texture agents modify the consistency of industrial products such as seasonings, bread, cakes, or dairy-based desserts. They emulsify, thicken, stabilize, or gelify them. The most commonly used texture agent is gelatine, which is extracted from beef or pork tendons, ligaments, bones, and skin. It gives elasticity to soft candies, for example, and a smooth or homogenous texture to desserts made from dairy. Anti-caking agents prevent lumps in flour, cereal-based preparations, and powdered milk, among others. The role of antioxidants is to limit the oxidation of food, which would change its colour or cause fats to go rancid. The addition of antioxidants to agricultural food products affects seasonings, cereals, and fatty products, among others. The addition of colouring agents makes it possible to enhance the colour of foods that are sometimes affected by processing, or to present a more attractively coloured food to the consumer. Colouring agents are incorporated into a wide range of products, particularly biscuits, drinks, confectionery, preserves, and yoghurts. Preservatives prevent the development of mould in industrially processed products like biscuits, sweets, deli meats, cakes, and ready meals. Artificial sweeteners add sweetness to a product with few or no additional calories. Flavour enhancers enhance the smell and taste of a product, and they are used in products like vacuum-packed sausages, crisps, ready meals, sauces, and soups.

These additives offer very practical solutions for the preservation and presentation of food products. However, depending on the doses, some pose a health risk. Indeed, some additives are likely to delay growth in children and cause allergies, problems in the intestinal flora, migraines, behavioural disorders, and even cancers or other diseases. For example, glutamate, which is used in crisps, stimulate the taste buds and brain, so that it is difficult to stop eating them even though they contain about 40% of fat.

Scientific research conducted at Nanyang Technological University in Singapore in 2018 presents new opportunities for preserving food products using flavonoids. These are secondary plant metabolites that are naturally present in almost all vegetables and fruits. Their effectiveness in preserving food is much greater than that of their artificial equivalents. Tests show that when incorporated into baker's yeast, flavonoids considerably strengthen two major qualities for food preservation: antioxidant and antimicrobial effects. Another comparison of the use of artificial preservatives versus flavonoids in the processing of meat and fruit juices at room temperature shows that the former prevent the growth of bacteria for six hours, while the latter render the food products bacteria-free for two days. The use of flavonoids would avoid, for example, the use of sodium nitrates and nitrites that are associated with certain types of cancer. A development like this documents the constant search for progress to improve nutrition.

Vacuum Cooking: Guaranteed to Preserve Flavour

In the 1960s, food manufacturers used vacuum trays or bags for processed cooked products, such as ham or sausages, in order to preserve them. In 1968, the mechanic Jacques Gatineau applied this process to cooked vegetables, especially sterilized potatoes sterilized in the same splastic packaging they would be sold. However, the types of plastic that were used initially produced carcinogens when heated around fat. Therefore, since the early 1970s, more sophisticated plastics such as polyamide, polyethylene, or polypropylene have been used as they do not pose any problems for the food or consumers. Complete vacuuming extends the shelf life of foods such as berries and liquid products while hindering germ proliferation. Oxygen plays a major role in the proliferation of bacteria. The process that is used for the extraction of air eliminates oxygen and the microorganisms that usually develop in its presence. Thus, without oxygen, food remains edible for longer, for a period that is between two to seven times longer than otherwise. The decompression process increases the shelf life and flavour of foods such as steamed red beets and cheeses, and protects them from contamination. Perishable foods remain fresher, while dry products, which are protected from moisture, benefit from an optimal packaging system.

While it is usually used for practical reasons by chefs in restaurants or caterers, vacuum cooking can also be useful in private homes for family use. Vacuum cooking has three applications: first, the use of packaging to keep food fresh in its raw state, when very fresh products are essential; second, to preserve cooked meals — for example, this grants a chef flexibility for preparing a banquet with variable schedules, and also during staff shortages — and third, the vacuum cooking of food, in a bag, which helps to concentrate and to enhance flavours as well as to prevent drying out. Thus, in addition to the logistical aspects of successfully preparing food for a large number of guests, vacuum catering preserves the flavour of food, in particular keeping meat tender and juicy. Indeed, the necessary use of low temperatures ensures that the highest amount of nutritients in the food, like vitamins and minerals, are kept intact.

However, vacuum cooking has some constraints. It requires expensive equipment such as complete vacuum machines or temperature-regulated water tanks; it also depends on trained personnel to use this equipment; and great precision is required in the regulation of cooking temperatures. The number of containers or plastic packaging used is high. Further, starch-based preparations may separate or become liquid. It is then necessary to use modified starches for stabilization. Additionally, chlorophyll is not resistant to vacuum cooking so that green beans turn a khaki colour, which is not very appetizing for consumers. Without rigour in its use, this conservation process presents health risks. This explains why there is some reluctance regarding its implementation, even though it makes it possible to offer mouthwatering menus.

Food Preservation through Ionization

Since the 1980s, ionization, also referred to in the United States as 'irradiation' or 'cold pasteurization', has been developed in the agro-food industry. From a technical perspective, this involves exposing food to electromagnetic radiation waves that are shorter than waves in infrared and ultraviolet radiation. The removal of electrons from the products' top layers of atoms and molecules leads to the formation of ions, hence the term 'ionization'. The sources of this ionizing radiation are cobalt 60 and cesium 137. With the ionizing rays (beta, gamma, and X rays), this process reduces the number of microorganisms on the packaging or on the surface of the food. While killing any insects that may be present, this process slows down the ripening of fruits and vegetables in order to preserve them better. The dosage of rays varies from low for the purpose of eliminating insects, to high for inactivating viruses.

The practice of ionization can easily be applied to the treatment of dried aromatic herbs, spices, or vegetable condiments, because it preserves taste and aroma. For relatively small volumes, the process is also used for dairy products (such as camembert, caseinate or casein), egg whites, condiments made from the allium family (garlic, shallots, onions), and potatoes to prevent their germination, as well as dried or dehydrated vegetables, dried fruit, rice flour, cereal flakes and grains, peeled frozen shrimps and meat including poultry offal. Ionization removes pathogenic bacteria, such as salmonella, from animal products. It avoids the use of pesticides and chemical preservative additives. Increasing the shelf life of food products makes it easier to transport them over longer distances.

However, ionization causes the loss of water-soluble vitamins (B12 and C) and fat-soluble vitamins (A, E, and K). Moreover, unlike sterilization, it does not remove toxins from certain bacteria. Also, even though ionized foods do not become radioactive, their composition changes and new molecules appear. Since this process is relatively new, it is further not possible to know all its potential consequences for humans.

Today, human beings are generally able to feed themselves, except in the case of major disasters, wars, or policies deliberately implemented by particular political regimes in order to starve a population. However, this does not mean that food is necessarily abundant and varied for all. Since the nineteenth century, and except for the situations just mentioned, significant progress has been made in diversifying and improving the conservation of food products in order to avoid food shortages and famines. However, economic and financial realities encourage the pursuit of the greatest possible benefit from technological innovations, while advances in the quality of packaging explain why some products are being sent to more and more distant locations (Marache, Meyzie, 2015). This competes with, or even penalizes, local agricultural production in favour of off-season fruits and vegetables. As for organic farming, heated greenhouses or shipping from distant countries complicates the definition of products as acutally organic. Laws to regulate these new types of farming and the markets are at risk of being bypassed. Today, fast food is spreading throughout the world with its standardised offering of fruit and vegetables, homogenized flavours, and uniform menus. The food industry derives a financial interest from the situation. However, in the face of this trend, people continue to prepare meals with local characteristics, and there is no substitute for culinary know-how in their preparation. For a healthy diet, the future will need to find a balance between high-quality agriculture and good hygiene for the preservation of products. In order for this to happen, people will have to accept to pay a higher price, but will that be possible for everybody?

References

Flandrin, Jean-Louis and Massimo Montanari, eds, *Histoire de l'alimentation* (Paris: Fayard, 1996).

Freedman, Paul, ed., *Food. The History of Taste* (Oakland, CA: University of California Press, 2007).

Hachez-Leroy, Florence, *Menaces sur l'alimentation. Emballages, colorants et autres contaminants alimentaires, XIXe–XXIe siècles* (Tours: Presses universitaires François-Rabelais, 2019).

Klein, Jakob A. and James L. Watson, eds, *The Handbook of Food and Anthropology* (London and New York, NY: Bloomsbury Academic, 2016).

Laloux, Ludovic, *La Betterave à sucre: essor agricole et industrialisation rurale. Réalités françaises et perspectives internationales: de Marggraf à Berlin (1747) à la conférence de Londres (1937)* (Valenciennes: Presses universitaires de Valenciennes, 2020).

Marache, Corinne and Philippe Meyzie, *Les Produits de terroir. L'empreinte de la ville* (Rennes and Tours: Presses universitaires de Rennes and Presses universitaires François–Rabelais, 2015).

Meyzie, Philippe, *L'Alimentation en Europe à l'époque moderne: 1500–1850* (Paris: Armand Colin, 2010).

Pitte, Jean-Robert, *Le bon vin. Entre terroir, savoir-faire et savoir-boire* (Paris: CNRS Éditions, 2010).

Shapiro, Howard-Yana, *Great Moments in Chocolate History. With 20 Classic Recipes from Around the World* (Washington, DC: National Geographic, 2015).

Valenze, Deborah, *Milk. A Local and Global History* (New Haven: Yale University Press, 2012).

FABIEN KNITTEL

Agricultural Techniques

Four Examples of Transformations from the Nineteenth to the Beginning of the Twenty-First Century

Rural areas worldwide are extremely diverse, especially when it comes to agricultural techniques. Since the nineteenth century, the world's agriculture has gone from stability to change, agriculatural land being seen by some as 'shifting terristories' ('terres mouvantes') (Moriceau, 2002: 13–18). The prominent developments of this shift become clear with an investigation of some of the main agrarian techniques.

The history of agricultural techniques on a global scale since the nineteenth century is that of the growing chasm in productivity between on the one hand western agriculture which was first mechanized and then motorized and is now industrialized (Fitzgerald, 2003: 10–32) and on the other hand the manual, food-producing agriculture that still dominates in Asia, Africa, and South America. Thus this is a history of extreme technical discrepancies between these two opposing types of agriculture. Marcel Mazoyer and Laurence Roudart (2002: 597) have shown that, at the end of the twentieth century, the equipment used on large farms was out of reach for manual farmers, by calculating that it would take thirty centuries for a manual farmer on a small farm in the countries of the Global South to acquire the full set of machinery used in the Global North.

It is not possible within the scope of this chapter to discuss all agricultural techniques. I have therefore chosen to outline some of the most important ones focussing on four examples whose analysis is particuarly relevant when looking at agriculture on a global scale and over a long time period, from the nineteenth to the beginning of the twenty-first century. In this article, I will first focus on field preparation and tillage techniques (manual, mechanized, and motorized), and then on some harvesting techniques, including the use of scythe or sickle, as well as on haymaking and threshing. Then, selected milk processing techniques will be analysed using the example of dairy cattle farming. Finally, some manuring techniques will be discussed together with the use of fertilisers.

Global History of Techniques, ed. by Guillaume Carnino, Liliane Hilaire-Pérez and Jérôme Lamy, Global Matters Culture Matérielles, histoire des techniques et circulation globale des savoirs, 9 (Turnhout: Brepols, 2024), pp. 403–414
BREPOLS PUBLISHERS 10.1484/M.TECHNE-EB.5.129820

Fig. 1. Dombasle Plough with (1885). Th. Barrau, *Simples notions sur l'agriculture*, Paris, Hachette et Cie, 1885.

Other topics such as field preparation techniques using fire (Sigaut, 1975), the slash-and-burn technique, grape harvest techniques, grain conservation techniques (rice storage, cereal storage, particularly for wheat, etc.), or irrigation techniques (Gourou, 1984) will not be discussed at length in the scope of this article.

Tillage, Ploughing, and Field Preparation

In pre-industrial agriculture, tillage is essential to preparing fields for planting and improving soil quality for crop production. This type of ploughing can be done with various tools, e.g. manual tools, such as a spade or hoe, or with tools that are towed by animals for instance, such as araires — swing ploughs -— and ploughs. The first attempts at weeding with sulphuric acid, a task mostly carried out with a plough, were done before the 1930s, and this made it possible for farmers to slow down the growth of weeds by burying or even eradicating them. At the same time, the plough turned and loosened the soil. In temperate climates, ploughing also made it possible to bury residues left on the fields from the previous harvest as well as promoting organic fertilization (mainly by ploughing under manure) (Figure 1). The main purpose of this ploughing was to improve the soil structure, to facilitate the mineralization of organic matter, increase nitrogen levels, and assist the root growth of cultivated plants. In addition to ploughing, it was necessary to use a harrow and roller. Although the plough was the typical tillage instrument in the nineteenth century, some farmers, particularly those working with sandy soils, still used the swing plough on the eve of the First World War. Archaic as this may seem, any choice of tillage technique is related to the type of soil being tilled: the swing plough is better suited to working on light sandy soils, while the plough is more efficient in heavy clay soils (Haudricourt, Delamarre, 2000: 39–51, 520, 537).

Before the Second World War, the selection of farming equipment and the shape of some parts, such as the straight or helical shape of the mouldboard (i.e. the blade of the plough) was as much the farmer's choice as it depended on the local resources at his disposal, i.e. the equipment offered by wheelwrights and saddlers. Each wheelwright designed his ploughs and other agricultural equipment based on models published in the agricultural press, or simply according to local traditions that were passed down from father to son. In Western and Northern Europe as well as in North America, metal was a well-established material, and from the early years of the twentieth century onwards, iron and cast iron were increasingly used for the manufacture of agricultural tools, ploughs, and harrows. Tools made entirely from metal were produced, whereas previously only the working parts, such as the ploughshare (i.e. the section of the plough that cuts the soil horizontally) and the mouldboard, were made of metal.

While mechanized agriculture in the temperate zones of Europe and North America uses manual ploughing increasingly rarely, and only in horticulture, this is not the case in other parts of the world, where hoe work is still widespread. Tillage machines are towed with the help of animal power, by horses and often also oxen, although the use of the latter has been considered dated in Europe and North America since the early nineteenth century. However, because of the cheap and practical labour provided by oxen up to the end of the twentieth century, it was easier to use them to draw cartwheels. Oxen and horses were therefore still essential to European and North American agriculture until the middle of the twentieth century, as were oxen in Asia and horses in South America. Shortly before WWI the Veuve Faivre company in Champlitte (Haute-Saône department), France, depicted a mower pulled by oxen on some of its advertising posters.

In the rice fields of Southeast Asia, the land is worked with a swing plough drawn by one buffalo or two oxen. Fields (which are often very fragmented) need to be ploughed three times in a row before the plots are harrowed and levelled. This explains the persistetn use of the hoe, i.e. the manual preparation of soil, as there are areas where the swing plough can not be used, for example in the corners of the rice fields. In Lower Casamance (Senegal), the Diolas turn over the soil with the help of *kayudo,* a kind of spade, as Pierre Gourou explains (1984: 151 –54). In Madagascar, Vietnam, Colombia, and parts of Indonesia, the soil of rice fields is prepared by trampling oxen. In the absence of tillage tools, this is a specific feature of Malagasy rice fields, and the technique is practiced across the island. There are two main stages to it: the oxen are first made to trample the fields in order to bury the weeds and, a few weeks later, they are brought back to trample the flooded soil into a uniform mud. Once the rice had been transplanted, weed control is either done by hand by women, or with a hoe, a mainly male technique. In Japan, farmers use small rotary machines to harvest rice, which makes harvesting easier but requires the rice plants to be transplanted in very regular straight lines.

The second half of the nineteenth century saw the mechanization of tillage operations, while mechanized ploughing did not become widespread in the so-called developed countries until after 1945. The first steam ploughing tests, using a plough towed by a steam engine, which was a precursor to the tractor, were carried out between 1850 and 1860, but motorization was still marginal, for example, in France in 1914. It was not until the end of the Second World War that the tractor gradually became more widespread.

Types of ploughing also differed according to local contexts. From the second half of the nineteenth century onwards, the Brabant plough — a plough with two ploughshares and two reversible mouldboards — made it possible for farmers to avoid having to physically drive the plough around at the end of the plot instead reversing the moudboards. This caused the spread of flat ploughing, also known as planking. But in some parts of western France, ploughing is still done in ridges and furrows: alternately turned slices of soil leaning against each other formed curved mouldboard that encouraged drainage in areas that tended to be too wet (this is also called ridge ploughing). Here, tillage also created drainage (Bourrigaud, Sigaut, 2007: 21–27).

In the canton of Gobo, northern Cameroon, the Musey plow in ridges using a hoe called a *bananga*. They cultivat groundnuts, sorghum, and, above all, cotton. In neighbouring Chad, the Ngambay, use the same type of hoe, which they call *pangal*. However, during the 1980s, agricultural advisors sought to impose the use of ploughs instead of banangas on the Musey to ensure a more intensive cultivation of cotton via flat ploughing. The *bananga* was considered outdated, because lines obtained are less regular than those of the plough. However, ridges were more efficient for the cultuivation cotton, since they ensured the drainage of plots and slowed down soil erosion (Seignobos, Marzouk, Sigaut, 2000: 53 –68). This competition between the hoe and the plough in Africa raised the question of what was to be considered 'good ploughing'.

In Europe, the question of 'good ploughing' was discussed as early as the nineteenth century. In many agricultural experiments (such as that carried out in Lunéville, southern Lorraine, on 9 February 1833), 'good ploughing' was defined as being regular in terms of spacing and depth. The strip of land had to be completely turned over and the work done by draught animals (which were often two horses) and a ploughman without undue exhaustion. The shape of the lines driven into the ground was the main criterion that defined 'good ploughing', and the consequences of the traction work on humans and animals, i.e. their (low levels of) fatigue, another.

Ploughing was not the only way to prepare the seedbed. Other techniques used in the nineteenth and twentieth centuries to prepare fields include the use of fire as part of the slash-and-burn technique — a type of farming that involves cutting and burning vegetation before planting more — which is done in various regions, from Scotland to Indonesia (Sigaut, 1975: 43–44).

The mechanization of field preparation work, which was characteristic of agriculture in the nineteenth-century process of industrialization, was followed by motorized mechanization in the following century.

The processes of motorization and industrialization in Europe and North America, which took place at the same time as the Green Revolution in the Global South, developed in parallel with opposing dynamics such as the growing practice of no-till farming. Technological transformations are therefore not straightforward.

In the early twentieth century, major farms in North America and later in Latin America quickly adopted motorization and a high level of mechanization. Central to this development is the tractor . In Africa and Asia, by contrast, manual cultivation has remained the dominant practice, and today tractors are present on only 10% of African and Far Eastern farms, and only on the largest farms (Mazoyer, Roudart, 2002: 592). While tractors were gradually introduced to most farms in the Global North during the twentieth century, animal traction is still not the norm on nearly 85% of farms in sub-Saharan Africa.

In order to transform agriculture in 'developing countries' (as they were still referred to in the 1960s and the 1970s), agricultural research in the Global North (i.e. in 'developed countries') promoted the introduction of a Green Revolution in the Global south (Dagognet, 1973). However, the incentive to use high-yielding varieties of market-driven crops harmed subsitence crops, and ultimately only benefitted a very small part of African and Asian agriculture. Only farms situated in the regions that were already the most fertile made the huge cash invenstiments worthwhile. With the standardised cultivation methods it promoted, the Green Revolution ignored the complexity of agricultural production systems in the Global South. Another consequence has been the development of crops for export (e.g. cotton and groundnuts) at the expense of traditional food crops. Local balances were disrupted in favour of a global market for agricultural raw materials that benefitted mainly the economies of the Global North. The sol-called 'Green Revolution' also negated the expertise and experience of farmers who had developed complex production systems that combined cereal, tree, and livestock farming and ensured the maintenance of soil fertility (Mazoyer, Roudart, 2002: 649).

At the end of the twentieth century and the beginning of the twenty-first, many farmers adopted direct seeding, a technique also known as 'no-till farming'. This technique, which consisted of cultivating the plots without turning over the soil at all, was not new and did not emerge with the notion of sustainable agriculture in the 1990s, but had been implemented in the United States since the 1930s. In this technique, crop residues are left on the field to decompose gradually before new seedlings are planted. This technique helps to preserve the soil by preventing excessive soil erosion. However, weeds multiply when fields are left unploughed, so that weed control becomes a problem.

Some Thoughts on the Progressive Mechanization of the Main Harvesting Operations

Until the First World War, harvesting, especially of crops, was mainly done by hand in most European countries. The scythe replaced the sickle when, as was often the case on large farms, more straw was to be harvested, because the scythe cut at ground level. This straw was essential for the bedding of livestock or draught animals and also, in part, as their fodder. The choice of harvesting technique was not always a response to a technical difficulty or the result of a search for the most innovative tool (Sigaut, 2012), but rather was linked to a number of causes, including technical, social, and even cultural factors. For a long time, many preferred using the sickle rather than the scythe, even though the latter represented significant technical progress. But the sickle allowed farmers to 'saw' wheat and limited ginning, which reduced the spread of weeds and promoted regrowth. The very widespread use of mechanical harvesters and mowers was due to social rather than technical reasons: the lack of labourers during and after the First World War led to the use of these machines in cereal harvests. Where labour was scarce or expensive, mechanization happened quite

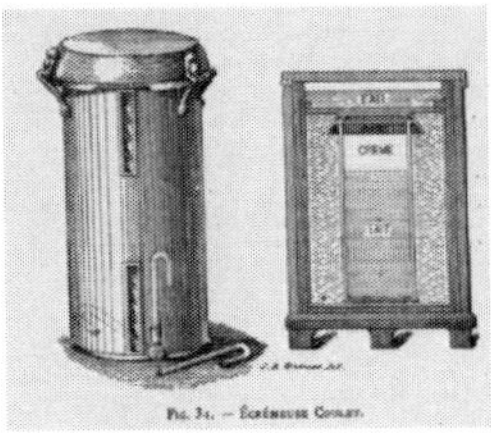

Fig. 2. Skimmer Cooley. René Lezé, *Les Industries du lait* (Paris, Firmin-Didot, 1904), p. 217. BnF/Gallica.

rapidly. It is also important to note that agricultural work, and perhaps even more so the harvesting operations, were organised according to age and gender. The men cut the wheat, while the women followed and gathered sheaves. The mechanization of harvesting went against the traditional gendered division of labour.

From the middle of the nineteenth century onwards, mowing and harvesting crops were operations that could be carried out mechanically, particularly on large capitalist farms. However, the mechanization of harvesting did not result in the disappearance of manual harvesting. Both techniques coexisted until at least the middle of the twentieth century, and in some areas until the 1970s.

In the United States, mechanical mowers and harvesters were in use from the mid-nineteenth century onwards. The latter were imported from across the Atlantic and sold in Western Europe. For example, in Haute-Saône, the company Veuve Faivre mostly sold McCormick machines. The widely distributed advertising posters promoted the machines' effectiveness. They were robust and easy to use. These advertisements represented an increasingly popular means of promoting technological progress in the agricultural press of the twentieth century.

Grain threshing was also highly mechanized. Mechanization happened quickly here, because threshing was done in a set place and did not have to be mobile. From the 1840s, steam-operated threshing machines were used on large farms, while on small farms, threshing machines were driven by animal traction. The operation was easy to mechanize. Further, mechanical threshing produced superior results in comparison with threshing by hand (for example with the flail). However, this transformation of threshing operations via their mechanization and then motorization required significant financial investments that were not always possible for small and medium-sized family farms, especially in the countries of the Global South.

Some Transformations of Dairy Techniques (Based on the Example of Cattle Breeding)

In the nineteenth century and through to the middle of the twentieth century, there were generally three models of dairy farms in the Western world (Vatin, 1996). The model in place in eastern France (and especially in the French-Swiss Jura Mountains) which was based on the cooperative production model (Knittel, 2021); the Danish industrial model (Henriksen, Lampe, Sharp, 2012: 770 –88), which was also used in western France and was done in conjunction with the production of butter and soft cheese; and the final model was that of urban dairy units. Here we discuss the technical aspects of the Jura Mountains' cooperatives known as 'fruitières', which provide a fairly broad overview of the methods used in traditional dairies before their mechanization and industrialization (Figure 2).

Fruitières are 'mutual milk lending associations' (Lullin, 1811: 9–10) that allowed associated farmers to make cheese for sale. *Fruitiers* produce a hard-paste, long life cheese of the vachelin and then of Gruyère types, which, since they can be kept for a long time, can be sold far from the production center, often in urban areas. The cheese was made in a common cauldron at the home of the person who had the *tour*, a privilege given to the farmer who had delivered the largest amount of milk. It was the principle of the so-called '*fruitière à petit carnet*'. In the nineteenth century, production in a dedicated chalet (sometimes referred to as a 'dairy', especially in Switzerland) became essential. This was the '*fruitière à grand carnet*' production model . In the fruitière model each member is paid for according to the milk they deliver. Once the milk is collected, it is placed in different containers. In summer, because milk storage is difficult, small cheeses have to be made. In winter it is possible to use the milk kept from a previous milking to make a larger cheese that keeps well, like the Comté. Dozens of litres of milk are required to produce cheese wheels with a minimum weight of about 15 kg. Technically, it was therefore necessary to pool together the small quantities of milk produced by individual families many of whom only had small herds, sometimes just the one cow, to make cheese that could be put on the market. While the cheeses produced in this manner were reserved for sale, farmers produced soft cheeses of the Munster type for their own consumption, for example in the region known as *Vosges comtoises*.

The motorization and industrialization of agriculture did not only concern soil tillage implements. Dairy cow breeding and milk processing techniques were also mechanized and industrialized, mainly in Europe and North America, but also in some agro-industrial sectors in the Global South. However, the chronology here is different to that soil preparation as most time transformations happened later in the twentieth century.

At the beginning of the twentieth century, an average dairy cow produced about 2,000 litres of milk a year. With the development of agronomy and animal husbandry, and the selection of specific bovine species there was a significant increase in milk production per head. Today, on mechanized farms in Europe and North America, a cow can produce up to 10,000 litres of milk a year. Breed selection was also intended to have animals that were adapted to mechanization, and then to milking automation. Milking machines require cows with teats that were neither too long nor too short. Mechanization has led to a standardization of practices as well that of of animals, even to the point that programmes have now been implemented to preserve domestic breeds that are on the verge of extinction. This is the case in eastern France since the 1990s with for the Vosges cow for instance.

Milk powder was developed at the beginning of the twentieth century and made it possible for milk to be widely distributed throughout the world, however, with important consequences for the agro-industrial dairy sector.

In the countries of the Global South, rapidly perishable foodstuffs like fresh milk, but also eggs and some fruits and vegetables, had escaped agro-export specialization (Mazoyer, Roudart, 2002: 598 –99). These staple foods were mainly sold in rapidly growing cities and produced in peri-urban farms. However, with milk powder the fragile local dairy agriculture was hit by competition. Milk powdering, an industrial process developed in Germany in the 1930s, was a good way to improve a series the by-products of the dairy industry while making milk a storable product with a good volume/price ratio (Vatin, 1996: 24). In the second half of the twentieth century, the milk market became more international and created a divide in traditional agriculture. For example, in Sahelian Africa, two parallel forms of milk processing started to coexist: an industrial sector that used imported milk powder, and a traditional sector based on small family cattle breeding.

Manure and Fertiliser

There were many techniques to fertilize soil before the development of agricultural chemistry. These techniques aimed to improve the soil and to create organic fertilisers. Many manuring techniques exist. Farmers can use organic soil improvers and/or fertilisers in varied proportions. Fertilisers are '*tout ce qui, mêlé à la terre arable, augmente ou rétablit la fécondité du sol en fournissant matières organiques et minérales*' ('anything that, when mixed with topsoil, increases or restores soil fertility by providing organic and mineral matter') (Lachiver, 2006: 533). Traditionnaly fertilisers thus included farmyard manure (i.e. yard and field manure), green manure (which was dug under during ploughing) and what in France was referred to as *colombine* (droppings from dovecotes and aviaries). Sheep manure was effective in manuring soil because it is very rich in nitrogenous material. The technique of marling, a soil amendment that corrects the acidity of cultivated soils with lime, was added to these organic fertilisers. In Normandy, kelp, marl, and mill sludge were also used as soil amendments. The contents of urban draining pits (known in Europe as 'powder') was also used as an amendment. It consisted of dried human excrements which are powdered and scattered onto agricultural plots, as was the case in some Chinese rice fields during the nineteenth and early twentieth centuries for instance.

In 1816, in France, manure provided only half of the fertilising requirements (Boulaine, 1995: 201 –08). The situation was the same in Galicia during the nineteenth century (Ballesteros, 2004: 144 –45). Therefore, farmers had to be supplied with organic fertilisers, or otherwise they were likely to see the fertility of the soil they cultivated inexorably reduced. Until the 1860s, the quantities of organic fertilisers produced on farms were insufficient to improve the fertility of cultivated soils, and therefore additional fertilizing agents had to be used. Sometimes it came from cities such as when the straw from large farms in the Paris Basin or Hainaut were sold to the city in return for which farmers were given the manure produced by horses in the city. Around Paris, cesspools produced manure used for market gardening before being used on large farms. In the Nantes region, what was known as the *noir animal*, a by-product of sugar production in the Caribbean was also used. It consisted of a residue of powdered animal bones used as filter in sugar production and was therefore rich in phosphoric acid that helped improve the acidic soils in western France.

In nineteenth-century Asia and well into the twentieth century, rice field manuring was often insufficient. In India, for example, animal manure was used to kindle fires. Therefore, the rice nursery technique was used, in which the germinated rice was sown into heavily smoked plots before the young plants were transplanted into the other plots of the rice field.

From the middle of the nineteenth century onwards, with the rise of agricultural chemistry, in addition to organic fertilization, chemical fertilisers became necessary to supplement the soil with nitrogen, potash, and phosphoric acid. The creation of the first agronomic centers, in some German states and then in France in particular, created a better knowledge of the effects of agricultural chemical fertilisers while at the same time establishing a body to control fertilisers and prevent fraud. The end of the nineteenth and the beginning of the twentieth century saw an increased use of artificial fertilisers, which led to a significant increase in yields per hectare. If the effects of phosphates (in the form of *noir animal*) on the granitic subsoil of the West had first been noted, superphosphates which could be used on all types of soils, were adopted on a grand scale at the beginning of the twentieth century, as was the use dephosphorus slag, which came from the residue of the Gilchrist-Thomas melting process. Around the turn of the twentieth century, phosphate fertilisers became essential for agriculture. Potash and nitrogen were introduced, but the scientific progress that brought about their full use came later. Nitrogen fertilisers required synthetic ammonia, which was first manufactured in 1913. Another element in increased productivity is also plant biology following the work of Pasteur and his disciples.

In the twentieth century, agronomists quickly discovered that phosphorus was a non-renewable resource. Therefore, animal and human excreta continue to be used or, if previously abandoned, they are now resumed, to be able to carry on using phosphate fertilisers. Potash and synthetic nitrogen were first used after 1914, mainly in Europe and North America.

The chemistry and microbiology of the nineteenth century was replaced by the genetics of the twentieth century (Bonneuil, Thomas, 2009), leading to the development of transgenic seeds, or 'GMOs', in the global North, which raised questions about food safety issues, especially health crises, such as the mad cow disease (bovine spongiform encephalopathy) in England and France in the 1990s. This led to the rejection of the globalized agri-food industry in favour of a more local approach to agricultural production, sometimes resulting in a return to traditional techniques. The industrialization of agriculture and livestock farming has had harmful consequences in terms of water pollution, turning consumers against chemical fertilizers and encouraging the development of organic or sustainable agriculture. As a result at the beginning of the twenty-first century, there is therefore a multiplicity of ways and techniques in use globally, from the most rudimentary and manual to the most sophisticated and scientific, with, for example, the genetic selection of certain cultivated plants. This takes into account the characteristic rural pluriactivity of family farming, which is still very widespread.

Since end of the nineteenth century, the global context of growing mechanization the and increasing use of chemicals in agriculture has led rural areas in the West to become more connected, thus improving the supply of distant mineral fertilisers and developing new commercial opportunities for them within the framework of specialized regional agrarian subsystems. This has increased the agricultural labour productivity in animal-drawn grain systems tenfold (Mazoyer, Roudart, 2002: 589–91). But everywhere, rich farmers who run large capitalist farms have coexisted with small farmers who run family farms and practise manual farming, like many rice farmers in Southeast Asia still do today.

References

Atkins, Peter J., *Liquid Materialities: a History of Milk, Science and the Law* (Farnham: Ashgate Publishing, 2010).

Bairoch, Paul, *L'Agriculture des pays développés (1800 à nos jours): production, productivité, rendements* (Paris: Économica, 1999).

Ballesteros, Antonio L., *Entre el vapor y el arado romano. Elites, intituciones y difusión del cambio técnico en la agricultura. Córdoba, 1780–1870* (Córdoba: Servicio de Publicaciones de la Universidad de Córdoba, Grupo de Historia Social Agraria, 2004).

Bonneuil, Christophe and Frédéric Thomas, *Gènes, pouvoirs et profits. Recherche publique et régimes de production des savoirs de Mendel aux OGM* (Paris: Quae-ECLM, 2009).

Boulaine, Jean, 'Quatre siècles de fertilisation', *Étude et Gestion des Sols*, 1995.2, 201–08 and 1995.3, 219–26.

Bourrigaud, René, *Le Développement agricole au XIXe siècle en Loire-Atlantique. Essai sur l'histoire des techniques et des institutions* (Nantes: Centre d'histoire du travail, 1994).

——— and François Sigaut, eds, *Nous labourons* (Nantes: Centre d'histoire du travail, 2007).

Campbell, Bruce and Mark Overton, *Land, Labour and Livestock. Historical Studies in European Agricultural Productivity* (Manchester: Manchester University Press, 1991).

Chouquer, Gérard and Marie-Claude Maurel, eds, *Les Mutations récentes du foncier et des agricultures en Europe* (Besançon: Presses universitaires de Franche-Comté / Guangzhou: Sun Yat-Sen University Press, 2019).

Dagognet, François, *Des Révolutions vertes. Histoire et principes de l'agronomie* (Paris: Hermann, 1973).

Emerson, W. W., Roy D. Bond, and A. R. Dexter, eds, *Modification of Soil Structure* (Chichester and New York, NY: John Wiley and Sons, 1978).

Fitzgerald, Deborah, *Every Farm a Factory. The Industrial Ideal in American Agriculture* (New Haven, CT: Yale University Press, 2003).

Gourou, Pierre, *Riz et civilisation* (Paris: Fayard, 1984).

Haudricourt, André Georges and Marielle Jean-Brunhes Delamarre, *L'Homme et la charrue à travers le monde* (Paris: éd. Renaissance du Livre, 2000).

Henriksen, Ingrid, Markus Lampe, and Paul R. Sharp, 'The Strange Birth of Liberal Denmark: Danish Trade Protection and the Growth of the Dairy Industry Since the Mid-nineteenth Century', *Economic History Review*, 65.2 (2012), 770–88.

Herment, Laurent, ed., *Histoire rurale de l'Europe, XVIe–XXe siècle* (Paris: EHESS, 2019).

Jean-Pierre, Jessenne, *Les Campagnes françaises. Entre mythe et histoire XVIIIe–XXIe siècles* (Paris: Armand Colin, 2006).

Knittel Fabien, *Agronomie et techniques laitières. Le cas des fruitières de l'arc jurassien 1790–1914* (Paris: Garnier, 2021).

Lachiver, Marcel, *Dictionnaire du monde rural: les mots du passé* (Paris: Fayard, 2006).

Lullin, Charles, *Des associations rurales pour la fabrication du lait, connues en Suisse sous le nom de fruitières* (Paris/Genève: J.J. Paschoud éd., 1811).

Mazoyer, Marcel, and Laurence Roudart, *Histoire des agricultures du monde, du néolithique à la crise contemporaine* (Paris: Seuil, 2002).

Moriceau, Jean-Marc, *Terres mouvantes. Les campagnes françaises du féodalisme à la mondialisation, 1150–1850* (Paris: Fayard, 2002).

Paillet, Antoine, *Archéologie de l'agriculture moderne* (Paris: Errance, 2005).

Seignobos, Charles, Yasmine Marzouk, and François Sigaut, eds, *Outils aratoires en Afrique. Innovations, normes et traces* (Paris: Karthala-IRD, 2000).

Sigaut, François, *L'Agriculture et le feu: le rôle et la place du feu dans les techniques de préparation du champ de l'ancienne agriculture européenne* (Paris and The Hague: Mouton, 1975).

———, *Comment Homo devint faber* (Paris: CNRS, 2012).

Stanziani, Alessandro, 'Les Paysans et la terre', in *Histoire du monde au XIXe siècle*, ed. by Pierre Singaravélou and Sylvain Venayre (Paris: Fayard, 2019), pp. 46–64.

van Bavel, Bas J. P. and Erik Thoen, eds, *Land Productivity and Agro-systems in the North Sea Area. Middle Ages – 20th Century. Elements for Comparison*, Corn Publications (Turnhout: Brepols, 1999).

Vatin, François, *Le Lait et la raison marchande. Essais de sociologie économique*, (Rennes: Presses Universitaires de Rennes, 1996).

Yante, Jean-Marie, ed., *Le Machinisme agricole, 1850–1950* (Université de Louvain-La-Neuve: *Cahiers Temps, Espaces et Sociétés*, n° 1, ajouter: 2010).

VINCENT KUITENBROUWER

Global Telecommunication Technologies from the Nineteenth to the Twenty-First Centuries

Electrifying Information

In August 1858 the inhabitants of New York celebrated the opening of the first intercontinental undersea telegraph cable, connecting the East Coast of North America with Western Europe. Inspired by the electrifying atmosphere in the city, a local artist made the lithograph 'Torchlight procession around the world', depicting a diverse parade of people from across the world holding a telegraph cable that connected them in a promise of global 'liberty' (Müller, 2016) (Figure 1). Although a few months later this first transatlantic cable broke (and was not replaced until 1866), the symbolism of the 1858 moment suggests that, like no other technology, the development of telecommunication has contributed to globalization in the modern age. Since the Industrial Revolution, and especially since the invention of electricity, contemporaries have hailed new communication technologies as great innovations that would transform the world they lived in. Such great predictions came true in the sense that, from the mid-nineteenth century onwards, various global infrastructures emerged which enabled human beings from all parts of the world to communicate directly with each other. The increasing simultaneity of these connections was attributed to the modern belief that technological developments would cause, in the words of Karl Marx, the 'annihilation of space by time' (Marx, 1857: 525).

The great transformative power attributed to global telecommunication technologies is also a major theme in the scholarly debate about media history. Canonic authors such as Marshall McLuhan considered technological innovation a key factor in the history of human communication, even more important than the content of media — a thought that was captured in McLuhan's famous aphorism 'the medium is the message' (McLuhan, 1964: 7). Additionally, McLuhan saw a clear, linear process throughout the centuries as communication technologies enabled ever closer connections between humans worldwide. This process, in his (again famous) words, culminated in the twentieth century in the development of 'electric media', which meant that the world had become a 'global village' (McLuhan, 1962: 31). Although they do consider technology to be an important factor, in their *A Social History of the Media*, Peter Burke and Asa Briggs have criticized McLuhan's work as deterministic. They call for a more historicized approach which takes into account the interplay

Global History of Techniques, ed. by Guillaume Carnino, Liliane Hilaire-Pérez and Jérôme Lamy, Global Matters Culture Matérielles, histoire des techniques et circulation globale des savoirs, 9 (Turnhout: Brepols, 2024), pp. 415–428
BREPOLS PUBLISHERS 10.1484/M.TECHNE-EB.5.129821

Fig. 1. 'Torchlight procession around the world. Dedicated to young America', lithograph by A. Weingärtner (c. 1858). Allegorical scenes related to the laying of the Atlantic telegraph cable. Library of Congress, https://www.loc.gov/resource/pga.08527/.

between political, economic, and cultural factors to analyse the impact of new media technologies on human societies of the past (Briggs, Burke, 2002: 5, 12).

This chapter draws on this historicized approach and considers the development of various communication technologies in the modern age in their historical contexts. The following sections analyse various telecommunication networks powered by electricity: wired telegraphy, wireless radio, satellite television, and the internet. As the anecdote about New York in 1858 shows, the introduction of each of these technologies gave rise to prophesies about the transformative powers of 'new media'. Edgerton's *Shock of the Old* provides a useful critique of the tendency of human beings to see technological developments in their own time as unique innovations (Edgerton, 2006). Further, despite the utopian visions of the unification of humankind through communication networks, contemporaries also were aware of the fact that these networks could be used to create division in times of war. As will be shown, discussions about telecommunication technologies in the modern age must be seen within the geopolitical contexts in which they emerged and developed. Throughout most of the period under discussion, contemporaries were aware of this connection between technology and power and therefore sought to regulate the use of emerging telecommunication networks.

Another theme in this chapter is that the subsequent rise of various media technologies in the modern age did not mean that they simply supplanted existing ones in an ongoing path of progress. On the contrary, 'new' media technologies often supplement 'old' ones and are entwined in complex patterns of interaction that form the basis of what can be called 'media ecologies' (Fuller, 2005: 2–5). In this respect, it is important to also take into account geography and the ways in which technologies have been diffused from the centres where they were developed to other parts of the world. In this respect this chapter has certain limits, as the main focus of scholarship on the history of telecommunication technology which informs this chapter has been on Western countries that have been the main sites of innovation. It is important, however, to realize that media ecologies varied (and vary) significantly around the world, and therefore the impact of telecommunication technologies has been diverse.

1850 to 1920: the Age of Wired Communication

The decades after the Industrial Revolution saw great transformations of global communication infrastructures, mainly as a result of the development of new sources of mechanical power. Steam power mainly had an impact on mobility patterns and the transport of physical carriers of information, such as letters, newspapers, and books. Steamships and steam trains enabled faster transport of these materials. At the beginning of the nineteenth century, the reach of these technologies was mainly limited to some developed parts of the Western world. But in the decades after 1850 there was a great expansion of both shipping and railway networks to all parts of the globe as part of the process of Western imperial expansion that intensified at the time. Also, the main continental and intercontinental routes in the West were increasingly supplemented with a finer grid of infrastructures that reached previously isolated communities as well. Pushed by this logistic transformation on global and local levels, the demand for information grew both in the Western and in the non-Western world.

The growth of direct information transfer was facilitated by the invention and mastering of electricity, which powered a technology that greatly increased global connectivity. In the early decades of the nineteenth century various inventors discovered that electromagnetism could be used to transfer signals via wires, but it was difficult to control this process and to turn it into a workable communication system. In the 1830s the American painter and inventor Samuel Morse achieved a breakthrough when he succeeded in creating a three-mile wired connection and a device that could be used to tap signals representing letters from the alphabet according to a coherent code. This invention, known as the Morse code, enabled the introduction of telegraphy as a commercial viable tool of communication. The first application of this technology was in the railway sector: telegraph lines along the tracks enabled better contacts between the operators of railway routes, who, as a result, could improve their schedules (Briggs, Burke, 2002: 134). Telegraphy also proved a welcome supplement to existing media technologies in other ways, particularly the press. Forward-looking journalists eagerly adopted telegraph news in their quest for scoops that would attract readers and boost sales numbers of their publications (Briggs, Burke, 2002: 192–93).

The growing global demand for information prompted inventors, mainly based in the United States and Great Britain, to look beyond the existing landlines and find ways to make telegraph connections that could cross seas as well. In the 1840s they developed methods to coat the fragile telegraph wires with rubber to create cables that could be lowered onto the seabed by crews on steamships. In 1851 the first underwater telegraph connection was constructed between Great Britain and France, crossing the North Sea Channel. In the years that followed, other lines between the British Isles and continental Europe were inaugurated. Visionaries also wanted to cross the Atlantic Ocean, but that proved to be a task of a much taller order. The first connection between Western Europe and Northern America, running from Ireland to Newfoundland, only functioned for a month when the cable broke beyond repair in September 1858. It took almost a decade before a durable transatlantic telegraph connection came into being in 1866.

The crossing of the Atlantic heralded a golden age of undersea telegraph connections: by 1874 there were 30,000 miles of submarine cables; in 1880 the mileage had increased to 100,000. The advent of the undersea telegraph cables caused much contemporary enthusiasm, celebrating the harmonious unification of mankind. In similar a vein, late-twentieth-century historians have described the intercontinental telegraph network as the 'Victorian internet', because it created, for the first time in history, a connected global communication system (Standage, 1998: 205–07). Others have criticized this terminology as an anachronism, because the wired telegraph system was difficult to access at the time. One important obstacle were the high prices that were charged per word by the operators of telegraph lines, which meant that, in practice, telegraph communication was available to a small social elite only (Potter, 2007). Moreover, telegraphy can been seen as a prominent 'tool of empire' used by Western powers to expand their influence over non-Western societies (Headrick, 1981: 157–64). From this perspective, wired telecommunication was not simply a unifying technology, it also acted as a catalyst for global inequality.

The most important organization in global telegraph communication at the end of the nineteenth century was the British news agency Reuters. The company, founded in London by Jewish immigrants from Germany in 1851, specialized in long distance communication. From the start it invested greatly in developing new telegraph routes that connected Britain with its expanding empire. As a result, Reuters acquired a government-sanctioned telegraph monopoly within the British sphere of influence, creating a single global telegraph network called the 'All-Red Route'. This colonial business model was implemented on a larger scale in 1870, when Reuters reached a deal with its main competitors on the European continent, Havas in France and Wolff in Germany. Echoing geopolitical practice in the age of imperialism, they divided the world into zones in which each of them could ensure their monopolies on wired information exchange, forming a hegemonic cartel in telegraphy communication (Silberstein-Loeb, 2014). In addition, in 1865 the International Telegraphy Union (ITU) was founded, an international administrative body in which official representatives of various countries took seat. At the time, post offices played an important role in telegraphy concessions, so they potentially had much power over news agencies. But in practice, at the end of the nineteenth century, they worked to ensure that the telegraph cartel could operate the way it did, rubberstamping regulations and high tariffs that benefitted it (Standage, 1998: 111).

This wired global telegraphy system functioned well for the Western powers during the last decades of the nineteenth century, as it facilitated their imperial projects in various parts of the world. The quick transfer of information gave colonial army officers a tactical advantage over their adversaries, and it provided businessmen with data to ensure they could gain substantial profits from colonial trade. Aditionally, telegraphy provided newspapers in the metropoles with the latest news about adventures in colonial peripheries that whipped up popular support for imperialism at home. However, at the beginning of the twentieth century, the telegraphy cartel started to show weaknesses. The main reason was the growing geopolitical tension between Great Britain and Germany. German opinion, and policy makers, were worried by the powerful position of Reuters, which had the largest sphere of influence by far. In nationalistic publications, often laced with antisemitism, Reuters was demonized as a powerful monster that flooded the world with misinformation (Tworek, 2015). The mistrust was mutual, as was illustrated when the First World War erupted: one of the British Navy's first acts of war was to cut German undersea cables worldwide. In the years that followed, the British firmly controlled wired global communication networks. That system, however, was already challenged at the time by the advent of a new technology: the wireless radio.

1920 to 1960: the Age of Wireless Radio

At the end of the nineteenth century, experiments with electronic telecommunication technologies reached a new level when the German scientist Heinrich Hertz invented a way to create electro-magnetic radio waves that could travel through air without a wire conducting them (1886). Despite the great promise of this invention, it proved to be difficult to cross a distance with this technology. In 1896, the Italian-born scientist and businessman Giuseppe Marconi made a breakthrough when he succeeded to transfer radio signals in Morse code from a transmission station on the shores of the Bristol Channel to a passing ship miles away. In 1899 he managed to connect stations in France and Britain across the North Sea Channel, and in 1903 he successfully sent a telegram across the Atlantic Ocean. At the time, his contemporaries in Britain thought that Marconi's technology would be mainly useful for military and maritime communication. The Germans, however, also saw wireless technology as a way to circumvent the powerful position of Reuters in wired communication and started to experiment with long-distance radio telegraphy in the 1900s. Spurred by this demand, the Telefunken company became a serious player in the radio market, and it provided the hardware for a transcontinental network of German wireless relay stations, as far as the Americas, operated by the news agency Transocean, which was inaugurated in 1913 (Tworek, 2010).

During the First World War, the Germans continued with wireless transmissions from the radio station at Nauen, which was one of the most powerful transmitters in the world at the time, but on a small scale. Engineers in other countries also took an interest in long-distance radio technology as a result of the wartime restrictions on British cables, for example in the Netherlands, which consequentially was virtually cut off from the Dutch East Indies, because the only existing cable to that colony was operated by Reuters. In 1916 a colonial engineer, who was on leave, proposed to construct a direct long-wave radio connection and received generous financial backing from the government in The Hague. Back in the Indies, he constructed a vast radio station in the foothills of Bandung in Java, which was matched by a 'radio cathedral' near the Dutch village of Kootwijk. In 1923 the first global radio telegraphy service was inaugurated between these two stations (Kuitenbrouwer, 2015: 30). In the 1930s, radio telegraphy received a new impulse with the introduction of the German *Hellschreiber* system, a machine that could translate radio signals directly into typographic texts. Downside to radio telegraphy was that the signals could be picked up even by those for whom the messages were not meant, which required for confidential information to be coded (Tworek, 2019).

In the interwar years, the greatest promise of radio-technology was the transfer of sound, which enabled the best-known application of the wireless: broadcasting. For this type of communication, which required modulation of waves, medium- and short-wave frequencies proved to be more suitable than long-wave frequencies. This also made it more difficult to cross long distances with broadcasting, but experiments showed that radio waves could travel around the world by reflecting them off the ionosphere. Again it was a Dutch experiment that caused a breakthrough in 1927, when a signal transmitted from the radio laboratory of the Philips company in Eindhoven was picked up in Java. After this feat, broadcasters in several countries started with international radio programming via the short wave. Like wired telecommunication had been previously, radio was a tool of empire at the time, and in these years the British were again on the forefront with the British Broadcasting Corporation (BBC) starting its famous Empire Service in 1932. The legendary director Lord Reith thought that the BBC had a 'civilizing mission', which meant that he set high standards for his organization in terms of journalism and neutrality. At the time, the main audience was conceived to be 'the lonely listener in the bush', i.e. white colonials who remained in touch with their home country via the wireless (Potter, 2012: 63–4, 93–4).

In the late 1920s, radio waves were used for various forms of telecommunication in addition to broadcasting and telegraphy, for example also for aviation and maritime communication. This situation meant that the ether became crowded, and that increased the risk of signal interference, making them unintelligible. The Geneva-based ITU (which changed its name in 1932, replacing Telegraphy with Telecommunication) found a new role for itself by addressing this problem, and from the late 1920s onwards it organised a series of international conferences to chart wave bands on a global level. Based on this schedule the organization (in cooperation with the Europe-based International Broadcasting Union) allocated frequencies to organizations for specific use. This initiative spurred optimism that emerged in the 1920s about radio as a medium to promote international harmony. In the 1930s, the League of Nations tried to promote such idealism with various initiatives, such as 'Broadcasting in the cause of Peace', which were transmitted and relayed around the world by various powerful stations in Western countries. The League even experimented with its own broadcasting station: Radio des Nations (Lommers, 2012).

In the late 1930s, however, the optimism about radio as a promotor of international peace was shattered. The rising fascist regimes in Germany and Italy used radio to spread propaganda beyond their borders. In Britain and France, there was particular worry about Arab-language broadcasts to Northern Africa and the Middle East, which the fascists used to stir up popular dissent against the colonial rule there. In reaction, the BBC started to broadcast in various languages other than English and inaugurated its World Service in the late 1930s. During the Second World War, these tensions in the ether escalated into a full 'war of words' between Allied and Axis broadcasters. Both sides invested greatly in broadcasting to mobilize support, inside and outside their spheres of influence, for their war effort. In addition to transparent forms of broadcasting, in which the stations clearly revealed their identity, and which was known as 'white propaganda', the warring parties engaged in clandestine uses of radio waves that violated the ITU conventions, often as part of operations of intelligence agencies. This so-called 'black propaganda' included various techniques, such as the spread of disinformation using pseudonyms or mimicking existing station by broadcasting on their frequencies (Briggs, 1995).

After the end of the Second World War there was a clear sense that the international harmony in the ether should be restored, and in 1947 the ITU became part of the United Nations. The emerging tensions between the two superpowers during the Cold War, however, prevented the development of the ether as a unified harmonious space. In Europe, which became the most apparent fault line between West and East, a propaganda war started in the late 1940s between the United States (via stations such as Voice of America and Radio Free Europe) and the Soviet Union (via Radio Moscow) (Badenoch, Fickers, 2013). In other parts of the world decolonization caused much dynamism in cross-border broadcasting. Many nations that came into existence after the end of the European colonial empires set up their own international radio stations for geopolitical purposes (Bronfman, 2016; Brennan, 2010). The bare fact that the global ether was so crowded by 1960 shows that wireless radio broadcasting was the most prominent medium for telecommunication at the time. In the decades that followed this position was challenged by other media emerging from new technologies, but radio remained important.

1960 to 1995: the Age of Satellite Television

Around 1960 the main challenge to radio as a mass medium came from television — a medium that, in addition to the ear, also caught the eye. Media making use of moving images were not new: film emerged around the same time as wireless radio, during the last decade of the nineteenth century. Mobility was a restricting factor in audio-visual communication, however, as it proved difficult to make it available to people in their homes. In the first four decades of the twentieth century, people had to go to cinemas to see moving pictures, whereas from the 1920s onwards they could receive radio broadcasts at home via receiving sets. In the 1930s, various technology companies successfully experimented with broadcasting moving images via ultra high frequencies, a technique that was named television, but the receiving sets were too complicated and too expensive for the consumer market at the time. That changed after the Second World War, when television sets were taken into mass production, and in Western countries in the 1950s, TV quickly grew to be a mass medium (Briggs, Burke, 2002: 234–35).

The quick rise of television did not mean it instantly killed the radio star, on the contrary. The 1960s also saw a great boom in radio listening, which was primarily due to the introduction of the battery-charged transistor radio receiving set. This device made radio portable and could also easily be fitted in automobiles, which meant that this medium, in terms of mobility, kept its edge over television, as at the time, TV sets were impossible to carry around and had to be plugged into electricity sockets to operate. In the West, the large-scale introduction of FM technology in the 1970s, which improved the reception and sound quality of mobile devices, gave another impulse to the popularity of radio broadcasting (Briggs, Burke, 2002: 230). In developing countries in Africa, transistor radio sets became popular too, as they were relatively cheap. These countries also often lacked an electricity grid to power television consumption on a large scale (Brennan, 2013). In many ways, radio remained a staple in the media diet of consumers around the world, and radio broadcasting remained a cornerstone of global telecommunication.

In the 1960s, television entered the realm of global telecommunication, too, mainly as a result of the advent of satellite technology. This development was a result of the Cold War competition between the United States and the Soviet Union, who tried to outdo each other in outer space in order to gain national prestige and military advantage. But the Space Race also had more peaceful consequences when it appeared that artificial moons were well-suited for long-distance communication, an idea that was first formulated by the science fiction writer Arthur Clarke in 1945. In 1962 the American Telephone and Telegraph company (AT&T) launched the first experimental communication satellite (Telstar 1) that, equipped with a simple relay transmitter, succeed to pass on the first transatlantic television broadcast (Briggs, Burke, 2002: 190). As before, the ITU noticed the great potential significance of this technological development early on, but also saw a possible bottleneck as it considered the geostationary orbit around the earth, the zone in which communication satellites are able to operate, a limited natural resource. As for the conceptualization and management of the global ether, the ITU organised a series of conferences from 1964 onwards, during which it set up rules for the allocation of orbital slots. Afterwards the organization remained responsible for monitoring these agreements and provided permits for new satellites.

The development of communication satellites took great flight in the 1970s and 1980s as a result of improvements in both aeronautic and communication technologies. Particularly digital forms of transmission were important, as these allowed for data transfer on a far bigger scale than electromagnetic waves. Spurred by these developments, hundreds of artificial moons for telecommunication were launched into orbit. These satellites took over parts of the existing functions of wireless stations, such as long-distance telephony, but above all, they facilitated a rise in long-distance television transmissions. From the late 1960s onwards, both the private cable television companies in the United States and the state broadcaster in the Soviet Union started to use satellite transmission to reach isolated parts of their networks (Evans, 2016; Parsons, 2008). In the early 1980s, the development of home satellitedishes enabled television broadcasts to be send via satellite directly to consumers, but this technology never became a successful business model because of the great size of the devices and their high cost (Briggs, Burke, 2002: 296). Eventually, satellites mainly facilitated the global dissemination of cable network television, which really caught on around fall of the Berlin Wall.

The breakthrough of international television can be seen as an important triumph of American soft power, leading to more homogenized visions of modernity as across the world large audiences watched television channels from the USA. The news channel Cable News Network (CNN), for example, became a successful global brand name with its live coverage of Operation Desert Storm, the American-led military operation to liberate Kuwait from Iraqi occupation in 1991 (Parsons, 2008: 480–83). Despite this prominent example, it would be wrong to think that satellite television simply facilitated the 'Americanization' of the world. In the 1990s, various countries increasingly used satellite communication to transmit their own news and cultural programmes, leading to a diverse supply of channels that could be watched via home satellite dishes. In Western countries, members of a number of migrant communities watched broadcasts from their countries of origin, contributing to their lingering feelings of loyalty and connection to their former homes (Mügge, 2010: 50). This tension between homogenization and heterogeneity would become even more magnified with the advent of the internet after 1995.

1995 the 2010: the Age of the Internet

The global breakthrough of telecommunication using information technology (IT) was the result of two developments in the late 1980s and the early 1990s. First, innovators such as Steve Jobs and Bill Gates succeeded in creating the personal computer (PC), a relatively small and cheap model with user-friendly software that made it possible for large groups of consumers to use computers at their workplace or at home. Secondly, innovators in different parts of the world, such as the CERN laboratories in Switzerland, started to make connections between computers at their institutions, which enabled information transfer between these devices. When they decided to link up all of these networks, the 'world wide web' came into being, also known as the 'internet' (a word that refers to the interconnection of different information networks) (Briggs, Burke, 2002: 309–10). The internet enabled data transfer in unprecedented quantities, which made it possible for IT companies to take over, combine, and expand the functions of earlier telecommunication technologies — something that scholars have referred to as 'convergence' (Briggs, Burke, 2002: 267–68). Although this did not mean that older media were simply abolished (paper media, radio, and television still exist today) the advent of the internet revolutionized telecommunication and gave rise to various hybrid media forms such as e-mail, digital newspapers, and podcasts.

As in previous eras, the ITU took it upon itself to regulate the internet where possible. The organization mainly focussed on the physical infrastructure that facilitated the global internet connection. Initially this infrastructure mainly consisted of communication satellites, which the ITU already monitored. In the twenty-first century, the ITU managed the construction of transoceanic glass-fibre cables, which became the major information highways around the world, largely using the same routes as the telegraph cables of the late nineteenth century. However, the internet created another, digital space that was more difficult to regulate, as it lay beyond the realm of the physical world: a virtual reality, which is also known as cyberspace. This concept initially captured the imagination of intellectuals who started thinking about what a digital future might look like. In the mid-1980s, pessimism dominated in the cyberpunk movement. William Gibson's iconic novel *Neuromancer* (1984) describes a dystopia in which a powerful artificial intelligence enslaves people in the real world (Gibson, 1984). By contrast, other thinkers and innovators hailed the internet as a liberating force for humankind. This utopian vision of virtual reality, captured in *A Declaration of Independence from Cyberspace* by J. P. Barlow, promised a world without boundaries and with complete individual freedom (Barlow, 1996).

In the 1990s, the latter, utopian narrative about the 'free internet' became mainstream in the Western world. Driving this process was the close relationship between American technology companies, based in Silicon Valley, and the federal US government in Washington, DC. The IT sector greatly benefitted from the policies promoting market deregulation that were initiated by successive US administrations. Additionally, there was a close but largely hidden connection between Silicon Valley and government agencies, such as the CIA, that sought to implement the latest technologies to protect national security. These two factors laid the groundwork for a boom of the IT sector in the first decade for the twenty-first century (Weiss, 2014). Technological innovations by manufacturers made computers so small that they were easy to carry around, such as laptops, tablets, and smart phones. Additionally, telecom providers created new wireless digital networks with wide coverage, so that people could access the World Wide Web almost everywhere, at any time. Also, the digital infrastructure of cyberspace evolved with the development of search engines (Google) and social media platforms (Facebook, Instagram). The latter services, which were free to subscribe to, seemed to fulfil the promise of the internet as a liberating force, because they enabled people to access and share information on an unprecedented level in human history. In Africa, IT technologies have arguably had the biggest impact during the last two decades: the majority of people on the continent had been cut off from global analogue networks, and the 'leapfrog technology' of the internet enabled many of them to connect to the world like never before (*The Economist*, 2017: 1).

These transformations also had dark sides, and since 2010 there have been increasing worries about the current state of the internet. The lack of regulation has given enormous powers to IT companies who gather vast amounts of information about those who make use of their services, feeding it into artificial intelligence systems that are capable of manipulating human behaviour. In the Western world, where an increasing number of consumer devices are connected to the world wide web (a process known as the Internet of Things), there are growing fears that such algorithm-based technologies infringe too much on the private lives of individuals. Further, information leaks have revealed that government bodies use the internet for large-scale intelligence gathering, combing through the huge data flows of IT companies in an apparent breach of privacy laws. Additionally, governments employ hackers to steal data and disrupt computer systems of adversaries. Some countries even seem to sanction the spread of disinformation via social media platforms, known as 'fake news', to destabilize other countries. The increasing worries about the potential disruptive effects of the internet in the real world have prompted calls for the regulation of cyberspace. Looking at the history of modern telecommunication, such calls make perfect sense. Throughout the nineteenth and twentieth centuries, technology has been an important factor in connecting people around the world, but in the end, people decide how they make use of these possibilities.

References

Badenoch, Alexander and Andreas Fickers, eds, *Airy Curtains in the European Ether: Broadcasting and the Cold War* (Baden-Baden: Nomos, 2013).

Barlow, John P., *A Declaration of Independence from Cyberspace* 1996, <https://www.eff.org/nl/cyberspace-independence> [accessed 27 December 2018].

Brennan, James, 'Communications and Media in African History', in *The Oxford Handbook of Modern African History*, ed. by John Parker and Richard Reid (Oxford: Oxford University Press, 2013), pp. 492–510.

———, 'Radio Cairo and the Decolonization of East Africa, 1953–1964', in *Making a World After Empire. The Bandung Moment and its Political Afterlives*, ed. by Christopher J. Lee (Athens, OH: Ohio University Press, 2010), pp. 173–95.

Briggs, Asa, *The War of Words: the History of Broadcasting in the United Kingdom*, vol. III (Oxford: Oxford University Press, 1995).

——— and Peter Burke, *A Social History of the Media. From Gutenberg to the Internet* (Cambridge: Polity Press, 2002).

Bronfman, Alejandra, *Isles of Noise. Sonic Media in the Caribbean* (Chapel Hill, NC: University of North Carolina Press, 2016).

Edgerton David, *The Shock of the Old: Technology and Global History since 1900* (Oxford: Oxford University Press, 2006).

Evans, Christine, *Between Truth and Time. A History of Soviet Central Television* (New Haven, CT: Yale University Press, 2016).

Fuller, Matthew, *Media Ecologies: Materialist Energies in Art and Technoculture* (Cambridge, MA: MIT Press, 2005).

Gibson, William, *Neuromancer* (New York, NY: Ace Books, 1984).

Headrick, Daniel R., *The Tools of Empire. Technology and European Imperialism in the Nineteenth Century* (Oxford: Oxford University Press, 1981).

Kuitenbrouwer, J. J. Vincent, 'The Dutch East Indies during the First World War and the Birth of Colonial Radio', *World History Bulletin*, 31.1 (2015), 28–31.

Lommers, Suzanne, *On Air: Interwar Projects for Radio Broadcasting* (Amsterdam: Amsterdam University Press, 2012).

Marx, Karl, 'Notebook V — The Chapter on Capital' 1857, <https://www.marxists.org/archive/marx/works/1857/grundrisse/ch10.htm#p. 525> [accessed 27 December 2018].

McLuhan, Marshall, *The Guthenberg Galaxy: the Making of the Typographical Man*, 2nd ed. (Toronto: University of Toronto Press, 1962).

———, *Understanding Media: the Extension of Man* (New York, NY and London: Mentor, 1964).

Mügge, Liza, *Beyond Dutch Borders: Transnational Politics Among Colonial Migrants, Guest Workers and the Second Generation* (Amsterdam: Amsterdam University Press, 2010).

Müller, Simone, *Wiring the World: the Social and Cultural Creation of Global Telegraph Networks* (New York: Columbia University Press, 2016).

Parsons, Patrick, *Blue Skies. A History of Cable Television* (Philadephia: Temple University Press, 2008).

Potter, Simon J., *Broadcasting Empire. The BBC and the British World, 1922–1970* (Oxford: Oxford University Press, 2012).

———, 'Webs, Networks, and Systems: Globalization and the Mass Media in the Nineteenth- and Twentieth-century British Empire', *Journal of British Studies*, 46.3 (2007), 621–46.

Silberstein-Loeb, Jonathan, *The International Distribution of News. The Associated Press, Press Association and Reuters, 1848–1947* (Cambridge: Cambridge University Press, 2014).

Standage, Tom, *The Victorian Internet. The Remarkable Story of the Telegraph and the Nineteenth Century's On-line Pioneers* (New York, NY: Walker and Co, 1998).

[The Economist], 'Technology in Africa: the Leapfrog Model', Special Report, *The Economist*, 11 November 2017, pp. 1–2.

Tworek, Heidi, 'Magic Connections: German News Agencies and Global News Networks, 1905–1945', *Enterprise & Society*, 15.4 (2015), 672–86.

———, *News from Germany. The Competition to Control World Communications, 1900–1945* (Cambridge, MA: Harvard University Press, 2019).

———, '"The Path to Freedom"? Transocean and German Wireless Telegraphy, 1914–1922', *Historical Social Research*, 35.1 (2010), 209–33.

Weiss, Linda, *America Inc.? Innovation and Enterprise in the National Security State* (Ithaca, NY: Cornell University Press, 2014).

GUILLAUME CARNINO

From Calculators to Data Centres

The History of Industrial Computing and Quantification

Historians sometimes trace the beginnings of computer science back to the Jacquard machine, or even to recipes on Egyptian papyri that are supposed to constitute the beginnings of algorithms (Essinger, 2004). In reality, as is almost always the case in the history of technology, what is significant is not the invention of a process, since the vast majority of patents are consigned to the dustbin of history (Baudry, 2015), but rather the passage to scale, e.g. the deployment in large territories and in dedicated professional universes. This chapter studies the history of computers and computing from this angle. What can be observed when the focus is no longer on the first appearance of a given device or machine, but rather on the entire socio-technical frame? In the same way that the *Annales* school was built against the history of great kings and famous battles (Burguière, 2006), it is important to radically detach oneself from the event-driven history of techniques in order to carry out long-term investigations which no longer focus on anecdotes relating to a particular discovery or invention, but attempt to pinpoint scalability issues. On the one hand, this is in line with a historiographical movement that insists on the importance of the old and the established to the detriment of innovation (Edgerton, 2007; Benoît, 2020). On the other hand it seems to align directly with recent works in industrial history (Fressoz, 2012; Jarrige, 2014; Hilaire-Pérez, Zakharova, 2016; van der Vleuten, Oldenziel, Davids, 2017), which describe scale as one of the important factors in industry. The argument presented in this chapter is that of a gradual transformation marked by breakthroughs that are not qualitative, but quantitative. More precisely, it is generally beyond a certain quantitative threshold that an evolution produces a qualitatively different universe. Where one classically opposes difference in nature and degree, it is rather necessary to see a continuity between the two: when a certain degree of transformation is reached, a divergence in nature occurs between two technical trajectories. This question also refers to the longstanding debate among historians between rupture and continuity: any phenomenon is generally described in terms of emergence or filiation. Industrial history allows us to understand that this debate (even though it is generally led by historians of technology) is hardly relevant, in the sense that a technique always combines novelty and continuity: beyond a certain threshold, any mass dissemination of an existing technology produces a new universe.

Global History of Techniques, ed. by Guillaume Carnino, Liliane Hilaire-Pérez and Jérôme Lamy, Global Matters Culture Matérielles, histoire des techniques et circulation globale des savoirs, 9 (Turnhout: Brepols, 2024), pp. 429–444
BREPOLS PUBLISHERS 10.1484/M.TECHNE-EB.5.129822

Emergence is very often only the consequence of a reorganization and deployment of the existing: new things are an aggregation of old ones. In other words, the history of computing (and of industry in general) is a history of deployment rather than invention.

The first part of this chapter will deal with the history of computer science in terms of mass computing, in order to grasp the emergence of this specific need for industrial quantification. The second part of the argument will focus on the deployment of the first computers and the development of hardware and software. The last part of the presentation will be devoted to the recent growth of networks (netware) and big data (dataware).

From Human to Mechanical Computing

The history of human computing is well documented today. From ancient Mesopotamia to the nineteenth-century Paris Observatory and its 'ladies of the sky chart', astronomical work always required a great deal of human computation (Grier, 2005). As industrialization unfolded, the intensity and amount of computation increased considerably, and some sectors hired women, especially black women (Shetterly, 2016), on a massive scale, for tasks that were often considered subordinate and poorly paid. NACA (the predecessor of NASA), as well as Bletchley Park (where the British centre for deciphering coded messages sent by the Axis countries was located) and the Manhattan Project, made intensive use of these calculators trained in mathematics (Evans, 2018).

The classic protagonists of the history of the computer are thus all witnesses of or direct actors in these large-scale calculation processes: Alan Turing, who anticipated the possibility of a universal calculator with his eponymous machine, worked in the British deciphering office (Smith, 2011); John von Neumann, who sketched out the conventional structure (which still bears his name) of today's computers, realized the importance of mass calculation as part of the Manhattan Project, when numerical solutions to partial differential equations had to be found (Ndiaye, 2001); Norbert Wiener, the founder of cybernetics and promoter of the feedback concept, worked in antiaircraft warfare (Galison, 1994); etc. While the abacus is an old and proven instrument (Cumin, Hossenlopp, 1994), tabulators were organizing machines for the advanced industrialization (Lazare, Mounier-Kuhn, 2016) and became very popular during the Second World War. Calculation gradually moved from craftsmanship, with its tried and tested methods, to industry, with its procession of regulations and optimizations. While mental arithmetic had long been taught as a professional skill, a large industrial market of computing emerged gradually and permanently reconfigured accounting practices (Gardey, 2008).

The industrial history of the IBM firm is exemplary here (Black, 2001). Between 1884 and 1887, Herman Hollerith, an American engineer at the US Census Bureau who managed the population census, filed patents for punched card statistical machines (or tabulators): the challenge was to shorten data processing times in this period of demographic growth. These tabulators read the punch cards produced in data entry and verification workshops and processed their contents according to a fixed program in order to produce a summary output on the card, which, in turn, could be reused downstream. The main function of these machines was therefore to count tabulated information, e.g. to quantify categorizations. After 1920, these tabulators even had an electrical connection board that allowed the processing program to be modulated. Based on its success, Hollerith founded the Tabulating Machine Company in 1896, which merged with other companies in 1911 to form the Computing-Tabulating-Recording Company, renamed the International Business Machines Corporation (IBM) in 1924. These mergers allowed for a diversification of the firm's activities, the core business of which remained organizational processing: under the supervision of its new president, Thomas Watson, the mechanographic workshops using IBM machines were deployed on all continents, particularly in banks and administrations, where the cards allowed for a more detailed and, above all, much larger management of customer accounts (Mounier-Kuhn, 2013). It was in Germany, after Hitler's seizure of power, that Dehomag (an acronym for the *Deutsche Hollerith Maschinen Gesellschaft*, the local branch of IBM) proved to be the most profitable of the subsidiaries: Dehomag then provided the technological solution allowing the massive task of identification and registration of Jewish people, communists, homosexuals, prostitutes, Sinti and Roma people ('gyspies'), and other groups. The Dehomag company's slogan, 'Oversight with Hollerith punch cards' ('*Übersicht mit Hollerith Lochkarten*'), praised the advertisements revealed by Edwin Black's meticulous archival work. This important market was supervised by Watson himself, who ensured that the customer was satisfied with these mechanographic procedures, for which each application required a long period of development by IBM engineers. Thus, each deportee was assigned a identification number consisting of three and six digits, the first three relating to the camp (001 for Auschwitz, 002 for Buchenwald, 003 for Dachau, etc.), and the next six identifying the inmate (tattooed on the arm).

Although Hitler decorated Watson with the highest distinction reserved for a non-German, the IBM logo soon disappeared from the tabulators in order to preserve the illusion of an entirely German company. As Black concludes, IBM's mechanography was at the origin not of the Shoah as such, but of the massive administration of the Holocaust as an industrialized death machine. IBM's tabulators allowed the partial automation of the processing of information in order to rationalize and thus increase the speed of the genocidal organization.

At the same time, the Allies also optimized their processing of military information, as demonstrated by the role of Patrick Blackett and his collaborators in the development of the founding algorithms for operational research and, ultimately, computer science, aimed at optimizing the management of convoys in the context of the Battle of the Atlantic (Budiansky, 2013).

Moreover, the mechanographic filiation of the early computer science is so strong that even though it became possible to work without punch cards — particularly after the launch of the hard disk by IBM in 1957 (the year the firm deployed the Fortran language) — and to use other peripherals for machine input and output, these small punch cards were so deeply rooted in organizational cultures that it was difficult to strip them of their prime position, as physical workshops concentrated these practices and cohorts of punchers had been trained for this purpose. From the end of the Second World War to the 1960s, the mainframes sold by IBM were thus powered by a language dedicated to the machine, and programmers were therefore always trained ad hoc. These gigantic computers were automatic mechanographic machines into which punch cards were entered and which output the result of the electronic data processing to the printer.

In addition to the close connections between military and technological developments, these early days of computing show that new processes and devices had emerged to break through the limits, particularly in terms of time, hitherto imposed by the critical masses that had to be processed. Producing data, e.g. formatting information to make it quantifiable and therefore processable in series, was at the heart of computing in the early days of hardware. Although the existing limits were often clearly identified, the rebound effect produced by the mass production of the technical solutions proposed was rarely anticipated. Howard Aiken, who designed the Mark I electromechanical computer at Harvard during the war, even wrote to Jay Forrester, who was responsible for the Whirlwind project (which gave birth to the first digital computer), that 'there will never be enough problems, enough work, to occupy more than one or two of these computers' (Edwards, 2004: 231). As is so often the case, the solution to a problem produced by scaling up creates a new world and its attendant complications.

The Hardware and Software Foundations of Modern Computation

Initially, IBM mainframes produced organizational results transmitted to operators, who then reacted accordingly. This was how the US Department of Defense's SAGE (Semi-Automatic Ground Environment) project worked to protect US territory against strikes by Russian bombers carrying nuclear warheads. The operator, in constant liaison with various detection systems that were designed to provide the status of air traffic, pointed at the enemy aircraft on his screen with an infrared 'gun' so that its coordinates could be transmitted to allied interceptors (Edwards, 1996).

The work carried out by these devices gradually raised the question of the interoperability of information processing systems and highlighted the need to increase the technological flexibility of computers. The fragility and weight of analogue computers was a real obstacle to their integration in certain devices (such as missile radio guidance), and this problem prompted the US Army's massive investments in the stabilization of transistor technology between 1947 and 1957, which finally reached the same degree of mastery as for vacuum tubes (Collett, 2003). This progressive technological replacement (from tubes to transistors) initiated the race to miniaturization and thus of the misnamed Moore's law which, far from being a physical fact, is a political and economic roadmap for the industrial world (Loeve, 2015). Three socio-technical phenomena then operated in parallel.

On the one hand, while the long-term and costly investment in training programmers appeared crucial, the pace of hardware innovation quickly rendered their skills obsolete, since they were only trained on a specific machine. Finding a way to sustain these educational investments therefore quickly became a goal in its own right for the administrators of these early IT systems. A group of IT experts then emerged which became institutionalized by being masculinized, and whose competence gradually merged with the holding of legitimate power (Ensmerger, 2010). At its limit, a convergence took place in terms of managerial skills, as in the British case, between the emergence of computer interoperability and that of the contemporary state (Agar, 2003).

On the other hand, the specific nature of each system made it difficult to coordinate technical skills within the market — a market in which IBM wanted to remain the sole player, but which was gradually opening up, as shown by the sustained competition from Bull in Europe (Mounier-Kuhn, 2013). Within these battles in skills, where the number of analyst-programmers could by far not satisfy the needs of the firms wishing to hire them, the need to standardize the use of automatic computing machines therefore gradually emerged. The development of programming languages, actual instances of the standardization of algorithmic writing, thus marked the beginning of the reign of software on organizational chains.

Finally, since the cost (in time as well as in salary) of transmitting information from the output of computers to subsequent operators was increasing, the idea of directly operating other machines via computational outputs appeared to be a logical evolution of existing devices. David Noble has shown that computerization then crossed the path of mechanization in production (Noble, 1984): since Taylor and Ford, automation had made it possible to increase the profits derived from the investment of industrial capital; at the same time, scientists and technicians trained during the Second World War were looking for a fruitful outlet for their newly developed technological skills. Thus, while mechanical automation seemed to be showing its limitations, electronic computation offered the hope of overcoming these technical locks.

The development of software marked the emergence of an instrumentation language that made it possible to describe — and therefore to execute — material processes: the deployment of numerically controlled machines provided the link between informational calculation and material action. Industrial production was then recomposed by the introduction of these machines, which were named CNC (for computer numerical control): from the 1960s to the early 1980s, perforated ribbons were used on which the machining instructions were encoded; between 1970 and 1985, this legacy of the punch cards was gradually replaced by magnetic strips. At the end of the century, thanks to advances in connectivity, the preference was to transmit instructions directly to the computer. Therefore, the various machines (boring machines, milling machines, drilling machines, etc.) were gradually centralized in numerically controlled machining centres.

Alain Jeantet and Henri Tiger described this succession of human replacements and the suffering caused by this dispossession of the technique hitherto incorporated by the workers in detail:

> During the first phase, [...] the automatic principle entered the workshop on tiptoe with machines whose settings remained manual while the machining operations were carried out automatically according to a program preset by the operator [...]. Their operating autonomy, e.g. their power over the machining process, remained intact [...]. The same cannot be said of the [...] second phase [...]. These machines [...] offered no programming possibilities of their own. This was therefore carried out outside the workshop by a technician from the Methods Department [...]. This led to a profound disqualification of the work of the operators assigned to them and the total divestment of their operating autonomy [...]. Things changed with the arrival on the market of second-generation numerically controlled machine tools, known as CNC (numerical controls with integrated computers) and the development of [...] computer-assisted programming [...]: the Methods Office, with its computer skills and the training provided by CNC suppliers, developed the programs, prescribed the tools, etc., in short, it concentrated all the qualified work within it (Jeantet, Tiger, 1988: 38–40, all translations [from the French] are the author's own).

In the same movement, computer-aided design (CAD) gradually replaced plans and drawing by hand and fed directly downstream to the numerically controlled machines: now the stated ambition was to process directly 'from file to factory'.

There was a desire to universalize operational commands through software instrumentation: the term 'command' then had the dual meaning of operational formalization and injunction. Computer symbols were the vehicle of a language that functioned as a standard, aiming to describe a sequence of operations, formerly carried out manually, that were now the prerogative of digital machines. Although mechanical automation was relatively reliable, it remained far too inflexible, since changing a procedure often meant physically modifying the workings of the machine: numerical control allowed great flexibility in the sequence of operations, which made it possible to envisage replacing the worker's know-how with its algorithmic integration into the control. The worker often initially remained the operator-programmer, but her knack and know-how were integrated and formalized in the computer code (CFDT, 1977).This deskilling initially mainly took place in the industrial sector, because the automation of services, although conceivable, required that part of the work be shifted to customers, as is shown by the debates about automatic supermarket checkouts (technically feasible at a very early stage, but commercially problematic, since the customer was initially frightened of doing unpaid work that was once carried out by paid staff at the shop's exit: Bernard, 2012).

The development of programming languages thus perpetuated this trend of putting information into calculation to make existing know-how commensurable in order to automate the execution of productive tasks. It was therefore no coincidence that IBM joined forces with Microsoft in 1981to develop the IBM PC (personal computer), based on an architecture known as 'fat client': the term 'fat' then meant that the terminals had the computing and storage power that would enable the development of office automation (as the encoding of elementary computing and writing operations) and, more generally, of all types of software that could be installed and used on a computer without having access to a server. The market was then optimal, and the virtual monopoly of Microsoft's Windows operating system embodied the versatility of small computer units that could perform various operations simply by modelling the software.

From Networks to Data Centres

SUN microsystems (an acronym for Stanford University Network) was founded in 1982 and developed the Network File System (NFS) protocol two years later. In the 1990s, the firm achieved a leading position in the IT market by massively mobilizing and promoting the TCP/IP standard implemented on a Unix layer. While organizational calculations were still performed on interoperable workstations, the major evolution consisted of connecting these machines to each other. From these interconnections emerged networks, and then networks of networks: what we know today as the internet. This global reorganization had important geopolitical implications regarding the nature and localization of root name servers, as was the case for the DNS (Benhamou, 2006). It was the era of the so-called 'three-tier' architectures: at the lowest level there is the data storage layer, then there is the layer where functional rules and algorithms are performed on remote servers, and finally there is the presentation layer — the layer that the users see on their screens, usually executed by a web browser. The network organization is a structuring instance, with data and algorithms managed on the server side (e.g. remotely) while the formatting is done on the client side — hence the 'thin client' and 'client/server' models (Pillou, Caillerez, 2016).

Organizational states were still optimized and physical machines still digitally operated, but this architecture authorized a real-time connection between them. Controlling optimization and instrumentation remotely became possible because infrastructures (fibre optics, satellites, etc.) allowed flows to transit at the speed of the electron, thus linking distant places and making temporalities commensurable. For example, one could perform calculations on the rise and fall of oil prices in New York, Tokyo, and London and compare them in real time, in order to instrument an automatic reaction to this calculation — a reaction which would very soon be implemented, as shown by the large-scale establishment of trading rooms in the world's major banks: while the back offices managed the calculations, the front offices were linked to the stock exchanges, and at the end of the chain, the accounting departments sent the results back to the finance departments. To put it another way, the deployment of digital networks made it possible to work on remote data that was provided in real time by the marketplaces, a market that became globalized at the same time. As the economist Marc Levinson stated: globalization is the meeting of the internet and the container — that is, the conjunction of a real-time communication medium and an industrial means of transport (Levinson, 2006).

As far as human work was concerned, networking largely reinvented practices, particularly in the implementation of workflows (Head, 2005; Head, 2014): these ERPs (enterprise resource planning, i.e. software deployed on a company's internal network to enable it to interweave its main functional components of production management, commercial management, logistics, human resources, accounting, and management control) made it possible to optimize the processing and instrumentation of geographically distant tasks. At Renault, for example, the workflow made it possible to match the execution times of the Guyancourt Technocentre with those of the subcontractors in Mumbai. The best-known integrated management software package is SAP (*Systeme, Anwendungen und Produkte in der Datenverarbeitung*), founded by five German IBM employees and developed as early as the 1980s. In 1992, the company abandoned the mainframe masterplan to migrate to a client/server architecture, enabling it to extend its ERP model to a large number of companies worldwide. SAP was listed on the Frankfurt stock exchange in 1988 and on the New York stock exchange in 1998. The arrival of workstations running on ethernet under Unix with a TCP/IP link revolutionized the world of information technology, because this type of infrastructure provided as many services as the mainframes of the major manufacturers, or more, and the manufacturers had to review their competitive offering in line with these new technologies.

What we now call 'big data' (or sometimes 'dataware') did not so much come from an increase in hardware performance as from a decrease in the cost of obtaining data, which fell drastically from the beginning of the twenty-first century onwards: big data is, above all, the possibility of obtaining considerable amounts of data for cheap. These digital traces are a product of the very nature of computing: since the temporal scale of electronic movements is in principle inaccessible to human beings, developers use logs (text files from operational archives) to be able to retrace *a posteriori* the main stages of processing carried out by machines, and thus to be able to detect errors (Carnino, 2007). Then a virtuous (or vicious) circle began: as more and more information and operations were digitized and computers were now interconnected, it cost almost nothing to collect these logs to obtain huge amounts of data. This had a knock-on effect on the globalized computer system itself: the more we digitized, the more the potential of mass computerization became apparent. And after a certain time, via a threshold effect, a new qualitative break occurred, bringing with it new possibilities. Therefore, in astronomy or biology the status of data changed during the 2000s: data was no longer expensive and rare, but free and potentially plethoric (Strasser, 2012, 2019).

However, this explosion in the amount of data implied a proportional multiplication of the physical storage infrastructure and the energy consumption that went with it (Carnino, Marquet, 2018). From the very beginning of computing, the size of computers had meant that industrial constraints in terms of rack space had to be managed, which are very similar to those experienced by today's data centres. When the IAS system was tested in 1951 in order to carry out with a calculation (of sixty days!) that was essential for the construction of the first thermonuclear bomb, the summer humidity in Princeton caused the air conditioning units to freeze — an incident that is in many respects comparable to the rainfall in the Facebook data centre in Prineville in August 2011, when the temperature difference coupled with the high circumstantial environmental humidity resulted in the formation of actual precipitation inside the buildings. Therefore, the early IT machines (which were considered critical infrastructures, mainly for military reasons) invariably raised the question of how their direct environment could be secured to avoid breakdowns, which almost always led to the construction of a specific building with its own security criteria: the American ENIAC (Electronic Numerical Integrator and Computer) of 1946 and the Soviet MINSK and EMBESM of 1959, the AN/FSQ–7 (which was used after 1958 for the US SAGE air defence project), the BNP IBM 705s of 1957, and the SABRE air ticket reservation system of 1960 were all built across entire floors, or even buildings (Redmond, Smith, 1980; Winkler, Webster, 1997; Balodis, Opmane, 2012; Mounier-Kuhn, 2013).

During the 1950s, researchers at the Stanford Research Institute created an Electronic Recording Method of Accounting computer processing system (ERMA) for banks, which was released to the public in 1955, tested on real bank accounts in 1956, and implemented for the Bank of America in 1959 (Balodis, Opmane, 2012). In the 1960s, most international banks acquired massive computers to support account management. In England, Barclays opened its first banking computer centre in 1961, where a massive IBM 1401 was installed. Over the years, these computing machines took up more and more space and were grouped together in huge Data Processing Centres (DPCs). However, staff strikes in the 1970s highlighted the fragility of these spaces, which were critical to bank operations: banks then began to spread out their information processing sites and circulate copies of computer tapes from one site to another. Some sites, such as the CTI (Centre de Traitement de l'Information) in Dinand (France), were subject to security measures that were the beginnings of today's data centres: 'powered by three generators in the event of an [electrical] power cut, with oil tanks and the necessary cooling by water drawn from the water table, it could operate in complete autonomy for several weeks' (Mounier-Kuhn, 2013: 107). Over the years, the spaces had become increasingly standardised: 'a move was made to separate operators and technicians' (Mounier-Kuhn, 2013: 106).

As these gigantic mainframes developed, the first local area networks (LANs) appeared, the first of which — ARCnet (Attached Resource Computer NETwork) — was established at Chase Manhattan Bank in 1977 to interconnect 255 machines. From the beginning of the 1990s onwards, the old mainframe rooms were colonized by these smaller servers and sometimes renamed data centres. The arrival of the World Wide Web in the 1990s led to the entanglement of these intranet networks with the nascent internet, which, in return, implied the multiplication of machine rooms, essential for hosting an ever greater amount of data and sites. As the digital economy developed, companies had to keep their IT infrastructure operational day and night, every day of the year. While some mainframe and telecom network spaces had retained robust power or air conditioning systems (Mastertel, Cineminers, 2019), most computer rooms in the 1990s were often tinkered.

However, these movements to rationalize infrastructure did not take into account the concentration of sites around cities. Very often, private data centres — whether they belonged to big retail companies, banks, or one of the big names on the internet — were located in the countryside, taking advantage of acres of inexpensive land. In the 1990s, with the development of the internet, new players in urban real estate emerged. The internet is a network of networks whose operation is based, among other things, on the spaces in which the networks interconnect. For reasons of economy, rather than developing in-house skills in computer science and data management, many companies used hosting companies to maintain servers and ensure the quality of the connection to the networks. These hosting companies tended to be located on the same premises as telecommunication operators, so that they could guarantee the highest possible data throughput for their customers. With the multiplication of servers in the 1990s, these spaces, which were often located in cities, became known as 'telecom hotels'. To avoid overdependence on one specific operator, third-party players developed a market called 'neutral collocation', which guaranteed customers equal access to all networks (Moriset, 2003). The issues of network access and data exchange speed contributed to the concentration of infrastructures in urban areas, with the various players often located close to each other. Moreover, since the largest operators needed a vast amount of computing power, they were gradually able to rent it. This was the development of what professionals today call IaaS (Infrastructure as a Service), which corresponds roughly to the emergence of the idea of the cloud in the general public — e.g. offshore computing 'in the cloud' (but in reality entirely physical and energy consuming). Contrary to some popular ideas, the cloud is only at the margin constituted by the multitude of personal computers (as is the case with the SETI@home research programme: Korpela et al., 2001). As a Parisian data centre manager states: 'the cloud is just someone else's data centre' (Carnino, Marquet, 2019: 37).

The data centre entangles two stories, that of information storage spaces that gradually became ultra-rationalized, and that of the networking of machines requiring multiple telecom accesses. In addition to security requirements, there were also expectations in terms of economic gains thanks to the concentration of infrastructures. The data centre market is therefore divided into several modes, from cloud computing services to hosting the information systems of corporate customers, making large-scale data interoperability possible.

Having access to vast amounts of information about real-time operations now makes it possible to work not, as previously, on samples, but on exhaustive sets of data available in the digital universe. The processing limitations imposed by the materiality of the data are disappearing, and thus there is a shift from sampling practices (the basis of quantitative science and surveys) to integral profiling (the basis of new emerging economic models) (Collomb, 2013). Distributors of Candy Crush Saga, a best-selling smartphone game adapted in multiple spin-offs, modify the playful performance and difficulty to the specificities of individual users in order to increase their interest (and addiction factor) for the application (Delbecq, 2014).

Big data pushes the de-semanticization specific to the digital world (which transforms the real into asemantic code: Bachimont, 2010) to its limit: mass processing of information concerning temporally correlated operations makes it possible to produce a commensurability of the signifiers themselves, and thus to generate what are akin to meaningful semantic connections without recourse to specific knowledge. Making words commensurable — that is, making language computable — opens the way to what Frédéric Kaplan calls linguistic capitalism (Kaplan, 2011): Google sells nothing but linguistic paths, which constitute a form of orientation within an infinite number of syntagmatically produced associative relationships, or, to put it more simply, links to signifieds generated from calculations on signifiers. The word exchange (the term 'ski' is sold to advertisers at a premium price in winter, and the expression 'sun cream' is auctioned off in summer) is at the basis of Google's economic model, which sells automated 'syntagmatic associations' that emulate semantic correlations. Mass aggregation of linguistic data has made an old capitalist dream possible: that of monetizing common language.

In 2010, there were 35 million IT technicians worldwide (Ensmerger, 2010: 10). Computation has now permeated every aspect of our lives, and our relationship to the world is marked by digital technologies. The four major phases of development in computing, hardware, software, netware, and dataware were mass transitions in scale, each allowing new technical and socio-economic relationships to emerge. As the historian of technology Thomas P. Hughes noticed concerning the electrification of Western society, the implementation of large technical systems required the involvement of actors of such power that only the military and banking institutions were generally able to provide it (Hughes, 1983, 1989). In return, the solidification of the military and economic interests that these computer systems brought about led to a profound intertwining between the initial objectives of their patrons and the structure of contemporary relations: while the army (at least since Napoleonic wars) dreamed of being able to know and control real-time information from the battlefield (which encompassed the entire rear production complex as industrialization progressed), digital technologies have brought about the ideal of real-time management of all aspects of life, whether it is waiting for a bus or a pizza delivery man, the duration of a traffic jam or the time remaining in the viewing of a video (Delfour, 2011); where the contemporary economy was built in opposition to the old practices of exchange in kind and interpersonal giving and receiving, a task which required a difficult education of populations and merchants in the rigours of monetary exchange and bank accounts (Strasser, 1989), digital technologies have finally made the emergence of *homo œconomicus* possible. This once-fictional being, supposed to minimize costs, maximize profits, and optimize capital in the manner of a company, is now partially real in a world where everyone has to work on their image in order to improve ratings on social networks. Thinking about the history of techniques and technologies over their long period of deployment, and not in terms of inventions, invites us to go beyond the individual illusion of novelty to integrate these recompositions into a fully-fledged anthropological narrative.

References

Agar, Jon, *The Government Machine: a Revolutionary History of the Computer* (Cambridge, MA: The MIT Press, 2003).

Bachimont, Bruno, *Le Sens de la technique. Le numérique et le calcul* (Paris: Les Belles Lettres, 2010).

Balodis, Rihards and Inara Opmane, 'History of Data Centre Development', in *Reflections on the History of Computing: Preserving Memories and Sharing Stories*, ed. by Arthur Tatnall (Berlin: Springer, 2012), pp. 180–203.

Baudry, Jérôme, 'The Order of Technological Knowledge. Crafting a New Language for Technology in France, 1750–1850' (unpublished doctoral dissertation, Harvard University, 2015).

Benhamou, Bernard, 'Organiser l'architecture de l'internet', *Esprit*, 5 (2006), 154–68.

Benoît, Serge, *D'eau et de feu: forges et énergie hydraulique, xviiie-xxe siècle. Une histoire singulière de l'industrialisation française* (Rennes: Presses universitaires de Rennes, 2020).

Bernard, Sophie, *Travail et automatisation des services. La fin des caissières?* (Toulouse: Octarès, 2012).

Black, Edwin, *IBM and the Holocaust: the Strategic Alliance Between Nazi Germany and America's Most Powerful Corporation* (New York, NY: Crown Publishers, 2001).

Budiansky, Stephen, *Blackett's War: the Men Who Defeated the Nazi U-Boats and Brought Science to the Art of Warfare* (New York, NY: Knopf, 2013).

Burguière, André, *L'École des Annales. Une histoire intellectuelle* (Paris: Odile Jacob, 2006).

Carnino, Guillaume, 'Rêve numérique ou cauchemar informatique', in *La Tyrannie technologique. Critique de la société numérique*, ed. by Cédric Biagini and Guillaume Carnino (Paris: L'Échappée, 2007), pp. 105–47.

——— and Clément Marquet, 'Les Datacenters enfoncent le cloud: enjeux politiques et impacts environnementaux d'internet', *Zilsel*, 3.1 (2018), 19–62.

——— and Marquet Clément, 'Du mythe de l'automatisation au savoir-faire des petites mains: une histoire des datacenters par la panne', *Artefact. Techniques, histoire et sciences humaines*, 11 (2019) 161–88.

CFDT, *Les Dégâts du progrès. Les travailleurs face au changement technique* (Paris: Seuil, 1977).

Collett, John Peter, 'The History of Electronics: From Vacuum Tubes to Transistors', in *Companion to Science in the Twentieth Century*, ed. by Dominique Pestre and John Krige (London: Routledge, 2003), pp. 253–74.

Collomb, Cléo, 'Penser la trace numérique: quels paradigmes, quelles conséquences?' (2013) <http://www.academia.edu/5454165/_Penser_la_trace_numerique_quels_paradigmes_quelles_consequences_seminaire_sur_le_paradigme_de_la_trace_Homo_Ludens_UQAM> [accessed 30 November 2022].

Cumin, Jean and Jean Hossenlopp, *Le Boulier* (Paris: Chiron, 1994).

Delbecq, Denis, 'Succès de Candy Crush Saga. Pourquoi ce jeu engendre-t-il une telle addiction?', *Science & vie*, 1161 (2014), 36–39.

Delfour, Jean-Jacques, *Télé, bagnole et autres prothèses du sujet moderne* (Toulouse: Érès, 2011).

Deluermoz, Quentin and François Jarrige, 'Écrire l'histoire avec les animaux', *Revue d'histoire du XIXᵉ siècle*, 54 (2017), 15–29.

Edgerton, David, *The Shock of the Old: Technology and Global History since 1900* (New York, NY: Oxford University Press, 2007).

Edwards, Paul N., *The Closed World. Computers and the Politics of Discourse in Cold War America* (Cambridge, MA: The MIT Press, 1996).

———, 'Construire le monde clos: l'ordinateur, la bombe et le discours politique de la guerre froide', in *Les Sciences pour la guerre, 1940–1960*, ed. by Dominique Pestre and Amy Dahan (Paris: EHESS, 2004), pp. 223–49.

Ensmenger, Nathan, *The Computer Boys Take Over: Computers, Programmers, and the Politics of Technical Expertise* (Cambridge, MA and London: The MIT Press, 2010).

Essinger, James, *Jacquard's Web: How a Hand-loom Led to the Birth of the Information Age* (Oxford and New York, NY: Oxford University Press, 2004).

Evans, Claire L., *Broad Band. The Untold Story of the Women who Made the Internet* (New York, NY: Penguin, 2018).

Fressoz, Jean-Baptiste, *L'Apocalypse joyeuse. Une histoire du risque technologique* (Paris: Seuil, 2012).

Galison, Peter, 'The Ontology of the Enemy: Norbert Wiener and the Cybernetic Vision', *Critical Inquiry*, 21.1 (1994), 228–66.

Gardey, Delphine, *Écrire, calculer, classer. Comment une révolution de papier a transformé les sociétés contemporaines (1800–1940)* (Paris: La Découverte, 2008).

Grier, David Alan, *When Computers Were Human* (Princeton, NJ: Princeton University Press, 2005).

Head, Simon, *Mindless. Why Smarter Machines are Making Dumber Humans* (New York, NY: Ingram, 2014).

———, *The New Ruthless Economy. Work and Power in the Digital Age* (New York, NY: Oxford University Press, 2005).

———, *Mindless. Why Smarter Machines are Making Dumber Humans* (New York: Ingram, 2014).

Hilaire-Pérez, Liliane and Larissa Zakharova, eds, *Les Techniques et la globalisation au XXᵉ siècle* (Rennes: PUR, 2016).

Hughes, Thomas P., *American Genesis. A Century of Invention and Technological Enthusiasm, 1870–1970* (New York, NY: Viking, 1989).

———, *Networks of Power. Electrification in Western Society, 1880–1930* (Baltimore, MD: Johns Hopkins University Press, 1983).

———, *American Genesis. A Century of Invention and Technological Enthusiasm, 1870–1970* (New York: Viking, 1989).

Jarrige, François, *Technocritiques. Du refus des machines à la contestation des technosciences* (Paris: La Découverte, 2014).

Jeantet, Alain and Henri Tiger, *Des manivelles au clavier: familles ouvrières en mouvement* (Paris: Syros, 1988).

Kaplan, Frédéric, 'Quand les mots valent de l'or. Vers le capitalisme linguistique', in *Le Monde diplomatique*, 692 (2011), 28.

Korpela, Eric, Dan Werthimer, David Anderson, Jeff Cobb, and Matt Lebofsky, 'SETI@home — Massively Distributed Computing for SETI', *Computing in Science & Engineering*, 3(1), (2001), 78–83.

Lazare, Emmanuel and Pierre Mounier-Kuhn, *Histoire illustrée de l'informatique* (Paris: EDP Sciences, 2016).

Levinson, Marc, *The Box. How the Shipping Container Made the World Smaller and the World Economy Bigger* (Princeton, NJ: Princeton University Press, 2006).

Loeve, Sacha, 'La Loi de Moore: enquête critique sur l'économie d'une promesse', in *Sciences et technologies émergentes: pourquoi tant de promesses?*, ed. by Marc Audétat, Gaïa Barazzetti, Gabriel Dorthe, Claude Joseph, Alain Kaufmann, and Dominique Vinck (Paris: Hermann, 2015), pp. 91–113.

Mastertel & Cineminers, *The Internet Architecture. Data Center* <https://www.youtube.com/watch?v=hg5zWgokNkI> [accessed 30 November 2022].

Moriset, Bruno, 'Les Forteresses de l'économie numérique. Des immeubles intelligents aux hôtels de télécommunications', *Géocarrefour*, 78.4 (2003), 375–88.

Mounier-Kuhn, Pierre, *Mémoires vives. 50 ans d'informatique chez BNP Paribas* (Paris: BNP Parisbas, 2013).

Ndiaye, Pap, *Du Nylon et des bombes. Du Pont de Nemours, le marché et l'état américain, 1900–1970* (Paris: Belin, 2001).

Noble, David, *Forces of Production. A Social History of Industrial Automation* (New York, NY: Knopf, 1984).

Pillou, Jean-François and Pascal Caillerez, *Tout sur les systèmes d'information* (Paris: Dunod, 2016).

Redmond, Kent C. and Thomas M. Smith, *Project Whirlwind* (Bedford, MA: Digital Press, 1980).

Shetterly, Margot Lee, *Hidden Figures: the American Dream and the Untold Story of the Black Women Mathematicians who Helped Win the Space Race* (New York, NY: HarperCollins, 2016).

Smith, Michael, *The Secrets of Station X: How the Bletchley Park Codebreakers Helped Win the War* (London: Biteback Publishing, 2011).

Strasser, Bruno J., *Collecting Experiments. Making Big Sata Biology* (Chicago, IL: University of Chicago Press, 2019).

———, 'Collecting Nature: Practices, Styles, and Narratives', *Osiris*, 27 (2012), 303–40.

———, *Collecting experiments. Making big data biology* (Chicago: UCP, 2019).

Strasser, Susan, *Satisfaction Guaranteed: the Making of the American Mass Market* (New York, NY: Pantheon, 1989).

van der Vleuten, Eric, Ruth Oldenziel, and Mila Davids, *Engineering the Future, Understanding the Past: a Social History of Technology* (Amsterdam: Amsterdam University Press, 2017).

Winkler, David F. and Julie L. Webster, *Searching the Skies: the Legacy of the United States Cold War Defense Radar Program* (Washington, DC: Headquarters Air Combat Command, 1997).

SABINE EFFOSSE, LAURE QUENNOUËLLE-CORRE

From Banking Computerization to FinTech

The Digital Revolution in Banking (1950s–1990s)

Like many sectors, the modern banking system has been significantly impacted by technological innovations, leading the transformation of customer services and the emergence of new professional roles and business segments. Without going all the way back to the origins of the banking system, the Second Industrial Revolution can undoubtedly be considered an important step in the creation of contemporary banks and capital markets. Indeed, the concentration of the means of production that accompanied the industrial revolution and the need for sometime heavy investment, such as for the construction of railways, required significant capital. In France, from the Second French Empire of Napoleon III onwards, when merchant bankers and small investment banks were not able to meet these demands, commercial banks and large investment banks were created (among them the Comptoir d'escompte de Paris, the Société générale, Crédit Lyonnais, Banque de Paris et des Pays-Bas, and Crédit mobilier). These new institutions were often listed, joint-stock companies which were able to raise vast amounts of capital to lend it to firms and individuals. In addition to traditional self-financing, they provided different forms of financing for the industry, including short- and long-term credit, the discounting of commercial drafts, the issuing of bank drafts, the launching of bonds and shares on the stock market, and the subscription of securities by investors. At the time French savings were relatively abundant and available (the savings rate was between 8% and 13% from 1890 to 1914) and could easily be mobilized thanks to the many banking branches that opened throughout France at the end of the century (1592 branches were opened for the three major banks, i.e. Crédit Lyonnais, Société générale and Comptoir national d'escompte de Paris in 1912). These connections between savers and companies via banks were made possible by the gradual evolution of the financing system of the French economy which developed throughout the nineteenth century. In 1900, 40% of the money supply consisted of gold and silver coins, 27% of bank notes, and 33% of bank deposits (Braudel, Labrousse, 1993). With the increasingly important role of banks and their growth — as well as that of debtors — the credit relationship became less personal and informal, and more formal and varied. The entire economic and financial context encouraged the development of banking services for customers, from current accounts to securities services and safe

Global History of Techniques, ed. by Guillaume Carnino, Liliane Hilaire-Pérez and Jérôme Lamy, Global Matters Culture Matérielles, histoire des techniques et circulation globale des savoirs, 9 (Turnhout: Brepols, 2024), pp. 445–454
BREPOLS PUBLISHERS 10.1484/M.TECHNE-EB.5.129823

deposit box rental. Brokers were also active in this golden age of the stock exchange, which witnessed the emergence of a form of 'financial democracy'.

Naturally, new technological developments in the banking and finance sectors accompanied this banking revolution. The telegraph and international telephone lines gave rise to the practice of arbitrage on the 1890s' financial markets for instance, while the development of unit record equipment in banks from the end of the nineteenth century onwards was no less influential. Clearly various new technologies have helped this sector to develop its activities from private companies individuals while reducing its running costs. Even more important than the telephone and the telegraph, the ticker tape, which emerged in the United States in 1867, was the ancestor of digital electronics. It made it possible to transmit information on share prices via telegraph lines, so that price and volume information could be monitored remotely, in real time, and accessible to all (Robb, 2017; Preda, 2003).

This modern banking and financial system changed little until the second half of the twentieth century, when financial globalization was accompanied by a diversification of financial products and services, particularly due to new calculation and accounting techniques. After a period of stagnation, which was followed by a period of strong state control between 1914 and 1950, the return of growth caused banks to bounce back, especially with the deregulation of the financial sector in the 1980s and 1990s.

The quantitative and qualitative leap of information and communication technologies (ICTs) in the second half of the twentieth century constituted a veritable industrial revolution that led to the establishment of the new information economy. Since the beginning of the twenty-first century, the term 'FinTech' — a portmanteau of 'finance' and 'technology' — has been used extensively in the banking and finance sector. While its definition is still unclear, the term is used to describe either the recent and rapidly expanding sector of start-ups that offer digital technologies for financial services, or traditional banking and financial companies that apply technological innovations to their businesses.

Our focus here is on the origins of FinTech, that is, innovations related to IT and the use of digital technologies in the banking and financial sector from the 1950s to the 1990s. Indeed, in these decades the IT revolution manifested itself in a spectacular way in the banking and finance sectors.

As for retail banks, they faced a major challenge from the 1950s onwards as they needed to manage increasing numbers of customers and the diversification of products on offer while keeping costs under control. To satisfy the financing demand of companies engaged in industrial modernization and the development of individual customers, the first computers needed to be introduced (the word *ordinateur* first appeared in French in 1955); since these had the capacity to process more and more information automatically and logically, it was now possible to cope with the growth in the sector while generating productivity gains. The early supercomputers, which limited the use of manual bank tranzation postings, were replaced by more powerful computers in the 1970s. The arrival of microcomputers and their connection to the telephone network ushered in a new banking era, epitomized by ATMs in particular

(Bátiz-Lazo, 2018). In 1983, the launch of Minitel in France paved the way for remote banking, an experience offered more widely by the internet from the late 1990s (Carpenter, 2011). In the financial sector, securities services and portfolio management were automatized in the early 1970s, while electronic trading combined with securitization grew exponentially in France and other European countries in the 1980s, taking inspiration from innovations that came from across the Atlantic.

This chapter therefore aims to analyze the transformative influence of the IT revolution on the products and services provided by banks, with a particular emphasis on France within a global context. What has been the impact of IT on the economic model (in terms of productivity, efficiency, security)? What impact has this revolution had not only on the banking profession and customer relations, but also on interbank relationships and the emergence of the contemporary banking system? Two questions will be explored. The first will focus on the consequences of the use of IT for processing transactions with a growing number of customers and for communicating with them. The second will deal with interbank transactions and the evolution of capital market interventions.

The Introduction of Computer Engineering in Banks: a Production Tool at the Service of large-scale Operations

Calculating, especially calculating accurately, is one of the banker's highest priorities. In the inter-war years, the first electric accounting machines were introduced to banks — first the American machines by IBM (founded in 1924), then the European machines by Bull (founded in 1931) — in order to reduce the manual work of bank transaction postings. Mechanical processing using punched cards made the collection, sorting, and classification of customer account data (like cash and securities, the calculation of premiums, and the execution of stock market orders) both easier and more reliable. As for industry, the mechanization of tasks was a major challenge for banks as service companies that employed an increasing number of staff. While certainly useful, unit record equipment reached its limits with the increase in banking customers from the late 1950s onwards, and even more so at the turn of the 1960s and the 1970s. The rise in living standards, the use of credit, and changes in regulations (e.g. the mandatory direct, monthly payment of wages, the increasing use of cheques or bank transfers) led to a vast increase in account openings and transactions. The increased management capabilities provided by unit record equipment soon was no longer sufficient. Profitability was in decline. Consequently, most major banks chose to introduce computers as early as 1957. Born out of military research from during and after the Second World War relating to the demands of ballistics and cryptography, these computers had a significantly higher processing power, and more memory and storage capacities, thanks to the use of electronics. Computer centres equipped with the IBM 1401 in particular appeared in American and European banks in the early 1960s — the Banque nationale de Paris, for example, used these computers from 1957 — which removed the backlog in processing an increasingly large mass of data while accelerating processing speeds (Mounier-Kuhn, 2013). Automation made it possible to increase productivity, which was an essential objective for banks distributing new financial products such as consumer credit, which typically granted relatively small amounts of credit (to finance vacuum cleaners, refrigerators, etc.), or personal savings accounts.

Computer manufacturers — IBM and Bull to name the main ones — played an essential role in the progressive computerization of banks and even in the diversification of their services. Initiated by IBM to reduce the cost of the early computers, the practice of leasing enabled banks to become equipped with information techniques in the 1960s, and this also became a commercial activity in its own right that was offered to companies to modernize their equipment.

The computerization of tasks also led to a standardization of banking practices which facilitated the exchange of information. A typical example is the processing of cheques when a huge number was in circulation. The United States, where ten billion cheques were issued in 1959, was a pioneer in this field and paved the way for automation by opting for the optical reading of cheques with Burroughs machines. Europe, however, followed the French proposal and chose magnetic reading. Before automation, cheques represented 55% of the operations in French banks, and according to an estimate for Crédit Lyonnais, their processing required between 750,000 and 1 million working hours per year. Following tests conducted by Bull at Crédit Lyonnais, the standard based on a particular code (the CMC7, a magnetic ink character recognition code with fine vertical stripes, which is the forerunner of the current bar code) was adopted by several European countries in 1963. This work on standards and uniformity prior to the computer processing of interbank transactions — France extended the CMC7 standard to all bank and postal cheques in 1965 — continued with the development of the 'universal payment certificate' (*titre universel de paiement*) a few years later, which was followed by the Basic Bank Account Number.

These advances in automation, including the reliability and speed of information processing, were made possible by the arrival of increasingly powerful computers on the market. In the late 1960s, the supercomputers of the 1950s were replaced by smaller and faster third-generation computers, in which peripheral devices increasingly supplemented the CPUs (central processing units). Their gradual connection to the telephone network transformed what now became a real IT system in banks. Concentrated in or even confined to dedicated centres (i.e. technical and computer centres) at first, IT was gradually decentralized due to several factors: first, there were technical reasons, such as the development of terminals that could work with computers remotely and the evolution of physical data storage media thanks to the switch to magnetic tapes; second, there were security concerns, for example with regard to the risk of fire; and third, there were also social reasons, such as when the French strikes of 1974 and 1977 paralysed several processing centres. At the end of the 1970s, the arrival of microcomputers in bank branches marked a new stage in the use of IT in banking, which was now more focussed on providing personalized services to customers.

From Telematics to Digital technologies: Banking IT at the Customer's Service

Following the model of large-scale food distribution (such as Carrefour or Walmart) and in a socio-economic context favourable to the 'consumer', banks took advantage of improved data transmission to introduce self-service banking.

Connected to local microcomputers, the first automatic teller machines (ATMs) appeared in France in 1972. They were accessible to customers who had a bank card with a magnetic stripe — a standard feature adopted by institutions after the launch of a new French payment method, the Carte bleue, in 1967. Their use was limited to cash withdrawals with amounts being capped on each card. A few years later, in 1974, the Frenchman Roland Moreno, who was the manager of the firm Innovatron and passionate about electronics, introduced a new technique about to know a bright future: the chip card. Safer than the magnetic stripe, which invited fraud on a grand scale, this innovation, which was later used in mobile phones in the form of SIM cards, was subject to many patents and aroused great interest from banks. Following negotiations leading to the adoption of this technical solution, they implemented a single chip card payment system in the mid-1980s. This agreement marked a turning point on two levels. On the one hand, the distribution of automated teller machines (ATMs), which were now connected directly to the banks' computer network and accessible twenty-four hours a day, seven days a week, provided a wider range of services as part of the commercial strategy of self-service banking that was implemented by banking institutions at the time. On the other hand, based on the American model, it contributed to the use of bank cards as a growing means of payment in France. In an effort to limit the use of cheques, which were more expensive to process, banks made it easier for retailers to install point-of-sale terminals. This strategy proved successful: within ten years, from 1979 to 1989, card payments in France increased from 3% to 22% of total transactions (Mounier-Kuhn, 2013). In 1983, the launch of Minitel, based on the development of Transpac, a packet switching network, heralded a new phase in the bank-and-customer relationship in France: the era of remote or home banking. A predecessor of online banking, which emerged at the end of the 1990s with the World Wide Web, remote banking developed in parallel with the computer equipment used in businesses and households. This equipment developed relatively fast. More than half of French small and medium companies for which access to cash flow in real time is crucial, were equipped with computers in 1985. In 1987, as the range of distance services like bill payment expanded, 25% of the exchanges between the Banque nationale de Paris and its customers were carried out by telematics (a portmanteau word between 'telecommunications' and 'informatics'). With this change, new professions emerged, for example in customer relations and staff training needs grew accordingly. Companies in IT services have been instrumental in the computerization of banks since the 1970s and have supported this development both through the installation and maintenance of computers and software and in terms of helping staff with getting used to the interface which was not always very user-friendly initially. By the early 1990s, the 'IT revolution' in banks was complete. The period of introduction IT had also been a time when IT was not always well understood because it was implemented by a small group of engineers using 'coded' language who were not directely part of the bank. By the early 1990s, this induction phase was over. Gradually most of the IT departments that were in charge of computerization passed into the hands of banking professionals, which showed both how increasingly accessible the field was and how common its use had become. IT, which was essential to the transformation of retail banking activities thanks to the speed of information exchange it facilitated, was also decisive in the development of financial activities.

From Telex to High Frequency Trading: Information and Communication Networks in the Financial Sector

There are four main forms of IT use in stock market transactions, which have developed at different time periods over time as a result of technological innovations: Information (telegraph, telephone, ticker, in the nineteenth century: arbitration between financial centres although letter exchanges dominated); the routing of stock market orders (fixing the price between supply and demand at the auction, with physical presence on the floor. This ceased at the end of the twentieth century); the clearing between operators, settlemend and delivery (accounting entries, which in the twentieth century became digital), and for commercial banks securities management (which has been automated sinced the beginning of the twentieth century). Banks, which have always had a major role in the purchase and sale of securities, have taken part in all these technological developments. But to what extent has the organization of their work flows changed over time as a result. After the Second World War, French banks used unit record equipment more intensely, in particular punched cards. But in the securities department — which used punched cards extensively — the transition from unit record equipment to computerization occurred much later than in other departments, as this was not a priority compared to payroll or the management of customers' accounts, of loans, and of relations with banking partners. Similarly, computers were gradually introduced for financial analysis. This began in 1965 in the United States, followed by Great Britain, and later by France. It was only in the early 1970s that securities accounting and portfolio management benefitted from significant IT resources to develop the quality of customer service and help the decision-making process of asset managers while reducing running costs.

As a rule, before the 1960s and 1970s, the switching network was mainly used between professionals in the field, e.g. partner banks, stockbrokers, brokers, or institutional investors. The latest innovation in this context was the creation of SWIFT (the Society for Worldwide Interbank Financial Telecommunication) in Brussels in 1973. Linked to the boom in foreign exchange and Eurodollar transactions since the 1960s and the resurgence of a financial globalization, this banking network was among the oldest that connected computers. Since then, this type of professional network has developed in many sectors. They are managed by computer manufacturers, administrative services, private companies, or universities. A cooperative company intended to 'provide its members with a private communication system for the transmission of international payments, statements and other international banking messages' SWIFT became operational in 1977, connecting more than 500 banks in Europe and North America, including some fifty French companies. At that time, the system only covered fifteen countries. SWIFT was financed by the participating banks, who could then improve their productivity and foreign currency cash management. This interbank cooperation has allowed the transfer of international funds via secure message switching.

In this period, within the context of deregulation and the development of financial markets, innovations mainly came from the United States. In 1971, with the creation of the NASDAQ (National Association of Securities Dealers Automated Quotations), Wall Street benefitted from an electronic market well before any other sector did. Unlike conventional stock exchanges, NASDAQ did not have a trading floor, but trading in stock options was carried out through decentralized transactions between investors, who bought and sold through computer screens and telephones (Alt, Beck, Smith, 2018). The use of telematics networks for order routing began at the New York Stock Exchange in 1976, while the first complete quotation automation system was installed at the Toronto Stock Exchange in 1977 via CATS (Computer Assisted Trading System). Subsequently, with the development of Bloomberg terminals in 1980, investment bank traders connected to one another within a computer network, long before other professionals did.

The Boom of the 1980s and the 1990s: Reorganization and Change of Scale

Over the next two decades, the financial markets of developed countries saw their ecosystems profoundly transformed by financial and technological innovations. Deregulation and financial globalization led to the development of derivatives markets and securitization. As a consequence, NICTs (new information and communication technologies) contributed to the banks' networking and the acceleration of the flow of information on a global scale. The share of international financial transactions in banking activities and income increased, as banks benefitted greatly from banking disintermediation. The bank intermediation rate (i.e. the credit financing of the economy) fell from 80% in 1979 to 48% in 1987. Conversely, financing through capital markets increased and became more international (Quennouëlle-Corre, 2015).

In France, the beginning of financialization in the 1980s led to the growth of market activities in the banking sector. The transition from auction to electronic quotation required the introduction of continuous trading which needed to be supported by computerization. But since not all stockbrokers opted for the same IT system and the Compagnie des agents de change baulked at the cost of this investment, Paris, as a financial centre, lagged behind in this department. It was not until 1986 that the Canadian CATS system was introduced to the Paris Stock Exchange when Continuous Assisted Quotation (CAC) was introduced (Muniesa, 2003). Its implementation, recommended by the Pérouse report of 1980, was laborious, but stockbrokers vying with each other on the floor of the Paris 'Palais Brongniart' were replaced by a fully automated trading system, in which an algorithm is used to order, match, and then execute electronically transmitted buy and sell orders.

At the national level, the dematerialization of securities had progressed in 1949 with the creation of Sicovam (*Société interprofessionnelle de compensation en valeurs mobilières*), which organised the clearing of securities between French and foreign brokers. As an interprofessional securities clearing company, Sicovam is the heir to the French *Caisse centrale de virement des titres,* which was supposed to prevent Germany from taking over the securities held in France during the occupation. In 1990, the company was entrusted with the centralization of the automated securities settlement system (RELIT which stands for *RÈglement et Livraison de Titres*). From 1997, the clearing house 'Clearing 21' was fully electronic. With the disappearance of auctions came the rise of the banks' trading rooms. The boom in financial innovations (especially the futures markets) allowed banks to intervene directly in the trade and the negotiation of these products. Banks thus became major players in MATIF (Marché à Terme International de France, or Futures Market for International Financial Instruments in France) that was created in 1985 in France.

A new organization of stock exchange services was then set up, in which banking institutions generally had a dual structure for derivatives markets: the front office or 'trading room' where traders took positions on the markets, and the back office, which registered, confirmed, and monitored the front office's operations. As a result, professional banking networks were set up, and the computer systems of all banks were aligned with international standards. In 1984, for example, Paribas set up a centralized data management system for swap contracts made worldwide. In terms of employment in banking services, new profiles emerged such as traders appeared for which mathematicians and engineers were employed (including graduates from the French *École polytechnique*), while careers in banking IT became increasingly complex (Mounier-Kuhn, 2013).

In the 1990s, NICTs took a leap forward. The spectacular increase in computer calculating power of computers and their reduced costs, as well as the use of mathematical formulas to price derivative products, have offered traders around the world new opportunities which have led to the development of market finance. The combination of mathematical formulas, of data imput, of computers able to do the necessary calculations and of electronic networks that connect computers internationally has led to the development of 'quantitative finance' and the exponential development of 'high frequency trading'. This potentially explosive combination reduces transaction costs and accelerates transaction speed, but also increases price volatility and market instability. The scale of the October 1987 crash, which was partly caused by investors' automated programmes, revealed the difficulty of using computer-based arbitrage programmes for futures and options markets. Official reports (like the Brady report or the BIS report) reveal that automated portfolio hedging mechanisms played a significant role in the speed, and probably the amplification, of price collapse. Whether they are seen as 'teething troubles' of the new technological and financial developments, or as the foreboding signs of issues to come, the 1987 episode is, in any case, indicative of the changes that occurred in what would be known as the FinTech sector.

The IT revolution which took place over the 1960–1980 period in banks made it possible to 'industrialize' the material tasks of a rapidly expanding market — as almost all households in Western countries opened a bank account and used banking services. It also prompted the development of commercial and financial activities thanks to an increasing speed in the execution and transmission of information. Both technological revolutions in the sector — for financial services and for banking services took place at different times. On the one hand in the 1980s, the banks' financial services acted as pioneers, turning the traditional activities of banks into highly profitable operations. At the same time, telematics paved the way for the digital revolution of the second half of the 1990s (Schafer, Thierry, 2016).

The arrival of the smartphone in 2007, when Apple presented the first generation iPhone in the United States, and the rise of social media and the harvesting of data they enable, marked another decisive turning point in the emergence of financial technologies. Emerging from the 2007/2008 financial crisis, FinTechs, in the strictest sense of the term, have several distinctive characteristics (Lee, Shin, 2018). Driven by the widespread use of smartphones, their business model is based on the exploration of new niche markets with high profitability potential such as means of payment (40% of FinTech worldwide) or crowdfunding. Targeting a young, urban, affluent, and connected clientele (i.e. millennials) or small and medium sized companies whose financing needs are not entirely met by traditional banks, they offer personalized services focussed on customer experience. Finally, FinTechs benefit from a regulatory environment that encourages competition and thus the arrival of new players. Notably, the adoption of SEPA (Single Euro Payments Area) by the European Commission in April 2007 broke the banking monopoly on payments; it was enforced in France in November 2009 with the Payment Services Directive, PSD1, which was supplemented with PSD2 in November 2015.

While traditional banks first perceived the emergence of FinTech as a threat because means of payment represented nearly 25% of their income, the challenge that these new players represent has enabled traditional banks to speed up their digital transformation and to diversify their activities by taking stakes in the capital or even buying up fast-growing start-up companies. Some French examples are BNP Paribas, who in 2017 bought *Compte Nickel* (a branch-free company that allows just anyone to open an account from cornerstore shops such as tobacconists), and *Crédit Mutuel Arkea*, who bought stakes in Leetchi (an online crowdfunding start-up created in 2009). Thus, traditional banks and start-ups coexist in a complementary way in terms of their activities and resources more than they compete against each other. However, the arrival of digital companies in financial services, such as Alipay (China) or Apple Pay (United States), poses a much more delicate challenge: in addition to their financial power, they are able to cheaply harvest data without having to set up a specific infrastructure to do so. With algorithms, they also have the means to exploit this enormously profitable resource now essential to banking and financial activities.

References

Alt, Rainer, Roman Beck, and Martin T. Smith, 'FinTech and the Transformation of the Financial Industry', *Electronic Markets*, 28 (2018), 235–43.

Bátiz-Lazo, Bernardo, 'Between Novelty and Fashion — Risk Management and the Adoption of Computers in Retail Banking', in Schönhärl Korinna (ed.), *Decision Taking, Confidence and Risk Management in Banks from Early Modernity to the 20th Century* (Switzerland: Palgrave Macmillan, Springer, 2017), 189–08.

———, *Cash & Dash: How ATMs and Computers Changed Banking* (Oxford: Oxford University Press, 2018).

———, Juan Carles Maixé-Altés, and Paul Thomes, 'In Digital We Trust: the Computerization of Retail Finance in Western Europe and North America', in Bernardo Bátiz-Lazo, Juan Carles Maixé-Altés, and Paul Thomes (eds), *Technological Innovation in Retail Finance. International Historical Perspectives* (London: Routledge, 2011), 1–14.

Bonin, Hubert, 'The Development of Accounting Machines in French Banks from the 1920s to the 1960s', *Accounting Business and Financial History*, 14.3 (2004), 257–76.

Carpenter, Marie, *La Bataille des télécoms. Vers une France numérique* (Paris: Economica, 2011).

Effosse, Sabine, 'El cheque en Francia: el lento ascenso de un medio de pago de masas (1918–1975)', *Revista de la Historia de la Economia y de la Empresa*, 11 (2017), 77–94.

Lee, In and Yong Jae Shin 'Fintech: Ecosystem, Business Models, Investment Decisions and Challenges', *Business Horizons*, 62 (2018), 35–46.

Llorca, Matthieu, 'Les Banques aux prises avec les FinTechs', *L'Économie politique*, 75.3 (2017), 43–58.

Mounier-Kuhn, Pierre, *Mémoires vives. 50 ans d'informatique chez BNP Paribas* (Paris: BNP, 2013).

Muniesa, Fabian, *Des marchés comme algorithmes: sociologie de la cotation électronique à la Bourse de Paris* (Paris: École Nationale Supérieure des Mines de Paris, 2003).

Preda, Alex, 'Les Hommes de la Bourse et leurs instruments merveilleux', *Réseaux*, 122 (2003), 137–65.

Quennouëlle-Corre, Laure, *La Place financière de Paris au XXe siècle. Des ambitions contraries* (Paris: CHEFF/IGPDE, 2015).

Robb, George, *Ladies of the Ticker, Women and Wall Street from the Gilded Age to the Great Depression* (Champaign, IL: Illinois University Press, 2017).

Schafer, Valérie and Benjamin G. Thierry, 'The "Web of Pros" in the 1990s: the Professional Acclimation of the World Wide Web in France', *New Media & Society*, 18.7 (2016), 1143–58.

DOMINIQUE BARJOT

Civil Engineering and Construction

Two Centuries of Technical Progress

From the beginnings of the British Industrial Revolution to today, construction and, more specifically, civil engineering, have seen considerable and multifaceted technological progress. Today's industrialized powers have only emerged thanks to this progress. The object here is not, and cannot be, to discuss the whole history of civil engineering, but rather to understand its mechanisms. Therefore, I will focus on the construction of major structures, road works, and the structural work of the building and cement industry. I will adopt an externalist approach, emphasizing both the role of social factors and actors (engineers, entrepreneurs, companies, and public authorities) and the economic, social, and cultural impact of technology.

In this context, three preliminary remarks are necessary. First, the innovator is not always the inventor or his company, and what applies to individuals or firms is also true for nations. Second, in construction, two models have historically challenged the relationship between the owner, the engineer (or architect) and the developer (or contractor): the Anglo-American model is organised around the engineer, who plans and designs and closely supervises the construction, while the contractor only performs the tasks (hence the global prevalence of American engineering); by contrast, in the continental model, which is very much used, for example, in Germany, France, and Spain, contractors have a more active role and, above all, companies with autonomous design capacities, on which their ability to generate profits is based. Finally, this study focuses on three important periods (the nineteenth century, the first half of the twentieth century, and the third quarter of the twentieth century), before concluding with the great changes that are taking place today for the benefit of emerging nations.

The Origin of Technical Progress: Engineers and Contractors

In nineteenth-century Western Europe and North America, the Industrial Revolution led to the rise of two key figures, i.e. the engineer and the contractor. As has been shown for the English Industrial Revolution, the boundaries between these two professions were not always as clear as they are today: in France for instance, state engineers who graduated from *grandes écoles* very much differed from civil engineers

Global History of Techniques, ed. by Guillaume Carnino, Liliane Hilaire-Pérez and Jérôme Lamy, Global Matters Culture Matérielles, histoire des techniques et circulation globale des savoirs, 9 (Turnhout: Brepols, 2024), pp. 455–480
BREPOLS PUBLISHERS 10.1484/M.TECHNE-EB.5.129824

who had often trained on the job, despite the development of the *Écoles des arts et métiers* and then of the *École centrale.* Engineers were key, their ability to invent and innovate was particularly important in civil engineering and the design and construction of bridges, ports, and tunnels. They knew how to take advantage of new technical opportunities, like the development of new materials, the introduction of electricity, or the adaptation of roads to the growing importance of cars.

Engineers

Historians of science and technology as well as researchers in the social sciences have discussed the different types of engineers that exist in different countries for a long time: engineers in England and Scotland are trained on the job but open to advanced science and reluctant to set up on their own as consulting engineers; engineers in Germany, may have trained at a Gewerbeinstitut (one of institutes established throughout Germany and in some Central European countries on the model of the Gewerbeinstitut Berlin, to train specialists for the manufacturing industries), but are still versed in abstract science such as done in universities, and commonly have a PhD; while French engineers graduated from a *grande école* and trained for employment in the state service or directly for private industry. Historians have also explored how the United States adopted or adapted most of these European models.

In nineteenth-century France, the engineers of the École nationale des Ponts et chaussées were very highly regarded. Created by Colbert, the school was under the authority of the Controller-General of Finances until the Revolution, and under Napoleon they were reorganised into a non-military strictly regulated body that was attached to the Ministry of the Interior. The Napoleonic period marked the starting point of the rising importance of engineers as they were tasked with planning and designing the infrastructures required by the Continental Blockade and the building of a great European empire. After taking an active part in the 1820 Becquey Plan, which had been influenced by Saint-Simonianism, they gained influence under the July Monarchy, hence the creation of a Ministry of Public Works. Under the Second Empire, this ministry gained full autonomy, reinforced in 1855 by the creation of a division of public works at the Council of State (Conseil d'état). This autonomy was further strengthened by the Republic, especially for the purposes of the Freycinet Plan. Considered the 'elite of the elite', they benefitted from unlimited leave during which they gained positions in the railway companies in France, Algeria, and abroad. The election of one of these engineers, Sadi Carnot, to the presidency of the French Republic confirmed their political importance.

The prestige of the Ponts et chaussées engineers declined from the 1880s to 1900 due to the failure of the Freycinet Plan, the financial crash of the Union générale, and, above all, the Panama scandals, followed in 1895 by the collapse of the Bouzey Dam. This resulted in the loss of expertise and, for a certain period, the threat of the dissolution of the school, but also in renewal, with the creation of an entry pathway into the school for works managers. This innovation was prompted, among other things, by their technical skills: Paul Le Gavrian, the author of pioneering work on the use of tar for paving roads, became the secretary general of an international permanent road association. *Ponts et Chaussées* construction overseers played a decisive role in three areas, i.e. the development of waterways, the construction of the Parisian underground railway (the Paris Métro), and the construction of bridges.

From the 1860s, engineers started being increasingly hired in private companies. After a plateau in the last twently years of the XIXth century, this trend accelerated until 1914. The engineers included an increasing number of *École polytechnique* as well as Ponts et chaussées graduates. These engineers were competing with engineers who had graduated from the Ecole Centrale de Paris although the latter tended to have careers in the industrial sector rather than in public works. By contrast, the engineers trained at the Écoles des arts et métiers had stronger positions than the *centraliens*. Often from a middle-class or upper-working-class background, the '*gadz'arts*' were very close to the elite workers on which the public works companies' success depended.

Between 1883 and 1889, there was a strong competitive climate between France and England. This rivalry was based on two large projects in the two countries at time: the Firth of Forth bridge in Great Britain and the Eiffel Tower in France. At the time, commentators noted that the French workers were more efficient than the British ones because they were better trained for difficult jobs, and also because they were, above all, assemblers, that is, very skilled workers who had perfected the art of assembling factory-made parts on site; the British by contrast, continued to use carpenters' techniques and finished the pieces which were supplied to them unfinished. Like Jean Compagnon (1837–1900) who was a metal carpenter, these assemblers constituted a group of elite workers. But they were not an exception. Elite workers were also numerous among the mechanics and masons, who were often of Limousin or Piedmontese origin, or the independent workers employed at railway yards.

Major construction work (Bridges, Ports, and Tunnels) as sites of Innovation

There were also elite workers in the building industry particular in France and Britain, but many fewer than in Germany or, even more so, the United States.

Technical progress remained modest in the building industry. As Abel Châtelain, Gabriel Désert, and Michel Lescure have shown, the industry was dominated by craft culture and on-the-job training. Building contractors were often connected to the world of farming. The passage from the farming to building was easy, except in England, which was already very urbanized. This is what happened in the Limousin region until 1914 and in the Piedmont region until the 1930s. The building trade also drew its strength in France from the culture of "compagnonage", the guild-like structure acting as a sector-specific right and regulation organization and an apprenticeship institution ensuring training by the elders. When "companions" lost ground under the Second Empire, the unions sought, above all, to defend the construction workers' immediate and material interests, neglecting the question of apprenticeship.

A training crisis ensued in the second half of the nineteenth century, which prevented the generation of a skilled workforce that the profession needed.

Productivity gains slowed down. Admittedly, from 1815 to 1882, there were many technical innovations, including Roger's mortar machine of 1843, Nepveu's crane of 1851, and, following the British model, the use of travelling cranes on rails for transporting materials. The most important development was probably the general use of the tarpaulin, which, from 1860, made it possible to continue work throughout winter. The impact of these transformations, however, was limited. In contrast to Germany, France was slow to adopt the system of the general contractor who was in charge of coordinating the on-site subcontracting of firms belonging to the various trades. This was mainly due to the resistance of the jobbers to this idea; these were craftsmen who were usually specialized in a trade but did not own a yard or workshop, and who derived their living from the difference between the lump sum received from the contractor and the wages paid out to workers. This practice also hindered the introduction of Taylorism that was advocated by people such as Auguste Vuillemin, one of the founders of the International Federation of Building and Public Works.

In civil engineering, Britain had taken the lead in the eighteenth century thanks to a combination of factors. The country had witnessed an early financial revolution (with shareholding companies financing canals, toll roads, or turnpike roads). It also had a highly skilled workforce and venturesome entrepreneurs. This combination of factors led to the construction of the railways. Financed by powerful public companies, innovative entrepreneurs such as Thomas Brassey or Morton Peto built a prime railway network, thanks to the labour of manual workers hired locally from among agricultural workers, but also thanks to the navvies who were specilialised workers who moved from job to job.

In France as in Germany, much more so than in Great Britain, the weakness of transport infrastructures required the use of new materials and processes allowing the construction of increasingly important and complex projects. Around 1840, the internal navigation network in France suffered from its obvious outdatedness. The Ponts et chaussées engineers introduced the first steam dredgers. To improve the navigation on the Seine, they called upon experienced mechanics such as Alphonse Castor, the inventor of the bucket dredgers worked by a single elinde (or mast), thanks to which the cost of work was reduced by two thirds. Castor's success sparked competition from Alphonse Couvreux. The dredging techniques evolved further in the construction of the Suez Canal: the Ponts et chaussées engineer Paul Borel and the polytechnician Alexandre Lavalley carried out most of the dredging works of the maritime canal with the most powerful machine that existed at the time. Designed in France, the budket dredgers soon superseded the spoon dredgers developed by the British and Americans. Under the leadership of Hildevert Hersent, Jules Goüin, and Louis Coiseau, bucket dredges were further developed until the end of the century. However, from the 1890s onwards, they faced increasing competition from suction dredgers using centrifugal pumping. Invented by Frenchman Ernest Bazin (1826–1898) in 1869, the suction dredgers were rapidly adopted especially in the United States, Belgium, and, above all, the Netherlands. In these countries, as well as in Britain, some manufacturers specialized in their production, which ensured the worldwide success of the contracting company Ackermans & Van Haaren of Antwerp.

In 1860, Alphonse Couvreux had created the bucket chain excavator, a type of dredge mounted on rails. Successfully used in the Suez canal and for dredging the Danube at Vienna in 1875, it then came to be more generally used in major earthworks. Perfected by Joseph Allard for the digging of the Tancarville Canal in 1884, the excavator was of great value for the works done in Antwerp and Terneuzen for the work done on the Meuse, for the digging of basins in Bremen and and the German North Sea Canal, for the works on the Manchester Ship Canal, and even for work in Japan. Although outclassed in Panama, its American competitor, the spoon digger, gave birth to the steam shovel. French engineers also played an important role in the development of 'titans', which are powerful traction machines for the construction of deep-sea ports. Their inventors, the Dussaud brothers, used them in Marseille, Port Said, Trieste, Smyrna, and Genoa. They faced competition from Alphonse Castor, of Hildevert Hersent and Joseph Lesueur, the builders of the port of Philippeville, in Algeria, and Alexandre Lavalley. Between 1883 and 1898, these materials were further improved by *Bartissol and Duparchy* at Leixoes in Portugal, by Felix Allard, Louis Coiseau, Jean Cousin, and Abel Couvreux at Hyest, near Bruges.

New construction methods appeared. For the use of compressed air, Jacques Triger invented a system of tubular foundations in 1841, which was adopted by the British and then reintroduced into France by *Castor and Jacquelot* in 1854; and a system of compressed-air caisson foundations was invented by Henri Joyant, Emile Vuignier, and Édouard Fleur-Saint-Denis, three engineers at the Eastern Railways (*Chemins de fer de l'Est*). The latter was successfully used for the first time by Hildevert Hersent for the construction of the Kehl Bridge in 1859–1860. Because it allowed underwater work of a large number of workers thanks to Gustave Eiffel, this system was quickly imposed in France, followed by Germany and the United States, where Washington Roebling used it for the foundations of the Brooklyn Bridge. The technique then spread very quickly thanks to *Fives-Lille* and Louis Coiseau.

Harbour engineering benefitted from this progress, especially for the construction of large refit basins (for instance in the docks of Missiessy at Toulon built by by Hildevert Hersent from 1877 to 1882, and the large basins of Toulon built by *Fougerolle, Daydé*, and *Grands travaux de Marseille*, from 1911 to 1921). Compressed air led to a significant overhaul of the construction techniques for wharf walls, notably because of the competition between Hildevert Hersent and Conrad Zschokke from Switzerland. Although the former introduced the technique in Antwerp in 1878, which was improved in Lisbon from 1888 to 1892, Zschokke offered a new process, substituting vaults for straight lintels, in Bordeaux from 1892 to 1895. The success of compressed air continued further with the progressive replacement of iron by reinforced concrete.

In addition to canals, bridges and ports, many tunnels were built by construction companies among which Marc Seguin, Germain Sommeiller (the inventor of the compressed-air locomotive and of the first known mechanical perforator as well as the main builder of the Mont-Cenis tunnel), and Louis Favre, who dug the Saint

Gotthard tunnel. The digging process became increasingly complex, with the most noteworthy work done between 1826 and 1849 on the tunnel under the Thames by the British engineer Marc Isambard Brunel, the inventor of the tunnelling shield (*bouclier*). It was used by *Couvreux and Hersent* in the early 1880s for the new Antwerp refit basins. The technology was used for the construction of the sewers and the construction of the Paris metro. At that time, many French engineers were very highly regarded internationally, among them Émile Dollot, the builder of a large part of the underground sewers of Mexico City; the *Fougerolle brothers*; and, above all, Léon Chagnaud. Chagnaud had some impressive achievements: first, with the Fougerolle brothers, he built the sewer collector of Clichy outside Paris; and second, on his own, the underground railway between the Austerlitz and Orsay stations, the section of the Paris Métro running between the Turbigo and Europe stations, the tunnel going under the river Seine between Châtelet and Saint-Michel, the Lötschberg tunnel in Switzerland, the Rove tunnel connecting Marseille to the Rhône and the Étang de Berre, for a log time the largest underground structure ever built.

New Technical Opportunities

Throughout Europe, and especially in France, a number of new technical opportunities emerged with the economic recovery of the early twentieth century. The great depression of 1880 to 1890 accelerated the search for cheaper materials. On the continent, much more so than in English-speaking countries, a large-scale transition from iron to concrete took place.

At the beginning of the 1880s, iron was the main material in civil engineering, because it made it possible to produce large-scale structures at fairly low prices. Used in large quantities, together with glass for the buildings housing the universal exhibitions, iron became essential in terms of rail infrastructure. In Great Britain, Germany, the United States, and also in France, a powerful industry of metal structures was established. In France, the largest companies (*Schneider et Cie, Cail, Fives-Lille, Ernest Goüin et Cie,* and later the *Société de construction des Batignolles*) came into competition with people from the locksmithing world, like Henri Moisant, Gustave Eiffel, and then Ferdinand Arnodin (the designer of ferries and the first suspension bridges with vertical lines and oblique stays), and Henri Daydé.

Despite their inventiveness, the international competition became tougher. The British and Americans had, much to their advantage, used steel in the construction of bridges, such as the St. Louis bridge built by James Buchanan Eads between 1867 and 1874 or, later, the Firth of Forth bridge, which were both made from steel while the Eiffel Tower was still being made from puddled iron. At the beginning of the twentieth century, the French companies' delayed adoption of steel became all the more serious as the Germans had also made great progress in the area. In France, a number of obstacles stood in the way of steel being more widely used: there were issues with the construction of suspension bridges, which were too weak to withstand the wind, and there was also the high cost of coal mining and therefore of steel, builders' resistance, and the rise of reinforced concrete.

Reinforced concrete, a new material that was both more resistant to traction than ordinary concrete and cheaper than iron or steel, was the result of a series of experiments. After Joseph-Louis Lambot's first experiments in 1848 and those of Joseph Monier a year later, a succession of experiments quickly discovered multiple uses for the new material: such as in François Coignet's reinforced concrete building in 1852, Lambot's iron and steel association in 1855 and Monier's reinforced concrete tank in 1872. After being granted a series of patents between 1867 and 1873, Monier exported much abroad: he struck a partnership with Schüster of Vienna, Austria-Hungary in 1879, while the German contractor Wayss acquired all the Monier patents in 1886. Over the next thirty years, the material was further developed as a result of the fierce competition between François Hennebique and Edmond Coignet, two ingenious construction businessmen active internationally. Hennebique won, and his technology was adopted all over the world: in Germany, the United States, Russia, as well as in Great Britain and Sweden.

After the material met success at the Universal Exhibition of 1900, the Union of builders in reinforced concrete (Chambre syndicale des constructeurs en ciment armé) was founded in in France in 1903. Among its members were Simon Boussiron, the founder the company of the same name in 1899 and the engineer behind the building of different large-span structures. Most importantly, he introduced the use of reinforced concrete for thinner slabs that were both strong and water-reistant. The new material was promoted by the Ponts et chaussées engineers, notably by Armand Considère, Charles Rabut, and especially the latter's students Albert Caquot and Eugène Freyssinet.

It is in this new context that the production of artificial or portland cement from clay and limestone heated at high temperature developed. Invented in 1840 in Great Britain, the process was not adopted in France until 1898, later than in Germany or Belgium. Thanks to large investments, the French cement companies perfected the production process, adopting a wet method to produce clinker, and then, from this, a homogeneous powder, artificial Portland cement. This was done by *Ciments français*, who had introduced portland cement to France in 1846, *Ciments Vicat* (which had been founded in 1856 to exploit the vast quarries around Grenoble), *Ciments d'Origny*, and especially the Pavin de Lafarge company, already the fourth largest company of its kind in Europe, which produced portland cement later than its competitors because of the importance of its existing lime kiln infrastructures. Nevertheless, under the direction of brothers Joseph and Auguste Pavin de Lafarge, the company undertook a major campaign of research, development, and diversification. On the eve of the First World War, the cement industry was able not only to provide traditional products (such as binders or blocks, which were cheaper than cut stone), but also to help with the construction of large structures.

At the beginning of the twentieth century, electricity was available in North America as well as in Europe, and a growth sector for the economy. Although Great Britain and France had fallen behind, from 1907, France also built large power plants to lower the costs and thereby develop the three major uses of electricity — lighting, power and electric traction (ie railways, tramways etc.). Further, the country showed an early interest in hydroelectric power or 'white coal': taking advantage of a long tradition, France was the European leader in hydropower, only surpassed by the United States. A certain number of companies played a decisive role in setting up infrastructures, including Thévenot, contractor of the *Compagnie du Midi*, Edmond Bartissol, the Fougerolle brothers, Alexandre Giros, and Louis Loucheur, as well as the *Société des grands travaux de Marseille* headed by Charles Rebuffel.

Roads presented another major sector of technical innovation. In 1914, France had the second largest number of cars in Europe after the United Kingdom, thanks to the competitiveness of its car industry, which had long dominated the sector worldwide, and thanks to the enthusiasm of national inventors. It benefitted from high domestic demand, stimulated by the short lifespan of vehicles, and the significant drop in prices, as well as the development of commercial networks and active promotion of automobiles in the press (see for instance the magazine *Auto*), at shows, through competitions, and through the activities of the *Automobile club de France*. In 1913, France was also the world's largest exporter of automobiles.

On the eve of the First World War however, France had failed to capitalize on its first graderoad network. Established the seventeenth century, it had benefitted from major restoration and maintenance during the Restoration period, and the July Monarchy. Despite the development of railroads, the effort was maintained under the Second Empire and the Third Republic. After the failure of the Freycinet Plan however, from 1883, French roads fell victim to financial cuts. In 1913, France was now the developed country where, for the main roads, the annual credit per kilometer was the weakest. In a context marked by the success of motor racing and the tragic episode of the Paris-Madrid race in 1903, the public authorities decided to tar the roads and the *Automobile club de France* created the race track of Auvergne in 1905, which was specially designed for competitions. French engineers, just as their British and Germans counterparts, mainly used their national techniques for tarrring the roads.

At the time, France had the edge as far as machinery was concerned thanks to the steamroller, which had been invented by Jean-Nicolas Fortin in 1830 and and later been fitted with a steam engine by Gaëtan Brun, and thanks to the hot tar spreading machine designed by Jules Lassailly in 1910. The adoption of the latter firmly established the use of tar, which had been an option since Dr Ernest Guglielminetti's research. Thanks to Émile Deutsch de la Meurthe, and the the Lassailly et Bichebois company, tar became the essential material for road surfacing. But the complexity of the problems that needed to be solved led French engineers and contractors increasingly to use foreign materials and machinery. Firmin Deschiron introduced the first American steamer to France, and Gaëtan Brun (1874–1923) the first motorized tar spreader. New coating techniques were introduced; G. Brun used cement in Grenoble from 1876 onwards, and bitumen emulsion, which had first been tested in Germany in 1900 and used there as early as 1904, was introduced in France in 1912.

The first road companies date back to the 1880s, the period when hydrocarbon binders were put on the market. At the time, the need to improve major urban centres and in particular circulation led to the creation of asphalt paving companies, such as the *Société des mines de bitume et asphalte du centre*, or *the Société des pavages et asphaltes de Paris*. Outside city centers the concentration of investment on maintenance supported the creation of rolling companies specialised in rolling asphalt with trained drivers and maintenance workshops. At the beginning of the twentieth century a third type of company emerged. Tarring companies were sometimes connected to companies specialised in rolling, but most often came from those specialized in the production and distribution of asphalt.

The First Half of the Twentieth Century: Contracting Markets and New Technological Opportunities

Between the two world wars, in public works and also construction, new requirements came to light as a result of the following major technological developments: the cement industry experienced remarkable productivity gains, so that its prices became stable while the prices of cast iron and steel increased significantly; related to this, reinforced concrete was substituted for steel in the construction of major structures; dams, power stations, and electrical networks; and the road industry experienced growth as a result of the combined effect of motorization and the search for uses for petroleum by-products.

Reinforced Concrete: a Material Adapted to its Time

While the First World War affected the construction industry in Europe and France on the one hand, and building materials on the other, the significant destruction caused by the war opened up considerable opportunities for the construction sector across Europe, in particular in Belgium, Central Europe, and France. In France, priority was given to the reconstruction of infrastructure until 1924, under the aegis of the Minister of Public Works, that is, the engineers of the *Ponts et chaussées*. For instance, Gaston Pigeaud, who was the director of the *Service central d'études techniques* (central service for technical studies), encouraged the standardization of processes and norms particularly for the reconstruction of bridges. In the absence of a similarly competent personnel, the Ministry of Industrial Reconstruction and Liberated Regions (Ministère de la reconstitution industrielle et des régions libérées) settled on the identical and costly reconstruction of the destroyed buildings. Certain large, modern companies were created, such as the *Société nationale de construction*, which was controlled by the Rothschild family, and the *Société nouvelle de construction et de travaux*, a joint subsidiary of *Paribas* and *Chemins de fer et entreprises*, the Belgian leader in the construction sector. Nevertheless, the French building contractors struggled to meet the challenge, especially as their counterparts in civil engineering won the most significant projects. It is true that during the inter war period, and then during the Second World War, the economic downturn was not conducive to technical innovation, contrary to what we know happened in the United States. during the period of prosperity from 1919 to 1929, or in Britain, where the 1930s saw the establishment of a powerful construction industry combining national presence (through a network of agencies) with the development of original techniques of prefabrication. In France, innovation was limited, despite the remarkable initiatives of some architects (Le Corbusier) and companies (Michelin, Solvay).

By contrast, the production of cement increased sharply in the 1920s, a period characterized by innovation in the manufacturing process with the development of the 'wet' process around 1925, a technology that dominated production until the 1950s. Heavy on energy consumption, it was nonetheless put to many different uses with concrete becoming the prime material of construction for individual and collective housing. Between 1920 and 1930, the use of cement multiplied two-and-a-half times in both France and Germany. Between 1920 and 1932, the European cement industry embarked on an impressive effort to adapt its operations: using portland cement rather than lime, developing of rotary kilns, and increasing both the number and size of factories. The cement industry was very much divided into two with on the one hand the proponents of a stronghold-type strategy, focussed on the local or regional market and on other large companies such as *the S. A. des chaux des ciments et du Teil, Poliet et Chausson* (who were the French or perhaps even European leader in the field), *Ciments français, Ciments Vicat, and Ciments d'Origny* in France; the Associated Portland Cement Company (Blue Circle) in Great Britain; *Dyckerhoff* and *Heidelberger Zement* in Germany; *Holderbank* in Switzerland; and *Italcementi, Unicem,* and *Buzzi* in Italy. Deeply affected by the global crisis, the cement industry reacted in Wetern Europe by forming national cartels and conglomerates. At the international level, however, the United States didn't follow suit even though cartel behaviour still characterizes each state.

Reinforced concrete was instrumental in the rise of the cement industry because it met the needs of the time. Its use was encouraged after the First World War by engineers like Albert Caquot (for use in multiple arches) and Eugène Freyssinet (who introduced mechanical vibration). Above all, the post-war reconstruction combined with motorization opened up considerable opportunities for the material across North America, Western Europe, and their colonial empires. The context led to the construction of structures using increasingly varied technologies (e.g. master beams with triangulation, cantilever multiple beams under pavement, bow-strings with suspended apron bow, with articulated or recessed arches, etc.). At the same time, engineers pushed the possibilities of material ever further, and for this, the Swiss engineer Henry Lossier was a main influence, as were A. Caquot, E. Freyssinet, and Simon Boussiron.

Reinforced concrete techniques progressed rapidly in Switzerland, in particular with Robert Maillart; in Italy with Pier Luigi Nervi; in Germany with Eugen Dyckerhoff, Franz Anton Dischinger, and Ulrich Finsterwalder, all leaders in the design offices of *Dyckerhoff und Widmann*; and in France, under the influence of both inventive engineers and building companies open to new construction techniques. Among these were Albert Caquot, who was behind major theoretical work and had designed impressive buildings as well as contributed to innovation in aviation, the design of machine tools, marine engineering, and the harnessing of tidal electricity and also

Simon Boussiron (who had built the biggest hangars of Bourget, Bizerte, Cherbourg, and Marignane), Alexis Brice, and Maurice Tricon. Nevertheless, the most significant achievements came out of partnerships between one or two talented businessmen and an important engineer, as was the case for Ferdinand Fourré, Fernand Rhodes, and Henry Lossier at the famous Polytechnicum of Zurich, and for Claude Limousin and E. Freyssinet, who built bridges and exceptional far-reaching structures.

Freyssinet was also influential in prestressing, which caused a 'revolution in the art of building' (Eugène Feyssinet, quoted in Fargeot, Montagnon and Xercavins, 2004, p. 127). His first patent for it, which he filed in 1928, defined the process for the first time. Unlike with reinforced concrete, resistance in prestressed concrete rests in the concrete itself which has to be compressed in order to eliminate dangerous traction and cracking. This discovery led him to break away from C. Limousin. This could have led to his ruin if the architect Urbain Cassan had not asked him in 1936 to save the harbor terminal of Le Havre, which threatened to collapse. Thanks to the development of prestressing by adherent wire, E. Freyssinet was able to avert the imminent catastrophe permanently. As early as 1937, while Freyssinet successfully applied his method to the construction of foundation boxes, it was also adopted by the German company *Wayss und Freytag*. Most importantly, a successful partnership between Freyssynet and Edmé Campenon ensued — Campenon was the founder of *Entreprises Campenon Bernard* — leading to the construction of the 44 kilometres of forced pipes in Algeria before the Second World War, of the deep-water foundation boxes of the port of Brest, and of the Algerian dam of Beni Bahdel.

A Prime Use: Power plants

Until 1934, French dams and hydroelectric power stations offered important outlets for public works companies. In France as across Western Europe and North America, investments in electric energy grew rapidly. Three factors contributed to this: first, the search for economies of scale, with the construction of ever more powerful power stations; second, the energy of private investors, and in this case, of the major production groups as well as those involved in transport and distribution, and of the railway companies; and third, state support, especially since the 1938 three billion plan.

Over time, hydroelectric developments became increasingly powerful and complex. Thermal power plants made spectacular progress, in particular via the use of reinforced concrete. Hydropower plants had to adapt by increasing the power of their facilities and regulating their production. As interconnection of the electricity transmission and distribution networks developed, which was at its most advanced in France and in Switzerland, investments in the civil engineering of power stations grew strongly until 1935. This was a response to geographical constraints, but also to the necessity of reducing the quantity of concrete: while the French engineers continued to build gravity dams, the Americans developed dams made from movable materials, particularly on the Mississippi, while the British and especially the Italians developed buttress dams.

In France, in the late 1920s, engineers opted for the solution of arch dams, taking advantage of what had been done in the United States. Good examples are Jean Résal and his theory of the active vault, or André Coyne. Having become an international expert in this field, Coyne invented the system of vibrating wire strain meters, introduced prestressing to the construction of dams, and built, with *Entreprises Léon Ballot*, the large vaulted dams of Marèges and L'Aigle. These innovations fostered emulation with the rockfill Ghrib Dam in Algeria and the cut of the Rhône at Génissiat by the Swiss engineer Marcel Cochard, done on behalf of *Entreprises Léon Chagnaud*, power plants and dams along the Durance and the Rhine by the *Société des GTM (Grands Travaux de Marseille)*, and a new type of weight dams by the *ECB (Entreprises Campenon Bernard)* (Le Chambon). In France in particular, electrification was also accompanied by extensive work to interconnect the electricity transmission and distribution networks. This work was carried out at the initiative of the production, transmission, distribution, or railway companies, but also supported by the state. This work encouraged the emergence of a new sector, that of electrical installers, a sector for which big companies competed. These companies showed a great capacity for innovation that would prevail globally after the Second World War.

A Leading Market: Road Works

As for water dams, the inter-war period opened up much larger markets for road works than those that had existed before the war. The state played a key role here: not only did the budget allocations increase rapidly, but also, the French Road Administration defined an effective road policy in the 1930s. This had four objectives: covering a significant part of the local authorities' expenses, the improvement of the existing network, the regulation of traffic, and the development of research (the research department of the *l'École nationale des ponts et chaussées* was bolstered up). This policy was supported by powerful professional organizations, organised by trade and which, in 1936, formed the Union des syndicats des industries routières de France (union of road work industries), which was more socially minded than the *Syndicat des Entrepreneurs de Travaux Publics de France* or than the *Fédération Nationale du Bâtiment et des Travaux Publics*. However, the state's investment efforts were hampered by the high financial committment. This led to delays in the construction of motorways (the Autoroute de l'Ouest project (Western motorway) did not start until 1939), while the great achievements of Italy, Germany, and the United States heightened the competition between companies, forcing them to increase their productivity rapidly to preserve their margins. This resulted in accelerated progress for road construction technologies.

Between 1918 and 1939, road works underwent a profound technical change. Public commissioners' focus — obtaing the best quality at the lowest prices — met the suppliers' interest in finding uses for the by-products of the coal, metal and oil industries and that of construction companies in securing regular business thanks to their technical expertise. Because of the rigidity of the public procurement system, innovation was a necessary condition for growth if markets were to be found and profits to be made. Towards the end of the 1920s, the previously accepted methods of compacting changed profoundly. At the end of the First World War, John Loudon MacAdam's idea, namely the invention of tarmac, and the use of the compressor network were still prevalent. In 1925 the diesel or semi-diesel compacting roller which had been developed simultaneously in Germany and Great Britain started being used in France. Two years later, the adoption of the heavy oil engine marked the beginning of the French production of compactors. Material progress was accompanied by theoretical advances.

The most significant progress related to road surfacing. From 1926 onwards, as a result of research on adhesion and bonding, surfacing processes progressed greatly. Companies developed bituminous concrete, which was tested on road tracks in Britain, the United States, Germany, and France, especially roads with heavy vehicle traffic and motor racing circuits. The less expensive superficial surfacing was more successful on paved roads, and tar gradually lost preeminence in favour of bitumen emulsions and fluidized bitumen. In the 1930s, gravelled surfaces were made by high-performance mobile construction sites. If the first large-capacity sprinkler appeared as early as 1932, in 1939 mechanical gravelling was often more expensive than the manual alternative.

This success of top surfacing was largely due to black products, especially on urban roads. In the early 1920s, there were seven types of urban road surfacing: stone paving, which was noisy and uncomfortable; wooden pavements, which were too slippery; asphalt, which was first compressed, and then cast, and was also slippery; cement concretes, which were tough, but expensive and could crack; brick, which was hardwearing, but became increasingly slippery with time; metal coatings, which were rather prone to wear and slippery and porphyry cast asphalt. In the end, the latter prevailed in 1923, despite the growing competition for bituminous concrete. Outside the cities, there were hurdles to the development of concretes or carpets of asphalt namely the archaism of quarry exploitation and the limited use of blast furnace slag until the implementation of the Dawes Plan and its introduction of in-kind benefits. The high cost of cement and concrete coatings made tar the most widely used material. In 1925 tar produced in France, imported from England or supplied by Germany, formed the basis of binders. Due to the British General Strike of 1926, the price rose so steeply that bitumen was at a clear advantage and became an indispensable material for road surfacing. The first known use, in Saint-Germain-en-Laye, dates back to 1902. Before the First World War, a natural bitumen — called Trinidad — was generally used for bitumen-and-tar hot-mix asphalt pavements, bitumen on its own being prohibitively expensive. In 1913, all main application techniques were established, including emulsion, hot spreading, surface coating, and asphalt layering.

In 1923, several oil companies joined together to import bitumen to France from the United States, Mexico, Great Britain, and the Netherlands, made possible by the use of iron barrels instead of wooden casks. Bitumen thus became the basic material for road surfacing. Because they were easier to use and lower in cost and could be be spread even in wet weather conditions, anionic solutions were widely used. However, since they turned slippery very quickly, first limestone and later siliceous gravel was used instead of sand, but both without much success. The result was intense research into the discovery of cut-backs, which had been developed in the United States and had been perfected in 1931 in France, followed in 1932 by adhesive dopes using heavy-metal soaps in order to reduce the slipperiness of superficial coatings based on bitumen emulsion, and in 1938 by double surfacing; a year later, cationic emulsions followed. They would revolutionize road engineering after the Second World War.

The use of new, more economical materials was accompanied by a rapid mechanization of production processes. In the inter-war period, a series of new machines appeared, most of them imported from the United States. The increasing use of coating asphalt mats gave rise to finishers, materials specifically designed to put these in place. Introduced in 1911, it was a Warren machine that first applied bithulite, a thick bituminous concrete. From 1913, it became established, as new patents for bituminous concrete were taken. At the same time, more powerful materials appeared that were of American, Swiss, or British origin, and later of French origin. One of the reasons for the large influx of foreign material from 1926 onwards was the development of tarmacadam; henceforth, the *Barber-Greene Company* was number one for the application of hydrocarbon coatings. On a smaller scale, the construction of concrete roads in the north of France used Lakewood or Koehring mobile concrete mixers, which also came from America. At the end of the 1930s, motorway worksites in Germany and airfields in Great Britain paved the way for a second generation of machines, i.e. graders, multi-blade machines, and especially travel plants, the last suitable for building temporary aerodromes in sand asphalt at very low prices. Barber-Greene's innovations would play an important role during the Second World War, although the Germans had also developed quality finishers.

The most impressive progress was in earthmoving and soil stabilization equipment. Early machinery was developed in the United States from agricultural equipment used in the Great Plains. The First World War accelerated progress as it clearly demonstrated the superiority of the diesel engine. The excavator was introduced by Émile and René Razel in France in 1922, together with *Caterpillar machines*. The Razels were also the first to use bulldozers, which had first been developed in the United States in 1925 and were equipped with a diesel engine in 1930. Scrapers, however, which were first towed and then pushed and were already in use in the United States, did not arrive in Europe until after the Second World War. But even before the war, a powerful road industry was already established in France, dominated by an oligopoly of companies including the *Société routière Colas*, a subsidiary of *Shell-France*, and a promoter of the use of the cold bitumen emulsion. A new era had begun, that of automobiles and planes, of the domination of oil.

Acceleration of Growth and Technical Progress (1945–1974)

The period after the Second World War was characterized by an unprecedented need for reconstruction, the international context of the Cold War and of the rise of the Third World, as well as by phenomena of increased mass consumption, urban explosion, the rapid growth of new means of transport, and vast energy needs, all of which presented challenges for civil engineering and construction. The thirty years that followed the Second World War saw an acceleration of technical progress. This was followed by a new era that started in the last thirthy years of the twentieth century and continues to this day marked by the growing power of developing countries thanks to large-scale technology transfers. This period has been characterized by an in-depth renewal of civil engineering techniques as well as of building materials and construction.

The In-depth Renewal of Civil Engineering Techniques: the Example of Prestressing

During the post war period, new types of materials were developed. In underground work, jumbo drills were adopted, which were powerful mobile platforms of simultaneous front-end perforation, as well as drilling machines that cut or burst the ground. The machines either operated independently from the machines that were used to pick up the cuttings or in conjunction with them. The first type was preferred for use when the ground was hard, the second in homogeneous and friable soil. Even more spectacular was the progress in maritime works. New processes for the construction of deep seaports emerged. Above all, the need for offshore oil exploration led to the creation of offshore, self-raising or semi-submersible platforms, which made it possible to build structures with foundations at a great depth. Taking advantage of their lift and the fact that they were not subject to nautical constraints, they were sometimes used for the works themselves, and sometimes as means of derocking, containment, and to build docking equipment. Maritime works saw the introduction of yet other innovations.

Technical progress often came from the development of unusual processes. Diaphragm walls made it possible to create large vertical walls from reinforced concrete, from the ground level up and at great depth, even in non-coherent or aquiferous environments. This technique proved to very useful in meeting the needs of underground urbanism. In the 1960s France saw the spread of the process of the mechanically stabilized earth, a groundbreaking innovation for soil consolidation. At the same time, engineers increasingly made use of special glues, including thermoplastic resins (to improve the adhesion of concrete of different ages) and polyester- or epoxy-based products, which were first developed in the United States of America. They were the result of research conducted on synthetic oil conducted during the 1930s and the Second World War. They made it possible to join precast concrete elements to concrete, to reinforce existing structures with reinforced concrete, to assemble metallic frameworks in specific ways, and to build frameworks from laminated wood.

Among these innovations, prestressing was undoubtedly one of the most fundamental innovations. Benefitting from favourable conditions, its application to concrete had spectacular results. Since the 1930s, science had made considerable progress in Germany, but also in France, particularly due to Pierre Chambaud, Pierre Lebelle, and Henry Lossier. Because of its nuclear programme, France was a pioneer in heavy concrete, although it was not as advanced in the use of lightweight concrete.

Although it was too dependent on public research coming from scattered bodies and institutions, the building and civil engineering industry nevertheless founded the *Centre scientifique et technique du bâtiment* (scientific and technical building centre) in 1948; in 1935 the profession had created an *Institut technique du bâtiment et des travaux publics* (technical institute of building and public works), followed in 1953 by the *Centre expérimental du bâtiment et des travaux publics* (experimental centre of building and public works). Due to the combined needs for reconstruction and motorization, France restored, rebuilt, or built 8,000 bridges in total between 1945 and 1966. Finally, it was necessary to respond to new security requirements as well as new concerns about aesthetics whilst the building of motorways meant increasingly complex road work.

Prestressing offered the opportunity to create larger works at a much lower cost and in increasingly shorter time periods. Between 1945 and 1970, the construction of bridges was one of the specialties in civil engineering that progressed most. Four major innovations stood out in this period, three of German origin and one French, with the latter developing the fastest. The first was the adoption of the self-launching hanger. Introduced in Germany in 1966, this technical solution was then used by the *Grands Travaux de Marseille* to set up nine successive viaducts along the A8 motorway between Roquebrune, France and the Italian border. But this process was quite expensive, so that prefabricated beams tended to be more popular. For very long structures, it is possible to prefabricate the beams before transporting them on site. Also German in origin, this technique was used in France for the construction of the boulevard Périphérique (or Paris ring road), but it was particularly important in Italy.

Invented in 1962 by the *Campenon Bernard* company in France, the technique of prefabricated voussoirs, supported by corbels and assembled with cables supplanted the previous method, and the voussoirs could even be cast on the spot. During the 1960s, prefabricated segmental corbelled construction became the most prevalent construction method for large structures using floating or deck-mounted lifting equipment. At the time, the construction of prestressed concrete structures progressed anew with the German pushing technique borrowed from metal construction. This authorised the assembly, along one given axis, of a prefabricated structure put in place by the addition of successive elements. It was introduced in France by the *Compagnie industrielle de travaux*, a subsidiary of the *Empain-Schneider* group.

These new processes were underpinned by industrialization, i.e. standardization and prefabrication. Standardization, based on both an internal and external repetition, is illustrated by the Champlain Bridge in Montreal. The development of standard structures and the automatic calculation of the engineering of bridges ensured a relative preponderance of GTMs, as they followed Dywidag processes (*Dywidag* stands for *Dyckerhoff und Widmann Aktien Gesellschaft*), and especially of the Campenon Bernard company on the French and international markets. The industrialization of methods increasingly relied on prefabrication. Here, ECB (Entreprises Campenon Bernard) played a pioneering role at the end of the 1930s with the creation of the Oued Fodda factory, whose model was adopted in France and throughout Europe.

Some workshops and factories became veritable laboratories in which new solutions were explored.

In the aftermath of the Second World War, prestressing had gained many supporters abroad, especially in Germany, Switzerland, and Sweden. Therefore, the *Association Française de Précontrainte* was formed in 1949, and an International Federation in 1952. In France, beyond Freyssinet, others like Louis-Pierre Brice and André Coyne adopted prestressing. In the aftermath of the liberation, some of the leaders in reinforced concrete, like the *Société des constructions Edmond Coignet,* took to prestressing, first using the Freyssinet processes, and then, from 1952 onwards, those developed by the *Établissements Sainrapt et Brice* and later, in 1965, those developed by *CITRA (Compagnie Industrielle de Travaux)*. Some companies did not hesitate to acquire foreign processes: GTMs perfected their own processes in technical cooperation with Dywidag, and in 1964, *SGE (Société Générale d'Entreprises)* cooperated with Philipp Holzmann. One of the most striking examples, however, is the Boussiron process for prestressing concrete. Despite the fact that a specific process had been invented, the company secured exclusive rights to the Swiss BBR-V process, which was used in building the bridges of La Voulte and Tancarville, as early as 1954. Under the leadership of the renowned engineer Nicolas Esquillan, the Boussiron company became one of the main leaders in the European market.

Together with *Dywidag* (number 2 in the market), the *Entreprises Campenon Bernard* (number 1) were leaders in the field. Although hard hit by the death of its founder in 1962, the Campnenon Bernard company remained at the forefront of innovation thanks to the remarkable team of engineers that E. Freyssinet and Edme Campenon had gathered around them. Jean Muller, the inventor of construction by means of voussoirs joined together with glue, developed the technique of building structures by successive corbellings, and his company surpassed the German competition for the construction of bridges by temporary bracing. *ECB* was instrumental in reconstruction. In North Africa, mainly in Algeria, the company built prestressed concrete runways at the Algiers-Maison Blanche aerodrome, hydraulic plants and dams: such as Erraguène on the Oued Djen-Djen river, which had multiple vaults and was the site of numerous innovations. Then retreating to the metropolis, the company focused on areas where it gained a competitive advantage from its mastery of prestressing.

The parent company of a dynamic group, Entreprises Campenon Bernard was active in Morocco, but also in Latin America, i.e. in Brazil, Uruguay, Colombia, and Venezuela, from 1947 to 1952. In 1953 they also settled in Iran. Also active in building and road works, *ECB* continued in the 1960s and 1970s to build extraordinary structures in France (including the Choisy-le-Roi Bridge — the first in the world to be made with a corbelled apron and with prefabricated epoxy-bonded voussoirs — as well as the Oléron bridge and the Brotonne Bridge, which is the world's largest prestressed cable-stayed bridge) and abroad. Other than bridges, there was the tidal power plant of La Rance, but also hydroelectric plants on the Durance, Rhine, and Rhône, the Idriss I dam in Morocco, and also sea structures, the building of water treatment and storage facilities, and prestressed concrete enclosures for nuclear power plants. Abroad, independently or via the *Compagnie de constructions internationales* (CCI), ECB built the Reza Chah Kabir Dam in Iran, one of the largest dams in the world (in association with *SGE, GTM, CITRA* and *Dragages TP*).

Campenon Bernard's successes were largely due to the engineering advances made by the *Société technique pour l'utilisation de la précontrainte* (STUP). Created in 1943 to make use of and develop the Freyssinet patents, it became a world leader: in the 1960s, 70% of the prestressing processes used in the world derived from the Freyssinet system, and only 30% from Dywidag's. With its exceptional engineers such as Pierre Lebelle and Yves Guyon, ECB expanded the uses for prestressing and designed pioneering works including the Underground Basilica at Lourdes, the 38-kilometre-long Lakde Pontchartraint Causeway Bridge in Louisiana, the Gladesville Bridge in Sydney, the Chesapeake Bay Bridge tunnel in Virginia, which is 19.6 kilometres long, the Brasilia road platform in pre-stressed concrete, the Oldbury nuclear power station in England, and the Tokyo-Haneda monorail in Japan. ECB's processes were also used elsewhere, in Germany, Belgium, and the United States. The company never ceased to innovate and had a major role in the training of French and foreign engineers.

Two Privileged Areas of Innovation: Dams and Roads

Dams and roads were also major areas of technical innovation. In France, *Électricité de France* encouraged contractors in this direction. Concreting techniques had made great progress. By the early 1950s, the cement industry was able to provide products of consistent quality. The mixing of artificial cement and slag on the site made it possible to produce cheaper and better-quality concrete. Above all, the use of contraction joints, which solved the problem of thermal shrinkage, and construction site laboratories became widespread. As the average size of the dams increased, the site facilities developed much further.

Between 1945 and 1974, techniques for arch dams constantly developed, partly due to the increased scientific control of construction. This progress benefitted from competition between three main national traditions: that of the Americans, the French, and the Swiss, while Italy also became influential later. However, with the introduction of computer science, both approaches (computation on the one hand and experimentation on reduced models on the other) were reconciled: in 1968, the engineering of the Vouglans dam for instance was born out of both.

The arch dam was so successful in France, in Italy, Switzerland, and Austria because it matched the conditions of construction in Europe: in the first two decades following the Second World War, demand increased to a level that required that the materials were saved at all costs. There was also a relative lack of unskilled workers while a specialized workforce did exist. The American model, which focused on the construction of massive structures, which prioritized economies of scales and the implemntation of industrialised construction techniques, opposed the French or Italian models, based on research on the optimization of materials and on the creation of more elegant structures.

Arch dams best suited narrow sites. When sites were too narrow it was necessary to build the hydroelectric plant and weirs underground, into the supporting rock, which was a very expensive solution. At the Aigle dam on the Dordogne, André Coyne proposed with the *Entreprises Léon Ballot* the solution of an arch-weight structure whose body was incorporated the power plant and ski-jump shaped weir projected the waters of the reservoir far. This technology was then adopted by the SGE before being disseminated abroad, in Portugal and Chile, an initiative of the engineering firm *Coyne and Bellier*. French engineers and contractors also promoted double-curved vaults abroad, a method that was adopted for the first time by A. Coyne and *Entreprise Fougerolle* in Salamonde in Portugal, and in Mozambique and Iran for two of the largest dams of the world at the time. A. Coyne was among those who contributed most to the adoption of vaulted dams in wide valleys, notably from 1956 to 1959 in Kariba, on the Zambezi in Rhodesia, when he worked as consulting engineer for the Italian company *Impregilo*. The Dumez company went even further in the Hendrik Verwoerd dam, in South Africa, built from 1970 to 1972. The vaulted dam provided savings of 40 to 50% for thin vaults compared to a dam of equivalent weight. However, with the widening of the valley the cost advantage of vaulted dams was reduced, and there was resistance to the construction of a gravity dam. However, engineers gradually promoted the construction of buttress dams. Moreover, they increasingly employed multiple vaults. Invented by Caquot in the 1920s, these were very popular after the Second World War, and the technology was exported abroad.

In Western Europe, concrete dams were prevalent for a long time, in France, Italy, and Switzerland, as well as in Germany. But from 1970 onwards, dams made from movable materials became more prevalent, in particular because their cost per m^3 was twenty times lower. Thanks to advances in soil mechanics and the systematic use of modern compaction methods from the United States, French companies acquired world-renowned expertise, either individually or together with the Italians. The same was the case for rockfill dams. Like all dams made from soft materials these caused difficulties, and required the ancillary structures to be distinct from the main structure. From 1960 to 1964, A. Coyne solved the problem while the *Compagnie française d'entreprises* constructed the Jatiluhur Dam in Java, which had a freestanding concrete tower combining a central spillway and a circular dam-hydroelectrice plant integrated at the periphery. This expertise in building backfilled dams allowed the CCI to build two of the largest dams in the world, the Keban Dam on the Euphrates in Turkey (1967 to 1973), and the Tarbela Dam in Pakistan (1968 to 1976), which, due to its dimensions (mass of materials mobilised), held the world record for dams with a mixed profile (earth and rockfill) and an inclined core.

Between 1944 and 1974, in all Western countries, road works expanded, accompanied by rapid technological progress. In France, for example, the conditions were favourable because of a late but powerful activity, demand, and a high level of integration and of concentration, as well as the vigorous competition that existed with an oligopoly of big companies. These firms undertook a major effort in research and development, which was conducted in conjunction with oil or cement companies and shaped by research conducted by the road administration. Around the *Laboratoire central des ponts et chaussées*, or LCPĆ (central laboratory for roads and bridges), which had been created in 1949, a network of sixteen regional research centers was set up to meet the needs of the *Direction départementale de l'équipement*. Since 1970, LCPC has cooperated closely with *SETRA (Service d'Etudes Techniques des routes et autoroutes)*, the professional organization for road works and aerodromes, in a context of advanced international cooperation.

Under the dual impetus of LCPC and private research, knowledges concerning road construction made much progress, particularly in terms of soil compaction and stabilization. Nevertheless, while the stabilized gravel had the advantage of reducing costs, from 1958 the United States' cement gravel surpassed them, and these were introduced in France in 1962 to 1963. From 1965 onwards, engineers focussed on surfacing, as maintenance had mechanized and rationalized. The increased efficiency of maintenance was largely due to the improved quality of the materials that were produced at ever lower cost. Although the increasing use of slag cement encouraged the use of tarmacadam and then bituminous concrete, black bitumen-based products remained prevalent: the arrival of crude oil from the Middle East encouraged the use of bituminous cements and cationic emulsions. By contrast, by 1973 the use of tar declined with the use of bitumen-tar, whereas poured asphalt was used in cities and on motorways.

While earthmoving equipment saw the most dramatic changes, new equipment for the surfacing of roads appeared, including, first, travel plants, and then slip forms which could create almost continuous pavement, as well as finishers and self propelled gravellers. The asphalt plant in particular was introduced by the American expeditionary force in 1943 to 1944. Two techniques — continuous coating and discontinuous coating — appeared in competition with each other. In France, the state opted for continuous coating in 1954. Around that time, German and French manufacturers became rivals as the plants continued to progress thanks to their adoption of electricity and automation.

The Construction Sector: Reinforced Dualism

In contrast to their German and British counterparts, the French construction industries had little technical integration. In Germany, this integration was strong upstream, in the UK downstream. Moreover, French structures were blatanted divided between obvious strong points and important zones of weakness. In the glass industry, the Compagnie de Saint-Gobain was a world leader ready to face the French competition and strengthened in 1970 by its exemplary merger with *Pont-à-Mousson*. Despite initially lagging behind for the adoption of flood glass, the company quickly caught up with its global competitors. Its European leadership contrasted with the persistent archaism of those industries which were still largely composed of small and medium-sized companies, like those working with wood, brick and quarry materials, quarries having often been purchased by French or foreign companies that came from outside the sector.

The cement industry, for its part, was increasingly dominated by a small oligopoly of large companies, i.e. *Lafarge, Poliet and Chausson, Ciments français, Vicat,* and *Ciments d'Origny.* In the aftermath of the Second World War, they avoided nationalization and were given priority within the Monnet Plan. A rapid growth in production ensued as a result of the power of demand and the technological revolution linked to the development of the 'dry track' process, which created a more homogenised cement and saved a substantial amount of energy. The 1960s saw new technological changes linked to the development of ready-mixed concrete (a US invention that was successfully developed by the British *Ready Mix Concrete company RMC,* and then introduced in continental Europe) as well as to prefabrication and progress in automation. As a result, costs decreased and new improvements were made. In this regard, the Ciments Lafarge and the Swiss group Holderbank had a growing advantage over German, British, French, and Italian competitors, while North America attracted an increasing amount of investments. In 1971, thanks to a strong Canadian presence, Lafarge became the second main North American producer, while in Europe, it became a close rival to the European plaster leader, *British Plaster Board (BPB).*

The building industry also underwent profound changes. Despite retaining its tradition of craftsmanship and manpower and the division into specialties or trades, related to structural work and then finishing, it experienced strong productivity gains. In France, the years from 1953 to 1973 constituted a prosperous period because of a strong demand for housing that was first supported by the state and then, from 1963 onwards, increasingly driven by the private sector. Until 1975, mainly collective housing was built, and afterwards, the focus shifted to individual homes. The year 1963 marked a turning point as well as the end of the idea of heavy prefabrication, whereby a whole building was constructed and then installed on pre-built foundations. It was supplanted by formwork, which allowed prefabrication on site, and was referred to as 'light prefabrication'.

In a sector often dominated by banking or industrial capital, however, important family enterprises continued to be influential both for the structural and for the finishing work, as were construction cooperatives. These competed with companies working with reinforced concrete and also, and especially, with public works. New companies burst onto the market, like *Raymond Camus et Cie,* whose processes were a worldwide success, or *the Société des Maisons Phénix* which, adopting the American home builders' formula, produced individual houses prefabricated in a factory and in bulk. During the 1960s, two leading companies emerged, *Bouygues* and the *Société auxiliaire d'entreprises* (SAE), which, after they had established themselves as leaders on the French market, precipitated the decline of heavy prefabrication in favour of a strategy that prioritised workers' skill and productivity and later gained a very strong global influence in the 1970s and 1980s.

From 1980 to Today: New Progress and the Increased Role of Emerging New Powers

Since the 1980s, technical progress has not stopped in the construction trades, as shown by, for example, the cement industry and civil engineering. At the same time, Western technological supremacy is being undermined by the rise of new global players, ie multinationals from the major emerging countries.

Continued Progress of Techniques

Thanks to advances in information technology, civil engineering continued to experience important technical progress in the 1980s and 2000s. In fact, within thirty years, contemporary concretes have undergone significant changes in their constituents, their composition, their characterization methods, and their performances. The development of the use of plasticizing or super-plasticizing admixtures made it possible to isolate different properties of concretes, where previously determined only according to the proportion of cement being used. Many constructions were made using High Performance concrete, which has been known since the late 1980s. At the end of the twentieth century, two new types of concrete appeared: ultra-high-performance fibre-reinforced concrete, which was developed by *Lafarge* and *Bouygues*, and self-placing concrete, which was developed by the Japanese in the late 1980s.

These developments impacted the use of prestressed concretes. Much faster than pouring in place, the use of glued conjugated segments has allowed for the construction of large structures. *Campenon Bernard's* introduction of the process of continuous corbelling from pile to pile made the construction of the Brotonne Bridge in 1977 possible, which was fitted with a prestressed concrete deck cast on site with symmetrical corbels. Since 1978 and promoted by Jean Muller, self-launching hangers have become essential for medium-span structures built by assembling, span by span, prefabricated segments, either glued or not, which are prestressed by cables outside the concrete. Finally, the use of sliding teflon supports has made it possible to build, thanks to the pushing technique, structures of constant height and average span, but of great length.

The most significant innovation continues to be that of glueing combined voussoirs with epoxy resin glue. Five methods of installation are used: the voussoirs can be built by corbelling to the beam, by corbelling to the portal, by crane, by full span on hanger from underneath or assembled gradually as the work progresses. These methods continue to evolve. Other factors contribute to this, such as the strengthening of lifting equipment, and the development of calculation means and materials. This progress is all the more necessary as prestressed concrete is increasingly competing with steel or mixed structures (e.g. the French pleated web process for bridge decks combining steel, concrete and, external prestressing, which was invented in 1986).

Technical progress has not been limited to large structures. It also concerns underground construction, foundations, and, to a lesser extent, the building sector. There are three categories of tunnels: open-cast tunnels, cast-in-place tunnels, and tunnels built from prefabricated walls, according to the Berlin or Paris methods; explosive digging combined with the use of compressed air; mechanized digging by means of point attack machines, adapted for soils of equal consistency and soft rocks, or tunneling shields, or tunnel boring machines. Open-front shields have been replaced with those with a confined front, which are adapted to soft ground or aquifers. Finally, tunnel boring machines have become more and more common.

Special foundations have also undergone a remarkable technical evolution. In fact, contemporary urbanism seems inseparable from underground works and foundations. Benefitting from advances in soil mechanics, from the recording and automatic processing of drilling parameters, as well as from the increasing power of machinery, the processes have diversified. In addition to foundations on beaten piles, on bars or on micropiles, there are now molded walls, which have been introduced from Italy, vibro-flotation, vibro-substitution, liquid and then solid cement injections, jet grounting, and the technology of mechanically stabilized earth. These technical advances have had a marked impact on both construction and civil engineering projects.

Is Today's Technological Supremacy of the West Threatened?

The abovementioned state-of-the-art techniques have ensured the successful international export of French, German, Italian, Swiss, Dutch, and Belgian civil engineering and construction companies, particularly in the context of the two oil crises and in response to the needs of the new industrialized countries. Their activity developed greatly until 1983. Afterwards, the shrinking revenues of OPEC countries and the Latin American bankruptcies led to the limitation of their overseas activities, followed by a recovery that was slowed down by the Asian crisis of 1997 and then revived at the turn of the millennium. Since the 1970s, European companies have faced increasing competition from Japanese and South Korean firms, particularly in Southeast Asia or the Middle East, while Spanish companies have now come to the fore.

These new companies grew further between 2000 and 2010, and in 2017, there were five Chinese firms among the fifteen largest global construction companies ranked by their turnover, including four in the lead, while their European counterparts came in fifth, seventh, eighth, eleventh, and fourteenth position. The highest-ranked American company was in thirteenth position, and the most successful Japanese company in fifteenth. A similar development has affected the cement industry, albeit more recently: in 2016, out of the top ten global groups ranked by turnover, five were Chinese, one Mexican (*Cemex*), one Taiwanese, and one Indian, contrasted with three European groups, including one Irish (*CRH*), one Swiss-French (*Lafarge Holcim*) and one German-Italian (*Heidelberg Italcementi*). But assessed in terms of production capacity, Chinese companies ranked first, third, fifth, sixth and eighth, Lafarge Holcim second, Heidelberg Italcementi fourth, amd CRH disappeared from the list, since its assets mainly consist of aggregate quarries.

Follwing the American model, European civil engineering and building companies have developed their engineering activities. In the 2010s, North American firms continued to dominate the world market: *Bechtel* (first position), *Fluor* (second),

KBR (ninth), and *Foster Wheeler* (tenth). European companiess occupy the third (*Saipem*, Italy), fourth (*Balfour Beatty*, UK), and sixth (*Technip*, France) positions. Nevertheless, the two largest French construction groups have conquered strategic positions. Since 1990, Bouygues has controled the Swiss *Vorspann System Losinger*, the world leader in concrete prestressing. This breakthrough from Bouygues led Vinci in 2008 to merge its subsidiary *Freyssinet International* (formerly STUP) with Solétanche Bachy, the world leader in soil treatment, and to acquire, *Entrepose Contracting* and *Cégelec*, the European leader in electricity engineering, in 2010, and then *Spie Capag* a year later. Thus, in the mid-2010s, the Vinci group rose to the level of *Bechtel*, just behind *Halliburton*, the new global leader.

Nevertheless, the game is not over yet. In the cement sector, the Cemex group has achieved a level of technology comparable to that of *Lafarge Holcim*, which it exceeds in terms of productivity and profitability, while the Chinese company Anhui Conch outperforms all its major global competitors in terms of profitability. Even in the engineering sector, the emerging countries have generated formidable competitors with the Brazilian firm *Odebrecht*, albeit at the heart of a huge scandal, the Russian company *Stroygazconsulting*, and especially, *Samsung Engineering*, respectively the fifth, seventh, and eighth in the world. More importantly, China alone now accounts for almost half of the global construction market. In fact, the end of the twentieth and the beginning of the twenty-first century have been marked, in the construction industry as in construction materials, by gigantic transfers of technology.

References

'100 ans de route en France (1926–1976)', *Revue générale des routes et des aérodromes*, special issue, (December 2000).

Bardet, Jean-Pierre, Chaunu, Pierre, Désert, Gabriel, Gouhier, Pierre, Neveux Hugues, *Le bâtiment. Enquête d'histoire économique, XIVe- XIXe siècles. Maisons rurales et urbaines dans la France traditionnelle* (1971) (Paris : EHESS, 2002).

Barjot, Dominique, 'The Americanization of the European Cement Industry: Lafarge in Comparative Perspective, from Fashion to a Structural Change', *Business and Economic History On-Line*, 7 (2009), 1–17.

———, ed., 'The Construction Industry in the 20th Century: an International Interfirm Comparison', *Revue française d'histoire économique — The French Economic History Review*, 1, special issue, (2014).

———, 'From Tournon to Tancarville. The Contribution of French Civil Engineering to Suspension Bridge Construction, 1824–1959', *History and Technology*, 6 (1988), 177–201.

———, *La Grande entreprise française de travaux publics (1883–1974)* (Paris: Economica, 2006).

———, *Travaux publics de France. Un siècle d'entrepreneurs et d'entreprises* (Paris: Presses de l'École des Ponts et Chaussées, 1993).

———, 'Why Was the World Construction Industry Dominated by European Leaders?', *Construction History. The International Journal of the Construction History Society*, 28.3 (2013), 89–114.

——— and Jacques Dureuil, eds, *150 ans de génie civil: une histoire de centraliens* (Paris: PUPS, 2008).

——— and Monique Trédé, eds, *Réseaux électriques et installateurs des origines à nos jours* (Paris: PUF, 1995).

Berthonnet, Arnaud, *Un siècle de construction routière. Une histoire d'entrepreneurs* (Paris: RGRA-USIRF, 2005).

Châtelain, Abel, *Les* migrants *temporaires en France de 1800 à* 1914 (Lille: Publications de l'Université
de Lille III, 1976).

Culture Technique, special issue, *Génie civil* (26), (1992).

Derry, Thomas K. and Trevor I. Williams, *A Short History of Technology from the Earliest Times to A.D. 1900* (New York, NY: Dover Publications, Inc., 1993).

Fargeot, Bernard, Montagnon, Jean and Xercavins, Pierre, eds., *Eugène Freyssinet. Une révolution dans l'art de construire* (Paris: Association Eugène Freyssinet, 2004).

Goldberg, Serge, ed., *1900–2000. Un siècle de constructions* (Paris: Moniteur, 1999).

Guillerme, André and Jean-François Crola, *Histoire des métiers du bâtiment aux XIXe et XXe siècles* (Paris: Plan construction et Architecture, 1991).

Jacobs, David and Anthony Neuville, *Bridges, Canals and Tunnels. The Engineering Conquest of America* (New York, NY: American Heritage, 1968).

Lemoine, Bertrand, *L'Architecture du fer. France: XIXe siècle* (Seyssel: Champ Vallon, 1986).

Lescure, Michel, *Immobilier et bâtiment en France (1820–1980): histoire d'une filière* (Paris: Hatier, 1983).

Martini, Manuela, *Bâtiment en famille. Mutations et petite entreprise en banlieue parisienne au XXe siècle* (Paris: CNRS Éditions, 2016).

Picon, Antoine, ed., *L'Art de l'ingénieur: constructeur, entrepreneur, inventeur* (Paris: Le Moniteur, Centre Georges Pompidou, 1997).

Pierre, Francis and Heinz-Herbert Cots, *The History of Road Building Equipment* (Southfield: AKHL Group Publications, 1998).

Voldman, Danièle, *La Reconstruction des villes françaises de 1940 à 1954. Histoire d'une politique* (Paris: L'Harmattan, 1997).

West, Graham, *Innovation and the Rise of the Tunneling Industry* (Cambridge: Cambridge University Press, 1988).

AUDREY MILLET

The Factory of Desire

Textiles and Techniques (Nineteenth to Twenty-First Centuries)

Despite technological improvements during the eighteenth century, the textile industry has long failed to meet the growing demand. Attempts to recycle Napoleon's army uniforms before the retreat from the Russian campaign confirm that there were supply difficulties even then (Watson, 1991). In the mid-nineteenth century, looms, which had remained mostly mechanized, still did not produce cloth quickly enough and in sufficient quantites. Additionally, raw materials, such as cotton, were expensive to import to Europe. While industrialization, mechanization, chemical processes, and the concentration of labour in factories changed textile production, another major change relied on increasing globalization, as transport improved and colonization expanded. As the economist Pietra Rivoli shows by retracing the life cycle of a cheap T-shirt in a globalized economy, within two centuries, the textile industry was widely transformed (Rivoli, 2005). Today, a Texan cotton fibre can be sold to a Chinese manufacturer who will sping the yarn, weave the fabric, and manufacture the T-shirt. After printing, the T-shirt is sent back to the US, where it is sold for $6. Simultaneously, technical developments made it possible to change scale. From the machine ginning cotton bales saturated with chemicals to the dyes that made it easier to hold prints, the global textile industry is defined by abundance and waste.

Giorgio Riello's work on cotton in particular has somewhat revised the traditional narrative (Riello, 2013). Kenneth Pomeranz dates the 'great divergence' to the late eighteenth century and considers it exceptional since it was concentrated on a short period (one century at the most) that coincided with the industrialization of the British Isles and Europe, while Riello suggests a longer process took place with Europe as a late benficiary (Pomeranz, 2000). The divergence between Asia and Europe is thus clearly linked to their convergence. The historiographical debate has so far schematically opposed the defenders of a 'Western exception' (Landes, 1969: 1–40) to those who interpret Europe's divergence as the product of historical contingency (Parthasarathi, 2009). For Riello, it is unlikely and unthinkable that a continent like Europe, which had little knowledge about the cultivation, processing, and consumption of cotton, could specialize in it. Before the eighteenth century, India dominated cotton production and trade in and around the Indian Ocean, and only later did north-western Europe dominate on a global scale. Numerous interactions

Global History of Techniques, ed. by Guillaume Carnino, Liliane Hilaire-Pérez and Jérôme Lamy, Global Matters Culture Matérielles, histoire des techniques et circulation globale des savoirs, 9 (Turnhout: Brepols, 2024), pp. 481–494
BREPOLS PUBLISHERS 10.1484/M.TECHNE-EB.5.129825

between Europe and other continents after 1500 contributed to establishing this future European domination, and therefore, technologies, products and raw materials, socio-cultural changes, and their integration into the Western socio-economic system are important.

An article interested in textile techniques should first and foremost study the Schumpeterian pattern of innovation. As early as the mid-nineteenth century, entrepreneurs saw clothing as a particularly promising market. To meet consumer demand and sell their products, they sought to produce larger quantities at a lower cost (Lipovetsky, 2006). The change in fashion — colours, styles, or shapes — was calculated to stimulate sales. The fashion industry developed around eight traditional 'seasons': spring, summer I, summer II, autumn, autumn-winter transition, winter I, winter II, and holidays (Birnbaum, 2005; Anguelov, 2015: 1–21). But today, the most fashion conscious customers look for new products, and there are twenty-four distinct 'seasons', including new classifications such as 'back-to-school', 'prom', and even 'wedding' (Jimenez, Kolsun, 2014). Nevertheless, this competitive schedule has certain consequences (Lipovetsky, 2017). The labourers that work in manufacturing are often exploited in workshops with deplorable wages and sanitary conditions. The workers' misery is the dark side of an industry that presents itself as avant-garde, creative, innovative, and responsible. Today, behind advanced techniques lie marketing techniques that promote the quality of products made in sweatshops and built in obsolescence.

The purpose of this article is to trace the major technological transformations of the textile industry, an industry increasingly connected to innovative sectors. The first part will present the characteristics of textile industrialization in the nineteenth century, when machines gradually replaced the skilled hand of the worker, and when transfers between sectors yielded their first concrete results. But production methods also exploited the workers until their death. Until the 1960s, techniques were oriented towards performance with synthetic fabrics that corresponded to new lifestyles and the development of recreational sports. This dynamic continued into the early twenty-first century: manufacturers sought to exploit high-tech sectors to offer intelligent fabrics. However, since the 1980s, a debate between old and modern techniques has criticized polluting techniques and the homogenizzation of cultures (Jarrige, 2014). Somewhat continuing from the working conditions of the nineteenth century, the textile industry continues to be the most unequal sector in the world.

The Industrialization of the Textile Sector

The textile industry has been one of the main drivers of the industrialization and the globalization of the economy. A change in morals, belief in progress, and the advent of the nineteenth-century consumer society accompanied the technical transformations that became key to low-cost fabric production (Tortora, Marcketti, 2015).

As technical innovation increased productivity, the price of clothing decreased and stimulated demand. The cotton ginning machine, patented in 1794, allowed the southern regions of the United States to increase gross annual production from 2 million to 85 million pounds within sixteen years (Cipolla, Birdsall, 1979; Tortora, Marcketti, 2015). Steam power used in textile mills also contributed to falling prices. The realization of the patterns was facilitated by the Lyon weaver Joseph Marie Jacquard's loom, and his followers mechanized the series of operations that constituted weaving. More than 10,000 looms were sold between 1801 and 1811, although they were not widely used until the 1860s (Cipolla, Birdsall, 1979). To be precise, these looms made it possible to imitate Indian cashmere shawls. In the nineteenth century, mechanization gradually made it possible to create substitute products, thereby increasing the connections between the textile and clothing sectors, while trade in goods and technology transfers were facilitated (Verley, 1997b). At the same time, the invention of the sewing machine had a direct impact on the making of clothing at home, on the expansion of subcontracted work, and on mechanization in factories (Cordillot, 2018; Godfrey, 1982). These developments concerned both women and men. Following the developments of Barthélémy Thimonnier, Walter Hunt, Elias Howe, and Isaac Singer, Nathan Wheeler and Allen B. Wilson marketed an efficient sewing machine with interchangeable parts, which was operated by both foot and hand. It specialized tasks and limited the need for skilled labour. As early as the 1870s, steam sewing machines were used to produce an entire wardrobe. In practice, the manufacturing time for a coat was reduced from six to three days (Godfrey, 1982: 87). The American Union Army's vast demand for uniforms (1.5 million per year) made ready-to-wear clothes and machine manufacture essential to the American Civil War. Army statistics were then collected by manufacturers to standardize civilian clothing. The speed of machine manufacture and the standardised measurements made clothes both available and affordable. The growing needs of individuals were being met (Tortora, Marcketti, 2015).

Mechanization presented a real economic incentive, because automating textile manufacturing provided a means to accelerate the production of fabrics and sell them at lower prices. In the meantime, the textile industry moved closer to other sectors to transfer, adapt, and adopt their techniques. These profoundly changed the sector as well as the daily lives of consumers (Tortora, 2015). Raincoats, zippers, and trainers revealed more of these technical inter-trade exchanges than the jeans produced since the sixteenth century (Muthu, 2017). The chemical industry inspired Charles Macintosh (1766–1843), who made the first raincoat from cotton and Indian rubber. In the 1820s, the coat was still coarse and heavy, it did not smell good, and the stiff fabric deformed the seams. Four years later, Macintosh invented a sealing process. By 1900, most raincoats were referred to as 'Mackintoshs' (Keers, 1987). Inspired by metallurgy, the zipper was introduced in 1890 and adopted *en masse* in the 1930s. It was a Swedish electrical engineer, Gideon Sundback, who, after many tests, invented the modern metal zipper and the machines for its mass production in 1913. The Hookless Fastener Company then commercialized the device, which was practical and reliable. The zipper reflected the entrepreneurial and innovative spirit of American industrialists at the beginning of the twentieth century. It replaced buttons and hooks after the designer Elsa Schiaparelli adopted it in her fashion shows, which were great publicity events for manufacturers (Friedel, 1994). However, the product best representing the alliance of innovative industrial sectors was the trainer. To manufacture a sports shoe, it was necessary to assemble small pieces of canvas that followed the three-dimensional shape of the foot, which made the manufacturing process more complex. The sewing machine could not yet replace the experienced and skilled hands that folded, shaped, and sewed the parts together and attached them to the sole. Jan Matzeliger solved this problem with a machine that attached the sole to the canvas within just one minute. Patented in 1883, it made it possible to manufacture hundreds of pairs of shoes a day (Tenner, 2000). This shoe was chosen by individuals across many social classes for its comfort and popularity in the United States (Coye, 1986; Vanderbilt, 1998; Hendrickson, 2000). Vulcanized rubber offered flexibility and comfort. The rubber industry, including the British company Dunlop, immediately took over the manufacture and marketing of trainers. Worn by athletes at the 1900 Paris Olympic Games, by the military, and by American college students, they gained much popularity. It was the Marquis Mills Converse (1861–1931) who had the idea of producing the 'All-Star Converse' shoe and of associating a famous basketball player who signed it 'Chuck Taylor'. As advertising agencies multiplied in the United States, Mills Converse understood that marketing was an accelerator of success (Heard, 2003). However, behind these innovations, which sprung from the meeting of innovative sectors, lay the realities of manufacturing.

At the beginning of the Industrial Revolution, the United States did not seem to be the most promising industrial centre or a future supplier of cotton. The cotton production boom that followed was staggering. During the first half of the nineteenth century, American production increased by a factor of 25. And with the outbreak of the Civil War, the South produced about two thirds of the world's total cotton, mainly for export (Rivoli, 2005). In 1860, cotton accounted for about half of the value of all American exports. This success was based on the work of slaves on the plantations. Productivity was, therefore, not only the result of technical innovations but also due to human captivity. This extended beyond the cotton fields to the sweatshops. Sweatshops are often perceived as particularly horrible work environments, morally and politically distinct from other lawful forms of low-paid work that seem acceptable by comparison (Hoskins, 2014). Size does not matter here. An employee who produces more value for the company than the firm pays him or her is being exploited. From the nineteenth century onwards, the development of the textile industry worked with aggressive pricing strategies. The pricing policy led to a reduction in expenses. Excessive capitalism has worsened the conditions of production. Friedrich Engels explained that 'the history of the proletariat in England begins with the invention of the steam engine and of machinery for working cotton'. Engels suggested that the Industrial Revolution had led to 'a rapid fall in the price of all manufactured commodities, prosperity of commerce and manufacture, [and] the conquest of nearly all the unprotected foreign markets'. But the new 'national wealth' did not benefit everyone (Engels, 2008: 1, 7). Rural depopulation intensified demographic pressure on industrial cities, which could not accommodate new inhabitants properly. The controversies around the English Factory Act of 1833 illustrate how difficult the situation was. Reducing children's working time would lead to an increase in employers' expenses, and employers also sub-contracted labour to workers whose activities were not regulated. The poor sanitary conditions in workshops led to serious diseases, including tuberculosis, which was referred to as 'tailors' disease'. The development of the garment industry was punctuated by human disasters. On 25 March 1911, a fire broke out at the Triangle Shirtwaist Factory in New York. The situation was dramatic: there was no emergency exit for workers, the workshop was overcrowded, the doors locked or blocked, and the windows closed with iron bars. The death toll was 146, and seventy-one people were injured. As the inescapable logic of the low-cost market required that costs be reduced as much as possible, manufacturers cut back on safety and working conditions to 'balance' the accounts (Hoskins, 2014).

Mechanization made it possible to produce fabrics at a lower cost, allowing a broadening of the customer base. The textile industry's connections to other sectors, such as the chemical industry, made it possible to have technical transfers to meet consumer demands, such as comfort and well-being. Nevertheless, production and manufacturing still relied on human labour, which was under increasing pressure. The exploitation of labourers enabled industrialists to remain competitive in a globalized economy seeking to lower prices, whatever the human cost. These dynamics continued throughout the first half of twentieth century.

The Future in Synthetic Form

Everyday techniques, social changes, and economic events, such as the depression of the 1930s that sounded the death knell for beaded dresses, stimulated manufacturers and buyers. In the early twentieth century, the increasing availability of ready-to-wear clothing of all kinds encouraged mass consumption (Tortora, Marcketti, 2015). Innovations of the period were largely connected to the increasingly strong links between the textile industry and chemistry. Various processes made fabric dyes more saturated and more reliable, but manufacturers were mainly interested in fibres. Cotton, linen, wool, and silk had long been the only available materials. The industry then sought to manufacture a synthetic fibre that would present a combination of low cost, strength, flexibility, and easy care.

Of the early synthetic fibres, nylon is particularly characteristic of the use of petrochemicals in the manufacture of clothing (Bide, Collier, Tortora, 2000). From parachutes to women's stockings, it seemed to be the answer to all needs. The formula for nylon was created in 1931 by the American chemist Wallace Carothers at DuPont, who named it '66'. Seven years later, Paul Schlack, a German chemist at IG Farben, developed another type of fibre, 'nylon 6'. Produced by British Nylon Inc, nylon heralded the era of synthetic fibres that would revolutionize the global textile industry. DuPont began commercial production in 1939, focussing mainly on the stockings officially presented at the San Francisco Exhibition. These immediately convinced thanks to their durability, look, and low cost. As Pap Ndiaye explains, during the Second World War, nylon was part of the war effort and was used in the production of parachutes, ponchos, tyres, and tents (Ndiaye, 2001). DuPont was able to meet the needs of the period thanks to the technical culture of a team of chemists competing within their workplace. After 1945, nylon was used to manufacture underwear for thousands of women. The 'miracle fibre' (Kadolph, Langford, 2002: 12) dominated until the 1980s, when it was finally replaced by polyester (Tortora, Marcketti, 2015).

Nylon's promises triggered the production of other fibres (Handley, 1999). For manufacturers, synthetics were the future (Tortora, Marcketti, 2015). The market was promising. Deprived of its rubber supply from Southeast Asia during the Second World War, the United States launched research programmes to find a synthetic substitute. Polymer, like nylon, was the key to its manufacture. The melt spinning process was traditional: the polymer was melted and spun. Nevertheless, the yarns either broke during the process or during their transformation into fabric. Manufacturers then created a bicomponent fibre: two types of polymers were melted and spun together, for example nylon and polyester. Microfibre came out of the combination of molecular science and textile industry. Each microfibre had a diameter of less than 0.00009 inches, about one-tenth the diameter of a silk fibre. Later, its diameter was reduced to less than 0.0025 millimetres. Modacrylics (1949), acrylics (1950), polyesters (1953), triacetate (1954), and spandex (1959) were crucial developments. Microfibre technology profoundly altered the textile industry. Due to their small size, microfibres were flexible and dense, easy to fold, and moisture resistant. These properties were particularly fitting for coats, sportswear, and underwear. But the manufacturing process was still complex and expensive (Hongu, Phillips, 1990).

With the introduction of new lifestyles, including an increase in the number of women working outside the home, the widespread use of public transport, and easier travel, consumers and manufacturers became increasingly preoccupied with fabric care. Mixing natural fibres, like wool or cotton, with polyester increased the fabric's resistance to creasing and staining, and made washing them easier. The significant growth in synthetic fibre production even forced the US Congress to pass a law on textile labelling to indicate the percentage of natural and synthetic fibres for the benefit of consumers, who could no longer identify the fabric (Tortora, Marcketti, 2015).

Social changes like the practice of sports — first by members of the wealthy classses, then in international competitions, and finally at schools — required the development of increasingly complex technologies. Watkins explains that technology must serve sports performance (Watkins, 2010). By taking on the manufacture of sports shoes, ski clothing, and swimwear, the textile industry contributed to the attractiveness of sports competitions and its economy (Tortora, 2015). In Europe and North America, competitive swimming started in 1883 at the first world championship, which was held in New York (McComb, 2004). In 1905, Australian swimmer Annette Kellerman wore a one-piece jersey that she coated with lard to make sliding easier. Manufacturers then looked for suitable textiles to increase the speed of swimmers. Produced by Dunlop, latex, a rubber yarn spun with cotton, silk, wool or nylon, covered the jerseys and then the corsets. In 1959, elastomeric fibres — called spandex in the United States and elastane in Europe — replaced nylon because they had the quality of stretching from 500% to 600% without breaking and recovering their shape after stretching. As for spandex, it is recommended for its resistance to weathering due to sunlight and chlorine (Collier, Bide, Tortora, 2009). These innovations profoundly changed the construction of swimwear (Poli, 1995; Martin, Koda, 1990).

The textile industry and sports joined forces to foster the visibility of international competitions and the development of a specific sports economy, within the context of globalization. Some manufacturers entirely dedicated themselves to this lucrative business. Fabrics had to become 'smart'. The sports coverage in the media and the search for athlete performance were essential components of the technological change in the textile sector. First applied to the sports sector alone, innovations then spread to everyday clothing. Throughout the second half of the twentieth century, manufacturers invested in advanced techniques and sought to reduce the prices of finished products by lowering labour costs.

From Intelligent Textiles to Empowering the Textile Industry

The so-called 'high-tech' fabrics are the result of scientific and technical advances. The latest textiles derive from the fields of chemistry, computer science, aerospace engineering, and the automotive industry. In constant connection with the fashion industry, the textile industry projects an image of rapid change and prospective reflection on production methods, on materials used and their integration into the environment. As technology became part of everyday life, including the space race in the early 1960s, it increased its influence on the manufacture of fabrics. Some of the textiles used on catwalk in the 1970s had been created in laboratory experiments. Production methods, techniques used for garment sizing, and the development of new materials influenced the clothing industry. Consumers' daily lives were made easier. But this was not the case for textile workers in Asia or Europe whose daily realities constitute the dark side of the textile industry.

High-tech fibres, textiles, and finishes are designed for a specific function or appearance. Specialists believe that 'the greatest potential for the future lies in experimental fashion using advanced synthetics to promote new aesthetics and methods of clothing construction' (Braddock, O'Mahony, 1999b: 12). Most synthetic products imitate natural fibres, but they are required to be stronger and more durable even when they are light, transparent, or elastic. 'Techno-nature' textiles, a mixture of natural fibres and synthetic materials, combine appearance and performance (Braddock, O'Mahony, 1999a). While techno-materials emerged alongside traditional manufacturing methods such as weaving and knitting, experimentation with various finishing techniques for non-woven fabrics introduced new aesthetic options. This was the case for Tyvek, a non-woven fabric made of high-density polyethylene (HDPE) fibres with a thickness of between 0.5 and 10 μm, which was developed by DuPont de

Nemours. After extrusion, the fibres are randomly and non-directionally arranged and then consolidated under pressure with a thermal process. Tyveck was first marketed in the form of sheets or rolls of various sizes, both non-toxic and recyclable, in the United States, Luxembourg, and Malaysia in 1967. In 1999, fashion designer Hussein Chalayan seized upon the non-woven fabric to make dresses. Nevertheless, most research concerns fabric finishing (Quinn, 2002). Fabrics can be covered with microencapsulated substances, such as vitamins, perfumes, insect repellents, or bacteriostats. When the small capsules burst, the substance is released onto the skin. Outlast® phase change technology, developed by NASA, has been used by manufacturers to adapt fabrics to temperature changes. Outlast® is a highly innovative technological fabric: it absorbs and distributes excess heat before storing it. When the environment cools, the fabric solidifies and transmits the stored heat to the wearer. Techno-materials are therefore intended to improve the wearer's well-being and hygiene on a daily basis. They are developed in close collaboration with applied research laboratories (Braddock, O'Mahony, 1999b). A finish radically transforms the visual and tactile qualities of a fabric as well as its performance, including resistance to staining and wind. Thermoplastic fibres can also be hot moulded to create permanent three-dimensional surfaces (Quinn, 2002). The sector's competitiveness is confirmed by its rapid integration into the mass market and the creation of laboratories dedicated to textile research and technologies.

Techniques developed for competitive sports have been rapidly integrated into the traditional clothing industry, while an increasingly connected global population has adopted a constantly evolving electronic lifestyle (Watkins, 1995). Laboratories fully dedicated to this type of research today include Media, Starlab, Charmed Technology, and International Fashion Machines at the Massachusetts Institute of Technology. The objective of these groups is to develop prototypes of portable electronic products and to explore the necessary synergy between IT, textiles, healthcare, and military defence. Innovations must be marketable and user friendly. The industry accompanies the social and economic changes that have also transformed it. Yet it often has a bad reputation.

In international sports competitions, some fabrics have been accused of promoting 'technological doping' by improving buoyancy. These include the Fastskin swimsuit by Speedo, which was developed with the help of research into sharks' swimming movements and skin structure, especially those species known for their speed. Although athletes wore these for the 2000 Olympic Games, FINA (the International Federation of Amateur Swimming) banned the fabric ten years later. Technical advances are so rapid that swimmers are required to present their swimsuits to the appropriate authorities twelve months before the competition. FINA's list of prohibited models is updated regularly. The performance culture that has underpinned contemporary society since the nineteenth century is being reassessed.

Many designers criticize the excessive power of technology in the clothing industry. The strength of the Japanese textile industry lies in the very combination of new technologies with traditional crafts such as pleating, shibori, and reserve dyeing (Iwamoto, Rice, Barton, 1983). Rei Kawakubo, Issey Miyake, and Yohji Yamamoto were the leaders in this trend that brought together the creative textile industry and the fashion sector. They called for a more balanced perspective of tradition and modernity. While advocating a return to working by hand and using less sophisticated materials, they actually criticise an entire economic and social system based on the race for technical innovations. Critics prefer imperfection and individuality to the uniformity created by the rapid evolution of technology.

The textile and clothing industries' lack of interest in environmental and social issues is regularly reported in the media (Braddock, O'Mahony, 2002). There are many debates concerning 'green fibres'. This is the case for tencel made from wood chips, whose production creates more pollution than its 'equivalent' viscose, an artificial fibre derived from cellulose. Manufacturers of bamboo fabrics, which is advertised as a highly environmentally-friendly natural material, were called out by the Federal Trade Commission (FTC) and fined for violating the Product Identification Act that had been adopted by Congress in 1960. Bamboo is a plant that consumes a lot of water, and the 'bamboo' label is considered as misleading as an advertisement that manipulates environmentally-concerned customers. Animal advocates, who are increasingly influential with the clothing industry, encourage scientific research to find alternatives to fur, crocodile skin, coral, and ivory (Renbourn, Rees, 1972).

Finally, the human cost of most of the world's textile and clothing industries is largely underestimated or hidden. I will emphasize three points here. First, technical advances and mechanization have not served to improve the lives of workers. Brand conglomerates relocated their production to the poorest countries in the world, where workers do not even receive a living wage. Their daily lives are punctuated with industrial disasters; in 2013, the Rana Plaza accident in Dhaka, Bangladesh, which caused 1100 deaths, echoed the deadly fire at the New York shirt factory a century earlier (Hoskins, 2014). Southeast Asia and, more recently, sub-Saharan Africa have become symbols of the modern slavery that feeds both the clothing economy and consumer desire. Simultaneously, global commodity chains have increased global subcontracting, for instance for sports shoes or in the clothing industry in China, where large companies produce for Western clients (Guiheux, 2019). Second, there are health disasters happening today: infertility, cancers, and autism are on the rise in countries producing textiles and clothing (Adams et al., 2012; Hoskins, 2014; Nanda et al., 2016: 44). Pesticides, dyes, and the lack of worker protection are responsible for the ongoing poisoning of workers (Masse-Stamberger, 2014). Third and finally, far from witnessing a shift in health disasters from West to East, we can see them spread across the globe. Indeed, the industrialized countries are also affected. From the female retailer who became sterile after contracting chromium poisoning from clothing to workers poisoned by phosphine in port warehouses and to the consumer suffering burns after being in contact with DFMU, an anti-mould chemical placed in shoe boxes in a small white bag, the list of victims of the textile industry continues to grow (Greenpeace, 2011; Hoskins, 2014; Struillou, 2015). Using techniques oriented towards economic performance and not improving working conditions or comsumer safety, the textile industry has become a veritable poison that feeds a regime of global inequalities.

In conclusion, the growth of the textile industry since the nineteenth century has made it possible to make clothing that is more comfortable, adapted to different types of activities, and cheaper for the consumer. However, techniques have changed the way in which fabrics are conceived, and they now need to anticipate consumer desires. The misleading nature of needs and the growing demand, the well-established discourse of companies' communication units and the difficulties of traceability in the manufacturing chain conceal the dangers faced by workers and consumers. Chemical advances, technical innovations, and mechanization regularly impede social welfare. Behind innovation, progress, and industrial organizations, it is the economic performance of business conglomerates that influences the choice of location for work and the choice of components. In short, the ever-growing consumption of fabrics has created the illusion of democratization in the sector. The quest for low prices that has driven technical research since the nineteenth century, which has been praised by advertising agencies, has been to the detriment of the worker and the consumer. Additionally, the transformations of the textile industry have made it the second most polluting sector after the oil industry. Finally, by relying on the least developed countries, both industrially and in terms of social protection, companies bypass human rights by introducing modern slavery in plain sight in order to feed hyperconsumerism (Lipovetsky, 2017).

References

Adams, James B., Audhya, Tapan, McDonough-Means, Sharon, Rubin, Robert A., Quig, David, Geis, Elizabeth, Gehn, Eva, Loresto, Melissa, Mitchell, Jessica, Atwood, Sharon, Barnhouse, Suzanne, Lee, Wondra, 'Toxicological Status of Children with Autism vs. Neurotypical Children and the Association with Autism Severity', *Biological Trace Element Research*, 151 (2012), 171–80.

Anguelov, Nikolay, *The Dirty Side of the Garment Industry: Fast Fashion and Its Negative Impact on Environment and Society* (New York, NY: CRC Press, 2015).

Bide, Martin, Billie J. Collier, and Phyllis G. Tortora, *Understanding Textiles* (New York, NY: Macmillan, 2000).

Birnbaum, David, *Birnbaum's Global Guide to Winning the Great Garnment War* (Hong Kong: Third Horizon Press, 2005).

Braddock, Sarah E. and Mary O'Mahony, *Sportstech: Revolutionary Fabrics, Fashion and Design* (London: Thames and Hudson, 2002).

——— and ———, *Techno Textiles: Revolutionary Fabrics for Fashion and Design* (London: Thames and Hudson, 1999a).

——— and ———, *Techno Textiles 2: Revolutionary Fabrics for Fashion and Design* (London: Thames and Hudson, 1999b).

Breward, Christopher, *The Culture of Fashion* (Manchester: Manchester University Press, 1995).

Cipolla, Carlo M. and Derek Birdsall, *The Technology of Man: A Visual History* (London: Wildwood House, 1979).

Collier, Billie J., Martin Bide, and Phyllis G.Tortora, *Understanding Textiles* (Upper Saddle River, N. J.: Pearson Education, 2009)

Cordillot, Michel, ed., *La Révolution de la machine à coudre*, special issue, *Les Cahiers d'Adiamos 89*, 16 (2018).

Coye, Dale, 'The Sneakers/Tennis Shoes Boundary', *American Speech*, 61 (1986), 366–69.

Engels, Friedrich, *The Condition of the Working-Class in England in 1844* (New York, NY: Cosimo Classics, 2008 [1887] transl. by Florence K. Wischnewetzky).

Friedel, Robert, *Zipper: an Exploration in Novelty* (New York, NY: W. W. Norton, 1994).

Godfrey, Frank P., *An International History of the Sewing Machine* (London: Robert Hale, 1982).

Greenpeace, *Dirty Laundry 2: Hung Out to Dry* (August, 2011).

Guiheux, Gilles, 'La Sous-traitance en Chine contemporaine: des chaines de valeurs globales aux réseaux d'entreprises familiales. Subordination et collaboration', *Revue de Synthèse*, special issue, *Pratiques du travail au forfait. Métiers, techniques et sous-traitance dans une perspective euro-asiatique, XVIIIe–XXIe siècle*, ed. by Manuela Martini, Liliane Hilaire-Pérez, and Giorgio Riello, 140 (2019), 239–57.

Handley, Susannah, *Nylon: the Manmade Fashion Revolution. A Celebration of Design from Art Silk to Nylon and Thinking Fibres* (London: Bloomsbury, 1999).

Heard, Neal, *Sneakers: Over 300 Classics from Rare Vintage to the Latest Designs* (London: Carlton Books, 2003).

Hendrickson, Robert, *Facts on File Encyclopedia of Word and Phrase Origins* (New York, NY: Facts on File, 2000).

Hongu, Tatsuya and Glyn O. Phillips, *New Fibres* (Chichester: Ellis Horwood, 1990).

Hoskins, Tansy E., *Stitched Up: the Anti-capitalist Book of Fashion* (London: Pluto Press, 2014).

Iwamoto, Wada Yoshiko, Mary Kellogg Rice, and Jane Barton, *Shibori: the Inventive Art of Japanese Shaped Resist Dyeing* (Tokyo: Kodansha International, 1983).

Jarrige, François, *Techno-critiques. Du refus des machines à la contestation des technosciences* (Paris: La Découverte, 2014).

Jimenez, Guillermo C. and Barbara Kolsun, *Fashion Law: a Guide for Designers, Fashion Executives and Attorneys* (New York, NY: Bloomsbury, 2014).

Kadolph, Sara J. and Anna L. Langford, *Textiles* (NewYork, NY: Prentice-Hall, 2002).

Keers, Paul, *A Gentleman's Wardrobe* (London: Weidenfield and Nicolson, 1987).

Landes, David S., *The Unbound Prometheus: Technological Change and Industrial Development in Western Europe from 1750 to the Present* (Cambridge and New York, NY: Press Syndicate of the University of Cambridge, 1969).

Lipovetsky, Gilles, *Le Bonheur paradoxal: essai sur la société d'hyperconsommation* (Paris: Gallimard, 2006).

———, *Plaire et toucher. Essai sur la société de séduction* (Paris: Gallimard, 2017).

Martin, Richard and Harold Koda, *Splash! A History of Swimwear* (New York, NY: Rizzoli International, 1990).

Masse-Stamberger, Benjamin, 'Bangladesh: les forçats du cuir de Hazaribagh', *L'Express*, 03 08 2014 <https://lexpansion.lexpress.fr/actualite-economique/bangladesh-les-forcats-du-cuir-de-hazaribagh_1512039.html> [accessed 03 04 2023].

McComb, David, *Sports: an Illustrated History* (New York, NY: 2004).

Muthu, Subramanian Senthilkannan, ed., *Sustainability in Denim* (Cambridge: Woodhead Publishing, 2017).

Nanda, Shikha, Kumar, Manoj, Kumar, Ashok, Behar, Sunny, Nanda, Sawal, 'Malwa Region, the Focal Point of Cancer Cases in Punjab: a Review Study', *International Journal of Current Research in Multidisciplinary*, 1.3 (2016), 41–45.

Ndiaye, Pap, *Du nylon et des bombes: Du Pont de Nemours, le marché et l'état américain, 1900–1970* (Paris: Belin, 2001).

Parthasarathi, Prasannan and Giorgio Riello, *The Spinning World: a Global History of Cotton Textiles, 1200–1850* (Oxford: Oxford University Press, 2009).

Poli, Doretta, *Beachwear and Bathing-costume* (Modena: Zanfi Editori, 1995).

Pomeranz, Kenneth, *The Great Divergence: China, Europe, and the Making of the Modern World Economy* (Princeton, NJ: Princeton University Press, 2000).

Probert, Christina, *Swimwear in Vogue Since 1910* (New York, NY: Abbeville Press, 1981).

Quinn, Bradley, *Techno Fashion* (Oxford and New York, NY: Berg, 2002).

Renbourn, Edward Tobias and William Howard Rees, *Materials and Clothing in Health and Disease* (London: H. K. Lewis and Company, 1972).

Riello, Giorgio, *Cotton: the Fabric that Made the Modern World* (Cambridge: Cambridge University Press, 2013).

Rivoli, Pietra, *The Travels of a T-Shirt in the Global Economy* (Hoboken, NJ: John Wiley & Sons, 2005).

Tenner, Edward, 'Lasting Impressions: An Ancient Craft's Surprising Legacy in Harvard's Museums and Laboratories', *Harvard Magazine*, 103 (2000), 37.

Tortora, Phyllis G., *Dress, Fashion and Technology: from Prehistory to the Present* (New York, NY: Bloomsbury, 2015).

——— and Sara B. Marcketti, *Survey of Historic Costume* (New York, NY: Bloomsbury, 2015).

Struillou, Yves, 'Circulaire N° DGT/CT2/2015/160 du 7 mai 2015 relative à la prévention et à la protection des travailleurs contre les risques chimiques dans les conteneurs et autres contenants de marchandises' (2015) <http://circulaire.legifrance.gouv.fr/pdf/2015/05/cir_39649.pdf> [accessed 03 04 2023].

Vanderbilt, Tom, *The Sneaker Book: Anatomy of an Industry and an Icon* (New York, NY: The New Press, 1998).

Verley, Patrick, *L'Échelle du monde. Essai sur l'industrialisation de l'Occident* (Paris: Gallimard, 1997a).

———, *La Révolution industrielle* (Paris: Gallimard, 1997b).

Watkins, Susan M., *Clothing: the Portable Environment* (Ames, IA: Iowa State University Press, 1995).

———, 'Functional Dress', in *Berg Encyclopedia of World Dress and Fashion*, ed. by Roach Mary Ellen and Joanne Eicher (Oxford: Berg, 2010: 117).

Watson, James, *Textiles and the Environment* (London: Business International, 1991).

PART 3

Connections

JÉRÔME LAMY

Introduction

No technique emerges, is transmitted or deployed in isolation. Techniques are the combined results of practices, political and economic circumstances, cultural choices, material opportunities, and converging social forces. This interdependency between contingencies and structures results in a multitude of intersections, overlaps, and connections. This section therefore intends to explore the interconnections of techniques and the social fabric. We want to look at these issues on a global scale as practices, uses, objects, and methods in a variety of fields of human activity circulate. In each major field of practice, different types of techniques have emerged from encounters between specific frameworks and the possibilities offered by the wider technological context.

The chapters follow the twists and turns of how a range of techniques variously connected to social structures along the nineteenth and twentieth centuries. In the process, the section explores some of the key global transformations witnessed in these two centuries.

Liliane Hilaire-Pérez's historiographical chapter unpacks the various meanings of the term 'technology'. Initially conceived of in the early modern period and eighteenth century as including all practices in all their diversity, the nineteenth century concept of technology, particularly in England, became centred on machines. Later a more procedural rationale prevailed where technology as understood not only as machinery but as encompassing all the processes and practices necessary to implement them. In addition to outlining conflicting definitions of the term, Hilaire-Pérez draws attention to the lively, mostly French, research tradition that has considered technology as part of the humanities. This position is both a methodological choice (to capture all forms of human practice) and a heuristic one (to avoid considering technology as an end in itself). This conceptual discussion allows us to foreground 'technique' (rather than 'technology') at the heart of different types of human activity, while the following chapters investigate the main technical developments witnessed in the nineteenth and twentieth centuries.

The laboratory, which has been a special place for experimentation since the beginning of the early modern period, is defined and shaped by various structures and conventions — be they social, epistemological, or political, while being naturally also at the centre of technical questions. Joris Mercelis explores the ways in which laboratories as places specifically devoted to science have integrated technical components (notably through instrumentation) as well as produced new techniques (as, for example, the optical equipment that came out of Fraunhofer's laboratory).

Due to the rise of scientific practices involving a large number of researchers, engineers, and technicians, large-scale organizational procedures and a vast deployment of technologies, a truly Big Science emerged after the Cold War. Olof Hallonsten highlights the many transformations that have affected Big Science since the 1950s, including the development of nuclear and particle physics, the increasingly diverse uses for large particle accelerators, and the emergence of combined instrumental structures.

Spreading beyond laboratories, techniques come into contact with the regulatory frameworks of human activities, starting with that of the law. Legal formats are a good indicator of transformations; throughout the last two centuries, patents in particular have reflected the development of industrial capitalism as well as the evolution of national and international legislations. Alessandro Nuvolari and Caterina Sganga investigate how the different phases of the industrial revolution were shaped by laws. Starting from a consideration of the co-evolution of techniques and legal frameworks, they analyse the role of patents in the practice of invention during the First Industrial Revolution. Patents then formed part of a system that was not strictly codified by law. In contrast, the Second Industrial Revolution was characterized by the rise of a powerful industrial sector and patents became the norm, including in international regulations. Finally, the Third Industrial Revolution was marked by the emergence of standardization and its framing by law.

One of the reasons technology was so present socially in the nineteenth and twentieth centuries was because it was disseminated via specific educational practices. Stéphane Lembré shows that whatever the particular national context, there was a significant broadening of technological education in Europe — be it in terms of its audience, the areas it covered or the professions involved. He first insists on the increasing diversity of educational spaces with specialized schools emerging as early as the eighteenth century. Economic requirements (particularly in terms of the quality of the workforce), the control of workers, changes in the professional structure, as well as various political agendas, all contributed to recasting technological education.

The question of power and how it is supported by technology is central to understanding the history of armed conflict. Patrice Bret devotes a chapter to the relationship between war and technology. He maps the historiography of what has been a very rich field of research, especially since the 1980s. The relationship between war, power, and technology informs global developments, which are based particularly on innovation, secrecy, and cooperation.

Work was one of the areas most influenced by technological developments in the nineteenth and twentieth centuries, as Patrick Fridenson shows. Modern management developed internationally out of the contributions of different types of actors. Frederick W. Taylor's ideas and methods were initially used on production lines. Aiming to apply science to workshops and, later, offices, Taylor's methods were among the most efficient ways of using skilled workers' know-how and stimulating semiskilled workers' and labourers' productivity. Fridenson demonstrates that they enabled engineers and technicians, who were gathered in special offices, to identify the best worker, the best tool, and the best way to do each task. They were also part of methods that came from the gun and ammunition sector. Thus, the goal of Taylor's methods was to standardize preparation and micromanage work processes while individualizing wages. From America, Taylorism spread to Europe and Asia thanks to consultants, international conventions, and national institutions. It was also introduced to the Soviet Union by communist leaders and activists, and then implemented more rigidly in Communist China. Taylorism differs from Fordism and Toyotism, although it is at the base of both. Although Taylorism has been at the centre of criticism ever since its introduction, it has proved remarkably resilient. Despite renewed criticism in the 1970s, it has survived to this day with automation, its focus being always to extract knowledge about work processes in order to control the cost of labour and administration. However, even automation cannot eradicate the resurgence of autonomy among the workforce.

For the past two centuries, techniques have been at the centre of many controversies as François Jarrige points out. Different moments can be stressed: first, the rejection of the cult of machinery at the beginning of the nineteenth century; then, with industrialization, the easing off of tensions. Later, in the twentieth century, came a growing critique of technology in the wake of the condemnation of the consumer society; and more recently, technoscepticism motivated by environmental considerations.

One of the reasons technology has been criticized so harshly is because it is perceived as being conducive to capitalist and destructive changes in living conditions. This confrontation between technology and the future of humankind is at the heart of technological transformations affecting the body, as Cecilia Calheiros points out in her chapter. In the ninetenth century, hygienist policies and germ theory paved the way for biopolitics and its articulation of social anxieties and industrial imperatives. At the same time, eugenics drew on the racist ideological matrix to implement so called regenerative technologies. In the twentieth century, self fashioning techniques as well as sport and fitness manifest the emergence of new demands on individuals. Overcoming biological boundaries prepared the political and ideological sphere for biomedicalization and transhumanism.

The entangled dynamics linking techniques and the body are also central to gender, as Virginie Julliard shows. She identifies three moments in the history of the intersections between gender and technology: the 1970s saw a critique of the underrepresentation of women in the literature; the 1980s corresponded to the rising importance granted to users of technology and the emergence of the subversive figure of the cyborg; and finally, the 1990s and 2000s, which aimed at the de-reification of the body and the gendered distribution of technological practices (and in particular digital practices). Julliard concludes by thinking about feminist technology as a means to redefine the tenets of patriarchal culture.

Martine Beugnet and Mathieu Duplay explore the foundations of the techniques of audiovisual culture. The reproduction of sound and moving images was the result of social, cultural, political but also sensual and aesthetic transformations. Throughout the nineteenth and twentieth centuries, immersive technologies which represented technological feats were often driven by aesthetics.

In the nineteenth and twentieth centuries, technology started penetrating every aspect of the everyday. Gil Bartholeyns and Manuel Charpy analyse the industrial, commercial, and economic changes that influenced the widespread circulation of technological objects. The objects that accessorize contemporary consumption, from plastic to serial, standardised consumer goods, are also technological objects of sorts.

Soraya Boudia and Nathalie Jas attempt writing a global history of the links between technology and pollution. They note parallels in the development of industry and the spreading of pollution over two centuries. Deconstructing this process, Boudia and Jas analyse the various ways pollution has been managed in history. Depending on time periods and contexts, a variety of resources and tools are deployed from control and monitoring to regulations and law.

If technologies impact the environment and contribute to its destruction, they also sometimes involve recycling and repair. Thierry Bonnot and Michel Letté identify a multiplicity of cycles in the technical treatment of objects and materials. They show, once again, that moral and ethical considerations are part and parcel of practices of recycling and reuse.

The cross-cutting issues addressed in this section of the book aim to help readers grasp the complex issues in play in the globalization of techniques. This is not a question of choosing top-down or bottom-up approaches, but of considering techniques as the combined result of practices, objects, innovations, regulations, cultural tastes, economic choices, and political orientations.

LILIANE HILAIRE-PÉREZ

Technology as a Human Science

In an article published in 1964 entitled 'La Technologie, science humaine', the ethnologist André-Georges Haudricourt offered a definition of the science of techniques, or rather of the science of human activities that has today been all but forgotten (Haudricourt, 1964). Trained as an agronomist, Haudricourt explained that he had naturally turned to linguistics and technology as a social science. Echoing Marcel Mauss's definition of technique as 'traditional effective action', Haudricourt defines techniques as efficient means of action shaped by shared experience (Mauss, 1973, 75). For Haudricourt, each discipline being defined by a specific point of view taken on phenonmena, technology can be seen as 'the humanities' perspective on material objects' (Haudricourt, 1964, 28; Haudricourt, 1987, 37).[1] Haudricourt uses the example of a table: for mathematicians a table is considered as a surface and shape in space; for physicists it's defined by its mass and its density; biologists may want to study the age of the tree and the tree species; and 'from the point of view of the humanities, it is the origin and function of the table for human societies that is being studied'. This in essence meant understanding 'technology as a human science'. He concludes: '[i]t is clear that for a manufactured object it is the human point of view, its manufacture and use by humankind that are essential, and that if technology is to be a science, it is as a science of human activities'.

In 1987, François Sigaut edited a series of essays by Haudricourt in a book entitled *La Technologie, science humaine*, among them the 1964 essay, with a long introduction titled 'Préface. Haudricourt et la technologie', in which he proposed to historicize the concept of technology as a human science (Haudricourt, 1987). Sigaut, who was also an agronomist working on techniques as an ethnologist, was a staunch supporter of this forgotten meaning of technology at a time when '*technologie*' had become 'some kind of scientific, or pedantic superlative' for technique (Sigaut, 1994, 52). For Sigaut, '*technologie*' had been derived from the English to refer to science-based industrial applications, and could be seen as sort of paradigm of the myth of progress and modernity, which we might also call 'techno-sciences' today. In his introduction

1 Unless otherwise stated, all quotations from non-English sources have been translated into English by the author of the present article.

Global History of Techniques, ed. by Guillaume Carnino, Liliane Hilaire-Pérez and Jérôme Lamy, Global Matters Culture Matérielles, histoire des techniques et circulation globale des savoirs, 9 (Turnhout: Brepols, 2024), pp. 501–516
BREPOLS PUBLISHERS
10.1484/M.TECHNE-EB.5.129826

to Haudricourt's essays, Sigaut recalled the long-term history of the concept of technology since Antiquity and highlighted its polysemy.

True to an inclusive and universalist ideal, 'technology' as the science of techniques encompassed the diversity of techniques. This vision was present in pre-modern encyclopaedias just as it is in the anthropological and ethnolograghical studies promoted by Mauss, who thought of techniques as manifestations of the kaleidoscopic variations that made various societies (Sibeud, 2004; Schlanger, 1991). 'Do not neglect any detail [...]. Not only must everything be described, but an in-depth analysis must be carried out, this is what will mark out a good observer, where their particular sociological talent will be revealed' (Mauss, 1947, 7).

The other meaning of 'technology' as a science of the arts was more abstract in nature. Over the centuries, especially since the Renaissance, and more clearly since the eighteenth century, a scientific approach to technology had been constructed in various publications. Treatises entitled '*réductions en art*' (reductions into art), which aimed to rationalize the diversity of practices into principles of action (Dubourg-Glatigny, Vérin, 2008), paved the way for attempts at conceptualizing various trades according to their operative principles. A good example of this attempt at a general technology was *Entwurf der algemeinen Technologie* written by the cameralist Johann Beckmann in 1806 (Beckmann, 1806; Carnino, Hilaire-Pérez, Hoock, 2017). Sigaut explains that this fundamental text on technology was brought out of oblivion by Jacques Guillerme and Jan Sebestik in their seminal 1968 essay 'Les Commencements de la technologie' (Guillerme, Sebestik, 2007). Guillerme and Sebestik wrote a history of the short-lived attempt at a general, theoretical technology. These treatises were only published until the mid-nineteenth century, and in all languages, 'technology' henceforth referred to science-based industrial techniques, but no longer to the social science of techniques, the word thus shifted from referring to the *study* of techniques to the techniques themselves (Schatzberg, 2006, 2018a). In the process it lost its epistemological anchorage in social sciences to become aligned with applied sciences. However, while the conceptual meaning of the term 'technology' was challenged, as a science of operative principles, the importance of describing and collecting techniques in their diversity continued throughout the nineteenth century, for intance in technical museums. This may explain why Mauss advocated for 'technology as a human science' in and after the inter-war period (Mauss, 1948), and why the idea of technology as human science spread more widely in a context of innovation in the social sciences.

The Ambition of a General Technology of the Arts

In the early modern period, the growing interest in techniques by various countries, cities, and local administrations for the purposes of war, prestige, or (increasingly) economic growth and socio-economic reforms was matched by an enormous effort to collect and codify techniques. For instance, the *Description des arts et metiers* supervised by the Académie des sciences was part of a growing practice of state-funded enterprises carried out in France aiming at recording technological information; among these, the '*enquête du Régent*' of 1718 to 1719, a vast survey of natural resources and their use, set up a pattern for turning technology into statecraft (Demeulenaere-Douyère, Sturdy, 2008). At the same time, a sustained governmental attempt to disseminate printed technical information was underway, echoing a growing public interest in technological knowledge in print.

In the rest of Europe, the growing dissemination of technological information in print, its connection with state policies and learned circles, and the international circulation of these works underpinned the emergence of a new speciality, that of '*technologie*'. This new discipline was connected to cameralists in the German states and to liberal reformers in France who made efforts to produce inventories of their countries' riches as a first step towards economic improvement (Meyer, Popplow, Bayerl, 2006; Garner, 2005). At this point, the different arts were subsumed under the general term 'technology', effecting cross-classification of all human activities. The term '*technique*' (or its equivalent in other Euroepan languages) was not in use at the time except as an exception, and an innovative way, Christian Wolff explained in 1728: '[a]lthough it has been neglected so far, a philosophy of the arts is also possible. You can call it "Technique" or "Technology" ('*Eam Technicam aut Technologicam appelare posses*') (Wolff, 1728: Chapter 3, Paragraph 71; Camolezi, 2021). The Wolffian idea of a philosophy of the arts as a conceptualization of the rules governing the use of tools and instruments flourished one generation later (Guillerme, Sebestik, 2007: 63). In his 1806 treatise, Johann Beckmann explained that 'a wide range of trades — whatever material they use or the finished product — have to achieve work by following the same purpose: the work of a joiner with a plane, the polishing of glass, the calendaring of linen fabrics are activities that derive from the same operative purpose, that of surfacing bodies' (Beckmann, 1806, 464; Carnino, Hilaire-Pérez, Hoock, 2017, 17). '*Technologie*' was meant to reduce the diversity of practices into distinct operations that could be transferred from one trade to another and then, to encourage invention, be considered as an art of association and a synthetic and analogical activity of the mind. '*Technologie*', in France, quite differently from the German cameralist perspective, was praised as a tool for enlightening artisans, and during the French Revolution, it was intended to enable their emancipation from the guilds which supposedly prevented them from connecting techniques across trades (Mertens, 2002).

This corresponded to an era of 'civic technology'. Joost Mertens stressed the part played here by Louis-Sébastien Lenormand, the author of the *Dictionnaire technologique* (published beween 1822 and 1835 in twenty-two volumes) and other works, who was instrumental in the rise of technological magazines (Mertens, 2002). As a newly appointed professor of technology, Lenormand thought that the mission of '*technologie*' was to promote the education of the working classes: his goal was to encourage workers to compare and transpose processes between the trades so as to foster innovation. In his *Dictionnaire technologique*, Lenormand provided a dual system of references to techniques, which enabled the reader to compare and cross-reference different trades, either by analogy (printing, for instance, could be related to books, wallpapers, printed cotton, or earthenware), or via related trades (for example, watchmaking was connected to polishing, gilding, making springs, etc.). These classifications were rather different from the institutional division of the trades in guilds (although skills that carried across different trades and their relations to each other formed connections, especially in the subcontracting networks) (Hilaire-Pérez, 2013). Lenormand's classifications opened up what Hélène Vérin has called '*un espace de la technique*' (a technical space) (Vérin, 1993), i.e. a space independent from the technical knowledge of the guilds, trades, but also of academies. This may explain why Lenormand claimed that 'technology' as a concept had a civic virtue. This he connected to the Tennis Court Oath (*Serment du Jeu de Paume*) of 20 June 1789 that had transformed the Estates-General into the National Assembly, and led to the birth of the Nation and the desire for social harmony, leading to the abolition of the Ancien Régime and its hierarchical society ('*la société de corps et d'ordres*') (Mertens, 2004).

In England, research has focused on the history of machinery following Maxine Berg's work on industrial science and engineering, and on what Andrew Ure called 'the philosophy of manufacture' or 'industrial economy' (Berg, 1980; Ure, 1835, 1836). In a seminal essay, Mertens compared Lenormand's dictionary to Ure's *Dictionary of Arts, Manufactures, and Mines* (Ure, 1839) and found that, while Lenormand proposed an inclusive view of human activities with an emphasis on equipment and actors, Ure listed only processes, and that the purpose of his *Dictionary* was to describe the transformation of materials by mechanical or chemical means into 'general objects of exchangeable value' (Mertens, 2010, 281). 'Technology' thus became the science of the 'operative industry' (Ure, 1839, iii quoted in Mertens, 2010, 280), which was intended to lower the cost of production. It was this 'technology' that Karl Marx denounced in *Das Kapital*, this 'new modern science of technology' which fundamentally aims to break down production into distinct constituent movements and 'discovered the few main fundamental forms of motion, which, despite the diversity of the instruments used, are necessarily taken by every productive action of the human body; just as the science of mechanics sees in the most complicated machinery nothing but the continual repetition of the simple mechanical powers' (Marx, 1887, vol. 2, 492). This approach put an end to the ambition of a general technology of the arts and initiated a radical shift in the meaning of the term 'technology'.

From the Science of the Arts to the Science of Processes

In 1855, George Wilson, a Scottish chemist who was also the director of the Industrial Museum in Edinburgh, became the Regius Professor of Technology at the University of Edinburgh. In the same year he delivered a lecture entitled 'What is Technology?' (Wilson, 1855; Anderson, 1992). Wilson felt the need to explain his position. He began by recalling that '[t]he Title of the Chair, also, is as novel as its creation, and the term "Technology," by which it is distinguished, is so unfamiliar to English ears, and so inexpressive to English minds, that I must, at the very outset, explain what the branch of knowledge is which I am called upon to profess'. He then argued that the term 'technology' was better known on the Continent, and that it referred to a discipline which was taught in Germany, where this notion was associated with Beckmann's legacy. He stated that '[t]he word "Technology" literally signifies the Science of the Arts, or a Discourse or Dissertation of these'. Hence, '[i]ts object is not the Art itself, i.e., the *practice* of Art, but the principles which guide or underlie Art'. Technology was then, the science of the 'Industrial Arts' or the 'Theory of the Arts', and its goal was to study the principles by which 'the artist secures his ends' (Wilson, 1855, 6). The Edinburgh Chair of Technology was discontinued in 1859 with Wilson's premature death. The classes he had initiated at the Industrial Museum were suspended, the museum's acquisitions took a new direction towards the decorative arts, and the museum evolved into the Museum of Science and Art — today's National Museum of Scotland.

From the nineteenth century onwards, the meaning of 'technology' shifted from referring to the science of the arts to the industrial applications of science, and from the study and theory of techniques to technique itself, as industrial engineering developed alongside mechanization and the heavy chemical industry. Technology became the science of processes, both mechanical and chemical. It was no longer the science of purposeful action. In 1855, this change was yet to come. Wilson quoted Friedrich Ludwig Knapp's *Chemical Technology, or, Chemistry Applied to the Arts and Manufactures* (1848), in which 'technology' was understood as 'the systematic definition λσγσς (*logos*) of the rational principles upon which all processes employed in the arts τεχηες (*technes*) are based' (Knapp, 1848, 1), which differed from the broader meaning of the science of action. As applied sciences became more established, both 'useful knowledge' and the idea that the world of industrial arts was based upon efficient action became marginalized (Mokyr, 2002, 2).

In France, from the Revolution onwards, the 1791 foundation of the *brevet d'invention,* the French system of patents of invention governed by the natural law theory, was connected to the renewed importance of scientists in the production of technological knowledge. Jérôme Baudry has shown that this trend was echoed in the newly created *Conservatoire des arts et métiers* published *Description des machines et proceeds.* In the publication, which set out designing norms and specifications, increasing importance was given to formalism, excluding any mention of tasks, uses, and work contexts. This was at odds with the encyclopaedic, 'inclusive' type of technology that at this point still coexisted with it (Baudry, 2014, 2017; Mertens, 2002). This ambiguity was also expressed by Gérard-Joseph Christian, the director of the *Conservatoire des arts et métiers,* in his *Plan de technonomie* of 1819 (Christian, 1819). While the *Conservatoire des arts et métiers* was important for technology as the science of the arts, in the spirit of the encyclopaedic and humanistic definition of techniques (Corcy, Hilaire-Pérez, 2019), and while Christian still placed an emphasis on skills and product economy, he also aimed at a unified theory of industrial operations, and he embodied the shift in technological thinking towards machine theory. His classification of the trades was not intended to provide a rational description of the arts, but rather to analyse the possibilities of substituting manual operations with appropriate machinery (Hilaire-Pérez, 2013). Christian defined 'industrial work itself' as including all production (whether artisanal, or made in manufactories and factories) and all 'transformation of raw materials and preparation of various organic combinations' (Christian, 1819, 43). He thus aligned technology with a science of processes which was highly specialized, and no longer a universalist science of human activities.

'*Technologie*' as a discipline developed at the same time as production changed from artisanal to industrial (Guillerme, Sebestik, 2007). 'This was because technological invention and innovation no longer depended on systematic descriptions of arts and manufacturing, but were increasingly inspired by the physical and chemical sciences', Mertens explained in his essay devoted to Léon Lalanne (Mertens, 2011, 108; see also Barron, 2021). Lalanne was a polytechnician who, while he included the study of artisanal crafts and trades in his definition of technology, in 1840 considered technology as an 'introduction to industrial mechanics' and the 'science of processes by which man acts on the forces and raw materials provided by organic and inorganic nature, to appropriate these forces and materials to his needs' (Lalanne, 1840, 7; Mertens, 2011; Hilaire-Pérez, forthcoming). The codification of techniques gradually moved towards science as the latter became applied to industry: artisanal technical culture faded away before that of the industrial entrepreneur, whose role was well established in the legal, economic, scientific, and technical spheres.

Critics of machinery and industrial capitalism provided some alternatives (Jarrige, 2016). The organicist metaphor of a functional society among the Saint-Simonians and the early socialists, for example, was full of hope regarding the potentiality of machines. Mechanization did not necessarily involve a dehumanisation. Machines could enhance communication, increase the amount of information and interaction and even provide auto-regulation (Hilaire-Pérez, 2016). Underpinning this was the 'optimization of technical systems and networks' (Tresch, 2016, 91). The socialists' reflections on techniques laid the foundations for its integration into the emerging social sciences, building a socialist technology in its own right, i.e. 'an epistemology that recognized the continuity between human work and that of engines' (Tresch, 2016, 92). This development of machines was necessary to imagine the place of human beings in a technological society which, in socialist thought, was seen as a society of communication, information, and collaboration.

The transition was much more radical in the United States, as Eric Schatzberg has shown (Schatzberg, 2006, 2018a, 2018b). While several attempts were made in the nineteenth century to define technology as the science of the arts — such as with Jacob Bigelow's *Elements of Technology* in 1829, with the foundation of the Massachusetts Institute of Technology in 1860 (which may have been due to Bigelow's influence upon its founder William Barton Rogers), or with John Wesley Powell's initiative in the Anthropological Society of Washington in the 1890s — this definition of technology never toold hold in America. Franz Boas's 'attack on the dominant evolutionary approach in anthropology, and in particular on the link between the useful arts and the level of civilisation' led to the disappearance of technology as a science in anthropology and, more generally, in the social sciences (Schatzberg, 2018b, 208–09). Instead, a transfer occurred and the German term '*Technik*' subsumed

'*Technologie*' and evinced a new meaning, that of industrial technology (Schatzberg, 2006). The broker of this transfer was the sociologist Thorstein Veblen. Veblen was drawing from the criticism of industrial capitalism and the defence of workers' skills as materialized into machines. In Marxist theory this criticism emphasized the idea of a 'community's collective technological knowledge' (Schatzberg, 2006, 504). Veblen combined 'the older definition of technology as a field of knowledge with the German concept of *Technik* as a set of material practices', 'the state of industrial arts', hence 'moving its meaning away from a field of study and closer to material practices and craft knowledge', 'while remaining firmly independent of science'. Instead, Veblen's legacy in the social sciences in America led to the connection of technology with 'a deterministic concept of material progress' as well as with applied science, blurring the boundaries between science and technology. It was only in the late 1930s that technology became 'a significant category in the social sciences', but it never showed any ambition to study human activity: 'no major field of *technology* ever emerged in the English-language social sciences', and the situation was not much different in Germany at the beginning of the twentieth century (Schatzberg, 2018b, 4).

Nevertheless, this failure essentially concerned academic disciplines and scholarly publications by prominent members of the scientific community. New research tends to balance the disappearance of this meaning of technology by integrating other actors into the narrative and by investigating their impact on the renewed interest in technology as a science of human action in the twentieth century.

The Multiple Lives of Technology

Even if technology is commonly understood today as applied science, the older meaning of the term has persisted and sometimes resurfaces unexpectedly. In 1852, Augustus Henry Lane Fox Pitt-Rivers proposed a typology of the many objects he had brought back from his colonial travels and then gathered at Oxford in the museum that still bears his name today (Bontems, 2016; de Beaune, 2013). Facing the difficulty of trying to classify objects, he considered a morphological classification, aligned with biological evolution. The objects were divided into large classes corresponding to their functions and exhibited accordingly. The objects in each class were grouped together according to their family resemblance, ranging from the simple to the most complex (de Beaune, 2013). This was fundamental for the development of the concept of a long-term technical lineage. This notion was central to the French philosopher Gilbert Simondon, who analysed artefacts grouped in families constituted by similar operating principles (Guffroy, Bontems, 2018).

But some historians have warned against the construction of a conceptual history of technology that would suggest continuity where contexts are varied and meanings multiple (Loeve, Deldicque, 2018). One solution is to identify the contexts of the different meanings given to technology as the science of the arts.

Long-lasting interest in technology was closely connected, for instance, with the prominent place occupied by traditional techniques in industrial exhibitions, both in

Europe and beyond. While novelty, among other things, defines the modernity of products and knowledge, this was matched by a growing interest in retrospectives on 'archaic' objects that highlighted the progress of humanity to glorify Western achievements. The process of classifying objects via their methodical comparison underpinned this and underlined the rise of the human sciences. The 1867 inauguration of the museum of national antiquities (*Musée des antiquités nationales*) in Saint-Germain-en-Laye coincided with the exhibition of a gallery of the history of labour (*Galerie de l'histoire du travail*) on the Champ de Mars, whose catalogue mentioned the need to 'facilitate the comparison of the products of human labour at different times and among different peoples' to promote both 'the practice of the arts' and 'the study of their history' (Riviale, 2010, 68). The comparison of products and processes, the very cornerstone of technological science, was established as a method for the organization of human sciences into disciplines.

This context has been further explored by recent research. A doctoral thesis on François-Edmond Pâris's work at the Louvre Marine Museum at the end of the nineteenth century has shown how his visual discernment, his taste in drawings, and his penchant for comparison led him to reproduce the boats that he had observed during his distant travels. He then exhibited models in the museum, contributing to a nautical science that was supposed to 'gain knowledge from the study of distant nautical objects in time and in space' (Barron, 2019, 211). In the 1880s, following the foundation of the Trocadero Museum, Pâris's activity reached its peak, and he had the ambition to create a universal museum of techniques that would bring together traditional navigation, shipping in general, industry, and civil engineering. At the same time, the 1889 *Exposition rétrospective du travail et des sciences anthropologiques* (retrospective exhibition of work and the anthropological sciences) sought to retrace the evolution of tools across all human activities, which suggests that 'the retrospective Exhibition was based on an intellectual construction very different from the criteria adopted by industrialists, manufacturers and builders to organize their stands' (Corcy, 2014, 66). Skill and craftsmanship gained recognition. The recent study of railway collections at the *Conservatoire des arts et métiers* reveals that the institution's choices were based on a rationale of quality and incremental improvement, while emphasizing the cross-fertilization between sectors that helped innovation in railways. The eclecticism of the *Conservatoire* made it the ideal place for highlighting the interrelationships, convergences, and analogies that allowed a new sector to be constituted. While the *Conservatoire* had become an ardent defender of applied sciences since its 1819 training reform (Fox, 1992), the institution maintained — through its collections — an inclusive vision of techniques.

The connections remain to be investigated between 'theoretical technology' and the 'descriptive technology' of collectors, which informed ethnography in the twentieth century and the utopia of describing the entire breadth of 'lived and pragmatic techniques' as envisaged by Marcel Mauss (Schlanger, 1991, 118). Sigaut pointed out that 'theoretical technology' was a cul-de-sac since it tended to reduce itself to a kinematic study of mechanisms, i.e. an algebra of movement, through a mathematization that made it impossible to understand the intellectual and material resources involved in techniques. Through the collections, technology reconnected with the descriptive trend and gained a new role in the rise of social sciences, at least in the French school of technology; in England, by contrast, a divorce between anthropology and museums occurred, leading to the promotion of studies in material culture which did not necessarily connect with the social sciences of technology (Coupaye, Douny, 2009).

Academic courses in France began with Marcel Mauss in 1926 (Vatin, 2004). Mauss was related to Emile Durkheim and a student of Alfred Espinas (a former colleague of Durkheim), who taught social economy at the Sorbonne and founded praxeology as the study of efficiency in acting (Iribarren, 2021). Mauss developed the investigation of objects within their context, associated with tasks, and then, beyond the artefacts, the human value of techniques (Vatin, 2004). The impact on his students was enormous, among them Haudricourt, who stated that '[i]t could be said that the material civilization of a given human group was not so much the set of objects it manufactures or uses as the set of its traditional and technically effective muscular movements' (Haudricourt, 1987, 39). Another student was André Leroi-Gourhan. An ethnologist and paleoanthropologist in charge of collections, he insisted on the need for a scientific study of techniques, not only for the purpose of gathering knowledge about technical tasks, but also in order to understand the principles of actions that materialized in technical actions (Leroi-Gourhan, 1943, 2004). Leroi-Gourhan inspired the philosopher Simondon in his search for genetic principles to techniques 'it was no longer a question of classifying machines according to functional genera and species, but according to the invention that gave rise to the abstract technical idea which is reactivated in its successive materializations'(Bontems, 2016). Such connections between the philosophy of techniques and anthropology are specific to the French context (Collomb, 2021; Loeve, Guchet, Bensaude-Vincent, 2018; Schlanger, 2023).

This leads us to 'theoretical technology'. Although this dimension of technology may have been a cul-de-sac at the end of the nineteenth century, several attempts to establish theoretical technology beyond collections were made that illustrate the multiple paths taken by technology as a human science. One attempt was undertaken by Jacques Lafitte, an engineer and architect, who was in favour of a 'technomorphic' society as well as being the author of *Réflexions sur la science des machines* (Lafitte, 1932), in which he suggested the creation of a new science called 'mechanology' (Lafitte, 1932, 33–34). According to Lafitte, the science of machines, far from being the study of mechanical movements inside machines, aims to study machines as phenomena in their own rights whose genesis is located in society 'since the causes that determine the variations in the organization of machines can only be found in the slow variations of human activity, these causes that mechanology seeks, and whose ultimate goal is to determine this science, are of a social order' (Lafitte, 1932, 109; Carrozzini, 2009). In short, 'mechanology is a social science', that is 'the science of bodies organised by man'. Therefore Lafitte claims in the *Revue de synthèse* in 1933: '[t]here are few objects whose full knowledge presupposes, to such an extent, the simultaneous application of all scientific disciplines. The knowledge of techniques, and therefore of machines, presupposes an comprehensive synthesis of human knowledge and faculties' (cited in Le Roux, 2015, 558).

A member of the *Centre d'études cybernétiques* (Centre for Cybernetic Studies) in the 1950s, Lafitte shared the hopes of this new discipline for a 'general theory of machines' (Le Roux, 2015, 207). Developing after the Second World War, cybernetics fostered an analogical paradigm that drew from the feedback model of automatons and spread to engineering sciences, molecular biology, and the social sciences. The transfer of the concept of feedback was the foundation of the discipline's goals according to Norton Wiener (Wiener, 1948). This was the meaning assigned to it by Simondon, who knew and appreciated Wiener. For Simondon 'cybernetics' generally encompassed all categories of interscientific operating relationships. It could be considered the ideal place 'for the interdisciplinary development of feedback models' (Le Roux, 2015, 29). It was illustrative of the role that technique played in the interdisciplinary dynamics that developed between the 1950s and the 1970s but could not be integrated into the academic world.

While Lafitte as well as Wiener and Simondon put their hopes in techniques and their conceptual dimension through technology, other critical approaches developed. They were much more radical than the former socialist ones. German scholars understood technique 'as an application of abstract rationality and calculation' (Bontems, 2009, 47), opening up the possibility of the now well-known phenomenon of technophobia (Bontems, 2009). In France, although critical approaches also developed — for instance those of Jacques Ellul (Ellul, 1988) and Jean-Jacques Salomon (Salomon, 1992), who denounced technological discourses as rhetorical devices that legitimized technical imperialism — a long-term 'critical technophile' school of thought was established that extended from Jacques Lafitte and his hope for a 'liberating machininism' (Carrozzini, 2009, 29) to cyberneticians and Gilbert Simondon, and to 'technology as a human science' in the tradition of Mauss and Haudricourt. Although the meaning assigned to '*technologie*' by Simondon (i.e. that of the reticular and transductive dimension of techniques) and that which it assumes in Haudricourt (i.e. that of the science of techniques) do not overlap, they share similar beliefs in technological humanism.

It is necessary to stress the uniqueness of the French context. Technology as a human science initiated a relationship between technology and social sciences. In the 1930s, Leroi-Gourhan took part in compiling *L'Encyclopédie française*, a project launched by Lucien Febvre, the co-founder of the école des *Annales*, who planned an encyclopaedia that was neither organised alphabetically nor by discipline, but rather through transversal approaches. This was implemented in *L'outillage mental*, the first volume of Leroi-Gourhan's *Encyclopédie* (de Beaune, 2011). Apart from developing a classification of technical actions, Leroi-Gourhan introduced the idea of publishing data in folder form so that pages could be moved around, added or amended to reflect his own techniques of writing, registering, identifying, etc. This suggests a mutual connection between the attempt by Leroi-Gourhan to derive abstract principles of action and the cognitive means he used to do so. The *logos* of *techné* was mirrored in the *techné* of *logos*. It could be argued that the renewed interest for technology as a human science, and more broadly for the operative schemes at work in human activities, has had an impact on the social sciences in terms of allowing them to identify their own operative principles. In *Les Formes de l'expérience*, Bernard Lepetit, echoing Marc Bloch, does not hesitate to qualify history as 'a technique (a craft) based upon the manipulation (of archives, of series, of scales, of hypotheses...) and experimentation'. He adds that in history 'practice defines relevance' (Lepetit, 1995, 13). This mirroring between the 'cultural turn' in the history of technology and the 'pragmatic turn' in the social sciences, both converging in the identification of operative schemes, is one of the most recent outcomes of the efforts to conceptualize technology.

References

Anderson, Robert G. W., '"What is Technology?": Education through Museums in the Mid-nineteenth Century', *British Journal for the History of Science*, 35 (1992), 169–84.

Barron, Géraldine, *Edmond Pâris et l'art naval. Des pirogues aux cuirassés* (Toulouse: Presses Universitaires du Midi, 2019).

Barthélémy, Jean-Hugues, *Simondon* (Paris: Les Belles Lettres, 2014).

Baudry, Jérôme, 'Écrire et dessiner l'invention: les brevets et la technologie en France et aux États-Unis', in *La Technologie générale. Johann Beckmann Entwurf der algemeinen Technologie / Projet de technologie générale (1806)*, ed. by Guillaume Carnino, Liliane Hilaire-Pérez, and Jochen Hoock (Rennes: Presses universitaires de Rennes 2017), pp. 177–97.

———, 'Une histoire de la propriété intellectuelle. Les brevets d'invention en France, 1791–1844: acteurs, catégories, pratiques' (unpublished doctoral thesis, Paris, EHESS, 2014).

Beckmann, Johann, *Entwurf der algemeinen Technologie, Vorrath kleiner Anmerkungen über mancherley gelehrte Gegenstände*, Vol. 3 (Göttingen: Johann Friedrich Röwer, 1806).

Berg, Maxine, *The Machinery Question and the Making of Political Economy 1815–1848* (Cambridge: Cambridge University Press, 1980).

Bontems, Vincent, 'Actualité d'une philosophie des machines. Gilbert Simondon, les hadrons et les nanotechnologies', in *Les Machines: objets de connaissance*, ed. by Ronan Le Roux, Vincent Bontems, and Paul Braffort, special issue, *Revue de Synthèse*, 130.1 (2009), 37–66.

———, 'Le Progrès des lignées techniques', *Implications philosophiques. Espace de recherche et de diffusion* (2016) <http://www.Implications-philosophiques.Org/Actualite/Une/Le-progres-des-lignees-techniques/> [accessed 20 May 2020].

Camolezi, Marcos, 'Technique, Technics, Technik, substantifs: les mots à travers les dictionnaires du XIXe siècle', *Artefact. Techniques, histoire et sciences humaines*, 15 (2021), 61–106.

Carnino, Guillaume, Liliane Hilaire-Pérez, and Jochen Hoock, *La Technologie générale. Johann Beckmann Entwurf der algemeinen Technologie / Projet de technologie générale (1806)* (Rennes: Presses universitaires de Rennes, 2017).

Carrozzini, Giovanni, 'Gilbert Simondon et Jacques Lafitte: les deux discours de la "culture technique"', *Cahiers Simondon*, 1 (2009), 25–46.

Charpy, Manuel, 'Les "Techniques archaïques": produits d'un autre temps et produits artisanaux dans les Expositions universelles', in *Les Expositions universelles en France au XIXe siècle. Techniques, publics, patrimoines*, ed. by Anne-Laure Carré, Marie-Sophie Corcy, Christiane Demeulenaere-Douyère, and Liliane Hilaire-Pérez (Paris: CNRS Éditions, 2012), pp. 279–99.

Christian, Gérard Joseph, *Vues sur le système général des opérations industrielles, ou Plan de technonomie* (Paris: Mme Muzard & Mme Courcier, 1819).

Collomb, Cléo, 'Cartographier les réseaux d'une philosophie franco-italienne des techniques. Compte rendu de deux ouvrages collectifs', *Artefact. Techniques, histoire et sciences humaines*, 14 (2021), 467–75.

Corcy, Marie-Sophie, 'La Muséification des galeries du Conservatoire des arts et métiers: le cas de l'exposition rétrospective du travail et des sciences anthropologiques à l'Exposition universelle de 1889', in *Les Expositions universelles. Les identités à l'épreuve de la modernité*, ed. by Christiane Demeulenaere-Douyère and Liliane Hilaire-Pérez (Rennes: Presses universitaires de Rennes, 2014), pp. 59–76.

——— and Liliane Hilaire-Pérez, 'Le Premier catalogue des collections du Conservatoire des Arts et Métiers (1818). Jalon d'une pensée technologique' (unpublished paper, 2019).

Coupaye, Ludovic and Laurence Douny, 'Dans la trajectoire des choses', *Techniques & Culture*, 52–53 (2009), 12–39.

de Beaune, Sophie, 'La Genèse de la technologie comparée chez André Leroi-Gourhan. Introduction à son article "L'homme et la nature", paru dans *L'Encyclopédie française* en 1936', *Documents pour l'histoire des techniques*, 20 (2011), 197–223.

———, 'Le Musée comme lieu d'administration de la preuve. Genèse et destin de deux collections du XIXe siècle', *Gradhiva*, 18.2 (2013), 166–99.

Demeulenaere-Douyère, Christiane and David J. Sturdy, *L'Enquête du régent, 1716–1718. Sciences, techniques et politique dans la France préindustrielle* (Turnhout: Brepols, 2008).

Dubourg-Glatigny, Pascal and Hélène Vérin, *Réduire en art. La technologie de la Renaissance aux Lumières* (Paris: Éditions de la MSH, 2008).

Dufaux, Lionel, *L'Amphithéâtre, la galerie et le rail. Le Conservatoire des arts et métiers, ses collections et le chemin de fer au XIXe siècle* (Rennes: Presses universitaires de Rennes, 2017).

Ellul, Jacques, *Le Bluff technologique* (Paris: Hachette, 1988).

Espinas, Alfred, *Les Origines de la technologie: étude sociologique* (Paris: Alcan, 1897).

Fox, Robert, 'Un enseignement pour une nouvelle ère: le Conservatoire des arts et métiers, 1815–1830', *Cahiers d'histoire du CNAM*, 1 (1992), 75–92.

Garner, Guillaume, *État, économie, territoire en Allemagne. L'espace dans le caméralisme et l'économie politique, 1740–1820* (Paris: Éditions de l'EHESS, 2005).

Guffroy, Yohann and Vincent Bontems, 'La Mécanologie: une lignée technologique francophone?', *Artefact. Techniques, histoire et sciences humaines*, 8 (2018), 255–78.

Guillerme, Jacques and Jan Sebestik, 'Les Commencements de la technologie', *Thalès*, XII (1968), 1–72 (repr. in *Documents pour l'histoire des techniques*, 14 (2007), 49–122).

Haudricourt, André-Georges, 'La Technologie, science humaine', *La Pensée*, 115 (1964), 28–35.

———, *La Technologie, science humaine. Recherches d'histoire et d'ethnologie des techniques*, ed. by François Sigaut (Paris: Éditions de la MSH, 1987).

Hilaire-Pérez, Liliane, 'La "Philosophie de la technologie" de Léon Lalanne: micro-histoire d'un concept', in *Léon Lalanne. Un ingénieur entre la science, la politique et l'industrie au XIXe siècle*, ed. by Guillaume Carnino and François Jarrige (Dijon: Presses universitaires de Franche-Comté, forthcoming).

———, *La Pièce et le geste. Artisans, marchands et savoir technique à Londres au XVIIIe siècle* (Paris: Albin Michel, 2013).

———, 'Postface. "Sauver l'objet technique": une utopie socialiste?', in *Dompter Prométhée. Technologies et socialismes à l'âge romantique (1820–1870)*, ed. by François Jarrige (Besançon: Presses universitaires de Franche-Comté, 2016), pp. 251–74.

Iribarren, Leopoldo, 'La Technologie: paradigmes grecs d'une science sociale française' *Artefact. Techniques, histoire et sciences humaines*, 15 (2021), 175–200.

Jarrige, François, ed., *Dompter Prométhée. Technologies et socialismes à l'âge romantique (1820–1870)* (Besançon: Presses universitaires de Franche-Comté, 2016).

Knapp, Ludwig, *Chemical Technology, or, Chemistry Applied to the Arts and Manufactures* (London: Hyppolite Bailliere Publisher, 1848).

Lafitte, Jacques, *Réflexions sur la science des machines* (Paris: Bloud & Gay, 1932).

Lalanne, Léon, *Essai philosophique sur la technologie* (Paris: Bourgogne et Martinet, 1840).

Lepetit, Bernard, 'Histoire des pratiques, pratique de l'histoire', in *Les Formes de l'expérience. Une autre histoire sociale*, ed. by Bernard Lepetit (Paris: Albin Michel, 1995), pp. 9–22.

Leroi-Gourhan, André, *Le Geste et la parole*, vol. II, *La Mémoire et les rythmes* (1964) (Paris: Albin Michel 2004).

———, *L'Homme et la matière* (Paris: Albin Michel, 1943).

Le Roux, Ronan, *Histoire de la cybernétique en France (1948–1970)* (Paris: Garnier, 2015).

———, 'L'Impossible constitution d'une théorie générale des machines? La cybernétique dans la France des années 1950', in *Les Machines: objets de connaissance*, ed. by Ronan Le Roux, Vincent Bontems, and Paul Braffort, special issue, *Revue de Synthèse*, 130.1 (2009), 5–36.

———, 'Pierre Ducassé et la revue *Structure et évolution des techniques* (1948–1964)', *Documents pour l'histoire des techniques*, 20 (2011), 119–34.

Loeve, Sacha and Timothée Deldicque, 'Les Malentendus de la technologie', *Artefact. Techniques, histoire et sciences humaines*, 8 (2018), 213–54.

———, Xavier Guchet, and Bernadette Bensaude-Vincent, eds, *French Philosophy of Technology: Classical Readings and Contemporary Approaches* (Dordrecht: Springer, 2018).

Marx, Karl. *Capital: A Critical Analysis of Capitalist Production* (London: Sonneschein, 1887).

Mauss, Marcel, *Manuel d'ethnographie* (Paris: Payot, 1947).

———, 'Techniques of the body', *Economy and Society*, 2 (1973), 70–88, originally 'Les techniques du corps', *Journal de psychologie*, 32 (1935), 271–93.

———, 'Les Techniques et la technologie', 1948, *Le Travail et les techniques*, *Journal de psychologie* (71–8), repr. in *Revue du Mauss*, 23 (2004), 434–50.

Mertens, Joost, 'Le Déclin de la technologie générale: Léon Lalanne et l'ascendance de la science des machines', *Documents pour l'histoire des techniques*, 20 (2011), 107–17.

———, 'Éclairer les arts. Eugène Julia de Fontenelle (1780–1842), ses manuels Roret et la pénétration des sciences appliquées dans les arts et manufactures', *Documents pour l'histoire des techniques*, 18 (2009), 95–112.

———, 'The Mere Handicrafts: Ure's Dictionary (1839–1853) Compared with the Dictionnaire Technologique (1822–1835)', *Documents pour l'histoire des techniques*, 19 (2010), 277–85.

———, 'Technology as the Science of the Industrial Arts: Louis-Sebastien Lenormand (1757–1837) and the Popularization of Technology', *History and Technology*, 18.3 (2002), 203–31.

Meyer, Thorsten, Marcus Popplow and Günter Bayerl, *Technik, Arbeit und Umwelt in der Geschichte: Günter Bayerl zum 60. Geburtstag* (Münster: Waxmann, 2006).

Mokyr, Joel, *The Gifts of Athena: Historical Origins of the Knowledge Economy* (Princeton: Princeton University Press, 2002).

Riviale, Pascal, 'Entre exotisme et pragmatisme: l'Amérique Latine dans les premières Expositions universelles en France, 1855–1899', in *Exotiques expositions… Les Expositions Universelles et les cultures extra-européennes, France, 1855–1937*, ed. by Christiane Demeulenaere-Douyère (Paris: Somogy-Archives Nationales, 2010), pp. 64–75.

Saada, Anne, 'De quoi la technologie est-elle le nom?', *Revue d'histoire des sciences humaines*, 33 (2018), 267–74.

Salomon, Jean-Jacques, *Le Destin technologique* (Paris: Gallimard, 1992).

Schatzberg, Eric, 'Technik Comes to America: Changing Meanings of Technology Before 1930', *Technology & Culture*, 47.3 (2006), 486–512.

———, *Technology: Critical History of a Concept* (Chicago, IL: The University of Chicago Press, 2018a).

———, 'Why is There No Discipline of Technology in the Social Sciences?', *Artefact. Techniques, histoire et sciences humaines*, 8 (2018b), 193–214.

Schlanger, Nathan, 'Le Fait technique total. La raison pratique et les raisons de la pratique dans l'œuvre de Marcel Mauss', *Terrain*, 16 (1991), 114–30.

———, *L'Invention de la technologie. Une histoire intellectuelle avec André Leroi-Gourhan*, (Paris : PUF, 2023).

—, 'Une technologie engagée: Marcel Mauss et l'étude des techniques dans les sciences sociales', preface to Marcel Mauss, *Techniques, technologie et civilisation* (Paris: Presses universitaires de France, 2012), pp. 17–134.

Sibeud Emmanuelle, 'Marcel Mauss: "Projet de présentation d'un bureau d'ethnologie" (1913)', *Revue d'histoire des sciences humaines*, 10.1 (2004), 105–15.

Sigaut, François, 'La Formule de Mauss', *Techniques & Culture*, 40 (2003), 153–68.

———, 'Préface. Haudricourt et la technologie', in André-Georges Haudricourt, *La Technologie, science humaine. Recherches d'histoire et d'ethnologie des techniques*, ed. by François Sigaut (Paris: Éditions de la MSH, 1988), pp. 9–34.

———, 'La Technologie, une science humaine', in *L'empire des techniques* (Paris: Seuil, 1994) 51–61.

Tresch, John, 'Les Moteurs divins des Saint Simoniens : technologie, socialisme, panthéisme', in *Dompter Prométhée. Technologie et socialismes à l'âge romantique (1820–1870)*, ed. by François Jarrige (Besançon: Presses universitaires de Franche-Comté, 2016), pp. 30–45.

Ure, Andrew, *Dictionary of Arts, Manufactures, and Mines, Containing a Clear Exposition of Their Principles and Practice* (London: Longman, 1839).

———, *Philosophie des manufactures, ou économie industrielle de la fabrication du coton, de la laine, du lin et de la soie, avec la description des diverses machines employées dans les ateliers anglais* (Paris: Mathias, 1836).

———, *The Philosophy of Manufactures. An Exposition of the Scientific, Moral, and Commercial Economy of the Factory System of Great Britain* (London: Charles Knight, 1835).

Vatin, François, 'Mauss et la technologie', *Revue du Mauss*, 23 (2004), 418–33.

Vérin, Hélène, *La Gloire des ingénieurs. L'intelligence technique du XVIe au XVIIIe siècle* (Paris: Albin Michel, 1993).

Wiener, Norbert, *Cybernetics or Control and Communication in the Animal and the Machine* (New York: The Technology Press, John Wiley & Sons, Inc.,1948).

Wilson, George, *What is Technology? An Inaugural Lecture Delivered in the University of Edinburgh on November 7, 1855* (Edinburgh: Sutherland & Knox / London: Simpkin, Marshall & Co., 1855).

Wolff, Christian, 'Discursus praeliminaris de philosophia in genere', *Philosophia rationalis sive logica, methodo scientifica pertractata et ad usum scientiarum atque vitae aptata* (Frankfurt and Leipzig: Officina Libraria Rengeriana, 1728).

JORIS MERCELIS

Laboratory Techniques*

Over the past century and a half, the laboratory has been an iconic institution of science, and laboratory items such as the test tube and the lab coat have shaped popular perceptions of what it means to be a scientist. In the early 1800s hardly anyone could have anticipated these developments. Up to that time, the term 'laboratory' had referred almost exclusively to private spaces where chemical operations were carried out (Klein, 2008). These rooms or buildings were as likely to be used by apothecaries as by academic chemists and were carefully shunned by members of 'polite society' wary of chemical odours and potential hazards (Jackson, 2016: 299; Homburg, 1999: 5–9; Klein, 2007). It was only in the second half of the nineteenth century that laboratories started to proliferate in educational, industrial, and government/military settings, and that 'laboratory talk' became part of searches for scientific and cultural authority in not just the natural sciences but also the social sciences and many other fields (Owens, 2014). The late-nineteenth- and twentieth-century spread of laboratories occurred almost simultaneously in Euro-America and in other parts of the world, particularly in new laboratory sciences such as bacteriology and, to some extent, experimental psychology and nuclear physics (Escobar, 2014: 296, Leslie, 2015). As a result of its introduction into a rapidly growing list of disciplines and places, the laboratory became an extremely heterogeneous institution, and its modern history is almost impossible to fit into any overarching framework. There are also important empirical lacunae in the historiography on which generalizations could draw (and have drawn), partly because there has been a persistent tendency to neglect less prestigious types of laboratories, such as the middle and high school lab, the testing lab, and the industrial works lab (van Rooij, 2011; Gooday, 2008: 784–85, 788).

Although new laboratory histories continue to appear at a steady pace, lab studies are no longer as central to science and technology studies (STS) and the history of science as they were in the late twentieth century, when they provided seminal new perspectives on how scientific knowledge is produced (see Knorr Cetina, 1995

* I am grateful to the editors of this volume for their feedback and encouragement, and would also like to thank Anna Guagnini, Bill Leslie, and Alex Parry for sharing their thoughts on an earlier version of this chapter.

Global History of Techniques, ed. by Guillaume Carnino, Liliane Hilaire-Pérez and Jérôme Lamy, Global Matters Culture Matérielles, histoire des techniques et circulation globale des savoirs, 9 (Turnhout: Brepols, 2024), pp. 517–532
BREPOLS PUBLISHERS 10.1484/M.TECHNE-EB.5.129827

and Doing, 2008 for reviews of early research in STS, and James, 1989 for early contributions by historians). Several scholars have argued that laboratory studies did not live up to their potential (Doing, 2008; Kohler, 2008; Jackson, 2016), and there are naturally different viewpoints as to how the field could be revived and whether a revival is at all necessary (see also Vinck, 2007). Historians do not agree on whether the most promising approach would be to write: 1) inclusive histories that analyse mainstream laboratories along with various other spaces (e.g. kitchens and museums) and examine which ones of these could and could not obtain the status of a 'laboratory' (Gooday, 2008); 2) more conventional laboratory histories organised by discipline (Morris, 2015); 3) macro-studies interpreting laboratories as period-specific socio-cultural infrastructure (Kohler, 2008), or particularly as key sites of modernity (Schmidgen, 2021); or 4) micro-histories that integrate laboratories' 'institutional essence with what goes on within their walls' (Jackson, 2016: 304). Unfortunately, none of these suggestions draws significantly on research on laboratories outside of Europe and North America, and recent surveys of laboratory history likewise discuss such labs only very briefly, if at all (Jackson, 2016: 301–02; Marsch, 2003; Schmidgen, 2021; van Rooij, 2011: 444). Although the notion that laboratories are 'placeless' institutions whose locations hardly matter has long been debunked (Livingstone, 2003: 3), the relatively narrow geographic scope of this literature partly mirrors limitations of the available historiography. For instance, even many, if not most, of the laboratory histories that adopt a transnational perspective examine the influence of various 'model' labs from one European country — typically Germany — in another European country or in the United States (e.g. Diser, 2016; Kichigina, 2009; Rossiter, 1975; Vanpaemel, 2015 reviews additional literature on German models). However, multiple important studies on laboratories outside of Euro-America have come out since at least the 1980s (e.g. Visvanathan, 1984; Anderson, 1992), and it is notable that recent commentaries and surveys of lab history do not engage with the findings of this research.

The lens of the history of technique proposed by the editors of this volume is well suited for exploring the development of different types of laboratories from more global perspectives. Studying laboratory techniques, for example, raises questions about the nature and origins of the instruments, materials, and animals used in various laboratory spaces, and about the knowledge and know-how necessary to produce and handle them. As research on resource extraction, commodities, and international standard-setting has shown, global connections and dependencies may be revealed by addressing these themes (Soto Laveaga, 2009; Pretel, 2019; Declercq, 2019; Druglitrø, Kirk, 2014; also Roberts, 2016: 342–50). Moreover, much like studies of colonial laboratories (Chakrabarti, 2012) and research on the roles and knowledge of laboratory technicians (Iliffe, 2008; Hentschel, 2009), these connections can prompt new questions as to who and what should be included in laboratory histories. In this respect, it is important to also explore the permeability of the boundaries and borders (supposedly) separating laboratories from their surrounding environments and societies. This is another theme that has long been central to laboratory studies (Latour, 1983; Kohler, 2002), but it has not previously been examined from the perspective of 'technique' in the sense of artefacts and savoir-faire, and 'technology' as the science of the human use of techniques (Carnino, Hilaire-Pérez, Lamy, this volume).

This volume's editors' revisionist distinction between technology and technique is in some ways especially relevant to this chapter. After all, laboratory research and development (R&D) work in chemistry, physics, and engineering contributed to the twentieth-century equation of technology with applied science and, less directly, to the Euro-American-centred histories of technology associated with both this conceptualization and the more recent notion of 'technoscience' (cf. Carnino, Hilaire-Pérez, Lamy, this volume). Instead of approaching labs as sites where technoscience was being practiced and science-based artefacts were being produced, this chapter primarily interprets them as socio-cultural spaces shaped by the presence (or absence) and application of techniques. I suggest that such a perspective can help provide a more global understanding of the history of the modern laboratory, including of the political and economic functions and significance of lab techniques. To develop and illustrate these points, this chapter first explores the globalization of the lab in the decades before the First World War through the lens of bacteriology and tropical medicine, two related fields that stand out in the historiography of the laboratory in terms of the amount of research that has been done on labs outside of Europe and the United States. Afterwards I propose that even the histories of well-known physical science labs in late-nineteenth- and early-twentieth-century Euro-America can be fruitfully reinterpreted from more global and longer-term perspectives by focussing on the origins, use, and circulation of lab techniques. Both of the approaches followed in this chapter could also be applied to various other types of laboratories, and in the concluding section I emphasize that much work remains to be done.

Globalizing the Laboratory: Bacteriology and Tropical Medicine

Bacteriological laboratories and laboratory techniques were remarkably quick to 'globalize' in the wake of the much-celebrated discoveries of Louis Pasteur, Robert Koch, and their respective associates. Less than two years after Pasteur's July 1885 test of his rabies vaccine on Joseph Meister, for instance, a group of creole physicians in Cuba had formed their own rabies laboratory (Palmer, 2011). By the end of the nineteenth century, Pasteurian bacteriology labs had also been established in French Indochina, Algeria, and Tunisia, as well as in the British colonial cities of Sydney, Hong Kong, Poona, and Bombay, and a 1900 outbreak of yellow fever in Senegal would lead to a Pasteurian research mission to Rio de Janeiro (Löwy, 1990; Moulin, 1992; Guénel, 1999; Chakrabarti, 2012; Peckham, 2013). Pasteur's rival Robert Koch even personally travelled to several British and German colonies to study the etiology of cholera, malaria, and sleeping sickness (Gradmann, 2009: chapter 5), and a central theme in recent literature on tropical medicine indeed concerns the circulation of laboratory techniques across different empires (Neill, 2012; Mertens, 2014; Webel, 2019). Koch's bacteriological techniques had been adopted in US and Russian laboratories by the late 1880s (Gossel, 1992; Yefremenko, 1968), and his Institute of Infectious Diseases (founded in 1891) served as a model for the similarly named Japanese institute whose formation Kitasato Shibasaburō initiated in 1892 (Kim, 2014: 135). The reference to infectious diseases in the names of these establishments was appropriate, as disease outbreaks go a long way towards explaining the rapid global spread of bacteriological laboratories and techniques. More specifically, the prevalence of endemic and epidemic infectious diseases was seen as a defining feature of the heterogeneous set of regions in Asia, Africa, and Latin America that were often lumped together as the 'tropics', and the new laboratory science of bacteriology appeared to hold great promise for making those regions more hospitable to mostly white colonizers and/or to domestic elites.

What happened with microbiological techniques initially developed in Western and Central Europe when they moved to other world regions? How did the use and the politics of these techniques change in this process? For analytical purposes it can be useful to differentiate between the spaces in which laboratory operations were carried out, the tools and materials that were used, and the knowledge and skills necessary to handle these resources. Indeed, with respect to not just bacteriology and tropical medicine but also many other fields of study, this differentiation can serve to highlight distinct (though closely related) factors that have facilitated and hindered the global circulation of lab techniques.

To begin with, it is important to appreciate that, both in Europe and elsewhere, virtually any space could be converted into a bacteriological 'laboratory'. Thousands of American women who enrolled in early-twentieth-century courses in household bacteriology, for example, were expected to turn their 'kitchens into [bacteriological] laboratories and conduct some simple experiments' (Elliott, 1907: 5; Parry, 2020). Outside of the United States and Europe, too, many early bacteriological labs were of a makeshift nature (Harrison, 1994: 113). In Hong Kong in 1896, as a case in point, Swiss-French Pasteurian Alexandre Yersin famously identified the plague bacillus in a 'temporary bamboo structure — hastily erected in the field by Chinese workmen for 75 piastres' (Peckham, 2013: 123). Yersin had to use this facility because he was a techno-political outsider in this British colony, and he had reached the plague hospital where he was hoping to do his research shortly after Koch's accomplished disciple Kitasato Shibasaburō, who had been sent to Hong Kong by the Japanese government and had already been provided with a hospital room (Cunningham, 1992: 224–38; Velmet, 2020: 26–31). For both Yersin and Kitasato, it was crucial that they could refer to these temporary facilities as their bacteriological 'laboratories' as soon as they had installed their equipment. This was not because of the exclusivity of these spaces, for the boundaries between these labs and the surrounding city and field sites actually remained highly fluid (Peckham, 2013). However, as Andrew Cunningham (1992) has argued, bacteriologists aimed to redefine the identity of infectious diseases by attributing responsibility to germs, and (makeshift) labs were indispensable for this purpose (see also Latour, 1983). In the words of French imperial bacteriologist Louis Calmette, the tropical lab was also a space for 'the study of exotic pathologies according to modern methods' (quoted in Guénel, 1999: 3), and not just colonial researchers, but also domestic elites in colonial and formally independent societies could gain authority and prestige by associating themselves with laboratory discoveries and techniques.

Given that bacteriological techniques were supposed to be universally applicable, and that (makeshift) bacteriological laboratories could, in principle, be established virtually anywhere, it is important to consider which factors determined these labs' actual locations. Access to human research subjects and laboratory animals was often a major consideration, which helps explain why some early labs were sited in or close to hospitals and zoos (Guénel, 1999: 9; Chakrabarti, 2012: 122). In British India, however, several bacteriological labs were deliberately sited in relatively remote hilltop locations where such resources were less readily available than in urban centres. As Pratik Chakrabarti (2012: 62–68) has noted, one of the arguments used to justify this decision was the claim that Pasteurian labs would be most effective in climatological conditions similar to those in Paris, even though the Institut Pasteur at the time was establishing an international network of labs deliberately located in environments perceived as 'tropical'. In reality, the preference for laboratories in the mountains was driven by British imperial concerns about isolation and exclusivity, and the related desire to create spaces pleasant for white physicians and patients. British visitors, for example, were delighted with the climate in Kasauli, the city where India's first Pasteur Institute was established in 1900 (Figure 1), and appear to have experienced the hill station as a refuge. Indian rabies patients, in contrast, initially had great difficulties in trying to reach the remote site (Chakrabarti, 2012: 67–68).

While travelling bacteriologists using makeshift laboratories typically brought along their own instruments, other researchers grew frustrated about a lack of suitable equipment and resources. In a 1902 diary entry, for example, a British colonial medical officer studying sleeping sickness at the Ugandan side of Lake Victoria complained that he had 'dissected a good many mosquitoes & worked very hard, but up to now have found nothing of much importance, which was to be expected with my instruments & no room with light enough for the work. Gov[ernmen]t does not yet provide us with microscopes!' (quoted in Webel, 2019: 277). This was a telling comment given the centrality of microscopic methods to tropical medicine, a newly emerging field that drew on (among other things) laboratory techniques developed by microbe-hunting bacteriologists and parasitologists studying parasites and their vectors (Neill, 2012; Mertens, 2014). Still, this British doctor's grievance would have resonated with German sleeping sickness researchers working in the same area, who indeed envied their British colleagues for their relatively well-endowed laboratory in the city of Entebbe and appropriated techniques used in this lab (Webel, 2019). In Belgian Congo, too, a lack of microscopes was a recurring issue affecting sleeping sickness missions, including the ways in which African patients were treated (Mertens, 2014: 60, 177, 238).

In any case, obtaining microscopes, cell-culture dishes, micro-photographic apparatus, and other microbiological equipment was often less challenging than acquiring the tacit knowledge and skills necessary to isolate, characterize, and visualize microorganisms in ways that were consistent with the latest research standards (Löwy,

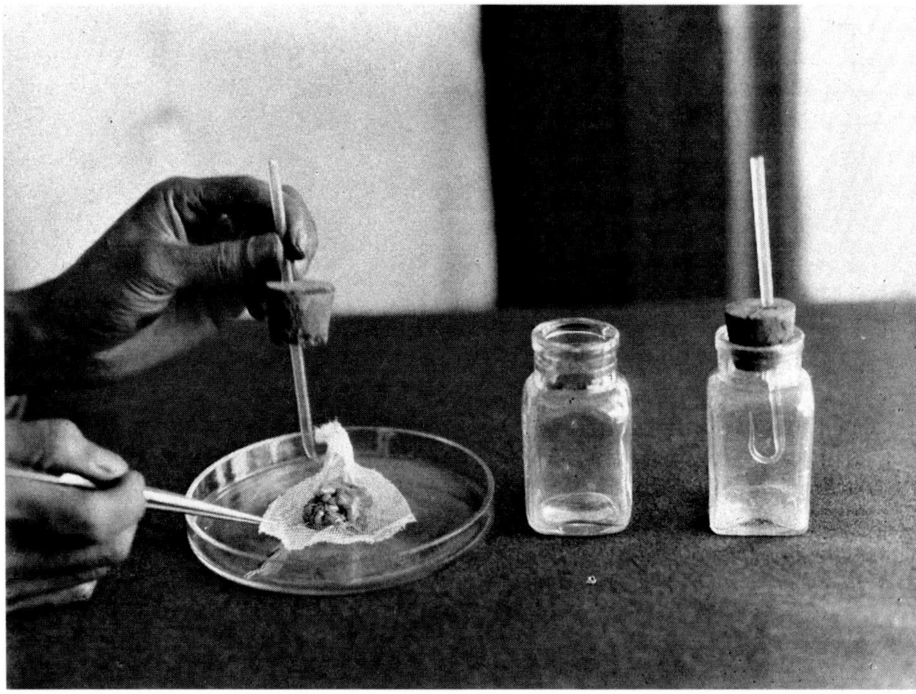

Fig. 1. The Pasteur Institute, Kasauli, India: one stage in the preparation of the rabies vaccine: a rabbit brain on a square of muslin. Photograph, *c.* 1910. Source: Wellcome Library, London. Wellcome Collection. Public Domain Mark.

1990). It was no coincidence that, in the second half of the 1880s and the early 1890s, many aspiring bacteriologists travelled to Western and Central Europe to participate in courses at Koch's Institute of Hygiene at the University of Berlin or in Emile Roux's 'Cours de Microbie technique' (Course on Microbe Research Technique) at the Pasteur Institute in Paris. By the early 1900s, a broader range of programmes had become available, partly due to the integration of bacteriology into tropical medicine (Neill, 2012: chapter 2). However, given that the command of laboratory techniques was central to bacteriology (Gossel, 1992: 287), and that state-of-the-art approaches to microbiological research had changed significantly by the start of the twentieth century, it continued to be important to have personal relationships with (perceived) leaders in the field — or at least with these individuals' collaborators or students. As Ilana Löwy (1990: 145) has emphasized, '[a] prolonged training in the few leading European centres, and later training with bacteriologists that had studied in these centres, were viewed as crucial steps in the proper transmission of bacteriological knowledge'. Consistent with this view, some self-taught Latin American bacteriologists were criticized for using inadequate techniques, and the findings of other autodidacts were simply ignored (Löwy, 1990; also Palmer, 2011: 449, 457).

At the same time, supposedly universalist microbiological techniques were not just imposed on local populations by researchers trained in Europe or the United

States. To be sure, as Deborah Neill (2012: 16) has noted in her analysis of the emergence and development of tropical medicine in the period between 1890 and 1930, the partial transition from clinical to laboratory medicine 'reinforced the marginalization of the broader social context' in which doctors and scientists were operating, as labs 'brought researchers inside, limiting the questions they might have asked about the larger environment'. Somewhat similarly, Warwick Anderson (1992: 519) has stressed that physicians at US government and military labs in the colonial Philippines examined Filipinos as residents of 'the idealized, reconstituted tropical region', rather than as inhabitants of a specific village or island. In addition, the colonial medical papers produced at the Manila Bureau of Science (founded in 1904) and the US Army Board for the Study of Tropical Diseases dehumanized Filipino patients by reducing their bodies to sources of data whose origins and methods of acquisition did not matter as long as they could be subjected to universalizing lab techniques (Anderson, 1992, 2006). Still, despite strong power asymmetries and the relatively impermeable boundaries of some laboratory spaces, colonial subjects' acts of resistance could and did affect the development of lab techniques and the nature of the research that was done in medical labs. For instance, as Mari Webel (2019) has observed in her research on sleeping sickness, not only lymph but also blood or cerebrospinal fluid could be examined under a laboratory microscope to confirm whether African patients in colonial isolation camps had been infected with parasites. These body fluids had to be extracted from large patient populations, however, and Webel (2019: 285) argues that physicians' choice to use cervical lymph-node punctures has to be understood 'in the context of uneven coercive power of European researchers and contingent compliance of their African subjects', who were likely to have opposed the use of methods that seemed more painful or invasive.

Colonial subjects were also relatively early to enter bacteriological and tropical medicine labs as technicians and even as doctors, a development that would facilitate later transitions to postcolonial laboratory medicine and global public health. Significantly, the nonwestern staff of late-nineteenth- and early-twentieth-century colonial laboratory spaces ranged from Indian snake charmers employed in a snake venom lab in Bombay and Vietnamese health professionals at the Pasteur Institute in Hanoi to African microscopists participating in sleeping sickness missions in Belgian Congo (Chakrabarti, 2012: 130; Mertens, 2014: 238; Guénel, 1999: 24–5). In Cuba, as mentioned earlier, Spanish colonial subjects even took the initiative to establish their own bacteriological lab (Palmer, 2011), and microbiological labs were also quick to appear in Brazil and other independent countries in Latin America. Indeed, not just in Latin America but also in Southeast Asia and Africa, national independence movements and postcolonial national elites have tended to enthusiastically embrace and appropriate Euro-American laboratory medicine, rather than to reject it as a biopolitical tool of empire (e.g. Anderson, 2006). Partly for this reason, the legacy of the late-nineteenth- and early-twentieth-century globalization of the bacteriological lab has remained very visible up to the present. Several of the French colonial

Fig. 2. A laboratory space at the Pasteur Institute in Saigon (Ho Chi Minh City) in 1962. Source: Wikimedia Commons (image in public domain).

Pasteur Institutes established in the decades before the First World War, for example, are among the laboratories that currently constitute the Institut Pasteur International Network (Moulin, 2016) (Figure 2).

Globalizing Laboratory Histories: From 'Technoscience' to Global Uses of Lab Techniques

Precisely when the mid- to late-nineteenth-century 'laboratory revolution' in European medicine (Cunningham, Williams, 1992) was becoming a global phenomenon, the rise of the modern chemical and electrical industries contributed to the emergence of a revised technology concept in the United States and beyond (Marx, 2010; Schatzberg, 2018: 72–73; also Hilaire-Pérez, this volume). In this process, too, laboratories played a significant role, as high-stakes inventions generated in chemical, physics, and engineering R&D labs made it easier to equate technology with applied science during the so-called Second Industrial Revolution. More recently, the late-nineteenth- and early-twentieth-century emergence of the corporate research lab has been identified as a major (supposed) starting point of a Euro-American-centred history of 'technoscience' (Channell, 2017; also Reich, 1983). However, if technology is redefined as the science of the human use of techniques, it becomes possible to evaluate the novelty and impact of the work done in not only industrial, but also academic and government labs in Europe and the United States from longer-term and more global perspectives.

As a case in point, Irving Langmuir's prize-winning use of the light bulb as a research instrument at the General Electric Research Laboratory thus becomes part of a long and institutionally diverse history of laboratory glassware and glassmaking that helps explain the global circulation (and non-circulation) of lab techniques in the physical sciences. Langmuir used lamps to study what Leonard Reich (1983: 210) has characterized as 'the natural phenomena of the artificial environment', and was famously able to solve both technical and scientific problems in this way. Less well-known is that Langmuir had been assigned a personal toolmaker who constructed the delicate apparatus that the later Nobel laureate desired to use in his research (Wise, 1985: 151–52), and that earlier chemists and physicists had likewise relied on the glassblowing skills of technicians and professional instrument makers (Gay, 2008: 55–56; Muller, 2011; Espahangizi, 2015). Organic chemists had even been quick to acquire these skills themselves in the years following the invention of Justus Liebig's *Kaliapparat* (potash apparatus), a relatively affordable and user-friendly carbon measurement device that contributed to the still limited internationalization of the chemical teaching-research laboratory in the first half of the nineteenth century (Jackson, 2006, 2015; Rocke, 2003). Although most available studies on the history of lab techniques in the nineteenth- and early-twentieth-century physical sciences have focussed on European and US cases, including on firms' role in producing and standardizing laboratory tools and materials and on the gradual emergence of generic research technique communities on both sides of the Atlantic (Gaudillière, Löwy, 1999; Shinn, Joerges, 2002), these developments were also highly pertinent from a more global perspective. In particular, while much of the laboratory glassware used by both academic/government physical scientists and industrial researchers like Langmuir was relatively easy to transport across long distances, the knowledge and skills needed to effectively use, standardize, or redesign sophisticated instruments were developed in heterogeneous networks of laboratory researchers, technicians, and firms that were hard to replicate. This was even the case for a science-oriented and relatively wealthy country like Meiji-era Japan, as Yoshiyuki Kikuchi (2013) has shown in a rare comprehensive analysis of the early history of the academic chemistry lab in a nonwestern society. Late-nineteenth-century Japanese chemists benefitted from their educational and professional relationships with British and US colleagues, yet the new laboratory-based pedagogy that they and their government sought to appropriate was inseparable from a glass-dominated material-scientific culture that was foreign to their country. Accordingly, Japanese chemists initially imported laboratory glassware from Europe and the United States and, at least in some cases, integrated glassblowing into their schools' curricula (Kikuchi, 2013: 125, 158).

A history-of-techniques perspective is also useful for appreciating the global ramifications of laboratory-based R&D activities like those that contributed to the equation of technology with applied science. Although Langmuir's employer and several other leading firms in the electrical industry were quick to globalize their manufacturing operations and, to some extent, R&D work, the laboratory practice of chemical synthesis was even more relevant in this respect. Chemists' eventual success in synthesizing camphor, for instance, was among the factors that helped end a turbulent history of camphor-related violence in Taiwan (Inkster, 2018). Ironically, however, the emergence of new knowledge-intensive industries in Europe and North

America during the Second Industrial Revolution had also indirectly contributed to this violence, as camphor's use in celluloid plastic and photographic film had made it much more profitable to control access to East Asian camphor trees. In other cases, moreover, the laboratory synthesis and development of new substances negatively affected the economies of Latin American, Asian, and/or African countries that had been exporting competing goods, although the replacement of (among other things) 'natural' dyes and rubber by synthetic varieties was a more gradual and less complete process than is often assumed (Pretel, 2019).

Finally, from a longer-term history-of-techniques viewpoint, it is worth noting that the emergence of the modern laboratory in the second half of the eighteenth century and the first half of the nineteenth century was a precondition for later lab investigations of scientists like Langmuir. The gradual removal of furnaces from laboratory spaces was one crucial development enabling the transition to the modern lab (Morris, 2015; Homburg, 1999). Furnaces had long been a defining feature of the laboratory, as the chemical operations carried out in early modern labs depended on the heat that they were capable of providing. As Peter Morris (2015: chapters 1–5) has explained, the emergence of pneumatic chemistry was among the factors that contributed to the disappearance of furnaces, for there was no need to study newly discovered 'airs' in their proximity. Other pertinent developments included the introduction of new analytical reagents and instruments (e.g. the blowpipe) and the adoption of steam-pipe heating systems and utilities that directly supplied water and gas to laboratory benches. Due to these and other changes, the lab could develop into a more comfortable and prestigious space that was no longer exclusively associated with heat-driven chemical techniques. Although the emergence of the modern lab is usually described as an exclusively European phenomenon, this development also had immediate consequences in other parts of the world. For instance, it contributed to the reorganization of slave economies in the Caribbean, where organic chemistry labs had started to appear on sugar plantations by the 1840s (Rood, 2017: 58). Cuban plantation holders' interest in laboratory knowledge can, in fact, be traced back to at least the late 1790s and early 1800s (Portuondo, 2003: 251–52), and elsewhere in Latin America, European-style chemical labs were actually being introduced in these years (Aceves, 1992).

The basic fact that laboratory techniques have been used outside of Europe and the United States since at least the start of nineteenth century runs counter to the persistent notion that surveys of modern lab history have to be Euro-American-centred due to the nature of the topic. As this chapter has tried to demonstrate, it is possible and important to offer more global perspectives on modern labs and lab techniques by paying more attention to African, Asian, and Latin American examples and developments. I have primarily illustrated this point by discussing the exceptionally well-studied cases of bacteriology and tropical medicine, while also drawing attention to global underpinnings and ramifications of late-nineteenth- and early-twentieth-century Euro-American R&D work that is usually analysed through a more geographically narrow lens. It speaks to the richness of lab history that various additional themes and strands of literature could have been considered to develop similar arguments, particularly but not exclusively with respect to Cold-War and contemporary laboratories. Historians of the life sciences, for example, have already started providing more global viewpoints on the multifaceted relations between the lab and the field in studies on botanical gardens and field stations, as well as on forests, mountains, national parks, and various other spaces that have been characterized as (natural) 'laboratories' at different times and places (e.g. De Bont, 2017; Raby, 2017; Tonn, 2019; Wille, 2019; see also Sivaramakrishnan, 2008: 45, 50; Heggie, 2013). Even the entire African continent has been approached as a lab (Tilley, 2011), and a better understanding of the metaphorical and conceptual histories of the 'laboratory' in a broad range of languages and cultures would certainly be helpful for researching more global histories of lab techniques. Given the extreme heterogeneity of both the modern lab and the societies and environments in which lab techniques have been used, the findings of such research will vary considerably, depending on the specific topics that are being examined. Studies on the global spread of commercial testing labs, for example, may not have much in common with analyses of the popularization of school labs of the type in which hundreds of millions of students have conducted basic exercises over the past century and a half. It is, however, clear that while at least some kinds of laboratories were already being globalized more than a century ago, much additional work remains to be carried out to create a more geographically and culturally inclusive historiography of the modern lab.

References

Aceves, Patricia, 'The First Chair of Chemistry in Mexico (1796–1810)', in *Science and Empires: Historical Studies About Scientific Development and European Expansion*, ed. by Patrick Petitjean, Catherine Jami, and Anne Marie Moulin (Dordrecht: Kluwer, 1992), pp. 137–46.

Anderson, Warwick, *Colonial Pathologies: American Tropical Medicine, Race, and Hygiene in the Philippines* (Durham, NC: Duke University Press, 2006).

———, '"Where Every Prospect Pleases and Only Man is Vile": Laboratory Medicine as Colonial Discourse', *Critical Inquiry*, 18 (1992), 506–29.

Chakrabarti, Pratik, *Bacteriology in British India: Laboratory Medicine and the Tropics* (Rochester, NY: University of Rochester Press, 2012).

Channell, David F., *A History of Technoscience: Erasing the Boundaries Between Science and Technology* (Abingdon: Routledge, 2017).

Cunningham, Andrew, 'Transforming Plague: the Laboratory and the Identity of Infectious Disease', in *The Laboratory Revolution in Medicine*, ed. by Andrew Cunningham and Percy Williams (Cambridge: Cambridge University Press, 1992), pp. 209–44.

———, and Percy Williams, eds, *The Laboratory Revolution in Medicine* (Cambridge: Cambridge University Press, 1992).

De Bont, Raf, 'A World Laboratory: Framing the Albert National Park', *Environmental History*, 22 (2017), 404–32.

Declercq, Robrecht, 'Natural Born Merchants: the Hudson Bay Company, Science and Canada's Final Fur Frontiers (1925–1931)', *Business History*, (2019) < [accessed 13 December 2019].10.1080/00076791.2019.1625331>

Diser, Lyvia, *Wetenschap op de proef: Laboratoria in het Belgisch overheidsbeleid, 1870–1940* (Leuven: Universitaire Pers Leuven, 2016).

Doing, Park, 'Give Me a Laboratory and I Will Raise a Discipline: The Past, Present, and Future Politics of Laboratory Studies in STS', in *The Handbook of Science and Technology Studies*, ed. by Edward J. Hackett, Olga Amsterdamska, Michael E. Lynch, and July Wajcman, 3rd ed., (Cambridge, MA: MIT Press, 2008), pp. 279–95.

Druglitrø, Tone and Robert G. W. Kirk, 'Building Transnational Bodies: Norway and the International Development of Laboratory Animal Science, *c.* 1956–1980', *Science in Context*, 27 (2014), 333–57.

Elliott, S. Maria, *Household Bacteriology* (Chicago, IL: American School of Home Economics, 1907).

Escobar, Rogelio, 'The Instruments in the First Psychological Laboratory in Mexico: Antecedents, Influence, and Methods', *History of Psychology*, 17 (2014), 296–311.

Espahangizi, Kijan, 'From *Topos* to *Oikos*: the Standardization of Glass Containers as Epistemic Boundaries in Modern Laboratory Research (1850–1900)', *Science in Context*, 28 (2015), 397–425.

Gaudillière, Jean-Paul and Ilana Löwy, eds, *The Invisible Industrialist: Manufactures and the Production of Scientific Knowledge* (Houndmills: Macmillan, 1998).

Gay, Hannah, 'Technical Assistance in the World of London Science, 1850–1900', *Notes and Records of the Royal Society*, 62 (2008), 51–75.

Gooday, Graeme, 'Placing or Replacing the Laboratory in the History of Science?' *Isis*, 99 (2008), 783–95.

Gossel, Patricia P., 'A Need for Standard Methods: the Case of American Bacteriology', in *The Right Tools for the Job: at Work in Twentieth-century Life Sciences*, ed. by Adele E. Clarke and Joan H. Fujimura (Princeton, NJ: Princeton University Press, 1992), pp. 287–311.

Gradmann, Christoph, *Laboratory Disease: Robert Koch's Medical Bacteriology* (Baltimore, MD: Johns Hopkins University Press, 2009).

Guénel, Annick, 'The Creation of the First Overseas Pasteur Institute, or the Beginning of Albert Calmette's Pastorian Career', *Medical History*, 43 (1999), 1–25.

Harrison, Mark, *Public Health in British India: Anglo-Indian Preventive Medicine 1859–1914* (Cambridge: Cambridge University Press, 1994).

Heggie, Vanessa, 'Experimental Physiology, Everest and Oxygen: From the Ghastly Kitchens to the Gasping Lung', *British Journal for the History of Science*, 46 (2013), 123–47.

Hentschel, Klaus, ed., *Unsichtbare Hände: zur Rolle von Laborassistenten, Mechanikern, Zeichnern u. a. Amanuenses in der physikalischen Forschungs- und Entwicklungsarbeit* (Diepholz: Verlag für Geschichte der Naturwissenschaften und der Technik, 2009).

Homburg, Ernst, 'The Rise of Analytical Chemistry and Its Consequences for the Development of the German Chemical Profession (1780–1860)', *Ambix*, 46 (1999), 1–32.

Iliffe, Rob, 'Technicians', *Notes and Records of the Royal Society*, 62 (2008), 3–16.

Inkster, Ian, 'Indigenous Resistance and the Technological Imperative: From Chemistry in Birmingham to Camphor Wars in Formosa, 1860s–1914', in *Technology and Globalisation: Networks of Experts in World History*, ed. by David Pretel and Linu Camprubi (Cham: Springer, 2018), pp. 41–74.

Jackson, Catherine M., 'Laboratory', in *A Companion to the History of Science*, ed. by Bernard Lightman (Chichester: Wiley, 2016), pp. 296–309.

———, 'Re-examining the Research School: August Wilhelm Hofmann and the Re-creation of a Liebigian Research School in London', *History of Science*, 44 (2006), 281–319.

———, 'The "Wonderful Properties of Glass": Liebig's Kaliapparat and the Practice of Chemistry in Glass', *Isis*, 106 (2015), 43–69.

James, Frank A. J. L., ed., *The Development of the Laboratory: Essays on the Place of Experiments in Industrial Civilization* (Basingstoke: Macmillan, 1989).

Kichigina, Galina, *The Imperial Laboratory: Experimental Physiology and Clinical Medicine in Post-Crimean Russia* (Amsterdam: Rodopi, 2009).

Kikuchi, Yoshiyuki, *Anglo-American Connections in Japanese Chemistry: the Lab as Contact Zone* (New York, NY: Palgrave Macmillan, 2013).

Kim, Hoi-eun, *Doctors of Empire: Medical and Cultural Encounters between Imperial Germany and Meiji Japan* (Toronto: University of Toronto Press, 2014).

Klein, Ursula, 'Apothecary's Shops, Laboratories and Chemical Manufacture in Eighteenth-century Germany', in *The Mindful Hand: Inquiry and Invention From the Late Renaissance to Early Industrialisation*, ed. by Lissa L. Roberts, Simon Schaffer, and Peter Dear (Amsterdam: Koninklijke Nederlandse Akademie van Wetenschappen, 2007), pp. 246–76.

———, 'The Laboratory Challenge: Some Revisions of the Standard View of Early Modern Experimentation', *Isis*, 99 (2008), 769–82.

Knorr Cetina, Karin, 'Laboratory Studies: the Cultural Approach to the Study of Science', in *Handbook of Science and Technology Studies*, ed. by Sheila Jasanoff, Gerald E. Marckle, and James C. Peterson (Thousand Oaks, CA: Sage, 1995), pp. 140–67.

Kohler, Robert E., 'Lab History: Reflections', *Isis*, 99 (2008), 761–68.

———, *Landscapes and Labscapes: Exploring the Lab–Field Border in Biology* (Chicago, IL: University of Chicago Press, 2002).

Latour, Bruno, 'Give Me a Laboratory and I will Raise the World', in *Science Observed: Perspectives on the Social Study of Science*, ed. by Karin Knorr Cetina and Michael Mulkay (London: Sage, 1983), pp. 141–70.

Leslie, Stuart W., 'Atomic Structures: the Architecture of Nuclear Nationalism in India and Pakistan', *History and Technology*, 31 (2015), 220–42.

Livingstone, David N., *Putting Science in Its Place: Geographies of Scientific Knowledge* (Chicago, IL: University of Chicago Press, 2003).

Löwy, Ilana, 'Yellow Fever in Rio de Janeiro and the Pasteur Institute Mission (1901–1905): the Transfer of Science to the Periphery," *Medical History*, 34 (1990), 144–63.

Marsch, Ulrich, 'Laboratory, Industrial Research', in *The Oxford Companion to the History of Modern Science*, ed. by John L. Heilbron (Oxford: Oxford University Press, 2003), pp. 443–45.

Marx, Leo, 'Technology: the Emergence of a Hazardous Concept', *Technology and Culture*, 51 (2010), 561–77.

Mertens, Myriam, 'Chemical Compounds in the Congo: Pharmaceuticals and the "Crossed History" of Public Health in Belgian Africa (ca. 1905–1939)' (unpublished doctoral dissertation, Ghent University, 2014).

Morris, Peter J. T., *The Matter Factory: a History of the Chemistry Laboratory* (London: Reaktion Books, 2015).

Moulin, Anne-Marie, 'Les Instituts Pasteur du Maghreb: mémoires divisées, histoire transversale', *Dynamis*, 36 (2016), 341–62.

———, 'Patriarchal Science: the Network of the Overseas Pasteur Institutes', in *Science and Empires: Historical Studies About Scientific Development and European Expansion*, ed. by Patrick Petitjean, Catherine Jami, and Anne Marie Moulin (Dordrecht: Kluwer, 1992), pp. 307–22.

Müller, Falk, 'Johann Wilhelm Hittorf and the Material Culture of Nineteenth-century Gas Discharge Research', *The British Journal for the History of Science*, 44 (2011), 211–44.

Neill, Deborah, *Networks in Tropical Medicine: Internationalism, Colonialism, and the Rise of a Medical Specialty, 1890–1930* (Stanford, CA: Stanford University Press, 2012).

Owens, Robert B., '"Laboratory Talk" in US Sociology, 1890–1930: the Performance of Scientific Legitimacy', *Journal of the History of the Behavioral Sciences*, 50 (2014), 302–20.

Palmer, Steven, 'Beginnings of Cuban Bacteriology: Juan Santos Fernández, Medical Research, and the Search for Scientific Sovereignty, 1880–1920', *Hispanic American Historical Review*, 91. (2011), 445–68.

Parry, Alexander, 'Gardens of Dust', *Lady Science* (7 May 2020) <https://www.ladyscience.com/features/gardens-of-dust-household-bacteriology-2020> [accessed 31 August 2020].

Peckham, Robert, 'Matshed Laboratory: Colonies, Cultures, and Bacteriology', in *Imperial Contagions: Medicine, Hygiene, and Cultures of Planning in Asia*, ed. by Robert Peckham and David M. Pomfret (Hong Kong: Hong Kong University Press, 2013), pp. 123–47.

Portuondo, Maria M., 'Plantation Factories: Science and Technology in Late-eighteenth-century Cuba', *Technology and Culture*, 44 (2003), 231–57.

Pretel, David, 'Technology and the Fates of Three Caribbean Commodities', *History of Technology*, 34 (2019), 127–48.

Raby, Megan, *American Tropics: the Caribbean Roots of Biodiversity Science* (Chapel Hill, NC: University of North Carolina Press, 2017).

Reich, Leonard S., 'Irving Langmuir and the Pursuit of Science and Technology in the Corporate Environment', *Technology and Culture*, 24 (1983), 199–221.

Roberts, Lissa R., 'Exploring Global History Through the Lens of History of Chemistry: Materials, Identities and Governance', *History of Science*, 54 (2016), 335–61.

Rocke, Alan J., 'Origins and Spread of the "Giessen Model" in University Science', *Ambix*, 50 (2003), 90–115.

Rood, Daniel B., *The Reinvention of Atlantic Slavery: Technology, Labor, Race, and Capitalism in the Greater Caribbean* (Oxford: Oxford University Press, 2017).

Rossiter, Margaret W., *The Emergence of Agricultural Science: Justus Liebig and the Americans* (New Haven, CT: Yale University Press, 1975).

Schatzberg, Eric, *Technology: Critical History of a Concept* (Chicago, IL: University of Chicago Press, 2018).

Shinn, Terry and Bernward Joerges, 'The Transverse Science and Technology Culture: Dynamics and Roles of Research-technology', *Social Science Information*, 41 (2002), 207–51.

Schmidgen, Henning, 'The Laboratory,' *Encyclopedia of the History of Science* (April 2021) doi: [accessed 17 November 2022].10.34758/sz06-t975

Sivaramakrishnan, Kalyanakrishnan, 'Science, Environment and Empire History: Comparative Perspectives From Forests in Colonial India', *Environment and History*, 14 (2008), 41–65.

Soto Laveaga, Gabriela, *Jungle Laboratories: Mexican Peasants, National Projects, and the Making of the Pill* (Durham, NC: Duke University Press, 2009).

Tilley, Helen, *Africa as a Living Laboratory: Empire, Development, and the Problem of Scientific Knowledge, 1870–1950* (Chicago, IL: University of Chicago Press, 2011).

Tonn, Jenna, 'Laboratory of Domesticity: Gender, Race, and Science at the Bermuda Biological Station for Research, 1903–30', *History of Science*, 57 (2019), 231–59.

Vanpaemel, Geert, 'The German Model of Laboratory Science and the European Periphery (1860–1914)', in *Sciences in the Universities of Europe, Nineteenth and Twentieth Centuries*, ed. by Ana Simões, Maria Paula Diogo, and Kostas Gavroglu (Dordrecht: Springer, 2015), pp. 211–25.

van Rooij, Arjan, 'Knowledge, Money and Data: an Integrated Account of the Evolution of Eight Types of Laboratory', *The British Journal for the History of Science*, 44 (2011), 427–48.

Velmet, Aro, *Pasteur's Empire: Bacteriology and Politics in France, Its Colonies, and the World* (New York, NY: Oxford University Press, 2020).

Vinck, Dominique, 'Back to the Laboratory as a Knowledge Production Space', *Revue d'anthropologie des connaissances*, 1 (2007), 160–66.

Visvanathan, Shiv, *Organizing for Science: the Making of an Industrial Research Laboratory* (New Delhi: Oxford University Press, 1984).

Webel, Mari K., 'Trypanosomiasis, Tropical Medicine, and the Practices of Inter-colonial Research at Lake Victoria, 1902–07', *History and Technology*, 35 (2019), 266–92.

Wille, Robert-Jan, *Mannen van de microscoop: de laboratoriumbiologie op veldtocht in Nederland en Indië, 1840–1910* (Nijmegen: Vantilt, 2019).

Wise, George and Willis R. Whitney, *General Electric, and the Origins of US Industrial Research* (New York, NY: Columbia University Press, 1985).

Yefremenko, A. A., 'From the History of Russian-German Scientific Relationships: Russian Doctors at Bacteriological Courses organised by the Koch Institute — Koch and Russian Doctors', *Journal of Hygiene, Epidemiology, Microbiology and Immunology*, 12 (1968), 498–505.

OLOF HALLONSTEN

Big Science and Technological Transformations

Big Science was originally a Cold War phenomenon. The use of large technological apparatus for scientific research and the organization of scientific research projects in large teams with hierarchical command structures and industrial modes of organization was a product of the buildup of scientific and technological capacity with a military rationale in the 1940s and on. The origins were wartime efforts to harness basic science and technology for the development of weapons and other defence technologies, among which the most famous and celebrated one was the United States nuclear bomb project nicknamed the Manhattan Project. But the cycles of the Cold War, and the scientific, technological, social, economic, and political developments that happened in their shadow, transformed Big Science beyond its 1940s and 1950s incarnation. Between 1989 and 1991, the Cold War ended and with it the bipolar world order and the threat of mutually assured destruction (MAD). New ambitions and objectives, based on developments to renew industrial production, improved communications and information technology, treated and cured disease, and achieved a more sustainable economic development that took environmental concerns into account, rose to the top of the agendas of policymakers and science bureaucrats across the Northern Hemisphere. Today, Big Science is something radically different than in the 1950s. It is still around, but it has certainly transformed, organizationally as well as technologically. In this chapter, the transformation of Big Science is outlined and conceptualized chiefly from the perspective of technology, with the help of a summarizing review of previous studies in the sociology and history of science.

Origins

In January 1961, as his second term in office was drawing to a close, US President Dwight D. Eisenhower (1890–1969) gave a farewell address to the nation which has become famous most of all for the warning he therein issued, of the inadvertently detrimental effects on democracy and public life of what Eisenhower dubbed the 'military-industrial complex'. The concept referred to the huge US military defence

Global History of Techniques, ed. by Guillaume Carnino, Liliane Hilaire-Pérez and Jérôme Lamy, Global Matters Culture Matérielles, histoire des techniques et circulation globale des savoirs, 9 (Turnhout: Brepols, 2024), pp. 533–544
BREPOLS PUBLISHERS
10.1484/M.TECHNE-EB.5.129828

system, including its governmental branch (i.e. the military itself) and the defence industry, which at the time had grown to record proportions and was 'felt in every city, every statehouse, every office of the federal government' (Eisenhower, 1961). It has been suggested that the president originally also intended to include the scientific establishment in the 'complex' and that the concept should be extended to 'military-industrial-scientific complex' or 'military-industrial-academic complex' (Giroux, 2007) in order to account for the vast influence of science, academically rooted, on the defence-related bureaucratic buildup in the US, and the negative influence it had not only on democratic institutions like local and federal government and its branches, but also scientific practice and autonomy. Later the same year, physicist and director of the Oak Ridge National Laboratory Alvin Weinberg (1915–2006) published an article entitled 'Impact of Large-scale Science on the United States' in which he warned about the bureaucratization of science and the negative effects of Big Science — 'the huge rockets, the high-energy accelerators, the high-flux research reactors' — claiming that its 'triple diseases' of too much administration, too much money, and too much (journalistic) attention were threatening core academic values of creativity and curiosity (Weinberg, 1961: 161).

By then, the buildup of Big Science in the United States and elsewhere had been occurring on a broad front for over fifteen years. The efforts during the Second World War of making use of the advances in fundamental physics for weapons development — pursued with varying successes in Nazi Germany, the United Kingdom, and the United States — got a spectacular and horrifying showcase demonstration on 6 August 1945, when the first nuclear bomb was dropped on Hiroshima. Technologies like the radar, cryptology, synthetic materials, and mass-produced penicillin had also advanced vastly during the war (Pestre, 2003: 70), and once it came to an end, major technological capacity and vast human capital in science and engineering was ready for utilization for peacetime, although not necessarily peaceful, purposes. The bipolar world order that was established in the late 1940s built on superpower competition, most of all, of course, in nuclear armament, but also in several other areas. In most larger countries, the United States and the Soviet Union, but also France, the United Kingdom, and eventually West Germany, Big Science became a natural ingredient of the postwar mobilization and the continued process of the modernization of society. Nuclear energy was the future and the central technology of this mobilization in science, but large programmes were also initiated in advanced materials research, space exploration, and electrical engineering.

Physics 'had a blank check from the military' in the 1950s (Pais, 1986: 19) and could expand not least in Big Science. The wartime use of particle accelerators and nuclear reactors in weapons development continued, and the peaceful exploitation of nuclear energy and research into the inner structure of the atom was not separated from weapons development until 1953 and the launch of the Atoms for Peace programme, which gave non-nuclear powers access to reactor technology and fuel and spurred the launch of nuclear and particle physics programmes across the Northern Hemisphere (Hewlett, Holl, 1989: 209 ff.) (Figure 1). Particle physics had broken off as a subdiscipline of its own, beside nuclear and solid-state physics

(Martin, 2015), and soon became the most prestigious peaceful research branch of Atoms for Peace, with accelerator construction projects at several locations across the United States, the Soviet Union, and Europe, including also West Germany once the Allied ban on nuclear research was lifted in 1955. In a sense, particle physics became a separate league of superpower competition, with increasingly larger accelerators built especially in the United States and the Soviet Union in the first two postwar decades. The 'grand scales and easily graspable logic' of accelerator-based particle physics (or high energy physics) made it a perfect peaceful art of competition; the larger (and costlier) the accelerator, the better (Hallonsten, 2016a: 46; Greenberg, 1999: 218–19). In Europe, emphasis on the peaceful use of nuclear energy led to vast investments in the collaborative European Organization for Nuclear Research (Conseil Européen pour la Recherche Nucléaire, CERN), which became one of the global hotspots for particle physics (Pestre, Krige, 1992). In the United States (and the Soviet Union), the 'blank check from the military' (Pais, 1986: 19) could easily be extended to support the built-in expansion logic of particle physics. In both cases, the logic was simple: in order to continue the exploration of the constituent particles and forces of the atom, whether the purpose was peaceful or had ultimately military ends, particles had to be collided (and thus forced to disintegrate) at higher and higher energies; and the higher energies demanded larger accelerators. The wartime cyclotron at Berkeley, used by Ernest Lawrence (1901–1958) and associates to produce radioactive isotopes, was circular, had a circumference of 4.5 metres, and cost about half a million dollars to build. The Cosmotron at Brookhaven National Laboratory on Long Island, starting operation in 1952, had a 72-metre circumference and cost 4 million dollars, the 1954 Bevatron at Berkeley measured 120 metres and cost 10 million dollars, and so on. Gradually expanding in size and cost, in the late 1960s the accelerators needed to continue the exploration of the smallest building blocks of the universe were several kilometres long and cost several hundred million dollars, requiring entirely new institutes to be set up, far away from urban areas and university campuses, on previous farmlands (Hallonsten, 2016a: 46; Hoddeson, Kolb, 2000: 308). But by then, of course, society had started to change abruptly, and with it both politics and science.

Change

In the two to three decades following immediately after the Second World War, economic growth seemed almost unstoppable, and there was little halt to the investments in science, technological development, (higher) education, and other improvements of society across the Northern Hemisphere. Big Science, in the original sense, flourished during this period. In the 1960s and 1970s, however, a range of social, political, and economic changes took place that cohered with transformations within and between scientific fields and the foundations for technological innovation.

Particle physics would continue to be the star of the show for another two decades, but its growth was halted by budgetary austerity, the need to also prioritize

Fig. 1. 'Tandem Accelerator Laboratory at the Institute for Nuclear and Particle Physics'. Dimitris Sideridis/Wikimedia Commons.

other (costly) areas of science, and the growth of the discipline itself, which became unsustainable. When the Fermi National Accelerator Laboratory (Fermilab) was established outside Chicago in the late 1960s, to a cost of 400 million dollars, it became the national flagship facility for particle physics, and other national laboratories had to shelve their plans for major new particle accelerator projects; it simply did not make sense to duplicate efforts across the US anymore. The labs in Argonne, Berkeley, Brookhaven, and Oak Ridge had to find new profile areas other than particle physics and new major facilities around which to build their mission portfolio (Westfall, 2008). Fermilab also inaugurated the transition of particle physics towards what has been called 'megascience' (Hoddeson Kolb, Westfall, 2008), and which had two parts: first, a growth in the size of the experimental equipment necessary to do new particle physics research to sizes that necessitated experimental runs to be extended by several years (to motivate the cost) and include large and heterogeneous teams of hundreds and eventually thousands of researchers and engineers, and second, the monopolization of budgets by the larger and most advanced labs. In the United States, the latter development concentrated particle physics research to Fermilab and Stanford, and in Europe, the major CERN upgrade of 1972 to 1976 made it impossible for smaller countries to maintain national experimental particle physics programmes and concentrated the efforts on CERN (Krige, 1996). On both continents, the transition of particle physics into 'megascience' left previous accelerator labs with a mission vacuum and unused talent pools that could be redirected to efforts to put accelerators to work for different purposes (Hallonsten, 2016a: 62 ff.).

Meanwhile, the distrust towards government and especially the military, from the social movements of the 1960s, and the superpower détente after the Cuban Missile Crisis of 1962, had altered the foundations for the 'military-industrial-scientific complex' (Smith, 1990: 71–72). When the economic crisis hit, in the mid-1970s, reprioritizations and reform of the governance structures of science and technology ensued, with bureaucratization taking hold together with more direct efforts by policymakers to steer and direct investments in science and technology, with the intention to boost innovation-based economic growth and strengthen national competitiveness. Particle physics, whose contributions to science were utterly esoteric but could be shown to have important secondary effects through high-tech procurement, still carried prestige not least in the competition between the superpowers. It now had to learn to coexist with other prioritizations, most of all, of course, in the life sciences and in materials science, where keys to the solving of grand challenges seemed to lie, and which had a clearer role to play in the improvement of daily life for people through technological innovation. But life science and materials science lacked the imaginative lure that particle physics had, and in the 1980s, it was still possible to get another big particle accelerator project funded, the Superconducting Super Collider (SSC), which was approved by the US Congress in 1988, at a cost estimate of 3 billion dollars. Five years later, however, as kilometres of tunnel had already been dug on site in Texas, several billion dollars had been spent, and the final price tag had been raised to over 10 billion dollars, Congress terminated the project (Riordan, Hoddess, Kolb, 2015). By then, the Cold War had ended, and a new generation of leaders were on the rise; Bill Clinton (1946-) had embarked on a programme to double the appropriations to the National Institutes of Health (NIH) and strengthen the technological capability of the US in materials and life science, including also the Human Genome Project and, eventually, the National Nanotechnology Initiative (NNI) (Kevles, 1997; Johnson, 2004). The days of Big Science were not over, but it was clear that the landscape had changed dramatically.

Neutrons and X-rays

The advanced study of materials, within solid-state physics, chemistry, and engineering, had received its fair share of funding and attention during and after the Second World War but nonetheless remained in the shadow of nuclear and particle physics.

Wilhelm Conrad Röntgen (1845–1923) had discovered X-rays in 1895, and their penetrating capability was soon utilized in medical applications. In the first half of the twentieth century, X-ray diffraction and X-ray spectroscopy were developed into laboratory techniques for the study of materials, including biomaterials, on molecular and atomic level, and it was soon realized that the quality of the results depended strongly on the intensity of the X rays used. A similar use of neutrons, one of the three constituent particles of atoms, which was discovered by James Chadwick (1891–1974) in 1932, was explored during the Second World War, when reactor construction took full speed and the waste product of nuclear fission, namely neutrons, was abundantly available at the weapons labs in Germany, the United Kingdom, and the United States.

The use of X-rays and other radiation (ultraviolet, infrared) and of neutrons for studies of materials remained on the margins of the postwar buildup of scientific and technological capacity within the 'military-industrial-scientific complex', but breakthroughs in the 1950s aroused interest in wider communities in solid-state physicists, chemists, and also biologists. In 1955, experimental use of neutrons for advanced studies of materials was taking place across the United States and Europe, in virtually every laboratory equipped with a nuclear reactor, and with the US national labs in Argonne, Brookhaven, and Oak Ridge as the leading institutes (Bacon, 1986: 52–119).

Extracting high-intensity radiation (including X-rays) from particle accelerators was a theoretical possibility predicted in a sense by James Clerk Maxwell (1831–1879) in his famous equations of the mid-nineteenth century, in which he postulated that particles of some energy and speed would emit radiation if their trajectory was bent. For particle physicists in the 1950s, such emitted radiation was an energy loss and thus a compromise on the performance of their accelerators and experiments, and 'synchrotron radiation' (as it was called due to its first detection from a synchrotron accelerator in 1947) was meticulously documented and studied by them. Thus, by the early 1960s, the quality of the radiation as well known, and advances in the field of optics and vacuum enabled the extraction of the radiation and its use in diffraction and spectroscopy experiments. Advances in accelerator technology that significantly improved the potential of extracting focussed radiation in continuous beams, increased the viability of routine experimental use of synchrotron radiation. A major breakthrough came in the early 1970s, when materials scientists at Stanford University connected a beam port to a newly built storage ring at the Stanford Linear Accelerator Center (SLAC), one of the US national labs dedicated to particle physics research, and were able to conduct both spectroscopy and crystallography experiments with a quality of results that was many orders greater than what could be achieved with benchtop X-ray sources (Hallonsten, 2015).

The transition of particle physics to 'megascience' that occurred in the 1970s (see above) meant a restructuring of Big Science and the organizations that hosted it. In smaller European countries, national accelerator laboratories were simply closed when CERN expanded to absorb most of national particle physics budgets (e.g. Hallonsten, 2011, 2016a: 88 ff.), and across the US system of national laboratories, a time of crisis ensued. Not only were the 1970s a time of relative decline in the overall US spending on science and technology, and especially Big Science (Hallonsten, Heinze, 2012), but several of the labs that had previously counted on being able to design and build accelerators, and thus renew their mission through participation in the advancement of nuclear and particle physics, now had to look for new projects in order not to stand without a main mission and *raison d'être* (Westfall, 2008; Hallonsten, 2016a: 125 ff.). Fortunately, the cadres of accelerator physicists and engineers that had been educated and kept busy by service to particle physics could be redeployed to other tasks, and in both Europe and the United States, their new partners became materials scientists with ambitions to further the initial success of neutron and synchrotron radiation and fulfil the arguably major potential of both techniques.

Today, several decades later, synchrotron radiation and neutrons are experimental resources used by tens of thousands of scientists across the world within a broad range of disciplines, including physics, engineering, chemistry, biology, and medicine, but also in cross-disciplinary fields, such as materials science, life science, and environmental science, and in remote areas like archaeology and cultural studies. Particle physics is still around, but experimentally conducted only at one place in the world, namely at CERN in Geneva, where in 2012 the Higgs Boson was finally detected by teams of several thousands of scientists and engineers. It could easily be argued that particle physics, because of the global concentration of resources to one hotspot and the growth in teams beyond hundreds of people, is nowadays less of a Big Science and should instead be called 'megascience' or something even larger. Restructuring of global scientific communities and the innovation systems where research environments of the academic and private sectors join forces to develop technologies and advance science for the greater benefit of humanity has also, to some extent, meant the end of those broad and mission-oriented programmes that characterized the Cold War era and made up the science part of the 'military-industrial-scientific complex'. But Big Science has survived through a transformation that has made the use of large accelerator complexes more attuned to today's needs of sustaining interdisciplinary and international R&D efforts to contribute to solving grand challenges (Hallonsten, 2016a).

Technological Evolution

The growth in size of particle accelerators from the mid-1940s and onwards was briefly mentioned in an earlier section. The largest machine ever proposed, the Superconducting Super Collider, which would have been large enough to encircle Manhattan Island (87 kilometres in circumference), was never built, and although particle physicists still dream of a next big machine that would dwarf all existing accelerator complexes (including the 27-kilometre Large Hadron Collider at CERN), it seems the expansion built into the scientific discipline of particle physics ultimately caused its relative decline, or at least the halt of its expansion ambitions. Today, Big Science is significantly smaller than the gargantuan Superconducting Super Collider. The neutron and synchrotron radiation facilities that are the current leaders in the development of these techniques use accelerators that are merely a few hundred metres long.

Most importantly, though, the 'transformed Big Science' that dominates today (for a comprehensive elaboration of this concept and its various parts, see Hallonsten, 2016a) is small science on big machines — i.e. facilities used in scientific projects undertaken by user groups of only a few people, and the 'big machines' that are used are small compared to megascience. The facilities as such are complex and varied, providing a wide range of instruments that can be used in parallel and that are optimized for different types of experiments, and that can also be substituted and upgraded, while all are dependent on the central infrastructure (accelerator, reactor) which remains big enough to warrant collaborative efforts and central governmental investments. Also, when we move beyond particle accelerators and looking at other so-called research infrastructures deemed strategically important for Europe to invest in, many of them cost more than 100 million Euros to build, but very few reach sums over 1 billion Euros (Hallonsten, 2020).

A convenient way of organizing the history of neutron and synchrotron radiation facilities, which stretches back to the 1940s and 1960s respectively and has been characterized by dramatic improvement in the performance of experiments and expansion of use across scientific communities, is to sort the facilities into generations. In the case of neutrons, the reactors built during the Second World War make up the first generation; these were constructed for weapons development and only used in an auxiliary function for other purposes. In the early 1950s, as technological development had advanced and weapons development was a routine activity at the US national labs and their counterparts elsewhere, reactors were custom-built for production of fissile material, for electric power production, or for research, with the last being better equipped for some work to optimize the production of neutrons. These are the second generation, and some reactors in this category are designed specifically to increase the quality of the neutrons and thus to enhance the experimental conditions for neutron diffraction studies. Due to this development, neutron use in the mid-1960s was 'a well established branch of international science' with users mostly in condensed matter physics and chemistry, but also biology (Bacon, 1986: 5). Third-generation neutron sources, i.e. reactor facilities purpose-built for neutron production and use, and fourth-generation neutron sources, i.e. accelerator-based so-called 'spallation sources', saw the light of day in the 1970s and 1980s respectively (Hallonsten, 2016a: 72–73).

Synchrotron radiation sources are similarly sorted into generations, but since this use of the radiation unavoidably emitted by particles in motion began later than neutron use and was more closely tied to the technological and scientific developments in particle physics, the categorization is somewhat different. The first-generation synchrotron radiation sources were the labs built at particle physics facilities in the 1960s and 1970s, called 'parasitic' by some (Hallonsten, 2015: 219). As results came out and the great potential of the technique was demonstrated, plans were soon made for purpose-built facilities, and these make up the second generation, built and operated mainly in the late 1970s and early 1980s. A technological innovation called 'insertion devices', which enables the manipulation of the particles in the accelerator so that they take several turns in straight sections of the accelerator and thus emit significantly more radiation, instead of merely turning once in every corner or bent of the accelerator, revolutionized synchrotron radiation and inaugurated the third generations — purpose-built facilities optimized for the use of such insertion devices. These were designed and built in the late 1980s and onwards, and constituted the most significant breakthrough for synchrotron radiation, with user numbers multiplying in the 1990s, and not least with users from biology and medicine (including the drug industry) entering in large numbers (Hallonsten, Heinze, 2015). A fourth generation is sometimes also mentioned in this succession, namely free electron lasers which use linear accelerators to produce laser radiation for experimental use similar to synchrotron radiation use, but since they are not synchrotron radiation facilities in the traditional sense, their place in this chronology is contested; furthermore, they did not replace third-generation synchrotron radiation facilities but complement them with new experimental opportunities on the basis of dramatically improved performance on some specific parameters (Hallonsten, 2016a: 253 ff.).

Contemporary Big Science facilities that are big machines used for small science, are extremely complex entities, both technologically and organizationally, which means that the user-friendliness and efficiency of operations that are demanded of these facilities today — since they serve mainly as a resources that dynamic and rather volatile user communities make use of — depends upon a refinement of both technological and organizational arrangements that has been achieved in a complex process not unlike what sociologist of science John Law dubbed 'heterogeneous engineering', meaning that 'bits and pieces, social, natural, physical, or economic, are interrelated and keep each other in place' (Law, 1987: 231).

A user facility, serving hundreds or thousands of scientists every year who apply for access in competition and make visits of only a few days to conduct experiments, obviously needs a well-oiled organization and a robust technological system. Accelerator operation and development, instrument operation and development, scheduling of user visits, sample preparation facilities, data storage and management, and taking care of administration and finances are obviously very diverse activities that all require their own professionals and apt organizational structures to be successful. And success in this regard is measured not by traditional indicators of output, such as productivity (publications) or efficiency of operations (limited downtime), since these are oversimplified and only partially capture the dynamics of user facilities whose primary missions are to serve a broad and transient community of visiting scientists (Hallonsten, 2013, 2016b). Success is broader and cannot be simply measured, but depends on the whole operation of the very complex technological systems and organizations that these contemporary Big Science facilities operate.

The emergence and growth of Big Science in the early Cold War era, and its transformation towards the end of the twentieth century, was a largely evolutionary technological development process rather than one characterized by discontinuous breakthroughs. Some specific innovations, of course, stand out, particularly the early accelerator and reactor development before and during the Second World War, which set off the development of Big Science, and later the insertion device technology in the case of synchrotron radiation facilities and the move from reactor-based to accelerator-based (spallation) neutron sources. But it is important to note the gradual changes and how technology quite evidently developed incrementally; accelerator complexes with experimental use in the sciences are extremely complex entities, and hence their refinement is a continuous process of inventing new technology on the micro level, improvement of the stability of operations of existing technical solutions, and adaptation to new demands from science, and by extension, society.

The transition of particle physics to 'megascience', and what this brought in the longer perspective in terms of redrawn maps of resource distribution and scientific use of large accelerator complexes, did not happen through a radical invention but by a gradual and reciprocal adaptation of technology and its scientific use to new conditions. The political and organizational aspect should not be underestimated; it was not a crisis of confidence in the particle accelerators per se that caused the demise of the Superconducting Super Collider and the reorientation (transformation) of Big Science, but changes in the politics and organization of and around (Big) science. Similarly, the real breakthrough for neutron and synchrotron radiation facilities as highly sophisticated and complex technological resources for experimental science was not technological, but mainly organizational. It came with the reorientation of the organizations and their stated purposes to this goal: to serve users and function as central nodes in the complex and broad networks of users. Europe entered an era of global leadership in neutron instrumentation and its use with the start of user operation in 1972 of the Institute Laue Langevin (ILL) in Grenoble, built jointly by France and West Germany, and with the United Kingdom joining as a member in 1974. The ILL was located in Grenoble which, together with Munich, formed the 'neutron axis' of Europe (Pestre, 1997: 138), and as the first-ever purpose-built neutron scattering facility it was designed as a service facility for all of (Western) Europe's current and future neutron users, with technical solutions and organizational arrangements deliberately implemented to put service to the user community first, which made the ILL and its users all over Europe take over the global lead in neutron scattering from the United States, which still holds (Bacon, 1986: 7; Rush, 2015: 140). The ILL served as a role model not least for the European Synchrotron Radiation Facility (ESRF), located next to it in Grenoble and opening to scientific use in 1994. The ESRF is organised as a European collaboration and, similar to the ILL, has become a global leader in synchrotron radiation mainly due to its heavy focus on user service which runs through the whole organization, from accelerator operation to administration, over scheduling of experimental time and instrument development (Hallonsten, 2013). Key to the breakthrough of synchrotron radiation in science — which took the technique 'from esoteric endeavor to mainstream activity' (Hallonsten, 2016a: 76) and was manifested especially in the construction of several new and dedicated synchrotron radiation facilities across Europe and the United States in the early 2000s — is their user orientation, which has lowered the thresholds significantly also for user groups less experienced in high-tech instrumentation and big facilities (e.g. biology, environmental science, cultural studies) (Hallonsten, Heinze, 2015). It was certainly enabled by technological innovations like insertion devices and the most recent advances in accelerator technology, not to mention vacuum, optics, handling and storage of data, sample preparation, and so on, but all these technical components need to be refined and deployed to create a coherent whole that caters to the need of a broad and dynamic user community. Such is the logic of the (technologically) transformed Big Science.

References

Bacon, George Edward, ed., *Fifty Years of Neutron Diffraction. The Advent of Neutron Scattering* (Bristol: Adam Hilger, 1986).

Eisenhower, Dwight, Military-Industrial Complex Speech, (1961) <https://avalon.law.yale.edu/20th_century/eisenhower001.asp> [accessed 21 November 2019].

Giroux, Henry A., *The University in Chains: Confronting the Military-industrial-academic Complex* (London: Paradigm Publishers, 2007).

Greenberg, Daniel S., *The Politics of Pure Science*, 2nd ed. (Chicago, IL: The University of Chicago Press, 1999).

Hallonsten, Olof, *Big Science Transformed. Science, Politics and Organization in Europe and the United States* (Cham: Palgrave Macmillan, 2016a).

———, 'Growing Big Science in a Small Country: MAX-lab and the Swedish Research Policy System', *Historical Studies in the Natural Sciences*, 41.2 (2011), 179–215.

———, 'Introducing Facilitymetrics: a First Review and Analysis of Commonly Used Measures of Scientific Leadership Among Synchrotron Radiation Facilities Worldwide', *Scientometrics*, 96.2 (2013), 497–513.

———, 'Is There an 'Iron Law' of Big Science?', in *Big Science and Research Infrastructures in Europe*, ed. by Katharina C. Cramer and Olof Hallonsten (Cheltenham: Edward Elgar Publishing, 2020), pp. 217–31.

———, 'The Parasites: Synchrotron Radiation at SLAC, 1972–1992', *Historical Studies in the Natural Sciences*, 45.2 (2015), 217–72.

———, 'Use and Productivity of Contemporary, Multidisciplinary Big Science', *Research Evaluation*, 25.4 (2016b), 486–95.

——— and Thomas Heinze, 'Formation and Expansion of a New Organizational Field in Experimental Science', *Science and Public Policy*, 42.6 (2015), 841–54.

——— and Heinze, Thomas, 'Institutional Persistence Through Gradual Adaptation: Analysis of National Laboratories in the USA and Germany', *Science and Public Policy*, 39 (2012), 450–63.

Hewlett, Richard G. and Jack M. Holl, *Atoms for Peace and War, 1953–1961* (Berkeley, CA: University of California Press, 1989).

Hoddeson, Lillian and Adrienne W. Kolb, 'The Superconducting Super Collider's Frontier Outpost, 1983–1988'; *Minerva*, 38 (2000), 271–310.

———, ———, and Catherine Westfall, *Fermilab: Physics, the Frontier & Megascience* (Chicago, IL: The University of Chicago Press, 2008).

Johnson, Ann, 'The End of Pure Science: Science Policy from Bayh-Dole to the NNI', in *Discovering the Nanoscale*, ed. by David Baird, Alfred Nordmann, and Joachim Schummer (Amsterdam: IOS Press, 2004), pp. 217–30.

Kevles, Daniel J., 'Big Science and Big Politics in the United States: Reflections on the Death of the SSC and the Life of the Human Genome Project', *Historical Studies in the Physical Sciences*, 27.2 (1997), 269–97.

Krige, John, ed., *History of CERN*, vol. III (Amsterdam: North-Holland, 1996).

Law, John, 'On the Social Explanation of Technical Change: the Case of the Portuguese Maritime Expansion', *Technology and Culture*, 28.2 (1987), 227–52.

Martin, Joseph D., 'Fundamental Disputations: the Philosophical Debates that Governed American Physics, 1939–1993', *Historical Studies in the Natural Sciences*, 45.5 (2015), 703–57.

Pais, Abraham, *Inward Bound: Of Matter and Forces in the Physical World* (Oxford: Oxford University Press, 1986).

Pestre, Dominique, 'Prehistory of the Franco-German Laue-Langevin Institute', in *History of European Scientific and Technological Cooperation*, ed. by John Krige and Luca Guzzetti (Luxembourg: Office for Official Publications of the European Communities, 1997), pp. 137–43.

———, 'Science, Political Power and the State', in *Companion to Science in the Twentieth Century*, ed. by John Krige and Dominique Pestre (London: Routledge, 2003), pp. 61–76.

——— and John Krige, 'Some Thoughts on the Early History of CERN', in *Big Science: the Growth of Large-Scale Research*, ed. by Peter Galison and Bruce Hevly (Stanford, CA: Stanford University Press, 1992), pp. 37–60.

Riordan, Michael, Lillian Hoddess and Adrienne W. Kolb, *Tunnel Visions. The Rise and Fall of the Superconducting Super Collider* (Chicago, IL: The University of Chicago Press, 2015).

Rush, John J., 'US Neutron Facility Development in the Last Half-century: a Cautionary Tale', *Physics in Perspective*, 17 (2015), 135–55.

Smith, Bruce L. R., *American Science Policy since World War II* (Washington, DC: Brookings Institution, 1990).

Weinberg, Alvin M., 'Impact of Large-scale Science on the United States', *Science*, 134 (1961), 161–64.

Westfall, Catherine, 'Surviving the Squeeze: National Laboratories in the 1970s and 1980s', *Historical Studies in the Natural Sciences*, 38.4 (2008), 475–78.

ALESSANDRO NUVOLARI, CATERINA SGANGA

Technique and Law

Virtually all historians recognize that the Industrial Revolution was a fundamental turning point in the history of humanity. Before the eighteenth century, economic growth was inconsistent and sporadic, so that stagnation in living standards was the defining feature of economic life. Underlying the development of this 'new' economic regime, which Nobel laureate Simon Kuznets termed 'Modern Economic Growth' (Kuznets, 1966), was a radical transformation in the process of technological change (Landes, 1969). Before the Industrial Revolution, innovations and changes in technological practices were, by and large, the results of accidental discoveries or improvements arising from processes of learning-by-doing and learning-by-using. The Industrial Revolution witnessed the development of a radically new approach to inventive activities characterized by the systematic search for innovations and improvements. A. N. Whitehead (1929: 120) pointedly labelled this transformation the 'invention of the method of invention'. This method involved the systematic utilization, in the struggle for inventions and improvements of existing technologies, of the existing stock of knowledge about natural phenomena. Moreover, the sustained effort in inventive activities prompted an incremental accumulation of the knowledge of the natural world, which, in its turn, fed back on technological developments. It is worth noting that, as Kuznets has aptly remarked, the stock of knowledge exploited in the search for innovations is much broader than academic science, for it encompasses all the knowledge of natural phenomena that is potentially useful for the development and improvement of technologies (Mokyr, 2002). In this context, innovations became a sort of regular steady flow constantly reconfiguring the structure of the production system.

Against this general background, economic historians have traditionally found it useful to adopt a periodization that considers the period that runs from 1750 to date as comprising three major historical phases: first, the First Industrial Revolution, spanning roughly the period from 1760 to 1880, characterized by the mechanization of many production processes and by the emergence of capital goods industries that focussed on the production of machines and machine tools (Rosenberg, 1963); second, the Second Industrial Revolution, approximately spanning the period from 1880 to 1970, characterized by the emergence of firms with laboratories carrying out systematic research activities (corporate R&D) and of large firms involved in the mass

Global History of Techniques, ed. by Guillaume Carnino, Liliane Hilaire-Pérez and Jérôme Lamy, Global Matters Culture Matérielles, histoire des techniques et circulation globale des savoirs, 9 (Turnhout: Brepols, 2024), pp. 545–562
BREPOLS PUBLISHERS 10.1484/M.TECHNE-EB.5.129829

production of consumer durables (Chandler, 1990); and third, the Third Industrial Revolution, characterized by the development of information and communications technologies (microelectronics, software, and the internet; see Isaacson, 2014). This periodization must be understood just as a broad frame of reference that can be useful as a descriptive device to summarize long-run technological trends. However, it is important to recognize that the long-run technological change is actually an evolutionary process characterized to a large degree by continuities and path dependence. This means that new technological systems will not completely 'substitute' previous technologies, as a naïve formulation of the notion of 'industrial revolution' may imply. Rather, at any given point in time, the structure of the production system will be constituted by 'layers' of different technologies with a strong overall continuity. For instance, the emergence of information and communications technologies does not imply that the large-scale technological system of electricity is displaced, but, more precisely, that information and communications technologies will co-evolve interacting with the technical system of electricity. In this way, the overall structure of the production system will gradually become more articulated and complex.

Clearly, these technological revolutions have been necessarily associated with a transformation of the socio-institutional system. Innovation and technical changes are inherently disruptive. Furthermore, the social benefits and costs related with the application of some new technologies may become apparent only after significant delays. Accordingly, the accommodation of new technologies in the broader social system may require significant revisions in the existing legal and regulatory frameworks. However, changes in the legal and regulatory framework, in turn, may affect inventive activities (e.g., the ban of a pollutant substance may stimulate the search for production processes using alternative materials). For these reasons, many social scientists argue that a proper understanding of the connection between long-run technical change and the transformation of the socio-institutional framework (comprising the legal system) must be based on a co-evolutionary perspective (Freeman, Louca, 2001).

This is the perspective that we will take in this chapter. We shall focus on three main themes that offer an insightful vantage point to study the connection between technical change and legal system, and that have important implications also for contemporary policy discussions. We shall begin by discussing the connection between inventive activities and the development of patent systems taking place during the First Industrial Revolution. We shall then move to the Second Industrial Revolution and the shift from independent inventors to corporate R&D laboratories. Finally, we will consider the tension between standards and patents during the Third Industrial Revolution.

Industrialization and the Patent System

In a major work of synthesis on long-run economic growth, Nobel laureate Douglass North noticed:

> The failure to develop systematic property rights in innovation up to fairly recent modern times was a major source of the slow pace of technological change [...][;] a systematic set of incentives to encourage technological change was established only with the patent system (North, 1981: 164–65).

North links the origins of the Industrial Revolution in Britain explicitly with the creation and the progressive operationalization of the English patent system, following the enactment of the Statute of Monopolies (1624). In the Statute, patents for inventions became an exception to the general prohibition of all monopoly privileges granted by the Crown. However, the roots of modern patent systems are to be found in earlier developments. Their origins lie in the proprietary and exclusionary attitudes towards craft knowledge that had emerged within the guild systems (Long, 1991). Early modern technology was, to a major extent, a matter of 'expert' knowledge and 'tacit' skills. The guild system represented the fundamental institutional set-up in which this kind of knowledge was stored and transmitted to the younger generations of apprentices. In this context, the diffusion of technology was the result of the international migration of skilled workers (Cipolla, 1972). The state of continuous military and economic competition between Renaissance states and polities made them keenly aware of the key role played by the acquisition of skilled workers from abroad in the process of international technology transfer. Accordingly, from the late Middle Ages, all over Europe, laws and regulations restricting the emigration of skilled craftsmen were introduced and refined. Concomitantly, there were continuous attempts of suborning skilled workers from other countries by offering them different kinds of incentives. One of the most common was the granting of special 'privileges' for the importation of new technologies, consisting in the exclusive right to use the technology in question for a limited amount of time (Belfanti, 2006).

In 1474 the Venetian government enacted a specific statute that provided a systematic codification of these previously ad hoc procedures. This law comprised, in an embryonic form, several fundamental traits of modern patent legislations. In particular, the law prescribed that the grants of the Statute, rather than being subject to the discretionary power of the government, depended on the applicants' ability to fulfil a specific set of criteria. First, the inventions in question should have resulted 'useful' for the State when implemented; second, and most importantly, they should not already be known within the Republic. In this way, the Venetian law inaugurated the two fundamental criteria of utility and novelty for the patentability of inventions that would later be taken up by all subsequent patent legislations.

The Venetian model of protection and exclusive privileges spread quickly to other states such as Spain, Holland, France, and England. This is hardly surprising, since, within the mercantilistic model of political economy, import substitution represented a key policy goal for avoiding outflows of bullion in favour of rival states — hence the cultivation of domestic technological capabilities that could enhance the competitiveness of domestic industries and thereby increase exports and reduce imports. In this context, privileges of invention by encouraging the immigration of foreign craftsmen represented a very effective tool of technology policy (Hilaire-Pérez, 2000).

In England, the Statute of Monopolies of 1624 had the objective to curtail abuses in granting privileges of monopoly which disrupted established trades. As we have already mentioned, in the Statute, patents for inventions became the only exception to the prohibition of monopolies. The Statute also introduced a fixed term of fourteen years for the duration of the exclusive right (corresponding to twice the common term of apprenticeship).[1]

Another key element for the functioning of early patent systems was the description of the invention. In the case of ad-hoc privileges, the inventor was actually requested to put his invention in practice. In the case of patents, the patentees' obligations concerning the disclosure of innovations remained relatively ambiguous throughout the sixteenth and seventeenth centuries. Bottomley (2014) has noticed a significant shift in the English practice from the early eighteenth century, when law officers started requiring patentees to submit a complete and rather detailed description of the inventions. This approach resulted in a reconfiguration of the rationale for patent grants, where patents were conceived as a substantially quid-pro-quo exchange between the inventor and the government ('patent bargain'). In other words, in exchange of a complete disclosure of the invention to the public, the inventor was given an exclusive right for a limited period to the use of the invention (Biagioli, 2011). This new rationale was explicitly elaborated in the IP clause of the US constitution (art. 1, section 8, § 8) and early patent legislation. Furthermore, in the United States, the fees that inventors were required to pay were relatively modest, in order to maximize access to the system for the largest segment of the population possible (Khan, 1999).

This new conceptualization of the function(s) of the system provided the foundation for the modern understanding of patents as 'property rights'. In this case, the first legal formulation is the French patent law enacted in 1791, which explicitly recognized property right bestowed by the patent grant as a natural right of the individual. At least from a formal point of view, the French patent law of 1791 may be seen as a watershed marking the definitive shift of patents from Ancien Régime privileges to intellectual property rights (Galvez-Behar, 2008; Baudry, 2020). Here, however, the change in the legal framework was mostly a byproduct of the philosophical and political approach followed by the French legislator after the Revolution. To break the spell of the Ancien Régime and affirm the independence of individuals from the tyranny of the State and the feudal order, property rights were conferred to citizens in order to provide them with the means to exercise their freedoms, develop their personality, and have a decent standard of living. In the legal vocabulary of the French Revolution, 'property' was the external manifestation of individual liberty and was granted not only on lands and material goods, but on every asset having a commercial value and/or being generated by the work of an individual (Gambaro, 2011). Inventors were granted property rights on their innovations, as much as authors were granted property rights over their creations, on the basis of natural law arguments and not of any utilitarian rationale. It was only due to a clear 'heterogony

1 The duration of protection contemplated in the Venetian patent law was ten years.

of ends' that the effect of the shift towards a property label empowered the tool insomuch as to transform it into an innovation policy instrument.

Interestingly, it is not possible to trace a similar sequence of events in China. In the Chinese context, the connection between the notions of authorship, ownership of knowledge, and a possible role for the State in granting and enforcing exclusive rights remained extremely hazy. This cultural outlook, giving paramount importance to the respect of an established canon in many creative domains, accounts for the lack of development of intellectual property laws in the Chinese empire (Alford, 1995).

What was the impact of the emergence and consolidation of patent legislation on the rate of technological progress? Economic historians have discussed the matter at length, and the debate does not show signs of imminent closure (MacLeod, Nuvolari, 2016). Clearly, the assessment is extremely difficult, and it also entails, to a great extent, counterfactual speculation (what would have been the rate of technical change without the patent system?). To be sure, several technological breakthroughs, such as James Watt's separate condenser, Richard Arkwright's water frame, and Edmund Cartwright's power loom, were patented. Furthermore, when the history of these inventions is examined in detail, it turns out that the prospect returns ensured by the patents were an incentive that their inventors took, in fact, into account. On the other hand, it is also possible to point to other technological breakthroughs, such as Samuel Crompton's spinning mule or Henry Maudslay's screw-cutting lathe, which were not patented. Meisenzahl and Mokyr (2012) have constructed a sample of 759 notable British inventors distilled from biographical dictionaries and other specialized sources. Of these, 60% took at least one patent. On the other hand, Petra Moser's research on the inventions presented at the Crystal Palace exhibition in 1851 (Moser, 2005) shows that only a relatively minor share of exhibits was patented (11% for the British and 14% for the American case). On reflection, this variety in quantitative appraisals of the propensity to patent inventions during the Industrial Revolution should not be too surprising, since it is simply a consequence of the nature of the different samples used, which can capture the elusive 'universe' of inventive activities in this period only in a partial and imperfect manner. Be this as it may, this research suggests that a significant segment of inventive activities took place outside the coverage of patent protection. This, in turn, indicates that inventors could probably reap economic returns from their inventive activities by means of appropriability strategies that did not contemplate the use of patents.

Interestingly, in this period it is also possible to find cases in which communities of inventors, rather than taking patents, preferred to carry out inventive activities practicing a very 'open' approach characterized by the systematic sharing of technical knowledge. Allen (1983) noticed this type of behaviour (which he termed 'collective invention') in the iron industry of Cleveland (UK) in the second half of the nineteenth century. Nuvolari (2004) has documented another significant case of collective invention during the nineteenth century. In this case, the technology developed by means of knowledge sharing was the steam pumping engines that were used to drain the copper and tin mines of Cornwall. In Cornwall, knowledge sharing activities revolved around a monthly journal called *Lean's Engine Reporter*, which reported on the technical characteristics and the performance of the engines at work in the Cornish mines (Figure 1). The systematic sharing of this type of data permitted a timely identification of the most fruitful trajectories of technical advance. At a close reading, it is, indeed, possible to find several other cases of collective invention during both the nineteenth and twentieth centuries (Bessen, Nuvolari, 2016). Overall, this line of research highlights that major industrial technologies such as blast furnaces and steam engines could be effectively developed outside the purview of the patent system. This form of organization of inventive activities seems to have a modern counterpart in the open-source approach to software production. Here, the first program developed may or may not be covered by a patent, but the open-source philosophy approach is enforced through licence agreements which implement a 'no-right-reserved' approach, leaving developers free to build upon the software upon the obligation not to file for a patent for the derivative invention (Carver, 2005).

Finally, when assessing the role of the patent system, one has also to consider the detrimental effect exerted on inventive activities by 'blocking patents', that is, broad patents that completely foreclose inventive activities in certain specific domains. Kanefsky (1978) and Torrens (1982) argue that Boulton and Watt's separate condenser patent obstructed technical progress in steam engineering, because it gave to the two partners the opportunity to interfere with the development of the compound steam engine of Jonathan Hornblower. It is not far-fetched to imagine that without Watt's patent, the rate of technical change in steam engineering would have been significantly faster.

These considerations illustrate the difficulties of pinning down the specific role played by patent systems in the development of the Industrial Revolution, mostly engendered by the complex set of connections that linked the patent system to inventive activities in this period. Furthermore, there is evidence that a significant share of inventive activities took place outside the coverage of the patent system. In this context, the main conclusion that one could draw is that patents were not the fundamental institutional prerequisite for the Industrial Revolution, as hinted at by North (1981). Rather, one might suggest more soberly that the patent system was just one component of a much broader and articulated innovation system, constituted both by formal and informal institutions, which gradually emerged and consolidated between the sixteenth and the nineteenth centuries (MacLeod, Nuvolari, 2012, 2016). In this respect, it is worth noting that throughout the nineteenth century, countries such as the United States and Germany, either *de jure* or *de facto*, accorded a very ineffective protection to foreign patentees. This discrimination provided some significant opportunities for the adoption and development of foreign technologies and, in this way, for accelerating the industrialization process. An interesting example of this kind of technology transfer, possibly instigated by discrimination approach towards foreign patentees, is the case of importation of spinning and weaving machines from England to the United States during the first half of the nineteenth century (Moser, 2021).

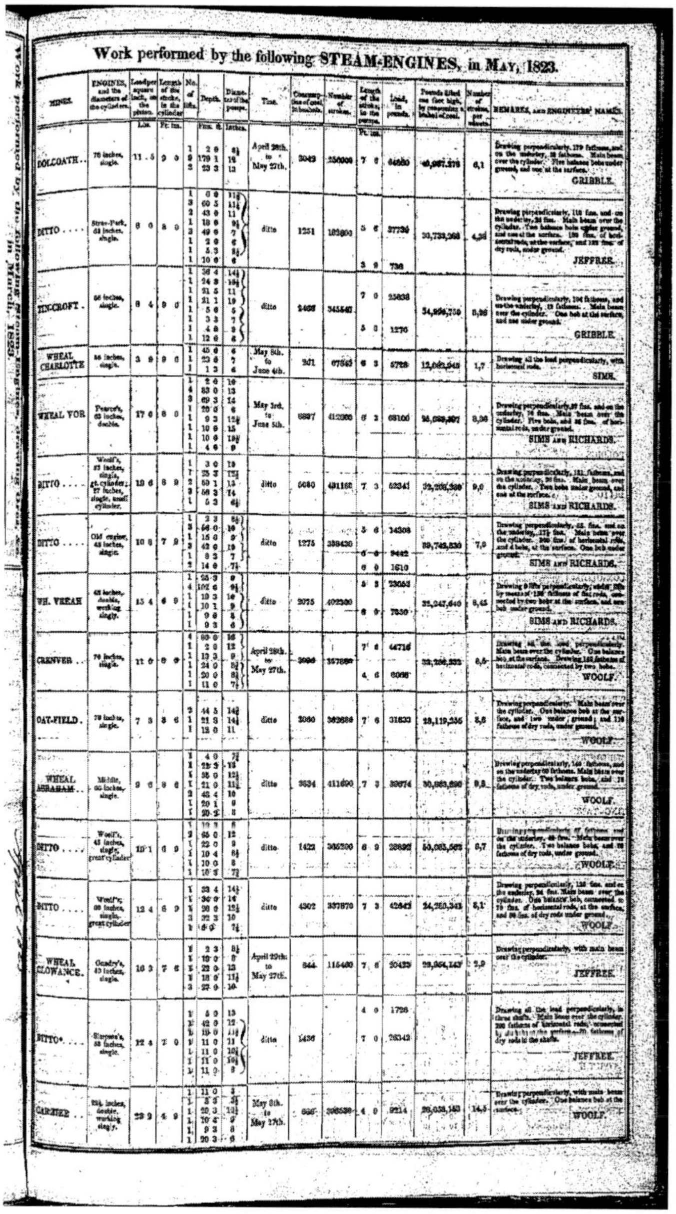

Work performed by the following STEAM-ENGINES, in MAY, 1823.

Fig. 1. Page from the *Lean's Engine Reporter* reporting the performance of the steam engines at work in the mines of Cornwall.

The Industrialization of Invention

The development of the 'technological systems' of the Second Industrial Revolution determined a significant transformation in the organization of inventive activities. During the First Industrial Revolution, the search for innovations was undertaken by independent inventors or, in some cases, by communities of independent inventors engaging in knowledge sharing activities as in collective inventions settings. The search for inventions in the newly emerging fields of chemicals and electricity required the use of major pieces of experimental equipment, the integration of different technical skills and scientific competences, and considerable development times. Accordingly, leading firms active in these industries established dedicated research laboratories where inventive activities were carried out, employing teams of professional scientists and engineers (Tunzelmann, 1995). Historian of science J. D. Bernal (1953) has pointedly described this historical process as 'the industrialization of invention'.[2]

From the perspective of the interaction between law and technology, it is important to highlight that the Second Industrial Revolution prompted a significant reconfiguration of the legal framework of the patent system. In cultural terms, there was a collision between the organizational set-up of the corporate R&D laboratories and the romantic perspective of the 'heroic' inventor, which was at the base of the individualistic perspective on patents that marked the transition from privileges to property rights discussed in the previous section.

The emergence of industrial research was coupled with the increasing use of employment contracts that assigned ownership and control of the patents of the inventor-employees to the corporate employer (Fisk, 2009). Under US patent law, corporations could be assignees of patents, but they could not be patentees, the latter being a status available only to natural persons. This was in line with the notion that an invention was a sort of 'single' event made by one individual (or also a group of individuals), in certain place and at a specific time. While this notion could serve well enough to adjudicate contentions in a world of competing individual inventors, it was not very effective when innovations were the outcome of large-scale research activities involving teams of scientists and engineers with prolonged development times. Following the diffusion of new employment contracts, the number of US patents assigned at the time of issuance to companies increased dramatically, from 12% in 1885 to more than three-quarters in 1950 (Fisk, 2009). As Fisk (2009) has shown, the emergence of corporate R&D also called for a corresponding change in the legal doctrine, with reverberations also on the more general cultural outlook on the inventor and his or her societal role. The romantic view of the inventor as a creator inspired by flashes of genius was dislodged in favour of that of a professional working methodically in a specialized working environment. Progressively, the corporate employers were increasingly successful in claiming before courts the ownership of their employees' inventions in a wider set of circumstances (Fisk, 2009).

2 It is interesting to note that recent research in economic history, even if still recognizing the historical significance of the emergence of corporate research and development, has pointed out that independent inventors did not completely disappear, and that some significant breakthroughs, also in this period, were developed by independents who could work with more freedom outside the bureaucratized context of the corporate R&D laboratory (Nicholas, 2010).

The other major development of this historical phase is the gradual development of a patent system with an international dimension. Patent systems had evolved from the Old Regimes privileges maintaining a quintessential national scope. Accordingly, national interests played a very important role in the functioning of the systems throughout the nineteenth century. In more general terms, considerations of national interest took the form of the discrimination of foreign patentees *de jure* or *de facto* (by means of higher patent fees, more restrictive conditions for granting and working the patent, etc.). However, the increasing globalization of the world economy in the second half of the nineteenth century prompted efforts for establishing some degree of harmonization. The first result was the Paris International Convention for the Protection of Industrial Property of 1883. The Convention established two important principles. The first is the principle of 'national treatment', which states that foreign inventors should be treated by each national patent system in exactly the same way as 'domestic' inventors. The second principle is the 'priority rule', which establishes that a 'priority claim' in one State member of the Paris Union must be recognized in all others. More specifically, when a patent is applied for in one member state, the applicant has a fixed period in which he can apply a patent for the same invention in the other member states. In this case, the protection of the subsequent applications dates from the original application (Guellec, van Pottelsberghe de la Potterie, 2007).

The original version of the Paris Convention did not contain any substantive-law definitions. The reasons of this omission were twofold. On the one hand, contracting parties believed that the national treatment and priority principles were enough to align their legal systems to their cross-border innovation policies. On the other hand, national patent laws did not feature enough points of convergence to facilitate an agreement on minimum common rules (Gervais, 2002). However, this approach gradually changed throughout the numerous revisions the Convention was subject to from 1883 to 1967. The number of contracting parties increased, and so did the number of substantive provisions, which were directed to set a minimum standard of protection and shared definitions on terms of protection and rights conferred. Significantly, the Paris Convention also marked a strengthening of patent protection, mostly through the curtailment of working requirements, which led to the exclusion of patent forfeiture for failure to work the invention. Forfeiture was also excluded in case of the patentee's importation, into the country where the patent was granted, of patented goods manufactured in other countries. The two provisions were introduced to limit the policy space left to contracting parties to implement protectionist measures, which often consisted in forcing foreign inventors and/or firms to establish manufacturing facilities in the country in order to maintain patent protection (Ricketson, 2015). The road towards a supranational standardization of patent laws and the abatement of national borders was already paved.

As Gervais correctly noted, the process launched by the Paris Convention was destined to be carried even further as soon as major IP-exporting countries realized the key role played by intellectual property in international trade (Gervais, 2002). Unsurprisingly, in fact, there has been a much stronger push since the 1980s, mostly due to the United States, towards the implementation of a stricter global intellectual property regime. The TRIPS Agreement (1994) can be regarded as a further stage in this process of global harmonization of IPR systems. Deeply embedded into the Uruguay Round negotiations, the Agreement was approved in Marrakesh in 1994 as a key pillar of the new WTO rules, and marked a substantial shift compared to the GATT, where intellectual property featured only in side provisions and mostly as an obstacle to international trade. With the TRIPs Agreement, IPRs were formally introduced within the framework of the WTO negotiations. No WTO member may opt out from the Agreement; at the same time, failure to comply with the stricter standards of IPRs protection defined by the TRIPs can result in a proceeding before the WTO Dispute Settlement Body and in subsequent trade retaliatory measures. Compared to the much lighter standardization impressed by the Paris Convention, the free adhesion to it, and the lack of real enforcement measures against non-compliance, the Agreement could not but result in a historical increase of pressure towards stronger IPR regimes worldwide (Ullrich, 2016).

Such a shift in the international IP law framework is responsible for a drastic change of conditions for countries that have approached the various phases of the industrial revolutions with some decades of delay. While today's developed countries embarked in their processes of industrialization during the nineteenth century with a relatively wide range of opportunities for finding configurations of their patent law that could accommodate their national interest, the room for developing countries for autonomous policy making in this crucial domain is now severely restricted (Chang, 2002; Gervais 2005; Yu, 2006, Helfer, 2004). The impact of such a compression of national sovereignty and discretion in the realm of innovation policies mostly descend from the regulatory matters on which the WTO decided to intervene. The TRIPs Agreement, in fact, provides a much more pervasive standardization of substantive patent rules. Suffice to say that Article 27 TRIPs prohibits discrimination between technological sectors and obliges WTO member states to grant patents for inventions in all fields of technology, with very limited and narrowly described exceptions. This made it impossible for national legislators to exclude specific industry sectors from patent protection on the basis of broader considerations of national industrial policies or other public policy goals (Dinwoodie, Dreyfuss, 2006). Similarly, the Agreement standardised the rights conferred in the case of product and process patents (Article 28), the main requirements of patentability (Articles 27 and 29), focussing particularly on disclosure, and the term of protection (Article 33), setting it to a minimum of twenty years from the filing date. And, to limit ever more the possibility for national legislators to depart from the high level of protection set by the WTO, Articles 30 and 31 circumscribe the conditions upon which member states may introduce exceptions to patent exploitation (i.e. free uses without authorization) and provide for compulsory licences. Unsurprisingly, this reduced the ability of developing countries to devise instruments directed to balance patent law against other public policy goals, ranging from the protection of health to systemic economic stimuli going beyond the mere provision of antitrust rules (Matthews, 2015; Ho, 2011). Last, but not least important, the TRIPs Agreement represented the first international act introducing provisions with regard to IP enforcement. Under Articles

41 to 49 TRIPs, WTO member states are now obliged to introduce in their legal systems provisions regarding evidence production and protection, damage calculation, injunctions, rightholders' right of information, and a set of other remedies to create an effective deterrent to infringement, such as the withdrawal from the channels of commerce and the destruction of infringing goods, without compenzation. As a consequence, costly enforcement systems and the heavy economic load they impose on national budgets have been also forced upon developing countries, increasing — if at all possible — the hardship of the side effects triggered by the supranational standardization of IP (and especially patent) regimes (Correa, 2020).

The TRIPs Agreement does not mark the end of the story of the headlong rush towards a pervasive international standardization of patent laws and the irreversible process of strengthening patent protection worldwide. Basically, every bilateral investment treaty (BIT) and multilateral trade agreement (MTA) concluded from 1996 onwards contains so-called 'TRIPs-plus' provisions, which are directed to increase the protection offered to IPRs by the 'receiving' (often developing) states beyond the TRIPs minimum standard, in order to reach levels similar to, if not higher than, those characterizing the legal systems of main international trade players, such as the US and the EU (Grosse Ruse-Khan, 2011; Drexl, 2016; Correa, 2004; Drahos, 2003). Along the same lines, a number of IP-exporter countries have tried to conceive additional instruments directed to establish international standards for IPRs enforcement. A chief example was the Anti-Counterfeiting Trade Agreement (ACTA), a proposed multilateral treaty which aimed at establishing an international legal framework to combat and prevent the commercialization of counterfeited patented goods and services, the parallel import of generic medicines, and the spread of copyright piracy.[3] ACTA also provided for the creation of an ad-hoc governing body outside the WTO, the WIPO, and the UN. The new Agreement needed the ratification of six countries to enter into force, a target that was never achieved, predominantly due to the large public outcry triggered by the secrecy that characterized its negotiation and the strong opposition of EU member states, and culminated in the rejection of its ratification by the European Parliament in July 2012 (Blakeney, 2012).

Technological Systems and Standards

The emergence of large-scale complex technological systems with many interacting components and spanning different engineering fields provided a major impetus towards the establishment of guidelines that could ensure the standardization of components, and thereby the compatibility and functioning of the system. This was a gradual process. Professional engineering societies at national level formed committees that outlined technical standards for industrial equipment such as steam boilers,

3 At the signature in October 2011, ACTA was supported by Australia, Canada, Japan, Morocco, New Zealand, Singapore, South Korea, and the United States. In 2012 these countries were joined by Mexico, the European Union, and twenty-two of its member states.

screw threads, and steel rails. This process culminated in the constitution of the Engineering Standards Committee in Britain (1901). The first general standard organization at a national level, this organization embodied the principles of operation that will characterize private, voluntary standard setting throughout the nineteenth century, the most important ones being, first, the involvement of a wide range of stakeholders; second, the voluntary compliance to the standard; and third, the continuous revision of standards in order to incorporate technical advances (Yates, Murphy, 2019). The international process of standardization received further impetus after the Second World War with the creation of the International Organization for Standardization (ISO). One of the main successes of this institution was the establishment of successful standard for shipping containers (Yates, Murphy, 2019). The development of information and communications technologies witnessed a further discontinuity in standard setting processes. In this domain, standards were the outcome of consortia of companies, which were less inclusive with respect to stakeholders, but could enjoy faster decision processes (Yates, Murphy, 2019).

The history of standardization is fascinating, because it points out that technological development is quintessentially a socially embedded process, requiring the development of effective mechanisms of coordination among different social actors. In this respect, it is also interesting to note that there is an inherent tension between standards and patents. Patents provide an incentive to inventive activities by ensuring (for a limited period of time) exclusionary rights on the use of innovations; standards aim to foster technological adoption instead (but also technical progress) by ensuring access to a technology to all the stakeholders involved. This tension is, indeed, a leitmotif in the history of several ICT domains, from software systems to computer networks and mobile phones.

Standardization also created new challenges for patent law. In fact, when a standard is developed and a standard-setting organization (SSO), constituted by producers belonging to a specific industry sector, adopts it in one or more consensus meetings, that particular piece of technology will be uniformly adopted across all competing products and services and their component, in order to ensure compatibility and interoperability (Gerardin, 2017). Should the standard be protected by one or more patent, the rightholder immediately becomes the owner of a key tollgate to operate in an entire industry sector, with clear risks of patent hold-up. It comes at no surprise, then, that against the inertia of international and national legal sources, SSOs have developed internal IPRs policies which, first, incentivize patent holders on standards to disclose the existence of the patent(s) before the standard is officially adopted; second, they request that, once the standard is adopted and the patent becomes a standard-essential patent (SEP), rightholders license them to producers according to fair, reasonable, and non-discriminatory (FRAND) terms. SSOs policies do not create legislative but only contractual obligations, thus their enforceability is quite limited. This is the reason why SSOs react to rightholders' refusal to comply with their policies by redesigning the standard in order to circumvent the contested patent(s) (Ginsburg, Gerardin, Klovers, 2018).

National regulatory responses, by contrast, have been slow and timid. The EU Commission first intervened in 1992 with a soft-law instrument, the "Communication on IPRs and standardization", which tried to address the challenges raised by the activities of SSOs in a twofold manner. First, it requested SSO to respect general EU antitrust rules, stating that

> abuse of a dominant position by a standard-making body and its members could manifest itself by the activities of imposing unfair purchasing prices (i.e. royalty rates to right holders) or selling prices, (rates including royalties for the use of standards) or other unfair trading conditions (EU Commission, 1992, § 5.1.6).

Second, it reported the main general principles of SSOs' IP policies, requiring the disclosure of patents on potential standards and the use of FRAND terms in the licensing of SEPs. A number of cases dealt with by the EU Commission have provided additional guidance on the interplay between antitrust law and SEPs, often finding an abuse of dominant position under Article 101(1) of the Treaty on the Functioning of the European Union (TFEU) in the case of violation of FRAND commitments or improper use of injunctions to enforce SEPs against producers.

Building on its precedents, the Commission also published a set of Guidelines on horizontal cooperation agreement, focussing on hold-ups of producers caused by a refusal to licence SEPs or by the imposition of excessive royalties (EU Commission 2011). The Guidelines adopted a safe-harbour approach, excluding an agreement from the scope of Article 101 TFEU if a number of requirements are met, among them the good-faith disclosure of patents which are destined to become SEPs, the transparency of the standard setting process, and the FRAND commitment. More recently, the Court of Justice of the EU also intervened with a landmark decision in the case of Huawei versus ZTE (015), which determined the conditions upon which an SEP owner could file for and obtain an injunction against a producer without committing abuse of dominant position.

A similar path was followed in the United States, where the US Patents and Trademarks Office (USPTO), the US Department of Justice, and the National Institute of Standards and Technology (NIST) recently published a joint 'Policy Statement on Remedies for SEPs Subject to Voluntary F/RAND Commitments' (December 2019), which represents the follow-up of a similar statement enacted in 2013 and follows the lines traced by the EU (Ginsburg, Gerardin, Klovers, 2018; Gerardin, 2017). No supranational intervention is envisioned to date — a circumstance that leaves such an important sector in a state of great territorial fragmentation and uncertainty as to what the future may hold.

Technical knowledge is a peculiar good which is partly public and partly private (Nelson, 1989). Economists of innovation, for example, tend to focus on the private dimension, highlighting investment in inventive activities (R&D) and the advantages that innovating firms enjoy in the competitive processes. On the other hand, sociologists and historians are more alert to the public dimension of technology, stressing the role that professional communities of scientists, engineers, etc. have in the accumulation, storage, and diffusion of technical knowledge. Interestingly, in this perspective, the relationship between law and technology since the eighteenth century has been characterized by a systematic attempt to expand the private dimension of technology by articulating the notion of intellectual property rights, in particular in the form of patent systems. This expansion of the private dimension of technology has been based on the assumption that the 'propertization' of knowledge by providing incentives to innovators would result in more rapid rates of technical progress. As highlighted above, however, historical evidence suggests that the proof of connection between knowledge propertization and innovation is mixed. In fact, during the nineteenth century, in many Western countries, there was a significant policy debate on the possible abolition of patent systems (Machlup, Penrose, 1950). As a matter of fact, the impact of the patent systems on inventive activities, and relatedly, on economic growth, is still a controversial issue among economists and law scholars. Many contemporary discussions on patent policy mirror the arguments of the nineteenth century patent question studied by Machlup and Penrose (1950) rather closely. On the one hand, advocates of strong patent systems insist on the indispensable incentive effect generated by the propertization of knowledge. On the other hand, we have scholars who claim that the social costs of patent systems by far outweigh their direct and indirect benefits, and that the time is ripe for an outright abolition of the system (Boldrin, Levine, 2008). Interestingly, today we also have significant evidence suggesting that a significant amount of inventive activities takes place outside the coverage of patent protection (Fontana et al., 2013). This finding should suggest caution in embarking on policy reforms resulting in stricter patent regimes.

Yet, since the 1990s, there has been a policy push for further expanding the private part of technology by means of an extension of patents to new domains (software programs, business models, etc.) and by strengthening and harmonizing patent systems at international level (the TRIPS agreement of 1994). Since, as we have seen, there are important reasons for maintaining a proper balance between the private and the public part of technology (Nelson, 1989), one might be concerned that the sudden shift towards an increased propertization of technical knowledge may have swung the pendulum too far.

References

Alford, William P., *To Steal a Book is an Elegant Offense. Intellectual Property in Chinese Civilization* (Stanford, CA: Stanford University Press, 1995).

Allen, Robert C., 'Collective Invention', *Journal of Economic Behaviour and Organization*, 4 (1983), 1–24.

Baudry, Jérôme, 'A Politics of Intellectual Property: the French Revolution and the Creation of a Patent System', *Technology & Culture*, 61.4 (2020), 1017–44.

Belfanti, Carlo M., 'Between Mercantilism and the Market: Privileges for Invention in Early Modern Europe', *Journal of Institutional Economics*, 2 (2006), 319–38.

Bernal, John D., *Science and Industry in the Nineteenth Century* (London: Routledge, 1953).

Bessen, James and Allesandro Nuvolari, 'Knowledge Sharing Among Inventors: Some Historical Perspectives', in *Revolutionizing Innovation. Users, Communities and Open Innovation*, ed. by Dietmar Harhoff and Karim R. Lakhani (Cambridge, MA: MIT Press, 2016), pp. 135–55.

Biagioli, Mario, 'Patent Specification and Political Representation: How Patents Became Rights', in *Making and Unmaking Intellectual Property*, ed. by Mario Biagioli, Peter Jaszi, and Martha Woodmansee (Chicago, IL: University of Chicago Press, 2011), pp. 25–40.

Blakeney, Michael, *Intellectual Property Enforcement: a Commentary on the Anti-counterfeiting Trade Agreement (ACTA)* (Cheltenham and Northampton: Edward Elgar Publishing, 2012).

Boldrin, Michele and David K. Levine, *Against Intellectual Monopoly* (Cambridge: Cambridge University Press, 2008).

Bottomley, Sean, *The British Patent System During the Industrial Revolution 1700–1852* (Cambridge: Cambridge University Press, 2014).

Carver, Brian W., 'Share and Share Alike: Understanding and Enforcing Open Source and Free Software Licenses', *Berkeley Technology Law Journal*, 20 (2005), 443–81.

Chandler, Alfred D., *Scale and Scope. The Dynamics of Industrial Capitalism* (Cambridge, MA: Belknap Press of Harvard University Press, 1990).

Chang, Ha-Joon, *Kicking Away the Ladder. Development Strategy in Historical Perspective* (London: Anthem Press, 2002).

Cipolla, Carlo M., 'The Diffusion of Innovations in Early Modern Europe', *Comparative Studies in Society and History*, 14 (1972), 46–52.

Correa, Carlo M., 'Investment Protection in Bilateral and Free Trade Agreements: Implications for the Granting of Compulsory Licenses', *Michigan Journal of International Law*, 26.1 (2004), 331–53.

———, *Trade-related Aspects of Intellectual Property Rights. A Commentary on the TRIPs Agreement*, 2nd ed. (Oxford: Oxford University Press, 2020).

Dinwoodie, Graeme B. and Rochelle C. Dreyfuss, 'Diversifying Without Discriminating: Complying with the Mandates of the TRIPS Agreement', *Michigan Telecommunications & Technology Law Review*, 13.2 (2006), 445–56.

Drahos, Peter, 'Expanding Intellectual Property's Empire: the Role of FTAs' (2003) <http://www.grain.org/rights_files/drahos-fta-2003-en.pdf> [accessed 27 September 2020].

Drexl, Josef, 'The Concept of Trade-relatedness of Intellectual Property Rights in Times of Post-TRIPS Bilateralism', in *TRIPS plus 20. From Trade Rules to Market Principles*, ed. by Hanns Ullrich, Reto M. Hilty, Matthias Lamping, and Josef Drexl (Heidelberg: Springer, 2016), pp. 53–83.

EU Commission, Communication, 'Intellectual Property Rights and Standardization', *COM(92) 445 final*, Brussels, 27 October 1992.

———, Communication, 'Guidelines on the Applicability of Article 101 of the Treaty on the Functioning of the European Union to Horizontal Co-operation Agreements', OJ C11/1, 14 January 2011.

Fisk, Caroline, *Working Knowledge. Employee Innovation and the Rise of Corporate Intellectual Property, 1800–1930* (Chapel Hill, NC: The University of North Carolina Press, 2009).

Fontana, Robert, Alessandro Nuvolari, Hiroshi Shimizu, and Andrea Vezzulli, 'Reassessing Patent Propensity: Evidence from a Data-set of R&D Awards, 1977–2004', *Research Policy*, 42 (2013), 1780–92.

Freeman, Christopher and Francisco Louca, *As Time Goes By. From the Industrial Revolutions to the Information Revolution* (Oxford: Oxford University Press, 2001).

Galvez-Behar, Gabriel, *La République des inventeurs. Propriété et organisation de l'innovation en France (1791–1922)*, (Rennes: Presses Universitaires de Rennes, 2008).

Gambaro, Antonio, 'Property Rights in Comparative Perspective: Why Property is So Ancient and Durable', *Tulane European & Civil Law Forum*, 26 (2011), 205–39.

Gerardin, Damien, 'European Union Competition Law, Intellectual Property Law and Standardization?', in *Cambridge Handbook on Technical Standardization Law*, ed. by Jorge L. Contreras (Cambridge: Cambridge University Press, 2017), pp. 78–94.

Gervais, Daniel J., 'Intellectual Property, Trade & Development: the State of Play', *Fordham Law Review*, 74 (2005), 505–35.

———, 'The Internationalization of Intellectual Property: New Challenges from the Very Old and the Very New', *Fordham Intellectual Property Media & Entertainment Law Journal*, 12 (2002), 929–90.

Ginsburg, Douglas H., Damien Gerardin, and Keith Klovers, 'Antitrust and Intellectual Property in the United States and the European Union', in *The Interplay Between Competititon Law and Intellectual Property. An International Perspective*, ed. by Gabriella Muscolo and Marina Tavassi (The Hague: Kluwer Law International, 2018), pp. 92–119.

Grosse Ruse-Khan, Henning, 'The International Law Relation between TRIPS and Subsequent TRIPS-plus Free Trade Agreements: Towards Safeguarding TRIPS Flexibilities?', *Journal of Intellectual Property Law*, 18.2 (2011), 325–65.

Guellec, Dominique and Bruno von Pottelsberghe de la Potterie, *The Economics of the European Patent System* (Oxford: Oxford University Press, 2007).

Helfer, Laurence R., 'Regime Shifting: the TRIPs Agreement and New Dynamics of International Intellectual Property Lawmaking', *Yale Journal of International Law*, 29.1 (2004), 1–83.

Hilaire-Pérez, Liliane, *L'Invention technique au siècle des Lumières* (Paris: Albin Michel, 2000).

Ho, Cynthia M., *Access to Medicine in the Global Economy: International Agreements on Patents and Related Rights* (Oxford: Oxford University Press, 2011).

Isaacson, Walter, *The Innovators* (New York, NY: Simon & Schuster, 2014).

Kanefsky, John W., *Boulton and Watt and the Development of the Steam Engine: a Reassessment*, Exeter Research Group, Discussion Paper 1 ([Exeter]: Exeter Research Group, 1978).

Khan, Zorina B., *The Democratization of Invention. Patents and Copyrights in American Economic Development, 1790–1920* (Cambridge: Cambridge University Press, 2005).

Kuznets, Simon, *Modern Economic Growth. Rate, Structure and Spread* (New Haven, CT: Yale University Press, 1966).

Landes, David S., *The Unbound Prometheus. Technological and Industrial Development in Western Europe from 1750 to the Present* (Cambridge: Cambridge University Press, 1969).

Long, Pamela O., 'Invention, Authorship, "Intellectual Property" and the Origins of Patents: Notes Towards a Conceptual History', *Technology & Culture*, 32 (1991), 846–84.

Machlup, Fritz and Edith Penrose, 'The Patent Controversy in the Nineteenth Century', *Journal of Economic History*, 10 (1950), 1–29.

MacLeod, Christine and Alessandro Nuvolari, 'Inventive Activities, Patents and Early Industrialisation: a Synthesis of Research Issues', *Rivista di Storia Economica*, 32 (2016), 77–107.

——— and ———, 'Technological Change', in *The Oxford Handbook of the Ancien Régime*, ed. by William Doyle (Oxford: Oxford University Press, 2012), pp. 448–66.

Matthews, Duncan, 'The Right to Health and Patents', in *Research Handbook on Human Rights and Intellectual Property*, ed. by Christophe Geiger (Cheltenham and Northampton: Edward Elgar Publishing, 2015), pp. 496–512.

Meisenzahl, Ralf and Joel Mokyr, 'The Rate and Direction of Invention During the British Industrial Revolution: Incentives and Institutions', in *The Rate and Direction of Inventive Activity Revisited*, ed. by Josh Lerner and Scott Stern (Chicago, IL: The University of Chicago Press, 2012), pp. 443–82.

Mokyr, Joel, *The Gifts of Athena. Historical Origins of the Knowledge Economy* (Princeton, NJ: Princeton University Press, 2002).

Moser, Petra, 'How Do Patent Laws Influence Innovation? Evidence from Nineteenth-century World's Fairs', *American Economic Review*, 95 (2005), 1214–36.

———, *Pirates and Patents* (Princeton, NJ: Princeton University Press, 2021).

Nelson, Richard R., 'What is Private and What is Public About Technology?', *Science, Technology and Human Values*, 14 (1989), 229–41.

Nicholas, Tom, 'The Role of Independent Invention in US Technological Development, 1880–1930', *Journal of Economic History*, 70 (2010), 57–82.

North, Douglass C., *Structure and Change in Economic History* (New York, NY: W. W. Norton, 1981).

Nuvolari, Alessandro, 'Collective Invention During the British Industrial Revolution: the Case of the Cornish Pumping Engine', *Cambridge Journal of Economics*, 28 (2004), 347–63.

Ricketson, Sam, *The Paris Convention for the Protection of Industrial Property. A Commentary* (Oxford: Oxford University Press, 2015).

Rosenberg, Nathan, 'Capital Goods, Technology and Economic Growth', *Oxford Economic Papers*, 15 (1963), 217–27.

Torrens, Hugh, 'New Light on the Hornblower and Winwood Compound Steam Engine', *Journal of the Trevithick Society*, 3 (1982), 21–41.

Tunzelmann, G. Nicholas von, *Technology and Industrial Progress* (Cheltenham: Edward Elgar Publishing, 1995).

Ullrich, Hanns, 'The Political Foundations of TRIPS Revisited', in *TRIPS plus 20. From Trade Rules to Market Principles*, ed. by Hanns Ullrich, Reto M. Hilty, Matthias Lamping, and Josef Drexl (Heidelberg: Springer, 2016), pp. 85–129.

USPTO, US Department of Justice, NIST, 'Policy Statement on Remedies for Standards-essential Patents Subject to Voluntary F/RAND Commitments' (December 2019) <https://www.uspto.gov/sites/default/files/documents/SEP%20policy%20statement%20signed.pdf> [accessed dd Month yyyy].

Whitehead, Alfred N., *Science and the Modern World* (Cambridge: Cambridge University Press, 1929).

Yates, JoAnne and Craig Murphy, *Engineering Rules. Global Standard Setting since 1980* (Baltimore, MD: Johns Hopkins University Press, 2019).

Yu, Peter K., 'TRIPs and Its Discontents', *Marquette Intellectual Property Law Review*, 10 (2006), 369–410.

STÉPHANE LEMBRÉ

Technical Training and Education since the Nineteenth Century

At the turn of twentieth century numerous rewards were attributed to technical educational institutions in world exhibitions, a sign of the diversity of institutions, their distribution throughout the world, and the general interest for technical training, as well as the belief in rapid and limitless technological progress (Dittrich, 2010). These prizes encouraged international economic competition but also gave rise to a fear of industrial decline, since many observers saw training and education as the very source of economic competitiveness (Edwards, 2008). Visitors to the exhibitions could admire machinery and participate in this 'publicization of technology' (Hilaire-Pérez, 2012, 13), in keeping with the institution of technological education that took place at the time. Nevertheless, these exhibitions, like all competitions featuring technical excellence up to the Worldskills Competitions at the beginning of the twenty-first century, sometimes obscure the importance of technical training.

Following different rhythms and taking different shapes, the design of vocational training courses can be studied on a global scale from the nineteenth to the twenty-first centuries. Although remarkable institutions were inherited from previous centuries, the last two centuries have witnessed a multiplication of educational institutions as well as the widening of the types of students, the geographical areas, and the professions involved. These developments manifest a stronger focus on education highlighting the importance of studying the history of vocational and technical education if one is to understand technology. To do so we need to move beyond a restrictive vision of technical education as being mainly male and related to industries.

The Multiplication of Sites, Audiences, and Objects

The appearance of higher scientific and technical training in the eighteenth century shaped the landscape of the great European engineering schools for the following centuries. The *École polytechnique* and *École des ponts et chaussées,* which were connected to France's major state bodies, were sources of inspiration abroad, for example in the United States (Angulo, 2012). Their evolution in the nineteenth century raises the question of what exactly are engineers: are they pragmatic technicians and project

Global History of Techniques, ed. by Guillaume Carnino, Liliane Hilaire-Pérez and Jérôme Lamy, Global Matters Culture Matérielles, histoire des techniques et circulation globale des savoirs, 9 (Turnhout: Brepols, 2024), pp. 563–580
BREPOLS PUBLISHERS 10.1484/M.TECHNE-EB.5.129830

managers or are they scientists mastering the most recent theoretical issues? As a scientific and research institution, the *École polytechnique* moved away from the teaching of techniques (Shinn, 1980), a job done by applied schools like the fact-focused *École des ponts et chaussées*, although this division was not straightforward and in itself created questions regarding the end purpose and the contents of lessons taught (Picon, 1992). Engineering schools specialized and grew in number. In 1913 to 1914, Russia — a barely industrialized country at the time — had fifteen state engineering schools for industry, which accommodated 23,500 students, in addition to offering courses in agronomy and forestry and having naval and military schools (Aref'ev, Aref'ev, 2013). European capitals and major cities established schools either through private initiatives, like, for example, the Montefiore Institute for electrical engineers in Liège (1883), or at the initiative of municipalities or states, as in the creation of the certificate of accountant, which encouraged the learning of accounting techniques in France (1927), or the establishment of the Institute of technical and accounting expertise at the Conservatoire national des arts et métiers (1931). Countries under European domination were also among those that founded educational institutions at a very early date: the Thomason College of Civil Engineering in the small town of Roorkee, Northern India, was founded in 1848 for students of both European and local origins. In the 1850s, other colleges were founded in Calcutta, Madras, and Poona, but these had some difficulty attracting British students (Cuddy, Mansell, 1994). Higher education in the commercial trades also developed. This was a mainly European movement, as some foundations illustrate, notably those of the *Hooger Handelsgesticht* (higher institute of Antwerp) (1852) or the *École des Hautes études commerciales* (school of higher commercial studies) established in Paris (Sanderson, 1995; Meuleau, 1995), both of which prompted other institutions to be created. To regulate the diversity of engineering schools and to guarantee that courses provide high-quality education, it became necessary to precisely define what an engineering school was and restrict the use this descriptor to certain schools only (Grelon, 1986).

In the wake of these engineering schools of various fame and reputation, the training of a 'middle level' of foremen and workshop managers, heads of offices and team leaders requires a combination of professionnal experience and more subject-specific knowledge. Workers and craftsmen are mainly trained in the workplace, and young workers start with an apprenticeship, which is probably the oldest mode of technical training — defined as the acquisition of trade skills from an experienced professional, i.e. a master (De Munck, Kaplan, Soly, 2007) (Figure 1). The nineteenth-century transformations of apprenticeships are generally viewed as the sign of the decline of technical training, although this process took place over a longer period and can rarely be attributed to the evolutions of the guilds alone. Criticism focuses on the fact that the transmission of know-how was not centre place, with the apprentice being put to merely subsidiary and non-productive tasks (Lane, 1996; Hamilton, 2000; Crowston, Lemercier, 2019). The number of apprentices in the United Kingdom fell throughout the period, from 315,000 apprentices in 1925 to 218,000 in 1970 and fewer than 54,000 in 1990. However, apprenticeships continued to be important in countries where they were more highly regarded, for example, in Germany and

Fig. 1. Apprentice. Man and Boy Making Shoes, *c.* 1914. Painting by (Louis-?) Emile Adan (1839–1937), copyrighted by Braun & Co., N.Y. United States Library of Congress Prints and Photographs division, digital ID cph 3b27512

Switzerland. The German apprenticeship system, in which guilds play an important part in the management of training, has often been praised for its efficiency by other nations since the nineteenth century. Although there are great regional variations in its implementation, overall the apprenticeship system continues to be controlled by the trades (Baltic Sea Academy, 2015).

At all levels, the variety of local initiatives resulted in a myriad of training arrangements. Evening classes and short qualification courses for the skilled workforce were first devised in cities, for example, for craftsmen in Florence (Pellegrino, 2012) or at the beginning of the twentieth century, vocational courses set up for male and female workers by the French labour clubs (*bourses du travail*) (Hamelin, 2011). These courses were centres of professional and technical culture and are inseparable from the ideals of popular education. With time, there was a gradual shift from the interpersonal transmission of knowledge and know-how in the workplace, to its integration into dedicated institutions, courses, and schools. This transition developed in different places and for different trades. It was not a generalized process, but rather, training *in situ* coexisted with school education. Agricultural training for instance, where demonstrations in fields and on agricultural experiment stations are necessary, shows the limits of a purely school-based education. For example, the United States'

1887 Hatch Act provided funding to individual states for the development of experiment stations and for the dissemination of information related to agriculture to help farmers update their methods (Brewer, 2009). The increasing use of dual training shows *in situ* and school-based education are seen as complementary, as evidenced by the development of engineering education through apprenticeship. The choices that were made between the introduction specialist schools and the continuation of apprenticeship type training in the workplace have shaped technical training and education.

Rather than adopting a diffusionist vision of specialist training spreading from pioneer countries to the rest of the world, the focus here will be on the flow of people and ideas. The development of the training and teaching of techniques in colonial settings is marked by ambivalence: was the purpose to train a skilled workforce for the economic development of the territories under colonial rule, or to contribute to the economic growth of the colonizing powers? In the latter case, agricultural training was often prioritized and industrial training rather neglected, as for example in French West Africa, a choice that sometimes became a bone of contention (Gamble, 2017). However, in these colonies, this led to real, diverse, and sometimes ambitious achievements. Initiatives were not necessarily dependent on European presence, although they sometimes took inspiration from European models: the first agricultural school of the Ottoman Empire was created in 1848, following the French model of the agricultural school of Grignon which the ambassador Mustafa Resit Pasha had visited. Then the school of Selemiah in the vilayet of Damascus, which was located at the heart of a cereal growing area, opened in 1910. The French, whose mandate for Syria was created in 1920, criticized the agricultural education that was provided but kept the same site and tried to rejuvenate the education programme. This was a difficult task given the very low level of education of the local populations who were ill equiped to integrate agronomical ideas. Focussing on practical demonstrations and the power of teaching by example, the French civil servants reproduced a mode of technical training that had already been tested in metropolitan France, but here ran into problems due to the programme's cost and low student numbers; the agricultural school of Sélémieh closed in 1933 (Sakatni, 2016). In the English colony of Cape Town in South Africa agricultural training was the brainchild of Charles T. Loram, the author of *The Education of the South African Native* of 1917, he believed in the need for education to be adapted to the local context and to move towards hands on and agricultural training (Paterson, 2005).

Transfers of experience combined with endogenous developments. In the 1920s, educators in Kabete, Kenya, were traveling to rural communities to teach agricultural techniques to the locals, drawing inspiration from education programmes for black people in place in the southern states of the United States (Gould, 1989). This type of itinerant training scheme was particularly appreciated in Europe: agricultural domestic economy education for girls, which developed in Belgium at the end of the nineteenth century, was replicated in France both before and after the First World War (Lelorrain, 1992; Lembré, 2010).

Any study of these schemes should look jointly at local, national, and transnational initiatives. The diversity of schemes makes it difficult to identify national models, as programmes are often influenced and shaped by international surveys. Not only is the national context not always the most relevant for understanding differences, but national typologies conceal the constant transformations of training in reaction to technical changes as well as those of learning techniques.

The Difficulties of Learning Techniques

Trying to explain the development of technological training and education is tricky. Two deterministic pitfalls should be avoided, technological change and economic growth not being the only factors. The multiplication of places of training was a response to various challenges, among which adaptation to technical changes, both for boys and girls about to enter the professional world and for adults retraining in a different profession. If economic factors prompt the recourse to specialist training both within and outside professional organisations, they are not sufficient to explain the shape taken by training. The example of the Ansaldo factories in Genoa shows how much the quality of production depends on workers' technical training in various trades (mechanical fitters, boilermakers, and turners), and also emphasizes the need for a training policy that is closely related to production, recruitment, and pay policies (Dewerpe, 2017). An investigation of the relationship between the educational and economic spheres, however, makes it possible to study changes in technology education in early industrialized countries such as Belgium (Bertrams, 2006) and in formerly colonized countries that experienced a fast economic growth at the end of the twentieth century, like Singapore (Blackburn, 2017).

Other motivations are always present; they may be, in particular, social, political, related to trade unions, or religious beliefs. For example, social control by the elites is at the heart of the rise of vocational and technical education in Bavaria in the first half of the nineteenth century (Laufer, 2000). The legal age of employment, which increased as legislation progressively protected children, also raised the question of the role of technical education in the school curriculum. In Sweden, technical education in elementary school became increasingly important in the first quarter of the twentieth century, as industrialization progressed (Hallström, 2009).

Technology is never independent from the social and political contexts of its uses. The vocational training of Spanish workers in the 1920s and the *Universidades Laborales* (labour universities) set up by the Franco regime combined professional training and ideological indoctrination (Rico Gomez, 2014; Delgado-Granados, Ramirez-Macias, 2017). Similarly, the creation of training courses in rural trades and agriculture in Europe in the 1920s and 1930s took part in the wider context of trying to slow down rural exodus. In domestic economy schools, girls learned the modern methods for manufacturing milk and cheese, and also took classes in sewing, cutting, laundering, ironing, housekeeping, cleaning and childcare, in keeping with the idea that educated girls played a stabilizing social role in rural societies. Missionaries,

Fig. 2. Unley Girls Technical School, Unley, South Australia. State Government Photographer, *c.* 1945. The History Trust of South Australian, South Australian Government, GN10986.

who were key figures in the European domination of the world in the nineteenth and the first half of the twentieth centuries, favoured vocational education, and this had significant repercussions where they succeed to establish themselves in the long term: the Salesians gradually established a school network in Cairo and several other Egyptian cities in spite of the fact that there was competition from other educational options on offer (Turiano, 2016).

The gender and race distribution reveals the social and political goals attributed to the teaching of technology. Women were often simply considered good housekeepers capable of taking on household chores, regardless of the professional techniques learned and used by textile workers or women farmers. Shorthand and typing partially reinforced these stereotypes. The introduction of typewriters in the last third of the nineteenth century and their rapid adoption in offices created the figure of the female secretary (Gardey, 2001). The feminization of some professions involved casting certain qualities as naturally 'feminine', an equation which the training courses provided in numerous schools were intended to reveal and reinforce (Figure 2). By contrast, the new jobs occupied by women reinforced the assimilation of certain skills to a male sphere: while technical training had long been pursued by both girls and boys, industrial technical education was presented as suitable only for men, a bastion of masculinity and virility (Ward, 2015). While practical schools were accessible to black people in the southern states of the United States as early as the 1830s, they were carefully separated from the technical education schemes reserved to white students (Bradshaw, Bohan, 2013), but there, too, there was some ambiguity, since a few decades later, leaders including Frederick Douglass and Booker T. Washington encouraged students to attend these courses, which they considered as a path towards emancipation. With the economic crisis of the 1930s, the proportion of African Americans in technical schools of the South declined sharply. In a context where learning a trade guaranteed employment, having a place in a technical school became competitive which in effect excluded African Americans (Anderson, 1988).

The growth of technical education should also be seen in the context of the evolution of trades. Some sectors required specialized skills, among them watchmaking: following the foundation of the Geneva watchmaking school (1824), La Chaux-de-Fonds, a modest production centre, set up a watchmaking school in 1865 to 'promote the training of watchmakers destined to occupy technical management positions', and thus fostered technical excellence in the context of otherwise fragmentated apprenticeship schemes (Donzé, 2007, 51). In order to strengthen their place in the world markets, Japanese companies introduced watchmaking training a few decades later, first for engineers in imperial universities, then at the Koto Seimitsu Kogakko (precision industry high school) (founded in 1933), and especially within companies. After 1945, the Daini Seikosha watch factory (Hattori group) recruited engineers to design a new production system; in the 1950s, the in-house technical managers had graduated from technical colleges (Donzé, 2014). More generally, training schools were resource centres that needed to cater to different target audiences: daytime further instruction for adolescents who wished to continue training beyond elementary school and evening classes for employees and workers. In both cases, they also needed to offer different specialty topics and be fitted with appropriate equipment. The increasing offer of technical education could foster the development of an industrial city and vice versa. In Belfast, economic growth throughout the nineteenth century culminated in 1907 with the opening of the Belfast Municipal Technical Institute (McCloy, 2009), while China's reinvention since the economic reform programme of 1978 has coincided with the growth of vocational secondary education (Munch, Risler, 1987). By contrast, the emergence and exponential development of computer science has initiated the training of engineers on a high level, while computer manufacturers have invested in training all those who needed to do work for their customers. In 1961, the *École d'application de la Compagnie des machines Bull* in Paris (the applied school of the Bull machine company) trained 548 perforators and 385 operators, and three years later, another 690 and 799 respectively (Neumann, 2013; Mounier-Kuhn, 2010). The proliferation of computer engineering schools first took place at the end of the twentieth century in the wealthiest countries, but India also took its first steps in this direction in the 1970s, and by 2018, about 400,000 new computer scientists were trained in India annually thanks to numerous, sometimes very selective, technology institutes. Computer engineering training in India is characterized by its strong emphasis on practice rather than theory.

While the military sector considers technical education a necessity even in times of peace, any war effort will involve the production and improvement of arms, and even innovation (Hacker, 1993). The end of wars is often marked by an awareness of the necessity of technical training for the workforce: much legislation is then adopted, as was the case, for example, in the United States with the Smith-Hughes Act of 1917 or with the Astier law in France in 1919 which organised technical training in the two countries. In the twentieth century, international organizations became involved. This involvement encouraged international comparisons, with institutions such as the International Labour Office, for which the International Bureau of Technical Education — which organised international congresses in the 1930s (Matasci, 2012) — was responsible. ILO recommendations are important for understanding the rise of technical education and career guidance, for example in South America (Pronko, 2003). "Development" as an ideological construct also plays an important role here. UNESCO, like other international institutions, considers the training of a qualified workforce essential for the so-called 'developing' countries, which often witnessed a shortage of managerial staff following independence.

Institutions for technical education are sometimes drivers of technological change in themselves. In France, the École supérieure des travaux publics (i.e. an advanced school of public works), which had been founded by Léon Eyrolles, considered colonies a field of expansion, and exported courses, especially using over the mail formats, to promote the figure of the colonial engineer (Vacher, Guillerme, 2017). When thinking about the adaptation of education to the colonial setting, specific thought should be given to technical education in terms of its design and implementation, and more generally techniques should be contextualized within the relations between the colonizers and the colonized. Agricultural education provides a good example here. The American missionary Elmory Alvord, who arrived in Southern Rhodesia (now Zimbabwe) in 1919 to train Africans in modern farming methods, faced the Africans' refusal to use commercial fertilizers for fear of damaging soil fertility (Mungazi, Walker, 1998). This should not lead us to comparing 'traditional' techniques (deemed obsolete) and technical training (a herald of modernity), or to underestimating the technical skills of local populations, which are all too often considered as skills for survival, like women's crafts (Rogers, 2013).

Pedagogical Modalities: Between Theory and Practice

Training and teaching systems are usually characterized by the juxtaposition or complementarity of practical and theoretical learning, in the workplace or in a place dedicated to training. Studying their history requires the use of various sources, i.e. the surviving curricula, textbooks, and equipment. Like curricula, textbooks indicate what is supposed to be learned but fail to document the concrete ways in which that learning takes place. Many teachers and professionals produced technical books that connected technical changes to learning. One of the people were instrumental in the restructuring of the Japanese watch industry, Aoki Tamotsu (1882–1964), who was also the founding director of the Koto Seimitsu Kogakko (precision industry high school) in 1933, published one of the first watch technology manuals (Donzé, 2014).

The equipment at technical schools deserves attention. Workshops are places of learning equipped with tools and machines that are either nearly obsolete or, on the contrary, in keeping with professional work environments (Lambert, Lembré, 2017). In this respect, the *écoles d'arts et métiers* (arts and crafts schools) served as pedagogical models in France and beyond: in the nineteenth century, workshop training was dominant for training skilled workers in the wood and iron trades (Day, 1987, 2001). Universal machines like the lathe, a device that allowed the worker to rotate any work piece, using his hands or tools such as, for example, a file, were widely introduced. But the workshops also needed be equipped with more specialized machines, which were also more expensive. If local companies were able supply the machines, technical transfers were not always evident. The technical school of Khala Bavan, established in 1890 in the fast industrializing city of Baroda (British India), was equipped with machinery provided by English and German companies (Raina, Habib, 1991). Outside industrial contexts, the diversity of training sites — from training ships to hotel schools — brought about specific learning methods (Kennerley, 2000; Gindraux, 1993).

A specific pedagogical approach was necessary but delicate to implement. Schools were constantly adaptated to changes in production methods. In the first half of the nineteenth century, engineering replaced architecture at the École des ponts et chaussées, while mechanics and construction played increasingly important roles (Picon, 1992). The teaching of mechanics is emblematic in this respect: at the École polytechnique, the mechanics class was merged with training on machines in 1850 in order address the criticism that it was too theoretical; but it was in the practical schools that courses in applied mechanics were developed. The evolution of the concepts and methods being used shows, however, that the distinction between applied and rational mechanics did not always imply a distinction between theory and practice, or between the abstract and the concrete (Belhoste, 2003). Charles Dupin's mechanics course at the Conservatoire royal des arts et métiers (royal conservatory of arts and crafts) from 1820 onwards, and those of Jean Victor Poncelet in Metz, manifest the will to present practitioners with real-life examples while also providing them with a theoretical background.

New modes of energy production created new disciplines. The introduction of electricity in many trades led to various technical and scientific courses, both public and private, some taught at a high level of education others introduced into secondary education curricula. Showing a spirit comparable to that of neighbouring Germany, the Czech and Slovak regions began to teach electrical engineering from the mid-1880s onwards, as electrification was still in its infancy. At the Czech Technical University of Prague, electronics was first integrated into the physics course designed for the mechanical sector. Automation was incremental but common to electrotechnical institutes that were created around the world, with the engineering school in Porto Alegre (Brazil), the *Technische Hochschule* in Germany, and the North American land-grant universities (Efmertova, Grelon, Mikeš, 2016). In the mines, the electrification of underground works, which was sometimes delayed until after the Second World War, involved a specific training effort. Electromotive power replaced compressed air and paved the way for mechanized longwall mining with the combined use of powered hydraulic chocks, shearer and conveyor. The introduction of this new technology required on going training and upskilling for mining engineers. First disseminated by a few specialists, training extended to all mining staff (Conus, Escudier, 2004).

The dichotomy between theory and practice is reductive because it maintains an artificial disjunction between knowledge and know-how. Whereas the distinction between the classroom and the workshop (in schools or the workplace) may be convenient, it fails to capture the role occupied by either, and also neglects other ways learning and training take place. Migration, which was not limited to engineers in training, is a case in point. Labout migrants becoming "technical gyrovagues" of sorts and receiving training as they move is an age-old phenomenon (Steffens, 2003). The figure of the journeyman represents the top echelon of this training through migration. The training of journeymen, which always involves the making of a masterpiece — essential for them to become master craftsmen —, contributes to the strong group identity of journeymen conferred partly through this specific training path. The practice of the journeyman years reached its peak in the first half of the nineteenth century, and it was then challenged by big industry and machinery. Learning the skills of so-called 'manual' trades via travel across France and beyond continues to be at the heart of the associations that have informed the current legal forms of French vocational education (Icher, 2010).

Some actual educational models were developed. The teaching method promoted by Victor Della Vos at the Imperial Technical School of Moscow aimed to enable engineering students to progressively acquire mastery of the tools and techniques necessary for practising civil engineering or the mechanical or electrical trades. Presented at the World's Fair in Philadelphia in 1876, this method inspired American manual training schools (fifteen schools in 1894 and 51 in 1913, with more than 50,000 students enrolled) (Alix, 2017). Other educators such as Georg Kerschensteiner sought to mobilize practical intelligence and challenge the dichotomy between general education and vocational training. Approaching this from a very different perspective as he connected technical training with political projects, Marx aimed for a polytechnic education that would prevent workers from losing their skills to capitalism (Small, 1984). A century later, Gandhi wished to organize the entire Indian programme of teaching around the learning of craft techniques to reverse the traditional social hierarchy (Holzwarth, 2015).

The body of teachers stood behind the educators and intellectuals who were interested in the teaching of techniques in its different forms. In schools, former professionals whose reputation was based on their experience and know-how taught alongside people who had graduated from institutions dedicated to the training of teachers of technical education (Grollmann, Rauner, 2007). This was the case in the École normale de l'enseignement technique (National school for technical training), which had been established in Paris in 1912 and whose reptutation slowly increased until it became a model after which other schools were set up in Algeria, Greece, and Cameroon (Le Bot et al., 2013). At the *Tianjin Gongshang xueyuan* (institute of higher industrial and commercial studies), which operated from 1923 to 1951 and had more than 2000 students in 1949, the Jesuit teachers, who were for the most part engineers and scientists, were also authors of many scientific and literary publications (Dehoux-Dutilleux, 2018). In the workplace, the apprentice works close to the tutor. In the German system, the latter needed to finish educational training and pass an examination for accreditation, to then occupy the role of *Ausbilder* — technical competence alone was not sufficient. The long tradition of the professionalization of teachers in technical education was encouraged in the context of modernization policies at the global level (UNESCO, 1962). More generally, the early training of technicians required mobility, reproducing the academic pilgrimage of medieval origin. This took the form of training tours organised by the engineering schools and then internships and periods abroad as part of the courses. The enrollment of an increasing number of international students in engineering degrees in elite universities, especially in America, also contributes to the globalization of technical education.

National characteristics can also be seen. In this respect, traditional corporatism (in the form of guilds) has been instrumental in the birth of the German system of vocational training and its particular language. Despite its various evolutions through time, the German model of the *Beruf* can be said to be focussed on the learning of a trade. In this respect it differs from the French model where a more general approach to technical culture is favoured. The discrepancy is clear with regard to apprenticeships, which was more disrupted by liberalism and industrialization in France than it was in Germany (Schriewer, Harney, 1999; Deforge, 1979). These characteristics are changing however with the emergence of new issues, for example those relating to the social responsibility of the engineer, an aspect which confirms the importance of the humanities and social sciences in curricula (Colby, Sullivan, 2008; Didier, Derouet, 2013).

Despite vastly different contexts and realities, vocational training and elite higher education in science are both characterized by better quality training. Although there are discrepancies in the historiography, it is now possible to study training policies on a global scale. To do so requires one look at new sources however if one is to understand how techniques are learnt and acquired as well as their wider economic, social, and political implications.

References

Alix, Sébastien-Akira, *L'Éducation progressiste aux États-Unis: histoire, philosophie et pratiques, 1876–1919* (Grenoble: Presses universitaires de Grenoble, 2017).

Anderson, James, *The Education of Black People in the South, 1860–1935* (Chapel Hill, NC: University of North Carolina Press, 1988).

Angulo, A. J., 'The Polytechnic Comes to America: How French Approaches to Science Instruction Influenced Mid-nineteenth-Century American Higher Education', *History of Science*, 50.3 (2012), 315–38.

Aref'ev, A. L. and M. A. Aref'ev, 'Engineering and Technical Education in Russia, in Numbers', *Russian Social Science Review*, 54.5 (2013), 38–56.

Baltic Sea Academy, ed., *Work-based Learning around the Mare Balticum* (Hamburg: Baltic Sea Academy, 2015).

Belhoste, Bruno, *La Formation d'une technocratie. L'École polytechnique et ses élèves de la Révolution au Second Empire* (Paris: Belin, 2003).

———, Amy Dahan Dalmedico, and Antoine Picon, eds, *La Formation polytechnicienne 1794–1994* (Paris: Dunod, 1994).

Berner, Esther and Philipp Gonon, eds, *History of Vocational Education and Training in Europe* (Bern: Peter Lang, 2015).

Bertrams, Kenneth, *Universités & entreprises. Milieux académiques et industriels en Belgique 1880–1970* (Brussels: Le Cri édition, 2006).

Blackburn, Kevin, *Education, Industrialization and the End of Empire in Singapore* (New York, NY: Routledge, 2017).

Bradshaw, Lauren Yarnell and Bohan, Chara Haeussler, 'The Struggle for Industrial Education in the "Lowell of the South", Columbus, Georgia, 1850–1930', *American Educational History Journal*, 40.2 (2013), 189–206.

Brewer, Ernest W., 'The History of Career and Technical Education', in *Definitive Readings in the History, Philosophy, Practice and Theories of Career and Technical Education*, ed. by Victor C. X. Wang (Hangzhou: Zhejiang University Press, 2009), pp. 1–16.

Cohen, Yves and Klaus Manfrass, eds, *Frankreich und Deutschland: Forschung, Technologie und industrielle Entwicklung im 19. und 20. Jahrhundert* (Munich: C. H. Beck, 1990).

Colby, Anne and William M. Sullivan, 'Ethics Teaching in Undergraduate Engineering Education', *Journal of Engineering Education*, 97.3 (2008), 327–38.

Conus, Marie-France and Jean-Louis Escudier, 'Électrification, savoirs scientifiques et techniques dans l'industrie houillère (1880–1980)', in *Artisans, industrie. Nouvelles révolutions du Moyen Age à nos jours*, ed. by Natacha Coquery, Liliane Hilaire-Pérez, Line Sallmann, and Catherine Verna (Paris: ENS éditions 2004), pp. 237–47.

Crowston, Clare H. and Claire Lemercier, 'Surviving the End of the Guilds. Apprenticeship in eighteenth and nineteenth-century France', in *Apprenticeship in Early Modern Europe*, ed. by Maarten Prak and Patrick Wallis (Cambridge: Cambridge University Press, 2019), pp. 282–308

Cuddy, Brendan and TonyMansell, 'Engineers for India: The Royal Indian Engineering College at Cooper's Hill', *History of Education*, 23.1 (1994), 107–23.

Day, Charles R., *Education for the Industrial World: The Écoles d'arts et métiers and the Rise of French Industrial Engineering* (Cambridge, MA: Massachusetts Institute of Technology, 1987).

———, *Schools and Work. Technical and Vocational Education in France Since the Third Republic* (Montreal: McGill-Queen's University Press, 2001).

Deforge, Yves, 'The Specific Nature of Vocational Training in School', *European Journal of Education*, 14.4 (1979), 329–39.

Dehoux-Duthilleux, Corinne, 'Les hautes études industrielles et commerciales de Tianjin (Tianjin Gongshang Yueyuan 天津工工商学院), 1923–1951. Un exemple de l'action éducative des jésuites en Chine' (unpublished doctoral thesis, Bordeaux Montaigne University, 2018).

de Lavernette, Claire, *Le Droit à l'éducation et à la formation technique et professionnelle*, on OIDEL, <https://oidel.org/doc/WP/WP%2018.pdf>, accessed 09 January 2023.

Delgado-Granados, Patricia and Gonzalo Ramirez-Macias, 'Conveniencia o necessidad? La formacion de la clase obrera en las Universidades laborales franquistas (1955–1978)', *Historia Critica*, 63 (2017), 117–36.

De Munck, Bert, Steven L. Kaplan, and Hugo Soly, eds, *Learning on the Shop Floor. Historical Perspectives on Apprenticeship* (New York, NY: Berghahn Books, 2007).

d'Enfert, Renaud, 'L'Introduction du travail manuel dans les écoles primaires de garçons, 1880–1900', *Histoire de l'éducation*, 113 (2007), 31–67.

Dewerpe, Alain, *Les Mondes de l'industrie. L'Ansaldo, un capitalisme à l'italienne* (Paris: Éditions de l'EHESS, 2017).

Didier, Christelle and Antoine Derouet, 'Social Responsibility in French Engineering Education: a Historical and Sociological Analysis', *Science and Engineering Ethics*, 19.4 (2013), 1577–88.

Dittrich, Klaus, *Experts Going Transnational: Education at World Exhibitions During the Second Half of the Nineteenth Century* (unpublished PhD. thesis, University of Portsmouth, 2010).

Donzé, Pierre-Yves, *Les Patrons horlogers de La Chaux-de-Fonds. Dynamique sociale d'une élite industrielle (1840–1920)* (Neuchâtel: éditions Alphil-Presses universitaires suisses, 2007).

———, *'Rattraper et dépasser la Suisse'. Histoire de l'industrie horlogère japonaise de 1850 à nos jours* (Neuchâtel: éditions Alphil-Presses universitaires suisses, 2014).

Edwards, Anthony David, *The Role of International Exhibitions in Britain 1850–1910. Perceptions of Economic Decline and the Technical Education Issue* (Amherst, MA and New York, NY: Cambria Press, 2008).

Efmertova, Marcela and André Grelon, eds, with the collaboration of Jan Mikeš, *Des Ingénieurs pour un monde nouveau. Histoire des enseignements électrotechniques (Europe, Amériques), XIXe–XXe siècle* (Brussels: P. I. E. Peter Lang, 2016).

Fox, Robert and Anna Guagnini, eds, *Education, Technology and Industrial Performance in Europe, 1850–1939* (Cambridge: Cambridge University Press / Paris: Éditions de la MSH, 1993).

Gamble, Harry, *Contesting French West Africa. Battles Over Schools and the Colonial Order, 1900–1950* (Lincoln, NE: University of Nebraska Press, 2017).

Gardey, Delphine, *La Dactylographe et l'expéditionnaire: histoire des employés de bureau, 1890–1930* (Paris: Belin, 2001).

Gardinier, David E., 'Vocational and Technical Education in French Equatorial Africa (1842–1960)', *Proceedings of the Meeting of the French Colonial Historical Society*, 8 (1985), 113–23.

Gay, Hannah, 'Association and Practice: the City and Guilds of London Institute for the Advancement of Technical Education', *Annals of Science*, 57 (2000), 369–98.

Gindraux, Philippe, *L'Art et la manière. L'école hôtelière de Lausanne* (Lausanne: Payot Lausanne, 1993).

Gobe, Éric, ed., *L'Ingénieur moderne au Maghreb (XIXe–XXe siècles)* (Paris: Maisonneuve & Larose, 2004).

Gould, W. T. S., 'Technical Education and Migration in Tiriki, Western Kenya, 1902–1987', *African Affairs*, 88.351 (1989), 253–71.

Gouzévitch, Irina, André Grelon, and Anousheh Karvar, eds, *La Formation des ingénieurs en perspective. Modèles de référence et réseaux de médiation, XVIIIe–XXe siècles* (Rennes: Presses universitaires de Rennes, 2004).

Grelon, André, ed., *Les Ingénieurs de la crise: titre et profession entre les deux guerres* (Paris: Éditions de l'EHESS, 1986).

——— and Girolamo Ramunni, eds, *La Naissance de l'ingénieur-électricien. Origines et développement des formations nationales électrotechniques* (Paris: Association pour l'histoire de l'électricité en France, PUF, 1997).

Grollmann, Philipp and Felix Rauner, eds, *International Perspectives on Teachers and Lecturers in Technical and Vocational Education* (Dordrecht: Springer, 2007).

Hacker, Barton C., 'Engineering a New Order: Military Institutions, Technical Education, and the Rise of the Industrial State', *Technology and Culture*, 34.1 (1993), 1–27.

Hallström, Jonas, 'Technical Knowledge in a Technical Society: Elementary School Technology Education in Sweden, 1919–1928', *History of Education*, 38.4 (2009), 455–74.

Hamel, Thérèse, Michel Morisset, and Jacques Tondreau, *De la terre à l'école. Histoire de l'enseignement agricole au Québec 1926–1969* (Montreal: Éditions Hurtubise HMH, 2000).

Hamelin, David, 'Les Bourses du travail: entre éducation politique et formation professionnelle', *Le Mouvement Social*, 235 (2011), 23–37.

Hamilton, Gillian, 'The Decline of Apprenticeship in North America: Evidence from Montreal', *The Journal of Economic History*, 60.3 (2000), 627–64.

Heikkinen, Anja and Lorenz Lassnigg, eds, *Myths and Brands in Vocational Education* (Cambridge: Scholars Publishing, 2015).

Hilaire-Pérez, Liliane, 'Les Expositions universelles en France au XIXe siècle. Lignes de tension et lignes d'horizon dans le champ technologique à l'ère de l'industrialisation', in *Les Expositions universelles en France au XIXe siècle. Techniques, Publics, Patrimoines*, ed. by Anne-Laure Carré, Marie-Sophie Corcy, Christiane Demeulenaere-Douyère, and Liliane Hilaire-Pérez (Paris: CNRS éditions 2012), pp. 13–34.

Holzwarth, Simone, 'Gandhi and Nai Talim: Rural Craft Education for a New Village-Minded Social Order' (unpublished thesis, Humboldt University of Berlin, 2015).

Icher, François, *Les Compagnons du Tour de France* (Paris: La Martinière, 2010).

Jacoby, Daniel, 'The Transformation of Industrial Apprenticeship in the United States', *The Journal of Economic History*, 51.4 (1991), 887–910.

Jarrell, Richard A., *Educating the Neglected Majority. The Struggle for Agricultural and Technical Education in Nineteenth-century Ontario and Quebec* (Montreal: McGill-Queen's University Press, 2016).

Kennerley, Alston, 'Vocational Further Education and Training for British Merchant Navy Ratings: the National Sea Training Schools, 1942–1972', *History of Education*, 29.4 (2000), 301–27.

Kiprianos, Pandelis, 'Greek Technical-Vocational Education (1870–1940): Intentions and Failed Aspirations', *Paedagogica Historica*, 40.5 (2013), 664–76.

Kritikos, George, 'The Proliferation of Agricultural Schools: a Practical Education in Greece (1922–1932)', *Agricultural History*, 81.3 (2007), 358–80.

Lambert, Guy and Stéphane Lembré, eds, *Les Lieux de l'enseignement technique (XIXᵉ–XXᵉ siècles)*, special issue, *Histoire de l'éducation*, 147 (2017).

Lane, Joan, *Apprenticeship in England, 1600–1914* (London: UCL Press, 1996).

Larsen, Eirinn, 'The Masculine Foundation of Business Education: France and Norway in Comparison (1870s–1940s)', *Entreprises et histoire*, 65 (2011), 24–42.

Laufer, Ulrike, *Technik und Bildung: bürgerliche Initiativen und staatliche Reglementierung im beruflich-technischen Schulwesen Bayerns und der bayerischen Pfalz 1789–1848* (Mannheim: Palatium, 2000).

Le Bot, Florent, Virginie Albe, Gérard Bodé, and Élisabeth Chatel, eds, *L'ENS Cachan: le siècle d'une grande école pour les sciences, les techniques, la société* (Rennes: Presses universitaires de Rennes, 2013).

Lelorrain, Anne-Marie, 'La Vie quotidienne de l'enseignement agricole ménager 1913–1922', *Annales d'histoire des enseignements agricoles*, 4–5 (1992), 31–41.

Lembré, Stéphane, 'L'Expérience de l'enseignement agricole ambulant dans le nord de la France (1900–1939)', *Histoire et Sociétés Rurales*, 34 (2010), 149–80.

Locke, Robert R., *The End of the Practical Man. Entrepreneurship and Higher Education in Germany, France, and Great Britain, 1880–1940* (Greenwich, CT: JAI Press Inc., 1984).

McCloy, Don, *Creating Belfast. Technical Education and the Formation of a Great Industrial City, 1801–1921* (Dublin: Nonsuch Publishing, 2009).

Marco, Luc, Samuel Sponem, and Béatrice Touchelay, *La Fabrique des experts-comptables. Une histoire de l'INTEC 1931–2011* (Paris: L'Harmattan, 2011).

Matasci, Damanio, 'L'Éducation, terrain d'action internationale: le Bureau international de l'enseignement technique dans les années 1930', *Relations internationales*, 151 (2012), 37–48.

———, 'International Congresses of Education and the Circulation of Pedagogical Knowledge in Western Europe (1880–1914)', in *Shaping the Transnational Sphere: Experts, Networks, Issues, 1850–1930*, ed. by Davide Rodogno, Bernhard Struck, and Jakob Vogel (New York, NY: Berghahn Books, 2012), pp. 352–82.

Meuleau, Marc, 'From Inheritors to Managers: The École des hautes études commerciales and Business Firms', in *Management and Business in Britain and France. The Age of the Corporate Economy*, ed. by Youssef Cassis, François Crouzet, and Terry Gourvish (Oxford: Clarendon Press, 1995), pp. 128–46.

Mounier-Kuhn, Pierre-Éric, *L'Informatique en France de la Seconde Guerre mondiale au Plan calcul. L'émergence d'une science* (Paris: PUPS, 2010).

Münch, Joachim and Matthias Risler, *Vocational Training in the People's Republic of China. Structures, Problems, and Recommendations* (Berlin: European Centre of the Development of Vocational Training, 1987).

Mungazi, Dickson A. and L. Kay Walker, *Colonial Agriculture for Africans. Emory Alword's Policy in Zimbabwe* (New York, NY: Peter Lang, 1998).

Musiedlak, Didier, 'La Création de l'Université Luigi Bocconi et le développement de l'enseignement supérieur commercial en Europe (1896–1914)', *Mélanges de l'École française de Rome. Moyen Âge, Temps modernes*, 92.2 (1980), 625–62.

Nagy, Adrienn, 'Secondary Agricultural Vocational Training in Hungary in the 19th and 20th Centuries', in *Erziehung und Bildung in ländlichen Regionen — Rural Education*, ed. by Claudia Gerdenitsch and Johanna Hopfner (Frankfurt am Main: Peter Lang, 2011), pp. 129–41.

Neumann, Cédric, 'De la mécanographie à l'informatique. Les relations entre catégorisation des techniques, groupes professionnels et transformation des savoirs managériaux' (unpublished doctoral thesis, Université Paris-Ouest Nanterre-La Défense, 2013).

Nilsson, Anders, 'The Unknown Story: Vocational Education for Adults in Sweden 1918–1968', *History of Education*, 43.5 (2014), 615–34.

Oelsner, Veronica, '"Forging the Fatherland": Work and Vocational Education in Argentina During Peronism (1944–1955)', *Paedagogica Historica*,49.3 (2013), 382–401.

Paterson, Andrew, '"The Gospel of Work Does Not Save Souls": Conceptions of Industrial and Agricultural Education for Africans in the Cape Colony, 1890–1930', *History of Education Quarterly*, 45.3 (2005), 377–404.

Pellegrino, Anna, *La Città piu artigiana d'Italia. Firenze 1861–1929* (Milan: FrancoAngeli, 2012).

Picon, Antoine, *L'Invention de l'ingénieur moderne: l'École des ponts et chaussées, 1747–1851* (Paris: Presses de l'École nationale des ponts et chaussées, 1992).

Pronko, Marcela A., *Universidades del Trabajo en Argentina y Brasil: una historia de las propuestas de su creación, entre el mito y el olvido* (Montevideo: CINTERFOR, 2003).

Raina, Dhruv and S. Irfan Habib, 'Technical Institutes in Colonial India: Kala Bhavan, Baroda (1890–1990)', *Economic and Political Weekly*, 26.46 (1991), 2619–21 and 2623–24.

Rico Gómez, María Luisa, *La Formación profesional obrera en España durante la dictadura de Primo de Rivera y la Segunda República* (Madrid: Consejo Superior de Investigaciones Cientificas, 2014).

Rogers, Rebecca, *A Frenchwoman's Imperial Story: Madame Luce in Nineteenth-Century Algeria* (Stanford, CA: Stanford University Press, 2013).

Romano, Antonio, '"Cultura" y "trabajo" en las discusiones sobre la Ensenanza Secundaria e Industrial en la decada de 1930 en el Uruguay', *Revista Brasileira de Història da Educação*, 17.1 (2017), 7–35.

Sakatni, Mehdi, 'Les Mondes ruraux', in *Le Moyen-orient. Fin XIXe–XXe siècle*, ed. by Leyla Dakhli (Paris: Seuil, 2016), pp. 231–85.

Sanderson, Michael, 'French Influences on Technical and Managerial Education in England, 1870–1940', in *Management and Business in Britain and France. The Age of the Corporate Economy*, ed. by Youssef Cassis, François Crouzet, and Terry Gourvish (Oxford: Clarendon Press, 1995), pp. 128–46.

———, *The Missing Stratum. Technical School Education in England 1900–1990s* (London and Atlantic Highlands, NJ: The Athlone Press, 1994).

Sarkar, Suvobrata, 'Technical Content and Colonial Context: Situating Technical Knowledge in Nineteenth Century Bengal', *Social Scientist*, 38.1–2 (2010), 37–52.

Schriewer, Jürgen and Klaus Harney, 'Beruflichkeit versus culture technique. Contribution à la sémantique du travail en France et en Allemagne', in *Le Travail et la nation. Histoire croisée de la France et de l'Allemagne*, ed. by Bénédicte Zimmermann, Claude Didry, and Peter Wagner (Paris: Éditions de la MSH, 1999), pp. 107–46.

Shinn, Terry, *L'École polytechnique: 1794–1914* (Paris: Presses de la Fondation nationale des sciences politiques, 1980).

Small, Robin, 'The Concept of Polytechnical Education', *British Journal of Educational Studies*, 32.1 (1984), 27–44.

Steffens, Sven, 'Migrations ouvrières temporaires de formation professionnelle en Belgique (de l'Ancien Régime à 1914). Première approche et comparaison avec la France et l'Allemagne', in *Angles d'approches. Histoire économique et sociale de l'espace wallon et de ses marges (xve–xxe siècles)*, ed. by Véronique Fillieux, Laurent Honnoré, and Paul Servais (Louvain-la-Neuve: Bruylant-Academia, 2003), pp. 159–74.

Turiano, Annalaura, 'De la pastorale migratoire à la coopération technique. Missionnaires italiens en Egypte. Les salésiens et l'enseignement professionnel (1890–1970)' (unpublished doctoral dissertation, Aix-Marseille University, 2016).

UNESCO, *Recommendation Concerning Technical and Vocational Education Adopted by the General Conference at Its 12th Session* (Paris: UNESCO, 1962).

Vacher, Hélène and André Guillerme, eds, *L'Essor de l'École Eyrolles au XXe siècle: technologies, professions et territoires* (Paris: Garnier, 2017).

Walker, Martyn, *The Development of the Mechanics' Institute Movement in Britain and Beyond: Supporting Further Education for the Adult Working Class* (London: Routledge, 2017).

Ward, Michael R. M., *From Labouring to Learning. Working-Class Masculinities, Education and De-Industrialization* (Cham: Palgrave Macmillan, 2015).

Wong, Ting-Hong, 'Crossing the Binary Line: the Founding of the Polytechnic in Colonial Hong Kong', *History of Education*, 43.4 (2014), 524–41.

PATRICE BRET

War and Technology from 1800 to Globalization

Prometheus Unbound

As a military historian, General John Frederick Charles Fuller stressed the influence of mechanics on war (Fuller, 1942), and then the influence of armament on world history (Fuller, 1946).[1] As the foremost British tank expert of the Great War, the prophet of mechanized warfare and an influential military theorist, he had a significant impact on the German, Soviet, and US headquarters during and long after the Second World War. As soon as January 1919, Fuller first claimed in his *Weekly Tank Notes* that innovation could lead to victory ('tools, or weapons, if only the right ones can be discovered, form 99 percent of victory [...] for given the right tool any war can be won at the smallest cost') and that overstatement became a leitmotiv in every of his later books (e.g. Fuller, 1946: 31). Hiroshima seemed to prove him right. Today, however, technological determinism in war — 'long a concern of historians of technology and often, though perhaps only implicitly, historians of war' (Walton, 2019: 1) — and the ideology of technologization are questioned by historians of war, social constructivists, and historians of technology, political scientists, and strategists (Henrotin, 2013; Mahnken, 2008; Black, 2013; Lefeez, 2017; Walton, 2019). Obviously, the pairing of 'war and technology' can no longer be reduced to a mere technological innovation, even a technological breakthrough or technoscientific achievement. It now involves not only military technologies, their design, production, and (strategic and tactical) uses, but also their material and moral impacts on civil life as well as geopolitical, political, economic, social, and cultural issues.

More than any other field, the historiography of 'war and technology' derives from various professional and disciplinary backgrounds. After a look at the momentous academic turn of the 1980s and some later trends, this chapter surveys the steps and spaces of war-related technological change and power from a global perspective, from the Revolutionary and Napoleonic Wars up to the early twenty-first century, then its social interplay, dynamics, paradoxes, and routes between the civil and the military, through innovation, conservatism, secrecy, cooperation, and proliferation.[2]

1 Fuller first published the latter in the US *Army Ordnance* from July 1944 to July 1945, prior to adding a chapter on 'The Age of Atomic Energy' for his British edition in 1946.

2 Naval, air and space technologies are developed in dedicated chapters in this book.

Global History of Techniques, ed. by Guillaume Carnino, Liliane Hilaire-Pérez and Jérôme Lamy, Global Matters Culture Matérielles, histoire des techniques et circulation globale des savoirs, 9 (Turnhout: Brepols, 2024), pp. 581–612
BREPOLS PUBLISHERS 10.1484/M.TECHNE-EB.5.129831

A Short Survey of Historiographical Trends

There is a vast, long-standing literature on weaponry which is particularly well established in the anglophone world, alongside reference books on national armaments. Some have a broader scope on the twentieth-century armaments in a global world (Hogg, 1973; Hogg, Weeks, 2000). Others only go beyond a mere description to propose a 'social history' of the machine gun (Ellis, 1975) or to question techno-military notions such as 'firepower' and 'effectiveness' (Hughes, 1974), or even the notion of the 'weapon', which is expanded to 'almost anything with the remotest of military applications, from a motorcycle to a pair of binoculars' (Cleator, 1967: 15).

In the 1970s, the human and social sciences refreshed the objects of research and took the field into a fruitful debate about the place of war-related technologies in the rise of the West and Western imperialism. Since the 1980s, academic research has also enlarged the meaning of the 'military', now including the military-industrial complex, and that of 'technology' as

> all the ways in which humans use the materials and energy in the environment for their own ends, beyond what they can do with their bodies [...] [i.e.] not only artifacts and domesticated plants and animals but also the skills needed to use them and the systems in which they are embedded (Headrick, 2010: 3).

In 1993, Alex Roland analysed that 'historiographical revolution', which first came from early modern history (Roland, 1993) and later led to include techniques in the academic field of war studies.

The 1980s Historiographical Revolution

Responding to a complaint on the lack of attention to the relationship between technology and war (McNeill, 1963) and expounding on a pioneering study by Michael Roberts (Roberts, 1956), Geoffrey Parker emphasized the role of technology in the *Journal of Modern History* (1976), and then in his landmark book, *The Military Revolution*, to which Jeremy Black responded (Parker, 1988; Black, 1991). Meanwhile, his concept had been applied to the American Civil War and the wars won by Prussia over Denmark, Austria, and France for the German unification (1864–1871) (Headrick, 1981). William H. McNeill too argued that the Western industrialized states owed their supremacy over the world to their superior military technology and to the free market and command economy through centralization of power (McNeill, 1982).

Scrutinizing this new scholarship, A. Roland defined six broad trends. Most studies addressed the impact of technology on war more than the reverse. Most placed the dynamic interplay between war and technology in the context of the recently-translated universal history by Hans Delbrück in 1900–1920 (Delbrück, 1975–1985), either in a political and diplomatic (Pearton, 1982) or a societal approach (McNeill, 1982), and most shared the broader vision of the term 'military' (Headrick, 1981, 1991; McDougall, 1985). Many stressed the close links between technology, war, and finance which led to the military-industrial complex, which emerged in England during the naval arms race with Germany in the 1890s, and to modern command economies (McNeill, 1982), or even to a 'technological radicalism' (Sumida, 1989) or 'technological fanaticism' (Sherry, 1987) exacerbated by twentieth-century technocracy and state-managed technology (McDougall, 1985). Otherwise, Roland warned against two dangerous risks: 'a kind of presentism [that] moves and infects much of this literature' (Roland, 1993: 133–34) and some degree of technological determinism, especially in large sweeps of history written from the perspective of military history (O'Connell, 1989; van Creveld, 1989). To him, the only exception was an investigation into the nature of technological change through the social construction of intercontinental ballistic missile (ICBM) guidance, which was a new approach to weapons indeed (MacKenzie, 1990).

However, Roland partly overlooked the incipient trend of critical ethnography, anthropology, and science studies. For instance, C. Cohn brought a gender approach to her analysis of the mechanisms of the mind's militarization through the discourse of US 'defence intellectuals', in terms of abstraction, alienation, and masculine identities of supermen, that functioned 'as a gloss, an ideological curtain behind which the actual reasons for [the nuclear deployment] decisions hide' (Cohn, 1987: 716). To her, even 'technostrategic' language and strategic thinking are 'part of technology itself', through 'a way of looking at problems — formal, mathematical modelling, systems analysis, game theory, linear programming'. Supporting Lord Zuckerman's critical views on the nuclear weapons (Zuckerman, 1982), she argued that strategic thinking changed 'in direct response to technological changes, rather than political thinking, or some independent paradigms that might be isolated as "strategic"' (Cohn, 1987: 690). At the same time, an anthropologist examined the 'nuclear rites' of the Lawrence Livermore National Laboratory weapon designers facing anti-nuclear activists; he later portrayed the 'people of the bomb' of the American nuclear complex (Gusterson, 1996, 2004).

Post-Cold-War Historiographical Interplay

Pioneering reference collections of papers must also be mentioned as an introduction to the field. Their contribution to the field was innovative and outstanding. Through various historical approaches and topics, the authors had begun to study — mainly for the 'American experience' — the complexity and variety of the previously neglected relations between science, technology, and the military, as these titles witness: *Military Enterprise and Technological Change* (Smith, 1985), *Science, Technology, and the Military* (Mendelsohn, Smith, Weingart, 1988), *The Restructuring of Physical Sciences in Europe and the United States* (De Maria, Grilli, Sebastiani, 1989), and *Big Science: the Growth of Large-scale Research* (Galison, Hevly, 1992).

Beside monographies and other collected papers of that mainstream 'social shaping of technology' (often mentioned *infra*), two further collections with a broader geographical scope must be added: *National Military Establishments and the Advancement of Science and Technology* (Forman, Sánchez-Ron, 1996), and *Les Sciences pour la guerre*, which also focuses on organizational techniques (Dahan, Pestre, 2004). Otherwise, the post-Cold War period was also characterized by a slow cross-fertilization between academic research and the specialized studies produced in the military and engineering spheres.

One of the new trends arose from armament heritage groups founded either by local scholars or weapons engineers, as military sites closed after the end of the Cold War. Following the Fortress Study Group, which had been founded at Oxford (1975), the Gunpowder Mills Study Group produced local publications and articles in the *Industrial Archaeology Review* prior to becoming an international network (later as the Gunpowder and Explosives History Group) in the 1990s and holding symposia at meetings of the International Committee for the History of Technology (ICOHTEC) (Buchanan, 1996, 2006). Also interested in their heritage and history, French weapons engineers opened their scientific symposia to international academics, e.g. on instruments and experiments (A3P, 2001) or on rocket solid propellants (A3P, 2004). The Service Historique de la Défense (SHD – Defence Historical Service) jointly published a social history of the Second World War armament factories with a state worker union and academic historians (Albert, 2019). Archaeologists and historians worked on British explosives factories before and during the Cold War (Cockroft, 1999) or Japanese facilities such as the sea forts of the Kaiho artificial islands in Tokyo Bay or the Hario radar towers in the Nagasaki Prefecture, recently recognized as 'Important Cultural Property' by the Japanese Agency for Cultural Affairs (Masaaki, 2011). Museum curators, too, have investigated the history of weapons beyond their collections, e.g. a curator from the National Air and Space Museum (NASM), Washington, DC, who has worked on nineteenth-century rockets (Winter, 1990), or a team at the Paris Army Museum, who has researched tactical nuclear artillery (Sevrin, 2010). The social history of military technology was included as a dedicated as a section in ICOHTEC meetings around Barton Hacker (NASM), who founded *Vulcan. The Journal of the History of Military Technology* and told 'the life story' of American military technology in the 'Greenwood Technographies' series (Hacker, Vining, 2006). Separate papers presented at other congresses — of the Society for the History of Technology (SHOT), the Society for Military History, and the History of Science Society (HSS) — were also collected (Walton, 2005).

As disarmament began after the Cold War, practitioners and experts of the military-industrial-academic complex wrote retrospective accounts of their fields, e.g. former staff members of the Soviet bacteriological war programmes who reported on their activities (Alibek, 1999) and members the Office of Technology Assessment (OTA), who from 1972 to 1995 provided the US Congress with analyses of the complex scientific and technical issues on the American defence acquisition (Alic, 2007). Others turned to the history of their own field or country, like a British naval architect who published many popular reference books on naval technological developments (Brown 1992, 1993, 2003), or a French electronics engineer and industrial manager who authored a history of the radar (Blanchard 2004). In the late 1990s, an institution key to the French military-industrial complex — the Délégation Générale de l'armement (DGA-General Delegation for Armament), which had been created in 1961 —

also initiated reflections on its own actions. It encouraged academic researchers to join state weapons engineers in the newly founded Comité pour l'histoire de l'armement(CHARME-Committee for the History of Armament) and the Department of History at its Centre des hautes études de l'armement (Centre for Higher Studies on Armament). Beside specialized workshops on information and command systems, French naval arsenals (CHARME, SHM, 2002), or military optronics (CHARME, 2004), they organised international conferences on the armament in modern France (Pestre, 2003), the mobilization of nations in the era of total wars (1914 to 1945) (Barjot, 2012), and the European armaments cooperation after the Second World War (Hébert, Hamiot, 2004). They also supervised working groups of weapons engineers (Comité d'histoire de l'armement Terrestre (COMHART-Committee for the history of land armament; Comité d'histoire de l'aéronautique (COMAERO-Committee for the history of aeronautics) in order to help future academic work on the French research and development (R&D) programmes since 1945. By way of example, the COMAERO series includes volumes devoted to ballistic (Arnaud, 2004) and tactical missiles (Carpentier, 2004) and on-board techniques (aircraft electronics, equipment, landing gear, armament, etc.) as well as aeronautic institutions (research centres, flight test centres, training centres, industrial maintenance workshops, etc.). CHARME also supported academic research (Schiavon, 2014; Aucouturier, 2017) and acted as a guest editor for popular scientific magazines (CHARME, 2002).

Regardless of the author, most of the literature on the post-Cold War decades could not easily sustain an academic methodology as it was produced without access to classified sources or the capacity to reveal secret data. Historical essays by defence analysts working in government agencies were reviewed to prevent the disclosure of classified information, e.g., the CIA checked a book on the spread of nuclear weapons (Kroenig, 2010). They often reflected the theory of 'Revolution in Military Affairs' (RMA), which spread in the 1990s to 2000s among the American intellectual community. Postulating that the US was technologically invincible, RMA was actually rooted in the 1950s 'military-technical revolution' that was theorized in USSR following the impact of nuclear weapons on the conduct of conventional warfare. Soviet military analysts asserted that improvements in US dual high technologies (computers and computer networking, which was so far underdeveloped in USSR) would give the United States dominance of the battlefield thanks to precision-guided munitions and 'surgical strikes': '[v]isionaries spoke of "net-centric warfare", in which electronically networked forces would reconnoiter, communicate, and coordinate on the battlefield faster than the foes' (Roland, 2016: 108). Beyond historians who try to put modern warfare with drones and technological intelligence into perspective (Kreuzer, 2016), political scientists also investigate the field and assess how cyber-, robotic, and space weapons have already changed and are yet to change future war (Jordan et al., 2008; Rabkin, Yoo, 2017).

Lastly, 'war and technology' had also been questioned in an environmental approach. Whether CBN (chemical, biological, nuclear) or conventional weapons (high explosive or incendiary weapons, cluster munitions, landmines, etc.), twentieth-century military technologies caused environmental harm (deforestation, soil erosion, water pollution) as an aftermath of war. Apart from the indirect effects of warfare (explosive remnants on the battlefields, loss of biodiversity), weapons with intentional effects have brought long-term environmental disruption since the Second World War. Napalm was used massively by the US in Vietnam in the 1960s, along with 'Rainbow Herbicides'. These chemical defoliants had long-lasting toxic impacts. Delayed radioactive fallout from hundreds of nuclear tests has taken months or years to come down to earth all around the globe; carcinogenic effects are still observed in the former sites of nuclear experiments in Chinese Xinjiang and French Polynesia, or in Iraq, from the depleted uranium missiles of the first Gulf War (1990–1991) (Jha, 2018).

Steps and Spaces of War-related Technological Change and Power

Since 1800 to 1830, technological change has been so fast and rich that authors cannot agree on naming or even defining later periods (Table 1). After Fuller's echo of Lewis Mumford's periodization based on dominant energies and materials, they often reflect innovations and paradigmatic technologies, although they vary much and even combine fears related to the hubris of power (Cleator, 1967; Massie, 1991). Others would stress the late twentieth-century transition from the 'industrial age' to the 'information age' (Goldman, Eliason, 2003).

Such types of periodization hide concomitant situations in the world. Beyond innovation-centric items (penicillin, high-technology weapons), David Edgerton has argued that older 'invisible technologies' still dominate in modern wars, especially in guerrilla wars against colonial powers or in civil wars (like the Kalashnikov assault rifle or even the machete, which killed thirty-eight per cent of the victims in Rwanda). To him, that reveals 'a division, implicit but powerful, between things which belong to the realm of the military and those in the world of science and technology that are taken to be essentially civilian'. Thus, 'civilian technologies applied to war', rooted in the industrialization of war since the late nineteenth century, 'transformed twentieth-century war [...] by civilianising and totalising it', and globalizing it (Edgerton, 2006a: 138).

The linear technological evolution of Table 1 (p. 587) also ignores the dynamics of the military technology production and trade. By contrast, K. Krause defines four types of coexistent 'degrees of mastery' which could present a dynamic chart of those capacities at any time over the world (Table 2, p. 588).

To Krause, only a few states in Europe and the United States reached the most sophisticated step before the First World War, and all the types coexist in any historical period. Krause's last step meets M. Pearton's concept of the 'knowledgeable

3 The articles published in Pestre (2015) do not actually define or name 'ages', but the characterizations are theirs.

4 Fuller changed the name between the first American edition in July 1945 in *Army Ordnance* and the post-Hiroshima British edition (see footnote 1).

Table 1. The War & Technology Ages

<table>
<tr><th></th><th>Fuller 1946</th><th>Cleator 1967</th><th>van Creveld 1989</th><th>Black 2013</th><th>Pestre 2015[3]</th></tr>
<tr><td>1500–1830</td><td>'The Age of Gunpowder'</td><td>'Powder and Shot'</td><td>1500–1830 'Machines' (iron)</td><td>1490–1800 'Gunpowder Technology'</td><td>early science-based war technologies</td></tr>
<tr><td rowspan="4">1830–1945</td><td>'The Age of Steam'</td><td rowspan="2">'Mechanisation and Immobilisation'</td><td rowspan="4">1830–1945 'Systems' (steel)</td><td rowspan="2">1775–1945 'Firepower, Steamships, Railways, Telegraphs, Radio: Technologies of Killing, Logistics, Command, and Control'</td><td>1770–1914 industrial and managing innovation; technological experiments; early military-industrial complex</td></tr>
<tr><td>'The Age of Oil' (internal-combustion engine and radio)</td><td rowspan="2">1914–1918 total & techno-scientific war engineers' & chemists' war (war of machines, mass production of firepower, early R&D)</td></tr>
<tr><td rowspan="2">'The Age of Mechanisation' and 'The Age of Oil II'[4]</td><td rowspan="2">'Prelude to Armageddon'</td><td rowspan="3">1903–2013 'A New Sphere: Air Power' and 1910–2013 'The Internal Combustion Engine: The Technology of Decentralized Power'</td></tr>
<tr><td>1939–1945 total & techno-scientific war physicists' war (technological breaks: radar, atomic bomb)</td></tr>
<tr><td>post-1945</td><td>'The Age of Atomic Energy'</td><td>'Nuclear Nemesis'</td><td>'Automation' (sophisticated alloys, ceramics, & other synthetic materials)</td><td>technoscientific Cold War & RMA mathematicians' war (technological breaks and permanent R&D: missiles, computer science, information technology)</td></tr>
</table>

state', which is when a government takes up an incentive position within knowledge production as 'overall policy-maker' defining the purposes and providing the suitable services, workforce, and public investment to reach them (Pearton, 1982, p. 254). It

Table 2. Krause's War & Technology 'Degrees of Mastery'

I	'[technical] skills needed to operate and maintain the weapons'
II	'basic scientific and engineering skills needed to reproduce (or copy) technology'
III	'military and economic organization needed to use, adapt or refine the weapons for particular battlefield or marketplace conditions'
IV	'social, political and economic organization (or motivation) needed to produce new forms of technology and to advance the production frontier'

Source: (Krause, 1992: 19).

also meets the concept of the 'warfare state' that Edgerton has applied to Britain for its 'liberal militarism' in the 1920s to 1960s, although it is usually seen as a 'welfare state' (Edgerton, 2006b). To a certain extent, France had prefigured this: from the 1770s to the 1830s, it experimented and set up long lasting pioneering configurations of science-based and state-led military R&D, a techno-military *État Prométhée* a long time before a 'welfare state' (*État providence*) was implemented (Bret, 2002a).

Technological Change in Industrial Europe and the United States

The decades around 1800 marked the peak and subsequent decline of the early modern classic age of war technologies based on natural energies (animal and wind) and the proto-industry (watermills and blast furnaces). France set up the first steps to standardization and arms systems with the *plans-types* (model plan) of ships (Vérin, 1993; Llinares, 1994) and Gribeauval's artillery system, attempts of stamped flintlock, and the revolutionary *Atelier de perfectionnement des armes portatives* (Workshop for the Improvement of Light Firearms) and *Atelier de précision de la Grosse Artillerie* (Precision Workshop for Heavy Ordnance) run by *savants* (Alder, 1997; Bret, 2002a). Chemists were in charge of gunpowder manufacturing with Antoine Laurent Lavoisier's scientific management of the production (1775), and then of the 'revolutionary processes', Claude Louis Berthollet's 'muriate powder' (potassium chlorate powder), and shells and sulphur sublimation (Gillispie, 1992; Bret, 1992, 1993, 2002a). France also first experimented with Fulton's submarine and so-called 'torpedo' (sea mines), accelerated long-distance communication with Chappe's optical telegraphy, and introduced the third dimension into war by using hydrogen balloons on the battlefield (Langins, 1983; Bret, 1990, 1991). Many *ingénieurs savants*, first trained at the *École Polytechnique* (founded in 1794), were now engaged in research in state laboratories and committees, either military (artillery and engineering schools) or civilian (the mint, the Natural History Museum), especially in Paris (Gay-Lussac on alloys) and Metz (Poncelet on hydraulic mills) (Grattan-Guinness, 1993; Bret, 2002a). Then, until 1914, Europe and the United States gradually produced a continuous series of 'duelling technologies' applied to war (defensive counter improvements respond to offensive innovations and vice versa).

Innovations in small arms depended less on the iron industry and more on the artisans' or users' smartness. Thanks to a series of micro-inventions, the old flintlock musket was progressively replaced with percussion rifles and then with needle guns, which were easier to load and provided a longer range and better accuracy, alongside fewer shot failures due to wet powder: First, metallic caps or primers filled with silver or mercury fulminates were patented in France by the gunmaker Prélat (1818); the invention, also claimed by several others in Britain and America, followed the discovery of these substances by the Italian chemist Brugnatelli (1798) and the American officer Howard (1800). Second, the 'hollow bullet' expanded into the rifling grooves (its base being deformed by the detonation gases) and provided far superior ballistics (in terms of power and accuracy) than the usual round bullets. The British government refused Captain Norton's and the gunmaker Greener's earlier proposals before adopting an improved French version for the Enfield Pattern 1853 rifle-musket or 'Minié rifle', which was first experimented with in the eighth Kaffir or Xhosa War (South Africa). Despite their major improvements, those rifles kept a slow firing rate as they still were muzzle loaded. Third, breechloading was an European innovation when used with a central percussion cap at the base of a metallic cartridge in 'needle guns'. Despite having a shorter range than the Minié rifle, the first military needle gun ('Dreyse rifle') was adopted by Prussia as early as 1841. Being loaded faster, even by a soldier lying on the ground — it was to bring a key change in modern warfare after the Prussian victory over the Austrians at Sadowa (1866). Fourth, an other crucial innovation, in the United States, was repetition gunnery and automatic loading, e.g. in the Colt (first patented 1836) and Remington revolvers, and then the Winchester rifle, which were all suitable for use on the Western frontier and the American Indian Wars, and were adopted by the Turks against the Russians in 1878. Lastly, the further achievement of repeating weapons was the machine gun, which was first used in the American Civil War before it became a crucial collective arm in the colonial wars and then the First World War.

Most of the technologies, most of the technologies for war were based on — and participated in the development of — new processes of the so-called 'Industrial Revolutions' and the heavy industry (iron works and chemical industry). New energy sources (steam, electricity) and management procedures caused the production to be more focussed on state and private weapons manufactures and arsenals, which were located within large and mechanized buildings made from new materials (iron architecture), for example, the Harpers Ferry Arsenal and the Springfield Armory (USA), the Royal Small Arms Factory at Enfield and W.G. Armstrong & Co (UK), the *Manufacture d'armes de Saint-Étienne* and the *Établissements Schneider* (France), and the *Spandauer Arsenal* and the *Fried. Krupp AG* (Germany) (Smith, 1980; Bacher, Brun, Perrin, 2007). Steam and steel changed civilian and military industry, logistics and communications significantly (van Creveld, 1977; Wolmar, 2010). The war preparedness had a huge impact on the development of the industrial empires built on railways, artillery, battleships, and submarines by steel barons (Krupp, Schneider, etc.). As early as 1833, Prussian industrials and economists preached the use of railways to move troops. The German economist Friedrich List (1789–1846) already saw them as defensive weapons, but complete armies, with horses and artillery, were successfully sent to fight against the Polish (1846), Danish (1864), Austrians (1866), and the French (1870). Railways even became a technological arm of German diplomacy, and the *Bagdadbahn* (Berlin–Baghdad railway) was later completed with the Hejaz railway until Medina as a crucial axises to control the Ottoman Arab provinces.

One of the major issues of this military-industrial collaboration was the decline and demise of sailing and wooden ships, which were replaced with steam (see chapter 'Naval') and steel. After Waterloo, the use of the carronade (a powerful, short British navy gun) and Colonel Paixhans' systematic study of earlier experiments the French Artillery records led him to propose the development of a big gun to shoot shells into the hulls of enemy vessels: thus, a mere gunned brig could sink the biggest man-of-war ships (H.T. Adams, 1994; Bret, 2003). The 'Paixhans gun' prompted the navies to have the wooden warships armoured with iron, and then to build steel-hulled ships, as Bessemer steel was resistant to the shells. Ironclad warships appeared in France (1859) and Britain (1861), and the first naval battle of ironclads was fought at Hampton Roads during the American Civil War (1862), long after te construction of Fulton's *Demologos*, which was the first attempt at a paddle steamer with timber clad in New York (1815). Eventually, once the Italian naval architect Vittorio Cuniberti had developed the concept of an 'all-big-gun' warship in the early 1900s, it was applied in Britain, Japan, and the United States. The other key innovation had been the replacement of paddle wheels by propellers, which were experimented with in the early 1800s, and notably developed in Britain and the United States from the 1840s by Francis P. Smith and the Swedish-born mechanic John Ericsson. The result of all those evolutions was a collection of high-end technology in the most powerful HMS *Dreadnought* (1906), which revolutionized the navies after the Russo-Japanese War (1905). The first 'all-big-gun' ship able to shoot both torpedoes and long-range (13-kilometre) armour-piercing shells under a fire control system by electrical transmission, she was also the fastest battleship (21 knots) as the first powered by steam turbine. Her launch strengthened national military-industrial complexes and boosted the arms race, especially with the German *Kriegsmarine* (War Navy), and the development of submarines and naval mines (Sumida, 1989; Massie, 1991; Brown, 1992, 1993, 1997; Sondhaus, 2001).

Above all, the great Western powers were rather symmetrically armed, despite temporary dysbalances. Their technological dominance led to a new wave of colonization over traditional societies and well-structured countries unprepared to face asymmetrical war technologies.

Military Technology, Colonial Power, and Asymmetrical Wars

Since the sixteenth century, European imperialism has been supported by its dominance in war technologies, and its organizational, financial, and trade techniques, alongside its administrative technologies for funding and marshalling military resources (Hoffman, 2015). 'The West owes warfare for its rise to global power' (Dincecco, Onorato, 2017).

Before 1800, Western imperialism had covered thirty-five per cent of the world's land surface; one century later, Europe ruled over eighty-four per cent. After 1871, while European powers were at peace, their rivalry spread out around the globe. They fully remodelled the world's map, thanks to various sophisticated tools of penetration and conquest (steamboats, quinine, and trading). The German victory of 1871 pushed France to conquer more than nine million additional square kilometres in Africa, the Indian Ocean, and the Pacific and implement arsenals around the globe, in order to manage its new colonial empire (CHARME, SHM, 2002). With its most powerful naval network, the British empire 'ruled the waves': its new expansion surpassed twelve million square kilometres, while Germany and Belgium gained two million square kilometres each. The Russians had also conquered territories in the Ottoman and Persian empires, in the Caucasus region and Kazakhstan, and even in the confines of China from the 1770s to 1860s; it later gained two million square kilometres in Central Asia, too, while the United States conquered all of the American Indian territories.

With the sole exception of the Ethiopians' victory over the Italian troops at Adwa (1896), economic dominance, continuing technological progress, and the better organization and military training of the Western powers maintained asymmetric relations and wars to their benefit (Headrick, 1981, 2010; Hoffman, 2015). Even when commercial or diplomatic considerations caused the Europeans and Americans to provide the native people with enhanced technologies, they sustained their own superiority: as well as native-made Asian matchlocks had met European flintlocks, imported flintlocks and Guinea powder later met percussion guns, and then obsolescent percussion guns fought breechloading rifles, before a few British Maxim machineguns made thousands of those small modern arms obsolete in the Matabele and Mahdist wars in Rhodesia and Sudan (1890s). The business of arms trading and local adaptation by the native craftsmen who were able to repair guns and make ammunitions increased violence and colonial rule in Asia and Africa (Reid, 2012).

Western techno-military pressure was, indeed, so great that even formerly strong powers were forced to open their territories (Japan and Persia) or cede parts of them (the Mughal, Ottoman, and Chinese empires), despite their efforts to modernize their own military capacities. Rather than an Ottoman decline and technological inferiority, historical research now again stresses the administrative-bureaucratic reforms of the Western 'military revolution', and hence of its production capacity, scientific engineering, and state patronage to supply the ever-growing armies (Grant, 1999; Ágoston, 2005). The Ottoman response started after the Russian victories in the 1770s, with French experts in naval engineering, fortifications, gun founding, and artillery, but it was slowed by the janissaries' strong opposition to Selim III's 'new order' (*Nizam-ı cedid*, 1793) until Mahmud II massacred them (1826), paving the way of the state modernization of his successor's Tanzimat era from 1839 onwards (Firges, 2014). In Egypt, Mehmet Ali had already modernized that Ottoman province. After the display of Western military and administrative technologies during the French occupation of the country, he had adopted and adapted them. After slaughtering the Mamluks who refused his reforms (1811), he soon conquered Hijaz and Sudan.

After the destruction of his modern European-type fleet at Navarin by the French and British fleets fighting for the Greek independence (1827), he turned against the Sultan and even conquered Syria.

Despite cultural or corporatist resistance, modernization was mainly conceived as a hybrid European 'military' pattern, especially after Waterloo, when many European experts proposed their services on all continents, e.g. the Maratha states and Punjab in India, the Merina empire in Madagascar, the Regency of Tunis, and the pashalic of Baghdad among others. All of them were nevertheless won by more experienced armies, which often included native troops trained by the colonial powers to supplement their owns troops in their conquest. The East India Company (EIC) 'sepoys' took part in the British conquest of India, from the Mysore Wars and the Maratha Wars to the Sikh Wars. Once the whole peninsula was under British control, the EIC was seen as competition on the open market of arms and driven out of it by the British government in 1856. Instead of the percussion smoothbores of the sepoys, London provided them with the modern Enfield Pattern 1853 rifle-musket. The appreciation of this rifle by the native troops reveals the dramatic social and political consequences of a military technological transfer into another cultural context: the new 'greased cartridge' sparked the Great Mutiny of 1857 among the sepoys of the Bengal Army, which led to the end of the EIC and the proclamation of the Empire of India (Harding, 1997–1999).

Actually, only Japan brilliantly succeeded in appropriating Western technologies. Opening to foreign trade in the 1850s, it went through a fast transition after the Meiji reform (1867) before successfully turning new naval and military techniques against its neighbours and becoming an imperialist power. After its victory over China (1895), Japan conquered Formosa and 'protected' Korea prior to annexing it. Above all, after defeating the Boxer Rebellion alongside the Western powers, it even defeated Russia with its young fleet at Tsushima and in Manchuria (1905): Japan now controlled and soon annexed the Chinese province prior to occupying a large part of China in the 1930s as a prelude to the total war in the Second World War.

The post-1945 decolonization wars showed the limits of technology in asymmetric wars within a global geopolitical context. Despite the victory of the colonialized peoples, Western imperialism has continued via the global arms trade — challenged by the military supplies from the Soviet Bloc and new actors like Brazil — and the exploitation of raw materials, such as uranium in Africa (Hecht, 2012).

Total War and Mass Destruction Technologies in the Era of the Mass Army and After

From 1800 onwards, strictly military and dual-use technologies — both military and civilian — moved from artisanal and protoindustrial system towards the 'American System', and then mass production (Hounshell, 1984). The First World War dramatically changed the scale of war, which became total and global — and the Second World War even more. The 'global war' arose from a massive participation of the colonial empires, the Commonwealth, the United States, Japan, and other non-European belligerents. To W. Mills, three historical revolutions were at the roots of 'total war': the French Revolution with its legacy of the ideology and practice of 'the nation in arms'; the material and organizational technologies of mass production bequeathed by the Industrial Revolution; and the Prussian military managerial revolution (Mills, 1956). In addition to their demographic and moral components, total wars were won by the alliances that produced the largest technological and industrial production (a large number of tanks was more useful than titanic heavy railway artillery (Barjot, 2012; Geyer, Tooze, 2015). Others see the transport and communication technologies as 'the single most important factor' in 'the Era of the Mass Army', from the railway used to mobilize and feed millions of soldiers, until long-distance and precision weapons (missiles and aircraft) made such mass armies less desirable in the late twentieth century (Onorato, Scheve, Stasavage, 2014: 449).

In the First World War, as men were mobilized as soldiers, female and colonial manpower massively participated in the European war industry, and 'inescapably, the enemy's population became a target, for it embodied both the national will and the productive capacity of the state' (Roland, 2016: 81). Weapons and techniques had to be adapted to the new kind of war: millions of soldiers on immobile military fronts of trenches, barbed wires, and machineguns; unprecedented mass bombing of artillery; new aerial and submarine battlefields; and heavy logistics (trains, lorries, cargo ships). To Frédéric Guelton, the human factor is the essential constituent in the four types of adaptation: 'retro-innovation' (e.g. steel helmets, trenches, mines, and grenades), technological stagnation, the capacity for innovation in wartime, and the invention of new weapons, such as breaking chemical (gas) or mechanical (tanks) military technologies (Guelton, 2014).

Although despite their moral impact (Haber, 1986; Lepick, 1998; Johnson, 1990), war gases failed to be the expected breakthrough weapon, the First World War was a war of chemists. The urgency and large-scale demands brought along state-directed or state-led solutions in the production of explosives and pharmaceuticals, to which chemists from the Commonwealth also contributed (MacLeod, 1989, 1993; Cockroft, 1999; MacLeod, Johnson, 2007). Many other scientists also took part in the scientific and industrial mobilization (Rasmussen 2010; Winter, 2014; Jouanjus et al. 2018), including mathematicians, mainly for ballistics (Aubin, Goldstein, 2014), and physicists, who helped in the areas of firing control and tackling aerial, submarine, and mass bombing warfare (Hackmann, 1984; Aubin, Bret, 2003). During the war, the scientists often worked for war in the 'battlefield laboratory' (MacLeod, 2000), but mainly in their own university laboratories. After the war, several of them carried on working with military laboratories, for example on chemical and biological warfare in Le Bouchet (Lepick, 1997; Cole, 1997; Aucouturier, 2017). Even a pacifist like the physicist Paul Langevin worked on remote sensing in the laboratory that had recently been implemented by the Navy in Toulon (Lelong, 2001),

Total and global war reached a peak in the Second World War. An institutionalized military R&D and new technologies in various fields were added — not only material technologies (microwave radar, jet propulsion, proximity fuses, the bazooka and the *Panzerfaust*, guided and early cruise missiles, 'precision' bomb sights, acoustic torpedoes, computerized code breaking, and the atomic bomb), but also intellectual technologies (propaganda) based on state- or industry-controlled mass communication technologies (the press, wireless transmission). All were later enhanced both during the Cold War (intercontinental missiles, the hydrogen bomb, nuclear propulsion), and after (space warfare, cyber war). R&D in state-led laboratories was now carried out in the new superpowers and the old colonial powers (Gowing, 1964, 1974; Bud, Gummett, 1999; Edgerton, 2006b).

While aircraft carriers became the most iconic of conventional military strength as an extension of sea power (Horowitz, 2010), the nuclear bombing on Hiroshima and Nagasaki (6 and 9 August 1945) gave birth to the symbol and concept of 'weapons of mass destruction', an expression which was used three months later by in the Truman-Attlee-King 'Declaration on Atomic Bomb'. Evoking the dual use of nuclear energy, the U.S. President and the British and Canadian Prime ministers jointly proposed to promote its 'widest use for industrial and humanitarian purposes' and eliminate atomic weapons 'and all other major weapons adaptable to mass destruction'. UNO later developed the last phrase into 'atomic explosive weapons, radioactive material weapons, lethal chemical and biological weapons, and any [similar] weapons developed in the future', i.e. chemical, biological, and nuclear (CBN) weapons (Martin, 2004; Oren, Solomon, 2020).

R&D on and the use of other weapons of mass destruction often circumvented international treaties. Chemical weapons had been banned at La Haye (1899) and Geneva (1925), and again in 1992, when the Paris convention forbade their production and R&D, and the stocks had to be destroyed. However, Sarin gas was used by Iraqi and Syrian forces. Meanwhile, other chemical weapons had been developed. Invented at Harvard University in 1942, napalm was used by the US Air Force against Japanese and German military and civilian targets, and later against the Greek communists and in Korea. Above all, it was used in Vietnam by the French in the early 1950s, and then massively by the US in the 1960s, along with Rainbow Herbicides, which had first been used by the British against Malay communist insurgents in the 1950s. 'Agent Orange', the chemical defoliant most employed by the US in Vietnam in the next decade, was produced by Monsanto and Dow Chemical for the US Department of Defense (Jha, 2018).

Regarding biological weapons, their production and use was also banned in 1972. Soon after, however, the Soviets created Biopreparat, a huge complex for underground military chemical and biological weapons programmes, with some forty pharmaceutical laboratories and virology research centres and sixty thousand scientists and technicians (Alibek, 1999). In China, the Lop Nor nuclear site in Xinjiang was also used for biological experiments. Other civil sites hid military R&D on CBN weapons around the world, in Iraq, Israel, Pakistan, South Africa, Iran, etc. They often developed strategies using mobile facilities to evade international controls.

After the USSR developed the atomic bomb in 1949, the Korean War opened 'a period of heroic technological ventures' (Alic, 2007), while the US budgets for military R&D increased dramatically. As the Cold War arms race happened between the two blocs, the nuclear-powered submarine USS *Nautilus* (1955) represented a major technological break that made it possible to prevent a third world war. Now intercontinental bombers could be destroyed in their air bases and the ICBMs in their launch pads, but the new component of the strategic nuclear triad did not require frequent refuelling, and was hence undetectable by the enemy. Thus, the 'submarine-launched ballistic missile' (SLBM) became the *ultima ratio* of Cold War deterrence, and nuclear technology shaped the Long Peace for decades, as its balance prevented direct war between the superpowers (Gaddis, 1987). Britain (1952), France (1960), and China (1964) developed their own nuclear armament before the UNO treaty of non-proliferation (1968), which was not signed by France and China until 1992. Later, Israel, India, Pakistan, South Africa, North Korea, and Iran also developed nuclear deterrence or menace,, while several areas were officially declared free from nuclear weapons: the Antarctic (1959), Latin America and the Caribbean (1967), the South Pacific regions (1985), Southeast Asia (1995), Africa (1996), and Central Asia (2006).

Nuclear weapons evolved via the miniaturization and multiplication of nuclear heads, increased range-performance, and the accuracy of missiles. High-technologies of ballistic and antiballistic missile systems were developed within the US 'military-industrial-academic complex' (Leslie, 1993) to prevent Soviet nuclear strikes. Using vacuum tubes and digital computing (IBM computers), the US Air Force and MIT set up the 'Semi-Automatic Ground Environment' (SAGE) in the late 1950s to promote the aerial defence against atomic weapons. Both a contribution and a visionary response to the emergence of a 'closed world' (Edwards, 1996), or an issue of and a metaphor for total defence, in spite of the pretended results of its tests, SAGE was soon supplanted by ICBMs, and technically, by transistors and integrated circuits. A dream and a myth of automatic control, SAGE has contributed to the ideology of an active defence that is able to diminish the nuclear threat (Bruce-Briggs, 1988; Edwards, 1996). Six SAGE centres were still in use in 1986, when President Reagan ordered a new shield, the 'Strategic Defense Initiative' (SDI) which was another myth commonly known as the 'Star Wars program' that used a new array of the most recent technologies (lasers).

A key source of funding for advanced materials or supercomputing research in national, industrial, and academic laboratories, despite the doubts about its feasibility, SDI was stopped by President Clinton in 1993. Now the Soviet threat was over, and the RMA theory of a revolution leading to US technological invincibility was thought to have been amply demonstrated in the Gulf War. 'September 11, 2001' and the following wars once more increased the defence spending: military R&D even exceeded fifty billion dollars in 2003 and seventy billion dollars in 2007. In the six decades from 1945 to the early 2000s, the USA spent 3.6 trillion dollars on defence acquisitions — both military R&D and procurement outlays — which was more than one-third of the entire post-war US defence spending. However, research on new weapons prompted R&D spending on competing technologies. In the context of the arms race with USSR, innovation resulted from the competition within and among the US military services, with much duplication and wastage (Alic, 2007). Thus, alongside the successful early US high-technology programmes like the nuclear-powered submarine, many others were cancelled in the early 1960s after billion-dollar expenditures, for technical reasons or due to not yet having any military utility, like the nuclear-powered bomber and the Dyna-Soar spaceplane. Others only reached fruition after decades, like the stealth aircraft: first tested in 1981, the Lockheed F-117 Nighthawk fighter was only used from the Gulf War (1991) onwards, and the Northrop B-2 Spirit bomber (1989) — which was twenty times more expensive and thus the most expensive plane (2.2 billion dollars each, which is the equivalent of Central African Republic's nominal GDP today) — in Afghanistan, Iraq, and Libya.

Social Interplay in War and Technology: Paradoxes and Dynamics

Despite the opening of new war fields (air, space, and cyber) and the astonishing technological, political, and geopolitical change from the old 'gunpowder age', with flintlock and smoothbore artillery, to the new 'information age', i.e. technologization and drone surgical strikes, cultural and social factors have always played a prominent role in the dynamics of innovation and diffusion of war-related technologies.

Paradoxes about and Divergences in Military Technology Innovation

A technological advance has never guaranteed superiority. France was the first to develop shell guns and to build a steam-powered warship, an ironclad warship, a mechanically powered submarine, and a steel-hulled warship, but the French Navy did not gain superiority over the Royal Navy (Horowitz, 2010). An apparent paradox is that 'technology does indeed favour victory, but it does not guarantee it'. Hence, 'belief in technology — even high technology — as a military panacea is misplaced and dangerous' (Roland, 2016: 112). Vietnam and decolonization wars proved that technological superiority cannot ever guarantee victory. In the asymmetric wars of Afghanistan and Iraq, the strongest moral power of the technologically weakest belligerents again resisted an expensive display of sophisticated tools. In the 2010s, as the most industrialized high-tech Western army could be challenged by artisanal low-tech partisans with 'improvised explosive devices' (IEDs), the RMA theory fell out of favour. At the least, weaker powers can ambush stronger enemies, provoking casualties or destroying tanks with mere individual Molotov cocktails. Otherwise, they can divert enhanced artefacts from their regular functions so as to use military aircraft or even civilian aircraft as bombs, as the Japanese kamikaze pilots did at the end of the Second World War, or Al Qaida on 11 September 2001. In non-conventional wars — any guerrilla or Al-Qaida and Daesh terrorist attacks — even lorries or common knives can be used as weapons.

Interestingly, weaker belligerents' weapons were occasionally so successful that the winners adopted them. This eventually led to small changes in Western warfare. During the final decades of the eighteenth century, the British troops and their regular smoothbore muskets were challenged by both the American insurgents' Kentucky rifles, which had better accuracy and a longer range, and the Sultan of Mysore's artisanal rockets in India, which could mainly startle cavalry horses — or even, by chance, blow up a British ammunition wagon, as was the case in Seringapatam (1799). In the following decades, the British Army incorporated Riflemen and rocket troops into their troops as new elite forces. Hence, from the Napoleonic Wars onwards, the Congreve rocket system started a European revival of the most ancient firearm of Chinese origin, which had challenged artillery in fifteenth-century Europe and challenged it again throughout the nineteenth century. Despite the poor accuracy of black powder rockets, they had higher mobility and, in their European adaptation, a longer range than artillery for a few decades. After the bombing and rocketing of Copenhagen, and the subsequent capture of the Danish fleet by the British (1807), rockets became a weapon of local deterrence, so that Denmark also developed rockets. The defence of the Belts (Danish Straits) was now partly operated by light rocket boats that could send fire into the most powerful warships. Thus, even the weakest flotilla armed with adequate technology (mobility, range, firepower) would be considered a danger to a big warship or a fleet (Winter, 1990; Bret, 2002a; Werrett, 2009).

To Lewis Mumford, the main 'paradox of modern technics' was that 'war stimulates invention, but the army resists it!' (Mumford, 1955, p. 95). Regardless of the historical context, that has, indeed, often been the case, for a single innovation might endanger the fragile balance between a familiar technology and its social and tactical adaptation or routine.

It is not only independent inventors that needed to fight against the military establishment's conservatism for a long time, but even scientists had to convince the political power first, prior to implementing science-based technological innovations or breakthroughs in the army. During the Revolutionary and Napoleonic wars, France developed early R&D programmes on military ballooning, gunpowder manufacturing, and shell guns under the supervision of outstanding French chemists. Although Lavoisier and Berthollet had first failed to replace saltpetre with potassium chlorate in black powder (1788), the latter now supervised the secret production of powerful chlorate shells and incendiary bullets on behalf of the Comité de salut public (Committee of Public Safety (1794). Charles C. Gillispie even compared the programme to the Manhattan Project on the atomic bomb in the Second World War (Gillispie, 1992). Despite the government's orders, French sailors only used these secret weapons once or twice, if that, as they feared that the Royal Navy could soon produce them too, and then fully destroy the French navy (Bret, 2002a). Until 1810, although Congreve rockets had set fire to Copenhagen and the Rochefort fleet and were already in service or developed in Britain, Denmark, and Austria, Napoleon could not see any utility in rockets: trained as an artillerist, he could not accept their erratic accuracy, until he found that they could help conquer besieged towns in Spain (Bret, 2002a). From 1820 to 1836, the French Navy also refused to use Paixhans shell guns (see above). In 1828, conscious of Britain's global leadership in sea power, the Admiralty rejected steam navigation, which was seen as a menace, for use against the British naval predominance (Horowitz, 2010). A few decades later, submarines were thought to be 'ungentlemanly' (Edgerton, 2006b).

For various reasons, armies have resisted the replacement of flintlocks. Before the above-mentioned metallic cap, in Scotland, Reverend Forsyth first proposed percussion locks with a chlorate priming powder ignited the black powder charge. As the Board of Ordnance refused his invention, he patented it (1807) and produced sporting percussion guns (Neal, Back, 1969). The French Artillery expert General Cotty refused both the first needle rifle, which had been invented by Pauly's — he feared a dramatic increase of ammunition consumption due to its fast breechloading — and Champy's mechanical fast process of 'round powder' in the early 1810s (Bret, 1993). In 1830, the French government also cancelled the British technologies that had recently been introduced into French gunpowder manufacturing, as they had yielded a 'breaking powder' challenging the bronze ordnance since 1826 due to a misuse of the new carbonization process (Bret, 2000).

Most paradoxically, military conservatism could even be found among innovators. General Gribeauval introduced standardization into gunnery but rejected Lavoisier's proposal of faster gunpowder manufacturing processes in 1781, and despite experiments that yielded positive results, he asked to keep the 1684 regulations. Moreover, external factors like the rivalry between producers and users, or nationalistic behaviours, could interfere in the adoption or rejection of innovations. In the 1890s, despite dangerous explosions of 'smokeless powder' aboard battleships, Paul Vieille, an engineer of the Service des poudres et salpêtres (gunpowder and saltpetre service), refused to introduce a safer chemical stabilizer (diphenylamine) that had been claimed by the French Artillery: not only was he the inventor of the first colloidal powder (1884), but Nobel and the Germans added diphenylamine to their 'ballistite' powder (Bret, 2002b). Otherwise, when Nobel successively patented his ballistite in France, Britain, Italy, and Spain from 1887 to 1889, a commercial and scientific rationale led his choices (Bergman, 2011).

Even within the Pentagon, the largest provider and funder of R&D and innovations, the 'Old Guard' tended to resist reformers and innovators (Burton, 1993). Moreover, the military services might have divergent views on technological efficiency. In Vietnam, the US Air Force firmly believed in a technological solution to improve its kill ratios, while the US Navy developed new tactics and upgraded pilot training instead, reaching far higher kill ratios than the Air Force pilots (Mahnken, 2008, p. 96). Rather than employing advanced technology in such wars, the key to victory lies in finding workable strategies to enhance their efficient use.

Lastly, the most recent technologies (robotics, microminiaturization, nanotechnologies, artificial intelligence, drones, autonomous weapon systems), be they of military or civilian origin, 'will enter a world that is paradoxically more dangerous and less lethal than at any time in human history' (Roland, 2016: 114).

Dualities: Civil and Military Interplay and Dual-use Technologies

Deriving from previous civilian-military collusion on war and technology matters (army and navy staff, government officials, and managers in the arms industry), the military-industrial-complex was born in England during the naval arms race with

Germany in the 1890s (McNeill, 1982) and later evolved into a 'military-industrial-academic complex' (Leslie, 1993) or the 'science-technology-military-industrial-political complex' (Robert S. Cohen in Forman, Sánchez-Ron, 1996: viii). Since the Cold War, the permanent presence of war has even led to a generalized war culture. British statistics, then the Organization for European Economic Cooperation (OEEC, 1948) — later Organization for Economic Co-operation and Development (OECD, 1961) — counted 'qualified scientists and engineers' (QSE) among the war tools (Shapin, 2015), while the large-scale investments in American defence and aerospace R&D had a strong impact on science and societies. Not only have military commands in electronics, materials, and chemicals pushed on and sustained the R&D at civil laboratories' , for example at the prolific Bell Laboratories and at German and Swiss chemical and pharmaceutical laboratories with their thousands of patents, but many techniques and technologies first designed for war preparedness (materials science, systems analysis, reconnaissance satellites, think tanks) have found civilian applications in private and public sectors, including city planning and management (Light, 2005; Pestre, 2015). The core technologies of missiles were developed for military purposes, although von Braun shifted from civilian to military research in Germany, and then from military to civilian pursuits with the Apollo programme.

In fact, most technologies are 'dual-use' technologies: they have both military and civilian applications (e.g. steam engines, internal combustion engines), or they are used by the military in both weapons and non-weapons roles. These technologies include chariots and roads (e.g. Napoleon's roads in Dalmatia and Java, Hitler's and Mussolini's motorways, the Burma Road of the British), trains and railways (cf. the American Civil War, Franco-Prussian War), ships and aircraft; electric, electronic, and digital communication (telegraph, telephone, radio, television, the internet); steam and internal combustion engines, and nuclear power. Even firearms are dual-use technologies (military vs hunting or sporting weapons) and their innovations are often driven by civilian gunmakers or amateurs (see above on percussion and needle rifles). Thus, once the 'Old Guard' conservatism had been overcome, war was crucial in developing technological innovations that first appeared in the civil sphere, like aviation — both lighter-than-air (during the Revolutionary Wars) and heavier-than-air aviation (in the First World War). Apart from specific new purposes, the military opened vast markets and funds that could boost many civilian-born inventions and discoveries, not only for industrial products, such as DuPont's nylon or Rohm and Haas' Plexiglas boosted by the Second World War, but also in various other fields. In the health sector, Jenner's smallpox vaccine (1796) was soon developed massively and used by the military in European (Napoleon's *Grande armée*) and non-European countries (Mehmet Ali's Egyptian army), with positive impact on the civilian population. Later, quinine, a tool of European imperialism, and penicillin were developed. Fleming's discovery (1928) was first developed between 1941 and 1943, and then entered mass production to cover the military requirements (100 billion units per month in the summer of 1944) (Bickel, 1992; Gaudillière, 2002). Direct effects of war also led to surgical and medical developments such as maxillofacial surgery, protheses, radiosurgery, serotherapy, psychiatry, and the treatment of war neuroses or post-traumatic stress disorder.

Many civilian technologies arose from military sources, even in meteorology, at least in the United States. The US Army Signal Service, which was created in 1860, was active during the Civil War for transmissions and telegraphy under A. J. Myer. Later in charge of storm telegraphy and the weather service (1870), Myer promoted world cooperation at the first International Meteorological Congress in Vienna (1873), which led to the publication of the *Bulletin of International Simultaneous Observations* (1875) (Whitnah, 1961; Fleming, 2005). In the peaceful 1920s, the shape and extension of telephony were largely determined by 'a triple purpose in war: for military communications (tactical and strategic), for propaganda, and as a means of maintaining commercial and social stability' (Kragh, 1996: 60). Long used in diplomatic and military spheres, cryptography became a crucial weapon of war when the Tokumu Han ('special section') of the Japanese Imperial Navy broke Chinese codes in the 1930s, the British broke the German naval 'Enigma' code in the Second World War (Kahn, 1996), and hackers recently hacked the Pentagon. The major communication technology of the internet appeared in the military context of the SAGE anti-aircraft system prior to spreading across the academic world, and then throughout the world across businesses and in homes. In 1969, the first interconnected computers at UCLA, the Stanford Research Institute, in Santa Barbara, at the University of Utah, and at the Lincoln Laboratory (MIT) belonged to the Advanced Research Projects Agency Network (ARPANET). The military network and the academic network were separated in 1980.

Lastly, even nuclear energy is a dual-use technology. Developed by civilian fundamental physicists (including Oppenheimer and Fermi) and civilian chemical engineers at DuPont (Ndiaye, 2007), the Manhattan Project had military purposes, but also had civilian uses in mind. Otherwise, within the military field, nuclear energy is both weapon (bomb) and non-weapon (nuclear propulsion, although this has been hitherto applied to weapons, i.e. to submarines and aircraft carriers). Nevertheless, civilian nuclear energy had a strong social impact around the world, through reactor accidents (Three Mile Island, Chernobyl, Fukushima) and national or transnational protests (Gusterson, 1996; Tompkins, 2016). Moreover, as usual, a closed technological paradigm (Thomas P. Hughes' 'technological momentum') put an end to the open time of competing technologies. The early military choices now had a cascading effect of 'lock-in' or 'closure' which led to a path-dependent commercial technology in later civilian applications, such as the Westinghouse light water reactor supported by US Admiral Rickover's . Paradoxically, the only country to be a later target of a nuclear weapon dared to turn towards a civilian nuclear industry. Since the 1990s, and even after Fukushima (2011), the Japanese government and nuclear sector have sought to export Japanese nuclear power plants in the name of national strategy (import, appropriation, export). Renewing nationalist justifications linked to technoscientific advancement and national progress, the Japanese went as far as to use the memories of the 1945 atomic tragedies in their nuclear policy (Inoue, 2019).

Circulation: Secrecy, Mobility, Cooperation, and Proliferation

The archives relating to recent secret technologies are most often classified. Historical accounts have been written by members of research programmes, such as the biologist Kanatjan Alibekov, deputy director of Biopreparat, the Soviet agency working on infectious agents in the last decade of USSR and preparing genetically modified germs to assassinate individual targets, who later defected to the United States and then became Ken Alibek (Alibek, 1999). Historians under contract for state research might get access to classified documents and produce both unpublished confidential reports and published academic works, as did Olivier Lepick on the French biological warfare programme from 1919 to 1945, alongside his book on the chemical Great War (Lepick, 1997, 1998). Yet, independent investigation on the US and Iraqi biological and chemical warfare programmes in the 1970s to 1980s also revealed underground political and economic links and support given to the latter by US and European firms before the first Gulf War (Cole, 1997). Through the mobility of men, designs, and devices, secrecy can be shared (cooperation), broken (spying), or a mixture of the two.

Mobility, considered at any scale of space and time, accompanies R&D, as well as regular war technological production or innovation. In the 1790s, the probably largest manufacture worldwide, the Ichapur gunpowder factory in Bengal, employed some 2,000 seasonal workers that often came from distant districts (Lucassen, 2012). The above-mentioned breechloading rifle and the revival of the war rocket reveal international and interprofessional mobilities, either state-led or private. J. S. Pauly, a Swiss carriage maker, became a gunsmith in Paris, where he patented the first needle rifle after the ministry refused it. Among his workers was the Thuringian J. D. N. Dreyse, who later improved the system and provided the Prussian Army with an unexpected advantage over Austria (1866). Of Chinese origin, the rocket had challenged gunpowder artillery in fourteenth and fifteenth centuries; then European armies neglected it as long as bronze artillery dominated for the next three centuries. After the British troops faced Indian rockets in Mysore, William Congreve Jr. developed a 'rocket system' based on European protoindustrial standards at Woolwich Arsenal, and the British reused rockets during the Napoleonic wars, in the War of 1812 against the United States, and in the colonial wars in South Asia. European 'Congreve rockets' were even manufactured in British India. Considering the longer range and greater mobility of rockets on the battlefield, and given the lightness of their launch apparatuses, the Western great powers and smaller countries soon competed in an arms race for half a century, involving cooperation, spying, reverse engineering, and the hiring of technicians into foreign service. Diplomacy was also a way of proliferation. In Ottoman Albania, the British provided the Pasha of Berat with rockets against Ali Pasha of Janina, who was now under French influence (1809). Denmark refused to divulge its own highly secret rocket system to the Swedes prior to giving it to the French allies in 1813, and then also to the Austrians, in exchange for their support against the Swedish aims upon Norway (1815) (Bret, 2002a).

At any time, imported technologies required foreign technical advisors and skilled foreign workers. In many countries, war-related technologies participated in the establishment of industry, from the French roots of E. I. DuPont de Nemours' gunpowder plants (Stapleton, 1987) and Eli Whitney's alleged 'interchangeability of parts' in musketry (Woodbury, 1960; Smith, 1980), and the 'American system of manufacturing' of the 1800s 1850s (Smith, 1985), to the coproduction of American sophisticated weapons such as aircraft and missiles in Taiwan and South Korea from the 1970s onwards (Nolan, 1986). Military technicians also accompanied the Cold War deployment and activation of such American or Soviet weapons through Asia, Africa, and South America, and even into nuclear powers like France, as long that country belonged to the integrated military organization of NATO. Those technical and organizational transfers required cultural adaptations (Goldman, Eliason, 2003; Sevrin, 2010).

The end of wars increased mobility as specialists lost or defected from their posts. Not only did the Americans incorporate Werner von Braun and his staff of scientists and technicians working on the V2 at Pennemünde into their R&D as the core of their airspace programme, but they also welcomed the military Japanese unit specialized in bacteriological warfare despite the crimes of war they had committed in China. France welcomed German rocket technicians at Vernon and the Institut de Saint-Louis, and the USSR enrolled others. In 1989, as the Soviet power declined, Vladimir Pasechnik, the head of a key institute of Biopreparat, defected during a scientific mission in France. He joined Britain, denounced the Soviet underground R&D activities, and worked for the British Department of Health before forming his own company on biotechnologies. The chief scientist Alibek also defected in 1992. In peace or at war, war and technology continue to have a strong relationship, but the technological illusion in the military thought must be questioned (Lefeez, 2017).

References

A3P, *Instrumentation, expérimentation et expertise des matériaux énergétiques (poudres, explosifs et pyrotechnie) du XVIe siècle à nos jours* (Paris: Centre de recherches en histoire des sciences et des techniques-Association pour le patrimoine poudrier et pyrotechnique-Commissariat pour l'énergie atomique/Division des affaires militaires, 2001).

A3P, *Histoire des propergols solides au XXe siècle* (Paris: Centre de recherches en histoire des sciences et des techniques-Association pour le patrimoine poudrier et pyrotechnique-Commissariat pour l'énergie atomique/Division des affaires militaires, 2004).

Adams, Henry Thomas, 'L'Évolution du canon-obusier Paixhans et sa place dans la Marine au XIXe siècle' (unpublished doctoral dissertation, 2 vols, University Paris IV, 1994).

Adams, James, *The New Spies: Exploring the Frontiers of Espionage* (London: Hutchinson, 1994).

Ágoston, Gábor, *Guns for the Sultan: Military Power and the Weapons Industry in the Ottoman Empire* (Cambridge: Cambridge University Press, 2005).

Albert, Marie-Claude, Pierre Bugnet, Michel Cortay, Geneviève Cerisier-Millet, and Joseph Chotard, *Hommes et femmes dans les usines d'armement. Aux sources d'une conscience ouvrière éclatée, 1914–1919* (Châtellerault: Institut fédéral d'histoire sociale des travailleurs de l'Etat-Confédération Générale du travail and Service historique de la défense, 2019).

Alder, Kenneth, *Engineering the Revolution: Arms and Enlightenment in France, 1763–1815* (Princeton, NJ: Princeton University Press, 1997).

Alibek, Ken, with Stephen Handelman, *Biohazard: the Chilling True Story of the Largest Covert Biological Weapons Program in the World, Told From the Inside by the Man Who Ran It* (New York, NY: Random House, 1999).

Alic, John A., *Trillions for Military Technology: How the Pentagon Innovates and Why it Costs So Much* (Basingstoke: Palgrave Macmillan, 2007).

Aubin, David and Patrice Bret, eds, 'Le Sabre et l'éprouvette: L'invention d'une science de guerre, 1914–1939', *14–18. Aujourd'hui. Today. Heute*, 6 (2003), 39–172.

——— and Catherine Goldstein, eds, *The War of Guns and Mathematics: Mathematical Practices and Communities through World War I in France and its Western Allies* (Providence, RI: American Mathematical Society, 2014).

Aucouturier, Étienne, *Biological Warfare. Another French Connection. Translated by the author* (Paris: Éditions Matériologiques, 2020).

———, *La Guerre biologique. Aventures françaises* (Paris: Éditions Matériologiques, 2017).

Bacher, Bernard, Jean-François Brun, and Éric Perrin, *La Manufacture d'armes de Saint-Étienne. La révolution des machines, 1850–1870* (Saint-Étienne: Musée d'art et d'industrie, 2007).

Barjot, Dominique, ed., *Deux guerres totales 1914–1918, 1939–1945. La mobilisation de la nation* (Paris: Economica, 2012).

Bergman, Yoel, 'Alfred Nobel, Aniline and Diphenylamine', *ICON. Journal of the International Committee for the History of Technology*, 17 (2011), 57–67.

Bickel, Lennard, *Howard Florey. The Man Who Made Penicillin* (Melbourne: Melbourne University Press, 1972).

Black, Jeremy, *A Military Revolution? Military Change and European Society, 1550–1800* (Basingstoke: Macmillan, 1991).

———, *War and Technology* (Bloomington and Indianapolis, IN: Indiana University Press, 2013).

Blanchard, Yves, *Le Radar, 1904–2004. Histoire d'un siècle d'innovations techniques et opérationnelles* (Paris: Thalès, Ellipses, 2004).

Bret, Patrice, 'Le Dernier des procédés révolutionnaires: la fabrication et l'expertise de la poudre ronde (1795–1830)', *Annals of Science*, 50.4 (1993), 325–47.

———, *L'État, l'armée, la science. L'invention de la recherche publique en France, 1763–1830* (Rennes: Presses universitaires de Rennes, 2002a).

———, 'Genèse et légitimation patrimoniale d'une invention: les archives de l'Artillerie à l'origine d'une innovation cruciale dans la Marine au XIXe siècle', in *Les Chemins de la nouveauté: inventer, innover au regard de l'histoire*, ed. by Liliane Hilaire-Pérez and Anne-Françoise Garçon (Paris: Éditions du CTHS, 2003), pp. 385–410.

———, 'La Guerre des laboratoires: Poincaré, Le Chatelier et la Commission scientifique d'étude des poudres de guerre (1907–1908)', (unpublished paper online, Colloque Le Chatelier, 127e Congrès des sociétés historiques et scientifiques (CTHS), Nancy, April 2002b).

———, 'Gunpowder Manufacturing in Cairo From Bonaparte to Muhammad 'Alî: Adaptation, Innovation and Transfer of Technology (1798–1820)', in *Gunpowder, Explosives and the State: A Technological History*, ed. by Brenda J. Buchanan (Aldershot: Ashgate, 2006), pp. 206–38.

———, 'La Mise en place humaine et industrielle de l'aérostation militaire (1794–1796). Essai de géographie technologique d'une mobilisation', in *Révolution française 1988–1989* (Paris: Éditions du CTHS, 1991), pp. 439–66.

———, 'Quand la poudre défie le métal. L'affaire des poudres brisantes (1826–1830)', (unpublished paper online, Colloque 'Métal et poudre', Varaignes, October 2000)

———, 'Recherche scientifique, innovation technique et conception tactique d'une arme nouvelle: l'aérostation militaire (1793–1799)', in *Lazare Carnot ou le savant-citoyen (Colloque Carnot, Université Paris IV, janvier 1988)*, ed. by Jean-Paul Charnay (Paris: Presses de l'Université de Paris-Sorbonne, 1990), pp. 429–51.

———, 'Une tentative d'exploitation militaire de la recherche en chimie: Berthollet et la poudre de muriate oxygéné de potasse (1787–1794)', in *Lavoisier et la Révolution chimique*, ed. by Michelle Goupil (Palaiseau: SABIX-Ecole polytechnique, 1992), pp. 195–238.

Brown, David K., *The Eclipse of the Big Gun: the Warship, 1906–1945* (London: Conway Maritime Press, 1992).

———, *Paddle Warships: the Earliest Steam Powered Fighting Ships, 1815–1850* (London: Conway Maritime Press, 1993).

———, *Warrior to Dreadnought: Warship Development, 1860–1905* (London: Caxton Editions, 2003).

Bruce-Briggs, B., *The Shield of Faith* (New York, NY: Touchstone, 1988).

Buchanan, Brenda J., ed., *Gunpowder, Explosives and the State: a Technological History* (Aldershot: Ashgate, 2006).

———, ed., *Gunpowder. The History of an International Technology* (Bath: Bath University Press, 1996).

Bud, Robert and Philipp Gummett, eds, *Cold War, Hot Science, Applied Research in Britain's Defence Laboratories 1945–1990* (London: Harwood Academic Publishers, 1999).

Burton, James G., *The Pentagon Wars: Reformers Challenge the Old Guard* (Annapolis, MD: Naval Institute Press, 1993).

Carpentier, Jean, *Cent vingt ans d'innovations en aéronautique* (Paris: Hermann, 2011).

Carpentier, René, *Les Missiles tactiques* (Paris: CHEAr/DHAr, 2004).

CHARME, ed., 'La Science et la guerre. 400 ans d'histoire partagée', *La Recherche*, hors série, 7 (2002).

CHARME, ed., *L'Optronique militaire en France, 1945–1985* (Paris: CHARME-CHEAr/DHAr, 2004).

CHARME, SHM, eds, *Les Bases et les arsenaux français d'outre-mer, du Second Empire à nos jours* (Paris: Lavauzelle, 2002).

Cleator, P. E., *Weapons of War* (London: Robert Hale, 1967).

Cochet, François, *Armes en guerre* XIXe–XXIe *siècle. Mythes, symboles, réalités* (Paris: CNRS éditions, 2011).

Cockroft, Wayne D., *Dangerous Energy. The Archaeology of Gunpowder and Military Explosives Manufacture* (Northampton: English Heritage, 1999).

Coffey, Patrick, *American Arsenal: a Century of Waging War* (New York, NY: Oxford University Press, 2013).

Cohn, Carol, 'Sex and Death in the Rational World of Defense Intellectuals', *Signs: Journal of Women in Culture and Society*, 12.4 (1987), 687–718.

Cole, Leonard A., *The Eleventh Plague. The Politics of Biological and Chemical Warfare* (New York, NY: W. H. Freeman, 1997).

Dahan, Amy, and Dominique Pestre, Les Sciences pour la guerre, 1940–1960 (Paris, Éd. de l'EHESS, 2004).

Delbrück, Hans, *History of the Art of War Within the Framework of Political History*, 4 vols, translated from the German by Walter J. Renfroe (Westport, CT: Greenwood Press, 1975–1985).

De Maria, Michelangelo, Mario Grilli, and Fabio Sebastiani, eds, *The Restructuring of Physical Sciences in Europe and the United States* (Singapore: World Scientific, 1989).

Dincecco, Mark and Massimiliano G. Onorato, *From Warfare to Wealth: the Military Origins of Urban Prosperity in Europe* (Cambridge: Cambridge University Press, 2017).

Edgerton, David E., *Britain's War Machine: Weapons, Resources and Experts in the Second World War* (Oxford: Oxford University Press, 2011).

———, *England and the Aeroplane. Militarism, Modernity and Machines* (London: Penguin Books, 2013).

———, *The Shock of the Old: Technology and Global History since 1900* (London: Profile Books, 2006a).

———, *The Warfare State: Britain, 1920–1970* (Cambridge: Cambridge University Press, 2006b).

Edwards, Paul N., *The Closed World* (Cambridge, MA:, MIT Press, 1996).

Ellis, John, *The Social History of the Machine Gun* (New York, NY: Pantheon, 1975).

Firges, Pascal, 'Gunners for the Sultan. French Revolutionary Efforts to Modernize the Ottoman Military', in *Well-Connected Domains: Towards an Entangled Ottoman History.*, ed. by Pascal Firges, Tobias P. Graf, Christia Roth, and Gülay Tulasoğlu (Leiden: Brill, 2014), pp. 171–87.

Fleming, James R., 'Telegraphing the Weather: Military Meteorology, Strategy, and "Homeland Security" on the American Frontier in the 1870s', in *Instrumental in War: Science, Research, and Instruments Between Knowledge and the World*, ed. by Steven A. Walton (Leiden and Boston, MA: Brill, 2005), pp. 153–78.

Forman, Paul and José Sánchez-Ron, eds, *National Military Establishments and the Advancement of Science and Technology* (Dordrecht: Kluwer Academic Publishers, 1996).

Fuller, John Frederick Charles, *Machine Warfare. An Enquiry Into the Influences of Mechanics on the Art of War* (London: Hutchinson, 1942).

———, *Armament in History. The Influence of Armament on History from the Dawn of Classical Warfare to the End of the Second World War* (London: Eyre & Spottiswoode, 1946).

Gaddis, John Lewis, *The Long Peace: Inquiries into the History of Cold War* (New York, NY: Oxford University Press, 1987).

Galison, Peter and Bruce Hevly, eds, *Big Science: the Growth of Large-scale Research* (Stanford, CA: Stanford University Press, 1992).

Gaudillière, Jean-Paul, *Inventer la biomédecine. La France, l'Amérique et la production des savoirs du vivant* (Paris: La Découverte, 2002).

Geyer, Michael and Adam Tooze, eds, *The Cambridge History of the Second War*, vol. 3, *Total War: Economy, Society and Culture* (Cambridge: Cambridge University Press, 2015).

Gillispie, Charles Coulston, 'Science and Secret Weapons Development in Revolutionary France, 1792–1804: a Documentary History', *Historical Studies in Physical Sciences*, 23.1 (1992), 35–152.

Goldman, Emily O. and Leslie C. Eliason, eds, *The Diffusion of Military Technologies and Ideas* (Stanford, CA: Stanford University Press, 2003).

Gowing, Margaret, *Britain and Atomic Energy 1939–1945* (London: Macmillan Publishing, 1964).

———, *Independence and Deterrence: Britain and Atomic Energy 1945–52* (London: Macmillan Publishing, 1974).

Grant, Jonathan, 'Rethinking the Ottoman "Decline": Military Technology Diffusion in the Ottoman Empire, Fifteenth to Eighteenth Centuries', *Journal of World History*, 10.1 (1999), 179–201.

Grattan-Guinness, Ivor, 'The *ingénieur savant*, 1800–1830. A Neglected Figure in the History of French Mathematics and Science', *Science in Context*, 6.2 (1993), 405–33.

Guelton, Frédéric, 'Technology and armaments', in Winter, Jay, ed., *The Cambridge History of the First World War*, 3 vols (Cambridge: Cambridge University Press, 2014), vol. II, pp. 240–65.

Gusterson, Hugh, *Drone: Remote Control Warfare* (Minneapolis, MN: University of Minesotta Press, 2016).

———, *Nuclear Rites: A Weapons Laboratory at the End of the Cold War* (Berkeley, CA: University of California Press, 1996).

———, *People of the Bomb: Portraits of America's Nuclear Complex* (Minneapolis, MN: University of Minesotta Press, 2004).

Haber, Lutz F., *The Poisonous Cloud: Chemical Warfare in the First World War* (Oxford: Clarendon Press, 1986).

Hacker, Barton C. and Margaret Vining, *American Military Technology: the Life Story of a Technology* (Westport, CT: Greenwood Press, 2006).

Hackmann, Willem, *Seek and Strike: Sonar, Antisubmarine Warfare and the Royal Navy, 1914–54* (London: HMSO, 1984).

Harding, David F., *Smallarms of the East India Company 1600–1856*, 4 vols (London: Foresight Books, 1997–1999).

Headrick, Daniel R., *The Invisible Weapon: Telecommunications and International Politics, 1851–1945* (New York, NY: Oxford University Press, 1991).

———, *Power Over People: Technology, Environments, and Western Imperialism, 1400 to the Present* (Princeton, NJ: Princeton University Press, 2010).

———, *The Tools of Empire: Technology and European Imperialism in the Nineteenth Century* (New York, NY: Oxford University Press, 1981).

Hébert, Jean-Paul and Jean Hamiot, eds, *Histoire de la coopération européenne dans l'armement* (Paris: CNRS Éditions, 2004).

Hecht, Gabrielle, *Being Nuclear: Africans and the Global Uranium Trade* (Cambridge, MA: MIT Press, 2012).

Henrotin, Joseph, *La Technologie militaire en question. Le cas américain et ses conséquences en Europe* (2nd ed., Paris: Economica, 2013).

Hoffman, Philip R. T., *Why Did Europe Conquer the World?* (Princeton, NJ: Princeton University Press, 2015).

Hogg, Ian V., *The Guns, 1914–18* (London: Macmillan, 1973).

——— and John S. Weeks, *Military Small Arms of the Twentieth Century* (1973) (Iola: Krause, 2000).

Horowitz, Michael C., *The Diffusion of Military Power: Causes and Consequences for International Politics* (Princeton, NJ: Princeton University Press, 2010).

Hounshell, David A., *From the American System to Mass Production 1800–1932* (Baltimore, MD: Johns Hopkins University Press, 1984).

Hughes, Major-General B. P., *Firepower. Weapons Effectiveness on The Battlefield, 1630–1850* (London: Arms and Amour Press, 1974).

Inoue, Masatoshi, 'L'Invention du nucléaire "nippon": histoire du développement électronucléaire au Japon par le prisme du nationalisme technologique (de 1954 à nos jours)' (unpublished master's thesis, EHESS Paris, 2019).

Jha, Umesh Chandra, *Military Weapons and Environment* (New Delhi: KW Publishers, 2018).

Johnson, Jeffrey, *The Kaiser's Chemists: Science and Modernisation in Imperial Germany* (Chapel Hill, NC: University of North Carolina Press, 1990).

Jordan, David, James D. Kiras, David J. Londstale, Ian Spelle, and Dale Walton, *Understanding Modern Warfare* (Cambridge: Cambridge University Press, 2008).

Jouanjus, Alain, Patrick Simin, and Carl Trémoureux, eds, *Les Armes de la Grande Guerre. Histoire d'une révolution scientifique et industrielle* (Paris: Editions Pierre de Taillac et Ministère des Armées/Conseil général de l'armement et Direction générale de l'armement, 2018).

Kahn, David, *The Codebreakers. The Story of Secret Writing* (New York, NY: Scribner, 1996).

Kaspi, André, *Le Temps des américains, 1917–1918* (Paris: Publications de la Sorbonne, 1976).

Kragh, Helge, 'Telephone Technology and Its Interaction with Science and the Military, ca. 1900–1930', in *National Military Establishments and the Advancement of Science and Technology*, ed. by Paul Forman and José Sánchez-Ron (Dordrecht: Kluwer Academic Publishers, 1996), pp. 37–67.

Krause, Keith, *Arms and the State: Patterns of Military Production and Trade* (Cambridge: Cambridge University Press, 1992).

Kreuzer, Michael P., *Drones and the Future of Air Warfare: the Evolution of Remotely Piloted Aircraft* (New York, NY: Routledge, 2016).

Kroenig, Matthew, *Exporting the Bomb: Technology Transfer and the Spread of Nuclear Weapons* (Ithaca, NY: Cornell University Press, 2010).

Langensiepen, Berndt and Ahmet Güleryüz, *The Ottoman Steam Navy 1828–1923* (London: Conway Maritime Press, 1995).

Langins, Jānis, 'Hydrogen Production for Ballooning During the French Revolution: an Early Example of Chemical Process Development', *Annals of Science*, 40 (1983), 531–58.

Lefeez, Sophie, *L'illusion technologique dans la pensée militaire* (Paris: Éditions Nuvis, 2017).

Lelong, Benoît, 'Paul Langevin et la détection sous-marine, 1914–1929. Un physicien acteur de l'innovation industrielle et militaire', *Epistémologiques*, 1:3–4 (2001), 205–32.

Lepick, Olivier, *La Grande Guerre chimique, 1914–1918* (Paris: PUF, 1998).

———, 'Le Programme français de guerre biologique: 1919–1945', *Guerres mondiales et conflits contemporains*, 185 (1997), 29–54.

Leslie, Stuart W., *The Cold War and American Science: the Military-Industrial-Academic Complex at MIT and Stanford* (New York, NY: Columbia University Press, 1993).

Light, Jennifer S., *From Warfare to Welfare: Defense Intellectuals and Urban Problems in Cold War America* (Baltimore, MD: Johns Hopkins University Press, 2005).

Llinares, Sylviane, *Marine, propulsion et technique: l'évolution du système technologique du navire de guerre français au XVIIIe siècle*, 2 vols (Paris: Librairie de l'Inde, 1994).

Lorber, Azriel, *Ready for Battle: Technological Intelligence on the Battlefield* (London: Rowman & Littlefield, 2015).

Lucassen, Jan, 'Working at the Ichapur Gunpowder Factory in the 1790s', *Indian Historical Review*, 39 (2012), 19–56, 251–71.

MacKenzie, Donald, *Inventing Accuracy: a Historical Sociology of Nuclear Missile Guidance* (Cambridge, MA: MIT Press, 1990).

MacLeod, Roy M., 'The "Arsenal" in the Strand: Australian Chemists and the British Munitions Effort, 1916–1919', *Annals of Science*, 46 (1989), 45–67.

———, 'The Chemists Go to War: the Mobilization of Civilian Chemists and the British War Effort, 1914–1918', *Annals of Science*, 50 (1993), 455–81.

———, 'Sight and Sound on the Western Front: Surveyors, Scientists and the "Battlefield Laboratory", 1915–1918', *War and Society*, 18.1, (2000), 23–46.

——— and Jeffrey Allan Johnson, eds, *Frontline and Factory. Comparative Perspectives on the Chemical Industry at War, 1914–1924* (Dordrecht: Springer, 2007).

McDougall, Walter, *The Heavens and the Earth. A Political History of the Space Age* (New York, NY: Basic, 1985).McNeill, William H., *The Pursuit of Power: Technology, Armed Force, and Society Since A.D. 1000* (Chicago, IL: University of Chicago Press, 1982).

McNeill, William H., *The Rise of the West: a History of Human Community* (Chicago, IL: University of Chicago Press, 1963).

Mahnken, Thomas G., *Technology and the American Way of War since 1945* (New York, NY: Columbia University Press, 2008).

Martin, Suzan B., 'Weapons of Mass Destruction: a Brief Overview', in *Ethics and Weapons of Mass Destruction: Religious and Secular Perspectives*, ed. by Sohail H. Hashmi and Steven P. Lee (Cambridge: Cambridge University Press, 2004), pp. 16–42.

Masaaki, Okada, 'Interpretation of Defence Heritage in Japan: a Survey of Its "Neglect and Reuse" and Future Potential as Significant Landscape', *ICON. Journal of the International Committee for the History of Technology*, 17 (2011), 94–106.

Massie, Robert K., *Dreadnought: Britain, Germany and the Coming of the Great War* (New York, NY: Random House, 1991).

Mendelsohn, Everett, Merritt Roe Smith, and Peter Weingart, eds, *Science, Technology, and the Military*, 2 vols (Dordrecht: Kluwer, 1988).

Mills, Walter, *Arms and Men: A Study in American Military History* (New York, NY: Putnam, 1956).

Moore, Richard, *Nuclear Illusion, Nuclear Reality: Britain, the United States and Nuclear Weapons, 1958–64* (New York, NY: Palgrave Macmillan, 2010).

Mortal, Patrick, *Les Armuriers de l'État du Grand Siècle à la globalisation (1665–1989)* (Villeneuve d'Ascq: Presses Universitaires du Septentrion, 2007).

Mumford, Lewis, *Technics and Civilisation* (London: Routledge and Kegan Paul, 1955 – first published 1934).

Ndiaye, Pap, *Nylon and Bombs: DuPont and the March of Modern America* (Baltimore, MD: Johns Hopkins University Press, 2007).

Neal, William Keith and D. H. L. Back, *Forsyth & Co.: Patent Gunmakers* (London: Bell & Sons, 1969).

Nolan, Janne E., *Military Industry in Taiwan and South Korea* (London: Macmillan, 1986).

O'Connell, Robert, *Of Arms and Men: a History of War, Weapons, and Aggression* (New York, NY: Oxford University Press, 1989).

Onorato, Massimiliano Gaetano, Kenneth Scheve, and David Stasavage, 'Technology and the Era of the Mass Army', *Journal of Economic History*, 74 (2014), 449–81.

Oren, Ido, and Ty Solomon, 'Weapons of Mass Destruction': Historicizing the Concept, *History of Global Arms Transfer*, 10 (2020), pp. 121–44.

Panzac, Daniel, *La Marine ottomane, de l'apogée à la chute de l'Empire (1572–1923)* (Paris: CNRS Éditions, 2009).

Parker, Geoffrey, 'The "Military Revolution," 1560–1660--a Myth?', *Journal of Modern History*, Vol. 48, No. 2, pp. 195–214.

———, *The Military Revolution: Military Innovation and the Rise of the West, 1500–1800* (Cambridge: Cambridge University Press, 1988).

Pearton, Maurice, *The Knowledgeable State: Diplomacy, War and Technology since 1830* (London: Burnett Books, 1982).

Pestre, Dominique, ed., *Deux siècles d'histoire de l'armement en France. De Gribeauval à la force de frappe* (Paris: CNRS Éditions, 2005).

———, ed., *Histoire des sciences et des savoirs*, 3 vols (Paris: Seuil, 2015) — vol. 2, *Modernité et globalisation*, ed. by Kapil Rak and H. Otto Sibum — vol. 3, *Le Siècle des technosciences*, ed. by Christophe Bonneuil and Dominique Pestre.

Planel, Anne-Marie, *Du comptoir à la colonie. Histoire de la communauté française de Tunisie, 1814–1883* (Paris: IRMC, Riveneuve éditions, 2015).

Ploeger, Joanna S., 'The Art of Science at Fermi National Accelerator Laboratory: the Rhetoric of Aesthetics and Humanism in the National Laboratory System in the Late 1960s', *History and Technology*, 18.1 (2002), 23–49.

Pô, Jean-Damien, *Les Moyens de la puissance: les activités militaires du CEA, 1945–2000* (Paris: Fondation pour la Recherche Stratégique, Ellipses, 2001).

Rabkin, Jeremy and John Yoo, *Striking Power: How Cyber, Robots, and Space Weapons Change the Rules of War* (New York, NY: Encounter Books, 2017).

Rasmussen, Ann, 'Science and Technology', in John Horne, ed., *A Companion to World War I* (Chichester, Wiley-Blackwell, 2010, pp. 307–22).

Reid, Richard J., *Warfare in African History* (Cambridge: Cambridge University Press, 2012).

Roberts, Michael, *The Military Revolution, 1560–1660* (Belfast: Queen's University Press, 1956).

Roland, Alex, *The Military-Industrial Complex* (Washington, DC: American Historical Association, 2001).

———, 'Technology and War: the Historiographical Revolution of the 1980s', *Technology and Culture*, 34.1 (1993), 117–34.

———, *War and Technology: a Very Short Introduction* (New York, NY: Oxford University Press, 2016).

Shapin, Steven, "Figures de scientifiques", in *Histoire des sciences et des savoirs* — vol. 3, *Le Siècle des technosciences*, ed. by Christophe Bonneuil and Dominique Pestre (Paris: Seuil, 2015), pp. 27–45.

Schiavon, Martina, *Itinéraires de la precision. Géodésiens, artilleurs, savants et fabricants d'instruments de précision en France, 1870–1930* (Nancy: PUN — Editions universitaires de Lorraine, 2014).

Sevrin, Michel, ed., *Histoire de l'artillerie nucléaire de Terre française, 1959–1996, CERMA. Cahiers d'études et de recherches du musée de l'Armée*, hors série, 7 (2010).

Shaw, Stanford J., 'The Origins of Ottoman Military reform: the Nizam-I Cedid Army of Sultan Selim III', *Journal of Modern History*, 37.3 (1965), 291–306.

Sherry, Michael, *The Rise of American Air Power: the Creation of Armageddon* (New Haven, CT: Yale University Press, 1987).

Smith, Merritt Roe, *Harpers Ferry Armory and the New Technology: the Challenge of Change* (Ithaca, NY: Cornell University Press, 1980).

———, ed., *Military Enterprise and Technological Change. Perspectives on the American Experience* (Cambridge, MA: MIT Press, 1985).

Sondhaus, Lawrence, *Naval Warfare 1815–1914* (London/New York: Routledge 2001).

Stapleton, Darwin H., *The Transfer of Early Industrial Technologies to America* (Philadelphia, PA: American Philosophical Society, 1987).

Sumida, Jon Tetsuro, *In Defence of Naval Supremacy: Finance, Technology and British Naval Policy, 1889–1914* (Winchester, MA: Unwin Hyman, 1989).

Tompkins, Andrew, *Better Active Than Radioactive! Anti-Nuclear Protests in 1970s France and West Germany* (Oxford: Oxford University Press, 2016).

Touchet, Elisabeth de, *Quand les français armaient le Japon. La création de l'arsenal de Yokosuka, 1865 à 1882* (Rennes: Presses Universitaires de Rennes, 2003).

van Creveld, Martin, *Supplying War: Logistics from Wallenstein to Patton* (Cambridge: Cambridge University Press, 1977).

———, *Technology and War: From 2000 B.C. to the Present* (New York, NY: Free Press, 1989).

Vérin, Hélène, *La Gloire des ingénieurs. L'intelligence technique du xvi*^{e} *au xviii*^{e} *siècle* (Paris: Albin Michel, 1993).

Villatoux, Marie-Catherine and Patrick Facon, 'La Coopération franco-américaine en matière d'aéronautique, 1917, 1918', *Revue historique des armées* (2007), 33–45.

Walton, Steven A., ed., *Instrumental in War: Science, Research, and Instruments Between Knowledge and the World* (Leiden and Boston, MA: Brill, 2005).

———, ed., *Technological Determinism in War*, special issue, *Vulcan. The Journal of the History of Military Technology*, 7 (2019).

Weart, Spencer R., *Scientists in Power* (Cambridge, MA: Harvard University Press, 1979).

———, *La Grande aventure des atomistes français*, ([s.l.], Fayard, 1980).

Werrett, Simon, 'William Congreve's Rational Rockets', *Notes & Records of the Royal Society*, 63 (2009), 35–56.

Whitnah, D. R., *A History of the United States Weather Bureau* (Urbana, IL: University of Illinois Press, 1961).

Winter, F. H., *The First Golden Age of Rocketry* (Washington, DC: Smithsonian Institution Press, 1990).

Winter, Jay, ed., *The Cambridge History of the First World War*, 3 vols (Cambridge: Cambridge University Press, 2014).

———, and Annette Becker, ed., *La Première guerre mondiale*, 3 vols, (Paris: Fayard, 2014).

Wolmar, Christian, *Engines of War: How Wars Were Won & Lost on the Railways* (London: Atlantic Books, 2010).

Woodbury, Robert S., 'The Legend of Eli Whitney and the Interchangeability of Parts', *Technology and Culture*, 1 (1960), 235–51.

———, *Studies in the History of Machine Tools* (Cambridge, MA: The MIT Press, 1972).

Zimmerman, David, *Top Secret Exchange: the Tizard Mission and the Scientific War* (Montreal: Alan Sutton Publishing, McGill-Queen's University Press, 1996).

Zuckerman, Solly, *Nuclear Illusion and Reality* (London: Collins, 1982).

PATRICK FRIDENSON

Taylorism

Frederick Winslow Taylor (1856–1915) was not quite as original as he claimed. Several American managers, often engineers by training, had pioneered the systematic management of people, work flows, and information to improve the coordination and control of new, large firms before he did. Taylor, who was first an engineer and then became a consultant, went one step further by focusing on shop management and administrative processes. Louis Brandeis, a famous US lawyer, named the doctrine Taylor had devised 'scientific management', a term then adopted by Taylor. The doctrine spread worldwide, first in capitalist countries, and later also in communist countries. Very early on, it sparked controversy among workers, trade unions, and social scientists. Taylorism changed overtime through intellectual debate, partly due to Taylor's own followers, partly due to those who opposed his methods, but it has remained a matter of lively debate since its inception. Today Taylor's contribution to organizational theories, which focusses on coordination according to rules, laws, and principles, continues to be important.

Taylor's World

Taylor's life and discoveries are well known today (Nelson, 1980) thanks to his surviving papers which are kept in the archives of the Stevens Institute of Technology near New York, where he had studied by correspondence and obtained a degree in mechanical engineering in 1883. After his apprenticeship as a patternmaker and a machinist, Taylor entered a career in the steel industry. At Midvale Steel he progressed from machine shop labourer to chief engineer of the works. He later occupied another major position at Bethlehem Steel. From 1893 onwards, he also worked as an independent consultant, which became his main occupation after 1901. His contribution to technological innovation was important with forty-six patents granted, the most outstanding of which was his development, with Maunsell White, of a process for subjecting tool steel to increased cutting speeds which 'revolutionized metal cutting techniques and paved the way for mass production methods' (Neck and Bedeian, 1996, 21). His contributions to workshops management were developed from the journal he kept when he worked for Midvale. They aimed to improve

Global History of Techniques, ed. by Guillaume Carnino, Liliane Hilaire-Pérez and Jérôme Lamy, Global Matters Culture Matérielles, histoire des techniques et circulation globale des savoirs, 9 (Turnhout: Brepols, 2024), pp. 613–626
BREPOLS PUBLISHERS 10.1484/M.TECHNE-EB.5.129832

industrial efficiency by increasing the productivity of men and machines and reducing costs via the use of time studies, wage systems, forms, slide rules, and calculation techniques. Scientific management was thus one of the most thorough ways of getting around skilled workers' know-how and their potential ability to restrict output and instead of fostering the production of semiskilled workers and of labourers. Engineers and technicians in special offices were to decide on the best worker, the best tool, and the best way of doing each task. This was also part of the trend towards benchmarking, which had first been introduced for guns and ammunition. In this way the preparation, measurement, and control of work were standardised while the actual wages were tailored to the individual. Taylorism was gradually extended to offices, to public administration, and to the army.

Some women applied scientific management to their homes to do housework more efficiently. Their main proponent was Christine Frederick, who was the first woman to forge a career as a home efficiency expert, and later became an advertising consultant and a consumer advocate. She advocated for the standardization of tasks and trained women to become proficient homemakers. While promoting a new image of women, Frederick still maintained the gendered division of labour (Rutherford, 2003).

In industrial contexts Taylor had an acrimonious relationship with trade unions, despite his repeated assertions that the scientificity of his methods and the objectivity of his measures would elicit consent. At the Watertown Arsenal, skilled workers demonstrated that they knew more about their tasks than Taylor's men who were in charge of their reorganization, and were therefore able to suppress the introduction of his system in the arsenal (Aitken, 1960). Nevertheless, such setbacks deserve further examination. On the one hand, they did not really prevent the expansion of Taylorism to a minority of firms manufacturing large numbers of a limited set of products. On the other hand, an entire sector of American firms involved in customized or small-batch production at labour-intensive mid-sized factories continued to grow thanks to technologically adept owners and skilled workers who developed new work systems beyond Taylorism (Scranton, 1997).

The success of scientific management was intrinsically linked to Taylor's talent for networking (Figure 1). He certainly became an expert in promoting his methods and system through correspondence and lectures, articles and books, as well as events in the public sphere such as state or federal hearings and debates. Even strikes, like the extensive 1913 strikes against his system at Firestone (USA), Renault (France), and Bosch (Germany), aided his cause. Taylor's networks also proved to be a major asset. Among his many — male and female — followers he carefully picked a select few from all over America whom he used as proponents of his ideas, he used also the American Society of Mechanical Engineers as a sounding board, and communicated with companies interested in implementing scientific management, providing them with thorough reports and studies and proposals for reorganization. In 1910, albeit against his will, his US followers met to prepare the formation of a dedicated association, the Society to Promote the Science of Management (1911).

Date	Industry	Organization
10/18/1890	Engineering	Engineers Club of Philadelphia
11/1/1890	Mechanical Engineering	American Society of Mechanical Engineers
4/28/1899	Political and Social Science	American Academy of Political and Social Science
8/29/1900	Mechanics	Franklin Institute
9/7/1906	Industrial Education	National Society for the Promotion of Industrial Education
4/1/1907	Science	Carnegie Institute
10/17/1908	Education	Carnegie Foundation for the Advancement of Teaching
11/2/1910	Politics/Economics	National Tariff Commission Association
7/13/1911	Conservation	National Conservation Association
9/16/1911	Efficiency	National Society for Promoting Efficiency
9/21/1911	Business	Bureau of Business Research
3/6/1912	Philosophy	American Philosophical Society
3/13/1912	History	Historical Society of Pennsylvania
4/11/1912	Politics	American Association for Labor Legislation
10/19/1912	Economics	The Economic Club of Philadelphia
2/13/1913	Social	The New England Society of Pennsylvania
5/1/1913	Geography	National Geographic Society
5/14/1913	Scientific Management	Society to Promote the Science of Management

Fig. 1. F. W. Taylor's Expanding Social Networks. The dark colored organizations are close to Taylor's roots in mechanical engineering. Taylor resisted the society at the bottom, which he saw as a withdrawal into a private world of scientific managers, rather than influencing others. (c) Margaret 'Amy' DiGerolamo, 2015. Lee Vinsel, Taylor's World Pt. 2: F. W. Taylor's Expanding Social Networks, September 25, 2015

The internationalization of Taylorism consultancies followed rapidly, and in two ways. Taylor (and also his many competitors) 'spent short periods abroad proselytizing his system among fellow engineers', aided by exhibitions and by translations of his works into Italian, French, German, Russian, Lettish, Dutch, Spanish, and Japanese (Kipping, 2020, 140). Interessees from countries other than the US wanting to observe scientific management and its implementation in action went to the US and took the idea with them back to their home countries.

It would be wrong to look at the internationalization of Taylorism through a diffusionist lens. A small number of experts, businesses, and other organizations in each country became interested in Taylorism and put some of Taylor's ideas and methods into practice, not only according to their economic needs, but also in keeping with the roles of engineers, of science, and of labour in their home countries. France was the largest and most obvious nation where Taylorism took root (Fridenson, 1987). Taylor's work as a metallurgical engineer attracted the attention of a number of innovative engineers and scientists who had already wanted to increase the role of science in industry and organize the influx of labour and machinery in the workshops of new industries under the leadership of Henry Le Chatelier, an engineer who had become a leading professor of chemistry. Industrialists followed the lead of pioneering engineers. Taylorism was also adopted in much smaller countries. In Sweden it appeared as a solution to the labour problem created by mass emigration to the US. In this way, Swedish businesses profited from American experiences to overcome this population deficit. The Swedish Manufacturers Association published the translation of *Principles of Scientific Management* in 1913 (Runeby, 1978). In Great Britain, the foremost industrial power at the time, developments were less visible, partly because the engineering profession was not yet institutionalized. Contrary to older historical accounts, recent research has revealed that the leading engineering journals gave Taylorism 'a fair amount of attention, and much of it positive' (Whitston, 1997, 207), and the confectionery company Cadbury introduced many of Taylor's methods. Beyond Europe, the main interest in Taylorism came from Japan, from two different groups. One of these were engineers wishing to prevent or solve the numerous repair problems in railways, one of the early sectors of big business in Japan; they also aimed for import-substitution of goods and technologies that came from the West. As a result, Japanese railway repair shops, which, as elsewhere, were labour intensive and employed vast numbers of workers, seemed like a good place to start implementing Taylor's principles (Daito, 1989). The other group consisted of individuals like Yukochi Ueno and Toichiro Araki, who were eager to use American knowledge to increase the workers' productivity in factories.

The last development that Taylor and his followers faced was the rise of Fordism. The relationship between Taylorism and Fordism is still much debated. Arguably their inspirations were different, but both drew upon systematic management, and Taylorism created part of the preconditions for Fordism. Taylor focused on the workshop and on standardised tools, broke down workers' motions into small sequences at the initiative of two of his followers, Frank and Lilian Gilbreth — one of the great husband-and wife teams in engineering. He also retrained workers, regulated work routines, and motivated workers to do more than the norm by modifying piece-rate wages with bonuses. Ford focussed on the entire plant and established continuous-flow manufacturing, which paced the work and resulted in redefining and simplifying jobs, hiring large numbers of common labourers and setting a high fixed wage. 'Taylor saved time, Ford sped up time' (Nye, 2013, 34).

Recasting Taylorism from One World War to the Other

The First World War accelerated the dissemination of Taylorism in every country that was involved in the conflict. Taylorism offered a recipe for maximizing the quantities produced while using unskilled labour such as women and foreign workers. It lifted the opposition of trade unions, as most of their leaders embraced the idea of national defence. The war effort also meant that sciences gained unprecedented relevance, which gave some academics increased importance. It was also often linked to putting psychotechnics into practice (Leimgruber, 2001). Yet this period did not just witness the spread of Taylorism. Several areas of tension emerged. Among managers, criticism was levelled by French manager Henri Fayol, who had been promoting science-based programmes and the exploration of the unknown and had emphasised management by autonomous work teams and a search for social consent (Bertilorenzi, Dubruc, Passaqui, 2019). With regard to the impact of work on the human body a large group of ergonomists attempted to reduce the sources of fatigue associated with Taylorism. There was a backlash among workers against Taylorist control due to a desire for more autonomy, and this was one of the many causes of the mass strikes in Europe, the USA, and Canada from 1916 to the early 1920s as well as the demand for increased workers' control which contributed to revolutions in Central and Eastern Europe (Montgomery, 1987).

It was also during the First World War that Taylor's influence spread to China. Two Chinese citizens who studied in the US introduced Taylor's ideas to their country. The more important of the two was Mu Xiangyue, who translated Taylor's writings, created three cotton mills, and 'was an extraordinary activist in business and public affairs' (Morgan, 2006, 211). He applied Taylor's ideas to personnel in three ways: addressing his middle managers' lack of technical competency by requiring them to receive technical training; tackling the issue of wasteful labour by asking labour contractors to improve their technical competency and to report to him daily; and reducing the 'brutal treatment' of ordinary workers by establishing operating and disciplinary rules enforced by foremen, thus increasing productivity and quality (Morgan, 2006, 412).

The strikes of the post-war period only had lasting consequences in one major Italian firm, Ansaldo, a giant in the areas of armament and shipbuilding, where the skilled workers' resistance led management to discontinue the use of Taylorism (Dewerpe, 2017). Otherwise, the spread of Taylorism carried on unhampered.

In Western Europe, one key move was the controlled conversion of Germany's main science-based corporations, led by Siemens, to its principles (Homburg, 1991). Contrary to Taylor's views, European companies placed their planning rooms under the control of central management. European consulting firms emerged, focussing on technological and organizational change, among them OAB (Organisatie Advies Bureau), which was created by two young Dutch engineers in 1920. 'However, only seldom was anything changed in the content of the work' (Bloemen, 1988, 191). The domestic sciences movement that Christine Frederick had launched in the US found disciples among enterprising European women. In France, from 1922 onwards, the Paris journalist Paulette Bernège began to adapt Frederick's standards and methods to French homes in order to streamline housework. She founded the Institut d'organisation ménagère (institute for household management), was personally close to Taylorite engineers and employers and took part in various congresses on the scientific organization of work (Clarke, 2011). By contrast, in Germany, household reform (which was also inspired by Frederick) was embraced by both the bourgeois women's movement and by socialist women. The latter were particularly active in Frankfurt, and under the aegis of socialist architect Margarete Schütte-Lihotzky, encouraged the timing of domestic chores, the use of modern appliances and sleek furniture (Nolan, 1990).

In Eastern Europe and in Asia Taylorism continued to spread. The most striking case was Soviet Russia (Beissinger, 1988). Taylorism became an official doctrine

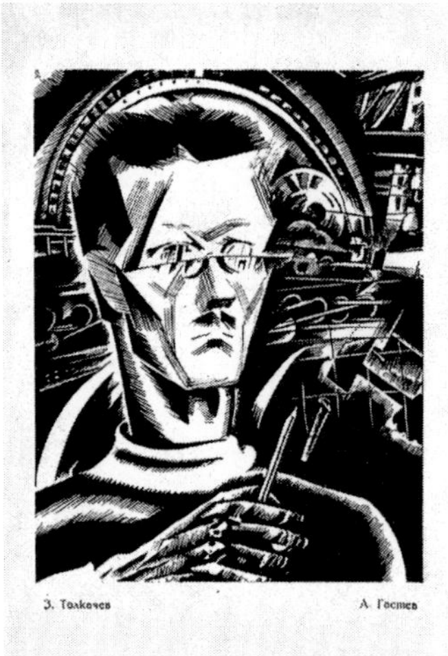

Fig. 2. Portrait of Gastev by Z. Tolkachev. From Gastev, *Revolt of Culture*, 1923.

under two main influences. First that of Lenin — who was initially critical of the doctrine in 1913 and 1914 — but who rapidly became convinced of the economic benefits of Taylorism and decreed it a national priority in April 1918. Additionally, there was the influence of the skilled worker and poet Aleksei Gastev, who had worked in France before the war and wanted both to encourage a new workers' culture and to improve the efficiency of labour in the whole country (Figure 2). In 1920 he founded the Central'nyj institut truda (Central Institute of Labour) which also developed laboratories (Koulaeva, 2021). However, in the 1930s, Stalin preferred the views of one of Gastev's critics, Platon Kerzhentsev, a left Communist, who implemented a version of Taylorism based on the principle of vertical authority: '[t]he totalitarian State of the 1930s was opposed to the creation of an anarchical network of socially engineered Cyborgs with liberated minds' (Smirnov, 2013, 106). In the late 1920s, a group of progressive engineers led by Peter Palchinsky went into the opposite direction and promoted 'upgrading the worker's knowledge to such a level that the primitive methods of Taylorism, designed for inexperienced workers, would not be necessary. Knowledgeable workers [...] would be masters of their work, not slaves of it' (Graham, 1996, 184). They were purged, and Palchinsky was executed in 1929, whereas Gastev was sentenced to death and shot in 1939.

Developments in Asia were far more incremental, less centralized, and less dramatic. In China there were few other experiments with Taylorism in the 1920s, but they multiplied during the modernization period between 1928 and 1937. In particular, in 1930 the nationalist government of China persuaded Shanghai industrialists to form the China Institute of Scientific Management. It emphasized three goals: the improvement of personnel administration and the cooperation between managers and the managed; the expansion of production skills and the continued effort to reduce waste; and the encouragement of the growth of national industry. The dissemination of Taylorism spread further. It often coexisted with traditional Chinese practices such as recruiting senior staff based on their place of origin or family affiliations. Therefore, Taylorism can be said to have undergone a process of acculturation, meeting with some resistance from contractors and foremen and also from part of the workforce. The interest in such changes extended beyond major 'industrialists and industry officials to reportage in the popular press' (Morgan, 2006, 422).

In Japan, pre-war developments expanded much more in the post-war period. While Ueno and Araki continued their effort promoting Taylorite writings and practised consultancy, two new groups appeared: mechanical scientists applying their knowledge to the metal industry, and Japanese employees working in Japanese-American joint ventures. The Japanese government supported this trend and created commissions on rationalization which became official forums for for sharing experience. An Industrial Efficiency Institute was privately founded in 1922 (Tsutsui, 1998).

In the West, the institutionalization of Taylorism also advanced. Each European country developed a national institute of Taylorism. The most important one was the German institute for rationalization (Reichskuratorium für Wirtschaftlichkeit or RKW), which was created in 1921 by the Ministry of Economy and is still very active today. In the Netherlands, engineers and accountants established the Dutch Institute for Efficiency in 1925 (Bloemen, 1988). Even a small country like Switzerland founded an equivalent institute for the French-speaking half of its population in 1928 (Leimgruber, 2001). Two countries saw the merger of a strictly Taylorite association with an association adhering to related but divergent management principles, a sign of the increased fluidity and openness of Taylorism. In France, Le Chatelier's *Conférence pour l'Organisation Française* merged in 1926 with Fayol's *Centre d'Études Administratives* into the Comité national de l'organisation française (CNOF, French national committee for organization). CNOF became both a centre for training managers in work rationalization and an office for improving the measure of elementary times, which became a distinct organization in 1938 (Moutet, 1997). In 1929 a department of domestic economy was added, led by Paulette Bernège. In the US, the association dedicated to Taylorism was renamed the 'Taylor Society' after Taylor's death; it soon coexisted with the Society of Industrial Engineers, which had been founded in 1917, and the two societies merged in 1936. Renamed 'Society for Advancement of Management', it is now the oldest of the professional societies of management in the US.

Additionally, dissemination efforts were organised at the international level, involving not only engineers, consultants, and business leaders, but also the new International Labour Organization, which had been created in Geneva in 1919. The efforts took two forms. First there was an annual international forum, in which small nations played a significant part. From 1924 to 1938, the *Comité International pour l'Organisation Scientifique* (CIOS) — or international committee for scientific management — organised annual congresses which focussed on innovations and experiments and were also attended by the new profession of accountants, who debated new methods like budget control. Then between 1927 and 1934, the CIOS was complemented by the Geneva-based International Management Institute. This private initiative was conceived and financed by the Boston department store entrepreneur, philanthropist, and social reformer Edward Filene and supported by the ILO (Cayet, 2010). It created a forum for discussions about the balance between economic modernization and social reform.

Within the US, the scientific management movement began to evolve in several ways in the 1920s. One wing of Taylor's followers reconciled with union leaders and endorsed the concept of collective bargaining. Further, the Taylorite movement developed closer relations with the personnel management movement which affected most major companies, not only in industry but also in services. Some Taylorites established links with the human relations movement to better understand the problems affecting workshops and to overcome them. With the Great Depression in 1929, the divide between the two factions of Taylorism deepened. One group helped manufacturers to find — sometimes drastic — ways to stay in business during economically difficult times. Another subscribed to the New Deal with « most non-academic leaders of scientific management » becoming « federal government officers » (Nelson, 1992, 3), applying their skills to government policy and to market regulation, whereas a few advocated collective bargaining or participatory management.

A lighter version of Taylorism was created in the US by the self-taught entrepreneur Charles Bedaux. Without being invested in organization restructure and training, it merely added to time study a wage system based on the number of points achieved in a given amount of time. Used in modern corporations like General Electric and Kodak, as well as car factories, textile mills, and coal mines, it often provoked confrontations with the workers, even in Fascist Italy. However, the example of the Pertusola mines in Italy shows that a typical implementation of the Bedaux system — which didn't divulge the references of its calculations and did not take work injuries into account — could be combined with a modelization of the body at work to optimize gestures, evaluate fatigue, and increase discipline and control (Sanna, 2019). After first hesitating about what stance it should take, the Taylor Society lashed out against the Bedaux system. In 1940, one of its former leaders called Bedaux no more than a 'time study merchant' (Thompson, 1940, 175). The Bedaux Company became one of the earliest multinational consultancy companies. By the mid-1930s, the firm 'employed several hundred consultants around the world' (Kipping, 2020, 140). It was 'poles apart' from most varieties of Taylorism, which were again called to arms in each nation when the Second World War broke out (Person, 1937, 71).

Moving towards Neo-Taylorism after 1945

In the period after the end of the Second World War Taylorism spread to emerging countries and experienced new developments in industrial countries.

Taylorism spread differently in emerging countries with low wages. While it spread gradually in Latin America (Vargas, 1984) it spread rapidly and extensively in Southeast Asia, as a resource that made it possible to increase the textile production for global markets. In Communist China, following Mao's death in 1976, Taylorism received official support from the regime. It was used as one of the keys to China's rise to being a global industrial power. At the 2015 international congress on industrial electronics and applications, a Chinese management scholar commemorated the centennial of Taylor's death and gave a warm tribute to his legacy (Zhang, 2015). In Soviet Russia, the late 1950s saw a rebirth of Taylorism 'on a technocratic basis' (Grachev, Rakitsky, 2013, 512).

In the West and in Japan Taylorist methods continued to become more complex, both in the workshop and in the office (Montmollin, Pastré, 1984). This was made possible by new types of machine tools and by the computer revolution. It was also supported by the growth of mostly American and West European multinational consultancy firms, to which Japanese firms were added later; this still left room for local consultants in complementary roles. It also relied on the intense collection of data and statistics on production processes, which in Japan led to the transformation of the control of shop floors (Wada, 2020). Among the most visible changes worldwide, time keepers were replaced by virtual stopwatches from the late 1970s onwards, and job evaluation, which had developed from the 1940s onwards, was dismissed in favour of new grading criteria that were simpler, less formal, and gave more room to performance and competencies. Meanwhile medicine, ergonomics, psychology, and psychiatry, often with the support of trade unions and public administration, stepped in to document, cure, or prevent the effects of Taylorism on bodies and minds. Such approaches were partly fuelled by the post-war, posthumous publication of the Italian Communist leader Antonio Gramsci's 1934 prison notebook. Its memorable passages on Taylorism highlighted the 'forced selection' of workers, the 'brutal' imposition of 'a new psychophysical nexus', and the 'crisis of adaptation' (Gramsci, 1992, 216).

Could Japan's Toyota production system be considered neo-Taylorist? Its emphasis on time seems to indicate so, and Toyota was able to create a one-minute cycle for workers on the automated and computerized assembly lines, which was then defined as the world standard of efficiency in the auto industry. However, this differed from original Taylorism because of its focus on quality, continuous improvement, and the participatory management at the workshop and office levels.

A new wave of criticism against Taylorism emerged in the West in the early 1970s. This had different origins. It was sparked by the 'blue-collar blues' at the modern, automated Lordstown factory of General Motors in the US in 1972 (Gooding, 1970). Taylorist managers told workers: '[l]eave your brain at the door' (Meyer, 2004). Criticism was also voiced by progressive industrialists in Sweden and in France, who emphasized features of Taylorism such as bureaucratization and partitioning as well as the failure of top-down command and control management systems, and some of them called for Taylorism to be turned 'upside down' (Dalle and Bounine-Cabalé, 1972). A new generation of sociologists, psychologists, and psychiatrists focussed on its negative impacts. Epitomizing this era, Harry Braverman, a Marxist metal worker-turned-editor, published *Labor and Monopoly Capital: the Degradation of Work in the Twentieth Century* in 1974, the English edition of which sold nearly one million copies. This critique had consequences (for instance, in France, the *Bureau des Temps Élémentaires* entered a long period of difficulties), and worldwide managers, engineers, and consultants introduced further changes to Taylorist methods.

Looking back at the period which followed, which was characterized by a worldwide decline of trade unions, it is possible to distinguish three types of policies: those which led to organizational forms that broke with Taylorist principles in terms of work management; those which brought about a reinforcement of Taylorism; and those which, despite important technological changes, simply reproduced the previous organization (Linhart, 1993). The first trend can be observed in the process industries, like the production of cement, petrochemicals, and steel.They were capital intensive, which implied strict safety regulations and high standards of accountability but also led to a decrease of the workforce, and resulted in a deeply segmented labour market. In these sectors, groups of polyvalent workers were established, with broader fields of action and skill levels, and tasks that included technical tasks defined in collaboration with office technicians. This was a way of enhancing workers' skills.

The second trend, which concerned a minority of industries only, aimed towards a tighter Taylorization of work situations. This was done in the clothing and textile industries, building trades, and public works. A heightened control of work rates was supplemented with the neo-Fordist introduction of automatic programmable conveyors. Here the multitasking of workers meant less specialization was required and went hand-in-hand with the implementation of human resource policies based on the exploitation of a workforce on precarious contracts.

The third trend was the continuation of former Taylorist practices. Although new technologies allowed for some preparation and design tasks to be transferred to the workshop, the role of the workshop was not really revalued, and the separation between design and execution persisted.

Outside industry and traditional services, Taylorist methods expanded into new areas of services: mass retail, fast food, call centres, internet-based commerce, and hospitals. 'Amazon's current work with robots, drones, and 'Mechanical Turk' brings neo-Taylorism to the warehouse, the skies, and even the 'digerati" (Duguid, 2015, 5).

What has the overall role of Taylorism been since 1945 in developed countries outside the process industries? (Lallement, 2007). The impact of information technologies and of automation has been twofold. On the one hand, in keeping with the Taylorist tradition, they have increased control and the precision of measures, and hence productivity. On the other hand, they have handed more responsibility to the local level, allowing for a greater variety of goods to be produced with a greater degree of flexibility. The consequences of IT on production have been a reduction of production waste and a tendency towards higher quality, contradicting the principle of programmed obsolescence. Its consequences on employment were initially favourable towards hiring immigrants and women, but this has not been the case lately given the recent levels of automation. The consequences on the health of the workforce have been identified as fatigue, stress, mental exhaustion, and, in the the worst cases, burnout. Yet field research yields a counterintuitive result: even in the most thorough or automated organizations, full control of work has not been entirely possible, and a discrepancy between the work that is assigned and that which is actually done can be observed. Pockets of autonomy either survive or are reconstituted by a fraction of the workers. Sociologists, anthropologists, ergonomists, and psychologists, as well as oral historians, have shown that whereas some workers may react to monotonous tasks with a restriction of output, alcohol, drugs, and absences, the search for autonomy remains a feature of working lives even in the more flexible neo-Taylorism, and periodically brings back the utopia of workplace democracy.

One question remains, that of the place and significance of Taylorism as one of the technologies of activity in the genealogy of the sciences and their relationship to technologies. Most scholars have connected Taylorism to the development of thermodynamics in the mid-nineteenth century and considered it a method of using bodily energy efficiently (Rabinbach, 1990). Some recent studies, by contrast, place its origins in the 1767 reflections of the Scottish philosopher Adam Ferguson, according to whom manufactures 'prosper most when the mind is least consulted; and where the workshop may [...] be considered as an engine, the parts of which are men' (Duguid, 2015, 5); and also in the 1799–1801 concept of French mechanics scholars, which reduced man to a machine. They suggest that it has persisted in spite of the many changes in both technologies and in industrial work since 1800. This legacy has indeed been appropriated by Taylor and Amar, as well as their many critics. However, overtime various engineers and social scientists have shown that work cannot be reduced to time and that, whenever it is not replaceable by machinery, there is always a delicate type of trade-off between quantity and quality, which calls for more complex metrics than Taylor's. Since the early twentieth century and following Armand Imbert, a number of ergonomists have attempted to obtain less general, more local measurements of work by combining mechanics with physiology. Therefore, the lasting distinction of Taylorism and ergonomics also relies on social and historical contrasts between different tools used for measuring work which convey value judgments on what is measured and how it is being measured in production-oriented activities (Moutet, 1997; Durrive, Saraceno, Vatin, 2019).

References

Aitken, Hugh G. J., *Scientific Management in Action: Taylorism at Watertown Arsenal, 1908–1915* (Cambridge, MA: Harvard University Press, 1960).

Beissinger, Mark R., *Scientific Management, Socialist Discipline and Soviet Power* (Cambridge, MA: Harvard University Press, 1988).

Bertilorenzi, Marco, Nadine Dubruc, and Jean-Philippe Passaqui, eds, *Henri Fayol. Les multiples facettes d'un manager* (Paris: Presses des Mines, 2019).

Bloemen, Erik S. A., *Scientific Management in Nederland 1900–1939* (Amsterdam: NEHA, 1988).

Cayet, Thomas, *Rationaliser le travail, organiser la production. Le Bureau international du travail et la modernisation économique durant l'entre-deux-guerres* (Rennes: Presses universitaires de Rennes, 2010).

Clarke, Jackie, *France in the Age of Organization. Factory, Home and Nation from the 1920s to Vichy* (New York, NY: Berghahn, 2011).

Dalle, François and Jean-Bounine-Cabalé, 'Le taylorisme à l'envers', *Le Monde* (20 June 1972).

Daito, Eisuke, 'Railways and Scientific Management in Japan 1907–30', *Business History*, 31.1 (1989), 1–28.

Dewerpe, Alain, *Les Mondes de l'industrie. L'Ansaldo, un capitalisme à l'italienne* (Paris: Éditions de l'EHESS, 2017).

Duguid, Paul, 'Making Sense of the Systems of Scientific Management', *Le Libellio d'AEGIS*, 11.4 (2015), 5–9.

Durrive, Barthélemy, Marco Saraceno, and François Vatin, 'Entretien avec François Vatin. Une sociologie historique de la mesure entre sociologie de l'activité et histoire des sciences', *e-Phaïstos*, 7.1 (2019) <http://journals.openedition.org/ephaistos/4518> [accessed 24 December 2020].

Fridenson, Patrick, 'Un tournant taylorien de la société française (1904–1918)', *Annales ESC*, 421 (1987), 1031–60.

Gooding, Judson, 'Blue-Collar Blues on the Assembly Line', *Fortune* (July 1970), 70–71.

Grachev, Mikhail and Boris Rakitsky, 'Historic Horizons of Frederick Taylor's Scientific Management', *Journal of Management History*, 19.4 (2013), 512–27.

Graham, Loren R., 'Palchinsky's Travels: a Russian Engineer's Adventures Among Gigantic Projects and Small Minds', *Proceedings of the American Philosophical Society*, 140.2 (1996), 175–85.

Gramsci, Antonio, 'Americanism and Fordism', in *Prison Notebooks*, edited and translated by Joseph A. Buttigieg with Antonio Callari (New York, NY: Columbia University Press, 1992), vol. 2, pp. 213–25.

Homburg, Heidrun, *Rationalisierung und Industriearbeit. Arbeitsmarkt — Management — Arbeiterschaft im Siemens-Konzern Berlin, 1900–1939* (Berlin: Haude und Spener, 1991).

Kipping, Matthias, 'Consultants and Internationalization', in *The Routledge Companion to the Makers of Global Business*, ed. by Teresa da Silva Lopes, Christina Lubinski, and Heidi J. S. Tworek (Abingdon and New York, NY: Routledge, 2020), pp. 138–55.

Koulaeva, Alexandra, 'Alexei Gastev, ingénieur du Nouveau Monde', *Entreprises et Histoire*, 103 (2021), 13–35

Lallement, Michel, *Le Travail. Une sociologie contemporaine* (Paris: Gallimard, 2007).

Leimgruber, Matthieu, *Taylorisme et management en Suisse romande (1917–1950)* (Lausanne: Éditions Antipodes, 2001).

Linhart, Danièle, 'À propos du post-taylorisme', *Sociologie du travail*, 35.1 (1993), 63–74.

Meyer, Stephen, 'The Degradation of Work Revisited: Workers and Technology in the American Auto Industry, 1900–2000', University of Michigan, 2004 http://www.autolife.umd.umich.edu/Labor/L_Overview/L_Overview8.htm [last accessed on 22 February 2023]

Montgomery, David, *The Fall of the House of Labor. The Workplace, the State, and American Labor Activism, 1865–1925* (Cambridge: Cambridge University Press, 1987).

Montmollin, Maurice de and Olivier Pastré, eds, *Le Taylorisme* (Paris: La Découverte, 1984).

Morgan, Stephen L., 'Transfer of Taylorist Ideas in China, 1910–1930s', *Journal of Management History*, 12.4 (2006), 408–24.

Moutet, Aimée, *Les Logiques de l'entreprise. La rationalisation dans l'industrie française de l'entre-deux-guerres* (Paris: Éditions de l'EHESS, 1997).

Neck, Christopher P. and Arthur G. Bedeian, 'Frederick W. Taylor, J. Maunsell White III, and the Matthew Effect. The rest of the story', *Journal of Management History*, 2.2 (1996), 20–25.

Nelson, Daniel, *Frederick Taylor and the Rise of Scientific Management* (Madison, WI: University of Wisconsin Press, 1980).

———, ed, *A Mental Revolution: Scientific Management since Taylor* (Columbus, OH: Ohio State University Press, 1992).

Nolan, Mary, '"Housework Made Easy": the Taylorized Housewife in Weimar Germany's Rationalized Economy', *Feminist Studies*, 16.3 (1990), 549–77.

Nye, David E., *America's Assembly Line* (Cambridge, MA: The MIT Press, 2013).

Person, Harlow S., 'The Bedaux System', *The New Republic*, 93 (November 24, 1937), 71.

Rabinbach, Anson, *The Human Motor. Energy, Fatigue, and the Origins of Modernity* (New York, NY: Basic Books, 1990).

Runeby, Nils, 'Americanisation, Taylorism and Social Integration', *Scandinavian Journal of History*, 3.1–4 (1978), 21–46.

Rutherford, Janice W., *Selling Mrs Consumer: Christine Fredrick and the Rise of Household Efficiency* (Athens, GA: University of Georgia Press, 2003).

Sanna, Francesca, 'Les Travaux sur la fatigue humaine à l'épreuve du terrain. Le cas du système Gennaper dans les mines de Pertusola (1930–1940)', *e-Phaïstos*, 7.1 (2019), <https://journals.openedition.org/ephaistos/4079> [accessed 24 December 2020].

Scranton, Philip, *Endless Novelty: Specialty Production and American Industrialization, 1865–1925* (Princeton, NJ: Princeton University Press, 1997).

Smirnov, Andrey, *Sound in Z: Experiments in Sound and Electronic Music in Early 20th-century Russia* (London: Koenig Books, 2013).

Thompson, Bertrand, 'The Taylor System in Europe', *Advanced Management*, 5.4 (1940), 171–76.

Tsutsui, William M., *Manufacturing Ideology: Scientific Management in Twentieth-century Japan* (Princeton, NJ: Princeton University Press, 1998).

Vargas, Nilton, 'Le taylorisme au Brésil', in *Le taylorisme*, ed. by Maurice de Montmollin and Olivier Pastré (Paris: La Découverte, 1984), pp. 155–66.

Wada, Kazuo, *The Evolution of the Toyota Production System* (Singapore: Springer, 2020).

Whitston, Kevin, 'The Reception of Scientific Management by British Engineers 1890–1914', *Business History Review*, 71.2 (1997), 207–29.

Zhang, Yangfei, 'The Principles of Scientific Management Spread Forever — to Commemorate the Principles of Scientific Management be Published by the 100th Anniversary', in *2015 AASRI International Conference on Industrial Electronics and Applications (IEA 2015)* (Amsterdam and Paris: Atlantis Press, 2015), pp. 600–02.

FRANÇOIS JARRIGE

Conflicts and Controversies Over Technologies

Disputes about genetically modified organisms (GMOs) and biotechnological products in Europe and Asia, protests against massive technical infrastructures deemed predatory, such as equipment related to nuclear, wind, or water power, high-speed railways, or airports — the current era is fraught with conflicts and controversies over the technical world (Fougier, 2011). According to some, including the well-known French doctor and businessman Laurent Alexandre, mistrust of new technologies is one of the evils plaguing contemporary societies. This mistrust may be a sign of a dangerous neo-Malthusianism, of self-hatred, or of an irrational fear of the inevitable transformations of the world. For Laurent Alexandre, such opposition translates to nothing more than techno-conservative views that put a brake on progress, and these controversies are the result of reactionary attitudes that might increase the distance between countries in the inevitable global competition. This short-term view, however, fails to capture the facts as it remains restricted to a straightforward and linear vision of technological development and of the conflicts it stirs up.

Technique and technology are categories difficult to delineate, as they are historically changing and variable, ranging from the science of arts and effective actions to the innovations put onto the market in any given period. However, any new technique, insofar as it acts against particular interests, marginalizes long-established practices and reshapes the relations between nature and society, and may generate conflicts and contradictory interpretations at the origin of heated controversies (Jarrige, 2014). Choosing a technique involves the rejection of others, and every technical choice in itself carries conflicts and controversies between competing actors, groups, and visions. One of the difficulties met with in the history of techniques arises from the fact that many innovations fail, are forgotten, or are never adopted. Besides, many techniques serve some interests at the expense of others, thus giving rise to divergent interpretations which may lead to the emergence of public conflicts. At a time of increased fascination with innovation, the persistence of an ancient technique is often interpreted as a manifestation of conservatism, although it may simply be the result of caution and of a choice in favour of an appropriate technique that has been tried and tested (Edgerton, 2007). Technologies have elicited a wide variety of attitudes across the world for the beginning of industrialization, and technical choices have often

Global History of Techniques, ed. by Guillaume Carnino, Liliane Hilaire-Pérez and Jérôme Lamy, Global Matters Culture Matérielles, histoire des techniques et circulation globale des savoirs, 9 (Turnhout: Brepols, 2024), pp. 627–642
BREPOLS PUBLISHERS 10.1484/M.TECHNE-EB.5.129833

taken place within the context of heated struggles that have kept changing according to the repertoires of action available and to the evolution of modes of production. Sustained by very diverse actors, practices, and narratives, the conflicts — from the disputes over incipient mechanization at the beginning of the nineteenth century to the current massive controversies over technosciences — have shaped the meanings of technology.

Controversies and Conflicts over Early Mechanization

At the turn of the nineteenth century, machines multiplied and mechanization boomed, accompanying the rise of industrialization. However, this process, which has long been called 'Industrial Revolution', went along with heated debates and conflicts. In opposition to the narrative of a pacifying industrialization at the origin of general enrichment and emancipation, historians insist on presence of debates, uncertainties, and conflicts accompanying the process. As early as the eighteenth century in England, and during the first half of the nineteenth century in many parts of Europe, communities of craftsmen and workers — as well as numerous intellectuals — raised doubts about the flood of machines in the workshops and factories. Sometimes they even went so far as to rebel and destroy them. The most spectacular troubles took place in England during the Luddite riots (1811–1813), which took their name from the mythical General Ned Ludd who led the revolt and whom, of course, the authorities failed to catch. In this context, a large uprising took place in the textile-producing regions, people protest during food riots and attacks against machines (Figure 1). In continental Europe, workers also banded together in small wool-producing centres in Languedoc or, in a more limited and more fragmented way, in very varied territories, such as in Paris, where printers destroyed the new mechanical presses in 1830. When machines arrived at Lodève, Saint-Pons, or Carcassonne, hundreds of workers gathered to oppose their impact and prevent their use. In 1830 in Verviers (Belgium), and again in Reims in 1848, workers took advantage of the revolutionary context to destroy machines accused of starving the people (Jarrige, 2009).

In the middle of the nineteenth century, small artisanal and domestic production was still dominant across the world. However, modern industry, with its vast system of powerful machinery, was spreading, threatening artisans and small producers' lifestyles and numerous local production systems. The independent artisan who masters a complex process and detailed know-how remained the norm. Confronted with an increasing number of attempts at mechanization, which they perceived as a threat to their liberty or job, or to the quality of the products that were the foundation of their identities, groups as diverse as printers, tailors, and milliners, pit sawyers and grain thrashers denounced the impact of large machinery, and even rose up against them whenever the crisis was too serious or the political context appropriate. These workers' protests did not just come from the lower sections of the world of labour, they also broke out in the most affected sectors and among elite workers, namely those who had sufficient resources to voice their complaints. Even if these revolts against the machines were put down rapidly, their memory persisted for a long time and helped shape discourse on popular ignorance and the need to justify and accompany innovations.

Fig. 1. Poster posted in Nottingham (26th March 1811). Reward offered for information leading to the arrest of machine breakers (poster reproduced in Brian Bailey, *The Luddite Rebellion* (New York: New York University Press, 1998), p. 17.

European economists and engineers dismissed those conflicts triggered by technological developments as the fruits of ignorance. For instance, Joseph Garnier noted in the entry '*Machines*' in the *Dictionnaire de l'économie politique* that it was essential to fight 'against the [workers'] prejudices that incite them to hate machines, to smash them, or to count on *illusionary means*' ('*contre les préjugés [des ouvriers] qui les incitent à haïr les machines, à les briser, ou à compter sur des moyens chimériques*') (Garnier, 1853: 120, all translations are the author's own unless indicated otherwise). Maxine Berg showed the centrality of the conflict over machinery in the construction of the British political economy during the first half of the nineteenth century. Far from being dominant or all-powerful, economists were attacked from all sides and compelled to justify profit, machinery, and the very possibility of continuous growth (Berg, 1980). David Ricardo settled the issue by distinguishing long-term benefits of machines from short-term adverse effects that might result from their use (Ricardo, 1817). In France, Jean-Baptiste Say attributed poverty to a military and aristocratic past. In his numerous texts on the subject, he depicted an ideal industrial society in which 'all that is purely mechanical action would be performed by machines. Such a nation would have all products, would enjoy all benefits'('*tout ce qui est action purement machinale serait exécutée par des machines. Une pareille nation aurait tous les produits, jouirait de toutes les utilités*') (Say, 1828, vol. I: 395).

While machinery was no longer a matter for debate among british economists, the controversy still existed among many social reformers, notably certain Tories and 'ultra-radicals' who feared the dissolution of former social hierarchies. Louis de Bonald and Louis Veuillot thus denounced the 'ill-considered development of machinery'('*développement inconsidéré des mécaniques*') (Lagrée, 1999: 34–45). In France, under the Restoration, the count de La Chabrière asked in a petition to the Parliament in 1817 'to prevent the multiplication of those machines, which one might call "infernal" if they were to be named after the effect they produce' ('*d'empêcher la multiplication de ces machines, qu'on pourrait appeler infernales si elles devaient être désignées par l'effet qu'elles produisent*') (Jarrige, 2014: 103). At the same time, the viscount De Saint-Chamans asserted that 'the machine is very harmful' ('*la machine est très nuisible*') and that it would be better to 'discourage their inventors and [...] restrain their use' ('*décourager leurs inventeurs et [...] en restreindre l'usage*') (Saint-Chamans, 1820: 263). Besides, the controversy was far from being settled at a time when the social question became more and more pressingand when the labour movement was awakening. In this context, the term *machinisme* ('machinery') appeared from the historian Jules Michelet's pen in *Le Peuple* (1846), to describe the 'servitude of the worker dependent on machines' ('*servitude de l'ouvrier dépendant des machines*') (Michelet, 1974: 98). The trope of the monster machine, the insatiable Moloch that devours humans, tended to spread among writers such as Honoré de Balzac, Alfred de Vigny, or Victor Hugo. In his poem 'Melancholia', which he finished in 1854, Hugo thus depicted the new tragic imagination of the industrial era associated with child labour: 'Stooped beneath the teeth of a sombre machine, | 'A hideous monster that chews who-knows-what in the shadows' ('*Accroupis sous les dents d'une machine sombre | Monstre hideux qui mâche on ne sait quoi dans l'ombre*') (Laforgue, 2003). The word *machinism* was immediately laden with a moral signification. It was a condemnation of the English-inspired industrial system and of the new social links based on the supremacy of the machine.

Resolution of Conflicts in the Era of Machinery (1860 to 1914)

From the 1860s onwards, industrialization settled in along with its cult of progress, its new technical systems, its education about machines, and its unprecedented confidence in science. Grievances were played down thanks to the establishment of a new mythology of innovation and an unprecedented belief in progress, or rather, a type of progress increasingly reduced to that of techniques (Carnino, 2015; Godin, 2015). Numerous conflicts and controversies over techniques remained, but they became less visible, as the culture of technical progress and its benefits became established, taught via multiple channels, including in schools, newspapers, and at European universal exhibitions. In 1867, the journalist and popularizer Louis Figuier published his *Merveilles de la science ou Description populaire des inventions modernes*, while Jules Verne became the symbol of a new literature of technological fantasy in which machines were primarily used for travel, communication, and daily comfort. They were the expression of human ingenuity, and savants and engineers asserted themselves as the heroes of modern times. As expressed by Michel Chevalier, a former Saint-Simonian who had become an economist close to the government, in this context, the 'war' against machines was futile, since 'but for the machines they hate, workers would not be free men, they would be slaves' ('*sans les machines qu'ils détestent, les ouvriers ne seraient pas des hommes libres, ils seraient des esclaves*') (Chevalier, 1848: 119). For the republicans and the radicals, all of whom were driven by an optimistic faith in a future built by technical progress, it was pointless to oppose machinery in this context. As the communist Étienne Cabet put it: 'it is not against the machines that one must struggle, it is the system of individualism and egoism that must be attacked, so as to shift to that of association, fraternity, and solidarity' (*'ce n'est pas contre les machines qu'il faut tourner ses efforts, c'est le système de l'individualisme et de l'égoïsme qu'il faut attaquer, pour y substituer celui de l'association, de la Fraternité et de la solidarité'*) (Cabet, 1844: 34). A few years later, in 1867, Karl Marx took up the same kind of analysis in *Das Kapital*, interpreting the progress of productive forces as the driver of history and teaching workers "to distinguish between machinery and its employment by capital, and to direct their attacks, not against the material instruments of production, but against the mode in which they are used" (Marx, 1906: 468). Social reformers and socialists looking for an emancipated future attempted to control technical change by having it serve as many people as possible. In this way, they tended to minimize conflicts and controversies over techniques and make them invisible by justifying the expansion of productive forces (Jarrige, 2016).

Notwithstanding, even if confidence in technical progress was growing, many conflicts still accompanied the adoption of new technical equipment on a daily basis, most specifically at the end of the nineteenth century, which witnessed the great depression and increased unemployment, though situations varied considerably from one area to the next. In the United States, where there was an insufficient number of workers, a massive amount of industrial equipment was adopted more rapidly and more pacifically than in Europe, where working communities were already established. In 1909, an eight-month-long strike thus took place in Hazebrouck, in northern France, against the introduction of large automated weaving looms imported from the United States (Benoit, 2009). During the same year a riot broke out in Concarneau, Brittany, against the machines that were to supplement the welders.

Outside Europe, tensions and conflicts existed too, for instance in the Ottoman Empire or in China, once the big Western productive machines arrived there. In Asia, nobody has illustrated the ambivalence about technique and the doubts it was raising better than Gandhi (1869–1948), who used a mere spindle to weave his own yarn for clothes and vehemently opposed industrial machinery of English origin, while also travelling by train and using a watch. According to Gandhi, the 'civilization of the machine' and modern industry had created an almost invisible, daily type of slavery which, in spite of the myth of abundance, impoverished whole sections of the Indian population. Political autonomy was therefore futile if it was not accompanied by a thorough questioning of modern industrial civilization and its commodities. Gandhi thus defended a local, self-sufficient craft industry within the framework of village autonomy and limited needs. Contrary to Indian neo-traditionalist currents defending the old Hindu civilization as intrinsically superior, and from nationalist modernizers seeking to mimic the West and turning its weapons against the colonial order, Gandhi imagined a third way based on a technical system adjusted to the needs of societies that was simple and easy for local populations to appropriate. For instance, he saw in the Singer sewing machine one of the rare artefacts created in the North that were truly useful to the peasant populations of the South (Gupta, 1994: 131).

The conflicts and controversies over the new industrial technical system were far from being absurd attempts to 'stop progress'. Instead, they were aimed at negotiating a change, which very often remained uncertain, by seeking out the best technical systems available. In practice, mechanization took place with great caution, and large machines presented as modern often did not work as their promoters had expected. They were expensive to maintain, required costly fuels, and the advantages they were supposed to provide were often not clear at all. Innovations presented by their promoters as advantageous were often criticized and deemed not very effective even by engineers, scientists, and manufacturers. A case in point are steam machines, although they were supposed to embody progress. Nineteenth-century societies were critics in the face of technological choices. They perceived the social, sanitary, and environmental threats that these choices introduced. Only the development of multiple social, legal, and cultural arrangements for disinhibition made it possible to render the technological changes accompanying the rise of industry acceptable, or even desirable (Fressoz, 2012). The engineer Edmond Vianne thus warned against the use of steam machines in farming, for it is 'a brutal force, always about to blow up' ('*une force brutale toujours prête à éclater*'), while in 1886 the agronomist André Sanson did not hesitate to assert that 'under certain specified circumstances [...] the steam machine is far behind the animal machine as an instrument for releasing energy' ('*dans certaines conditions déterminées [...] la machine à vapeur est fort en arrière de la machine animale, comme appareil à dégager de l'énergie*') (Vianne, 1866: 189; Sanson, 1886: 787). In England and in the British Empire, Andreas Malm, has shown that the introduction of steam-powered technologies was not the most rational choice — water-powered technologies were still available to produce energy — but a political choice in favour of power adjusted to the new ethos of the English capitalists. Water power involved both a geographical dispersion and a collective coordination to which manufacturers did not want to submit. Then again, the steam machine, although more costly and raising many doubts, offered a flexible, adaptable, and individualist source of energy which suited the ideology of English textile capitalism after 1830 (Malm, 2016).

More than simple opposition to the onward march of linear, abstract progress, conflicts and controversies over techniques can be interpreted as attempts to negotiate transformations, while at the same time preserving lifestyles and long-established qualifications and defending simple, flexible technologies. In the middle of the nineteenth century, artisans and craftsmen were still powerful, as they were capable of controlling the labour market and of imposing their views. Their opposition may have shaped investment choices at the local level, and it was not rare to come across manufacturers who did not make investments for fear of the workforce's reactions, or who chose to go and live in the country to avoid complaints and protests. In this context, the conflicts can be understood not as backward or reactionary actions, but rather as manifestations of the negotiated character of technical change, and the persistence of old tools and techniques shaped by a long process of adjustment to social and ecological constraints.

Muted Conflicts in an Age of Extremes?

The experience of the two world wars put an end to these debates for a while, imposing a new modernizing framework, both industrial and technological. Both world wars opened up a new ambiguous trajectory: on the one hand they were powerful drivers of innovation and technical modernization, on the other hand, the scale of destruction and wholesale massacres were more troubling than ever and led to the questioning of technical choices. In 1919, authors such as Paul Valéry thus observed that science had been 'dishonoured by the cruelty of its applications' ('déshonorée par la cruauté de ses applications') (Valéry, 2000: 405). During the interwar years, confidence and faith in technical progress coexisted with the most intense fears and warnings. For instance, during the grave crisis of the 1930s, George Orwell admitted a 'distaste for "progress" and machine-civilization' (Orwell, 2021, 150). At the same time, the French doctor Georges Duhamel wrote a great number of articles arguing against mechanization and its false promises (Duhamel, 1933). At the time, The United States embodied a disturbing dominant technological power. At the time, Gandhi's criticism of mechanization intrigued interwar Europe. This critique was part of his economic programme for the 'de-mechanization of the textile industry', which seemed to be utopic and unworkable. Faced with incomprehension, his critique gradually became more subtle. In October 1924, a journalist asked him, '[a]re you against all machines?' He answered, '[h]ow could I be? [...] [I am] against the thoughtless craze for machines, not against machines as such' (Gupta, 1994: 131). 'My object is not to destroy the machine but to impose limitations to it", that is, to control its usages so that it will affect neither natural environments nor the situation of the poorest (Gupta, 1994: 131–32). In fact, he developed a philosophy of the limits to and control of very-large-scale technology against the risk of loss of control, a theme that continually accompanied political theories in the contemporary period (Winner, 1978).

After 1945, the horrors of the war may have been interpreted as the result of a '*barbarie polytechnique*' ('polytechnical barbarity'), as the French writer Georges Bernanos put it in 1947 (Bernanos, 1999). The explosion of the two nuclear bombs in Japan drove Albert Camus, in his famous editorial in *Combat* on 8 August 1945, to condemn these events, which inspired in him with a feeling of horror and in which he saw the proof that 'the machine age has just reached its ultimate degree of savagery. It will be necessary, in the fairly near future, to choose between collective suicide or the intelligent use of scientific achievements' ('*La civilisation mécanique vient de parvenir à son dernier degré de sauvagerie. Il va falloir choisir, dans un avenir plus ou moins proche, entre le suicide collectif et l'utilisation intelligente des conquêtes scientifiques*') (Camus, 1950). Among intellectuals, the experience of war strengthened the tragic feeling that accompanied the development of the 'machinery civilization'. But the world had to be reconstructed, and confidence prevailed over doubts. The denunciation of 'anti-technicism' became commonplace while Europe made a commitment to fast-paced modernization, including the peasants who were slow to avail themselves of motorized equipment, or intellectuals rejected as ignorant of the technical world

(Mounier, 1949). The spectre of 'anti-mechanization' and of 'technophobia' — the latter term appeared at the time — justified campaigns for popular education and major programmes for having new techniques accepted. In 1954, the European Productivity Agency, concerned with the fact that 'the resistance to change can only delay and, in some cases, block progress and slow down economic expansion' ('*la résistance au changement ne peut que retarder et, dans certains cas, empêcher le progrès et ralentir l'expansion économique*'), ordered an investigation into workers' attitudes towards techniques in several European countries (Dofny, 1966). During the phase of intense modernization known as the Trente Glorieuses (the 'glorious thirty' years), as the motorization and multiplication of technical equipment accelerated, many intellectuals turned to the study of techniques. The philosopher Gilbert Simondon thus detected in his contemporaries a 'misoneism directed against machines' ('*misonéisme orienté contre les machines*') that was less related to fear of the new than to the refusal of the unknown and the 'other'. Opposition to new techniques would be a kind of anti-technical '*xénophobie*' (xenophobia) allegedly resulting from ignorance and the '*déshumanisation*' ('dehumanization') of objects, that is, from the concealment of the human reality they always contain (Simondon, 1958: 9–10).

In this context, conflicts and controversies over technological development waned and became less visible. They were taken over by peripheral figures in the intellectual field. It was at that time that, in Europe, Jacques Ellul (1912–1994) held that technique was 'the issue of the century' ('*l'enjeu du siècle*') and began to construct his critical analyses of the technological society, or what he would call the '*bluff technologique*' ('technological bluff') (Ellul, 1988). In the United States, the historian and urban planner Lewis Mumford attacked the 'myths of the machine' and the authoritarian technologies of the industrial era. In *Die Antiquiertheit des Menschen*, an all-encompassing critical picture of the industrial and nuclear society published in 1956, and thus at a time when technological consumerism was growing apace, the German philosopher Günther Anders developed an ambitious criticism of technology and its gigantism. For instance, he observed that

> 'today nothing discredits a man more rapidly than being suspected of criticizing the machines [...]. It is no surprise that the fear of this inevitable disgrace drives most critics to tone down their discourses and that the publication of a criticism of techniques has now become a matter of civic courage' ('*rien ne discrédite aujourd'hui plus promptement un homme que d'être soupçonné de critiquer les machines [...] Il n'est pas étonnant que la peur de cette inévitable disgrâce pousse la plupart des critiques à mettre une sourdine à leurs propos, et que la publication d'une critique de la technique soit devenue aujourd'hui une affaire de courage civique*') (Anders, 2002: 17).

This text, published during the Cold War and therefore in a period of intense technical transformations, had little impact at the time. The general public was discovering mass consumption and its string of innovations of all sorts, optimism prevailed after the war-induced privations. At the time, challenging new techniques and their promises, against the tide of history and progress, seemed unthinkable or was even the manifestation of visceral conservatism.

Conflicts over Techniques in the Ecological Age

Fig. 2. Front cover, *La Gueule ouverte*, 28, 20th November 1974. The alternative and ecological press of the 1970s challenged nuclear power.

This modernizing consensus has been breaking down since the end of the 1960s, inaugurating a renewal of conflicts and controversies over technology and resulting in an unprecedented politicization of technological issues and in much research on the relationships between the sciences, techniques, and societies (Quet, 2013; Bauer, 1995) (Figure 2). In the intellectual circles, technical choices have triggered many debates that are '*technocritiques*' (technocriticisms), a term the philosopher and engineer Jean-Pierre Dupuy used for the title of the collection he launched at the Éditions du Seuil in 1975. In most Western countries heated debates emerged over the issue of technosciences, their organization, and dangers. Even the unions sometimes challenged the meaning of technical progress, such as the Confédération française démocratique du travail (CFDT), a French Trade-Union which published *Les Dégats du progrès* in 1975, following a colloquium in which the impact of technical evolutions on workers had been discussed (CFDT, 1975). The rise of political ecology, too, contributed to re-politicizing the debates over techniques by questioning the industrialists' and authorities' choices, as well as their impact on the environment. Ten years later, soon after the Chernobyl catastrophe, Ulrich Beck formulated the theory that the rise of controversies was the manifestation of man's entry into a 'risk society'('*société du risque*'), characterized by an unprecedented reflexivity with regard to the effects of technical change (Beck, 1992).The 1970s were effectively marked by the multiplication of warnings about the dangers that artefacts and the modern technologies of the industrial era posed to human beings and their environment — the black tide in Brittany caused by the Torrey Canyon shipwreck of 1967 and the pollution of the Rhine River caused by the release of pesticides in 1969 may serve as examples here, creating news headlines that aroused people's indignation. In the Soviet republics as in the West, mobilizations grew and gave rise to vast campaigns against smog in the cities, pesticides, hydrocarbon or lead release, and against asbestos, which triggered a public health scandal. Initiated by scientists and writers, the first significant mobilization of public opinion in the USSR occurred at the very beginning of the 1960s and was directed against the risk of pollution to Lake Baikal posed by the massive industrial complex that had been built nearby. Movements of critical scientists also appeared alongside social mobilizations: the group Survivre et vivre was founded by three renowned mathematicians in Montréal in 1970 (Pessis, 2013), much influenced by the hippie movement, pacifism, and ecology. From August 1970 to June 1975, the group published the journal *Survivre... et vivre* whose goal was, as explained by the mathematician Alexandre Grothendieck, 'the fight for the survival of the human species, and even of life as a whole, as it is threatened by the growing ecological dysbalance caused by the indiscriminate use of science and technology' (*'la lutte pour la survie de l'espèce humaine, et même de la vie tout court, menacée par le déséquilibre écologique croissant causé par une utilisation indiscriminée de la science et de la technologie'* (Pessis, 2013: 21).

While in the West, the Club of Rome's report led to the criticism of the meaning of industrial techniques and the limits to growth, in Asia, Indian subalternist intellectuals and ecologists like Ashis Nandy developed 'a critical traditionalism' against the violence of European technical creations (Nandy, 1990). Young authors like Claude Alvarez challenged Western technologies and accused them of imposing ideologies of domination and of destroying local technical systems (Alvarez, 1980). The physicist Vandana Shiva set up a foundation for 'science, technology and ecology' in 1982 that led to the creation of the Navdanya movement in 1991, with a view to defending local crop varieties and farming techniques against the Green Revolution and, later on, biotechnologies. Vandana Shiva formulated a radical ecologist and feminist critique of the surge of techno-sciences and the violence of the Green revolution of the 1970s (Shiva, 1991).

The scale of technical infrastructures and the issue of pollution were at the heart of the debates over technical choices. Authors such as Ernst Schumacher and Ivan Illich argued for appropriate, convivial, clean technologies, rejecting industrial technologies as authoritarian and polluting. Again, the controversy concerned not so much technique in a general and abstract way but rather technical choices and their political and social basis. Then there were countless publications and experimentations: faced with the onslaught of consumer goods, many people turned to hand-made, Low energy tools, and some even thought that these 'soft technologies' were going to leave their niches and conquer the world. Democratic, appropriable, and soft techniques were set against massive technical systems. In 1975, in the United States, Edward Abbey published *The Monkey Wrench Gang*, a famous satire which takes the form of a road movie in the Wild West, and in which a gang of environmentalists wages a war against 'the machine' and sabotages equipment such as bridges, roads, and railways that have disturbed the desert (Abbey, 1975).

At the same time, new modes of managing controversies appeared: major economic actors and multinational corporations deployed a wide variety of tools in order to settle conflicts, impose to citizens their solutions and equipment, and avoid regulations interpreted as a brake on progress (Oreskes, Conway, 2010). Faced with the growing opposition to automobiles, car manufacturers continued to create new ways of defending themselves, promising, for instance, the advent of the clean, electric car and innovations that were supposed to solve the issue of pollution. Controversies and conflicts were continually settled and resolved by the multiplication of technological promises announcing a new technological world, a digital, dematerialised, clean, and more democratic that would solve the problems induced by former industrial technologies. During the 1980s and 1990s, a new technological progressivism thus appeared in parallel with the globalization of exchanges and the computerization of societies, which were both initially viewed as answers to the problems inherited from heavy, polluting industrial technologies. In the early twentieth century, however, more and more neighbours and communities grew to be opposed to massive technical projects which posed a threat to their way of life and natural environments. Today, while the study of techniques has become more complex, political speeches and programmes are still full of hope in technology. Innovation remains a magical word on which many rely to resolve social issues, the shortage in resources, or the climate crisis. All dissenting opinion and all sceptical points of view are rejected as unsuccessfully, or even dangerous.

Resistance to technological development, however, has become more dynamic than ever in the early twenty-first century. The rise of GMOs has been accompanied by massive protests and sabotage in Europe as well as in Southeast Asia (Bonneuil, 2010). Critical analyses of the damage, dead ends, and threats caused by new technologies have multiplied. Warnings are given by experts or citizens who investigate and endeavour to reveal the social and political issues of technologies. The French group Pièces et main d'œuvre, for example, has been fighting against the technoscience for about fifteen years by continually denouncing, for instance, 'the techno-totalitarianism and the "artificialization" of the world' (*'le techno-totalitarisme et l'artificialisation du monde'*) (PMO, 2008). In a world marked by the constant proliferation of technological promises, critical analyses are numerous, even though they have some difficulty making themselves heard. With regard to the new digital and robotic world, for instance, an increasing number of voices have been raised to contest advertising promises, among them that of the German psychiatrist Manfred Spitzer, who warns against 'digital dementia' in his best-selling book; further, the fear of a future without employment has become a genuine topic in its own right as such (Ford, 2017; Spitzer, 2012).

The early twenty-first century has been witnessing a proliferation of technoscientific controversies interpreted either as the manifestation of emerging societies that are more reflective and risk-conscious or as the manifestation of overcautious, ageing societies, allegedly constrasting with the 'old Europe' to the dynamic nations of the South. History shows that, far from being new or recent, these conflicts and controversies have continually shaped contemporary technologies in parallel with the evolutions of visions and collective-action répertoire. Far from expressing 'technophobe' or reactionary tendencies, as it is often depicted by observers, the study of these practices and discourses instead shows that they pertain to forms of socio-technical negotiations between groups with conflicting interests and visions. Thanks to such conflicts and controversies, debates over current trajectories still exist, forms of the social re-appropriation of techniques are still possible, and dynamics of the negotiated construction of future tools and infrastructures will be possible

(Hess, 2007). In the face of the rapid deployment of new technical trajectories and the multiplication of promises which are now blossoming more than ever, at a time of global socio-ecological crises (Huesemann, 2011), these conflicts remain indispensable defences and tools for domesticating changes that are not necessarily beneficial.

References

Abbey, Edward, *The Monkey Wrench Gang* (Philadelphia: Lippincott, 1975).

Alvarez, Claude, *Homo Faber: Technology and Culture in India, China and the West from 1500 to the Present Day* (The Hague: Nijhoff, 1980).

Anders, Günther, *L'Obsolescence de l'homme. Sur l'âme à l'époque de la deuxième révolution industrielle* (Paris: Encyclopédie des nuisances, 2002).

Bauer, Martin (ed.), *Resistance to New Nechnology. Nuclear Power, Information Technology and Biotechnology* (Cambridge: Cambridge University Press, 1995).

Beck, Ulrich, *Risk Society: Towards a New Modernity* (London, Sage, 1992).

Benoit, Serge, 'L'Introduction des métiers Northrop', in *Tisser l'histoire. L'industrie et ses patrons, XVIe–XXe siècles*, ed. by René Favier, Gérard Gayot, Jean-François Klein, Didier Terrier, and Denis Woronoff (Valenciennes: Presses universitaires de Valenciennes, 2009), pp. 85–96.

Berg, Maxine, *The Machinery Question and the Making of Political Economy* (Cambridge: Cambridge University Press, 1980).

Bernanos, Georges, *La France contre les robots* (Paris: Librairie générale française, 1999).

Bonneuil, Christophe, 'Saboter des champs transgéniques pour étendre le champ de la démocratie. Une histoire de la contestation radicale des *OGM* en France', in *Les Luddites en France. Résistance à l'industrialisation et à l'informatisation*, ed. by Cédric Biagini and Guillaume Carnino (Montreuil: L'Échappée, 2010), pp. 211–49.

Cabet, Étienne, *L'ouvrier, ses misères actuelles, leur cause et leur remède, son futur bonheur dans la communauté, moyens de l'établir* (Paris: Le Populaire, 1844).

Camus, Albert, 'Éditorial de *Combat*, 8 août 1945', in Albert Camus, *Actuelles 1: Écrits politiques* (Paris: Gallimard, 1950).

Carnino, Guillaume, *L'Invention de la science. La nouvelle religion de l'âge industriel* (Paris: Seuil, 2015).

Chevalier, Michel, *Lettres sur l'organisation du travail ou études sur les principales causes de la misère et sur les moyens proposés pour y remédier* (Paris: Capelle, 1848).

Confédération française démocratique du travail (CFDT), *Les dégâts du progrès. Les travailleurs face au progrès technique* (Paris: Seuil, 1977).

Dofny, Jacques, Claude Durand, Jean-Daniel Reynaud, and Alain Touraine, *Les Ouvriers et le progrès technique* (Paris: Armand Colin, 1966).

Duhamel, Georges, 'Sur la querelle du machinisme', *Revue de Paris*, 2.15 (1933), 721–52.

Edgerton, David E., *The Shock of the Old: Technology and Global History since 1900* (Oxford: Oxford University Press, 2007).

Ellul, Jacques, *La Technique ou l'Enjeu du siècle* (Paris: Économica, 1990 [1954]).

———, *Le Bluff technologique* (Paris: Hachette, 1988).

Ford, Martin, *L'Avènement des machines. Robots et intelligence artificielle: la menace d'un avenir sans emploi* (Paris: FYP éditions, 2017).

Fougier, Étienne, *Contester les technosciences: leurs raisons,* (Paris:Fondation pour l'innovation politique, 2011)].

Fressoz, Jean-Baptiste, *L'Apocalypse joyeuse. Une histoire du risque technologique* (Paris: Seuil, 2012).

Garnier, Joseph, 'Machines', in *Dictionnaire de l'économie politique*, vol. II, ed. by Charles Coquelin and Henri Guillaumin (Paris: Guillaumin, 1853), pp. 116–23.

Godin, Benoît, *Innovation Contested. The Idea of Innovation Over the Centuries* (London: Routledge, 2015).

Gupta, Shanti Swarup, *The Economic Philosophy of Mahatma Gandhi* (New Delhi: Concept, 1994).

Hess, David J., *Alternative Pathways in Science and Industry. Activism, Innovation and the Environment in an Era of Globalization* (London and Cambridge, MA: The MIT Press, 2007).

Huesemann, Michael Joyce, *Techno-Fix. Why Technology Won't Save Us or the Environment* (Gabriola Island: New Society Publishers, 2011).

Jarrige, François, *Au temps des tueuses de bras. Les bris de machines à l'aube de l'ère industrielle* (Rennes: Presses universitaires de Rennes, 2009).

———, ed., *Dompter Prométhée. Technologies et socialismes à l'âge romantique* (Besançon: Presses universitaires de Franche-Comté, 2016).

———, *Technocritiques. Du refus des machines à la contestation des technosciences* (Paris: La Découverte, 2014).

Laforgue, Pierre, « Machinisme et industrialisme, ou romantisme, modernité et mélancolie. Quelques jalons (1840–1870) », *Revue d'histoire littéraire de la France*, 1/2003 (Vol. 103), 63–92.

Lagrée, Michel, *La Bénédiction de Prométhée. Religion et technologie (xixe-xxe siècles)* (Paris: Fayard, 1999)

Malm, Andreas, *Fossil Capital: the Rise of Steam Power and the Roots of Global Warming* (London: Verso, 2016).

Marx, Karl, *Capital: A Critique of Political Economy* (New York: Modern Library, 1906).

Michelet, Jules, *Le Peuple* (Paris: Flammarion, 1974).

Mounier, Emmanuel, 'La Machine en accusation. Aux origines psycho-sociales de l'anti-technicisme', in *Industrialisation et technocratie*, ed. by Georges Gurvitch (Paris: Armand Colin, 1949), pp. 1–22.

Nandy, Ashis, *Science, Hegemony and Violence: A Requiem for Modernity* (Delhi-Oxford: Oxford University Press, 1990).

Oreskes, Naomi and Erik M. Conway, *Merchants of Doubt: How a Handful of Scientists Obscured the Truth on Issues from Tobacco Smoke to Global Warming* (London: Bloomsbury Press, 2010).

Orwell, George, *The Road to Wigan Pier* (Oxford: Oxford University Press, 2021).

Pessis, Céline, ed., *Survivre et vivre. Critique de la science, naissance de l'écologie* (Montreuil: L'Échappée, 2013).

PMO, *Terreur & Possession — Enquête sur la police des populations à l'ère technologique* (Paris, L'Échappée, 2008).

Quet, Mathieu, *Politiques du savoir. Sciences, technologies et participation en France dans les années 68* (Paris: Éditions des Archives contemporaines, 2013).

Ricardo, David, *On the Principles of Political Economy and Taxation* (London: John Murray, 1817).

Roy, Christian, 'Charbonneau et Ellul, dissidents du "progrès". Critiquer la technique face à un milieu chrétien gagné à la modernité', in *Une autre histoire des Trente Glorieuses, Modernisation, contestations et pollutions dans la France d'après-guerre,* ed. by Céline Pessis, Sezin Topçu, Christophe Bonneuil (Paris: La Découverte, 2013), pp. 283–302.

Saint-Chamans, Vicomte de, *Du système d'impôt fondé sur les principes de l'économie politique* (Paris: Le Normant, 1820).

Sanson, André, 'Le Moteur animé et le moteur à vapeur', *Revue scientifique,* 25 (1886), 787–89.

Say, Jean-Baptiste, *Cours complet d'économie politique pratique,* 6 vols (Paris: Rapilly, 1828).

Shiva, Vandana, *The Violence of the Green Revolution: Third World Agriculture, Ecology and Politics* (London: Zed Books, 1991).

Simondon, Gilbert, *Du mode d'existence des objets techniques* (Paris: Aubier, 1958).

Spitzer, Manfred, *Digitale Demenz: Wie wir uns und unsere Kinder um den Verstand bringen* (Droemer eBook, 2012).

Valéry, Paul, 'La crise de l'esprit (1919)', in *De l'Antiquité au XXe siècle : Anthologie critique et commentée,* ed. by Yves Hersant et Fabienne Durand-Bogaert (Paris: Robert Laffont, coll. Bouquins, 2000), pp. 405–14.

Vianne, Edmond, *La Culture économique par l'emploi raisonné des instruments, machines, outils, appareils et ustensiles usités dans la petite et la grande culture* (Paris: J. Rotschild,1866).

Winner, Langdon, *Autonomous Technology. Technics-out-of-control as a Theme in Political Thought* (Cambridge, MA: MIT Press, 1978).

CECILIA CALHEIROS

Body Techniques and the Search for Human Perfection

From the Regeneration of the Social Body to the Entrepreunarial Subject

Techniques to shape the body, make it beautiful, or enhance it are intimately linked to the increasing medicalization of contemporary western societies and the associated self-care involved in the construction of individuals. The social, political, and technical scope of the desire to change the human body can be explained in part by a new relationship with science that emerged during the eighteenth century. Man no longer marvels at the ingenuity of God's creation but uses human energy to transform nature (Le Breton, 2013). The intermingling of knowledge, and particularly of medical knowledge, and political power has contributed to the emergence of new ways of imagining bodies and new techniques for standardizing them. The boundaries between the normal and the pathological, between the acceptable and the unacceptable, between remedial and improvement started being redrawn. Initially this was essentially a question of correcting deformities (Vigarello, 1978; Stiker, 2013), but the techniques that then developed aimed to take charge of the body in order to free it from its biological constraints and to guide individuals towards the realization of their true selves (Bujon, Dourlens, Le Naour, 2014). As an expression of a societal trend, the use of techniques to improve the individual is as much a part of the growing standardization of appearances as it is of the redefinition of the body's uses. These techniques are, first of all, techniques of the body, as they train the body into becoming both an instrument for and the target of strategies aimed, in this instance, at improved performance (Mauss, 1936). However, this aim, when defined as an 'improvement', leads to a twofold movement. In their pursuit of changing the appearance of the body primarily through training, body modification techniques seem to be increasingly embodied and to rely on initiation and cooperation from the subject. Therefore, if for Mauss the body is man's first and most natural instrument, modern techniques of the governement of bodies that are simultaneously employed on, in, and with the body, also open up questions about the status of the individual and their body. They also question the boundaries of the body with its environment. This invites an analysis of the development of body modification techniques within the economic, political, and social contexts in which they have developed. More precisely,

Global History of Techniques, ed. by Guillaume Carnino, Liliane Hilaire-Pérez and Jérôme Lamy, Global Matters Culture Matérielles, histoire des techniques et circulation globale des savoirs, 9 (Turnhout: Brepols, 2024), pp. 643–668
BREPOLS PUBLISHERS 10.1484/M.TECHNE-EB.5.129834

the aim is to understand how the improvability of human beings is embedded in a variety of forms of biopolitical governance that have 'brought life and its mechanism into the realm of explicit calculation and has made the power of knowledge an agent of transformation of human life' (Foucault, 1979, 143). This applies first to the social body, and more particularly to subordinate bodies, and then it extends to individual bodies through growing levels of self-control a global phenomenon which is increasingly internalized. The question of how human life is governed so as to improve human bodies through the control, management, engineering, reshaping and modulation of the very vital capacities of human beings as living creatures (Rose, 2007, 3) is therefore at the heart of this chapter.

Historiography

In the second half of the twentieth century, the growing attraction for techniques of the body gave historians the feeling that they were exploring new territory. However, as Georges Vigarello (2004) and Michelle Perrot (1979) point out, this understanding of historiography is misleading. While cultural historians of modern societies have shown a strong interest in the body, this interest was also present before that time. Yet the approaches followed did not satisfy historians (Corbin, 2008). The investigation of the links between bodies and techniques, which had been initiated in part by Marc Bloch and Lucien Febvre (1935) between the two world wars, developed in the 1960s into a 'French school' which established the physical dimension of the individual and gave particular attention to techniques of the body (Memmi, Guillo, Martin, 2009). Yannick Ripa (2007) points out that while the body is a plural object that is part of social, cultural, political, economic, and military history, it was the history of medicine (Léonard 1978, 1986) and the history of women (Knibiehler, Fouquet, 1980; Badinter, 1980; Gélis, 1984; Shorter, 1984) which initially turned the body into an object of study in its own right. Rafael Mandressi underlines the dispersed nature of studies on the body, leading to what he calls a 'scattered body' (Mandressi, 2009). Further, the social significance of facial expression has been studied (Courtine, Haroche, 1988), as have the subjects of the rhetorics of tears (Vincent-Buffault, 1986), of skin in identity formation (Boëtsch et al., 2008; Ory, 2008), of hygiene practices in the context of rising notions of privacy (Vigarello, 1985; Csergo, Guerrand 2008), of the evolution of criteria for measuring beauty (Vigarello, 2004), of the sexualized body (Corbin, 1978; Sohn, 2011), the female body (Perrot, 1984), and the dead body (Ariès, 1977; Pouchelle, 1983; Vovelle, 1983; Carol, 2004). Whether it be an interest in body related practices or in representations of the body, this newly inventoried body (Feher, Nadaff, Tazi, 1989) manifests the existence of multiple, divided fields, a feature of the complexity inherent to the emergence of a new research area. Historians did not therefore discover the body as much as they uncovered its close connection to the construction of identity and explored the links between human existence and heightened self-awareness.

In sociology and anthropology, the body has long been regarded with mistrust, an epistemological mistrust inherited from the mind-body dichotomy that marks it as an illegitimate object of study (Memmi, Guillo, Martin, 2009: 12). Therefore, Marcel Mauss's famous 1934 conference on the 'techniques of the body' — in which he recommended analysing bodies as shaped by sociological, psychological, and biological factors — long remained unheeded. In addition to Mauss's post-war academic and social isolation, the cautious reception of this proposal more broadly reflected the perception of biological aspects in sociology and of material culture in anthropology (Memmi, Guillo, Martin, 2009: 23). Claude Levi-Strauss, following Mauss, made body techniques the starting point for all technical behaviour (Lévi-Strauss in Mauss, 1950), but it was not until the 1970s that they became an object of study in anthropology in their own right. The initial emphasis was placed on a cultural approach to techniques that highlighted the fundamental role of the body and then considered human action as intrinsically linked to the production of tools separate from the body (Leroi-Gourhan 1964, 1983; Haudricourt 1955, 1968). This approach challenges the distinction between nature and culture (Lévi-Strauss, 1987; Latour, 1991; Warnier, 1999; Descola, 2005) and invites us to think of human beings as unfinished (Leroi-Gourhan, 1964; Remotti, 2003; Diasio, 2012). Sociological work on body techniques increased considerably in the 1990s, on the one hand, because until the mid-twentieth century, the Durkheimian school — which then dominated French sociology — had distanced itself from biologizing approaches by explaining the social with the social. For the body to become an object of study in its own right, it had be dissociated from its biological dimension (Guillo, 2009).[1] On the other hand, until the 1970s, the body was essentially studied as part of the sociology of work, which presented it as material on which work inflicted pain and as a site of performance (Pillon, 2016).

The body has therefore gradually been constituted as an epistemic object illuminating social phenomena studied by a myriad of sub-disciplins (health, sports, etc.) and through various theoretical frameworks (structuralism, symbolic interactionism, etc.). It is sometimes presented as an instrument for reading the social world, reflecting the workings of reproduction and domination (Bourdieu, Elias) through which power is exercised (Foucault, Federici), or as a carrier of social meaning (Douglas, Goffman). Sometimes the body is the product of natural selection and the vehicle for mental activities for the supporters of cognitivism (Sperber, Conein). What has changed, therefore, is as much the status of the body and the place that corporality occupies in contemporary societies as the way in which historians, anthropologists, and sociologists have made the body visible by viewing it as an epistemic object. Nevertheless, these studies, which essentially focus on European bodies (Ripa, 2007), show a double tension: that between the individual and the social body, and that of the hard-working, productive body versus a fantasized, behaviour-inducing

1 This is why studies dealing with the body from before this period were marginalized and were only rediscovered later, see for instance Veblen Thorstein, *The Theory of the Leisure Class* (New York: The MacMillan Company, 1899).

body. As we will see, the history of the body intersects with the question of social, sometimes spiritual control, as well as with the repression of transgression. As Ripa puts it 'this conquest of oneself by oneself intersects with attempts at controlling bodies, which are always sites of power' (2007: 893).

From Social Medicine to Eugenic Body Sculpting

In the eighteenth century, medicine distanced itself from hippocratic and galenic approaches and gradually adopted a clinical approach (Faure, 1994; Sournia, 2004). Based on the concrete observation of the patient, it also benefitted from developments in the natural sciences (biology, chemistry, and physics), the improvement of instruments (to measure temperature, weight, heart rate, etc.) and surgeons' lessons in anatomical dissection (Rabier, 2010), which provided an increasingly refined knowledge of the functions of the human body.

In the nineteenth century, the Pasteurian revolution and the advances in microbiology, the development of hygienic measures (antisepsis, the isolation of patients, vaccination campaigns, etc.) and, more broadly, the framework for the production of new knowledge (anatomy, physiology, natural history, etc.) led to the emergence of new concepts of life[2], health, and disease, which in turn led to a change in the demands placed on medicine (Faure, 1994; Salomon-Bayet, 1994). The increasing role of such scientific tools and practices therefore coincided with the increasing institutionalization of medicine (medical training, the establishment of medical schools to accommodate ever-increasing numbers of students, laws governing the organization of medical professions, etc.) (Léonard, 1978), and its subdivision into specialisms (midwifery in hospitals, the creation of nursing schools, etc.) (Faure, 1994; Lagrée, Lebrun, 1994; Porter, 1995, 2002). The increasing demand for medical care called for the creation of new social forms and care structures. Therefore, hospitals, clinics, and private practices developed as the medical profession, wishing to establish a monopoly on medical care, gradually placed itself into a dominant position by banishing popular practices and making medical knowledge inaccessible to the general population (Freidson 1984; Abbott, 1988; Lagrée, Lebrun, 1994; Bourdelais, Faure, 2005). The shift from extractive to synthetic chemistry marked the transition from individual pharmacies to the pharmaceutical industry (Sournia, 2004; Bourdelais, Faure, 2005), the care sector moving to an altogether different scale. Alongside this process of the medicalization of society the state started to view health as a way to preserve the nation (Faure, 1994; Dessertine, Faure, 1994). The extension of medical jurisdiction, authority, and practices to areas that were traditionally not medical (sexuality, education, etc.) (Zola, 1972) allowed for the increased social control of populations (Imbault-Huart, 1994; Clarke et al., 2000, 2003, 2010). How did medicine, through its close relations with political power, become a legitimate

2 For instance, with microbiology and evolution theory, after Darwin life becomes defined as a conceptual unit: the natural system (Franklin, 2015).

authority for regulating bodies? How did medicine participate in the development of body techniques within a biopolitical framework?

As Pasteurian theory spread, so did representations of the microbe, an invisible danger (Salomon-Bayet, 1986). Then a shift in aetiologies happened: physical dirt carried moral implications which threatened society as a whole (Lowy, 2015). Social reformers turned the hygienic concerns of the previous century into a moral discourse (Seignan, 2010), health into a public affair and scientific truths into legal imperatives.[3] Public authorities considered doctors as the instruments of the future transformation and management of populations (Faure, 1994). As Alain Corbin points out, 'it was as if doctors were led to translate into a scientific language the fantasies that haunted the society of their times; but in so doing, they provided these fantasies with the scientific backing that turned imagined representations into scientific certainty' (Corbin, 1981, 146). This explains why the radical transformation of working conditions that happened with industrialization, the increased density of the population, the impoverishment of the working classes, and the rise of protest movements caused fears of a popular uprising among the bourgeoisie (Chevalier, 2002). The notion took hold that modern civilization lowered morals and weakened the population. With the upheaval created by this shift in paradigm, the question of the social body emerged in the first half of the nineteenth century. The need to control and treat the social body through measures aimed at the biological body (Jorland, 2010) became more urgent. Within this framework the political ambition of social medicine was deployed through the functions it was assigned: control, regulation, and the maintenance of public health. Therefore, between 1825 and 1840, social medicine and public health linked medicine and pharmacy with chemistry, statistics, political economy, and civil engineering. Moreover, the rise of hygienics made it possible to generate industrial profits that were compatible with urban sanitation (Fressoz, 2012). In the name of science, which was all-pervasive, the question of industrial pollution came behind that of the morality of workers (Jarrige, Le Roux, 2017). The physician Louis René Villermé (1989) is one example of how the emerging disciplines concerned with containing the deterioration of the physical and moral health of the working classes triggered the development of hygienist measures (Bourdelais, 2001). The shift from medical aetiology to hygienist investigation made workers responsible for their depradation and poor health, and this justified the control of populations and the administration of their lives. Hygienist concerns led to bodies being measured, and in the context of rationality and materialism of the times, to be increasingly examined, imagined, and mapped.

New fields of knowledge developed, including demography, epidemiology, anthropology, and statistics (Porter, 1986; Hacking, 1990; Desrosières, 1993; Schweber, 2006). One early question asked by these new body sciences concerned the origins of different skin colours. Then, physical anthropology based on anthropometric

3 As explained by Jacques Poirier in Lagrée and Lebrun (1994), during the 3rd Republic many doctors like Brouardel play an important political role encouraging hygienist measures for urban sewage systems, improved food hygiene, implementing policies against industrial pollution or the spread of diseases.

measurements, and particularly osteometric measurements, the characteristics of each 'race' were compared (Stepan, 1982). The doctor and anatomist Franz-Joseph Gall (1758–1828) created phrenology, the art of analysing an individual's moral and intellectual dispositions based on the study of their skull (Gall, 1838). Under the aegis of Paul Broca (1824–1880), the French founder of the *Société d'Anthropologie de Paris* (the Paris Anthropological Society), instruments to refine cranial measurements were developed and standardised (anthropometric compasses, etc.) (Blanckaert, 2009). The expansion of anthropometric measurements was also supported by the development of psychological and psychiatric studies on intelligence (Carson, 2007). Physiognomy further developed with the doctor and founder of criminology Cesare Lombroso's theories (1835–1909). Positivist uses of phrenological methods were developed, leading to science serving as instrument of social order when scientists started to try and identify what Lombroso called 'born criminal' (Lombroso, 1885, 20). The anatomist and mathematician Adolphe Quételet (1796–1874) used statistics to form the basis of social physics, which sought to measure the moral qualities of various nations (Quételet, 1844). His theory of the 'average man' inspired Francis Galton (1822–1911), who was an important figure in eugenics at the end of the nineteenth century (Galton, 2001).

Already present in the antiquity, eugenics resurfaced in Europe during the first third of the nineteenth century (Hecketsweiler, 2010) and then in North America, Scandinavia, Russia and South America, as well as in some African and Asian countries under colonial domination. Incorporating Darwin's theory of evolution and informed by theories of heredity, it was also linked to political concerns over the construction of the nation-state (Müller-Wille, 2015). Modernity was considered responsible for the physical, social, and moral decline of people. Encouraged by the idea that progress is achieved with the help of technology and science, the determined quantification of bodies and their defects justified the practices of bodily correction gradually established by medicine. The idea was, then, to reform society by encouraging the growth of the 'right' type of people. This would lead to 'improving' the genetic pool of the nation by revealing its 'beauty' through practices aimed at 'perfecting' or 'preserving' the species (Vigarello, 2004, 131). Through their willingness to combat what was seen as the degeneration of peoples, physicians contributed to the construction of medicine as an institution of social control (Faure, 1994; Zola, 1972; Conrad, 1992). They played an important role in the spread of eugenics and the development of an economy of human performance (Müller-Wille, 2015), making the body a place where a new form of government was exercised. In the inter-war period, medicine's normative dimension importantly made even more room for the politicization of science and the naturalization of politics, which gave rise to practices of experimentation on bodies aimed at improving their performance and appearance (Müller-Wille, 2015).

Various techniques have been used to preserve and regenerate the social body. First of all, the desire to curb the perceived degeneration of individuals by measuring deviations from the norm was achieved by correcting deformities. Vigarello talks about the advent of new bodywork through orthopaedics and orthopraxies. This type

Fig. 1. École normale de gymnastique et d'escrime (Joinville-le-Pont, Val-de-Marne). Flexibility exercises [overall movement]. 1906. Rol Agency.

of correction uses heavy iron devices (like boots and steel corsets) and machines intended to make up for disabilities (Vigarello, 1978). Already in use in the eighteenth century, these techniques were perfected by targeting specific parts of the body. While medicine humanized the abnormal by making it possible to educate the disabled body, this shift was also already one way of normalizing the body through the correction and control of posture (Stiker, 2013). With the industrial revolution, the corrective discipline imposed on bodies matched the desire to make humans more adaptable and better adapted to the environment.

Since the eighteenth century, physical exercise and gymnastics have been associated with civic virtues. Medical educators such as Guillaume Daignant, Jean Verdier, and Francisco Amoros thought that exercise had the potential of building up citizens, of restoring their morals, and of ensuring the perfectibility of man (Arnal, 2013). In hygienics, sporting activities were considered a means of combating the softening of modern man by providing benefits for the brain and strengthening muscles (Figure 1). Thus praised for its moral virtues, sport spread throughout society. In France, first private exercise rooms were opened, and then physical education at school came to be compulsory in the course of the twentieth century (Orthous, 2013). In the United States, exercise became instrumental to encouraging fertility and supported the perceived need for regeneration of the white middle class (Courtine, 1993). On both sides of the Atlantic, the female body, a medical object *par excellence* (Jordanova,

1989; Schiebinger, 1989), was understood as a sexual body and a reproductive body at the same time. Directly involving the 'body of the nation' (Gardey, 2015, 363), women's bodies were at the centre of medical and political concerns (birth campaigns, etc.).

Reconstructive and cosmetic surgery has also been a tool of regeneration (Jarrin, 2017). At the end of the nineteenth century, Brazil was influenced by the eugenicist imagination brought there by Arthur de Gobineau, a French proponent of the ideology of the Aryan race. He promoted the idea that Brazil was condemned to mixed-race generation 'polluted in blood and spirit and very ugly' (Jarrin, 2017, 29). First photography was used to make visible the bodies that needed to be 'cleaned and medicalized' (Jarrin, 2017, 35), then surgical 'correction' was legitimized via a political and moralist discourse seeking to normalize mixed-race and deformed bodies, the black bodies of the countryside, and homosexual bodies (the vaginal labia majora were said for for example to lead to 'abnormal' sexual practices) (Rodriguez Balanta, 2012). In the twenty-first century, the aesthetic hierarchy in Brazil is still being marked by an overly high valuation of white beauty and the idea of racial improvement. From this perspective and in the context of vast social inequalities in the country, beautification practices are strategies for climbing up the social ladder (Jarrin, 2017).

Self-sculpture and Beautification Techniques

The control of bodies was not carried out solely with a view to restoring the morality of the populations by controlling subaltern bodies. In the eighteenth century with the rise of subjectivity, the beautification of bodies became a more pressing concern. Attachment to both general and individual beauty occurred together with the development of self-care, the attention given to beauty, and fresh importance given to individuals' inner feelings and to their perception of their own bodies (Vigarello, 2004, 2014). Not only was identity no longer confined to the classic idea of consciousness, but in the twentieth century the newly imagined body became a site for a new biopolitical enterprise and self fashioning (Vigarello, 2004).

The aesthetic importance given to the body has included both the whole body and some of its individual parts. Many practices have contributed to reinventing the body by reshaping it. The appearance of the face changed at the end of the eighteenth century. The massive spread of toothbrushes in Europe and North America, the use of tooth cleaning and whitening powders and then, in the nineteenth century, the development of anaesthetics (novocaine, chloroform, etc.) improved techniques in dentistry (Sinatra, Jahr, Watkins-Pitchford, 2010; Handge, Nawathe, 2012). Old style expressionless faces gave way to wide smiles that revealed one's teeth. Previously considered a trait of the lower classes, this smile carried new meaning and gradually spread across society. This trend can be observed in Elisabeth Vigée-Le-Brun's self portrait, in the paintings of Ribera, Velasquez, and Hogarth, as well as it can been seen in everyday life (Jones, 2014). Smiling expressed sensitivity, participated in the redefinition of sociabilities, and became a marker of an individual's identity (Courtine, Haroche, 1988; Jones, 2014). Similarly, the invention of the stylised figure, or silhouette, contributed to the formation of a certain aesthetic normativity. Painters, fashion plate engravers and graphic artists making silhouettes (Honoré Daumier, Nadar) developed drawing techniques that focussed on the proportions of bodies and the differences in their outlines. Thus, if there was an individualization of appearances through the demand for personal portraits (Vigarello, 2014), these images still conformed to gendered aesthetics that, for example, magnified the aristocratic male body in the *Ancien Régime* or emphasized the bourgeois appearance in the nineteenth century — with the insistance on broad shoulders for example. Female bodies were characterized by significant gendered differences based in motherhood. During the Belle Époque, the aesthetic ideal was that of a slender woman with a prominent bosom, so that between 1870 and 1900, more than six million corsets were sold (de Saint Pol, 2010). The introduction of mirrors in homes reinforced the attention given to appearance by enabling individuals to have a precise impression of their silhouettes. At first reserved for homes of the wealthy — at a time when the body of the masses was still perceived as a working tool — mirrors spread to all homes in the twentieth century thanks to chemical technology that made it possible to produce them on a large scale. Later placed into the privacy of the bathroom, mirrors allowed the body to be transformed and became a tool of self fashioning.

Gradually, collective norms of beauty developed. The medical supervision of the body and the commodification of beauty all contributed to this development. Some beautification techniques spread rapidly, such as the use of make-up and powders. Mainly reserved for the wealthiest parts of society, beauty products became increasingly used across the classes in the nineteenth and twentieth centuries, as the cosmetics industry developed and the idea of changing one's appearance became morally more acceptable (Vigarello, 2004). Cosmetic techniques (manucure, skin care, hair care, etc.) mainly aimed at women became accessible to all[4] and their benefits were promoted on the radio and in magazines (Banner, 2006; Haiken, 1997). Once it became acceptable to expose one's body, it became scrutinized even more and its appearance subjected to normative expectations (Vigarello, 2004). From the 1930s onwards, the beauty market extended to other parts of the body (skin, hands, feet, legs, etc.), gradually transforming appearances and identifying bodily defects (cellulite, etc.) that had not existed previously. In the 1950s cosmetics and therapeutics grew closer to each other, and more dermatologists and chemists appeared in laboratories in the cosmetic industry (Bernuzzi de Sant'Anna, 2007). The space occupied by beauty treatments (from make-up and powder, to liposuction, plastic surgery, chemical peels and Botox injections) in the daily lives of twentieth-century women testifies to an individualization of behaviours inspired by the will to achieve the body's potential and even surpass it. Here too medical discourse is lending its support.

4 Haiken (1997) shows that in 1907, there were 36 000 hairdressers (as opposed to 18 000 in 1900 and only 9 000 in 1890); 25 0000 manicurists and 30 000 face massage and skin specialists.

The impact of physicians in hygienist considerations was predominantly visible in the act of making the desirable body the subject of public policy (de Saint Pol, 2010). Nevertheless, the medicalized society of the nineteenth century also obeyed a commercial rationale that sought to create a health market legitimized by scientific discourse (Faure, 1994). Body standards were thus justified in the name of health. A good example is body measurements and the idea of an ideal weight. From the end of the nineteenth century onwards, slimming diets appeared. The ideal weight was calculated in relation to the person's height, although the calculation method changed at different time periods. This mathematization of beauty, under the guise of medical necessity, was disseminated via magazines (de Saint Pol, 2010). Social historians Peter N. Stearns (1997) and Elizabeth Haiken (1997) refer to the 1920s as the period during which an obsession with beauty and slimming developed. In the United Kingdom and the United States, the transition from nineteenth-century society that valued inner beauty, which was a combination of character and morality, to a society that valued outer beauty, and therefore worldly beauty, can be seen as marking a major development in the triumph of consumer culture. The body replaced the soul by becoming an object of intramundane salvation. The new morality of self-improvement turned physical appearance into a tool of social distinction and ugliness into a sign of a lack of care (Haiken, 1997). Because science held the promise of metamorphosis, individuals were held responsible for their appearance. In the second half of the twentieth century, which much emphasized personal development, this trend reached a new peak. The body, seen as central in identity formation, became the primary foundation of existence. The body became a work in progress, the site of a project of self transformation and improvement, in which being and having became one and the same thing with the common goal of to curate the self in order to reach authenticity. This obligation to become the best version of oneself required a diversification of techniques, including personal development practices (meditation, etc.), cosmetics, and the hiring of professionals (personal trainer, life coach, etc.) who adapted exercises to the individual's objectives.

Sport is an indicator of the profound transformation that Western societies underwent. In the 1870–1880s, exercise derived from military training became a leisure activity for the middle classes, aimed at taking control of one's physical appearance and to shape one's body by developing muscle mass (Courtine, 1993). Initially, this promoted hygienic and virile beauty in men and harmonious women's bodies. Around the early 1920s, sport aimed at making the body adaptable to the world of work and leisure for both sexes (Vigarello, 1978). The athletic body promotes the need for freedom of movement by showing and valuing flexible, tension-free bodies. Gradually in addition to the civic dimension of sport and to the modelling of one's appearance, the idea of surpassing onelself through physical effort emerged (Viaud, 2013). Societies where sports play an important role rely on physiology, the progress of physics, biochemistry, and histology (Rieu, 2010) and systematize exercises to optimize health and performance. Therefore, body culture is part of a paradigm linking sports and medicine (Queval, 2008) that creates the concept of a healthy, slim, and fit body, an expression of the utopia of perfect health. Sport is supposed to protect against premature death and the prevailing expectation to do sports puts the individual in charge of his or her own body. The bodybuilding ethos of 'suffering while having fun' particularly applies to body management techniques, which oscillate between duty and pleasure. Rather than being an expression of narcissism, they are remnants of the conspicuous display of puritanism inherited from Victorian times. Courtine (1993) considers the self-inflicted labours of bodybuilders as the extension of the work ethics, putting effort at the heart of leisure time in the era of mass consumption. He sees in this quest for taut skin the fear of any physical signs of ageing.

Cosmetic surgery as a medical practice for aesthetic purposes, also became much more common and accessible in the twentieth century. The medical and surgical management of deformities and physical imperfection dates back to before the twentieth century, but cosmetic surgery was reinvented with the First World War (Haiken, 1997). Maxillofacial repair techniques such as taking facial imprints to prepare prostheses or bone and cartilage grafting that were developed for reconstructive surgery during the First World War were adopted by cosmetic surgery (Delaporte, 1996). Cosmetic surgery, often in contrast to reconstructive surgery, is related to individual wishes (Jacobson, 2000) and concerns individuals with a 'healthy body'. It is intended to empower the individual by opening up medical technical possibilities beyond medical care. Sometimes described as a psychosurgical act contributing to psychological well-being, cosmetic surgery reaffirms the role of medicine in managing human suffering and achieving individual fulfilment (Le Henaff, 2014). Nevertheless, other authors who adopt a structuralist perspective think that this self-creating body produces hyper-conformism on a beauty market that participates in the creation of the very 'inferiority complexes' that they are supposed to heal. Jarrin explains that self-fashioning involves a form of work that generates value on and through the body. However, the freedom to buy anything one desires (including youth and beauty) cannot be untied from the trend towards a globalization of beauty that values white bodies most highly. The example of Brazil, which is the third largest consumer of beauty products and the second largest in terms of plastic surgery worldwide, illustrates how beauty is used a biopolitical tool for reconfiguring subaltern bodies. Beauty, happiness, and health became symbolic assets in the 1990s, enabling individuals to achieve their full potential through technology, but it must be said that this resource is not accessible to the poor. Inherited from the eugenicist agenda of the nineteenth century, which concentrated inequalities of race, class, and gender onto and through the body, beauty generates an aesthetic but also social hierarchy, since in this context, ugliness becomes a marker of poverty. This trend can be observed particularly in skin whitening techniques or rhinoplasty, which aim to erase the 'ethnic' characteristics of individuals in order to allow them to access jobs that enable them to move up the social ladder. The desire to take control of one's appearance therefore primarily expresses a need for social conformity (Haiken, 1997).

The Biomedicalization of Life

Once epidemics (the plague, cholera, smallpox, etc.) had been controlled with the help of the biotechnologies of the first half of the twentieth century (penicillin, streptomycin, antibiotics, vaccines, etc.), medicine held the belief that any disease could be cured with an appropriate therapy (Lagrée, Lebrun, 1994). If the body had long been a fated body limited to the narrow realities of human existence and its deadly condition, the second half of the twentieth century, supported by growing materialism and the promise of technoscience, disrupted the individual's relationship to the world and to him- or herself by understanding the body as a mechanism that could be controlled (Queval, 2008). The power of medicine was strengthened by the transition from curative and preventive medicine to biomedicine, which seeks to establish a 'supernormal state, a modified condition supposed to meet our multiple demands: to be more beautiful, stronger, more intelligent, etc.' (Goffette, 2006: 9). Since the 1970s, with the emergence of new technical devices linked to NBIC convergence (nanotechnologies, biotechnologies, information technology, cognitive sciences), the medicalization of life has given way to its biomedicalization (Clarke et al., 2010). In other words, the aim was no longer to simply restore health or promote better health, but to take on the perfectibility of human existence. In the near future, the body would be saved from itself. This paradigm shift is in line with the WHO's proposed concept of health as 'a state of complete physical, mental and social well-being and not merely the absence of disease or infirmity' ([Anon.], 1946). As the boundary between the normal and the pathological became porous, it became difficult to determine whether medical techniques that modified the body were repairing or improving it (Goffi, Missa, 2011). Not only did the notion of disease become imprecise, but it extended to conditions that were not previously considered pathological. Control had shifted: it was no longer about controlling disease or disability, but rather about the creation of 'more life' by modifying and transforming the body (Clarke et al., 2003; Rose, 2007) and by hybridising the human and non-human to achieve desired properties for it.

Biological life had now become a branch of politics, what Nikolas Rose defines as the 'politics of life itself' which formed the backdrop of what Le Dévédec calls 'improvement society' (Rose, 2007, 3; Le Dévédec, 2015).[5]. It manifests particularly in improved medicine, also called 'human enhancement'. It concerns all extra-medical transformation techniques aimed at improving human, physical, or intellectual capacities in a sustainable or unsustainable way. This field of human modification uses emerging technologies such as molecular biology, gene therapy, pharmacogenetics, and transplantation medicine to break free from the constraints of biology. Transhumanism is bringing the paradigm of augmented humanity and self improvement to

5 Nikolas Rose defined the « politics of life itself » thus: « It is neither delimited by the poles of illness and death, nor focused on eliminating pathology to protect the destiny of the nation. Rather, it is concerned with our growing capacities to control, manage, engineer, reshape, and modulate the very vital capacities of human beings as living creatures. It is, I suggest, a politics of "life itself" » (Rose, 2007, 3)

new heights. This international movement claims the moral necessity to transcend the limits of humans to take them to the higher evolutionary stage of the posthuman. Underpinned by an ideological conception of the optimization of human life, this is a programme aiming at governing bodies by transforming them from within. Human enhancement, as a contemporary version of biopolitics and as a corollary of the biomedicalization of existence, is manifest through a set of body governance mechanisms that invite us to rethink the notion of health. Health is no longer considered a default but a prized possession, a capital to be exploited. It becomes both a social responsibility (Conrad, 1992) and an individual objective on which the attention is focused for the purpose of and through the realization of oneself. The power of the devices being used lies in the interdependence of the techniques of domination exercised over others in the name of well-being and wellness. They are deployed both at the collective level through public health policies (Bossy, Briatte, 2011) and at the individual level through permanent self-control.

This is particularly evident in the ways in which food and nutrition are used to 'optimize' health. Whether it is through public authorities' recommendations for 'healthy eating' (Coveney, 2006; Fournier, 2014) or market incentives encouraging consumers to buy certain products (neutraceuticals, superfoods, food supplements, etc.), the aim is to improve physical or intellectual performance. Some sport-related practices are also emblematic of the biomedicalization of society and the way in which the body is subject to the dual demand of well-being and performance (Queval, 2001). Results have been central to sports since the beginning of the twentieth century (Corbin, Courtine, Vigarello, 2006), leading to a level of rationalization. On the one hand, the efficiency of sports equipment has been increased; training sessions have become structured, and athletes' muscle mass improved. The use of steel instead of wood for instance made it possible to use stronger and lighter materials such as hollow iron tubes for certain gymnastics equipment or drawn steel for elastic fixed bars (Corbin, Courtine, Vigarello, 2006). On the other hand, technology shapes the body from head to toe, under the pervasive influence of the industrial society. After the First World War, regeneration was intertwined with performance through the development of systematized processes adapted to the individuals' objectives, such as rational sleep management and the use of massages for curative, preventive, or recovery purposes (Viaud, 2013). Efforts were geared towards understanding the way the body, and in particular the muscles, worked during exercise. Photography made it possible to observe bodies in action, and techniques were developed to rationalize objective progress in order to improve performance (via thermographs, chronometers, myographs, etc.) (Queval, 2008). In the twenty-first century, this 'culture of measurement' (Ajana, 2017) has led to the quantification of the self or the quantified self (Swan, 2013). It allows for a quantification of the self thanks to connected technology devices (applications) measuring sports performance (the time and distance covered, the heart rate, the number of steps taken in a day, calories burned, etc.) and monitoring the body through variables (weight lost, the ratio between muscle and fat, the amount of food consumed, etc.). This production of anthropometric statistics is the latest in a long history of body measurement (Crawford, Lingel, Karppi, 2015) and is based on a self-monitoring quantitative approach to health indices (de Saint Pol, 2010) that is nothing more than a contemporary version of biopower (Lupton, 2016). These applications objectify physical efforts and invite the individual to share their results with virtual communities. In competition with oneself as well as with others, one finds motivation and comparison in a form of self-improvement that is meant to be fun. Self-fashioning is also achieved by doping products that enhance performance and/or recovery, blurring the boundary between medical care, sports training, and doping (Brisonneau, Aubel, Ohl, 2008; Quet, Hubert, 2013). Studies on the increased intake of products highlight the standardization of meliorative practices such as using performance-enhancing drugs (William, Gabe, Martin, 2011; Steiner, 2016). Whether in private or professional life, and whether intended to improve cognitive skills (Coveney, 2006; Gabe, Williams, Martin, 2011) or sports performance, the use of doping products pathologizes the natural state while aiming for optimization using artificial substances.

In this new moral framework of self-transformation (Clarke *et al.*, 2010), the individual is therefore invited to perform 'intensive exploitation' on him- or herself. The individual, who has become his or her own project, is a neo-subject (or neoliberal subject) structured by entrepreneurial rationality and encouraged to constantly develop and surpass itself. With the biomedicalization of existence, self-improvement allows for the reimagining of identities and, more broadly, of the future of the human species as the very boundaries of humanity shift (Franklin, 2015). In the current era of 'project management', the desire for self-improvement is grounded in a precise historical moment when biotechnological possibilities shape and invite the production of subjectdhood and identity (Gardey, 2015). The biotechnological developments of the twentieth and twenty-first centuries, especially synthetic biology, have changed the concept of life, which is now seen as a complex set of physico-chemical processes that can be synthesized. In this context it is hardly surprising to see medicine — now aimed at improving rather than curing — be seen as turning the tables on the determinism of human condition with a promise of a new definition of the somatic and biological (Goffi, Missa, 2011). Bernice Hausman (1994) explains that the conditions that were necessary for the emergence of a demand for sex reassignment depended in part on advances in endocrinology and plastic surgery.

Many fields show possibilities for medical applications in human transformation, but the applications relate specifically to the control of living organisms and the mechanisms of life (from birth to old age). According to Adele Clarke and her colleagues (Clarke, 2003), new reproductive technologies such as surrogacy, medically assisted procreation or in vitro fertilization mark a new frontier of biomedicalization since they alter not only the mechanisms of life but the very facts of life. By adopting and reinforcing the gendered division of reproductive labour, these technologies increase the measures of control exerted over women (Angeloff, Gardey, 2015). At the other end of the spectrum of life, this is the case for anti-ageing medicine, which corresponds to the medicalization of the ageing process. This dates from the end of the nineteenth century (Haycock, 2008) but became more significant at the end of the twentieth century with the increase in life expectancy. In contrast to gerontology, anti-ageing medicine is a technique to improve their lives by delaying the effects of ageing such as erectile disfunction and hormonal disorders (Hauray, Dalgarrondo, 2014), or even by reversing the very process of senescence.

The biomedicalization of life is therefore also based on a rhetoric of promise. It posits that it is no longer enough to read and understand the mechanisms of life, but with a view of calculated life management, it is about being able to rewrite them. Stem cell research, for example, promises to pave the way for regenerative medicine (Fontaine, 2013). As for genetics, it promises the development of personalized medicine based on an analysis of information about human genes and genetic pathologies (Bourgain, Darlu, 2008; Shostak, Conrad, Horwitz, 2008). Underpinned by biopolitical projects for the optimization and control of plants, animals, and humans (Bonneuil, 2015), the history of genetics aims to reprogramme living beings and take control of the evolution of species. As early as the 1970s, with the birth of molecular biology and molecular genetics, a new conception of life with genes as the basis of life adapted the conceptions of the biological nature to that of genetic biology. Thus, with the discovery of genetic kinship, the adjective 'genetic' became synonymous with 'biological' (Franklin, 2015). In the 1980s, work on in vitro DNA amplification made it possible to copy a DNA fragment, and in the 1990s GMOs were created by transgenesis; both developments challenge genetic determinism with a new concept of DNA no longer as a programme but as a memory selectively mobilized by cell dynamics under the influence of the environment (epigenetics). The sequencing of the total human genome by the Human Genome Project in the early 2000s, and more recently the CRISPR CAS9 molecular scissors, have changed the relationship to innateness and lent unprecedented potential to the promise of writing the book of life. The implementation of a project of rationalization of life on a industrial level becomes possible. The decoding and then manipulation of the genome have ushered life into the realm of human management, on an unprecedented scale. In addition to producing knowledge about genes, genetics makes it possible to control the insides of bodies at the individual and collective levels. This new knowledge of the genome, especially the human genome, has social implications, not only through new types of knowledge and medical interventions, but also through forms of social categories, since the genetic approach to populations classifies abnormality by detecting and classifying sick bodies, bodies at risk, and bodies potentially compromized by their genes (Rabinow, 2000).

The invitation to take on the so-called perfectibility of humans, that is, to increase the power that each individual can have over their own body and appearance, refers to an understanding of health that makes risk and vulnerability increasingly less acceptable. In this context, the individual is not perceived as enhanced but rather as defective in terms of fragility and mortality. The dictates of well-being and good health, which are largely supported by a discourse positing humans as not adapted to their environment, do not therefore increase the number of superhumans but rather that of the sick in the making, who will have to be taken care of (Collin, 2007). Moreover, although biomedicine plays a central role in attempts to 'tailor-make' individuals, some of the current reconstructive medical technologies nevertheless invite

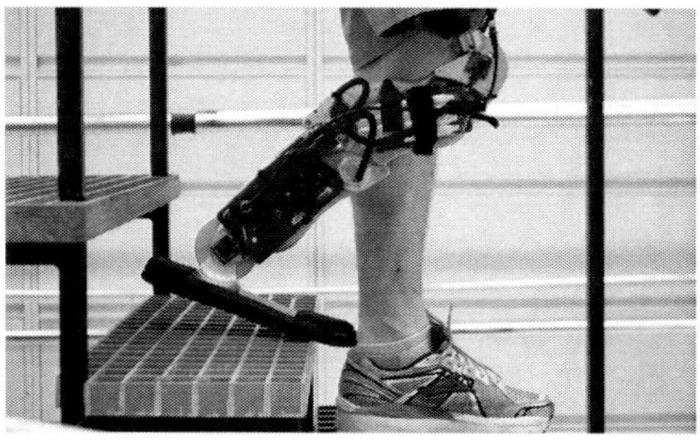

Fig. 2. Leg prosthesis. 2013. Romain Pernet.

the hyperbolic hopes of the promise for transhumanism to be qualified, particularly when it comes to self and identity.

Organ transplantation is a medical and technical act that supports the idea of a machine-like body whose parts can be exchanged. It allows for the line between life and death to be blurred. By exchanging its own organ for another one and turning itself into a 'time machine' the body becomes internally rejuvenated. However, the introduction of a healthy organ from another body leads the recepient to a painful (re)consideration of body awareness and image (Vaysse, 2007). Similarly, as techniques proposing partial substitutes for the body, prosthetics and organ transplantation change both body image and as a result subjectivity. The demand for prosthetic devices first originated in the army (Coker, 2014), which produces injured bodies and then seeks to care for them. In the nineteenth century, medical and orthopaedic techniques for mutilated bodies developed widely in the military, and prostheses were made, for example, in weapons factories (Nicogossian, 2012). At the beginning of the twenty-first century, the Defense Advanced Research Projects Agency (DARPA) created the *Revolutionizing Prosthethics* programme. This project aims to develop an arm prosthesis connected to the nervous system so that amputees can regain their motor functions and return to taking part in military operations (Dixon, 2006). The promise of reconfiguring the body with technosciences was first seriously considered with the figure of the perfected soldier (Robitaille, 2008) and the notion of the cyborg (cybernetic organism). This appeared in the 1960s in a military-industrial context and refers to an improved man who could survive in an extraterrestrial environment without necessarily relying on genetic modification (Dixon, 2006). Nevertheless, despite hyperbolic discourses overestimating the capacities of technology to produce an 'augmented' body, the reality is more modest. These technical extensions of the body through connected or orthopaedic prostheses, mainly concern elderly or sick people struggling to understand these new devices, which are often painful and difficult to handle (Jarrassé *et al.*, 2015). Additionally, the body fitted with prosthetics is above all an amputated body, wounded due to illness or accident, which often causes identity issues in individuals (Gourinat, Erhler, 2015). Furthermore, the prosthesis, especially a myoelectric prosthesis, requires some time for the individual to learn how to use it correctly. Only then is it accepted by the patient as a physical, functional, and symbolic substitute for the absent limb (Gourinat, 2014) (Figure 2).

In conclusion, the techniques of the body aimed at modifying the human being are part of a dialectical tension between autonomy and coercion. If the construction of the modern subject consists in its emancipation, it also takes place in and through forms of power that shape individual identities, particularly through the government of bodies. This rationalization of bodies radically embodies this new form of biopolitics characterized by a fusion between medicine, which has become biomedicine, and capitalism (Lafontaine, 2014). Thus, while the quest for beauty, power, or perfection certainly emanates from one's own will, it nevertheless responds to the invitation to be more than oneself, and ultimately reflects social expectations requiring individuals to be part of a productive regime, whatever this may be.

References

Abbott, Andrew, *The System of Professions. An Essay on the Division of Expert Labor* (Chicago, IL: The University of Chicago Press, 1988).

Ajana, Btihaj, 'Digital Health and the Biopolitics of the Quantified Self', *Digital Health*, 3 (2017), pp. 1–18.

Andrieu, Bernard, *Bronzage. Une petite histoire du soleil et de la peau* (Paris: CNRS Éditions, 2008).

———, *L'Éducation physique au xxe siècle: une histoire des pratiques* (Paris: Les cahiers Actio, 1993).

Angeloff, Tania and Delphine Gardey, 'Corps sous emprises. Biopolitique et sexualité au Nord et au Sud', *Travail, genre et sociétés*, 34 (2015), pp. 31–38.

[Anon.], 'Préambule à la Constitution de l'organisation mondiale de la santé', *Actes officiels de l'Organisation mondiale de la Santé*, 2.100 (1946), pp. 1–2.

Ariès, Paul, *L'Homme devant la mort. Le temps des gisants* (Paris: Le Seuil, 1977).

Arnal, Thierry, 'Les Rapports du physique et du moral en question au xixe siècle', in *L'Exercice corporel du xviiie siècle à nos jours, de la thérapeutique à la performance*, ed. by Grégory Quin and Anaïs Bohuon (Paris: Éditions Glyphe, 2013), pp. 41–60.

Arnold, Odile, *Le Corps et l'âme. La vie des religieuses au xixe siècle* (Paris: Le Seuil, 1984).

Badinter, Élisabeth, *L'Amour en plus. Histoire de l'amour maternel (xviiie–xxe siècle)* (Paris: Flammarion, 1980).

Banner, Lois, *American Beauty: a Social History. Through Two Centuries of the American Idea, Ideal, and Image of the Beautiful Woman* (Los Angeles, CA: Figueroa Press, 2006).

Bernuzzi de Sant'Anna, Denise, 'Fabriquer la beauté au Brésil', *Communications*, 81 (2007), pp. 121–34.

Blanckaert, Claude, *De la race a l'évolution: Paul Broca et l'anthropologie française (1850–1900)* (Paris: L'Harmattan, 2009).

Boëtsch, Gilles, Bernard Andrieu, David Le Breton, Nadine Pomarède, and Georges Vigarello, *La Peau. Enjeu de société* (Paris: CNRS Éditions, 2008).

Bonneuil, Christophe, 'Le siècle du gène', in *Histoire des sciences modernes*, vol. iii, *Le Siècle des technosciences*, ed. by Christophe Bonneuil and Dominique Pestre (Paris: Le Seuil, 2015), pp. 297–317.

Bossy, Thibault and François Briatte, 'Les Formes contemporaines de la biopolitique', *Revue Internationale de Politique Comparée*, 18.4 (2011), pp. 7–12.

Bourdelais, Patrice, ed., *Les Hygiénistes, enjeux, modèles et pratiques (xviiie–xxe siècles)* (Paris: Belin, 2001).

——— and Olivier Faure, *Les Nouvelles pratiques de santé, xviiie–xxe siècle* (Paris: Belin, 2005).

Bourgain, Catherine and Pierre Darlu, *ADN superstar ou superflic? Les citoyens face à une molécule envahissante* (Paris: Seuil, 2008).

Brisonneau, Christophe, Olivier Aubel, and Fabien Ohl, *L'Épreuve du dopage. Sociologie du cyclisme professionnel* (Paris: PUF, 2008).

Bujon, Thomas, Christine Dourlens, and Gwenola Le Naour, *Aux frontières de la médecine* (Paris: Éditions des archives contemporaines, 2014).

Carol, Anne, *Les Médecins et la mort, xixe–xxe siècle* (Paris: Aubier, 2004).

Carson, John, *The Measure of Merit: Talents, Intelligence, and Inequality in the French and American Republics (1750–1940)* (Princeton, NJ: Princeton University Press, 2007).

Chevalier, Louis, *Classes laborieuses et classes dangereuses à Paris pendant la première moitié du XIXe siècle* (Paris: Plon, 2002).

Clarke, Adele, Jennifer R. Fishman, Jennifer Ruth Fosket, Laura Mamo, and Janet K. Shim, 'Biomedicalization: Technoscientific Transformations of Health, Illness, and U.S. Biomedicine', *American Sociological Review*, 68.2 (2003), pp. 161–94.

———, Laura Mamo, Jennifer Ruth Fosket, Jennifer R. Fishman, and Janet K. Shim, *Technoscience, Health, and Illness in the U.S.* (Durham, NC: Duke University Press, 2010).

———, Jennifer R. Fishman, Jennifer Ruth Fosket, Laura Mamo, and Janet K. Shim, 'Technosciences et nouvelle biomédicalisation: racines occidentales, rhizomes mondiaux', *Sciences sociales et santé*, 18.2 (2000), pp. 11–42.

Coker, Christopher, 'Biotechnology and War: the New Challenge', *Australian Army Journal*, 2.1 (2014), pp. 125–40.

Collin, Johanne, 'Relations de sens et relations de fonction: risque et médicament', *Sociologie et sociétés*, 39.1 (2007), pp. 99–122.

Conrad, Peter, 'Medicalization and Social Control', *Annual Review of Sociology*, 18.1 (1992), pp. 209–32.

Corbin, Alain, 'Le Corps… enfin!', *L'Histoire*, 331 (2008), pp. 24–27.

———, *Les Filles de noce. Misère sexuelle et prostitution (XIXe et XXe siècles)* (Paris: Aubier, 1978).

———, 'L'Hérédosyphilis ou l'impossible rédemption, contribution à l'histoire de l'hérédité morbide', *Romantisme*, n°31, 1981. pp. 131–49.

———, Jean-Jacques Courtine, and Georges Vigarello, *Histoire du corps*, vol. III, *Les Mutations du regard, le XXe siècle* (Paris: Seuil, 2006).

Courtine, Jean-Jacques, 'Les Stakhanovistes du narcissisme', *Communications*, 56 (1993), pp. 225–51.

——— and Claudine Haroche, *Histoire du visage: exprimer et taire ses émotions XVIe–début XIXe siècle* (Paris and Marseille: Rivages, 1988).

Coveney, John, *Food, Morals and Meaning: the Pleasure and Anxiety of Eating* (New York, NY: Routledge, 2006).

Crawford, Kate, Jessica Lingel, and Terro Karppi, 'Our Metrics, Ourselves: a Hundred Years of Self-tracking from the Weight Scale to the Wrist Wearable Device', *European Journal of Cultural Studies*, 18.4–5 (2015), pp. 479–96.

Csergo, Julia and Roger-Henri Guerrand, *Le Confident des dames. Le bidet du XVIIIe au XXe siècle: histoire d'une intimité* (Paris: La Découverte, 2008).

Delaporte, Sophie, *Les Gueules cassées. Les blessés de la face de la Grande Guerre* (Paris: Éditions Noêsis, 1996).

de Saint Pol, Thibault, *Le Corps désirable. Hommes et femmes face à leur poids* (Paris: PUF, 2010).

Descola, Philippe, *Par-delà nature et culture* (Paris: Gallimard, 2005).

Desrosières, Alain, *La Politique des grands nombres. Histoire de la raison statistique* (Paris: La Découverte, 1993).

Dessertine, Dominique and Olivier Faure, *La Maladie. Entre libéralisme et solidarité, 1850–1940* (Paris: Mutualité Française, 1994).

Diasio, Nicoletta, 'Maillage des temps et gouvernement des corps dans la construction des rapports d'âge et de genre', *SociologieS* (2012), <http://sociologies.revues.org/4118> [accessed 15 November 2012]

Dixon, Bernard, 'From Broken to Bionic', *The New Scientist*, 192.2577 (2006), 58.

Faure, Olivier, *Histoire sociale de la médecine (XVIIIe–XXe siècles)* (Paris: Anthropos/Economica, 1994).

Febvre, Lucien, 'Réflexions sur l'histoire des techniques', *Annales d'histoire économique et sociale*, 7.36 (1935), pp. 531–35.

Federici, Silvia, *Caliban et la sorcière. Femmes, corps et accumulation primitive* (Geneva: Entremonde / Marseille: Senonevero, 2014).

Feher, Michel, Ramona Nadaff, and Nadia Tazi, *Fragments for a History of the Human Body*, vol. I (New York, NY: Zone, 1989).

Fontaine, Bernard, *Biotechnologies: quelles limites? Quo vadis homo sapiens?* (Paris: L'Harmattan, 2013).

Foucault, Michel, *The History of Sexuality*, vol. 1. *An Introduction* (London: Allen Lane, 1979).

———, '"Pouvoir et corps", Quel corps? n°2', in *Dits et Écrits*, vol. II (Paris: Gallimard, 2014), pp. 2–5.

Fournier, Tristan, 'Face à l'injonction diététique, un "relativisme nutritionnel" en France', *SociologieS* (2014) <http://sociologies.revues.org/4628>, [accessed 07 March 2014].

Franklin, Sarah, 'Foucault et les transformations du biopouvoir', in *Histoire des sciences modernes*, vol. III, *Le Siècle des technosciences*, ed. by Christophe Bonneuil and Dominique Christophe (Paris: Le Seuil, 2015), pp. 210–31.

Freidson, Eliot, *La Profession médicale* (Payot: Paris, 1984).

Fressoz, Jean-Baptiste, *L'Apocalypse joyeuse. Une histoire du risque technologique* (Paris: Seuil, 2012).

Gall, Franz Joseph, *Précis du système phrénologique du docteur Gall* (Paris: Garnier, 1838).

Galton, Francis, *Hereditary Genius: an Inquiry into Its Laws and Consequences* (Honolulu, HI: University Press of the Pacific, 2001).

Gardey, Delphine, 'Genre, corps et biomédecine', *Histoire des sciences modernes*, vol. III, *Le Siècle des technosciences*, ed. by Christophe Bonneuil, Dominique Pestre (Paris: Seuil, 2015), pp. 360–79.

Gélis, Jacques, *L'Arbre et le fruit: la naissance dans l'Occident moderne, XVIe–XIXe siècle* (Paris: Fayard, 1984).

Goffette, Jérôme, *Naissance de l'anthropotechnie. De la médecine au modelage de l'humain* (Paris: Vrin, 2006).

Goffi, Jean-Yves and Jean-Noël Missa, 'Amélioration de l'être humain', *Journal international de Bioéthique*, 22.3–4 (2011), pp. 13–14.

Gourinat, Valentine, 'Déstructuration et restructuration identitaire du corps prothétique', *Sociétés*, 125 (2014), pp. 127–35.

——— and Solange Ehrler, 'Du morcellement à la reconfiguration', *Revue des sciences sociales*, 53 (2015), pp. 100–07.

Guillo, Dominique, 'Les avatars de l'organisme dans les sciences sociales françaises', in *La Tentation du corps. Corporéité et sciences sociales*, Memmi, Dominique, Dominique Guillo, and Olivier Martin (Paris: Éditions de l'EHESS, 2009), pp. 221–43.

Hacking, Ian, *The Taming of Chance* (Cambridge: Cambridge University Press, 1990).

Haiken, Elizabeth, *Venus Envy: a History of Cosmetic Surgery* (Baltimore, MD: Johns Hopkins University Press, 1997).

Handge, Keshav T, and Abhay Nawathe, 'History of Dentistry', *Global Journal for Research Analysis*, 4.11 (2015), p. 36.

Haudricourt, André-George, 'La Technologie culturelle', in *Ethnologie générale*, ed. by Jean Poirier (Paris: Gallimard, 1968), pp. 731–822.

——— and Mariel Bruhnes Delamarre, *L'Homme et la charrue à travers le monde* (Paris: Gallimard, 1955).

Hauray, Boris, Dalgarrondo Sébastien, 'Soigner le vieillissement: sociologie d'une promesse médicale', in Aux frontières de la médecine, Bujon, Thomas, Christine Dourlens, and Gwenola Le Naour, (Paris: Éditions des archives contemporaines, 2014), pp. 231–48.

Hausman, Bernice, *Changing Sex: Transsexualism, Technology, and the Idea of Gender* (Durham, NC: Duke University Press, 1994).

Haycock, David Boyd, *Mortal Coil: a Short History of Living Longer* (New Haven, CT: Yale University Press, 2008).

Hecketsweiler, Philippe, *Histoire de la médecine, des malades, des médecins, des soins et de l'éthique biomédicale* (Paris: Ellipses, 2010).

Imbault-Huart, Marie-José, 'Où va l'histoire de la médecine?' in *Pour l'histoire de la médecine. Autour de l'oeuvre de Jacques*, Léonard Lagrée, Michel and François Lebrun, eds, (Rennes: Presses Universitaires de Rennes, 1994), pp. 71–82.

Jacobson, Nora, *Cleavage: Technology, Controversy, and the Ironies of the Man-made Breast* (New Brunswick, NJ: Rutgers University Press, 2000).

Jarassé, Nathanel and Marina Maestrutti, 'Robotic Prosthetics: Moving Beyond Technical Performance. (A Study of Socio-Anthropological and Cultural Phenomena Influencing the Appropriation of Technical Objects Interacting with the Body)', *IEEE Technology and Society Magazine*, 34.2 (2015), pp. 71–79.

Jarrige, François and Thomas Le Roux, *La Contamination du monde. Une histoire des pollutions à l'âge industriel* (Paris: Seuil, 2017).

Jarrin, Alvaro, *The Biopolitics of Beauty. Cosmetic Citizenship and Affective Capital in Brazil* (Oakland, CA: University of California Press, 2017).

Jones, Colin, *The Smile Revolution in Eighteenth Century Paris* (Oxford: Oxford University Press, 2014).

Jordanova, Ludmilla, *Sexual Visions. Images of Gender in Science and Medicine between the Eighteenth and Twentieth Centuries* (Madison, WI: The University of Wisconsin Press, 1989).

Jorland, Gérard, *Une société à soigner. Hygiène et salubrité publiques en France au XIXe siècle* (Paris: Gallimard, 2010).

Kleinman, Arthur, *The Illness Narratives. Suffering, Healing, and the Human Condition* (New York, NY: Basic Books, 1988).

Knibiehler, Yvonne and Catherine Fouquet, *L'Histoire des mères du Moyen Âge à nos jours* (Paris: Montalba, 1980).

Lafontaine, Céline, *Le Corps-marché. La marchandisation de la vie humaine à l'ère de la bioéconomie* (Paris: Seuil, 2014).

Lagrée, Michel and François Lebrun, eds, *Pour l'histoire de la médecine. Autour de l'œuvre de Jacques Léonard* (Rennes: Presses Universitaires de Rennes, 1994).

Latour, Bruno, *Nous n'avons jamais été modernes* (Paris: La Découverte, 1991).

Le Breton, David, *Anthropologie du corps et de la modernité* (Paris: PUF, 2013).

———, *La Chair à vif: usages médicaux et mondains du corps humain* (Paris: Métailié, 1993).

Le Dévédec, Nicolas, *La Société de l'amélioration. La perfectibilité humaine des Lumières au transhumanisme* (Paris: Liber, 2015).

Le Henaff, Yannick, 'La Chirurgie esthétique, l'authenticité médicale en question', in *Aux frontières de la médecine*, Bujon, Thomas, Christine Dourlens, and Gwenola Le Naour (Paris : Editions des archives contemporaines, 2014), pp. 195-208.

Léonard, Jacques, *Archives du corps. La santé au* XIXe *siècle* (Rennes: Ouest-France Éditons, 1986).

———, *La France médicale: médecins et malades au* XIXe *siècle* (Paris: Gallimard-Julliard, 1978).

Leroi-Gourhan, André, *Le Fil du temps. Ethnologie et préhistoire 1935–1970* (Paris: Odile Jacob, 1983).

———, *Le Geste et la Parole*, vol. I, *Technique et langage* (Paris: Albin Michel, 1964).

Lévi-Strauss, Claude, 'Introduction à l'œuvre de Marcel Mauss' in *Sociologie et anthropologie.* sous la direction de Mauss Marcel (Paris: PUF, 2013 [1950]), pp. IX–LII.

Lévi-Strauss, Claude, *Race et Histoire* (Paris: Folio essais, 1987 [1952]).

Lombroso, Cesare, *L'Homme criminel. Étude anthropologique et psychiatrique. Étude anthropologique et médico-légale,* (Paris: Ancienne Librairie Germer Baillière et Cie, Félix Alcan, 1895). http://classiques.uqac.ca/classiques/lombroso_cesare/homme_criminel_1895/homme_criminel_1895.html (consulté le 12/02/2023)

Lowy, Ilana, 'Les Microbes et les humains', in *Histoire des sciences et des savoirs*, vol. II, *Modernité et globalisation,* ed. by Kapil Raj and Otto Sibum (Paris: Seuil, 2015), pp. 223–41.

Lupton, Deborah, *The Quantified Self* (Cambridge: Polity Press, 2016).

Mandressi, Rafael, *Le Regard de l'anatomiste. Dissections et invention du corps en Occident* (Paris: Le Seuil, 2003).

———, 'Le corps et l'histoire. De l'oubli aux représentations', in *La Tentation du corps. Corporéité et sciences sociales,* Memmi, Dominique; Guillo, Dominique; Martin, Olivier (dir.), (Paris, Éditions de l'EHESS, 2009), pp. 143–69.

Mauss, Marcel, *Sociologie et anthropologie* (Paris: PUF, 1950).

———, 'Les Techniques du corps', *Journal de Psychologie,* 32.3–4 (1936), pp. 271–93.

Memmi, Dominique, Dominique Guillo, and Olivier Martin, *La Tentation du corps. Corporéité et sciences sociales* (Paris: Éditions de l'EHESS, 2009).

Müller-Wille, Staffan, 'Hérédité, race et eugénisme dans le long XIXe siècle', *in Histoire des sciences et techniques,* vol. II, *Modernité et globalisation,* ed. by Kapil Raj and Otto Sibum (Paris: Seuil, 2015), pp. 391–401.

Nicogossian, Judith, *Le Corps désirable. Hommes et femmes face à leur poids* (Paris: PUF, 2010).

———, 'La Prothèse de guerre: réparation du corps du soldat', *Corps,* 10 (2012), pp. 225–32.

Orthous, Marie-Hélène, 'Effets moraux des exercices du corps et science du cerveau au début du XIXe siècle', in *L'Exercice corporel du* XVIIIe *siècle à nos jours,* ed. by Grégory Quin and Anaïs Bohuon (Paris: Glyphe, 2013), pp. 19–40.

Ory, Pascal, *L'Invention du bronzage. Essai d'une histoire culturelle* (Paris: Complexe, 2008).
Perrot, Michèle, 'Le Corps a son histoire', *L'Histoire*, 8 (1979), p. 81.
Perrot, Philippe, *Le Travail des apparences ou les transformations du corps féminin, XVIIIe–XIXe siècles* (Paris: Seuil, 1984).
Pillon, Thierry, 'De quelques traces du corps en sociologie du travail', *Sociologie et Sociétés*, 48.1 (2016), pp. 83–96.
Porter, Roy, *Blood and Guts: a Short History of Medicine* (New York, NY and London: W. W. Norton & Company, 2002).
———, *Disease, Medicine and Society in England,1550–1860* (Cambridge: Cambridge University Press, 1995).
Porter, Theodore, *The Rise of Statistical Thinking (1820–1900)* (Princeton, NJ: Princeton University Press, 1986).
Pouchelle, Marie-Christine, *Corps et chirurgie à l'apogée du Moyen Âge* (Paris: Flammarion, 1983).
Quet, Mathieu and Mathieu Hubert, *Éthique du sport* (Paris: L'Âge d'Homme, 2013).
Quételet, Adolphe, *Sur l'homme et le développement de ses facultés ou Essai de physique sociale* (Paris: Fayard, 1991).
———, 'Sur l'appréciation des documents et en particulier sur l'appréciation des moyennes', *Bulletin de la Commission centrale de statistique (de Belgique)*, 2 (1844), pp. 205–86.
Queval, Isabelle, *Le Corps aujourd'hui* (Paris: Gallimard, 2008).
———, 'Le Dépassement de soi, figure du sport contemporain', *Le Débat*, 114 (2001), pp. 103–24.
Quin, Grégory and Anaïs Bohuon, eds, *L'Exercice corporel du XVIIIe siècle à nos jours, de la thérapeutique à la performance* (Paris: Éditions Glyphe, 2013).
Rabier, Christelle, 'La Disparition du barbier chirurgien. Analyse d'une mutation professionnelle au XVIIIe siècle', *Annales. Histoire, Sciences Sociales*, 65.3 (2010), pp. 679–711.
Rabinow, Paul, *Le Déchiffrage du génome, l'aventure française* (Paris: Odile Jacob, 2000).
Remotti, Francesco, 'De l'incomplétude', in *Figures de l'humain. Les représentations de l'anthropologie*, ed. by Francis Affergan, Silvana Borutti, Claude Calame et al. (Paris: Éditions de l'EHESS, 2003), pp. 19–74.
Rieu, Michel, 'La Santé par le sport: une longue histoire médicale', *La Revue pour l'histoire du CNRS*, 26 (2010), pp. 30–35.
Ripa, Yannick, 'L'Histoire du corps, un puzzle inachevé', *Revue historique*, 644.4 (2007), pp. 887–98.
Robitaille, Michèle, 'Le Cyborg contemporain. Quand les technosciences visent le remodelage du corps humain', *¿Interrogations?*, 7 (2008) <https://www.revue-interrogations.org/Le-cyborg-contemporain-Quand-les>, [accessed 1st December 2008].
Rodriguez Balanta, Beatriz Eugenia, 'Especimenes antropometricos y curiosidades pintorescas: la orquestracion fotografica del cuerpo negro (Brasil circa 1865)', *Revista Ciencias de la Salud*, 10.2 (2012), pp. 223–42.
Rose, Nikolas, *The Politic of Life Itself. Biomedicine, Power, and Subjectivity in the Twenty-First Century* (Princeton, NJ: Princeton University Press, 2007).
Salomon-Bayet, Claire, ed., *Pasteur et la révolution pastorienne* (Paris: Payot, 1986).

——— 'Histoire des sciences et histoire de la médecine', in *Pour l'histoire de la médecine. Autour de l'oeuvre de Jacques*, Léonard Lagrée, Michel and François Lebrun, eds. (Rennes: Presses Universitaires de Rennes, 1994), pp. 49–55.

Sawday, Jonathan, *The Body Emblazoned. Dissection and the Human Body in Renaissance Culture* (London: Routledge, 1995).

Schiebinger, Londa, *The Mind Has No Sex? Women in the Origins of Modern Science* (Cambridge, MA: Harvard University Press, 1989).

Schweber, Libby, *Disciplining Statistics: Demography and Vital Statistics in France and England (1830–1885)* (Durham, NC: Duke University Press, 2006).

Seignan, Gérard, 'L'Hygiène sociale au XIXe siècle: une physiologie morale', *Revue d'histoire du XIXe siècle*, 40 (2010), pp. 113–30.

Shorter, Edward, *Le Corps des femmes* (Paris: Seuil, 1984).

Shostak, Sara, Peter Conrad, and Allan Victor Horwitz, 'Sequencing and Its Consequences: Path Dependence and the Relationships between Genetics and Medicalization', *American Journal of Sociology*, 114.1 (2008), pp. 287–316.

Sinatra, Raymond, Jonathan Jahr, and Michael Watkins-Pitchford, *The Essence of Analgesia and Analgesics* (Cambridge: Cambrigde University Press, 2010).

Sohn, Anne-Marie, 'Le corps sexué', in *Histoire du Corps*, vol. III, *Les mutations du regard, le XXe siècle*, ed. by Alain Corbin, Jean-Jacques Courtine, and Georges Vigarello (Paris: Seuil, Paris, 2011), pp. 93–128.

Sournia, Jean-Charles, *Histoire de la médecine* (Paris: La Découverte, 2004).

Stearns, Peter W., *Fat History: Bodies and Beauty in the Modern West* (New York, NY: New York University Press, 1997).

Steiner, Pierre, *Le Dopage sans duperie. Essai sur le sport augmenté* (Paris: Les Belles Lettres, 2016).

Stepan, Nancy, *The Idea of Race in Science: Great Britain (1800–1960)* (London: Macmillan, 1982).

Stiker, Henri-Jacques, *Corps infirme et société* (Paris: Dunod, 2013).

Swan, Melanie, 'The Quantified Self: Fundamental Disruption in Big Data Science and Biological Discovery', *Big Data*, 1.2 (2013), pp. 85–99.

Vaysse, Jocelyne, 'Le Corps greffé', *Communications*, 81 (2007), pp. 225–35.

Veblen, Thorstein, *The Theory of the Leisure Class* (New York: The MacMillan Company, 1899).

Viaud, Jean-Baptiste, 'Réparer, régénérer, optimiser: la perfectibilité des corps au prisme de l'hétérogénéité des médecines des activités physiques et sportives dans l'entre-deux-guerres français', in *L'Exercice corporel du XVIIIe siècle à nos jours*, ed. by Grégory Quin and Anaïs Bohuon (Paris: Glyphe, 2013), pp. 181–204.

Vigarello, Georges, *Le Corps redressé. Histoire d'un pouvoir pédagogique* (Paris: J.-P. Delarge, 1978).

———, 'Gouverner et expliquer l'"impossible". La non-maîtrise dans les modèles anciens et modernes d'entretien de la santé', *Communications*, 56 (1993), 9–23.

———, *Histoire de la beauté. Le corps et l'art d'embellir de la Renaissance à nos jours* (Paris: Seuil, 2004).

———, *Le Propre et le sale: l'hygiène du corps depuis le Moyen Âge* (Paris: Seuil, 1985).

———, *Le Sentiment de soi. Histoire de la perception du corps* XVIe–XXe *siècle* (Paris: Seuil, 2014).

Villermé, Louis René, *Tableau de l'état physique et moral des ouvriers employés dans les manufactures de coton, de laine et de soie* (Paris: EDI–Études et documentation internationales, 1989).

Vincent-Buffault, Anne, *Histoire des larmes* XVIIIe–XIXe *siècles* (Paris: Rivages, 1986).

Vovelle, Michel, *La Mort et l'Occident de 1300 à nos jours* (Paris: Gallimard, 1983).

Warnier, Jean-Pierre, *Construire la culture matérielle. L'homme qui pensait avec ses doigts* (Paris: Presses Universitaires de France, 1999).

William, Simon, Jonathan Gabe, and Paul Martin, 'The Pharmaceuticalisation of Society? A Framework for Analysis', *Sociology of Health & Illness*, 33.5 (2011), 710–25.

Zola, Irving Kenneth, 'Medicine as an Institution of Social Control', *The Sociological Review*, 20.4 (1972), pp. 487–504.

VIRGINIE JULLIARD

Rethinking the Gender and Technology from a Feminist Perspective*

Technologies are an integral part of contemporary societies. We perceive, think, act and communicate through them, and all the more so today as they are 'increasingly close to [...] our bodies' (*'de plus en plus près [...] des corps'*) (Dalibert, 2016: 628 — unless otherwise stated, all translations are the author's) and as they act upon them, for example, through prostheses, implants, or reproductive technologies. Feminist studies on technology investigate its effects on women, the way in which technologies shape gender, and how technology and gender are mutually constitutive. They question the systematic association of hegemonic masculinity with technology and technical skills on one hand, and of femininity with technophobia and technical incompetence on the other (Gill, Grint, 1995; Wajcman, 2004). They thereby reveal the mythical character of these associations (see, for example, Sadie Plant's history of the link between women and computer simulation, assembly, and programming (1997)). These studies also reveal the way in which technologies shape gendered bodies, particularly with respect to recent developments in biomedical technologies and 'somatechnologies' (Lettow, 2011) or, in other words, 'technologies that act upon and within bodies' (*'les technologies qui agissent sur et interviennent dans les corps'*) (Dalibert, 2014: 631).

This chapter attempts to track the history of the way in which gender studies have addressed the topic of technology. While dividing this chapter into ten-year periods has a pragmatic advantage in terms of clearly presenting the different approaches and research, this division comes with caveats. Different approaches may be contemporaneous. Some may be challenged by different disciplinary fields, feminist theories, or specific contexts while persisting in others. Some technologies may be the preferred objects of study of some approaches or disciplinary fields — for example, the media are studied by both media studies in English speaking countries and in France by information and communication sciences. Given my disciplinary background, my particular focus is communicational objects, which are also technical artefacts (Jouët, 2007), such as the telephone and the French Minitel, and more widely the media,

* Translation by Gill Gladstone, Words in action.

Global History of Techniques, ed. by Guillaume Carnino, Liliane Hilaire-Pérez and Jérôme Lamy, Global Matters Culture Matérielles, histoire des techniques et circulation globale des savoirs, 9 (Turnhout: Brepols, 2024), pp. 669–686
BREPOLS PUBLISHERS
10.1484/M.TECHNE-EB.5.129835

including electronic media in the form of websites. In line with Teresa de Lauretis (1987), I consider media as 'technologies of gender', i.e. structures of power producing representations of gender that impact bodies, behaviours, and social relations. As early as the 1960s, feminists took an interest in the role that women's magazines and advertising played in constructing gender stereotypes and maintaining asymmetrical gender relations, among them Friedan (1963) in the United States, Mattelard (1970) in the South American context, and Sullerot (1966) and Dardigna (1974, 1978) in France. This chapter is less a comprehensive literature review, than it retraces a personal research path.

The 1970s: the Emergence of "Studies" and the Relations between Gender, Science, and Technologies

Since the mid-1970s, feminist scholarship has underlined the fact that women have been neglected by the history of science and technology (Harding, 1986). Scholars began to question the way in which science and technology shape social relations between men and women. Ecofeminism in particular looked into the relationship between humanity and technology and investigated male domination over nature and women (d'Eaubonne, 1974, 2018; Griffin, 1984; Daly, 1978; Merchant, 1990). The study of this domination analysed military, medical, and reproductive technologies in particular (on the latter, see especially Daly, 1978). These studies very often adopted a differentialist vision of feminism with a level of biological determinism (according to which men and women are supposedly programmed to perform different tasks). Gender determinism is often coupled with technological determinism, a perspective that dominated the philosophy of technology during the twentieth century (Dalibert, 2016). Technology was viewed as 'autonomous' (Winner, 1999) and, due to its increasing complexity, escaped the control of social systems. Determinism therefore sometimes acted as a conservative force that undermined the possibility of social change. Studies grounded in liberal feminism also considered technologies gender-neutral, but their research on gendered stereotypes led to initiatives encouraging and helping women to start using technologies (van Nostrand, 1993). While a negative vision of technology was dominant in feminist scholarship, some studies nonetheless took a more positive view, such as the work by Shulamith Firestone (1970), who thought that reproductive technologies could empower women. Whether they expressed a negative or positive view of technologies, these studies primarily dealt with the effects of technologies on women's lives. This was also key to more focused studies that addressed 'domestic' technologies (Bray, 2007; Wajcman, 2004), and which showed that 'mechanization of the domestic sphere has not substantively reduced the time that women spend on household tasks and chores, but has instead served to raise standards of cleanliness' (*'la mécanisation de la sphère domestique n'a pas substantiellement diminué le temps passé par les femmes à réaliser les tâches et les corvées ménagères, mais [...] elle a plutôt servi à élever les standards de propreté'*, Dalibert, 2016: 631). As for the domestic technologies introduced in the course of the nineteenth century, historians underline that this mechanization not only contributed to the dissolution of traditional spaces of feminine sociability (e.g., wash houses; Perrot, 1980), but also to the proletarianization of women (Perrot, 1983). Notably, this was the case during the first phase of the industrialization of textile production (1820 to 1830). Within the space of a few years, female wheel-spinners were replaced by spinning mills in which labour was 'divided up into several workshops and several operations' (*'parcellisé en plusieurs ateliers et plusieurs opérations'*) (Robert, 1980, cited in Perrot, 1983: 8). Although the spinning jenny required no physical strength, it was a highly complex machine, and its introduction led to the masculinization in the spinning trade. In other sectors, mechanization and proletarianization went hand in hand with feminization and the depreciation of traditionally male occupations. A notable case in point is typesetting. Mechanization only became cost-effective due to the feminization of labour — in the nineteenth century, women's wages were 50% lower than those of men. With the advent of the pianotype, the typesetter's job was redefined: the skills acquired during a lengthy apprenticeship were no longer necessary, and preference was given to women's 'manual dexterity' (Jarrige, 2007, 206). In all sectors, women found themselves relegated to implementation rather than decision-making tasks. Moreover, as women now worked outside the home, their morality came under scrutiny (in for instance the suspicion surrounding telephone operators, see Julliard, 2004).

The 1980s: a Growing Interest in Male and Female Users of Technology

In the 1980s, as Science and Technology studies (STS) flourished and the philosophy of technology took an empirical turn (Dalibert, 2016), technical determinism and the idea that technology was disconnected from the social disappeared. Studies adopting a constructivist approach showed how technologies result from choices made by 'groups and forces whose interests and purposes can be identified and demonstrated: capitalist employers, the industrial/military complex, the state' (Cockburn, Ormrod, 1993: 8). This approach is consistent with two major turning points, i.e. the shift from the 'study of science to the study of technology' on one hand, and the 'generalised *turn to the users*' (Dagiral, 2006: 197) on the other. Feminist STS scholars conceive technology and gender not as independent from each other, but as mutually constitutive (see also the position supported by Chabaud-Rychter and Gardey, 2002). This critique of the essentialism of both technology and gender calls intead for a more nuanced approach to categories as can be seen in the anthropology of science and technology, gender studies or indeed feminist STS. Moreover, these studies also argued that the interconnections between gender and technology may be observed from innovation down to production, distribution and use (Chabaud-Rychter, 1997a). Studies then began to focus on the normative power of technologies and the way in which they reveal, participate in, or reinforce power relations. This research developed the concept of the 'script' (Akrich, 1987) to shed light on the fact that technological artefacts can define a 'framework of action' (see below on the notion of 'gender script' proposed by van Oost, 2003). The script designates the prescriptions embedded in technological objects and which support, constrain, or shape the way in which these objects are used. The script stems from the notions designers of various technologies have of their intended users, including their competence, their reasons for using the object, and how they will use it (Wajcman, 1991). If technological objects materialise prescriptions, they also exclude some practices as well as users. Technological objects discriminate (Latour, 2009).

The 1980s also saw the birth of the cyborg, which prompted Donna Haraway (1985) to transcend binary categories (nature vs culture, biology vs technology, the material vs the immaterial). The cyborg is part of a broader movement that seeks to liberate bodies from gender binarism and decouple gender from heteronormative sexuality (Gardey, 2003). Yet while the figure of the cyborg 'embodies, even in its name which is the abbreviation of *cybernetic organism* [...] the intimacy of bodies and technologies', it above all invites reflection on the way that 'bodies are shaped by discourses, practices and power relationships arising from techno-sciences, rather than by technological objects themselves (Dalibert, 2016: 634).

Around the turn of the 1980s, feminist theorists emphasized the situatedness of scientific knowledge production introduced by experimental philosophy. In *Modest_Witness@Second_Millennium,* Donna Haraway (1997) investigated the relationship between the world of scientific knowledge production and gender relations. In her work on Robert Boyle's experiments in pneumatics she underlined that the mechanism for constructing (scientific) facts was underpinned by a social technology that established the conventions to be used in the ways philosophers of nature interacted; Boyle had invented a new aspect of masculinity: these philosophers were 'modest witnesses' (Haraway, 1997) — 'modesty' is not to be understood here in sense of a feminine quality. Scholars such as Donna Haraway (1988), Sandra Harding — alone (1986) and with Merril Hintikka (1983) — Helen Longino (1990), Evelyn Fox Keller (1996), and Linda Alcoff and Elizabeth Potter (1992) suggested that the embodied nature of intellectual work should be recognized rather than hidden, and that the production of scientific knowledge should be presented as 'founded on the incommensurability of the subject and object of knowledge' ('*fondée sur l'incommensurabilité du sujet et de l'objet de la connaissance*') (Gardey, 2006: 650). Standpoint feminism re-assessed 'the meanings of "objectivity" as a measuring category for what will be considered "scientific" or not — for example because it is putatively "biased"'(Bracke and Puig de la Bellacasa, 2009: 52). By contrast, it considered that strong objectivity implied '[taking] into account the position of those who are not "at home" in dominant positions and to value the position of the Other in order to look at our own situations in a more critical way' (Harding, 1991: 151). In this view, knowledge production is guaranteed by the confrontation of multiple viewpoints arising principally from marginal perspectives (Star, 1991).

The 1990s to 2000s: a Proliferation of Perspectives and Sectors

In the early 1990s, feminist scholarship that focussed on denaturalising the body came in the wake of critical studies investigating scientific knowledge production through a feminist prism: 'the meaning of the body appears, as always, in relation to a specific context and following perceptions and interpretations translated into the language of a given time period' (*'le corps apparaît, comme toujours, signifié dans un contexte particulier et suivant des perceptions et des interprétations traduites dans la langue d'une époque'*) (Gardey, 2006: 659). In her 1991 article, 'The Egg and the Sperm', anthropologist Emily Martin showed that gender stereotypes not only shaped scientific discourse — the meeting between egg and sperm was told as a fairy tale in which the egg has the role of the 'sleeping beauty' while the sperm is the 'heroic warrior' — but more importantly, that they brought their influence to bear on research insofar as they prevented reproductive biologists from envisaging the biochemical capacity of the egg. Feminist STS scholars also observed that in western societies the norm and ideas of universality were primarly represented by white, young, able-bodied, heterosexual, urban, middle- and upper-class male bodies, so that 'technological design may reproduce their interests and privileges to the detriment of bodies judged to be different, even deviant — in other words illegitimate bodies' (*"le design technologique risque de reproduire leurs intérêts et privilèges au détriment des corps jugés différents, voire déviants — des corps illégitimes"*, Dalibert, 2016: 630) (see also the case study in Oudshoorn, Rommes, Stienstra, 2004, which will be presented below).

In the course of the final decade of the twentieth century, feminist STS scholarship evolved under the growing influence of cultural studies (Dagiral, 2006). It specifically focussed on domestic technologies, as their design, production, distribution, and use is more likely to involve women. In their study on the microwave oven, Cynthia Cockburn and Susan Ormrod (1993) showed how the marketing for different target groups (British single men, then British mothers) impacted not only the gendered design of the object (its colour, functionalities, discourses around it, display in shops, etc.), but also women's participation in the aforementioned process. When the object was designed for mothers, women were involved as home economists alongside the engineers in the design phase, while on the sales floor, white-goods saleswomen took the place of brown-goods salesmen. Some studies took an interest in the technologies more likely to involve women in product design and marketing (Cockburn, Ormrod, 1993); others looked into technologies and production steps less likely to involve women, in order to establish how masculine culture permeated the design stage in particular, and how it contributed to producing an expectation of the skills that users were supposed to bring into play (Cockburn, 1985, 1991). Although many feminist studies adopted a constructivist approach that challenged the way in which society influenced technical objects, this combination of feminist and constructivist perspectives caused some problems (Dagiral, 2006). Cynthia Cockburn (1992), for example, underlined that Latourian Actor-Network Theory rendered women even less visible because attention was concentrated on technical objects and network-building rather than on human actors. She added that ANT tended to obscure power relations (see also Star, 1991).

The advent of the World Wide Web in the same decade led feminist scholarship to examine how technology constructed gendered social relations. Anna Cossetta (2012) identified three approaches to understanding the relationship between women and the Web during the 1990s. The first is through the prism of a feminist cyberutopia, as formulated in the cyberfeminist manifesto written by the Adelaide-based VNS Matrix collective and inspired by Donna Haraway's (1985) *Cyborg Manifesto*. It suggested that a parallel could be drawn between the autonomy of women and that of machines (Plant, 1997), as 'technological development afforded a glimpse of a liberation of identities' (*'le développement technologique laiss[ant] entrevoir une libération des identités'*) (Cossetta, 2012: 392). The second approach, referred to 'Women in Technology', observed that women used the Web less and investigated the reasons for this gender digital divide. The related literature highlighted that women's access to the Web was hampered not only by gender stereotypes, but also by the burdens of domestic chores (Kennedy, Wellman, Kleme, 2003), child care (Liff et al., 2004), and a limited access to technical devices (especially before home computers were common). As this approach viewed technologies as gender-neutral, it advocated for the development of women's skills to encourage them to make full use of these technologies. The third approach, named 'Women & Technology', encompassed studies that also considered technology to be gender neutral but contended that women were reluctant to master computer technology as they viewed it as culturally alien to them (Turkle, 1988). Her in-depth interviews with female computer programming students at Harvard and MIT revealed that they were less involved in programming than their male counterparts, as they rejected hacker culture and the stereotype attached to it, of a masculine figure in an intimate relation with machines, an image which then dominated computer science. What appeared to prevent these female students from viewing the computer as an 'intimate machine' was the value they attached to human relations. Anna Cossetta noted that, when the Web came to be seen as a male hunting ground that opened up the practice of cybersex, (in her words) 'digital gynaecea' appeared: 'groups for lesbians [...], magazine for geek girls or women hackers [...] and sites devoted to health and mutual support' (Cossetta, 2012: 394). The 1990s also saw a growing number of studies in France on the relationship between gender and technology. These focused more specifically on domestic technologies (Chabaud-Rychter, 1994, 1997a, 1997b), technical practices found in women's work (Gardey, 1998, 2001), and, more rarely, on the media, mostly in an international context (Mattelard, 1986).

In the 2000s, feminist scholars took an increasing interest in the mutually constitutive interactions of gender and technology. These studies generated greater insights into how the design of technical objects produces a feeling of technical incompetence in women as subordinate groups, particularly with respect to digital technology. Ellen van Oost's (2003) study on the design trajectories of Philips shavers for women and for men is noteworthy due to its development of the notion of the 'gender script' (following the concept of 'script') (Akrich, 1987). The idea of a gender script prompts a reflection on what — sometimes unwittingly — attributes a gendered identity to targeted users. Ellen van Oost's study emphasized the gendered dimension of the Philips shaver design, underlining that those products designed for women concealed the shaver's technical elements. The 'lipstick' shaver, for instance, which was produced in the late 1950s, had a name, shape, packaging, etc. that were closely associated with cosmetic products. The gendered dimension of the Philips shaver design became even more pronounced in 1967, when the company introduced separate production lines for men's and women's shavers. The products for men benefitted from technical innovations (e.g. additional microelectronic components), whereas the idea that women were technology-averse led the designers to mask the shaver's technical aspects and avoid innovation. Adjustable razor heads for men gave users more options and agency, while the shift from visible screws to a system that irreversibly locked the different parts of ladies' shavers in place limited women's options. Technical incompetence was thus created as feminine. As pointed out by Marion Coville, the design of lady shavers 'largely contributes to concealing the possibility for women to perceive themselves as being interested in technology or as technically competent' (*'participe largement à dissimuler la possibilité pour les femmes de se percevoir comme intéressées par les technologies ou comme ayant des compétences techniques'*) (Coville, 2017). In their study of an online information service developed by the City of Amsterdam, Nelly Oudshoorn, Els Rommes, and Marcelle Stienstra (2004) have shown that when most members of a design team are well-educated men with high technical competence, and when they take themselves as the point of reference for making design choices and use their own representations of the potential users of technical artefacts rather than referring to actual users (the I-methodology), they actually exclude certain users — first and foremost those who are the least competent as well as women. Although gender scripts do not directly dictate users' behaviour — users can always resist the script (Oudshoorn and Pinch, 2003) — they nonetheless produce frameworks for action.

In France, the 2000s saw the publication of an increasing number of studies on the links between information technology (Gardey, 2003) and communication technology (Jouët, 2003), as well as on the role of the media in the construction of gender (Bonnafous, 2003). Events happening in France at the time — namely debates on the civil solidarity pact (PACS) and gender parity — played a role (Julliard, 2009). Josiane Jouët (2003) pointed out that gender was seldom a central issue in the literature of the 'sociology of uses', even though it is a key element in the adoption of telematics or computing. One exception is the telephone, which was the subject of several studies (Quéré, Smoreda, 2000), some of them written from a gender perspective. In my study of the telephone in the French context during the period of manual switching, I describe how the economic model, the work organization, practices, and associated representations all interact to produce a low-performing and feminine communication object, along with a femininity that is either idle (users) or unproductive (switchboard operators). In other words, in France at the turn of the twentieth century, the telegraph and the telephone produced gender (Julliard, 2004). Covering the same period in the United States, Venus Green (2001) has shown that the work organization in the Bell Telephone Company, where segregation was rife, not only produced gender but also race; different jobs were proposed to white and black men and white and black women, and each group occupied a separate work space. In the French context, the production of gender and nationality interacted in a way that disqualified women in particular (Julliard, 2004). When a French woman married a foreigner, she automatically lost her French nationality and, as she was presumed to be less loyal to the institution and pose a threat to the confidentiality of conversations, she was obliged to leave public service. While the sociology of uses reduced gender to the sole dimension of feminine experience (Jouët, 2011: 82) without considering its relational dimension and the evolving masculine and feminine identities that developed in the relationship to information and communication technologies, Laurence Le Douarin's study (2004) showed that the advent of the home computer clearly revealed gender power relations. In her study involving twenty-seven families (i.e. heterosexual couples with children) in the Paris area, the author showed that the recognition of technical competence was linked to power relations not only in the couple but also between parents and children.

In the 2010s, different perspectives for exploring the relationship between gender and digital technology flourished. Several studies pointed out that although the digital divide between men and women tended to be narrowing — an increasing number of women used the internet and spent more time than men on social media — this quantitative catch-up can not conceal the reality of contrasting levels of appropriation of technologies which reflected distinct gendered skills resulting from men and women's different social integration. For example, Josiane Joüet (2011) asserted that women appropriated computers as soon as these became common and established as a means of communication (mainly with the rise of the Web). This was also one of Anna Cossetta's (2012) findings. In her view, women's use of the internet was linked to their social skills, while men's practices were more likely to lead to increased technical skills. Mindful to avoid a gender-based categorization, Josiane Joüet (2007) nonetheless reminded her readers of the heterogeneity of the 'male' and 'female' categories, which encompass extremely variable practices depending on age, social class, profession, or the type of couple (van Zoonen, 2002) of which men and women form part. Age remains particularly key as a discriminatory variable (Délias, 2019).

In her study on museum exhibitions featuring video games, Marion Coville (2016) has highlighted how masculine and feminine identies were differently constructed in relation to technological objects. Using the notion of 'gender script' (van Oost, 2003), the author examined 'the articulation between the gendered imaginary mobilised by designers and the design of technical objects' (*'l'articulation entre les représentations, genrées, que les concepteurs mobilisent et la conception des objets techniques'*) and underlined how exhibitions suggested not only a relationship between individuals and technical objects, but also 'a relationship between individuals by proposing an organised visit and a division of activities and skill sets between [male and female] visitors' (*'une relation entre les individus, en proposant une organisation de la visite et une répartition des activités et compétences entre les visiteur·se·s'*) (Coville, 2016: 244). In a study more specifically focussed on the hegemonic male figure of the video game creator, Marion Coville (2014) has suggested that the restrictive definition of video games makes any alternative forms of creativity invisible. It rules out girls' games, which are produced by the studios that female creators turn to, notably, because the video game industry fails to cater for female gamers. Queer games — where gamers who identify as queer or trans can share their experience — are often denied the status of games.

In my research I have looked into the new materialities of languages in the digital environment. By 'materiality', I mean the physicality of signs and media. In line with Monique Wittig (2010) and Stéphanie Kunert (2016), I establish a link between this materiality and the material oppression of individuals perceived as women, in keeping with the idea that language shapes material reality. Understanding the way in which digital writing tools contribute to renewing the signs through which gender is produced has been the main focus of my recent research. I have shown how gender has been constructed by new signs circulating on the Web. Examples include the stereotyped default male/female profile pictures and the different interests available to internet users on dating sites depending on what sex they declare which disregard other user-provided explanatory information or place of residence, age, or social class (Julliard, 2013). I have also studied the way in which gender has been naturalized by images circulating on the Web which present non-binary existences as monstrous, and I have explored how digital writing tools encourage an emotional attachment to the gender binary (Julliard 2018a, 2018b).

These studies are part of a broader movement in French academic scholarship which adopts an intersectional perspective (Hooks, 1990; Crenshaw, 1994; Hill Collins, 2000) to understand how media discourses depend on and produce articulations of specific power relations (notably in relation to gender, class, and race) (Bilge, 2009). In the information and communication sciences, Marion Dalibert (2013, 2014) has explored the way in which the media produce race through gender and the definition of masculinities and femininities specific to suburban areas.

The expanding scholarship on the relationship between gender and technology covers an increasingly diversified range of objects including 'somatechnologies'. In line with Donna Haraway's proposals (1985), the notion of 'prostheticity' has become widespread in research looking at the proximity between bodies and technologies (Jain, 1999; Smith, Morra, 2006). Nevertheless, feminist studies on disability consider Donna Haraway's conceptualization to be disconnected from the 'material and experienced realities of prosthetic bodies and other somatechnologies' ('réalités matérielles et vécues des corps munis de prothèses et autres somatechnologies', (Dalibert, 2016: 634). Haraway's conceptualization seems to suggest a normative view of (supposedly able) bodies that increases the invisibility of disabled bodies.[16] The academic interest in somatechnologies is justified as they direct attention to the 'trans-corporeality' of human existence. As Lucie Dalibert (2016: 635) asserts:

> [t]he concept of trans-corporeality highlights that 'the human is always intermeshed with the more-than-human world', and notably 'the often unpredictable and unwanted actions of human bodies, non-human creatures, ecological systems, chemical agents, and other actors' (Alaimo, 2010: 238).

> (*Le concept de transcorporalité met en relief 'les interconnections de nature matérielle [qui existent] entre la corporalité humaine et le monde plus-qu'humain', et notamment 'les actions souvent imprévisibles et non désirées [sur] les corps humains, des créatures non-humaines, des systèmes écologiques, des agents chimiques, et d'autres acteurs'* (Alaimo, 2010: 238)

This is also the case when a body fitted with a prosthesis triggers a security alarm and attracts attention to itself. Although openly living with somatechnologies at the sensory-kinetic level (embodiment) may prove delicate, viewing them as an intimate part of one's body and fully identifying with this technologically transformed body (incorporation) is particularly difficult in this framework (Dalibert, 2015, 2016).

Following the pioneering work of Lisa Nakamura (2015), Safiya Umoja Noble (2018), and Ruha Benjamin (2019), French research has investigated the discrimination perpetuated by algorithms (Jean, 2019; Grison, Julliard, 2021). This research highlights that gender, class, and race inequalities are reinforced by ignoring the discriminatory effects of the variables used (Amabile et al., 2020; Bertail et al., 2019) and of the bias in the collection and sorting of data, its quality (Zuiderveen Borgesius, 2018), and even the definition of economic and profitability objectives (Lambrecht, Tucker, 2019). The increased use of algorithms in moderating 'hate' content online — in the shape of legislation encouraged at national and European levels —, thus has the effect of making the discrimination of groups of people that Web companies see as 'protected categories' worse. Indeed, charities and NGOs observe an increase in the censorship of activists' accounts due to algorithms being unable to differentiate between actual insults and the ironic, militant reappropriation of insults by feminists and LGBT+ groups (for example 'dyke') (Grison et al., 2023).

Is There Such a Thing as Feminist Technology?

I wish to conclude this chapter with a line of research that questions what actually makes a technological artefact truly feminist, and not simply an object designed for women or presented as feminist for the sake of marketing (Layne, Vostral, Boyer, 2010). Such an object should be beneficial to women and broaden their ability to act (Gorenstein, 2010), or even ensure their autonomy, as in the case of reproductive techniques that open up the possibility of parenthood for female couples (Lewin, 1993). However, a technological artefact might benefit some women to the detriment of others, given the diversity of lives encompassed by the category of 'women', and it may well be perceived as feminist or anti-feminist depending on the feminist perspective taken (Layne, 2010). The menstrual suppression pill (Aengst, Layne, 2010) was rejected by radical feminists as being profoundly anti-feminist and met with approval from liberal feminists, who believed it gave women a broader choice of contraceptives. The attempt to qualify a technological artefact as either feminist or anti-feminist means not only investigating the complexity of the context in which the object is used, but also examining different feminist perspectives (Gorenstein, 2010). The specific culture-based form of patriarchy is a further contextual element to be taken into account (Meyer, 1987). On this point, the use of Wi-Fi by American and Japanese women is enlightening. In the American case, internet accessibility from diverse locations has enabled women to accomplish several domestic tasks concurrently (e.g., online shopping while taking their children to various activities). In Japan, it has allowed women who find it difficult to enter the labour market to trade anonymously online (with the double advantage of not being disqualified as a woman and being able to have a hand in the household budget without informing their husbands). In both cases, women's scope of action has increased, but still within the limits of the patriarchal system in which they live (Layne, 2010). Despite the difficulty of calling a technological artefact feminist, and referring back to Cynthia Cockburn's and Susan Ormrod's suggestion (1993) that it is no longer novelty that should be promoted but what will actually benefit consumers, Linda Layne (2010) has proposed that the design of feminist objects should be encouraged. With this in mind, she imagines not only setting up a classification system based on the benefits the objects have for women, much like the ratings of a building's energy efficiency in the United States, but also creating foundations that evaluate feminist technological innovations and publicize the most effective among them (see also Bronet, Layne, 2010).

References

Aengst, Jennifer and Linda L. Layne, 'The Need to Bleed? A Feminist Technology Assessment of Menstrual Birth Control Pills', in *Feminist Technology*, ed. by Linda L. Layne, Sharra L. Vostral, and Kate Boyer (Urbana, Chicago and Springfield, IL: University of Illinois Press, 2010), pp. 55–88.

Akrich, Madeleine, 'Comment décrire les objets techniques?', *Techniques et Culture*, 9 (1987), 49–64.

Albury, David and Joseph Schwartz, *Partial Progress: the Politics of Science and Technology* (London: Pluto Press, 1982).

Alcoff, Linda and Elizabeth Potter, eds, *Feminist Epistemologies* (New York, NY: Routledge, 1992).

Amabile, Arnai, Théophile Lenoir, Tanya Perelmuter, and Basile Thodoroff, *Algorithmes: contrôle des biais S.V.P.* (Paris: Institut Montaigne, 2020).

Benjamin, Ruha, *Race After Technology: Abolitionist Tools for the New Jim Code* (Cambridge: Polity Press, 2019).

Bertail, Patrice, David Bounie, Stephan Clémençon, and Patrick Waelbroeck, *Algorithmes: biais, discrimination et équité* (Paris: Télécom Paris Tech, Fondation Abeona, Institut Mines-Télécom, Institut Carnot, 2019).

Bilge, Sirma, 'Théorisations féministes de l'intersectionnalité', *Diogène*, 225 (2009), 70–88.

Bonnafous, Simone, 'Femme politique: une question de genre', *Réseaux*, 120 (2003), 119–25.

Bracke, Sarah, and Maria Puig de la Bellacasa, 'The Arena of Knowledge: Antigone and Feminist Standpoint Thinking', in *Doing Gender in Media, Art and Culture*, ed. by Rosemarie Buikema and Iris van der Tuin (London: Routledge, 2009), pp. 39–53.

Bray, Francesca, 'Gender and Technology', *Annual Review of Anthropology*, 36 (2007), 37–53.

Bronet, Frances and Linda L. Layne, 'Teaching Feminist Technology Design', in *Feminist Technology*, ed. by Linda L. Layne, Sharra L. Vostral, and Kate Boyer (Urbana, Chicago and Springfield, IL: University of Illinois Press, 2010), pp. 179–96.

Chabaud-Rychter, Danielle, 'Introduction au dossier Genre et techniques domestiques', *Cahiers du Gedisst (Groupe d'Études sur la Division Sociale et Sexuelle du Travail)*, 20 (1997a), 5–13.

———, 'L'Industriel et le domestique dans la conception d'appareils électroménagers', *Cahiers du Gedisst*, 20 (1997b), 63–96.

———, 'La Mise en forme des pratiques domestiques dans le travail de conception d'appareils électroménagers', *Sociétés contemporaines*, 17 (1994), 103–18.

——— and Delphine Gardey, eds, *L'Engendrement des choses. Des hommes, des femmes et des techniques* (Paris: Éditions des Archives Contemporaines, 2002).

Cockburn, Cynthia, *Brothers: Male Dominance and Technological Change* (New York, NY: Macmillan, 1985).

———, *In the Way of Women*, (Ithaca and New York, NY: ILR Press, 1991).

———, 'The Circuit of Technology: Gender, Identity and Power', in *Consuming Technologies. Media and Information in Domestic Spaces*, ed. by Roger Silversone and Eric Hirsch (London and New York, NY: Routledge, 1992), pp. 32–47.

——— and Susan Ormrod, *Gender and Technology in the Making* (London: Sage, 1993).

Cossetta, Anna, 'Que donnent les femmes sur le web?', *Revue du MAUSS*, 39 (2012), 391–404.

Coville, Marion, 'La Construction du jeu vidéo comme objet muséal. Le détournement d'un objet culturel et technique de son cadre d'usage initial et son adaptation au contexte muséal. Étude de cas dans un centre de sciences' (unpublished doctoral thesis, University of Paris-1-Pantheon-Sorbonne, 2016).

———, 'Créateurs de jeux vidéo et récits de vie: la formation d'une figure hégémonique', *Revue française des sciences de l'information et de la communication (RFSIC)*, 4 (2014) <http://journals.openedition.org/rfsic/763> [accessed 31 March 2023].

———, 'Feminist STS. E. Van Oost, la conception des objets techniques et la production de "scripts de genre"', *L'Exposition en jeu. Exposition, médiation et jeu vidéo* (2017) <https://expojeu.hypotheses.org/226> [accessed 31 March 2023].

Crenshaw, Kimberley W., 'Mapping the Margins: Intersectionality, Identity Politics, and Violence against Women of Color', in *The Public Nature of Private Violence*, ed. by Martha Albertson Fineman and Roxanne Mykitiuk (New York, NY: Routledge, 1994), pp. 93–118.

Dagiral, Éric, 'Genre et technologie. (Note critique)', *Terrains & travaux*, 10 (2006), 194–206.

Dalibert, Lucie, 'Living with Spinal Cord Stimulation: Doing Embodiment and Incorporation', *Science, Technology & Human Values*, 41.4 (2015), 1–25.

———, 'Technologie', in *Encyclopédie critique du genre*, ed. by Juliette Rennes (Paris: La Découverte, 2016), pp. 628–39.

———, 'Posthumanism and Somatechnologies. Exploring the Intimate Relations Between Bodies and Technologies' (unpublished doctoral thesis, University of Twente, 2014).

Dalibert, Marion, 'Authentification et légitimation d'un problème de société par les journalistes: les violences de genre en banlieue dans la médiatisation de Ni putes ni soumises', *Études de communication*, 40 (2013), 167–80.

———, 'Le marquage socio-discursif de la race par le genre. Les "roms", les Tunisiens, les Ukrainiens et les habitants des banlieues françaises dans les médias', *Revue Française des Sciences de l'Information et de la Communication*, 4 (2014) <http://rfsic.revues.org/743> [accessed 31 March 2023].

Daly, Mary, *Gyn/Ecology. The Metaethics of Radical Feminism* (London: Beacon Press, 1978).

Dardigna, Anne-Marie, *Femmes sur papier glacé* (Paris: Maspéro, 1974).

———, *La Presse 'féminine'. Fonction idéologique* (Paris: Maspéro, 1978).

d'Eaubonne, Françoise, *Écologie et féminisme, révolution ou mutation?* (Paris: Libre et solidaire éditeur, 2018.

———, *Le Féminisme ou la mort* (Paris: Pierre Horay Éditeur, 1974).

de Lauretis, Teresa, *Technologies of Gender: Essays on Theory, Film, and Fiction* (Bloomington, IN: Indiana University Press, 1987).

Délias, Lucie, 'L'Âge des usages. Usages sociaux des technologies numériques par les adultes âgéés et représentations du "bien-vieillir" connecté' (unpublished doctoral thesis, University Paris-3-Sorbonne Nouvelle, 2019).

Firestone, Shulamith, *The Dialectic of Sex. The Case for Feminist Revolution* (New York, NY: Bentam Books, 1970).

Fox Keller, Evelyn and Helen Longino, eds, *Feminism and Science* (Oxford: Oxford University Press, 1996).

Friedan, Betty, *The Feminine Mystique* (New York, NY: Norton and Co, 1963).
Gardey, Delphine, *La Dactylographe et l'expéditionnaire. Histoire des employés de bureau (1890–1930)* (Paris: Belin, 2001).
———, 'De la domination à l'action. Quel genre d'usage des technologies de l'information?', *Réseaux*, 120 (2003), 87–117.
———, 'Les Sciences et la construction des identités sexuées. Une revue critique', *Annales. Histoire, Sciences Sociales*, 61.3 (2006), 649–73.
———, 'La Standardisation d'une pratique technique: la dactylographie (1883–1930)', *Réseaux*, 87 (1998), 75–103.
Gill, Rosalind and Keith Grint, 'The Gender-technology Relation: Contemporary Theory and Research', in *The Gender-technology Relation. Contemporary Theory and Research*, ed. by Keith Grint and Rosalind Gill (London: Taylor and Francis, 1995), pp. 1–28.
Gorenstein, Shirley, 'What we now know about Feminist Technologes', in *Feminist Technology*, ed. by Linda L. Layne, Sharra L. Vostral, and Kate Boyer (Urbana, Chicago and Springfield, IL: University of Illinois Press, 2010), 203–14.
Green, Venus, *Race on the Line: Gender, Labor, and Technology in the Bell System, 1880–1980* (Durham, NC: Duke University Press, 2001).
Griffin, Susan, *Woman and Nature: the Roaring Inside Her* (London: The Women's Press, 1984).
Grison, Thibault and Virginie Julliard, 'Les Enjeux de la modération automatisée sur les réseaux sociaux numériques: les mobilisations LGBT contre la loi Avia', *Communication, technologie et développement*, 10 (2021) <http://journals.openedition.org/ctd/6049> [accessed 31 March 2023].
———, ———, Félix Alié and Victor Ecrement, 'La moderation abusive sur Twitter. Étude de cas sur l'invisibilisation des contenus LGBT et TDS en ligne', *Réseaux*, 237 (2023), 53–86.
Haraway, Donna J., 'A Manifesto for Cyborgs: Science, Technology and Socialist Feminism in the 1980s', *Socialist Review*, 80 (1985), 65–108.
———, *Modest_Witness@Second_Millennium. FemaleMan© _Meets_ OncoMouse: Feminism and Technoscience* (New York, NY and London: Routledge, 1997).
———, *Simians, Cyborgs, and Women. The Reinvention of Nature* (London: Free Association Books, 1991).
———, 'Situated Knowledges: the Science Question in Feminism and the Privilege of Partial Perspective', *Feminist Studies*, 14.3 (1988), 575–99.
Harding, Sandra, *The Science Question in Feminism* (Milton Keynes: Open University Press, 1986).
———, *Whose Science? Whose Knowledge? Thinking from Women's Lives* (Ithaca, NY: Cornell University Press, 1991).
——— and Merrill Hintikka, eds, *Discovering Reality: Feminist Perspectives on Epistemology, Metaphysics, Methodology, and Philosophy of Science* (Dordrecht: Reide, 1983).
Hill Collins, Patricia, *Black Feminist Thought. Knowledge, Consciousness, and the Politics of Empowerment* (London and New York, NY: Routledge, 2000).
hooks, bell, *Yearning. Race, Gender, and Cultural Politics* (Boston, MA: South End Press, 1990).

Jain, Sarah S., 'The Prosthetic Imagination: Enabling and Disabling the Prosthetic Trope', *Science, Technology & Human Values*, 24.1 (1999), 31–54.

Jarrige, François, 'Le Mauvais genre de la machine. Les ouvriers du livre et la composition mécanique (France, Angleterre, 1840–1880)', *Revue d'histoire moderne et contemporaine*, 54.1 (2007), 193–221.

Jean, Aurélien, *De l'autre côté de la machine: voyage d'une scientifique au pays des algorithmes* (Paris: Éditions de l'Observatoire, 2019).

Jouët, Josiane, 'Du genre et des objets communicationnels', *MEOTIC Conference* (Grenoble, 2007) <https://lesenjeux.univ-grenoble-alpes.fr/2007-meotic/Jouet/index.html> [accessed 31 March 2023].

———, 'Technologies de communication et genre. Des relations en construction', *Réseaux*, 120 (2003), 53–86.

———, 'Des Usages de la télématique aux *Internet Studies*', in *Communiquer à l'ère numérique*, ed. by Julie Denouël and Fabien Granjon (Paris: Presses des Mines, 2011), pp. 45–90.

Julliard, Virginie, 'La "Différence des sexes" sur Twitter: les conditions d'observabilité d'un engagement affectif et émotionnel' (habilitation thesis, University of Paris Est Créteil, 2018a).

———, 'Éléments pour une "sémiotique du genre"', *Communication & Langages*, 177 (2013), 61–76.

———, 'L'Idéologie raciste en appui aux discours antiféministes: les ressorts émotionnels de l'élargissement de l'opposition à la "théorie du genre" à l'école sur Twitter', *Les Cahiers du genre*, 65 (2018b), 17–39.

———, 'Pour une intégration du genre par les sciences de l'information et de la communication', *Questions de communication*, 16 (2009), 191–210.

———, 'Une "femme machine" au travail: la "demoiselle du téléphone"', *Quaderni*, 56 (2004), 23–32.

Kennedy, Tracy, Barry Wellman, and Kristine Kleme, 'Gendering the Digital Divide', *IT&Society*, 1.5 (2003), 72–96.

Kunert, Stéphanie, 'Monique Wittig: de la matérialité du langage', in *Matérialismes, culture et communication*, vol. II, *Cultural studies, théories féministes et décoloniales*, ed. by Maxime Cervulle, Nelly Quemener, and Florian Vörös (Paris: Presse des Mines, 2016), pp. 143–64.

Lambrecht, Anja and Catherine Tucker, 'Algorithmic Bias?' An Empirical Study into Apparent Gender-based Discrimination in the Display of STEM Career Ads', *Management Science*, 65.7 (2019), 2966–81.

Latour, Bruno, 'Where Are the Missing Masses? The Sociology of a Few Mundane Artifacts', in *Technology and Society. Building Our Sociotechnical Future*, ed. by Deborah G. Johnson and Jameson M. Wetmore (Cambridge, MA: The MIT Press, 2009), pp. 151–80.

Layne, Linda L., 'Introduction', in *Feminist Technology*, ed. by Linda L. Layne, Sharra L. Vostral, and Kate Boyer (Urbana, Chicago and Springfield, IL: University of Illinois Press, 2010), pp. 1–36.

———, Sharra L. Vostral, and Kate Boyer, eds, *Feminist Technology* (Urbana, Chicago and Springfield, IL: University of Illinois Press, 2010).

Le Douarin, Laurence, 'Hommes, femmes et micro-ordinateur: une idéologie des compétences', *Réseaux*, 123 (2004), 149–74.

Lettow, Susanne, 'Somatechnologies: Rethinking the Body in Philosophy of Technology', *Techné*, 15.2 (2011), 110–17.

Lewin, Ellen, *Lesbian Mothers: Accounts of Gender in American Culture* (Ithaca, NY: Cornell University Press, 1993).

Liff, Sonia A., Adrian Shepherd, Judy Wajcman, Ronald Rice, and Eszter Hargittai, 'An Evolving Gender Digital Divide?', *OII Internet Issue Brief*, 2 (2004) <https://papers.ssrn.com/sol3/papers.cfm?abstract_id=1308492> [accessed 31 March 2023].

Longino, Helen E., *Science as Social Knowledge: Values and Objectivity in Scientific Inquiry* (Princeton, NJ: Princeton University Press, 1990).

Martin, Emily, 'The Egg and the Sperm: How Science Has Constructed a Romance Based on Stereotypical Male-Female Roles" Signs, 3(1991), 485–501.

Mattelart, Michèle, 'El Nivel mitico de la prensa seudo-amorosa', *Cuadernos de la realidad national*, 3 (1970), 221–80.

———, 'Femmes et médias: retour sur une problématique', *Réseaux*, 120 (2003), 23–51.

———, 'Women and the Cultural Industries', in *Media, Culture and Society. A Critical Reader*, ed. by Richard Collins, James Curran, Nicholas Garnham, Paddy Scannell, Philipp Schlesinger, and Colin Sparks (London: Sage, 1986), pp. 63–81.

Merchant, Carolyn, *The Death of Nature. Women, Ecology and the Scientific Revolution* (San Francisco, CA: HarperOne, 1990).

Meyer, Donald, *Sex and Power: the Rise of Women in America, Russia, Sweden, and Italy.* (Middletown, CT: Wesleyan University Press, 1987).

Moser, Ingunn, 'De la normalisation aux cyborg studies: comment repenser le handicap', *Cahiers du Genre*, 38 (2005), 127–62.

Nakamura, Lisa, 'The Unwanted Labour of Social Media: Women of Colour Call out Culture as Venture Community Management', *New Formations: a Journal of Culture/Theory/Politics*, 86 (2015), 106–12.

Noble, Safiya Umoja, *Algorithms of Oppression: How Search Engines Reinforce Racism* (New York, NY: New York University Press, 2018).

Ormrod, Susan, 'Genre et pratiques discursives dans la création d'un nouveau mode culinaire', *Cahiers du Gedisst*, 20 (1987), 37–62.

Oudshoorn, Nelly and Trevor Pinch, eds, *How Users Matter: the Co-Construction of Users and Technologies* (Cambridge, MA: The MIT Press, 2003).

———, Els Rommes, and Marcelle Stienstra, 'Configuring the User as Everybody: Gender and Design Cultures in Information and Communication Technologies', *Science, Technology & Human Values*, 29.1 (2004), 30–63.

Perrot, Michelle, 'Femmes au lavoir', *Sorcières*, 19 (1980), 128–33.

———, 'Femmes et machines au XIXe siècle', *Romantisme*, 41 (1983), 5–18.

Plant, Sadie, *Zeroes and Ones: Digital Women and the New Technoculture* (New York, NY: Doubleday, 1997).

Quéré, Louis and Zbigniew Smoreda, 'Le Sexe du téléphone. Présentation', *Réseaux*, 103 (2000), 9–17.

Smith, Marquard and Joanne Morra, eds, *The Prosthetic Impulse. From a Posthuman Present to a Biocultural Future* (Cambridge, MA and London: The MIT Press, 2006).

Star, Susan L., 'Power, Technology and the Phenomenology of Conventions: On Being Allergic to Onions', in *A Sociology of Monsters: Essays on Power, Technology and Domination*, ed. by John Law (London: Routledge, 1991), pp. 21–57.

Sullerot, Évelyne, *La Presse féminine* (Paris: Armand Colin, 1966).

Turkle, Shirley 'Computational Reticence. Why Women Fear the Intimate Machine', in *Technology and Women's Voices. Keeping in Touch*, ed. by Cheris Kramarae (London: Routledge & Kegan Paul Ltd, 1988), pp. 41–61.

van Nostrand, Catherine, *Gender-responsible Leadership* (London: Sage, 1993).

van Oost, Ellen, 'Materialized Gender: How Shavers Configure the Users' Feminity and Masculinity', in *How Users Matter: The Co-construction of Users and Technologies*, ed. by Nelly Oudshoorn and Trevor Pinch (Cambridge, MA: The MIT Press, 2003), pp. 193–208.

van Zoonen, Lisbet, 'Gendering the Internet', *European Journal of Communication*, 17.1 (2002), 5–23.

Wajcman, Judy, *Feminism Confronts Technology* (University Park, PA: The Pennsylvania State University Press, 1991).

———, *Technofeminism* (Cambridge and Oxford: Polity Press, 2004).

Winner, Langdon, 'Do Artifacts Have Politics?', in *The Social Shaping of Technology*, ed. by Donald MacKenzie and Judy Wajcman (London: Open University Press, 1999), pp. 26–38.

Wittig, Monique, *Le Chantier littéraire* (Lyon: Presses universitaire de Lyon, 2010).

Zuiderveen Borgesius, Frederik, *Discrimination, intelligence artificielle et décisions algorithmiques* (Strasburg: Conseil de l'Europe, 2018).

MARTINE BEUGNET, MATHIEU DUPLAY

Technologies of Art and Culture

Film and Sound

The growth of cities and the rise of the assembly line, the railway, the automobile and the plane, the telephone and the telegraph, cinema and the phonograph: the emergence, with the Industrial Revolution, of modern ways of life, work, transport, and means of communication cannot be dissociated from the cultural and artistic production of their time. Modernity's artistic expressions and cultural practices did not merely grow out of technological, economic, and social transformations, but embraced, extended, and reflected on them (Charney, Schwartz, 1995: 10). Mobility, fragmentation, sensory excess, ephemerality: modernity's key attributes arguably find their 'fullest expression and combination' in the new, technologized forms of expression, amongst them time-based image and sound media that would later combine in sound film (Charney, Schwartz, 1995: 10). In effect, the evolution of the media forms that emerged at the turn of the nineteenth and the twentieth centuries continued to serve as a key marker of technological, social, and economic changes, including those ushered in by the advent of digital media.

The introduction of technologies of audio-visual reproduction had three characteristic effects on the conditions of production and reception of images and sounds: they extended the recording and representation of the visible and audible beyond the possibilities of ordinary human perception; they revolutionized the traditional understanding and experience of distance and time; and they normalised the practice of the copy, displacing uniqueness to the margins of audio-visual culture. More than a century after the invention of the phonograph, photography and cinema so radically altered the face of audio-visual culture, the advent of digital media not only renewed, but also intensified art and culture's technological turn.

Early on, theorists of modernity drew ambivalent conclusions, pointing to the gains, but also to the losses inherent in the fusion of art and technology (Frisby, 1986). Accordingly, the brief survey that follows accounts for some of modernity's enduring paradoxes as they manifest themselves through its emblematic media, the cinema and the phonograph.

In the early years of film, a number of key inventors such as Thomas Edison and William Kennedy-Laurie Dickson were interested in combining image and sound. Arguably, the first known example of film with synchronized sound, the Dickson Experimental Sound Film system dates from 1895. Furthermore, on the occasion

Global History of Techniques, ed. by Guillaume Carnino, Liliane Hilaire-Pérez and Jérôme Lamy, Global Matters Culture Matérielles, histoire des techniques et circulation globale des savoirs, 9 (Turnhout: Brepols, 2024), pp. 687–704
BREPOLS PUBLISHERS 10.1484/M.TECHNE-EB.5.129836

of the Universal Exhibition in Paris in 1900, a number of colour and sound films produced by the Frenchman Henri Lioret and Clément-Maurice Gratioulet's *Phono-Cinéma-Théâtre* were screened (featuring theatre and music-hall stars such as Sarah Bernhardt, Jean Coquelin, Réjane, Cléo de Mérode, Polin, Little Tich, and Footit and Chocolat).[1] Yet technological research in sound and in moving images also progressed independently, ultimately leading to their separate treatment. Therefore, at a time when intermedial experiments abounded across music, painting, poetry, literature, dance, and film (from the 'musicality' of Matisse's work, to the 'impressionism' of Debussy, and Viktor Eggeling and Hans Richter's 'painting in movement' and 'visual music'), in film and sound media, the devices that were most successfully developed and commercialized (i.e. the cinematograph, the phonograph, and the gramophone) dealt with either sound or image (Marcus, 2007). Although early on, film screenings were accompanied by live sound and music, it would take until the late 1920s for sound and image to be recorded and combined in one single medium. The history of technologies has accordingly tended to dissociate them, and our account follows suit. Yet it is of import to underline how, in choosing such a path for technical and commercial reasons, time-based sound and image technologies effectively worked to further de-naturalize perception, with significant repercussions. As far as sound was concerned, the principle of dissociation (of audio recording from its origin, the emitting body) was the source of a schizophrenic or spectral mode of representation (Figure 1). The initial disconnection of moving images and sound lent film similar ghostly qualities, but also kept film away from the lure of realism, thus arguably helping to establish it as an art form.[2]

Furthermore, if optics and sound technology helped spectators and listeners see and hear beyond what their human senses allowed, such forms of perception were also defined by a fundamental loss: that of presence or, more precisely, the co-presence of the object of the gaze or of the listening, and the spectator/listener. In turn, the massive democratization of access to culture facilitated by the digital display and circulation of images and sounds is marked by a power to dematerialize (its objects) and individualize (its target, consumer, or subject) that continues to trouble observers of our contemporary hyper-technologized culture.

---

1 Some of the films were screened again on the occasion of the symposium 'Cinéma 1900: magie & technologie' at the Cinémathèque française on 14 June 2019.

2 The disconnection of moving images and sound is illustrated by Maxim Gorky's classic likening of the Lumière films to a 'kingdom of shadows' in an essay originally published on 4 July 1896 in *Nizhegorodski listok* (this was translated by Leda Swan and reproduced in Jay Leyda, 1960: 407–09); see also the Soviet montageurs' manifesto for non-synchronized sound and image (Eisenstein, Pudovkin, Alexandrov, 1992: 317–18). René Clair also voiced doubts about synchronized sound as encouraging a merely illustrative use of images (Clair, 1951: 156, 173).

Fig. 1. Self-caricature of tenor Enrico Caruso making his first record (11 April 1902).

Cinema

The 'flow of life': this is the expression coined by Siegfried Kracauer to describe cinema's unique capacity to capture the elusive and the ephemeral — that which normally escapes our attention, but that the film lens records and re-presents with a wealth of detail that exceeds ordinary perception. Describing the films of the Lumière brothers, Kracauer pointed to recurrent motifs that belong to life's most elusive spectacles: crowds, smoke, waves, leaves (Kracauer, 1947 [2010]: 66–67). As a scientific tool and an as art form, film has the ability to show us the world as we do not normally see it: close-up and extreme close-up, blurred, speeded-up and slowed-down sequences are part of the techniques of recording and replaying that filmmakers experimented with from the beginning, seeking to take perception beyond the limits of what the human eye normally perceives.[3]

It is no wonder that trains became such an emblematic motif in early cinema. Train journeys are exemplary not only of the new mobility, but also of the paradoxical nature of vision in motion. On the one hand, to traverse space at a speed that exceeds unaided human mobility is to see more of the world. On the other, in so doing, one also loses access to part of the visible world: as the train speeds up, the foreground ineluctably becomes blurred, remaining in the margins of the traveller's vision as

3 'Without wanting to cultivate the paradox at all costs, one cannot help but stress one notable point that Etienne-Jules Marey (1830–1904) and Georges Méliès (1861–1938) have in common: neither of them saw the point of reproducing on the screen that which our eyes see in reality' (Malthête, 2002; see also Beugnet, 2017: 39–40).

she sets her eyes on the faraway landscape that can be seen distinctly. Photography and film, however, can record both the defined and the blurred, and already in the first decades of their invention, motion blur was used to create an impression of speed. Interestingly, the instruction manual for the *cinématographe*, published by the Lumière brothers in 1897, points to the best ways of using the shutter in cases where a certain amount of blurredness was sought, that is, to reproduce 'very fast movements, in which case blurred images contribute to a better illusion than more defined images that produce a halting recreation of movement' (Lumière, Lumière, 1897; translation by the present authors). A century later, creators of video games would come to the same conclusion and add blurred effects to initially defined images. None other than William Turner anticipated this through the older medium of painting: *Rain, Steam, and Speed — The Great Western* Railway (1844), his depiction of a train crossing the Maidenhead bridge, stands as an astonishing portent of the new foms of imaging that avant-garde filmmakers such as Walter Rutman (*Berlin*, 1927), Joris Ivens (*De Brug/The Bridge*, 1928), or Abel Gance (*La Roue*, 1923) would celebrate in their cinematographic treatment of similar motifs. Modernity, as exemplified by the railway, was synonymous with mobile vision and a capacity to saturate perception with a sensory overload that the cinema, more than any other medium, was able to harness to the grammar of its effects. In his classic depiction of a train journey to Normandy, Marcel Proust described the traveller crossing from one side of the carriage to the other, desperate to catch the landscapes passing through the right or left window: night, still, on one side, and dawn already breaking on the other (Proust, 1988 [1918]: 15–16). As a time-based medium whose configurations quickly developed around concepts of montage and transitions (cuts, wipes, fades, dissolves, and superimpositions), the cinema could offer the spectator that which escaped the train traveller: night and day edited together, alternating, overlapping, or sharing the screen. Cinematic montage could momentarily abolish spatial and temporal distance, not only bringing to the audiences of one part of the world lifelike images of places and people located thousands of kilometers away, but also, as in Dziga Vertov's famous example, thanks to the wonder of edited shots, having them greet each other from Chicago to Petrograd as if from across the street.[4]

But cinematic montage resonated with the fragmentation characteristic of modernity in broader terms: to some observers, the assemblage of a multiplicity of separate sequences of images into one film appeared like an allegory of industrial work patterns (Benjamin, 2010: 28–32). Even the contrast between the continuous movement of the projected image and the stillness of the spectator could recall the juxtaposition of the factory worker's fixed station with the uninterrupted movement of the assembly line. Hence, to Walter Benjamin, film as the experience of fragmentation (cuts, edits) offered itself as a sensory training ground for modern man, but also as the momentary revenge of the ordinary man. The painted portrait had been the privilege of the wealthy and socially established classes. Photography had started to democratize

4 'You are walking on a Chicago street today in 1923, but I make you nod to comrade Volodarsky, who is, in 1918, walking down a street in Petrograd; he acknowledges your greeting' (Vertov, 1978: 4).

access to representation. But it was cinema that established itself, both in its mode of recording and in that of reception, as the art of the masses (Benjamin, 2010). If the moving image had initially circulated simultaneously in the form of individual viewing apparatuses and as public projection (the Edison Kinetoscope versus the Lumière cinematograph), it is the latter that ultimately became the standard mode of film viewing. Where the modern, urban way of life heralded the reign of anonymity (a freedom to invent or reinvent oneself that chimed with new forms of solitude), going to the cinema offered an equally modern, and affordable, mode of socialization. Hence 'cinema' would come to describe the experience of watching films together, and even today it remains, next to the theatre and the opera, one of the more popular forms of urban spectacle. Although it rapidly integrated industrial models of production and distribution and shaped its commercial appeal on the star system, cinema was also the first art form that offered the crowd a vision of itself: populated with the anonymous faces of ordinary men and women, or brought to life by the appearance of the archetypal tramp-like figure of Chaplin, the cinematic space of representation originally yielded a power of collective identification that escaped neither the powers that were (intent on controlling it through censorship) not the propagandists (intent on exploiting it).[5]

As a primarily urban phenomenon, in the form of the large screen and the public projection of films, cinema exemplified modernity in one additional way: through the coexistence of massively divergent scales of representation that media archaeologist Erkki Huhtamo has called the 'Gulliverization' of urban spaces (Huhtamo, 2009: 20). Although this phenomenon is ancient (as with the disparity of size between large murals and miniature paintings from antiquity to the medieval era, for instance), in the nineteenth century such spectacular disproportions became a staple of modern urban life, best exemplified in the difference in scale between the enormous circular panorama or diorama paintings that predated the appearance of cinema screens and the tiny photographs whose circulation started to increase in the same period (Gunning, 1995).

Film's beginnings contributed to a systematic disruption scalar relations. Montage and framing (the editing together of close-up and long or wide-angle shots) produced an 'explosion' of the space of representation where the standard human body ceased to be the key reference and measure. The origination of this radical alteration in the representation of space through film is usually attributed to D.W.Griffith: "Griffith exploded filmic space; he stopped adjusting the point of view of the camera to that of the audience in front of the screen, thus radically changed the conditions of the cinematographic spectacle" (Bonitzer, 1982: 26, translation by the present authors). Special effects did the rest. Starting with James Williamson's *The Big Swallow* (1901), where a cameraman ends up falling into the massively enlarged mouth of the man he was filming, to the rise of the 'man of the future' who becomes invisible in the

5 See, for instance, John Burrows' study of the archives of the London County Council and of the Metropolitan Police (Burrows, 2004a, 2004b). See also Walter Benjamin's classic observations on propaganda (Benjamin, 2010: 34–35).

Incredible Shrinking Man (1957) and the endless remakes of King Kong's journey from the wilderness to the centre of New York, cinema never stopped creatively exploiting its power to destabilize our traditional, anthropomorphic sense of scale. After all, in its analogue form, such shifts in scale were part and parcel of the medium's material basis: destined to be enlarged and projected by a powerful beam of light, each of the individual frames recorded on film already contained, in the space of a few millimetres, a whole universe, recorded with photographic precision down to the most minute details.

To some extent, the evolution of the cinema in the twentieth and at the beginning of the twenty-first centuries can be read as the progressive intensification, rationalization, and diversification of the principles that determined its appearance as one of modernity's creations. The dual phenomenon of mobile vision and of 'Gulliverization' for instance, manifests itself anew with the advent of digital media and the dissemination of screens, from the massive advertising boards hung in city centres to the tiny mobile phones that most of us now carry though the same urban centres, and which function both as means of communication and as screening devices.

But such retrospective accounts have their limits: in terms of scale and screen size, for instance, television, and later the computer, offered themselves as alternative paths. Initially the medium of live transmission, television established a domestic form of spectatorship, based on a screen format and on filming techniques that rapidly aligned with the anthropomorphic scale (as exemplified by the 'talking heads', or close-ups on the commentators' faces as they seemingly address the spectator watching from her living room).

Undoutedly, the sensory overload characteristic of modernity's transformation of ordinary perception found one of its most pervasive expressions in cinema's sensational power of affect. Yet the ensuing evolution of the cinema as an immersive and creative medium bent on challenging ordinary perception does not follow a straight line of technological or artistic innovation. The advent of sound cinema in the late 1920s, and the need to precisely synchronize sound and image, brought the experimentation of the early avant-gardes to a halt: arguably, with the exception of its experimental branch, cinema never fully regained the power to innovate that characterized the filmmaking of its first three decades (Abel, 1988: 8–10; Belaygue, 1988). Nor does the evolution of the cinema necessarily find its logical conclusion in the appearance and diffusion of VR and 3D images: witness the multiple attempts and successive failures at launching and commercializing such technologies (Elsaesser, 2013).

Correspondingly, the discourses on obsolescence and the 'death' of cinema now tend to be considered in light of the complex, and often unforeseen, impact of technological and cultural evolutions. Ever since Louis Lumière's (sincere or calculating) labelling of the cinema as 'a medium without a future', its death has been proclaimed at various points in the twentieth and twenty-first centuries — most urgently at the time of the advent of television, then video, and finally digital media(Christie, 1994: 95).

The dissemination of television in the 1950s, and then the rise of video and the spread of VHS (Video Home System) in the 1970s had, indeed, supplemented cinema's traditional modes of production and consumption of moving images. And from the 1990s onwards, the digital modes of access, recording, archiving, and distributing (from DVD versions of films to platforms where they can uploaded, downloaded, or streamed) had a profound impact on film's creative and viewing practices. In addition to the spreading of a metaphorical discourse on the vulnerability and mortality of the cinema, a variety of theorizations attempted to account for digital media's absorption and transformation of cinema (Cherchi Usai, 2001). From 'media convergence' to 'remediation' or 'relocation', as discourses that tend to naturalize media obsolescence, such conceptualizations effectively mirror the commercial discourses of technological replacement that target consumers of equipment and services (Jenkins, 2006; Bolter, 1999; Casetti, Sampietro, 2012). To Evan Watkins, such theorizations are part of what he calls 'technoideology' (Watkins, 1993: 27; Rosen, 2001).

Yet historical precedents warrant a critical approach to media obsolescence, one that accounts for the coexistence of a diversity of formats as well as creative, archiving and viewing practices, rather than merely shadowing dominant discourses of media replacement. The ease of access and storage permitted by digital technology is irrefutable and unmatched: digitization allowed for the restoration and the wider circulation of films, but also the flourishing of a multitude of new creative artistic and amateur practices, including mash-ups, compilations, machinima, remixes, and other forms of foundfootage productions. At the same time, however, the accelerated rate of hardware and format discontinuance have rendered digital modes of archiving differently, but just as problematically impermanent as their analogue counterpart, forcing digital film archives to continually invest in data migration (National Research Council, 2000). Moreover, in contrast with the multiplicity of analogue formats (both professional and amateur), offering distinctive qualities of grain, contrast, and colour cast, digital modes of recording, storing, and postproduction have so far tended to draw film aesthetics towards a more standardised look and feel. Together with the dematerialized nature of digital film media — contrary to data stored on a hard disk, analogue film, like human skin, bore the marks of time — the production of a more homogeneous and timeless quality of image arguably weakens the medium's historicity: its capacity not only to record the realities of its time, but to account for the passing of time as well as evolving perceptions, sensibilities, and subjectivities.

As for the new 'freedom' (from set times and restricted choices) arguably permitted by domestic and mobile viewing, though aggressively advertised by a growing number of film-on-demand platforms, it remains largely dependent on catalogues of films that cater for the most consensual taste, promoting mainstream filmmaking rather than art and experimental practices, often reducing the latter to complete invisibility.

The naturalizing discourses on the evolution of watching practices not merely mirror commercial strategies, but also have aesthetic and political implications. To refer to film's original modes of circulation, the kinetoscope and other examples of 'peephole' viewing, is one way to relativize the import of cinema's main viewing *dispositif*: the projector, the darkened room, the audience. But watching a film on a large cinema screen continues to be a distinctive experience on at least two counts: on the one hand, the expanse of the cinema screen invites the gaze of the spectator to travel across and, to a certain extent, to lose itself in the image (rather than consume the smaller image at one glance, while the hand controls its appearance via remote control or tactile screen). On the other hand, at a time when the computer, electronic communication, and the internet encourage an increasingly individual mode of existence, the resilience of the cinema as a mode of collective spectatorship accounts for the ongoing need to engage with forms of socialization that are excluded from domestic viewing, as well as from the experience of immersive media and virtual reality. Cinema viewing, by contrast, implies the sharing of a space and the codes of film spectatorship, but most crucially, as a group, or as a crowd, the experience of watching together (Hanich, 2014, 2018).

Sound Technologies

The origins of sound recording technologies are difficult to situate precisely, and no strictly chronological account is capable of doing them justice. According to popular belief, the transformative moment occurred in 1877, when Thomas Alva Edison (1847–1931) tested a device of his own invention by recording the well-known nursery rhyme *Mary Had a Little Lamb* on tin foil wrapped around a metal cylinder and then playing it back (Gelatt, 1995). However, historians have long been aware that the poet and inventor Charles Cros (1842–1888) formulated a similar concept several months before Edison in a memoir submitted to the French *Académie des sciences*; strictly speaking, therefore, Edison did not invent the modern concept of sound recording, but was the first to put into practice an idea that had been taking shape for some time — considerably longer, in fact, than was generally understood prior to the beginning of the twenty-first century. In 2008, the First Sounds collective drew attention to the experiments conducted in the mid-nineteenth century by the French inventor Edouard-Léon Scott de Martinville (1817–1879). In 1857, Martinville patented a device he called the 'phonautograph'; it recorded sound vibrations in visible form on a cylinder coated with lampblack. His intention was to contribute to the nascent science of acoustics by visualizing sound waves, and the phonautograph was not capable of playback; however, this problem has been remedied with the aid of modern computer technology, and Martinville's 1860 recording of himself singing the French nursery rhyme *Au clair de la lune* is now considered as the earliest playable record of the human voice ever recovered.[6] In his pioneering study *The Audible Past* (2003), Jonathan Sterne has argued that devices such as Edison's gramophone and Martinville's phonautograph have their origins in even earlier developments, and that they are rooted in a distinctive sound culture whose progressive emergence dates

6 *First Sounds. The Phonautograms of Edouard-Léon Scott de Martinville* <http://www.firstsounds.org/sounds/scott.php> [accessed 8 April 2019].

back to the mid-eighteenth century (Sterne, 2003). Indeed, there are clear indications that the possibility of recording, in playable form, actual performances by famous musicians was already an object of intense speculation over a hundred years before Edison; a French Catholic priest, Marie Dominique Joseph Engramelle (1727–1805), described a device capable of recording keyboard music in his 1775 treatise *La Tonotechnie, ou l'art de noter les cylindres* (Engramelle, 1775).

While it is probably impossible to pinpoint the moment when sound recording was invented, it may be more helpful to focus on the epistemological transformations that made its current manifestations possible. To Engramelle, recording music did not entail capturing the actual sounds as they reached the listener's ears. Like later devices such as the player piano, his invention notated the way the musician handled his or her instrument, not what was heard by the audience; it recorded the key parameters governing sound production, not the sounds themselves, which Engramelle made no attempt to capture. Cros and Edison (and, to a lesser extent, Martinville) took a very different tack: they were interested in recording and recreating the actual sound vibrations perceived by the listeners present at the performance; this makes them pioneers of our current recording technologies, which seek to turn sound waves into something else — markings on soot-blackened paper, a groove on a shellac record, or a set of digital data — and to play them back again in audible form. Thus, the mid-nineteenth century did, in fact, mark an important turning point, as it witnessed a major conceptual shift without which our current practices could never have been imagined. This crucial transformation directly led to the emergence of what Jonathan Sterne calls the 'acousmatic' theory of sound recording, viz. the belief that it 'separate[s] a sound from its "source"' (Sterne, 2003: 25), and that it is therefore to be analysed in terms of audio 'reproduction' (thus raising the question of 'fidelity' to the supposed 'original'). Much of the history of sound reproduction in the twentieth and twenty-first centuries can be understood as the progressive unfolding of the implications inherent in this concept, which eventually came under critical scrutiny as musicians, theorists, and other cultural actors understood its limitations.

Roughly speaking, this history can be divided into four major phases, each characterized by the rise of a new recording technology which soon came to displace — if not replace altogether — its immediate predecessors. (1) Edison's phonograph recorded sound on wax cylinders, whereas the gramophone patented in 1887 by Emile Berliner (1851–1929) opted for the disc format which remained in use until the end of the twentieth century; however, both operated on a similar principle, as they mechanically converted sound waves into grooves by means of a stylus connected to a vibrating diaphragm (Millard, 2005 [1995]: 33). (2) This early phase, known as the 'acoustic' era, was brought to a close in 1925 by the invention of electrical recording, which made it possible to amplify, filter, and balance sounds; (3) the advent of magnetic tape in the immediate aftermath of the Second World War opened up further possibilities by allowing for various forms of editing and sound mixing such as multitrack recording (Millard, 2005 [1995]: 143, 346–52). Finally, starting in the mid-1970s, the introduction of digital recording (4) — which converts audible phenomena into sets of discrete numbers — broke with all previous practices, since earlier technologies invariably stored sounds as continuous 'analogue' signals. In turn, the digitization of sound recording has freed it from its dependence on specific, tangible modes of storage such as cylinders, records, or spools of magnetic tape, allowing recorded music to migrate freely from one device to another on the World Wide Web. These technological changes have all encountered various modes of resistance, and supposedly 'outdated' devices have tended to retain a minority, but by no means insignificant, following: the twenty-first century vogue of 'vinyl' (i.e. mostly analogue) records at a time when internet streaming has become one of the dominant modes of music consumption parallels the earlier attachment to pre-1945 recordings among music lovers nostalgic for their distinctive acoustic properties.

This clearly suggests that teleological approaches to the history of sound recording — understood as a narrative of linear 'progress' and increasing 'fidelity' — do justice neither to the actual experience of many listeners, nor to the complex temporality in which it unfolds. In fact, one could easily turn this optimistic tale on its head and argue that modern sound recording technology has long been associated with a nostalgic, melancholy outlook rather than the futuristic visions encouraged by the music industry. Despite the enthusiastic language used in advertising campaigns — such as the claim by the Edison Record label that 'comparison with the living artist reveals no difference', or the much later promise that the Sony compact disc 'turns your living room, car, and backpack into a concert hall' — there is a long-standing association between listening to recorded sounds and feelings of loss and mourning. Jonathan Sterne's observation that twentieth-century sound culture has its roots in Victorian attitudes towards death is borne out by the famous passage from James Joyce's *Ulysses* (1922) in which Leopold Bloom wishes that people could '[h]ave a gramophone in every grave or keep it in the house. After dinner on a Sunday. Put on poor old greatgrandfather. [...] Remind you of the voice like the photograph reminds you of the face'; as Jacques Derrida comments at the start of his 1987 book *Ulysse gramophone*, which takes its cue from Bloom's monologue, 'it is always too late with Joyce' ('il est toujours trop tard avec Joyce') (Joyce, 1997 [1922]: 109; Derrida, 1987: 15). From its inception, sound recording technology has been understood as a bulwark against oblivion, and the idea of recording the voices of the living so as to bequeath them to later generations is by no means to be treated as a purely literary fantasy, as it has given rise to well-documented practices dating back more than a hundred years. In addition to music, late-nineteenth-century engineers deliberately captured the voices of the great and famous in anticipation of their passing: thanks to them, we can now listen to Victorian and Edwardian-era luminaries including Robert Browning, Florence Nightingale, Theodore Roosevelt, and Pope Leo XIII; and a time capsule containing twenty-four gramophone records of famous *Belle Époque* singers was deposited in a vault at the Paris Opéra in 1907 by request of the French government in the hope of educating twenty-first century listeners about the history of musical interpretation. This backward-looking, elegiac understanding of sound recordings as 'traces' or 'remnants' of a lost past — or of a present understood to be transient and soon to be superseded — largely accounts for the urge to 'clean up' or 'remaster' old records in hopes of recreating the 'original' performances in all their glory, if necessary by making full use of technologies as yet unavailable at the time they were captured.

Interestingly, the history of recorded sound also provides several instances of the opposite process, whereby 'inferior' or 'low-fi' devices tend to prevail over earlier, and technically more advanced, technologies, as if a loss in sound quality were sometimes deemed preferable by the recording industry and the public at large. Notable examples include the 'wall of sound' promoted in the 1960s by producers Phil Spector (b. 1939) and Joe Meek (1929–1967), who turned their backs on the 1950s quest for 'high fidelity' and 'actively courted high infidelity', aware that 'the postwar generation would not merely tolerate "corrupted" sounds, but also embrace them' (Milner, 2009: 153–54). Other well-known instances include the choice of a low digital sampling rate that has left many audiophiles dissatisfied with the CD ever since the industry adopted it in 1980; the current prevalence of the highly compressed MP3 format also comes to mind: today, as Greg Milner observes, music is often 'heard through cheap ear-bud headphones or small, tinny computer speakers', and '[t]he high-fidelity ideal of re-creating the concert hall in the living room has little relevance when the living room is no longer the place where most recorded music is heard' (Milner, 2009: 354). These two phenomena — the urge to erect a fitting memorial to notable artists and individuals, and the conscious preference for technically inferior, but more easily marketable, recording technologies — can be seen as two sides of the same coin, since they both express a deep-seated ambivalence about the very idea of recorded sound. On the one hand, a recording is a 'reproduction' of a lost original, a belated — and, almost by definition, feeble — echo of a bygone age. On the other hand, it is irremediably different from its 'source', and it can therefore afford to renounce the illusion of absolute 'fidelity' in order to embrace what makes it an autonomous, self-sufficient creation.

This duality is particularly apparent in the treatment of recorded music. The urge to preserve audible traces of culturally significant musical performances is as old as the recording technologies themselves — indeed, it is probably one of the main reasons why they were developed at all, and Engramelle's 1775 treatise begins with the observation that we would be fortunate indeed if we could hear the music of Jean-Baptiste Lully (1632–1687) or Louis Marchand (1689–1732) as played by the composers (Engramelle, 1775: ii). The modern practice of making 'live' recordings of concerts began in earnest when Lionel Mapleson (1865–1937), the librarian of New York City's Metropolitan Opera House, purchased an Edison home phonograph and attempted to record famous singers from the prompter's box, and then — with significantly improved results — from the theatre's fly system. Between 1901 and 1903, Mapleson amassed a remarkable collection of audio documents; these include the only surviving recordings of tenor Jean de Reszke (1850–1925), often considered one of the greatest classical vocalists of all time. Initially pursued by enthusiastic amateurs, the practice was eventually adopted by recording companies, and when the famous Australian soprano Nellie Melba (1861–1931) gave her farewell concert in London in 1926, much of it was recorded and commercially issued by HMV. The advent of magnetic tape made recording live performances of long works much easier; in the 1940s, this allowed German engineers to capture wartime concerts by Wilhelm Furtwängler (1886–1954) and other leading artists, and the 1966 recording of Wagner's *Ring*, as conducted by Karl Böhm (1894–1981) at the Bayreuth Festival, is the first live account of this mammoth work to have been commissioned for commercial purposes. Similar practices were adopted at an early stage by folklorists and, later, ethnomusicologists in an effort to document and preserve musical styles current in non-Western cultures and among the 'uneducated' classes. In Europe, collectors such as Béla Bartok (1881–1945), Constantin Brailoiu (1893–1958), and Adolf Chybinsky (1880–1952) were early advocates of this approach as they recorded local 'folk' music in the spirit of cultural nationalism. Meanwhile, pioneering scholars such as Hugh Tracey (1903–1977) and Klaus Wachsmann (1907–1984) made the first recordings of traditional African music. Similar efforts took place in other parts of the world. In the United States, Alan Lomax (1915–2002) collected thousands of recordings for the Archive of American Folk Song at the Library of Congress; he is credited with popularizing musicians such as Robert Johnson (1911–1938), Woody Guthrie (1912–1967), and Pete Seeger (1919–2014). All these endeavours illustrate the 'acousmatic' theory of sound recording, which treats it as a 'reproduction' of a lost, or endangered, 'original' assumed to consist of a live performance captured, but in no way significantly altered, by the recording apparatus — hence its scientific function as a passive witness to musical practices that would presumably have existed without it.

Meanwhile, the notion that recordings can be approached as self-sufficient artificial constructs also gained ground. Whereas early acoustic records had been promoted as passive extensions of the artist's presence, this illusion was difficult to maintain when it turned out that some exceptionally 'phonogenic' voices sounded better on disc than in concert. An early example of this phenomenon is provided by the career of Italian soprano Amelita Galli-Curci (1882–1963), who made her most successful records as she was beginning to develop vocal problems that could not be disguised outside the studio; as a result, her British fans, who only knew her from her recordings, were disappointed when she toured the United Kingdom in 1924, an episode recalled in Nancy Mitford's 1945 novel *The Pursuit of Love*. Eventually, some musicians began to realize that this constituted a failing only to the extent that records were approached as more or less truthful recreations of live performances, and that the problem disappeared once the phonograph was treated as an autonomous sound source instead of a reproduction apparatus. With characteristic radicality, John Cage (1912–1992) made this clear in his early composition *Imaginary Landscape No. 1* (1939), scored for a small ensemble which includes two phono turntables used as musical instruments; the later *Imaginary Landscape No. 5* (1952), a collage of fragments from LPs recorded on tape, drives the point home as it dissuades the listener from identifying 'sources' and instead encourages him or her to enjoy the recorded sound for its own sake. Later examples of electronic music or *musique concrète*, such as *Poème électronique* (1958) by Edgard Varèse (1883–1965), illustrated the same principle as they demonstrated the musical potential of sound recording technologies once stripped of their mimetic function. Some sectors of the music industry took notice and began creating recordings that provided alternatives to actual concerts, instead of merely documenting them. When producer John Culshaw (1924–1980) was put in charge of the Decca recording (1958–1965) of Wagner's *Ring*, which was conducted by Georg Solti (1912–1997), he deliberately tried to make the sound as vivid as possible when reproduced on home audio equipment; in *Götterdämmerung*, the voice of tenor Wolfgang Windgassen (1914–1974), who sang Siegfried, was electronically altered to make him sound more like a baritone in the scene where he dresses up as Gunther, creating an effect that would have been impossible in the theatre. In recent years, the increasing sophistication of sound processing technologies and the frequent recourse to amplification and other modes of sound manipulation in public performances of 'popular' and, to a much lesser extent, 'classical' works have blurred the distinction between 'live' and 'recorded' music, accustoming audiences to hearing technologically mediated sounds even when attending a concert.

In the end, it could be argued that in the course of the late nineteenth, twentieth, and early twenty-first centuries, the most significant changes of all affected not recording technologies *per se*, but what Jonathan Sterne calls 'audile techniques', viz. the 'set of practices of listening' (Sterne, 2003: 23) to which people resort in the presence of technologically mediated music. As Wayne Koestenbaum asks in *The Queen's Throat*, what happens 'when Adelina Patti's voice emerges from the furniture' (Koestenbaum, 1993: 66)? Prior to the invention of modern sound recording, most music could exist only in the presence, or the close proximity, of the performer — 'acousmatic' sounds, let alone voices, were the exception, not the rule, and could be perceived only under specific conditions and for a limited length of time. In *The Tempest* (1611), Caliban's description of a magical island where the air is 'full of noises' and 'a thousand twangling instruments / Will hum about [his] ears' is the stuff of fantasy; so, for a long time, was the dream of a self-playing musical device — the Aeolian harp, a string instrument which produces eerie sounds when activated by the wind, exerted a powerful fascination on nineteenth-century culture, as evidenced by the writings of Percy Bysshe Shelley (1792–1822), Eduard Mörike (1804–1875), and Robert Louis Stevenson (1850–1894) (Shakespeare, 1997: 3087). The invention of the gramophone and, later, of the 'wireless telephone' or radio (patented in 1904) made the experience of listening to sounds issuing from an invisible source via an automated device available to a widening range of listeners, to such an extent that it is now one of the standard ways of experiencing music. This has resulted in a condition R. Murray Schafer (b. 1933) has called 'schizophonia', whereby musical sounds are forcibly separated from their 'natural' sources such as the human throat; this gives the modern 'soundscape' a uniquely pathological quality, leading Schafer and other writers to promote 'acoustic ecology' in the belief that humans should take responsibility for the sound environments they have created for themselves and, if possible, 'improve' them (Schafer, 1977: 90). Interestingly, Schafer's key distinction between 'hi-fi' and 'low-fi' environments — viz. between those where each individual sound is clearly perceived and those where all perspective disappears as undifferentiated 'noise' prevails — is itself couched in a terminology borrowed from the language of sound recording, establishing it as the norm even as Schafer seeks to alert his readers to its dangers (Schafer, 1977: 43).

Sound and Image

From the 1930s onwards, recorded sound and images finally came together in the form of the 'talkies'. But in the 'audiovisual contract' (to use the term coined by Michel Chion) that ensued, sound became accessory to the image (Chion, 1994). Contrapuntal uses of sound and image have remained largely limited to the realm of experimental filmmaking. In mainstream cinema, sound is secondary to the image, it fills the gaps, dramatizes, complexifies, and enriches the visual content, but it fits in with the patterns of mise-en-scène and visual editing. However, to many observers, the shift from analogue to electronic and digital media has unsettled the initial pattern. For Steven Shaviro, the advent of television, and then video, worked to overturn cinema's ocularcentrism (Shaviro, 2016). Based on the reduction and manipulation of sounds and images into similar data components, 'post-cinematic' aesthetics de-hierarchize the sound–image relationship: in music video for instance, the visual content alters in tune to the music and not the opposite. However, computer generated immersive environments, such as provided in VR, in turn, often favour 'real-time' experiences where sound remains secondary to the experience of the visual rendering of a virtual space.

The determining factors for the intensive technologizing of audio-visual culture are rarely solely technical, nor do they follow a 'natural', homogeneous pattern. Hence, in the case of the emergent immersive media, André Bazin's 1950s 'Myth of Total Cinema' retains some of its pertinence: in spite of the controversies raised by the pursuit of the fusion with the represented, in spite of the suspicion with which immersive technologies are met,[7] and the technical, aesthetic, and commercial failures that have occurred in their name nonwithstanding, the dream of total immersion continues to determine the development of audiovisual technologies (Bazin, 1992 [1962]: 23–27).

7 See the dossier on immersive technologies edited by Martine Beugnet and Lily Hibberd (Beugnet, Hibberd, 2020).

References

Abel, Richard, *French Film Theory and Criticism: a History/Anthology, 1907–1939* (Princeton, NJ: Princeton University Press, 1988).

Bazin, André, 'The Myth of Total Cinema', in *What is Cinema?*, trans. by Hugh Gray (Berkeley, CA, Los Angeles, CA, and London: University of California Press, 1962 [1992]), pp. 17–22.

Belaygue, Christian, ed., *Le Passage du muet au parlant* (Toulouse: Cinémathèque de Toulouse, 1988).

Benjamin, Walter, 'The Work of Art in the Age of Its Technological Reproducibility [First Version]', trans. by Michael W. Jennings, *Grey Room*, 39 (2010), 11–37.

Beugnet, Martine, *L'Attrait du flou* (Paris: Yellow Now, 2017).

———, and Lily Hibberd, 'Absorbed in Experience: New Perspectives on Immersive Media', *Screen*, 61.4 (2020), 586–93.

Bolter, David Jay and Richard Grusin, *Remediation. Understanding New Media* (Cambridge, MA: MIT Press, 1999).

Bonitzer, Pascal, *Le Champ aveugle* (Paris: Cahiers du cinéma, 1982).

Burrows, John, 'Penny Pleasures: Film Exhibition in London during the Nickelodeon Era, 1906–1914', *Film History*, 16.1 (2004a), 60–91.

———, 'Penny Pleasures II: Indencency, Anarchy and Junk Film in London's "Nickelodeons", 1906–1914', *Film History*, 16.2 (2004b), 172–97.

Casetti, Francesco and Sara Sampietro, 'With Eyes, With Hands. The Relocation of Cinema Into the iPhone', in *Moving Data — the iPhone and the Future of Media*, ed. by Pelle Snickars and Patrick Vonderau (New York, NY: Columbia University Press, 2012), pp. 19–32.

Charney, Leo and Vanessa R. Schwartz, *Cinema and the Invention of Modern Life* (Berkeley, CA: University of California Press, 1995).

Cherchi Usai, Paolo, *The Death of Cinema: History, Cultural Memory and the Digital Dark Age* (London: BFI, 2001).

Chion, Michel, *Audio-vision: Sound on Screen*, trans. by Claudia Gorbman (New York, NY: Columbia University Press, 1994).

Christie, Ian, *The Last Machine: Early Cinema and the Birth of the Modern World* (London: British Film Institute and BBC, 1994).

Clair, René, *Réflexion faite: notes pour servir à l'histoire de l'art cinématographique de 1920 à 1950* (Paris: Gallimard, 1951).

Derrida, Jacques, *Ulysse gramophone. Deux mots pour Joyce* (Paris: Galilée, 1987).

Eisenstein, Sergei, Vsevolold I.Pudovkin, and Grigori V. Alexandrov, 'A Statement (On Sound)', *in Film Theory and Criticism*, ed. by Gerald Mast, Marshall Cohen, and Leo Braudy (Oxford: Oxford University Press, 1992), pp. 315–17.

Elsaesser, Thomas, 'The "Return" of 3-D: On Some of the Logics and Genealogies of the Image in the Twenty-first Century', *Critical Inquiry*, 39 (2013), 217–46.

Engramelle, Marie Dominique Joseph, *La Tonotechnie ou l'art de noter les cylindres* (Paris. P. M. Delaguette, 1775).

First Sounds Collective, *The Phonautographs of Édouard-Léon Scott de Martinville.* <http://www.firstsounds.org/sounds/scott.php> [accessed 8 April 2019].

Frisby, David, *Fragments of Modernity: Theories of Modernity in the Work of Simmel, Kracauer, and Benjamin* (Cambridge, MA: MIT Press, 1986).

Gelatt, Roland, *The Fabulous Phonograph: From Edison to Stereo* (Philadelphia, PA: Lippincott, 1955).

Gunning, Tom, 'Tracing the Individual Body: Photography, Detectives, and Early Cinema', in *Cinema and the Invention of Modern Life*, ed. by Leo Charney and Vanessa R. Schwartz (Berkeley, CA: University of California Press, 1995), pp. 15–44.

Hanich, Julian, *The Audience Effect: On the Collective Cinema Experience* (Edinburgh: Edinburgh University Press, 2018).

———, 'Watching a Film With Others: Towards a Theory of Collective Spectatorship', *Screen* 55.3 (2014), 338–59.

Huhtamo, Erki, 'Messages on the Wall: an Archaeology of Public Media Displays', in *Urban Screens Reader*, ed. by Scott McQuire, Meredith Martin, and Sabine Niederer (Amsterdam: Institute of Network Cultures, 2009), pp. 15–28.

Jenkins, Henry, *Convergence Culture: Where Old and New Media Collide* (New York, NY: NYU Press, 2006).

Joyce, James, *Ulysses*, ed. by Danis Rose (London: Picador, (1922) [1997]).

Koestenbaum, Wayne, *The Queen's Throat. Opera, Homosexuality, and the Mystery of Desire* (London: GMP Publishers, 1993).

Kracauer, Siegfried, *Théorie du film. La rédemption de la réalité matérielle* (Paris: Flammarion, 1947 [2010]).

Leyda, Jay, *Kino: a History of the Russian and Soviet Film* (London: George Allen & Unwin, 1960).

Lumière, Auguste and Louis Lumière, 'Fiche technique et manuel d'emploi du cinématographe', *La Revue du siècle* (1897).

Malthête, Jacques, 'Marey et Méliès', *Bulletin de la Sémia*, 2 (2002) <http://www.artefake.com/marey-et-melies.html> [accessed 8 April 2019].

Marcus, Laura, *The Tenth Muse: Writing about Cinema in the Modernist Period* (New York, NY: Oxford University Press, 2007).

Millard, Andre, *America on Record. A History of Recorded Sound* (Cambridge: Cambridge University Press, 1995; second ed., 2005).

Milner, Greg, *Perfecting Sound Forever. The Story of Recorded Music* (London: Granta, 2009).

National Research Council, *The Digital Dilemma: Intellectual Property in the Information Age.* (Washington, DC: The National Academies Press, 2000).

Proust, Marcel, *À l'ombre des jeunes filles en fleurs*, in *À la recherche du temps perdu*, vol. II (Paris: Gallimard, Bibliothèque de *la Pléiade*, 1988 [1918]).

Rosen, Philip, 'Old and New: Image, Indexicality, and Historicity in the Digital Utopia', in *Change Mummified: Cinema, Historicity, Theory* (Minneapolis, MN: Minnesota University Press, 2001), pp. 301–49.

Schafer, R. Murray, *The Soundscape: Our Sonic Environment and the Tuning of the World* (New York, NY: Knopf, 1977).

Shakespeare, William, *The Tempest*, in *The Norton Shakespeare*, ed. by Stephen Grennblatt (New York: W. W. Norton, 1997).

Shaviro, Steven, 'Splitting the Atom: Post-cinematic Articulations of Sound and Vision', in *Post-cinema: Theorizing 21st-century Film*, ed. by Shane Denson and Julia Leyda (Falmer, Sussex: Reframe Books, 2016), pp. 362–97.

Sterne, Jonathan, *The Audible Past* (Durham, NC: Duke University Press, 2003).

Vertov, Dziga, 'Selected Writings', in *The Avant-garde Film: a Reader of Theory and Criticism*, ed. by P. Adams Sitney (New York, NY: Anthology Film Archives, 1978), pp. 1–13.

Watkins, Evan, *Throwaways: Work Culture and Consumer Education* (Redwood City, CA: Stanford University Press, 1993).

GIL BARTHOLEYNS, MANUEL CHARPY

Everyday Technologies

A Short History, from the Late Eighteenth to the Twenty-First Centuries

At the end of the 2000s, the design student Thomas Thwaites, in a bid to debunk the magical aura of everyday equipment, started manufacturing a toaster from scratch. A modest appliance if ever there was one, the toaster nevertheless combines all kinds of techniques at different levels. The undertaking was daunting: the young designer had to comb through abandoned copper and iron mines, recycle plastic because he could not find oil, etc. The result was an extravagantly costly but average toaster and more importantly a design experience whose narrative was intended to change our relationship with everyday objects (Thwaites, 2010).

Thwaites's project questions the role of techniques in everyday life and reveals how even the humblest of objects results from a series of elaborate knowledge related to materials, techniques, and know-how. Everyday life, and in particular the domestic space, is revealed as a place where many technologies are put to use, in this case the power grid, electrical resistance, and the electromagnet. In other words, contemporary industial societies are like a toaster: elaborate knowledge and unbelievably complex technologies have been accumulated, incredible savings in cost and time have been achieved all for a trivial end, and all governed by a concern for comfort and taste rather than necessity (Figure 1). This complexity vanishes with daily usage. Thwaites's toaster project is all the more revealing as the toaster, as Simondon noted as early as 1958, is one of those mysterious objects whose workings are hidden underneath a cover that is synonymous with comfort and modernity. Here, the black box becomes a laboratory for analysis. Finally, the object itself encapsulates ways of buying and preserving food, of eating, and, quite simply, of eating breakfast.

This could be a definition of everyday techniques: a way of bringing together a series of technologies, skills and practices and cultural phenomena in one object. There is no action during the day — eating, moving around, working, entertainment — that does not make use of a technical object or a certain type of know-how, whether it be toasting bread, hoovering the floor, starting a fire in a stove, operating a water pump, or harnessing horses. The flexibility with which the term 'technologies of everyday life' is used almost obscures the fact that the social sciences have made it into an object of global history.

Global History of Techniques, ed. by Guillaume Carnino, Liliane Hilaire-Pérez and Jérôme Lamy, Global Matters Culture Matérielles, histoire des techniques et circulation globale des savoirs, 9 (Turnhout: Brepols, 2024), pp. 705–728

BREPOLS PUBLISHERS 10.1484/M.TECHNE-EB.5.129837

Fig. 1. The first electric toaster, 4 heating elements, ceramic base. Model D-2, General Electric Company (Schenectady, NY), 1908, 17.5 × 20.4 × 10.1 cm, Creative Commons. Smithsonian's National Museum of American History.

Technologies as Culture

An interest in everyday life is not new and can be traced back to antiquarian histories of the daily life which started being popular from the second half of the eighteenth century and to the work of folklorists from the end of the nineteenth century, if we think of Alfred Franklin's *Vie privée d'autrefois* [The Private Life of Yesteryear] (1887–1902), James George Frazer's work between 1890 and 1915, and Arnold Van Gennep's *Manuel de folklore français contemporain* [A handbook of French contemporary folk life] (1937–1958). Yet the everyday only became a historiographical category in France with the emergence of the history of material culture, for which Fernand Braudel's work continues to be emblematic. *Les Structures du quotidien* (1967), the first part of *Civilisation matérielle,* deals with the topics of housing and consumption whose incremental development is seen as the hallmark of modernity, particularly in relation to luxury goods. This was the beginning of the reversal of the event-driven 'history from the top' model, a reversal which came full circle with the rise of social history in England, of Italian *microstoria* (microhistory), *Alltagsgeschichte* (history of the everyday) in Germany, and the history of sensibilities and historical anthropology in France in the wake of the *Annales* school.

A general movement towards investigating the working classes (Hoggart, 1957; Thompson, 1963) and their ways of living (Perrot, 1987) spread. It rose against a type of history that focussed on dramatic events and important figures, although the lives of the latter were also immersed in everyday experiences, e.g. dressing exquisitely, driving a car with a purring engine, hearing the clock strike tea time, or contemplating a photograph of a dead relative, are all actions involving everyday objects that tell us about these practices. Techniques are related to the experience of the self or of time, to ways of moving, of withdrawing to the privacy of a home made comfortable by artificial lighting, central heating and running water.

However, the history of techniques and materiality is often not present in these bottom-up approaches to history despite the fact they explore lived experience, living conditions, and the fundamental parameters of daily life, namely time and space. The fact technology is not mentioned in the vast *Histoire des mœurs* edited by Jean Poirier shows that the link was not yet obvious in 1990. And Braudel only integrates techniques through mentions of the 'great innovations' of gunpowder, printing, and deep-sea navigation. Moving from inventions to social and everyday uses requires yet another step.

Everyday life involves techniques, first of all, indirectly, through their inventions or improvements. In the early 1930s, seeking to anchor history more firmly to reality, Marc Bloch and Lucien Febvre embarked on writing a history of the mill, the carriage, and the forge. Innovation was seen the key to social transformation, the driving force of history. And when technical environments and objects were considered in terms of everyday experience it was to comment on their influence on a person's state of mind: 'the nights, which as we know were dimly lit', Marc Bloch writes, 'were darker, the cold, even in the halls [...] more rigorous. There is no instrument that allows us to weigh the influence that such an environment could exert on souls' (Bloch, 1939: 115).[7] Historical materialism remained for a long time the driving force of the historical rationale, and material culture was thought of, above all, as the response to the constraints of the environment — a catalogue of functional instruments and arrangements where technical innovation acted as a break in time that allowed for an improvement in living conditions.

Conversely histories of techniques and technologies seldom pay attention to the everyday. While Siegfried Giedion's *Mechanization Takes Command: a Contribution to Anonymous History* (1948) broke new ground by focussing on the way in which mechanization took over everyday life in the nineteenth century the home being the place of invention — so much so that Giedion identified the early nineteenth-century

7 Unless otherwise stated, all translated of works not published in English originally are the authors' own.

spit as the 'first automatic appliance in the home' — his history was, nonetheless, one in which machines rationalize the domestic space. In the monumental *History of Technology* (Singer et al., 1954–1984) and the 'general histories of technology' by Maurice Daumas (1962–1979) and Bertrand Gille (1978), everyday life only comes into play to the extent that technology (in agriculture, construction, etc.) is linked to people's livelihoods. They form a 'technical system', i.e. a coherent whole linked to the social and political system. Admittedly, these histories recognize that technical progress unsettles vernacular ways, but it almost always continues to be perceived as the Holy Grail of 'civilization', its widespread adoption in domestic spaces being the measure of this.

Social Uses and Everyday Life

If a new interest in everyday undoubtedly appeared in the 1970s it is undoubtedly because everyday life itself was changing. Edgar Morin and André Burguière noted this in Plozévet, a place analysed by dozens of researchers in the 1960s: the arrival of refrigerators changed the fabric of daily life and signalled changes, particularly with regard to the roles of women. 'It is time to look around us, to question the objects themselves, to untangle the chaos in domestic set up where the most recent often coexists in close promixity with the most archaic' writes Burguière (1975: 156). The fact that an entire generation saw daily life change — think of Jacques Tati's *Mon Oncle* (1958) — probably has something to do with this new focus. Following this new interest in rural ethnography, history started to focus on the role of technology in everyday life.

The shift in the 1970s and 1980s led to a critique of the traditional history of technology: it was no longer possible to study 'the relationship between technology and society' without 'first dealing with technologies that are in widespread use' (Edgerton, 1998: 816). Another development was influential: that of the history of the so-called 'private life', which is, by definition, linked to the everyday. Historically, everyday life in Europe witnessed a slow process of gradual 'privatization', primarily through the gradual separation of the home and the workplace from the late Middle Ages onwards, and later driven by the psychological need for a'shelter', although this happened much later than is usually thought. For instance, it was not until the first third of the nineteenth century that the word 'interior' (in Latin *interior*), the word used to describe the soul and the inner self, was applied to the domestic sphere. A whole series of techniques were then used to structure this interior and to reinforce 'private security', including complex locks, electric burglar alarms, which were introduced as early as the 1860s — speaking tubes and later electric intercom systems that further segregated between the home and the street. Acoustic techniques, bell systems to command servants at a distance, were refined to create increasingly silent spaces. Central heating and ventilation completed the comfortable home. During the second half of the twentieth century, mechanical ventilation, which became air-conditioning for offices and then homes, created spaces where, thanks to technology, people could dress and behave the same all year round. To have a room of one's own required a series of technical devices and if the middle classes did turn their homes into sanctuaries, they are too easily credited for this transformation given that the working classes also needed to rest and retire from the harshness of urban industrialization (Sennett, 1991).

With the domestic space as a lens, the study of innovation and the implementation of major technical systems, such as the railway or the electricity network, petrochemicals, and nuclear power loses its primacy. How are techniques used and by whom? How are they domesticated, appropriated, transformed? How do they change the delicate fabric of everyday life?

Technology still goes relatively unnoticed in works on the domestic space, but the everyday opens up a field of possibilities. The domestic space is no longer simply a space in which technologies first developed for public spaces — such as gas lighting, electric lighting, servants' bells, telephones, etc. — are adapted. It is a space where technologies are continually adaptated and reinvented, and often the very place where techniques from industry or the public sphere are taken up and become socially successful. The sewing machine, which had been invented for the garment industry, did not become popular until the Singer sewing machine, with its slogan 'A sewing machine in every family!', conquered the domestic market, first in the United States, France, and Russia, and then throughout the world (Godley, 2001: 1) (Figure 2). In fact, from the 1830s to the 1890s, all of daily life became mechanized, including transportation (by bicycle), personal hygiene (with heated showers), cooking (with autoclaves and gas stoves), communication (with the telephone), and even making and playing music (with the player piano — a mechanical invention if ever there was one — and the gramophone). The tens of thousands of patents registered in the nineteenth century show the domestic space was the focus of intense innovation efforts.

This also reminds us that everyday life involves complex systems or devices that combine a series of — sometimes quite simple — techniques. Taking a shower in the 1880s required wood or gas for heating the water, heating tanks, and very elaborate plumbing. This is also the case for water-closets or closed systems, such as the intercom systems used from the 1890s onwards to control building entrances from the upper floors (Figure 3).

Seen through the lens of the everyday, the history of technology does not allow any teleological interpretation. The history of electricity can no longer be simply the history of the advent of large-scale networks. Between the 1850s and the 1890s, many households in London, Paris, and Washington used electricity to light up rooms and operate doorbells remotely; batteries and accumulators permeated everyday life more than forty years before the networks, both urban and local, were established (Charpy, 2010). This set a notable precedent for the contemporary method (especially since the 2010s) of using lithium batteries in everyday objects, from toys to toothbrushes and scooters. This reversal of focus upsets hierarchies: small techniques become more important than larger ones — the percolator becomes more important than the aeroplane. This also leads to a reappraisal of the chronologies of innovation. The

Fig. 2. Lithographic advertisement for Singer sewing machines, British India, 1892. The slogan proclaims: 'Singer Sewing-Machines are known all over the world as being the best. The women of all nations prefer them. [...] Salesrooms in every city in the world'. Private collection (M. Charpy).

steam engine, which was characteristic of the industrial changes of the 1820s, can be said to only come to real fruition a century later when it was used in the domestic setting. Sometimes social diffusion only happens after one or two generations, as in the case of the technical objects used in the interwar period in North America and only transferred to Europe in the 1950s.

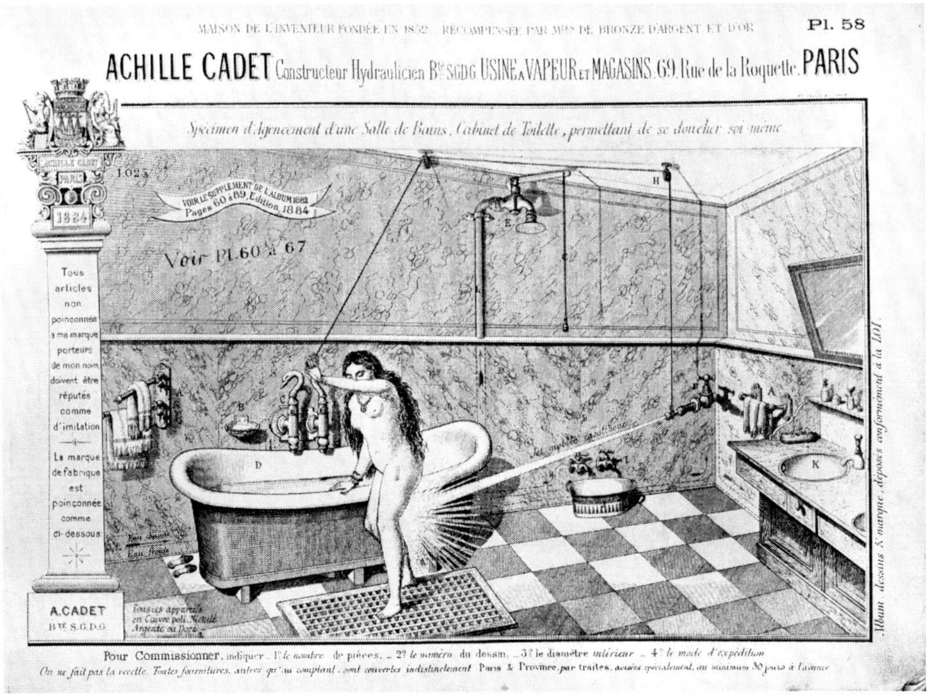

Fig. 3. Cadet. Presentation of arrangements for 'showering oneself', album by Achille Cadet, 69 rue de la Roquette, Paris, 1880s. Private collection (M. Charpy).

Seen through the lens of the everyday, new and old techniques coexist at any given point in time, and the latter often dominate. The everyday history of techniques is therefore cumulative, while innovation is more disruptive. Daily life proceeds less via transition than by accumulation: the radio was to be replaced by television, which in turn was supposed to come to an end with the arrival of the Internet. However, almost always the opposite happens. Techniques within the same ecosystem tend to become complementary of each other with each playing a specific, individual role. The convergence of the media and the emergence of the transmedial notion of 'content' is an illustration of this (Jenkins, 2006). The example of lighting in the nineteenth century is just as striking: its history should be presented as the successive decline of the candle and the oil lamp and the rise of gas and then electricity. But invention primarily focuses on old techniques, with stearic candles replacing tallow, the introduction of the hollow cylindrical wick and glass chimney, and oil being replaced by petroleum, a more efficient type of fuel. Thus, the lighting revolution, which happened alongside a growing desire to be able to see clearly at all hours of the day, seems to have been accomplished not so much through innovation and energy transition, but thanks to the 'global availability of fat' (Fressoz, 2020: 86, Zallen, 2019). But even then, implementations varied according to perceptions and social representations: gas was adopted quite quickly for lighting public spaces — 'shops are flooded with bright light' wrote Armand Husson in *Les Consommations de Paris* in 1875 (Husson, 1875: 14) — while the bourgeoisie, who had long been fond of candles with their naked flames, feared that gas would cause their decorations to fade and cause a risk of explosion or asphyxiation (Perrot, 1987). This social hierarachy is reflected in objects: one of the earliest batteries used in lighting appeared in the 1870s to remotely trigger a spark that ignited a paraffin lamp. At the end of the nineteenth century, gas lighting in flats coexisted with other lighting, allocated according to the time of day and the room, so that gas lighting was used in staircases, while mechanical Carcel oil lamps were used in living rooms, chandeliers were used for social occasions, and electric nightlights for reading (Charpy, 2010).

The Sociability of Techniques and Class Techniques

If the Ancien Régime period witnessed the 'birth of consumption' (Roche, 2000), the nineteenth century saw the mechanization of the domestic sphere, and primarily the proliferation of technical objects, both small and large, from doorbells to hydraulic water or pneumatic lifts (Charpy, 2011). The twentieth century saw the rise of normative goods as shown by the work of Henri Lefebvre (from 1947 onwards), Jean Baudrillard (1968), and Michel de Certeau (1980), a process that had already started much earlier (Brewer, Porter, 1993; Bartholeyns, 2021). The 'Americanization' of post-war European material culture is probably the best example of this normativity. In this reading the commodification of everyday life is viewed as alienating and consumers as engineers and bricoleurs of their lifestyles who respond to this commodification by singular appropriations of objects. This marks the beginning of techniques being looked at through the prism of their uses, cultural effects, and appropriations. The domestic space is now seen as an aesthetic and political arena, where social class dynamics are being played out.

By abandoning a heroic history of inventors and innovators, techniques and their uses reveal groups and categories of people who are often invisible: the poor, workers, children, women, etc. Technical devices reveal and shape the roles ascribed to certain groups or individuals by society. This is evident for women, who were forced, for example, to wear corsets (Steele, 2001) and, from the nineteenth century onwards, made to operate a range of household appliances — a term dating from the 1940s — which, far from liberating women, aligned them with the home (Delaunay, 1994). These machines also occupy their time, hence the success of sewing machines in Japan, for example, which became a means of controlling women's use of of their time in the twentieth century (Gordon, 2011). The same is true case for nineteenth-century workers whose time was controlled by time-keeping devices: the factory clock but also the adjustable mechanical morning alarm clock, developed in the late 1840s, which controlled the sleeping and waking of individuals.

The smallest technical object, however familiar it may be, involves social relations, which are often power relations. Does a simple Berlin key (a double-blade key that, after opening a door, needs to be pushed through and to the other side of the keyhole to lock the door again before it can be retrieved) not say something about the social obligation to lock collective doors, and thus about the relationship between private life and public space (Latour, 1993)? Does the presentation of a fully equipped kitchen at the 1959 American National Exhibition in Moscow, at a time when the Soviet Union had just successfully sent Sputnik into space, not say something about how the everyday day became the terrain of ideology and power, as exemplified by the famous 'Kitchen Debate' on ways of living between Khrushchev and Nixon (Colomina, 2002)?

The domestic space where the social reception of technology can take place. The shape objects is telling in this respect. Many assume historic forms in order to enter the private sphere. In the early days, lifts were sedan chairs with a velvet bench, and all devices related to entertainment, from the piano of the 1830s to the television of the 1950s, were disguised in peirod furniture. Enthusiasm for and resistance to certain technologies are perceptible in everyday life and show that even when an object is widely disseminated it may not be commonly used. In the 1960s, gas cookers, which are now common in kitchens and on wedding lists, did not enter rural kitchens, where the tradition of cooking on wood-burning stoves persevered (Perrot, 1981). An object may just be popular due to social pressure but not actually used — one only needs to think of the many exercise bikes or raclette grills that are discarded.

Technical choices confer distinction according to Pierre Bourdieu (1979). By merit of their prices and the fact that they are widely adopted, or not, technical objects are constant markers of social and cultural differentiation. Several decades after the advent of photographic prints on paper, the bourgeoisie still chose to use daguerreotypes, i.e. photographic images on silvered copper plates, a near-permanent medium well suited to the social aspirations of upper bourgeois families. Every technical object is a social programme which consumption materializes. This is exemplified by the elastic social capital of luxury goods: if they are too expensive, they do not enter the market, but if their price falls too much, they lose their power within the hierarchy. In this sense, the form of consumption reflects the form of society (Douglas, Isherwood, 1979).

Resistance to technology is also a matter of culture and imagination. Members of the aristocracy still refused to sit for photographs considered as vulgar around 1900 and rejected the cinema as a mechanical and childish entertainment (Pange, 1962). Members of the bourgeoisie presented themselves as technophiles, but for a long time considered it unimaginable to touch certain technical objects. Before they made high-rise buildings possible, lifts had a completely different function c.1900: in buildings of the middle classes, they only serviced the first floor (considered the elite floor) and until fairly late it was the concierge, a communal employee, who operated the ropes. The same social intermediation occurred throughout Europe for the telephone and electric lighting: for a long time, the bourgeoisie preferred to use an electric switch hidden under a carpet to summon a servant to light a paraffin lamp rather than switching on an electric light. The numerous instances where the rejection of appliances collides with local beliefs or religious prohibitions should also be mentioned here. The Amish community for instance uses gas washing machines because of their refusal to connect to the electricity grid (Kraybill, 2017).

Everyday life witnesses power relations and resistance because technology is so transformative. Sometimes the arrival of a new technical object consolidates hierarchies, but it can also redistribute power, status, change sociability and practices. The umbrella was originally a complex and luxurious object in the 1830s (Galizzia, 2018), but then quickly entered the sphere of ordinary objects and changed people's relationship to the weather as well as to the city. The arrival of the washing machine in the countryside in the 1960s led to the desertion of washhouses and the loss of these as social centres (Perrot, 1981). In India, the introduction of running water to some villages was similarly met with resistance, because it undermined the social interactions that took place around wells and fountains.

Techniques are now seldom studied outside their everyday implementation, and the everyday seldom studied without technology being taken into account. In the final decades of the twentieth century, there was even an emphasis in the literature on the use of technology and its societal impact, rather than on establishing technical genealogies. For example, digital technology and the mobile devices that revolutionized lifestyles on a global scale in the run-up to 2010 have been the subject of historical or genealogical analyses in the context of information culture (Manovich, 2001; Cardon, 2019; Nova, 2020) and in work addressing resulting conundrums such as attention (Crary, 2000, 2013) and distraction (Crawford, 2015). It is primarily as an everyday culture that the smartphone has been studied (Goggin, 2006; Pièces et Main d'Oeuvre, 2008), it is primarily through their cognitive and desocializing effects that social networks have been studied.

The Contemporary Mode of Existence of Technological Objects

Our daily experience of technological objects makes them familiar to us. However, it takes a whole range of domestication processes to do so, making it difficult to track innovation. It is necessary to learn how to operate these objects as well as to acquire the appropriate gestures and body postures their use requires. In other words, it is necessary to learn how to use them well in society, and sometimes how to maintain and repair them. For this purpose, 'demonstration salons' appeared from the end of the eighteenth century and proliferated in European cities from the middle of the following century onwards: saleswomen and salesmen, most often in mock interiors, demonstrated the use of gas of « machines amusantes » such as small hot air balloons, roasting machines, gas cookers, electric toys, hydrotherapy equipment, sewing machines, etc. Printed instructions which continued this development only appeared from the 1860s onwards; previously handwritten texts in the form of instruction manuals existed from the end of the eighteenth century, for example for toy aerostats (Hilaire-Pérez, Thébaud-Sorger, 2006). But these instructions were rare, generic and often not adapted to a particular object; and on the verso were handwritten notes explaining how the owner might repair the machine, including the constuction of replacement parts. Paradoxically, as soon as handling instructions became commonplace in the 1880s, parts were standardised and a network of repair shops was set up — among other things, for sewing machines, plumbing tools, weapons, and bicycles — which made it increasingly difficult to repair objects.

The question of the loss of quality is experienced by every contemporary user when an object breaks down less than a year after purchase, or a month after the end of the warranty. Characteristic of a new mode of existence for technical objects, obsolescence is experienced by users as technological depowerment. The introduction of electronic elements to mechanisms (in the age of the transistor), the transition from analogue to digital devices, and also the commercial impetus to prevent the repair of objects or even make it possible for the user to open them: all this has contributed to making technology inaccessible. Objects have also been covered for ergonomic and aesthetic reasons, following a long evolution from covering locomotives to pencil sharpeners and even toasters with aerodynamic shells (Loewy, 1951). In keeping with this lack of transparency, contemporary technical networks have become rather vulnerable due to their size and interdependence. Today, any breakdown, even a local disruption, often has catastrophic collective consequences. A flock of starlings on the tarmac or a suspicious piece of luggage are all it takes for tens of thousands of travellers to be grounded. In July 1977, it was lightning and, above all, a cascading series of breakdowns that plunged New York into the chaos of a blackout. Through the fear of a systemic collapse, technique has now become emblematic of the anthropocene (Bourrier, Nova, 2019).

The status of contemporary technical objects and devices results from a slow process of separation of design and use, production and consumption, the consequence of a framework of thought and practice in which technology is traditionally excluded from the realm of cultural values. It is not technology in itself but 'ignorance of the machine' which, for Simondon, is 'the strongest cause of alienation in the contemporary world' (2012: 10).

With regard to uses, the modern lack of quality involves a phenomenon of great historical import. After the age of proliferation and that of standardization, today seems to be the age of technical programming. As mentioned previously, our daily experience of technologies since the end of the twentieth century has been defined by a lack of quality, but this notation needs to be fleshed out. Technological objects break down. They are made to fail. Some are even disposable. Their use is programmed with their end in sight. It is, then, paradoxical, to say the least, that the concept of quality, which is specific to industrialization and its standardization, became central to businesses in the twentieth century. A new, so-to-speak improved definition of quality has emerged. Quality has moved from final quality control of the Taylorist type to 'total quality', which combines product quality, process quality, and customer relations (Weill, 2009). The economic programming of goods is not recent. Three centuries ago already, the Controller General Colbert scolded: '[i]f factories imposed, by dint of care, the superior quality of products, foreigners would find it advantageous to buy in France' Colbert's reprimand points out that quality was already considered as a production variable and no longer the ambition for any product which, in the fragmented production process that resulted from corporations, was to be subjected to inspection at each and every stage (for every material and process) before it was authorized onto the market, certified with official marks, as was the case in the medieval textile industry.

Consumption can be understood as resulting from the various ways utility, quality and fashion have combined, each factor having primacy at different time periods, in different social context and depending on the types of objects concerned. Around the 1900s, made-to-measure production was still dominant in women's clothing, which is partly explained by the guaranteed quality of handmade versus mass produced garments (Charpy, 2018). Even as early as the fifteenth century, a woman of high rank could tell her supplier that she only wanted the 'latest fabrics — unless a better quality was available. A chasm separates the premodern and the contemporary consumers who will buy a plastic mechanical pencil or the latest smartphone, even though the one they already have is still in perfect working order.

Technology has led to a form of everyday submission for which programmed or planned obsolescence, originally designed to revive the economy during the Great Depression of the 1930s (London, 1932; Latouche, 2013), is emblematic. Set against this logic, DIY has become one way reappropriating technical objects. This was already the case in the inter-war period, when small workshops on farms and working-class suburban housing offered the opportunity to repair machines on a daily basis as well as building machines, although these were sometimes unusable (Jarreau, 1985). The hippie movement also questioned technical objects, which were considered as tools of technicist and commercial control, by engaging with craftsmanship — for example, making a loom to be able to make one's own fabric (Chazalon, 2017; Rouvière, 2017). In 1968, the American biologist Stewart Brand proposed a programme of independent education and access to tools and parts in his *Whole Earth Catalog*, which was published intermittently until 1990s. Popular culture has its own repair and bricolage practices: from the 1940s onwards, tuning cars and motorbikes, which was popular in the United States particularly with Latin American immigrants, spread to Europe, where it is still common in the north, and to Japan in the 1970s and 1980s in the form of Bōsōzoku. In the French countryside, repairing, improving, and transforming one's moped was at the heart of juvenile, popular, and male culture for a long time.

In recent decades, the reappropriation of technical knowledge as part of the DIY movement and in repair cafés in cities is a striking development among a new group of the relatively well-off and those connected with the arts and design (Golsenne, Ribault, 2015). Matthew Crawford's account in *Shop Class as Soulcraft* (2009) of giving up his job at a renowned think tank to open a motorbike repair garage is emblematic of this regaining of control over techniques fostered by third places and decisions to lead alternative lives. Material subversion, reassembly, technical creolization, parallel economy practices: the low-tech and 'wild-tech' practices used out of necessity in poor countries and in DIY have entered alternative urban and rural lifestyles (Schumacher, 1973; Bihouix, 2014; Grimaud, Tastevin, Vidal, 2017).

In the twenty-first century, the mission is no longer to construct the world but to repair it, nor is it to make the everyday more technical, but to manage its waste (Furniss, Joulian, Tastevin, 2016). Our description of the toaster as civilisation must therefore be expanded. Just like General Electric's initial toaster, the D-12 model launched in 1908, the toaster as icon of contemporary civilization doesn't just constantly threaten to break down it is also an appliance whose components (nickel, chrome, zinc, etc.) are extracted from every continent, from New Caledonia to the Ural Mountains.

Globality, Differentiation, Hybridization

Unlike a history focussing on innovations and the deployment of techniques, as pointed out above, the history of everyday techniques is essentially global. But if the circulation of objects is evidence of different forms of acculturation, it is not necessarily synonymous with the spread of uses, in fact far from it. A whole range of types of acclimatization, adaptation or transformation occur between cultures, breaking with the great progressive narrative of the spread of Western goods and ways of life, what James Clifford has described as 'the pervasive dichotomy: absorption by the other or resistance to the other' (1988: 343). Do the Native Americans of the North who take up guns become Anglo-Americans? If colonial contexts provide an obviously political arena to study techniques, it is within the framework of a comparative approach to the history of their uses. A revision of the historical ranking of societies according to the level of their technical engagement or even to the presence or absence of Western techniques is also important (Adas, 1989).

A wide range of everyday objects can be seen as agents of colonization. The introduction of clocks and watches to schools, administrations, and workshops throughout the colonies in Africa was intented to establish a European conception of time (Nanni, 2012). The same applies to bicycles and sewing machines in India, which introduced new ways of travelling and dressing (Arnold, 2013). Here, technical objects aim to reform everyday life. Reappropriations, including sometimes unexpected ones, are nevertheless common. The example of photography is significant here: while ethnographers, anthropologists, and Europeans revelled in the African and Latin American 'natives'' apparent fear of the camera (Frazer, 1911), in fact, a large part of these populations adopted the technique as early as the 1860s. The middle classes of Liberia, Senegal, and Chile considered photography a way of establishing heir social status. At the same time, the complex folding umbrella became a symbol of power in Central Africa and was placed on the graves of chiefs, carefully torn to pieces to become unusable, underlining the appeal of the imported object for the local population, in the same way that Europeans collected ethnographic 'curiosities'. This type of appropriation happened in China between1900 and the 1920s, when the thermos flask initially used in European medicine became an everyday object for keeping tea hot, and torches became a popular accessory for urban dwellers who needed to walk on darker streets and alleyways beyond the main streets (Dikötter, 2007). The sewing machine and the accordion were successes thanks to their essential adaptability: the sewing machinde made it possible to create clothes including saris and gowns the accordion, despite essentially being the colonists' instrument, became a key instrument in the Congolese rumba, the Argentine chamamé, or the Martiniquan chouval bwa.

Notions such as the Creolization of technique (Edgerton 2006) and hybridization capture the complexity of techniques on the ground. A technique never arrives in a vacuum, there is always a preexiting technical tradition. Seen in the *longue durée*, material culture can always be seen as a process of hybridization: we evolve everywhere in 'an unprecedented overlay of traditions' (Clifford, 1988:9). In 1812, the Russian-American Company took men from Kodiak Island in Alaska to its colony of Fort Ross in California. At the foot of the Russian fort an interethnic village was established. The village had buildings and followed the hunting practices of Alaskan origin, while the domestic arrangements, which were managed by native women, were more specifically indebted to the Pomo and Miwok cultures. Food in the village inivolved foodstuffs that were not consumed by the native populations (seal, whale, scorpion fish, etc.), but cooked according to local techniques and traditions. Living together had produced a new culture.

But everyday techniques are also areas of passive or active resistance. Colonial rule gave rise to many instances of resistance. As early as the 1900s, settlers were flustered by seeing elegant Congolese showing off their non functional wristwatches. In the 1940s and 1950s, French sociologists in French Equatorial Africa were still perplexed by this (Chaumeton, 1947), without realizing that it was obviously a way of recasting the watch as a purely decorative jewel and refusing the general imposition of Western time. The Sawadeshi movement from the 1890s onwards should be interpreted in the same way, as a rejection of both imported English products and the ways of producing them while reviving the use of the traditional spinning wheel, the chakra, for making one's own clothes at home (Bayly, 1986).

Any history of material and technical imperialism is only valid if all forms of cultural dissemination are considered. This means looking at Europe's fascination with foreign countries and the transformation of Western daily life by products, techniques, and decorations, especially coming from the East. It is not surprising that since the 1840s, Europeans and Americans have been enjoying both objects made by 'superior' craftsmen such as the Indian shawl and the 'archaic' productions of various tribes from across the world who were seen as representatives of altogether different and supposedly inferior technical regimes. Universal Exhibitions dramatized the story of labour and industry as well as that of middle-class homes. A guitar or a bow from South America or West Africa, just like an obsolete spinning wheel now turned decorative object, told the grand narrative of technological progress spearheaded by Europe.

Globality and Subject formation

The globalization of the consumer goods market is another factor that shapes technology in everyday life and how the everyday is shaped by technology. The globalization of markets is the main feature of the twentieth and twenty-first centuries. In addition to creating a problem of variable quality (encapsulated by the derogatory use of the phrase 'made in China'), there is also a problem of quantity, visible in both in shops and in the home. 'It's a never ending battle…', 'we have too much stuff, 'It's overwhelming!' It is with these expressions of distress that the Netflix series *Tidying Up with Mary Kondo* opens. In it a series of families are being helped by a professional to cope with the clutter that is making their lives miserable. Alongside superfluous pairs of shoes and clothes, homes are cluttered with crates full of electric cables and obsolete electronics — an old hi-fi system, a now redundant household appliance, as well as cassettes, floppy disks, or DVDs kept because they form collections of sorts that are difficult to throw away. The pathological hoarding of objects, also known as Diogenes's syndrome, was given its scientific name, syllogomania, by English doctors in 1975. Syllogomania has been on the rise since the end of the twentieth century. In 2013, it came to be known as a type of obsessive-compulsive disorder (American Psychiatric Association, 2013). People's basements, garages, and attics are full. The shipping container, developed by Malcolm McLean in 1965, more than halved the cost of storing and handling bulky objects such as ovens, washing machines, and flat screen TVS. Between 2000 and 2017, the global traffic of bulk carriers and other tankers increased from 65 million containers to more than 175 million. In the United States, in 2017, the number of square meters of personal storage doubled in just two years, going from 18.5 million m^2 to 36 million m^2. Supply can barely keep up with demand. Europe showed similar developments in the same year, with an increase of more than 10% of rented storage space at an average monthly cost of 10 euros per square meter.

Houses in the Sun Belt States were and are typically built without cellars or attics, and it was here that the first furniture storage or self-storage companies were founded in the late 1960s. New homes are larger on average, but the problem grew rapidly in the 2010s due to larger numbers of single households (resulting from increased mobility, family restructuring, etc.) and the generational problem of what do with the possessions of baby boomers at the end of their lives. The predicament created by all this stuff is mirrored by the Black Friday frenzy leading to the purchase of ever more clothing and electronics. From destocking to total liquidation, saturation affects the entire distribution circuit from suppliers to consumers, who at the same time call for 'buy nothing days', take global action against waste, and favour tiny houses. If the everyday can be seen as a technique, it is a dysfunctional one. A life that until recently would have been considered a 'good' and 'satisfying' life by most has become ominous (Næss, 1989)

The reason the history of techniques is also that of lifestyles, it is because objects are no longer regarded as mere props, or even as the means for or the interface of social relations: they have agency. This shift in the status of objects is linked to the rise of the anthropology of objects brought about both by the biography of objects — the idea that things have a 'social life' and a life cycle (Kopytoff, 1986) — and actor-network theory (Kien, 2009), which considers society an arrangement of humans and non-humans that include artefacts. In other words: the fact that a society is disrupted if one object or technique is removed from it shows this object and technique to be social. No description of this society would be complete without this object, i.e. without calling on its agency (Latour, Lemonnier, 1994; Miller, 1997).

This means that not only are innovations and techniques socially constructed and constantly negotiated between a large number of groups and intermediaries (Akrich, 1994), but they are also part of the processes of subject formation. What individual changes occur in someone who has running water or electricity in their home before his or her neighbours do? What new sociabilities arise from the lighting of public places? What conception of the home, privacy or intimacy does technology satisfy or create when you think of devices for surveillance, locks, alarms, and doorbells, which multiplied in the 1860s if we go by the number of patents taken at the time? Historians of the eighteenth century were quicker to point out material sociability and how 'everyday things' contribute to subject-formation (Roche, 2000) than their nineteenth-century and twentieth-century counterparts. One reason for this is probably because they had account for a shift from scarcity to plenty and how this affected the everyday with production ranging from 'luxury to comfort' (Goubert, 1988).

As the everyday became an autonomous sphere of life, and both an extension and the locus of the self and its recognition, the world kept entering interiors, living rooms becoming boxes 'in the theater of the world' to use Walter Benjamin's expression. Curiosities from all continents entered European homes magnified by viewing machines, such as magic lanterns, home stereoscopes from the 1850s onwards, and slide projectors after the 1950s, etc. Technical objects have constantly tended to add new depth to the domestic space through the integration of images. The world had begun to enter interiors with images and their essentially religious subjects, but also in the form of landscapes, portraits of ancestors and genealogies, giving daily life a new temporalities. This developed as activities of contemplation and prayer traditionally carried out outside the home and alongside daily routines were brought home (Bartholeyns, Dittmar, Bourin, 2018). The omnipresence of the world's media really takes place with the radio, an object present in family homes from the 1930s onwards, which was followed by television in the 1950s. For Raymond Williams (1975), television solved the profound paradox of post-war Western societies, and particularly American society, which were both hyper-mobile and home-centred. It allowed the viewer to enjoy the world without leaving home. The history of the integration through technology of the world into homes has reached new levels with mobile and connected devices, although it also goes the other way round with the private being broadcast to the world. Whereas the domestic TV screen connected people, the new digital age creates social isolation. Yet nomadic technology also offers the user the comforting and always accessible feeling of being 'at home' everywhere and of finding oneself (Groening, 2010), because these objects give access to virtual spaces with strong identity and sentimental value.

Techniques of the Body, Techniques of the Self

The body is immersed in an increasingly personalized, everyday technical space, and this personal daily life is now experienced thanks to advanced technological devices which we hold in our hands. The history of daily life is here informed by two essential sources. Firstly, Marcel Mauss saw the body as 'man's first and most natural instrument', and everyday life in its entirety should be seen as made up of 'techniques of the body' (Mauss, 1973, 75). Sleeping is a technique, not merely because beds and bedding are artefacts and the expression of a certain material culture but also because we only fall asleep after we have learned to sleep while lying down or in an almost seated or even standing position, on a soft or hard surface, etc. And the search for comfort from the 1830s onwards called for articulated furniture, elasticated materials and springs, and rubber cushions.

The social construction of bodies depends on a series of techniques. Crinolines, corsets, and ties 'straighten' bodies; they are cultural orthopaedic devices (Vigarello, 1978). Zippers, busks, whalebone, wire and prostheses of all kinds are all mechanical devices of sorts. Bodies are not only alienated or freed by machines (Sennett, 2008); they form a whole with a class of artefacts — glasses, watches, helmets, etc. — that promote effective action in manual labour, or are used for dramatic feats of engineering. How did the Wright Brothers, Glenn Curtiss or Henri Farman fly aircrafts in the early 1910s, in front of an audience clutching to their canes and peering through their monocles, if not by swapping mere bodily strength for cognitive input and engineering? At any given moment, the body equips itself or is equipped with complex prosthetic additions to socialize or accomplish what it cannot do naturally (Bartholeyns, Joulian, 2017). Artefacts become integrated into bodies. They modify the body self image or produces a synthesis, so that with transportation devices one becomes 'one' with the device, an experience shared by all cyclists and car drivers (Warnier, 1999). And tools, on the farm, in the factory, and in urban workshops, seamlessly extend the arm, increasing its strength or precision in everyday actions. The replacement of craft activities with production lines and automation changes the ratio between subject and object but the process of integration is unchanged whether it be for a blacksmith and farrier or a steelworks maintenance worker.

Michel Foucault has shown body techniques to be techniques of the self, i.e. ways of mastering oneself and one's history, of becoming a fully-fledged subject. Foucault (1984) explored the interpretation of dreams, sexuality, and memory. How much time should one devote to oneself each day? What work should be done on oneself to live better? What truth does my life hold? We should add to these techniques of existence and self care the practice of meditation and yoga which has developed since the 1960s but but also the mobile phone which is a tool of self knowledge (Allard, 2010). In the second half of the nineteenth century, photography was a tool of self fashioning, a way of recording one's state at different stages of life (Charpy, 2011). Truth be told we are surrounded by instruments allowing us to become aware of ourselves and often to control ourselves. Measuring instruments are central. The transition from the clock to the pocket watch from the end of the eighteenth century internalised time and the increased collective control over individual time (Glennie and Thrift, 2009; Dequidt 2012; Granger, 2017). Thermometers, hygrometers, and barometers are still instruments that allow to track changes in the atmospheres and thus objectify bodily senzations. This is also the case for all devices relating to hygiene and the body, from showers and washbasins to washing machines, from electric shavers to hair dryers and bodybuilding machines, all of which are ways of combining social and personal motivations. Here once more, the unexpected arises at the heart of intimacy, especially for home-ground women. The telephone quickly became a means of communication that excaped social control, and the success since 1910 of massaging devices such as the Pulsoconn, which was initially hand-operated and then electric, in the United States and then in Europe, derived from its use as a vibrator (Mortas, 2020). In the privacy of the home, technological objects are always ambiguous.

Discipline, control, and the government of populations have been thought of as techniques in themselves, technologies of power that apply to all areas of life — medical, educational, and judicial. In particular, biopolitics as a technique meant to ensure the economic profitability of a population, permeates all aspects of its life: from birth practices to diseases and healthcare but also, on a daily basis, hygiene practices and living spaces. The study of these endemic realities increasingly requires the use of statistical counting techniques, census operations, standardization procedures, and security measures in addition to the 'old' spatial and material techniques of surveillance and discipline (Foucault, 2004). In this respect, between the eighteenth century and the end of the nineteenth century, everyday life became a political category specific to the liberal and technical world.

References

Adas, Michael, *Machines as Measure of Men: Science, Technology, and Ideologies of Western Dominance* (Ithaca, NY: Cornell University Press, 1989).

Akrich, Madeleine, 'Comment sortir de la dichotomie technique/société. Présentation des diverses sociologies de la technique', in *De la préhistoire aux missiles balistiques: l'intelligence sociale des techniques*, ed. by Bruno Latour and Pierre Lemonnier (Paris: La Découverte, 1994), pp. 105–31.

Allard, Laurence, *Mythologie du portable* (Paris: Le Cavalier bleu, 2010).

American Psychiatric Association, *DSM-5: diagnostic and statistical manual of mental disorders*, Fifth ed., Washington D.C. American Psychiatric Association, 2013.

Appadurai, Arjun, ed., *The Social Life of Things: Commodities in Cultural Perspectives* (Cambridge: Cambridge University Press, 1986).

Arnold, David, *Everyday Technology: Machines and the Making of India's Modernity* (Chicago, IL: University of Chicago Press, 2013).

Bartholeyns, Gil, Pierre-Olivier Dittmar, and Monique Bourin, 'La Domestication des images', in *Images de soi dans l'univers domestique*, ed. by Gil Bartholeyns, Pierre-Olivier Dittmar, and Monique Bourin (Rennes: Presses universitaires de Rennes, 2018), pp. 11–18.

———, Nicolas Govoroff, and Frédéric Joulian, eds, *Cultures matérielles. Anthologie raisonnée*, 2 vols (Paris: Éditions de la Maison de sciences de l'homme, 2011).

——— and Frédéric Joulian, eds, *Le Corps instrument* (Paris: Éditions de la Maison des sciences de l'homme, 2017).

Bartholeyns Gil, 'Choses, objets, produits', in *Une nouvelle histoire du Moyen Âge*, ed by Florian Mazel (Paris: Seuil, 2021), pp. 719–26.

Baudrillard, Jean, *Le Système des objets* (Paris: Gallimard, 1968).

Bayly, Christopher Alan, 'The Origins of Swadeshi (Home Industry): Cloth and Indian Society, 1700–1930', in *The Social Life of Things: Commodities in Cultural Perspectives*, ed. by Arjun Appadurai (Cambridge: Cambridge University Press, 1986), pp. 285–321.

Benjamin, Walter, *Paris, Capital of the 19th Century*, in *Selected Papers*, vol. III, 1935–1938 ed. by Howard Eiland and Michael W. Jennings trans. by Edmund Jephcott and Howard Eiland (Cambridge, MA and London: The Belknap Press of Harvard University Press, 2002).

Bihouix, Philippe, *L'Âge des low tech. Vers une civilisation techniquement soutenable* (Paris: Seuil, 2014).

Bloch, Marc, *La Société féodale* (Paris: Albin Michel, 1939).

Bourdieu, Pierre, *La Distinction. Critique sociale du jugement* (Paris: Éditions de Minuit, 1979).

Bourrier Mathilde and Nova Nicolas ed., *En cas de panne, Techniques & Culture*, 72 (2019).

Brand, Stewart , *Whole Earth Catalog* (1968–1998).

Braudel, Fernand, *Civilisation matérielle, économie et capitalisme, XVe–XVIIIe siècle*, vol. I, *Les Structures du quotidien* (Paris: Armand Colin, 1967).

Brewer, John and Roy Porter, *Consumption and the World of Goods* (London: Routledge, 1993).

Burguière, André, *Bretons de Plozévet* (Paris: Flammarion, 1975).

Cardon, Dominique, *Culture numérique* (Paris: Presses de Sciences Po, 2019).

Charpy, Manuel, 'Matières et mémoires. Usages des traces de soi et des siens dans une grande famille bourgeoise de la seconde moitié du XIXe siècle', *Revue du Nord*, 390 (2011), 395–432.

———, 'Piqué-machine et cousu-main. Succès et rejets de la machine à coudre dans la confection au XIXe siècle', *La Révolution de la machine à coudre*, ed. by Michel Cordillot, special issue, *Les Cahiers d'Adiamos* 89, 16 (2018), 67–112.

———, 'Le Théâtre des objets. Espaces privés, culture matérielle et identité sociale. Paris, 1830–1914' (unpublished doctoral thesis, University of Tours, 2010).

———, 'La broyeuse de cœur gras', *Artefact*, 11 2019, 109-41.

——— and François Jarrige, 'Penser le quotidien des techniques. Pratiques sociales, ordres et désordres techniques au XIXe siècle', *Revue d'histoire du XIXe siècle*, 45 (2012), 7–32.

Chaumeton, Josette, 'Note préliminaire sur les niveaux de vie dans les villes indigènes de Poto-Poto, Wenzé et Bacongo', Office de la recherche scientifique et technique Outre-Mer, 1947.

Chazalon, Élodie, 'À contre-mode? Du vêtement hippie et de son authenticité aux États-Unis, de 1967 à nos jours', *Modes Pratiques. Revue d'histoire de la mode et du vêtement*, 2 (2017), 309–27.

Clifford, James, *The Predicament of Culture: Twentieth-century Ethnography, Literature, and Art* (Cambridge, MA and London: Harvard University Press 1988).

Colomina, Beatriz, 'Enclosed by Images: Architecture in the Post-Sputnik Age', in *CTRL [SPACE]. Rhetorics of Surveillance from Bentham to Big Brother*, ed. by Thomas Y. Levin, Ursula Frohne, and Peter Wiebel (Karlsruhe and Cambridge, MA: ZKM–MIT Press, 2002), pp. 322–37.

Crary, Jonathan, *24/7: Late Capitalism and the Ends of Sleep* (London and New York: Verso, 2013).

———, *Suspensions of Perception: Attention, Spectacle and Modern Culture* (Cambridge, MA: The MIT Press, 2000).

Crawford, Matthew B., *Shop Class as Soulcraft. An Inquiry into the Value of Work* (London: Penguin, 2009).

———, *The World Beyond your Head: on Becoming an Individual in an Age of Distraction* (New York: Farrar, Straus and Giroux, 2015).

Daumas, Maurice, ed., *Histoire générale des techniques*, 5 vols. (Paris: Presses Universitaires de France, 1962–1979).

de Certeau, Michel, with Luce Giard and Pierre Mayol, *L'Invention du quotidien*, vol. I, *Arts de faire*; vol. II, *Habiter, cuisiner* (Paris: Union général d'editions, 1980).

Delaunay, Quynh, *Histoire de la machine à laver: un objet technique dans la société française* (Rennes: Presses universitaires de Rennes, 1994).

Dequidt, Marie-Agnès, 'Comment mesurer l'intériorisation du temps? (Paris, début XIXe siècle)', *Revue d'histoire du XIXe siècle*, 45 (2012), 69–81.

Dikötter, Frank, *Exotic Commodities: Modern Objects and Everyday Life in China* (New York, NY: Columbia University Press, 2007).

Douglas, Mary and Baron C. Isherwood, *The World of Goods: Towards an Anthropology of Consumption* (London: Allen Lane, 1979)

Edgerton, David E., 'De l'innovation aux usages. Dix thèses éclectiques sur l'histoire des techniques', *Annales. Histoire, Sciences Sociales*, 53.4–5 (1998), 815–37.

———, *The Shock of the Old. Technology and Global History since 1900* (London: Profile Books, 2006).

Foucault, Michel, *Histoire de la sexualité*, vol. III, *Le Souci de soi* (Paris: Gallimard, 1984).

———, *Sécurité, Territoire, Population. Cours au Collège de France. 1977–1979* (Paris: Seuil, Gallimard, 2004).

Franklin, Alfred, *La Vie privée d'autrefois. Arts et métiers, modes, mœurs, usages des Parisiens, du XII*[e] *au XVIII*[e] *siècle d'après des documents originaux ou inédits*, 27 vols (Paris: Plon, 1887–1902).

Frazer, James George, *Taboo and the Perils of the Soul* (London: Macmillan and Co, 1911).

Fressoz, Jean-Baptiste, 'Une histoire matérielle de la lumière', in *Face à la puissance. Une histoire des énergies alternatives à l'âge industriel*, ed. by François Jarrige and Alexis Vrignon (Paris: La Découverte, 2020), pp. 84–99.

Furniss, Jamie, Frédéric Joulian, and Yann-Philippe Tastevin, *Réparer le monde. Excès, reste et innovation* (Paris: Éditions de l'EHESS, 2016).

Gallizia, Steeve, 'La Mécanique des intempéries. Parapluies et ombrelles: évolutions et dérivés à travers les brevets d'invention', *Modes pratiques. Revue d'histoire de la mode et du vêtement*, 3 (2018), 250–63.

Gerritsen, Anne and Giorgio Riello, eds, *Writing Material Culture History* (London: Bloomsbury, 2014).

Gille, Bertrand, ed., *Histoire des techniques: technique et civilisations, technique et sciences* (Paris: Gallimard, 1978).

Giedion, Sigfried, *Mechanization Takes Command: A Contribution to Anonymous History* (Oxford: Oxford University Press, 1948).

Glennie, Paul and Nigel Thrift, *Shaping the Day: a History of Time Keeping in England and Wales 1300–1800* Oxford: Oxford University Press, 2009

Godley, Andrew, 'The Global Diffusion of the Sewing Machine, 1850–1914', *Research in Economic History*, 20 (2001), 1–46.

Goggin, Gerard, *Cell Phone Culture. Mobile Technology in Everyday Life* (New York, NY: Routledge, 2006).

Golsenne, Thomas and Patricia Ribault, eds, *Essais de bricologie. Ethnographie de l'art et design contemporain* (Paris: Éditions de l'EHESS, 2015).

Gordon, Andrew, *Fabricating Consumers: the Sewing Machine in Modern Japan* (Berkeley, CA: University of California Press, 2011).

Goubert, Jean-Pierre, ed., *Du luxe au confort* (Paris: Belin, 1988).

Granger, Christophe, 'Montre', in *Histoire du monde au XIX*[e] *siècle*, ed. by Pierre Singaravélou and Sylvain Venayre (Paris: Fayard, 2017), pp. 687–93.

Grimaud Emmanuel, Tastevin Yann-Philippe, Vidal Denis ed., *Low Tech? Wild Tech, Techniques & Culture*, 67 (2017).

Groening, Stephen, 'From "a Box in the Theater of the World" to "the World as your Living Room": Cellular Phones, Television and Mobile Privatization', *New Media & Society*, 12.8 (2010), 1331–47.

Hilaire-Pérez, Liliane and Marie Thébaud-Sorger, 'Les Techniques dans l'espace public. Publicité des inventions et littérature d'usage au XVIIIe siècle (France, Angleterre)', *Revue de synthèse*, 127.2 (2006), 393–428.

Hoggart, Richard, *The Uses of Literacy: Aspects of Working-class Life with Special References to Publications and Entertainments* (London: Chatto & Windus, 1957).

Husson, Armand, *Les Consommations de Paris, deuxième édition entièrement refondue,* (Paris: Hachette, 1875).

Jarreau, Philippe, *Du bricolage. Archéologie de la maison* (Paris: Centre de création industrielle, 1985).

Jenkins, Henry, *Convergence Culture: Where Old and New Media Collide* (New York, NY: New York University Press, 2006).

Kien, Grant, 'Actor-Network Theory. Translation as Material Culture', in *Material Culture and Technology in Everyday Life: Ethnographic Approaches,* ed. by Vannini Phillip (Bern: Peter Lang, 2009), pp. 27–44.

Kopytoff, Igor, 'The Cultural Biography of Things: Commoditization as Process', in *The Social Life of Things,* ed. by Arjun Appadurai (Cambridge: Cambridge University Press, 1986), p. 64–91.

Kraybill, Donald B., 'Out of fashion. Les Amish et la mode, entre le XVIe siècle et aujourd'hui', *Modes pratiques. Revue d'histoire du vêtement et de la mode,* 2 (2017), 71–89.

Landes, David S., *L'Heure qu'il est. Les horloges, la mesure du temps et la transformation du monde moderne* (Paris: Gallimard, 1983).

Latour, Bruno, *La Clef de Berlin et autres leçons d'un amateur de sciences* (Paris: La Découverte, 1993).

——— and Pierre Lemonnier, eds, *De la préhistoire aux missiles balistiques. L'intelligence sociale des techniques* (Pairs: La Découverte, 1994).

Lefebvre, Henri, *Critique de la vie quotidienne* (Paris: Grasset, 1947).

———, *Critique de la vie quotidienne,* vol. II, *Fondements d'une sociologie de la quotidienneté* (Paris: L'Arche, 1961).

———, *Critique de la vie quotidienne,* vol. III, *De la modernité au modernisme* (Paris: L'Arche, 1981).

Loewy, Raymond, *Never Leave Well Enough Alone* (New York, NY: Simon and Schuster, 1951).

London, Bernard, *L'Obsolescence planifiée, pour en finir avec la grande dépression* (1932), afterword by Serge Latouche, *Bernard London, ou la planification de l'obsolescence à des fins sociales* (Paris: Éditions B2, 2013).

Manovich, Lev, *The Language of New Media* (Cambridge, MA: The MIT Press, 2001).

Mauss, Marcel, 'Techniques of the Body' (1935), *Economy and Society,* 2 (1) (1973), 70–88.

Miller, Daniel, ed., *Material Cultures: Why Some Things Matter* (Chicago, IL: University of Chicago Press, 1997).

Mortas, Pauline, 'Sextoy', in *Le Magasin du monde. La mondialisation par les objets, du XVIIIe siècle à nos jours,* ed. by Pierre Singaravélou and Sylvain Venayre (Paris: Fayard, 2020), pp. 402–05.

Mosaddad, Seyed-Hashem, 'Le Produit bicylette en France (1883–1995): engouement, conception, concrétisation' (unpublished doctoral thesis, Sorbonne University, 2001.

Næss, Arne, *Ecology, Community and Lifestyle: Outline of an Ecology* (Cambridge: Cambridge University Press, 1989).

Nanni, Giordano, *The Colonisation of Time: Ritual, Routine, and Resistance in the British Empire* (Manchester: Manchester University Press, 2012).

Nova, Nicolas, *Smartphone. Une enquête anthropologique* (Geneva: Mêtis Presses, 2020).

Ogle, Vanessa, *The Global Transformation of Time, 1870–1950* (Cambridge, MA: Harvard University Press, 2015).

Pange, Pauline de, *Comment j'ai vu 1900* (Paris: Grasset, 1962).

Perrot, Martyne, 'Machines du dedans: les automatismes ménagers en milieu rural', *Culture Technique*, 3 (1981), 199–204.

Perrot, Michelle, 'Manières d'habiter', in *Histoire de la vie privée*, vol. IV, *De la Révolution à la Grange Guerre*, ed. by Michelle Perrot (Paris: Seuil, 1987), pp. 281–97.

Pièces et Main d'Œuvre, *Le Téléphone portable, gadget de destruction massive* (Paris: L'Échapée, 2008).

Roche, Daniel, *A History of Everyday Things: The Birth of Consumption in France, 1600–1800* (Cambridge: Cambridge University Press, 2000)

Rouvière, Catherine, 'Du refus de la mode à la création d'une "anti-mode". Les communautés néo-rurales en France', *Modes pratiques. Revue d'histoire de la mode et du vêtement*, 2 (2017), 347–71.

Schumacher, Ernst Friedrich, *Small Is Beautiful. A Study of Economics As If People Mattered* (London: Blond & Briggs, 1973).

Sennett, Richard, *The Conscience of the Eye. The Design and Social Life of Cities* (London: Faber & Faber, 1991).

———, *The Craftsman* (New Haven, CT: Yale University Press, 2008).

Simondon, Gilbert, *Du mode d'existence des objets techniques* (Paris: Aubier, 2012).

Singer, Charles, Eric John Holmyard, Alfred Rupert Hall, and Trevor I. Williams, *A History of Technology*, 8 vols (Oxford: Clarendon Press, 1954–1984).

Steele, Valérie, *The Corset: a Cultural History* (New Haven, CT: Yale University Press, 2001).

Thompson, Edward P., *The Making of the English Working Class* (Harmondsworth. Penguin Books, 1963).

Thwaites, Thomas, *The Toaster Project: or a Heroic Attempt to Build a Simple Electric Appliance* (Princeton, NJ: Princeton Architectural Press, 2010).

Van Dommelen, Peter, 'Colonial Matters. Material Culture and Postcolonial Theory in Colonial Situations', in *Handbook of Material Culture*, ed. by Christopher Tilley, Webb Keane, Susanne Küchler, Michael Rowlands, and Patricia Spyer (London, Thousands Oaks, CA, and New Delhi: Sage Publications, 2006), pp. 105–24.

Van Gennep, Arnold, *Manuel de folklore français contemporain* (Paris: Picard, 1937–1958).

Vigarello, Georges, *Le Corps redressé. Histoire d'un pouvoir pédagogique* (Paris: J.-P. Delarge, 1978)

Warnier, Jean-Pierre, *Construire la culture matérielle. L'homme qui pensait avec ses doigts* (Paris: PUF, 1999).

Weill, Michel, *Le Management de la qualité* (Paris: La Découverte, 2009).

Williams, Raymond, *Television. Technology and Cultural Form* (New York, NY: Schocken Books, 1975).

Zallen, Jeremy, *American Lucifers: The Dark History of Artificial Light, 1750–1865* (Chapel Hill: University of North Carolina Press, 2019).

SORAYA BOUDIA, NATHALIE JAS

Technology, Pollution, Environment

'To avoid the end of the world': this was the title of the editorial of a special issue of the French news weekly *Le Nouvel Observateur* devoted to the First Earth Summit, sponsored by the United Nations and held in Stockholm from 5 to 16 June 1972 (Le Nouvel Observateur, 1972). This summit, officially named the United Nations Conference on the Human Environment, assembled representatives from 132 countries, attesting to growing planetary awareness of how finite the earth's resources are and the extent of the social and environmental problems engendered by the Western industrial society (Speth, Haas, 2006). 'In the name of the progress that was the specificity and pride of man began the most enormous enterprise of destruction that any species has ever undertaken against the environment that supports life, and against life itself', declared Alain Hervé (1972, 4), author of the editorial. In this he picked up an argument heard in a myriad of environmental movements that had begun emerging in the 1960s, which often connected heightened ecological awareness with an engagement in a fight against Western capitalism and imperialism. Foremost among the denounced causes of this 'enterprise of destruction' was local- and global-scale pollution that had been exposed by studies on the presence of particular pollutants in the environment, notably radioactivity, organochlorine pesticides, and Polychlorinated Biphenyls (PCBs) Thereafter, even the areas farthest from human activity seem to have been defenceless against the flow of environmental contaminants, which many scientific fields and their new methods demonstrated could not only be carried over great distances, but could also affect living organisms insidiously, reaching their very DNA and the regulation of their fundamental biological functions (Craeger, 2014; Frickel, 2004; von Schwering, 2015). The Stockholm summit was a milestone and moment of public acknowledgement of two centuries of environmental pollution engendered by technological and industrial activities and their deleterious effects on all ecosystems and the human body. Today, more than forty years after this summit and despite the development of numerous national and international regulations, a considerable body of research demonstrates that the planetary expansion of pollution continues, although the scope and toxic consequences of this pollution are geographically and socially variable.

In the limited context of this chapter and without any pretense of exhaustivity, we will examine the ways in which global history can study the relationships between

Global History of Techniques, ed. by Guillaume Carnino, Liliane Hilaire-Pérez and Jérôme Lamy, Global Matters Culture Matérielles, histoire des techniques et circulation globale des savoirs, 9 (Turnhout: Brepols, 2024), pp. 729–742
BREPOLS PUBLISHERS 10.1484/M.TECHNE-EB.5.129838

technology, pollution, and the environment. We examine the theme from two particular approaches, the crossing and hybridization of which may help account for the accumulation and sedimentation of the layers of various pollutants and policies that are the foundation of the contemporary environmental crisis: the first is the history of the joint expansion of industrial production and pollution, and the second is the history of how pollution has been governed.

The Joint Expansion of Industrial Production and Pollution

It is difficult to fully grasp the nature of the relationships between technology, pollution, and the environment without also studying the development of an economy based in the ever greater proliferation and intensification of the production and consumption of goods. Environmental pollution and the constitution of what could be called a 'toxic world' are inextricably linked with the rapid growth of industrial activity since the late eighteenth century and its notable acceleration after 1945 (Boudia, Jas, 2019; Jarrige, Le Roux, 2020). Pollution is produced all along the chain, spanning the extraction of primary materials, the production and consumption of a variety of products and the waste products that ultimately result from them, and the rapid growth and geographical spread of these activities contributes almost 'naturally' to the spread and lasting establishment of pollution across space and over time. This configuration can be found in a wide range of industrial activities, including the mining sector, the metallurgical, chemical, and petrochemical industries, the nuclear industry, and the electronics industry.

The history of the development of the chemical industry provides an emblematic example of this joint expansion. The development of the inorganic chemical industry in England in the late eighteenth century was based on technological innovations that made it possible to produce sulfuric acid on a large scale. Sulfuric acid is necessary for the production of alkali and bleaching agents that are essential for the textile industry. Likewise, the capacity to produce great quantities of sulfuric acid was essential for the development of the glass industry, followed by the establishment of a prosperous superphosphate industry and the production of the aromatic nitrations needed to produce dyes for the textile industry. By the 1920s, the organic and inorganic chemical industries worked particularly in service of the textile industry, agriculture (superphosphates, nitrogen fertilizer, pesticides), and the armament industry (explosives, chemical weapons). By then they had established the foundations of their long-term development by being vital to a number of industrial sectors (including agriculture), developing production systems for mass production (continuous procedures, catalytic procedures, industrial electrochemistry, chemical engineering), and building very large companies with financial and research and development capacities capable of accompanying the transformation of other industrial sectors through chemistry (Ilgen, 1983; Murmann, 2003). Starting in the 1930s and accelerating after the Second World War and the surge in petrochemical and synthetic chemistry, the chemical industry had a hand in the emergence of other industrial sectors, particularly agribusiness, pharmaceuticals, and a variety of other sectors using materials produced through chemistry (nylon, polyester, resins, insulators, glass, metal alloys, concrete, paint, varnish, adhesives, plastics, and so on). Such industrial expansion resulted in the continual development of many kinds of pollution. This ame in part from the sheer mass of molecules produced by the chemical industry, extractive industries, and industries dependent on their products further down the chain that are toxic to ecosystems and the human body. A sizeable literature has documented this pollution and its effects on many scales (Fressoz, 2013; Le Roux, 2010; Rainhorn, 2019; Rosental, 2017; Zimmer, 2016).

These dynamics of global expansion and industrialization and the resulting contamination that began in the eighteenth century took a major leap forward after the Second World War, which had itself been a tremendous scientific, technical, and industrial catalyst (Dahan, Pestre, 2004; Oreskes, Krige, 2014). No part of the planet is now immune to these dynamics, as evidenced by the installation of offshore oil platforms. Of the technologies that emerged at the time, nuclear science is one of the most prominent, and its development is emblematic of the global proliferation of pollution (Brown, 2013; Hecht, 2012; Stawkowski, 2016). The perfection of the first nuclear bomb under the Manhattan Project stimulated research and development efforts involving the military, scientists, and major American companies. After the war, developing and testing atomic weapons rapidly became a major stake for several countries, while civilian uses for nuclear technology also began to take shape to produce the energy necessary for economic development, which was once again gaining speed under the effect of national and international voluntarist policies (Hecht, 1998). In the 1960s, military and civilian nuclear industries grew in countries of the Eastern Bloc as well as in the West, requiring the extraction and transformation of radioactive ore. This period produced a significant toxic legacy whose inventory is still incomplete: lands ravaged by old uranium mines in the United States, Africa, and Kazakhstan; plutonium production sites in Hanford (United States) and Mayak (in the former Soviet Union); thousands of tons of waste that will be extremely dangerous for millennia to come; and dozens of ageing nuclear reactors that fuel a growing concern over the possibility of major nuclear accidents or the cost of their dismantling in the future. This massive production of radioactive pollutants continues to this day, and in manifold ways. Despite major controversies, minimization, denial, and periodic dissimulation, there have been enough studies to confirm that all these activities have left a significant amount of radioactive contamination that will linger for thousands of years, and that these activities have particularly affected disadvantaged populations in Africa, several of the Pacific Islands, and elsewhere in the world, and are at the origin of hundreds of thousands of victims among civilian nuclear workers (including miners) and military personnel (such as soldiers who participated in atomic tests), in addition to populations subjected to nuclear bombs and radioactive fallout from major accidents like that at Chernobyl (Brown, 2019).

Lastly, the chemical and nuclear industries illustrate another important aspect of the joint expansion of industrialization and pollution: the development of industrial activities comes with, and leads to, a nearly continuous amassing of waste, further

Fig. 1. E-Waste recycling factory, Silverdale, Auckland. Judit Klein. https://www.flickr.com/photos/juditk/.

aggravating the pollution stemming from extraction and production processes. Since the nineteenth century, capitalism could only develop by accumulating increasingly greater quantities of leftovers resulting notably from the expansion of mass consumption and the flood of propaganda and marketing contesting the re-use of goods and establishing a 'disposable culture' favourable to mass production (Cohen, 2003; Strasser, 1999; Packard, 1960). The dizzying increase in the quantity of waste since the 1950s has resulted in a significant dispersal of potentially toxic products in the environment, with major consequences, i.e; soil pollution by heavy metals, PCBs, hydrocarbons, and plastics, and contamination of ecosystems, water, air, the entire food chain, and human populations (Boudia et al., 2021) (Figure 1).

Plastic provides a good illustration of the extent of waste distribution in the environment and the generalized pollution that results (Gabrys, Hawkins, Michael, 2013; Monsaingeon, 2017). Using figures from a variety of industrial sources, one important relevant study indicates that the production of plastic (resin and fibers) went from two megatonnes in 1950 to 380 megatonnes in 2015, amounting to a total of 7800 megatonnes, half of which was produced in only the last thirteen years (Geyer, Jambeck, Law, 2017). Its authors also estimate that 2500 megatonnes of plastic, meaning 30% of the mass produced since 1950, is still in use, and that accumulated primary and secondary waste (recycled products) stands at 6300 megatonnes, of which only 9% has been recycled, 12% was incinerated, and 79% is accumulating in dumps or nature.

In another study, the same authors explored the dispersal of plastic in the ocean and its disastrous impact on the marine environment (Jambeck et al., 2015). Such research fuels work by other researchers and activists who are concerned about a phenomenon causing a bit of a stir in the media: oceanic plastic pollution. Many media outlets have shown the formation of a 'plastic soup' in what has come to be called the Great Pacific Garbage Patch (also known as 'the seventh continent' or 'eighth continent'), described as a vast and growing patch of trash in the Pacific Ocean, composed of clumps of plastic bags, bottles, jugs, and other floating detritus. To this 'continent' in the northern Pacific are added four others of the same kind, a concentration of trash moved by oceanic marine currents and accumulating in the midst of gyres, which are great areas of swirling surface water. Since plastic waste can last hundreds of years and breaks down over time, it has made a sea filled with micro-particles of less than five millimetres in diameter in suspension to a depth of thirty metres. When ingested by wildlife, these micro-particles are indigestible and difficult to eliminate, so they accumulate in the bodies of fish, tortoises, birds, and other marine animals, eventually killing some. Moreover, these microparticles help fix persistent organic pollutants (POPs) such as DDT and PCBs, thus exacerbating their various deleterious effects on ecosystems.

This handful of examples shows the reach of industrial pollution. It often persists over a very long time, accumulating or forming sediment, or circulating and ending up in the water, air, soil, food chain, and bodies. Such pollution, that is ubiquitous and protean, also has many toxic qualities that effect the environment as well as human health (Homburg, Vaupel, 2019; UNEP, 2013). Although pollution became global over the second half of the twentieth century, it is still unequally distributed in social and geographical terms, and therefore some populations have paid, and continue to pay, a disproportionate price for the industrialization of the world (Fortun, 2001; Green Cross, Pure Earth, 2016; Nixon, 2011).

Ways of Governing Pollution

Industrial pollution is never a given: it is defined, delimited, and formatted with categories and paradigms that are built by a collection of scholarly communities, activist movements, and public administrations that configure how they are discussed and handled. Its framing and formatting can be historicized by examining the transformations of the scientific and technical practices which have contributed to its objectification, as well as the renewed visible or more confined mobilizations it has generated, and the reworked regulations and public policy it has been associated to.

Pollution and its deleterious effects were not ignored in the late eighteenth and nineteenth centuries. It prompted the mobilization of people whose health but also economic activities (especially farming and forestry) were affected, from workers to consumers. It prompted scientific research in a range of domains and disciplines, including analytical chemistry, human and veterinary medicine, public health, agronomy, forestry, and engineering, which was used to contribute to public or legal expertise in identifying sources of pollution, characterizing its effects, and developing tools for its regulation. The multiple conflicts surrounding pollution and its deleterious effects led industry and public authorities to develop several kinds of regulations. Notable among the emblematic regulations developed over the course of the nineteenth and into the early twentieth century, with timeframes varying according to country and the issue at hand, are those relative to dangerous and unhealthy industries, occupational illnesses and injuries, and poisonous substances (Bartrip, 2002; Massard-Guilbaud, 2010; Moses, 2018; Rosner, Markowitz, 1991)

These regulations were rooted in different rationales: trying to confine pollution to particular areas and populations using regulatory and technical means; promoting forms of controlled use for certain substances and uses; or organizing some kind of reparations in particular situations. These regulations were also intended to limit the search for responsibility and recourse to the courts by establishing an administrative management system that defined which pollution-related problems it would handle and how they should be managed, which made any other problems invisible and thus illegitimate. These regulations selecting the issues and the tools for managing them were inadequate for stamping out pollution and its detrimental effects. The period between the two world wars saw a great number of serious accidents, regular pollution, local mobilizations, and, in the United States, the first nationwide mobilization against the chemical industry and its pollution, which was led by engineers (Pujol, 2014; Ross, Amter, 2010; Whorton, 1974).

The explosion in the number of new substances and new products on the market at the end of the Second World War was accompanied by a discourse that industry and public authorities developed to be reassuring. Chemistry would improve everyday life, and pollution and its negative effects could be controlled. It was then that what we have termed 'governing through mastery' was developed (Boudia, Jas, 2019). This way of governing was based on prewar rationales and practices that sought to confine pollution and limit exposure through the adoption of norms and the concentration of dangerous industries in specific zones. It was based on the development of regulatory tools that required ever more scientific and technical knowledge. These tools were created in rapidly developing sectors such as the nuclear and agricultural industries. Positive lists of authorized substances, classifications according to use and/or hazardousness, specifications (regulatory definitions of the physiochemical properties of substances and the methods for establishing, measuring, and quantifying these substances), and labelling norms began to structure the regulations that were produced. Maximal exposure limits came to occupy a prominent place among the regulatory tools used in a government through mastery. Countless committees of experts in the United States and the burgeoning international organizations of the period were kept busy formalizing these values. The production of maximal exposure limit values was rooted in the conviction that it was possible to determine a dosage under which there would be no observable serious harmful effect, and crossing this dosage with exposure made it possible to set regulated thresholds (Boudia, Jas, 2013).

These regulatory products were not merely intended to master the harmful effects of hazardous substances; they were mainly intended to facilitate the industrial production and circulation of goods, especially given the extension of production and markets to a global scale that were not supposed to be constrained by excessively different technical and safety norms between countries. The established regulations only took control of over a limited number of substances, uses, and issues, so that most pollution was free of any kind of regulation — a state of affairs that was facilitated by the conviction, widespread in the scientific, industrial, and governmental communities, that the earth could absorb many of them without any serious damage.

For all these reasons, in the 1950s and gaining speed in the 1960s, visible pollution and scientists' alarms in public and closed circles challenged the claim that the rapid development of industry and consumer society could happen without serious or even irremediable damage (Boudia, 2007; Jas, 2013). The worldwide success of biologist Rachel Carson's book *Silent Spring*, published in 1962 and soon translated into a dozen languages, symbolizes the rising strength of a new environmentalism, rooted as much in everyday experience with pollution and its effects as in a critique of capitalism and imperialism, and a slew of new scientific findings would lead to a veritable 'toxicity crisis' in the 1970s (Vogel, 2012, 43 see also von Schwering, 2010).

This toxicity crisis emerged from discoveries of background pollution by substances like radioactive isotopes, DDT, and PCBs that had even reached areas where they had never been used, and from revelations of troubling toxic properties. As early as the late 1940s, some scientific research taken up by a few expert committees had concluded that insofar as especially dangerous carcinogenic substances were concerned, it was impossible to set a threshold under which no deleterious effect could be observed. How is it possible to maintain a discourse promising mastery of pollutants and their effect when some substances — some of which are widely used the world round — could potentially cause serious damage, even at low doses? This vital question was broadcast by numerous protests, which spread around the world with the founding of new international organizations like Greenpeace and Friends of the Earth. It was also fuelled in the 1960s and 1970s by the development of new fields of research and the renewal of the old by researchers studying pollutants and their potential effects on scales from the planetary to the microscopic. They would contribute descriptions of properties like persistence and the ability of certain pollutants to travel long distances redoubling concerns about the identification of potentially limitless toxicity, much-discussed carcinogenicity, mutagenicity, and reprotoxicity.

This deep and relentless contradiction between governing through mastery and the exposure of phenomena that defy mastery was handled in at least two ways: significant transformations in regulatory instruments and the development of a new mode of government that we describe as 'governing through risk' (Boudia, Jas, 2019). The 1970s were thus marked by the adoption of new legislation — the Clean Air Act of 1970, the Clean Water Act of 1972, and the Toxic Chemical Substances Act of 1976 in the United States, and the first environmental laws in Europe — and the creation of new agencies such as the EPA (Environmental Protection Agency) in the US and ministries of the environment in European countries. International spaces were not left out, and the 1970s were marked by a significant development in expertise. This expansion was partly due to the United Nations, which formed more new organizations and mechanisms to address the issue, such as the 1969 creation of the Scientific Committee on the Problems of the Environment (SCOPE) under the auspices of the International Council for Science (ISC) (which served as the scientific committee for the United Nations Environment Programme, UNEP), and the International Programme on Chemical Safety (IPCS), founded by the World Health Organization in 1980. Other organizations also invested in the issue, especially the Organization for Economic Co-operation and Development (OECD), which quickly became a key actor in the space of international expertise in the burgeoning field of dangerous chemical substances (Lönngren, 1992).

Despite being based on the reassuring discourse of public authorities and industry, these institutional and regulatory transformations were an inadequate response to these mobilisations and the issues they raised. It was henceforth less a matter of claiming mastery over these excesses than proposing a new social contract built on partial acknowledgement of risks and on new procedures for evaluation, management, and reparations. In the climate of controversy and political tension characteristic of the 1970s, several communities of experts became deeply interested in the issue of decision-making processes for risk-prone technologies and set out to define new procedures for acting in a controversial setting. 'Risk assessment' became the focus of work in several American federal agencies, which jointly began thinking about how they might define a cross-cutting methodology for evaluation and decision-making procedures for activities presenting technological, health/safety, or environmental risks, a preoccupation echoed in several international organizations. This work would ultimately culminate in the 1983 publication of a report, known as the 'Red Book' but formally titled *Risk Assessment in the Federal Government* that proposed a procedure that has since been adopted by most national and international institutions responsible for managing toxic substances. The procedure is based on the scientific methods of mathematical modelling and meant to be meticulous and applicable to a wide range of problems and situations (Boudia, 2014; Boudia, Demortain, 2014).

Governing through risk was designed to foster acceptance of a degree of environmental and health damage from industrial, technical, and scientific activities as a rationally determined necessity, but it failed to limit either pollution or protest. For one thing, bolstered by a fresh wave of mobilisation, scientific research set about signalling new forms of toxicity, once again leading to the creation of new categories that are simultaneously scientific, technical, regulatory, and political: 'endocrine disrupters' (EDs) in the 1990s to 2000s, 'persistent, bioaccumulative, and toxic' (PBT) in the 2000s, and 'substances of very high concern' from the European Union's REACH regulation would thus join the ranks of categories coined in the 1970s and 1980s, like 'carcinogens, mutagens, and reproductive toxicants'(CMR) and 'persistent organic pollutants' (Eckley, 1999; Scheringer, 2006). For another thing, the emergence of new problems stemming from the global expansion of industrial, technical, and scientific activities — moving highly polluting industrial activities from developed countries to the global South, industrial accidents like this that occurred in Bhopal (India), and the illegal circulation of dangerous products and toxic waste for example — came with the development of a variety of new mobilizations and international regulations. Many of these regulations were not very strict, and they often took the form of international conventions, like the Stockholm convention on POPs, the Basel convention on toxic waste, and the Rotterdam convention on hazardous chemicals (Paellemaerts, 2003; Pellow, 2007; Selin, 2010).

Starting in the 1990s, the proliferation of zones identified as very contaminated, either by particularly polluting industrial activity or by especially serious industrial accidents, led some groups of experts and national and international organizations to advocate what we have called 'governing through adaptation' (Boudia, Jas, 2019). This framing of problems is based on the recognitionthat there is permanent pollution that cannot be remedied (Brown, 2019; Langston, 2016; Topçu, 2013; Torny, 2013). This means that life in these highly polluted areas would henceforth be managed with a view to providing residents with recommendations for food, health and safety, and outdoor activities intended to limit their exposure and make everyone into their own risk manager. This implies vaunting supposed forms of empowerment while, in a great many cases, clearing public authorities and industry of responsibility for the pollution and its management. The area impacted by the Chernobyl accident was the first testing ground where tools for this way of governing were developed, and they are now used in many zones that are highly contaminated from routine activity or accident. They are also routinely used in less polluted areas to handle background pollution that is impossible to eradicate, such as atmospheric pollution, particularly severe peaks in pollution, major pollution of drinking water sources, and contamination of foodstuffs by toxic materials like PCBs and heavy metals.

Conclusion

Over the last two centuries, many regulatory systems have been built to address certain kinds of pollution and some of their effects. The production of regulations is such that we are now faced with a nearly illegible 'institutional millefeuille' operating at various scales and including regulations developed during different periods and with divergent aims (Jas, 2014). Not only did this proliferation of institutions, regulations, instruments, and tools fail to limit the pollution it was intended to control, it furthermore accompanied its spread and accumulation. As we have seen, this paradox is mainly explained by the recurring inability of dominant economic and political actors to conceive of forms of oversight and control that challenge forms of economic development dependent on massive industrialization, based on increasingly high-performing technologies for producing consumer goods that have never succeeded in limiting the production or emission of pollution, or remedying the result.

The development of technology and the related industrial expansion and associated pollution have been at the origin of irreversible environmental alterations of the planet. For instance, climate change, one of the most serious crises that humanity must now face, is rooted in these transformations deployed over the last two centuries, with regular accelerations from the 1950s onwards.

The environmental crisis at the heart of many current debates stressing the major risks to be faced by human societies from now on is therefore the product of a history in which the analysis of the interrelations of technology, pollution, and the environment is central. This history is as indispensible to informing debates over the exhaustion of resources and biodiversity, the carbonization of the economy and its impact on climate change, the toxification of ecosystems, and the acidification of oceans, as it is to discussions of the limitations of some of the technical and political solutions under consideration.

References

Bartrip, Peter W. J., *The Home Office and the Dangerous Trades: Regulating Occupational Disease in Victorian and Edwardian Britain* (Amsterdam and New York, NY: Editions Rodopi, 2002).

Boudia, Soraya, 'Global Regulation: Controlling and Accepting Radioactivity', *History and Technology*, 23.4 (2007), 389–406.

———, 'Managing Scientific and Political Uncertainty: Environmental Risk Assessment in a Historical Perspective', in *Powerless Science? Science and Politics in a Toxic World*, ed. by Soraya Boudia and Nathalie Jas (London and New York, NY: Berghahn Books, 2014), pp. 95–114.

———, Angela N. H. Creager, Scott Frickel, Emmanuelle Henry, Nathalie Jas, Carsten Reinhardt, and Jody R. Roberts, *Residues. Thinking Through Chemical Environments* (New Brunswick, NJ: Rutgers University Press, 2021).

——— and David Demortain, 'La Production d'un instrument générique de gouvernement. Le "livre rouge" de l'analyse des risques', *Gouvernement et Action publique*, 3.3 (2014), 33–53.

——— and Nathalie Jas, *Gouverner un monde toxique* (Versailles: Editions Quae, 2019).

——— and ———, eds, *Toxicants, Health and Regulation since 1945* (London and New York, NY: Pickering and Chatto, 2013).

Brown, Kate, *Manual for Survival: a Chernobyl Guide to the Future* (New York, NY: W. W. Norton and Company, 2019).

———, *Plutopia: Nuclear Families in Atomic Cities and the Great Soviet and American Plutonium Disasters* (Oxford: Oxford University Press, 2013).

Cohen, Lizabeth, *A Consumers' Republic: the Politics of Mass Consumption in Postwar America* (New York, NY: Knopf, 2003).

Craeger, Angela, 'The Political Life of Mutagens: a History of the Ames Test', in *Powerless Science? Science and Politics in a Toxic World*, ed. by Soraya Boudia and Nathalie Jas (London and New York, NY: Berghahn Books, 2014), pp. 46–65.

Dahan, Amy and Dominique Pestre, eds, *Les Sciences pour la guerre* (Paris: Editions de l'EHESS, 2004).

Eckley, Noelle, *Drawing Lessons About Science-policy Institutions: Persistent Organic Pollutants (POPs) Under the LRTAP Conventions*, ENRTP Discussion Paper E–99–11 (Cambridge, MA: Harvard University, 1999).

Fortun, Kim, *Advocacy after Bhopal. Environmentalism, Disaster, New Global Orders* (Chicago, IL: University of Chicago Press, 2001).

Fressoz, Jean-Baptiste, *L'Apocalypse joyeuse. Une histoire du risque technologique* (Paris: Le Seuil, 2013).

Frickel, Scott, *Chemical Consequences: Environmental Mutagens, Scientist Activism, and the Rise of Genetic Toxicology* (New Brunswick, NJ: Rutgers University Press, 2004).

Gabrys, Jennifer, Gay Hawkins, and Mike Michael, eds, *Accumulation: the Material Politics of Plastic* (New York, NY: Routledge, 2013).

Geyer, Roland, Jenna R. Jambeck, and Kara L. Law, 'Production, Use and Fate of All Plastics Ever Made', *Science Advances*, 3 (2017), e170078.

Green Cross and Pure Earth, *The Toxic Beneath our Feet*, (2016). <https://www.worstpolluted.org/docs/WorldsWorst2016.pdf> [accessed 28 November 2022].

Hecht, Gabrielle, *Being Nuclear: Africans and the Global Uranium Trade* (Cambridge, MA: The MIT Press, 2012).

———, *The Radiance of France: Nuclear Power and National Identity after World War II* (Cambridge, MA: The MIT Press, 1998).

Hervé, Alain, 'Editorial', *Le Nouvel Observateur*, special issue 'La dernière chance de la terre. Dossier spécial écologie', (1972), 4.

Homburg, Ernest and Elisabeth Vaupel, eds, *Hazardous Chemicals: Agents of Risk and Change, 1800–2000* (Oxford and New York, NY: Berghahn Publishers, 2019).

Ilgen, Thomas, '"Better Living Through Chemistry": the Chemical Industry in the World Economy', *International Organization*, 37.4 (1983), 647–80.

Jambeck, Jenna R., Roland Geyer, Chris Wilcox, Theodore R. Siegler, Miriam Perryman, Antony Andrady, and Kara L. Law, 'Plastic Waste Inputs From Land Into the Ocean', *Science*, 347.6223 (2015), 768–71.

Jarrige, François and Thomas Le Roux, *The Contamination of the Earth: a History of Pollutions in the Industrial Age* (Boston, MA: The MIT Press, 2020).

Jas, Nathalie, 'Adapting to Reality. The Emergence of an International Expertise on Food Additives and Contaminants in the 1950s and early 1960s', in *Toxicants, Health and Regulation since 1945*, ed. by Soraya Boudia and Nathalie Jas (London: Pickering and Chatto, 2013), pp. 47–69.

———, 'Gouverner les substances chimiques dangereuses dans les espaces internationaux', in *Le Gouvernement des technosciences. Gouverner le progrès et ses dégâts depuis 1945*, ed. by Dominique Pestre (Paris: La Découverte, 2014), pp. 31–63.

Langston, Nancy, 'La Convergence entre santé humaine et santé environnementale: le toxaphène dans le lac Supérieur', *Sciences Sociales et Santé*, 34.3 (2016), 103–23.

Le Nouvel Observateur, La dernière chance de la terre. Dossier spécial écologie', *Le Nouvel Observateur*, special issue, (1972)

Le Roux, Thomas, *Le Laboratoire des pollutions industrielles: Paris, 1770–1830* (Paris: Albin Michel, 2010).

Lönngren, Rune, *International Approaches to Chemicals Control. An Historical Overview* (Stockholm: Kemi, 1992).

Massard-Guilbaud, Geneviève, *Histoire de la pollution industrielle. France, 1789–1914* (Paris: Éditions de l'EHESS, 2010).

Monsaingeon, Baptiste, *Homo détritus* (Paris: Le Seuil, 2017).

Moses, Julia, *The First Modern Risk: Workplace Accidents and the Origins of European Social States* (Cambridge: Cambridge University Press, 2018).

Murmann, Johann P., 'Chemical Industries After 1850. An Article for the *Oxford Encyclopedia of Economic History*' (2002) <http://www.professor-murmann.net/murmann_oeeh.pdf> [accessed 28 November 2022].

Nixon, Rob, *Slow Violence and the Environmentalism of the Poor* (Cambridge, MA: Harvard University Press, 2011).

Oreskes, Naomie and John Krige, eds, *Science and Technology in the Global Cold War* (Cambridge, MA: MIT Press, 2014).

Packard, Vance, *The Waste Makers* (New York, NY: David McKay, 1960).

Paellemaerts, Marc, *Toxics and Transnational Law: International and European Regulation of Toxic Substances as Legal Symbolism* (Oxford and Portland, OR: Hart Publishing, 2003).

Pellow, David, *Resisting Global Toxics: Transnational Movements for Environmental Justice* (Cambridge, MA: The MIT Press, 2007).

Pujol, Hervé, 'L'Action des syndicats de défense contre les fumées des usines de Salsigne (1932–1950), exemple de judiciarisation d'un conflit environnemental', in *Tristes mines. Impacts environnementaux et sanitaires de l'industrie minière*, ed. by Hervé Pujol (Paris: Édition Les Études Hospitalières, 2014), pp. 183–208.

Rainhorn, Judith, *Poison légal. Une histoire sociale, politique et sanitaire de la céruse et du saturnisme professionnel (XIXe–XXe siècle)* (Paris: Presses de Sciences Po, 2019).

Rosental, Paul-André, ed., *Silicosis. A World History* (Baltimore, MD: John Hopkins University Press, 2017).

Rosner, David and Gerald Markowitz, *Deadly Dust: Silicosis and the Politics of Occupational Disease in Twentieth-century America* (Princeton, NJ: Princeton University Press, 1991).

Ross, Benjamin and Steven Amter, *The Polluters: the Making of Our Chemically Altered Environment*, (Oxford: Oxford University Press, 2010).

Scheringer, Martin, Sefab Boschen, and Korand Hungerbühler, 'Will We Know More or Less About Chemical Risks Under REACH?', *Chimia*, 60 (2006), 699–706.

Selin, Henrik, *Global Governance of Hazardous Chemicals* (Cambridge, MA: The MIT Press, 2010).

Speth, James F. and Peter M. Haas, *Global Environmental Governance* (Washington, DC: Island Press, 2006).

Stawkowski, Magdalena, '"I am a Radioactive Mutant": Emerging Biological Subjectivities at the Semipalatinsk Nuclear Test', *American Ethnologist*, 43.1 (2016), 144–57.

Strasser, Susan, *Waste and Want: A Social History of Trash* (New York: Metropolitan Book, Holt, 1999).

Topçu, Sezin, 'Chernobyl Empowerment? Exporting "Participatory Governance" to Contaminated Territories', in *Toxicants, Health and Regulation since 1945*, ed. by Soraya Boudia and Nathalie Jas (London and New York, NY: Pickering and Chatto, 2013), pp. 135–57.

Torny, Didier, 'Managing an Everlastingly Polluted World. Food Policies and Community Health Actions in the French West Indies', *in Toxicants, Health and Regulation since 1945*, ed. by Soraya Boudia and Nathalie Jas (London and New York, NY: Pickering and Chatto, 2013), pp. 117–34.

UNEP, *Global Chemical Outlook. Towards Sound Management ofCchemicals* (Nairobi: United Nations Environment Programme, 2013).

Vogel, Sarah, *Is It Safe? BPA and the Struggle to Define the Safety of Chemicals* (Berkley, CA: University of California Press, 2012).

von Schwering, Alexander, 'Low Dose Intoxication and a Crisis of Regulatory Models. Chemical Mutagens in the Deutsche Forschungsgemeinschaft', *Berichte zur Wissenschaftsgeschichte*, 33 (2010), 401–18.

———, 'Shaping Vulnerable Bodies at the Thin Boundary Between Environment and Organism: Skin, DNA Repair, and a Genealogy of DNA Care Strategies', *Science in Context*, 28.3 (2015), 427–64.

Whorton, James, *Before Silent Spring: Pesticides and Public Health in Pre-DDT America* (Princeton, NJ: Princeton University Press, 1974).

Zimmer, Alexis, *Brouillards toxiques. Vallée de la Meuse, 1930, contre-enquête* (Brussels: Zones Sensibles, 2016).

THIERRY BONNOT, MICHEL LETTÉ

Repair, Recycling, and Waste*

Given the vast range of practices, the widely varying quantitative and qualitative scale of materials, and their diverse temporalities — from extraction of resources, processing, and manufacture to use and reuse of objects before they become waste — it is not really possible to contemplate a global history of repair and recycling techniques without going into the fine detail of the nature of the operations involved and the environments in which they take place. Instead, our aim in this article is to outline the key stages in the development of some of the main lines of thinking and practice that make repair and recycling sociotechnical systems of waste management.

Yet such a history cannot ignore the nature of the substances and materials incorporated in the composition of objects, nor waste in its material form. Up until the early nineteenth century, artisanal and industrial economy was essentially organic (Barles, 2005). The manufacture of materials produced no or very little non-degradable waste. Systematic recovery was an integral part of a small, circular local economy (Guillerme, 2007). To cite one key example, animals were abundant sources of additional materials, providing substances that could be processed to produce objects and foodstuffs. Their final remains were rendered into tallow and lard. Their dung was incorporated into manure. Their urine was used in dyeing. The fluids that could not be used were drained into waterways and absorbed by the soil. The ultimate fumes from factories and workshops were taken away by the first wind. The final deposits, rubbish dumps and tips, provided a few more — admittedly limited — resources on which those excluded from consumer society, who were reduced to feeding and clothing themselves from its scraps, survived (Strasser, 1999).

The combined triumph of the mining and synthetic chemical industries at the turn of the twentieth century led to a major transformation (Homburg, Travis, Schröter, 2011; Bernhardt, Massard-Guilbaud, 2002). The coal economy and the kind of organization it fostered, the extraction and processing of metals at high temperatures, and the subsequent introduction of synthetic materials such as plastic into routine use meant that extractivism, productivism, consumerism, pollution, and waste management were inextricably linked (Le Roux, 2011; Amter, Ross, 2010). Ever increasing areas of land became saturated with the by-products of the mining industry, and

* Chapter translated by Rachel Gomme.

Global History of Techniques, ed. by Guillaume Carnino, Liliane Hilaire-Pérez and Jérôme Lamy, Global Matters Culture Matérielles, histoire des techniques et circulation globale des savoirs, 9 (Turnhout: Brepols, 2024), pp. 743–758
BREPOLS PUBLISHERS 10.1484/M.TECHNE-EB.5.129839

later with substances generated by synthetic chemistry and its disposable objects (Jarrige, Le Roux, 2017). Improvements in efficiency and technical performance led to a substantial increase in production and came to represent the promise of increased standards of living and comfort. Economic growth became established as the dogma of human progress, and rationalization was the means of achieving it. Waste now consisted of the excess substances generated by industrial production (Le Roux, Letté, 2013).

Homo Faber Entangled in Waste

Repair and recycling are two responses to one of the central questions for *Homo faber*: how to use objects or materials that are not quickly replaceable until they are completely worn out. These are ancient practices, but the problematization of them — in other words, constituting them as a subject of technical, political, economic, and legal consideration — is recent and inseparable from industrialization and mass consumerism (Lupton, 2011). It is the appearance of new materials, mainly plastics, and the proliferation of objects, particularly new technology, that pose a problem by generating waste that cannot be reused or can only be recycled in specialized industries (Russel, Vinsel, 2020).

The industrialization of production and consumption intensified over the nineteenth century with the rationalization of the world economy. Its corollary was the invention of waste (Barles, 2005), as it shifted from 'transitional state of matter' (Monsaingeon, 2017: 49) to a 'waste object [that] persists, takes up space, is potentially toxic and is not necessarily decomposing' (Larroque, 2013: 4). The production infrastructure of the heavy industry of this period required massive structures, in terms of both buildings and machine tools. The technical specialization of these structures substantially limited the potential for repurposing them when they ceased to be used for their original production. Their massive scale meant the costs of demolition and securing them were high, to the point where simply abandoning them seemed a possible solution. Thus arose the problem of industrial wasteland, which may offer precious archaeological remains and scientific resources (Fluck, 2017), and in some cases aestheticized cultural monuments (Tornatore, 2004), but is also a source of lasting pollution, whether visual, of the landscape, or other. Where products are concerned, waste derives both from the processing of materials and from the obsolescence of the manufactured products generated. Made up of composite, artificial, and, more recently, synthetic materials, these products are characterized by an increasingly long material existence compared to their ever shorter periods of use.

Throughout the world, whatever the cultural context and the industrial system and whether or not recycling facilities exist, the management of waste objects is a major socio-economic issue (Porter, 2002). These objects are circulated and exported. They are exchanged in the play of global trade, the international organization of labour and its power relations. At other times, they stay in the same place, breaking down at a pace of their own. There are countless legal rules and technics designed to confine and control them, and to make them safe. At the level of humanity, these questions arise from the entanglement of things and human societies. Archaeologist Ian Hodder describes this interdependence as a sort of material, social, and environmental trap in which human beings have become mired over centuries (Hodder, 2014). The model of exponential growth that was embarked on thousands of years ago, and that has been engaged in with particular intensity since the nineteenth century, still dominates. Hodder believes the movement is irreversible, notwithstanding the now widely recognized consequences for the climate.

Geographical Differences in the Culture of Waste

The distinction between what is waste to be disposed of and what can or should be recovered, repaired, or recycled partly derives from systems of cultural and social reference that have long been studied by anthropologists and archaeologists (Douglas, 1966; Thompson, 1979; Rathje, Murphy, 1992; Douny, 2007). Nobody accepts another's waste, which is to be removed and kept out of reach of the senses. Our own waste is discharge that needs to be hidden. It makes real the need to be separated from lost objects that are to be mourned. Rejects represent an economic loss when they generate additional production costs. For the manufacturer or industrialist, they are evidence that technical efficiency, if not lacking, can at least be improved. For the user, obsolescence and uselessness are a guarantee of waste that the waste object once again makes concrete. Hence the imperative to get rid of it as soon as possible, in order to return to calmly producing and consuming.

But the cultural aspect of our relation to the material world cannot be isolated from historical, technical, and economic realities: '[t]he richer a society is, the more it piles up waste and processes it industrially; the poorer it is, the more it recovers it and processes it manually' (Droulers, 1986: 62). While this observation is no longer new and needs to be qualified, it does express the reality of differentiated perceptions and practices with regard to waste on the global scale. This geographic and cultural contrast between rich and poor countries, which correlates with a global north/south division, has to be articulated with another historical and technical division in a way that generates certain paradoxes: although it would be logical for a technologically advanced society to move toward the dematerialization of certain goods and thus to produce less physical waste, the opposite occurs. The rich countries exploit the raw materials of the poor countries, particularly for ICT (information and communication technologies), and send back used manufactured products, along with their trail of pollution (Dobré, Flipo, Michot, 2013; Flipo, 2014; Bihouix, 2019). Yet the formal and informal practices of the composition and recomposition of materials, of repair and recycling, of appropriation and functional redeployment of commodities, limit the circulating mass of objects that are too rapidly rejected and thrown away. This is not enough, however, since their global accumulation continued to increase with the emergence of the consumer society. Since the Second World War, the systematic use of container shipping and the floating rubbish dump has reinforced the logic of a seemingly inevitably continuous expansion of waste (Veltz, 2017). A sort of planetary metabolism generates and digests this mass of substances, materials, and equipment in cycles of use and reuse, ending in the excreta of the hyperindustrial society (Alexander, Reno, 2012).

The informal economies of poor countries recycle some of the Western world's waste. They deploy extraordinary ingenuity in reappropriating and reinventing objects and materials, to the detriment of their immediate environment (Fanchette, 2016). In poorly industrialized countries and very unequal societies, necessity knows no law. Traditional recycling remains the rule, in the sense that it continues to predominate in a set of individual and collective practices determined by people's standard of living and their access to international distribution networks. At the level of practices, however, the reuse of common objects, whether repaired or not, remains a very widespread mode of appropriation and consumption. To take the example of plastic, bottles made of PET (polyethylene terephthalate) are recovered in some African countries for reuse on the petrol black market, particularly in Benin with 'kpayo', which is low-cost petrol smuggled in from Nigeria (Agossou, 2004). In Egypt, the same material generates a multiplicity of practices and raises critical economic and social issues (Desvaux, 2014), whereas in France, legislation enforced elimination of disposable PET disposable since 2021,[1] and henceforth requires companies to provide reusable cups with a deposit, in order to overcome the effects of this ban.[2] These different approaches to the same problem are each set within their own particular node of the problem of waste that affects all of the modern world, but it would be wrong to interpret them uniformly as a continuation of traditional uses.

The Great Recycling: From the Extraction of Materials to the Incineration of Objects

Before any repair or recycling comes the extraction of the resources needed to process materials and manufacture goods for sometimes transient uses. Each stage of this process involves an amount of fossil fuel, but also of substances extracted from underground, either directly in the form of materials that go to make up the finished product, or for the machines used in its manufacture and distribution (Bednik, 2015). Processing of materials is the first and largest source of the waste known as clinker, a term designating both the residues from the manufacture of substances at the start of the cycle and what remains after incineration at the end of the chain of production of consumer objects made from those substances. Removed as the residue from industrial roasting ovens and domestic and urban waste incinerators, clinker is stored to await either potential reuse, final rejection or permanent storage. The huge size of slag heaps, barren rubble heaps, or waste tips derived from mining and quarrying testify to the volume of this waste. The blast furnaces, steelworks, metallurgical plants,

1 https://www.economie.gouv.fr/cedef/interdiction-plastique-usage-unique [accessed 18 November 2022]
2 <https://gobeletsgreencup.fr/> [accessed 18 november 2022].

and roasting ovens associated with them generate clinker in their turn, which is symptomatic of the system. Since the nineteenth century this waste has primarily been used in public works, as backfill or levelling in civil engineering projects and for reclaiming land from the sea. Sometimes mixed with clay, it is used in road building and the development of public infrastructure, to reinforce carriageways and to surface paths, roads, and pavements. Its other regular use is in the manufacture of building materials, where it is ground and, with the addition of lime, compressed into bricks, shaped into tiles and various types of artificial stone, and incorporated into cements known as slag cements. Thus, clinker is one of the most widespread recycled materials, distributed throughout transport infrastructure and construction.

During the first phase of industrialization the majority of clinker comprised slag from steelworks, but with consumption taking over from production, this has now been replaced by clinker from incineration. The plumes of white smoke that can be seen today on the outskirts of large conurbations in Western countries usually derive from the generation of power or heat, or from plants processing urban and domestic waste, rather than from factories processing raw materials as they would have a century ago. This is the result of deindustrialization, which began in the late 1970s, and has been accompanied by the relocation of centres of slag production — coking, steelmaking, metalworking, and primary processing industries (Altena, Van der Linden, 2003). With the environmental turn that goes along with it, deindustrialization has paradoxically intensified people's refusal to continue putting up with the excess from any industrial activity that still continues locally, as well as that of the past whose consequences are no longer in doubt. Combating pollution, risk, and dangers to health has become a key issue, a focus of conflict (Pellow, 2007). Is this the debt, or the legacy of industrial history? It does not matter -waste must disappear or be confined by technical and legal measures, hence the desire to limit its production upstream, which repair and recycling promises to make possible.

Reuse: an Anthropological Necessity?

Today as in the past, common sense seems to determine the solution, whether that be reuse, recycling, or upcycling, an incorporation into new products, or intelligent recycling, the organization of a technical metabolic system in which everything, or almost everything, is assimilated within a cycle of energy and materials that it is hoped will be virtuous (Erkman, 2000). Thus, repair and redeployment for other purposes are broadly taken for granted, even if not yet systematically implemented. To recycle an object is to transform it into a raw material destined for a new function, for the manufacture of other goods or the provision of materials that can be resold. To reuse an object is to re-employ it for the same or a similar function, if necessary after repairing it. Extending its use — sometimes by adapting the way it works — means delaying the point when it becomes waste. To give an outline of a chronology, the global evolution of the problem can be summed up thus: in Europe, before the 'birth of consumption' (Roche, 1997) and before the industrial era, repair and recycling were integrated into the routine operations of everyday life, which were determined by the available resources. Depending on the relative abundance of a given material or foodstuff, societies would limit the use of them. They reused worn objects and recycled food and bodily waste. A useful object was not abandoned until it was absolutely impossible to use it for its original purpose. Archaeology offers evidence of repairs designed to prolong the use of objects despite physical deterioration,[3] and of multiple reuse, for example in architecture (Bernardi, Dessales 2017) and other domains (Figure 1). What cannot be repaired is recovered by 'freelancers', the rag-and-bone men who are key figures in the life of preindustrial cities (Florin, Garret, 2017; Droulers, 1986), to the point where they are now organised in the Global Alliance of Waste Pickers.[4] Collecting rubbish is once again becoming a job, and gleaning continues as a method of survival in some contexts (Silguy, 2009). From the nineteenth century onwards, these practices formed the basis for the artisanal and industrial waste recycling sectors, which are now concentrated in the hands of a few corporations that have built whole economic empires (Bertolini, 2008; Lorrain, 2001, 2002; Ortar, Anstett, 2017). Thus, waste materializes (shapes) huge global industrial and financial stakes in the world economy, working in support of the global pursuit of productivism and consumerism by other means. It also acts as the focus of local resistance to the constraints of Western capitalism, based as it is on the quest for perpetual economic growth.

Built-in Obsolescence of Products

Contemporary waste is set within a paradoxical temporality: the durability of industrial materials could have made it possible to slow down the rate of production by extending the life of products. But while materials take centuries to decompose, objects are now only used, or even usable, for a few years (Larroque, 2013), justifying, for industry, an exponential growth in the volume produced. Mass consumerism, as it has developed since the 1960s, is based on the principle of not repairing objects at the end of their life; the unusable object is not repaired but replaced by another with the same function. The technical and economic conditions of production have evolved in such a way that it becomes more expensive to repair an object or have it repaired, particularly in the case of electrical appliances, than to replace it with a new equivalent — even if, in theory, technical products remain 'repairable'. In industrialized countries, an object is now replaced not because it is no longer usable, but rather because another has come onto the market which is presented as being better (or more closely aligned with the image of modernity presented by industry, marketing, and the media). This change in the relationship to things represents a material revolution (Daumas, 2018) and correlates with the emergence of the consumer

3 Examples of ceramic repairs are widespread in the archaeological literature. *Kintsugi*, the repair of porcelain with gold lacquer in Japan, is a well-known example (see also Speranza, 2007).

4 <http://globalrec.org/> [accessed 18 november 2022].

Fig. 1. Archaeological excavation of an industrial ceramics dump, early twentieth century (Pouilloux, Saône-et-Loire, December 2011). © Bonnot, 2011.

society 'marked by the choice of disposable and perishable objects' (Jarrige, Le Roux, 2016). Zygmunt Bauman analyses this phenomenon through the notion of 'liquid life', noting the 'reversal of values attached respectively to duration and transience' (Bauman, 2005: 83) in consumerist society, which now dominates productivist society. 'For the expectations to be kept alive and for new hopes to promptly fill the void left by hopes already discredited and discarded, the road from shop to garbage bin needs to be short and the passage swift' (Bauman, 2005: 82). While objects from the first phase of industrialization, such as containers, laid claim to a durability and a potential for reuse that offered the possibility of a 'life' over centuries (Bonnot, 2002), twentieth- and twenty-first-century products have only a deliberately limited existence.

While the beginnings of built-in obsolescence can be identified in the 1950s, when industry began to think of 'the ready-to-throw-away as an engine of growth' (Monsaingeon, 2017: 143), it was theorized as early as 1932, in a pamphlet written by New York estate agent Bernard London. He invented a system of 'planned obsolescence' that would force consumers, on pain of taxation, to replace things once they had gone past a date fixed by law (London, 1932, 2012). According to London, shoes, houses, and any industrial or architectural product should have a limited life span. In order to avoid the practices of repair and saving that had become widespread following the 1929 Wall Street crash, he recommended that the law should fix a life duration for things. The aim at this point was to relaunch industrial growth by accelerating and controlling the production–consumption–disposal cycle, in order to avoid an obstruction in the economic system.

Today this way of thinking may seem absurd, not least because it assumes unlimited supplies of raw materials. Nevertheless, it can be seen as one of the first explicit theorizations of the logic of productivism and consumerism. Eighty years on, French law prohibits built-in obsolescence, defined as 'the set of techniques whereby a seller seeks deliberately to reduce the life of a product in order to increase its rate of replacement'.[5] The National Assembly has even discussed requiring products to be marked with the period of their expected useful life[6] — a move that would be the exact opposite of the allotted lifespan of things invented by London, since it is intended to sanction an excessively short period of use and encourage the 'reparability' of objects. These contrasting goals, which arose within a few decades of one another, shed light on the relation between human societies and technical objects. The comparison invites attention to the trajectory of the materials that make up the object, from extraction to transformation into their final form as waste. It is in this interval between extraction/processing and incineration/disappearance that the sociotechnical existence of objects and their constituent substances is negotiated, passing through the interplay of phases of repair and recycling. Thus, neither repair nor recycling can be assumed, since they relate to an incommensurable variety of economic, technological, social, and cultural contexts.

Countering Obsolescence in the Era of Hacking: Ethics, Value, and Morality

Practices of reuse are currently the focus of renewed attention, yet the rhetoric of sustainable development struggles to translate this into concrete action. It is worth noting that the difference in scale between repair, which is practiced at the level of the individual object, and recycling, which involves the mass collection of constituent materials, results in a different relationship to things and techniques. Thus, repair is often rooted in a system of artisanal work (Dant, 2010), as opposed to recycling, which deals with much larger quantities (Figure 2). There is, indeed, no comparison between the individual practices of keeping a compost bin at the bottom of the garden, or collective repair in a skillshare workshop, and those, for example, of the industrial sector, reprocessing fuel or substances derived from the dismantling of nuclear power stations. Yet the logic is basically the same, materials and objects are used to the limits of their potential in order to optimize reuse, with the declared

5 https://www.legifrance.gouv.fr/codes/article_lc/LEGIARTI000031053376/2015-08-19 [accessed 18 november 2022] This type of obsolescence differs from London's planned obsolescence in that it is fixed by industry rather than by the law, outside of official control and without the consumer knowing the end-date.

6 https://www.ecologique-solidaire.gouv.fr/duree-vie-des-produits-obsolescence-programmee.

Fig. 2. Recycling of porcelain ink bottles, transformed into lamp bases, c. 1960 (Palinges, Saône-et-Loire, 2018). © Bonnot, 2018.

aim of economic, social, cultural, political, and environmental optimization of use. The conflicts aroused by the practices of the nuclear industry nevertheless serve as a reminder that any disused object constitutes intrinsically unstable waste, in terms of both its physical and its socioeconomic materiality, depending on whether its existence in time and space is considered as a solution or a problem, a profit opportunity or a danger to the environment or health (Beaune, 1999; Harpet, 1999).

Resistance to the industrial and commercial logics of consumerism was initially organised at the level of non-profit organizations and citizen collectives. Originating in the Netherlands,[7] the 'repair café' movement offers one possible solution, offering a grassroots path of commitment toward 'deprogrammed obsolescence' (Goyon, 2016). Along with other similar initiatives, these approaches aim to combat organised waste. This kind of laboratory of technical citizenship offers participants a direct relationship with objects, demystifying and democratizing access to the black boxes of technology. Those organizing them seek to 'free' objects from intellectual property,

7 <https://www.repaircafe.org/fr/a-propos//> [accessed 18 novembre 2022].

and from their functions by insisting on recovery, reuse, and local short-circuit manufacture, all within the framework of a political, cultural, and social project (dissemination, facilitation, education). Repair, reuse, and diversion into alternative functions are thus embedded in a movement of resistance to the dominant model of mass consumption, with 'kludge' becoming a mode of appropriation of objects and techniques.

What might appear a game or a way of adapting to the technical and economic constraints of capitalism is now proclaimed a real political project. Ad hoc repair, previously confined to poor countries and the lower social strata of rich countries, becomes an act of resistance against the dominant system. The work of humanitarian and social recovery organizations such as refurbishing centres (Messal, 2014; Corteel, 2016) and the Chiffonniers d'Emmaüs[8] (Hoarau, 2009), as well as a number of similar initiatives (Anstett, Ortar, 2015), sit within a similar perspective.[9] In France the success of second-hand shops and flea markets, which trade commercially, is part of a different way of thinking (Sciardet, 2003). States and public bodies asserting a commitment to sustainable development also engage in recovery and reuse, through agencies such as ADEME (Agence de l'environnement et de la Maîtrise de l'Énergie), which offers tips on techniques and procedures designed to 'make your objects last' (ADEME 2019).

However, comparing these practices opens up a series of ethical questions on the values ascribed to things, objects, materials, and waste in different economic and social contexts. As Martine Droulers writes with reference to Brazil, 'burying without recycling is seen as a crime in poor countries' (Droulers, 1986: 68). Whether a material or an object is seen as waste or a resource is a matter axiological and moral judgment. In India, the recycling of second-hand clothes that Westerners donated to charities into blankets for humanitarian emergencies shows how moral values, material goods, and the market economy are interwoven, not without raising ethical problems around working conditions (Norris, 2012). The moral values proclaimed by those who promote recycling as a virtuous resistance against waste are diminished by their social practices. In France, non-profit organizations focussed on upcycling waste posit an analogy between the reuse of objects and return to employment for workers, despite the fact that 'only a small proportion of the objects actually acquires new value by being sold, [just as] only a small proportion of the workers said to be "returning to work" actually gain employment in the regular workplace' (Corteel, 2016: 118). Similarly, in China the work of repairing and reconditioning electrical equipment in most towns and large villages — despite the official prohibition on international trade in ICT devices, which are seen as waste in the West — forms an integral part of the consumerist industrial system, far from the illusion of a circular economy reducing the environmental impact of mass production to zero (Schulz, 2016). Repair

8 The 'ragpickers of Emmaüs', a charitable organization, was founded in 1949 to support homeless people that collected and recycled or repaired unwanted household goods.

9 For a regularly updated overview, see Elisabeth Anstett and Nathalie Ortar's blog 'La Deuxième vie des objets: anthropologie et sociologie des pratiques de récupération' <https://dvo.hypotheses.org/> [accessed 18 november 2022].

and recycling are thus never absolutely virtuous practices, but any ethical evaluation of them needs to take into account the entirety of the context, the circulation of materials, and the resulting effects.

Repair, recycling, and waste management are bound up together. In this short article we have attempted to convincereaders that they also point to the huge variety of sociotechnical relations in the never-ending cycle of processing of substances, materials, and equipment. On the basis of an entirely everyday object, it might (perhaps) be possible to envisage an analysis of all these links woven between the local and the global, past and future, nature and culture. Yet such an analysis would need to be set in the context of the forms of existence of the object in question, from extraction to the ultimate decomposition of materials. Between the workshop of a Parisian car dealer and the streets of Havana, the same carburettor surely has a different story to tell. The journeys of the two relate to widely divergent ways of thinking and practices, which nevertheless have some points of similarity. Would a global but contemporary history of techniques of repair, recycling, and waste management, then, be possible? Could it be something more than a list or a catalogue of local practices and dedicated international industries?

References

ADEME, 'Comment faire durer ses objets?', janvier 2019, online: https://librairie.ademe.fr/cadic/1207/fiche-comment-faire-durer-ses-objets.pdf?modal=false. [accessed 18 november 2022]

Agossou, Noukpo, 'Dynamique spatiale à Porto-Novo: les effets de la diffusion des produits pétroliers kpayo', *L'Espace géographique*, 33.3 (2004), 211–18.

Alexander, Catherine and Joshua Reno, eds, *Economies of Recycling: the Global Transformations of Materials, Values and Social Relations* (London and New York, NY: Zed Books, 2012).

Altena, Bert and Marcel Van Der Linden, eds, *De-industrialization: Social, Cultural, and Political Aspects* (Cambridge: Cambridge University Press, 2003).

Amter, Steven and Benjamin Ross, *The Polluters: the Making of Our Chemically Altered Environment* (Oxford: Oxford University Press, 2010).

Anstett, Elisabeth and Nathalie Orstar, eds, *La Deuxième vie des objets. Recyclage et récupération dans les sociétés modernes* (Paris: Éditions Pétra, 2015).

Barles, Sabine, *L'Invention des déchets urbains. France 1790–1970* (Seyssel: Champ-Vallon, 2005).

Bauman, Zygmunt, *Liquid Life* (Cambridge: Polity, 2005).

Beaune, Jean-Claude, *Le Déchet, le rebut, le rien* (Seyssel: Champ Vallon, 1999).

Bednik, Anna, *Extractivisme. Exploitation industrielle de la nature: logiques, conséquences, résistances* (Lyon: Éditions Le Passager clandestin, 2015).

Benelli, Nathalie, Delphine Corteel, Octave Debary, Bénédicte Florin, Stéphane Le Lay, and Sophie Rétif, *Que faire des restes? Le réemploi dans les sociétés d'accumulation* (Paris: Presses de Sciences Po, 2016).

Bernardi, Philippe and Hélène Dessales, 'Les Réemplois en architecture, entre Antiquité et Moyen Âge: introduction à l'école d'été (Atelier doctoral, Rome, 19–23 septembre 2016)', *Mélanges de l'École Française de Rome*, 129.1 (2017), 161–64.

Bernhardt, Christoph and Geneviève Massard-Guilbaud, eds, *Le Démon moderne. La pollution dans les sociétés urbaines et industrielles d'Europe* (Clermont-Ferrand: Presses Universitaires Blaise-Pascal, 2002).

Bertolini, Gérard, 'Le Marché des déchets: structures et acteurs, croissance, concentration et recompositions', *Annales des Mines. Responsabilité & Environment*, 45 (2007), 99–109.

———, 'Le Service public local des déchets ménagers à la recherche d'indicateurs de performance', *Réalités industrielles*, 1 (2008), 80–88.

Bihouix, Philippe, *Le Bonheur était pour demain. Les rêveries d'un ingénieur solitaire* (Paris: Seuil, 2019).

Bonnot, Thierry, 'Une manufacture de céramique vue à travers ses déchets: expérience pluridisciplinaire autour d'un dépotoir', *Artefact. Techniques, histoire et sciences humaines*, 6 (2017), 11–28.

———, *La Vie des objets* (Paris: Éditions de la MSH/Mission du Patrimoine Ethnologique, 2002).

Compagnon, Antoine, *Les Chiffonniers de Paris* (Paris: Gallimard, 2017).

Corteel, Delphine, 'Réemploi: le travail de requalification des déchets en question', *Mouvements*, 87 (2016), 107–18.

Dant, Tim, 'The Work of Repair: Gesture, Emotion and Sensual Knowledge', *Sociological Research Online*, 15.3 (2010), 97–118.

Daumas, Jean-Claude, *La Révolution matérielle. Une histoire de la consommation, France XIXe–XXIe siècle* (Paris: Flammarion, 2018).

Desvaux, Pierre, 'Tribulations de fin de vie d'une bouteille en plastique au Caire', *Géographie et cultures*, 91–92 (2014), https://journals.openedition.org/gc/3429 [accessed 18 november 2022].

Dobré, Michel, Fabrice Flipo, and Marion Michot, *La Face cachée du numérique. L'impact environnemental des nouvelles technologies* (Paris: L'Échappée, 2013).

Douglas, Mary, *Purity and Danger: an Analysis of Concepts of Pollution and Taboo* (London: Routledge and Kegan Paul, 1966).

Douny, Laurence, 'The Materiality of Domestic Waste. The Recycled Cosmology of the Dogon of Mali', *Journal of Material Culture*, 12.3 (2007), 309–31.

Droulers, Martine, 'Poubelles de pauvres. L'agglomération de Recife', *Les Annales de la recherche urbaine*, 31 (1986), 61–70.

Erkman, Suren, *Vers une écologie industrielle. Comment mettre en pratique le développement durable dans une société hyper-industrielle* (Paris: Fondation Charles-Léopold Mayer, 2000).

Fanchette, Sylvie, 'Papeterie et recyclage dans les villages de métier', *Techniques & Culture*, 65–66 (2016) <http://journals.openedition.org/tc/7954> [accessed 18 november 2022].

Flipo, Fabrice, 'Expansion des technologies de l'information et de la communication: vers l'abîme?', *Mouvements*, 79 (2014), 115–21.

Florin, Bénédicte and Pascal Garret, '"Ils n'ont qu'à bouffer leurs ordures!"', *Vacarme*, 81 (2017), 99–106.

Fluck, Pierre, *Manuel d'archéologie industrielle, archéologie et patrimoine* (Paris: Hermann, 2017).

Gonzalez-Lafaysse, Linda, 'Les Chiffonniers bordelais à la fin du XIXe siècle. Entre professionnalisation et stigmatisation', *Ethnologie française*, 40.3 (2010), 521–30.

Goyon, Marie, 'L'Obsolescence déprogrammée: prendre le parti des choses pour prendre le parti des hommes', *Techniques & Culture*, 65–66 (2016) <http://journals.openedition.org/tc/7983> [accessed 18 november 2022].

Guillerme, André, *La Naissance de l'industrie à Paris: entre sueurs et vapeurs, 1780–1830* (Seyssel: Champ Vallon, 2007).

Harpet, Cyrille, *Du déchet: philosophie des immondices. Corps, ville, industrie* (Paris, L'Harmattan, 1999).

Hoarau, François, 'La Communauté Emmaüs de Besançon: des sujets et des objets pris dans des réseaux d'actions sur les actions', in *Le Sujet contre les objets, tout contre. Ethnographies de cultures matérielles*, ed. by Marie-Pierre Julien and Céline Rosselin (Paris: Éditions du CTHS, 2009), pp. 97–106.

Hodder, Ian, 'The Entanglements of Humans and Things: a Long-term View', *New Literary History*, 45 (2014), 19–36.

Homburg, Ernst, Anthony S. Travis, and Harm G. Schröter, *The Chemical Industry in Europe, 1850–1914: Industrial Growth, Pollution and Professionalization* (Dordrecht: Springer, 2011).

Jarrige, François and Thomas Le Roux, *La Contamination du monde. Une histoire des pollutions à l'âge industriel* (Paris: Seuil, 2017).

——— and Thomas Le Roux, 'Le Rôle des déchets dans l'histoire', interview with Stéphane Le Lay *Mouvements*, 87 (2016), 59–68.

Larroque, Claire, 'Déchets et recyclage', *Pertinence*, 5 (2013), 'Cycle de vie', Éditions de l'UPA (Université Antonine, Liban) <https://www.academia.edu/2924088/D%C3%A9chets_et_Recyclage> [accessed 18 november 2022].

Le Roux, Thomas, *Le Laboratoire des pollutions industrielles. Paris, 1770–1830* (Paris: Albin-Michel, 2011).

——— and Michel Letté, eds, *Débordements industriels. Environnement, territoire et conflit, XVIIIe–XXIe siècle* (Rennes: Presses Universitaires de Rennes, 2013).

London, Bernard, *Ending the Depression through Planned Obsolescence* (New York, NY: 1932).

———, 'En finir avec la crise grâce à l'obsolescence planifiée', trans. by Christophe David, *Écologie et politique*, 44 (2012), 167–79.

Lorrain, Dominique, 'Le Secteur des déchets aux Etats-Unis (I)', *Flux*, 43 (2001), 73–84.

———, 'Waste management: les entreprises américaines de déchets (II)', *Flux*, 48–49 (2002), 107–21.

Lupton, Sylvie, *Économie des déchets. Une approche institutionnaliste* (Brussels: De Boeck, 2011).

Messal, Stéphanie, 'La Réanimation des objets mourants' (unpublished doctoral thesis, Aix-Marseille University, 2014) <https://tel.archives-ouvertes.fr/tel-01001683/document> [accessed 18 november 2022].

Monsaingeon, Baptiste, *Homo Detritus. Critique de la société du déchet* (Paris: Seuil, 2017).

Norris, Lucy, 'Economies of Moral Fibre? Recycling Charity Clothing Into Emergency Aid Blankets', *Journal of Material Culture*, 17.4 (2012), 389–404.

Ortar, Nathalie and Elisabeth Anstett, eds, *Jeux de pouvoir dans nos poubelles. Economies morales et politiques du recyclage au tournant du XXIe siècle* (Paris: Éditions Petra, 2017).

Pellow, David Naguib, *Resisting Global Toxics: Transnational Movements for Environmental Justice* (Cambridge, MA: MIT Press, 2007).

Porter, Richard C., *The Economics of Waste. Resources for the Future* (Washington, DC: Taylor & Francis Inc., 2002).

Rathje, William and Cullen Murphy, *Rubbish! The Archaeology of Garbage* (New York, NY: Harper Collins, 1992).

Roche, Daniel, *Histoire des choses banales. Naissance de la consommation, XVIIe–XIXe siècle* (Paris: Fayard, 1997).

Russel, Andrew L. and Lee Vinsel, *The Innovation Delusion: How Our Obsession with the New Has Disrupted the Work That Matters Most* (New York, NY: Currency, 2020).

Schulz, Yvan, '"Fin de vie" et renaissance clandestine en Chine du Sud', *Techniques & Culture*, 65–66 (2016), <http://journals.openedition.org/tc/7858> [accessed 18 november 2022].

Sciardet, Hervé, *Les Marchands de l'aube. Ethnographie et théorie du commerce aux Puces de Saint-Ouen* (Paris: Economica, 2003).

Silguy, Catherine de, *Histoire des hommes et de leurs ordures: du Moyen Âge à nos jours* (Paris: Le Cherche Midi, 2009).

Slade, Gile, *Made to Break: Technology and Obsolescence in America* (Cambridge, MA: Harvard University Press, 2006).

Speranza, Gaetano, *Objets blessés: la réparation en Afrique* (Paris: Musée du quai Branly and Milan: 5 continents, 2007).

Strasser, Susan, *Waste and Want: a Social History of Trash* (New York, NY: Metropolitan Books, 1999).

Thompson, Michael, *Rubbish Theory: the Creation and Destruction of Value* (Oxford: Oxford University Press, 1979).

Illustrations

8- D. Raina and N. Sarma, 'Framing the History of Technology in India'

Fig. 1. Passenger Train crossing the Dapoorie Viaduct (first railway bridge in India) from Bombay to Thane in 1855. Source: https://scroll.in/article/669579/the-long-history-of-indian-railways-told-through-images

Fig. 2. This photograph shows the India's first geo-stationary satellite Ariane Passenger Payload Experiment (APPLE) mounted on a bullock cart to provide a non-magnetic environment to figure out a communication problem. Source: https://www.thebetterindia.com/204852/india-isro-historic-pictures-thumba-mangalyaan-vikram-sarabhai-kalam-pslv/

10- J. Daheur, 'Central European Historical Writing on Technology, from 1945 to the Present Day'

Fig. 1. Exhibition hall in the National Museum of Technology in Warsaw (1964) Narodowe Archiwum Cyfrowe 3/51/0/10.4/409 https://www.szukajwarchiwach.gov.pl/ru/jednostka/-/jednostka/5987583/obiekty/558758

Fig. 2. Ceremony for the 200th anniversary of the opening of the Mining Academy in Banská Štiavnica (1964) Virtuálna Banská Štiavnica Fotka č. 7987. https://www.virtualna-banska-stiavnica.sk/hzizdibyeb.html/historicke-fotky/historicke-fotky/7987&k = 200-rokov-akademie

12- E. Kochetkova and J. Obertreis, 'The Achilles Heel of a Big Player'

Fig. 1.The Mosin rifle, a sample of 1891, «Марка» n° 1995, Michel n° 2212. https://ru.wikipedia.org/wiki/%D0%92%D0%B8%D0%BD%D1%82%D0%BE%D0%B2%D0%BA%D0%B0_%D0%9C%D0%BE%D1%81%D0%B8%D0%BD%D0%B0#/media/%D0%A4%D0%B0%D0%B9%D0%BB:RUSMARKA-1995.jpg

Fig. 2. Soviet Agat-9 computer, the last mass-produced model https://en.wikipedia.org/wiki/List_of_Soviet_computer_systems#/media/File:Agat-9_Black.jpg

13- J. Baudry, 'Technology in North America'

Fig. 1. *Steamboat Travel on the Hudson River* (1811–*c.* 1813), drawing by Pavel Petrovich Svinin, Collection of the Metropolitan Museum of Art, New York. https://www.metmuseum.org/art/collection/search/12715

Fig. 2. *Buzz Aldrin with Apollo 11 Lunar Module on the Moon* (1969), photograph by Neil Amstrong, Collection of the Metropolitan Museum of Art, New York. https://www.metmuseum.org/art/collection/search/786294

14- D.Pretel, 'Latin America in the Global History of Technology'

Fig. 1. Unloading Cane at a Modern Cuban Mill (*c.* 1900–1906). Source: Detroit Publishing Co, P. Unloading Cane at a Modern Mill. Cuba. [Between 1900 and 1906] [Photograph] Retrieved from the Library of Congress, https://www.loc.gov/item/2016798979/

Fig. 2. Workers building the Panama Canal (1903). Source: Harris & Ewing, photographer. Panama Canal, 1913. [Photograph] Retrieved from the Library of Congress, https://www.loc.gov/item/2016864862/

17- K. Troch, 'Prospecting, Extracting, and Post-mining Techniques in the Anthropocene Era'

Fig. 1. Diagram illustrating the hydraulic backfill process. Baijot, Achille and Adolphe Demeure, *Notes recueillies dans quelques mines du Bassin Rhénan-Westphalien, dans le Nord et dans le Pas-de-Calais sur le remblayage hydraulique,* dans *Revue universelle des mines, de la métallurgie, des travaux publics, des sciences et des arts appliqués à l'industrie,* 4th série, vil. 26, 1909, ill. 4.

20- J Lamy and A. Saint-Martin, 'Aeronautics and Aerospace'

Fig. 1. STS-70 Rollout. NASA on The Commons

21- A. Passalacqua, 'Mobility Techniques'

Fig. 1. Between technical transfer, advertising promotion and globalization stakes: Citroën-Kégresse caterpillar cars of the Citroën Central Asia Cruise, known as the Yellow Cruise, in circulation in Asia (1931) (*Le Monde colonial illustré,* 100, December 1931, gallica.bnf.fr/ Bibliothèque nationale de France).

22- G. Barron and J. Lajus, 'Sailing and Fishing'

Fig. 1. Le *Great Eastern*. Robert Charles Dudley, 1892. The Atlantic Telegraph Cable Fleet Assembled at Berehaven, 1866. New York, The Metropolitan Museum of Art, https://commons.wikimedia.org/wiki/File:The_Atlantic_Telegraph_Cable_Fleet_Assembled_at_Berehaven_(Southwest_Coast_of_Ireland)-_Ships,_the_Great_Eastern,_H.M.S._Terrible,_the_Alby,_the_Medway_and_the_William_Cory_MET_DP119840.jpg

Fig. 2. Beam-Trawl. Collins, J. W., 1889. 'Beam-Trawl Fishery of Great Britain, with Notes on Beam-Trawling in Other European Countries, etc., *Bulletin of the United States Fish Commission*, 7, 1887, Washington, DC, Government Printing Office, https://upload.wikimedia.org/wikipedia/commons/6/60/FMIB_37802_Towing_a_Beam-Trawl.jpeg

25- F. Knittel, 'Agricultural Techniques'

Fig. 1. Dombasle Plough (1885). Th. Barrau, *Simples notions sur l'agriculture*, Paris, Hachette et Cie, 1885.

Fig. 2. Skimmer Cooley. René Lezé, *Les Industries du lait* (Paris, Firmin-Didot, 1904), p. 217. Bibliothèque nationale de France) Les industries du lait (2e édition revue et augmentée) / par R. Lezé,... | Gallica (bnf.fr)

26- V. Kuitenbrouwer, 'Global Telecommunication Technologies from the Nineteenth to the Twenty-First Centuries'

Fig. 1. 'Torchlight procession around the world — Dedicated to young America', lithograph by A. Weingärtner (c. 1858). Allegorical scenes related to the laying of the Atlantic telegraph cable. Library of Congress, https://www.loc.gov/resource/pga.08527/ Publications older than 95 years can be published freely. See: https://www.loc.gov/rr/print/res/248_pga.html

33- J. Mercelis, 'Laboratory Techniques'

Fig. 1.The Pasteur Institute, Kasauli, India: one stage in the preparation of the rabies vaccine: a rabbit brain on a square of muslin. Photograph, *c.* 1910. Wellcome Library. https://commons.wikimedia.org/w/index.php?sort=relevance&search=Kasauli+pasteur+institute&title=Special:Search&profile=advanced&fulltext=1&advancedSearch-current=%7B%7D&ns0=1&ns6=1&ns12=1&ns14=1&ns100=1&ns106=1#/media/File:The_Pasteur_Institute,_Kasauli,_India._Wellcome_V0030195.jpg

Fig. 2. Research lab at the Saigon Pasteur Institute, 1961, source The-gioi Tu-do, author USIA https://commons.wikimedia.org/w/index.php?

search=Saigon+pasteur+institute&title=Special%3ASearch&go=Go&nso=1&ns6=1&ns12=1&ns14=1&ns100=1&ns106=1#/media/File:SaigonPasteurInstitute.jpg

34- O. Hallonsten, 'Big Science and Technological Transformations'

Fig. 1.'Tandem Accelerator Laboratory at the Institute for Nuclear and Particle Physics'. Dimitris Sideridis/Wikimedia Commons.

35- A. Nuvolarai and C. Sganga, 'Technique and Law'

Fig. 1. Page of *Lean's Engine Reporter* reporting the performance of the steam engines at work in the mines of Cornwall.

36- S. Lembré, 'Technical Training and Education since the Nineteenth Century'

Fig. 1. Apprentice. Man and Boy Making Shoes, *c.* 1914 painting by (Louis-?)Emile Adan (1839–1937), copyrighted by Braun & Co., N.Y. This image is available from the United States Library of Congress Prints and Photographs division under the digital ID cph.3b27512.

Fig. 2. Unley Girls Technical School, Unley, South Australia State Government Photographer, *c.* 1945 https://commons.wikimedia.org/wiki/File:Unley_Girls_Technical_School(GN10986).jpg?uselang = fr The History Trust of South Australian, South Australian Government, GN10986.

38- P. Fridenson, 'Taylorism'

Fig. 1. F. W. Taylor's Expanding Social Networks. The dark colored organizations are close to Taylor's roots in mechanical engineering. Taylor resisted the society at the bottom, which he saw as a withdrawal into a private world of scientific managers, rather than influencing others. (c) Margaret 'Amy' DiGerolamo, 2015. Lee Vinsel, Taylor's World Pt. 2: F. W. Taylor's Expanding Social Networks, September 25, 2015 http://leevinsel.com/blog/2015/9/25/taylors-world-pt-2-f-w-taylors-expanding-social-networks

Fig. 2. Portrait of Gastev by Z. Tolkachev. From Gastev, *Revolt of Culture*, 1923.

39- F. Jarrige,'Conflicts and Controversies Over Technologies'

Fig. 1. Poster posted in Nottingham (26 March 1811). Reward offered for information leading to the arrest of machine breakers (poster reproduced in Brian Bailey, *The Luddite Rebellion* (New York: New York University Press, 1998), p. 17.

Fig. 2. Front cover, *La Gueule ouverte*, 28, 20 November 1974. The alternative and ecological press of the 1970s challenged nuclear power.

40- C. Calheiros, 'Body Techniques and the Search for Human Perfection'

Fig. 1. École normale de gymnastique et d'escrime (Joinville-le-Pont, Val-de-Marne). Flexibility exercises [overall movement]. 1906. Rol Agency. https://gallica.bnf.fr/ark:/12148/btv1b53110699t?rk= 278971%3B2&fbclid=IwAR2Ck-ez8W3FpopAnIgDemXbA4bnEQxlimoQabQeRswIs_47sUL9NlTGS13k

Fig. 2. Leg prosthesis. 2013. Romain Pernet. https://commons.wikimedia.org/wiki/File:Jambe_bionique.png?uselang=fr

42- M. Beugnet and M. Duplay, 'Technologies of Art and Culture'

Fig. 1. Self-caricature of tenor Enrico Caruso making his first record (11 April 1902) https://commons.wikimedia.org/wiki/File:Caruso_gramophone_cartoon.jpg

43- G. Bartholeyns and M. Charpy, 'Everyday Technologies'

Fig. 1. First electric toaster, 4 heating elements, ceramic base. Model D-2, General Electric Company (Schenectady, NY), 1908, 17,5 × 20,4 × 10,1 cm.

Fig. 2. Lithographic advertising card for Singer sewing machines, British India, 1892. The slogan proclaims: 'Singer Sewing-Machines are known all over the world as being the best. The women off all nations prefer them. [...] Salesrooms in every city in the world'. Private collection (M. Charpy).

Fig. 3. Cadet. Presentation of arrangements for 'showering oneself', album by Achille Cadet, 69 rue de la Roquette, Paris, 1880s. Private collection (M. Charpy).

44- S. Boudia and N. Jas, 'Technology, Pollution, Environment'

Fig. 1. E-Waste recycling factory, Silverdale, Auckland. Judit Klein. www.flickr.com https://www.flickr.com/photos/juditk/

45- T. Bonnot and M. Letté, 'Repair, Recycling, and Waste'

Fig. 1. Archaeological excavation of an industrial ceramics dump, early twentieth century (Pouilloux, Saône-et-Loire, December 2011). © Bonnot, 2011.

Fig. 2. Recycling of porcelain ink bottles, transformed into lamp bases, *c.* 1960 (Palinges, Saône-et-Loire, 2018). © Bonnot, 2018.

Biographies of the authors

Larissa Zakharova — in memoriam

Larissa Zakharova was born in Leningrad in 1977. She died in 2019. She was Associate Professor in EHESS (École des hautes études en sciences sociales) (2010) and a member of the CERCEC research unit (Centre d'études des mondes russe, caucasien et centre-européen). Her research field was 'Tools of communication, governing patterns, social configurations in USSR, 1917–1991'. She was a specialist in Sovietism, which she studied from multiple angles. Her PhD thesis, published as *S'habiller à la soviétique. La mode et le Dégel en URSS* (Paris, CNRS Éditions, 2011), dealt with fashion in the Soviet Union. She also edited special issues of journals which revealed her growing impact in the history of technology: *Le Quotidien du communisme: pratiques et objets* for *Annales. Histoire, Sciences Sociales* (2013) and with Kristin Roth-Ey, *Communiquer en URSS et en Europe socialiste* for *Cahiers du monde russe* (2015). In 2011, she organised a symposium on globalization with Liliane Hilaire-Pérez, which was published as *Les Techniques et la globalisation au XX*[e] *siècle* in 2016. Larissa Zakharova was a brilliant young scholar. Before she died, she had completed her habilitation thesis, which has been published under the title *De Moscou aux terres les plus lointaines. Communications, politique et société en URSS* (Éditions de l'EHESS, 2020).

Dominique Barjot

Dominique Barjot was Emeritus Professor of Economic History at Sorbonne University and Professor at Renmin University of China from 2017 to 2020. In addition to being the former Deputy Director of SHS CNRS (Sciences humaines et sociales Centre national de la recherche scientifique) and a representative of France on the Standing Committee of the European Science Foundation, he was also the scientific Director of Humanities at the Ministry of Education, Higher Education and Research and the Director of the Graduate School of Modern and Contemporary History at Sorbonne University. Since 2017 he has been the President of the French Committee of Historical Sciences. He is the former president of the French Association of the historian economists and the author of many books and papers in peer-reviewed journals.

Geraldine Barron

Geraldine Barron is a French historian and a University Librarian at Université du Littoral Côte d'Opale. Her work focusses on maritime history and its scientific and technical aspects, like shipbuilding and navigation. Her publications include *Edmond Pâris et l'art naval: des pirogues aux cuirassés*, Histoire et Techniques (Presses universitaires du Midi, 2019) and 'Dévoiler la vapeur. La construction navale au XIXe siècle', in *L'Europe en transitions. Énergie, mobilité, communication, XVIe–XXA^{e} siècles*, Histoire nouvelle de l'Europe, ed. by Yves Bouvier, and Léonerd Laborie (Nouveau Monde éditions, 2016).

Gil Bartholeyns

Gil Bartholeyns teaches at the Humanities Department of the University of Lille. He has held the CNRS (Centre national de la recherche scientifique) Chair of Visual Studies and is in charge of the master and doctoral programmes in visual and material cultures. He is chief editor of the journal *Techniques & Culture*, member of the editorial board of *Terrain*, *Modes pratiques*, and *Entre-temps*, and the author of *Image et transgression* (Presses universitaires de France, 2008), *Politiques visuelles* (Presses du réel, 2016), *Le Hantement du monde* (Éditions Dehors, 2021), and *L'étrange et folle aventure du grille-pain, de la machine à coudre et des gens qui s'en servent* (Premier Parallèle, 2022) with Manuel Charpy.

Jérôme Baudry

Jérôme Baudry is Assistant Professor of History of Science and Technology at the Swiss Federal Institute of Technology in Lausanne (EPFL). His work focuses on intellectual property, on the textual and visual languages of technology, and on amateur practices in science and technology. He recently published 'A Politics of Intellectual Property: Creating a Patent System in Revolutionary France', *Technology and Culture* (vol. 61, no 4, 2020), 'Turning Crowds into Communities: The Collective of Online Citizen Science', *Social Studies of Science* (vol. 52, no, 2002), and he co-edited the volume *Produire du nouveau ? Arts, techniques, sciences en Europe* (1400-1900) at CNRS Éditions, 2022.

Yamina Bettahar

Yamina Bettahar is a sociologist and a historian of science, Associate Professor at the University of Lorraine, and a researcher at the Archives Henri Poincaré — Philosophie et Recherches sur les Sciences et les Technologies. She focusses on the socio-history of academic and scientific institutions (universities and 'grandes écoles') and the circulation of people and knowledge in the Euro-Maghreb space. She is Former responsible of international affairs and Fellow of JSPS foundation (Japan). She is fellowship of JSPS Foundation (Japan) and she teaches at Universities of Roma 3 (Italy), Hiroshima, Hitotsubashi and Kanazawa (Japan). She has co-edited *Les Universités au risque de l'histoire. Principes, configurations, modèles* (Presses universitaires de Nancy and Éditions universitaires de Lorraine, 2014) with Marie-Jeanne Choffel-Mailfert, co-authored 'Les Circulations scientifiques internationales depuis le début du xxe siècle. Nouvelles perspectives d'étude', *Philosophia Scientiæ* (2019) with Denis Guthleben, and edited *Universités et grandes écoles. Circulations étudiantes et scientifiques d'hier à aujourd'hui* (Presses universitaires de Nancy, 2020).

Martine Beugnet

Martine Beugnet is Professor in Visual Studies at the University of Paris. She is a member of the LARCA (Laboratoire de recherche sur les cultures anglophones) and a CNRS fellow (Centre national de la recherche scientifique). She has curated exhibitions and written articles on a wide range of film and media topics and authored several books on contemporary cinema, including *Claire Denis* (Manchester University Press, 2004), *Proust at the Movies* (Ashgate, 2005) (co-written with Marion Schmid), *Cinema and Sensation: French Film and the Art of Transgression* (Edinburgh University Press, 2007, 2012), and *L'Attrait du flou* (Yellow Now, 2017), on the history and aesthetics of blur in film. She co-edited the volume of collected essays *Indefinite Visions: Cinema and the Attractions of Uncertainty* with Allan Cameron and Arild Fetveit (Edinburgh University Press, 2017). Her new monograph, entitled *Le Cinéma et ses doubles. L'image de film à l'ère du foundfootage numérisé et des écrans de poche*, was published in 2021 (Bord de l'Eau Press). She co-directs, with Kriss Ravetto, the Edinburgh University Press book series Studies in Film and Intermediality, and is a member of the editorial board of the journal *NECSUS*.

Thierry Bonnot

Thierry Bonnot, a Research Fellow in anthropology at the French National Centre for Scientific Research (CNRS), is a member of the French Institute of Interdisciplinary Research on Social Issues (IRIS, Paris). Trained as a historian, he first worked at the Ecomuseum of the Urban Community at Le Creusot-Montceau (Saône-et-Loire department of France), for which he conducted several research projects. His research focusses on material culture and the construction of heritage. His most recent investigations are based on a multidisciplinary approach using the tools of history and archaeology as well as anthropology. He has organised an exhibition on an experiment carried out at an industrial waste dumping site near a disused ceramic production plant. He continues to work on the social status of objects, on the ways in which they are appropriated, and on how heritage sites are constructed. His publications include *La Vie des objets* (Maison des sciences de l'homme, 2002) and *L'Attachement aux choses* (CNRS Éditions, 2014).

Soraya Boudia

Soraya Boudia is a historian of science and Professor of Sociology at Université Paris Cité. She has worked extensively on the role of science and technology in politics and policy, particularly the transnational government of health and environmental risks. She has co-edited, with Nathalie Jas, *Toxicants, Health and Regulations Since 1945* (Pickering & Chatto, 2013), *Powerless Science? Science and Politics in a Toxic World* (Berghahn, 2014), *Gouverner un monde toxique* (Quae, 2019), with Emmanuel Henry, *La Mondialisation des risques. Une histoire politique et transnationale des risques sanitaires environnementaux* (Presses Universitaires de Rennes, 2015), and with Angela Creager, Scott Frickel, Emmanuel Henry, Nathalie Jas, Carsten Reinhardt, Jody Roberts, *Residues, Thinking Through Chemical Environments* (Rutgers University Press, 2021).

Francesca Bray

Francesca Bray is a historian and anthropologist of science, technology, and medicine specializing in China, and a global historian. Her books include the volume on *Agriculture* in Needham's series *Science and Civilization in China* (Cambridge University Press, 1984), *The Rice Economies* (Blackwell, 1986), *Technology and Gender: Fabrics of Power in Late Imperial China* (University of California Press, 1997), *Graphics and Text in the Production of Technical Knowledge in China* (Brill, 2007), *Rice: Global Histories and New Networks* (Cambridge University Press, 2015), and *Moving Crops and the Scales of History* (Yale University Press, 2023). She is Professor Emerita of Social Anthropology at the University of Edinburgh.

Patrice Bret

Dr Patrice Bret is Senior Researcher at Centre Alexandre Koyré and the Secretary General of the Lavoisier Committee at the Academy of Sciences in Paris. Trained in history, political sciences, and history of art (University of Provence and University of Paris I Panthéon-Sorbonne), as well as computing science (University Paris XIII Villetaneuse), he gained his PhD at the University of Paris I Panthéon-Sorbonne in 1994. He was a researcher in history of science and technology (CNRS and City of Science and Industry, 1995 to 2003) prior to joining the Centre Alexandre Koyré while also being the scientific head of the History of Armament Department at General Delegation for Armament (DGA) and the Institute for Strategical Research at the *École militaire* (IRSEM) of the Ministry of Defence (2003 to 2012). A large part of his own work focusses on the social and cultural history of scientific and technological institutions, networks, and practices of research and invention and deals with war, science, and technology in France (eighteenth to twentieth centuries), including exchanges with the Ottoman Empire, Egypt, and Latin America.

Cécilia Calheiros

Cécilia Calheiros is a doctor in sociology at the École des hautes études en sciences sociales (EHESS) in Paris, where she is a member of CéSor (Centre d'études en sciences sociales du religieux). Her research focusses on how techno-scientific environments create beliefs about self-transcendence and reconfiguration of the human. Located at the cross-roads of sociology of belief and Science and Technological Studies, her work re-examines the boundaries between knowledge and beliefs, between science and spirituality. Her thesis deals with how transhumanists appropriate science and technology to shape their actions, their identities, and their representations.

Guillaume Carnino

Guillaume Carnino, formerly a computer engineer, is Associate Professor in history of science and technology at the University of Technology of Compiègne. As a nineteenth-century historian, he is interested in the epistemological and industrial foundations of technology, and more broadly as a historian of the modern period, he works on the history of quantification and computer science, particularly data centres. He has published *L'Invention de la Science. La nouvelle religion de l'âge industriel* (Seuil, 2015) and co-edited, with Liliane Hilaire-Pérez and Aleksandra Kobiljski, *Histoire des techniques. Mondes, sociétés, cultures, XVIe–XVIIIe siècles* (PUF, 2016).

Manuel Charpy

Manuel Charpy is a researcher at the CNRS (Centre national de la recherche scientifique). A qualified teacher in applied arts and PhD in history, he focusses his research on the world of objects and pictures and their role in the construction of social identities in Europe, the United States, and West and Central Africa. He specializes in the history of material culture of the Parisian bourgeoisie, the history of the uses of portraiture, the history of advertising, the history of everyday techniques, and the history of clothing and fashion, including the trade and consumption of second-hand clothing, the birth of industrial clothing, and clothing in colonial situations, particularly in Central Africa. In 2015, he created (with Patrice Verdière) *Modes pratiques, a Journal of the History of Clothing and Fashion*. He teaches history in universities and schools of applied art and is the director of the InVisu laboratory

(CNRS / INHA) Centre national de la recherche scientifique / Institut national d'histoire de l'art).

Jawad Daheur

Jawad Daheur is a historian who works on the relations between human societies and their environment in Central Europe, with a special focus on the German-speaking areas and Poland. After completing his PhD at the University of Strasbourg in 2016, on German trade in the forests of the Vistula and Warta rivers basins (1840 to1914), he began a postdoctoral research project dedicated to the environmental drivers of poverty in Austrian Galicia (1873 to 1914) in 2017. His current field of research addresses the ecological impact of large-scale trade in the southern Baltic hinterland in the modern period.

Mathieu Duplay

Mathieu Duplay is Professor of American Literature at Université Paris Cité (LARCA, Laboratoire de recherche sur les cultures anglophones). His research focuses on the relationship between literature and other artistic fields, particularly music and drama. He is the author of numerous articles on the works of American musicians/writers such as John Cage, Samuel Barber, Stephen Sondheim, Steve Reich, and John Adams; his most recent book, a study of John Adams's stage works, will be issued in 2023 by the French publishing company Honoré Champion. In addition, Mathieu Duplay has co-edited a volume on the American philosopher Susanne K. Langer (*Vie, Symbole, Mouvement. Susanne K. Langer et la danse,* Mathieu Duplay & Anne Boissière [eds.], Grenoble, De l'Incidence, 2012). Mathieu Duplay currently serves as President of the French Association for American Studies (AFEA).

Marie Durand

Marie Durand is Associate Professor in Anthropology at Strasbourg University, member of the Interdisciplinary Laboratory in Cultural Studies (LinCS) and associate at the Centre de Recherche et de Documentation sur l'Océanie (CREDO) in Marseille. Her research interests include the anthropology of material culture and techniques, architecture and the built environment, as well as Pacific art and museum collections. She has been conducting fieldwork in Vanuatu since 2010, focussing especially on the Banks Islands in the northernmost province of the archipelago.

Sabine Effosse

Sabine Effosse is Professor of Contemporary History at Paris Nanterre University and a researcher at IDHES lab (Institutions et dynamiques historiques de l'économie et de la société). Her current research is focussed on credit policy, business history, and consumption and real estate finance in twentieth-century Europe. She has recently published 'Gendered Finance. Women and banks in the 20th century', *Quaderni Storici*, 166, 1/2021; *L'Entreprise post RSE vue par ses parties prenantes* (Paris, Institut de l'entreprise, 2020); 'La loi du 13 juillet 1965 et l'émancipation bancaire de la femme mariée', *Cahiers français* (2020); 'Consumer Credit as a Marketing Tool: the French Experience in European and Transatlantic Comparison, 1950s–1960s', in *Consumer Engineering, 1930–1970. Marketing from Planning Euphoria to the Limits of Growth*, ed. by Jan Logemann, Ingo Köhler, and Gary Cross (Palgrave McMillan, 2019).

Patrick Fridenson

Patrick Fridenson is Professor Emeritus of International Business and Labour History at the École des Hautes Études en Sciences Sociales (EHESS) in Paris. His research deals with the strategies, innovations, structures, labour, ethics, and performances of business enterprises in relation to consumers and to the regulatory and social environment in an international perspective. He is the author of *Histoire des usines Renault*, vol I (2nd ed., Éditions du Seuil, 1998), the co-author of *Ethical Capitalism. Shibusawa Eiichi and Business Leadership in Global Perspective* (University of Toronto Press, 2017), *Reimagining Business History* (Johns Hopkins University Press, 2013), *The Automobile Revolution* (University of North Carolina Press, 1982), *Thomson's First Century* (Campus Thomson, 1995) and the editor of *The French Home Front, 1914–1918* (Berg, 1992). He is a founding editor of the journal *Entreprises et Histoire*, and the former editor of the journal *Le Mouvement Social*. He is a board member of *Enterprise and Society*, *Business History Review*, *French Politics, Culture and Society*, and *Japanese Research in Business History*. He is a former president of the French Economic History Association and of the Business History Conference.

Joshua Grace

Joshua Grace is Associate Professor of African History in the Department of History at the University of South Carolina, USA. His research explores the intersections of technology, mobility, gender, and development in Tanzanian history. His book, *African Motors: Technology, Gender, and the History of Development in Tanzania* was published by Duke University Press in 2022. Grace's work has appeared in *Africa: Journal of the International African Institute* and in the volume, *Africa Every Day: Fun, Leisure, and Expressive Culture on the Continent*. His research has been supported by Fulbright Programme, the Andrew W. Mellon Foundation, and by the American Council for Learned Societies.

Florence Hachez-Leroy

Florence Hachez-Leroy, Associate Professor of Modern History at Artois University, Centre de recherche et d'études Histoire et société (CREHS-UR 4027), her work crosses the economic, technical, social, cultural and heritage aspects of entrepreneurial history and the history of modern industrial materials. She early studied the role of international cartels in the aluminium sector and their impact on the market. Her most recent book, *Menaces sur l'alimentation* (PUFR, 2019), focuses on environmental history, particularly the regulation of food additives and their relation to public health, in Europe and the United States. She is currently directing an ANR research program called "Aluminum, Architecture and Heritage, 20-21st century' (Archipal), 2019-2024.

Olof Hallonsten

Olof Hallonsten is a sociologist and historian of science. He teaches organizational sociology at the Lund University School of Economics and Management and conducts research on the politics and organization of historical and contemporary science, including the role of Big Science and technically advanced instrumentation in the development of science, economy, and society in the twentieth century and beyond. He is the author of several articles and books on these and similar topics.

Liliane Hilaire-Pérez

Liliane Hilaire-Pérez is Professor in Early Modern History at the University Paris Cité, Director of Research at the EHESS (École des hautes études en sciences sociales), and Senior Fellow of the Institut Universitaire de France. She is interested in the history of invention (*L'Invention au siècle des Lumières*, Albin Michel, 2000), in the global history of techniques (*Histoire des techniques. Mondes, sociétés, cultures XVIe–XVIIIe siècle*, with Guillaume Carnino, PUF, 2016), and in technological thought (*La Pièce et le geste. Artisans, marchands et savoirs techniques à Londres au XVIIIe siècle*, Albin Michel, 2013; *La Technologie générale. Johann Beckmann Entwurf der allgemeinen Technologie / Projet de technologie générale (1806)*, with Guillaume Carnino and Jochen Hoock, Presses universitaires de Rennes, 2017). She is co-editor of the journal *Artefact. Techniques, histoire et sciences humaines*.

François Jarrige

François Jarrige is Associate Professor in Contemporary History at the University of Burgundy (Dijon), where he is part of the LIR3S research unit. He works on the history of industrial society and investigates the conflicts, debates, and controversies that have shaped industrialisation in the West. He has notably published (with Thomas Le Roux) *The Contamination of the Earth: A History of Pollutions in the Industrial Age*

(MIT Press, 2020) and *Technocritiques. Du refus des machines à la contestation des technosciences* (La Découverte, 2014).

Nathalie Jas

Nathalie Jas is a STS scholar and a permanent researcher at the French National Institute for Agriculture Food and Environment (INRAE). Her current scholarly work analyses the transformation of the ways hazardous chemicals have been governed since 1945 in Europe, in international arenas and more recently in Africa. Her publications on these subjects include three co-edited books, three co-edited special issues and two co-authored books, the most of which being *Residues: Thinking Through Chemical Environments* (Rutgers University Press, 2021, co-authored with Soraya Boudia, Angela Creager, Scott Frickel, Emmanuel Henry and Jody Roberts).

Virginie Julliard

Virginie Julliard is Professor of Information and Communication Sciences at Sorbonne University/CELSA. She is interested in Media Production of Gender, Online Controversies and Digital Methods. Her current research focuses on the structuration of anti-gender movement in digital social networks and on online moderation techniques. She created and is head of the Center for Experimentation in Digital Methods for Human and Social Sciences (CERES) of the Faculté des Lettres of Sorbonne University. She has published *De la Presse à Internet, la parité en question* (Hermès Lavoisier, 2012).

Mahdi Khelfaoui

Mahdi Khelfaoui has been Professor at the Department of Human Sciences at the Université du Québec à Trois-Rivières since 2020. He holds a master's degree and a PhD in science, technology, and society from the University of Quebec in Montréal. His thesis focussed on the socio-historical trajectory of the MAPLE nuclear reactors project in Canada, and he is currently preparing a book on the history of the Canadian medical isotopes industry. He is also the associate editor of *Scientia Canadensis: Canadian Journal of the History of Science, Technology and Medicine.* His research interests are in the history and sociology of technology, the history of energy, and in economic and industrial history. His publications appeared in *Nonproliferation Review, Canadian Historical Review, Journal of Canadian Studies,* and *European Review of International Studies.* He is also active in the field of bibliometrics, where he applies quantitative tools and network analysis to the history and sociology of science. His bibliometric work has been published in journals such as *Scientometrics, Quantitative Science Studies, Physics in Perspective,* and *Research Evaluation.*

Fabien Knittel

Fabien Knittel is Associate Professor in Modern and Contemporary History at the University of Bourgogne-Franche-Comté (Besançon) and a researcher at the Centre Lucien Febvre (EA 2273). He is a specialist in the history of agronomy and of agricultural techniques of the nineteenth century. He published *Agronomie et innovation. Le cas Mathieu de Dombasle (1777–1843)* (Presses universitaires de Nancy, 2009). Fabien Knittel and Pascal Raggi are the co-editors of *Genre et techniques (XIXe–XXIe siècles)* (Presses universitaires de Rennes, 2013). He has recently defended his habiliation thesis with the unpublished essay entitled 'Le lait des agronomes (c. 1790–c. 1914)' (2019), which is forthcoming as La Fabrique du lait. Europe occidentale (moyen Âge-xxe siècle) (CNRS Éditions).

Aleksandra Kobiljski

Aleksandra Kobiljski is Researcher in Modern and Contemporary History at the French National Centre for Scientific Research (CNRS), specializing in modern Japan. She is the Principal Investigator of J-InnovaTech, a project funded by the European Research Council (ERC) aiming to forge the framework of a new consensus on the nature of Japan's early industry (2020–2025).

Elena Kochetkova

Elena Kochetkova is historian of the environment, technology, economy and state socialism. Currently, she is an Associate Professor in Modern European Economic History at the Department of Archeology, History, Cultural Studies and Religion at the University of Bergen. She served as a Secretary of the European Society of Environmental History in 2019–2021. Among her publications are articles in leading international journals, such as Technology and Culture, Environment and History, Contemporary European History, Journal of Contemporary History, among others. Her first monograph *The Green Power of Socialism: Wood, Forest, and the Making of Soviet Industrially Embedded Ecology* is forthcoming with the MIT Press in 2024.

Vincent Kuitenbrouwer

Vincent Kuitenbrouwer is Assistant Professor of the History of International Relations at the University of Amsterdam. He specializes in nineteenth- and twentieth-century imperial history and colonial media. He was the co-investigator of the project

'Connecting the Wireless World' which was funded by the Leverhulme Trust (2016–2019) and publishes on the history of Dutch global radio broadcasting.

Julia Lajus

Julia Lajus is currently a Visiting Associate Professor at Columbia University in the City of New York and Associate Researcher at the University of Helsinki. Previously she was a Head of Laboratory for Environmental and Technological History and Associate Professor at the Department of History, National Research University Higher School of Economics (HSE), St. Petersburg, Russia. In 2011-2015, she served as vice-president of the European Society of Environmental History. Her research focuses on history of field sciences and environmental history of natural resources, especially in marine and polar areas.

Ludovic Laloux

Ludovic Laloux is Professor of Modern History at the Polytechnic University of Hauts-de-France in Valenciennes. He is a member of the centre of multidisciplinary research in social science (CRISS Centre interdisciplinaire en sciences de la société). He also belongs to the scientific committee of the *Revue politique et parlementaire* and to the research group on 'European Academies of sciences' within the framework of the History and Epistemology Committee of the Academy of Sciences of France. His research focusses in particular on agriculture, agri-food, and food industry in the countryside. His habilitation thesis is entitled 'Sugar Beets: the Advancement of Agriculture and Rural Industrialization.'

Jérôme Lamy

Jérôme Lamy is a historian and sociologist of science, and researcher at the CNRS. His research focusses on the history of spatial activities, science materiality, and the circulation of concepts. He published *Faire de la sociologie historique des sciences et des techniques* (Hermann, 2018). He is the co-editor (with Arnaud Saint-Martin) of the academic journal *Zilsel*, which is devoted to the social studies of science and technology.

Stéphane Lembré

Stéphane Lembré is Professor of Contemporary History at the University of Lille and a researcher at the IRHiS (Institut de recherches historiques du Septentrion) lab. His work mainly focusses on the history of technical and vocational education and training in the nineteenth and twentieth centuries. Among his recent publications are *Histoire de l'enseignement technique* (La Découverte, 2016), 'Les Lieux du technique', *Histoire de l'éducation* (with Guy Lambert, 2017), *Histoire des élèves en France, vol II: Ordres, désordres et engagements 16e–20e siècles*, ed. with Jérôme Krop (2020) and *Le* *retour de l'apprentissage au XXe siècle. Comment la France a adopté l'alternance* (Presses des Mines, 2023).

Michel Letté

Michel Letté is Associate Professor at the French National Conservatory of Arts and Crafts (*Conservatoire national des arts et métiers, Cnam*) in Paris and a historian of techniques and environment at the research unit HT2S (History of Techno-sciences in Society). He is the head of the professional training programme in mediation in science and technology through cultural design. His research focusses on controversies as a mechanism for the informal mediation of the relations between science, technology, and society, on the history of environmental conflicts since industrialization, and on the history of popular education in science and technology.

Anaël Marrec

Anaël Marrec is a Post-doctoral Fellow at the Centre d'histoire sociale des Mondes contemporains (CHES) and an Associate Researcher at the François Viète Centre (University of Nantes). She defended a PhD thesis in Epistemology, History of science, and technology on the history of renewable energies in France (1880 to 1990) in 2018. Her work currently focusses on the history of energy in contemporary society, in its technical, political, and imagined perspectives. She is co-editor, with Sarah Claire, of a special issue of the journal *Socio-anthropology* on the theme 'Memory and Energy' (October 2020), and is the editor of a special issue on synergies and persistence in the history of energy techniques for the *Cahiers François Viète* (2022).

Joris Mercelis

Joris Mercelis is Assistant Professor at the Department of History of Science and Technology at Johns Hopkins University in Baltimore. His research interests include the economic history of science and technology and the histories of chemical and photographic technology, science, and business. He is the author of *Beyond Bakelite: Leo Baekeland and the Business of Science and Invention* (MIT Press, 2020) and a co-editor of special journal issues on academic entrepreneurship (*History and Technology and Management & Organizational History*, 2017) and on the history of women and gender in chemistry (*Ambix*, 2022).

Audrey Millet

Audrey Millet holds a PhD in history (University of Paris 8 and University of Neuchâtel). She is a specialist in the history of industry, knowledge, and know-how. Her fields of research interconnect history of technology, economic history, and social history, especially in textile. She has recently published *Vie et destin d'un dessinateur textile, d'après le journal d'Henri Lebert (1794–1862)* (Editions Champ

Vallon, 2018) and co-authored *Histoire des sciences et des techniques (1500–1789). Documents* (Atlande, 2016). She has published several articles in journals (*Revue d'histoire du 19ᵉ siècle, Artefact. Techniques, histoire et sciences humaines*). She is based at Oslo University.

Raphaël Morera

Raphaël Morera is a CNRS research fellow and co-director of the Centre de recherches historiques (CNRS-EHESS). As an environmental historian, he works specifically on the management of water and wetlands, considering both urban and rural territories. His research also focuses on contemporary history. In 2020, with Nicolas Lyon-Caen, he published *À vos poubelles citoyens ! Environnement urbain, salubrité publique et investissement civique (Paris, XVIᵉ- XVIIIᵉ siècle)* (Ceyzérieu, Champ Vallon).

Elisabeth Mortier

Elisabeth Mortier is a historian. She specializes in environmental history and the history of technology in the Middle East. Her doctoral research focusses on water uses in Palestine during the British Mandate (1920 to 1948). She has defended a PhD thesis on 'Les territoires de l'eau en Palestine rurale: une histoire environnementale, sociale et politique durant la domination britannique (1917–1947)' at Sorbonne Université (2022). She is the author of the article 'The heroic figure of the Zionist pioneer: The appropriation of the marshlands in Palestine (from the end of the nineteenth century to the 1950s)', *Études rurales*, 203–1, 2019. She has co-edited a dossier on 'L'irrigation au Moyen-Orient contemporain', *Artefact. Techniques, histoire et sciences humaines*, 12, 2020.

Alessandro Nuvolari

Alessandro Nuvolari is Professor of Economic History and Director of the Institute of Economics at Scuola Superiore Sant'Anna, Pisa. He was educated at Bocconi University, Milan and at Eindhoven University of Technology, where he received a PhD in economics. His main field of research is the role played by science and technology in the emergence and consolidation of modern economic growth with a particular focus on the Industrial Revolution in England and the early industrialization of Italy. He has also studied the connection between patents and inventive activity in both historical and contemporary contexts. He is the editor of *European Review of Economic History* (since 2022), associate editor of *Industrial and Corporate Change* (since 2015) and of the *Scandinavian Economic History Review* (since 2012), and a member of the scientific board of *Revue d'Economie Industrielle* (since 2015).

Julia Obertreis

Julia Obertreis was Professor of Modern and East European History at the Friedrich-Alexander-University Erlangen-Nuremberg (Germany). Her research areas included the imperial and colonial as well as environmental and technological history of the Russian Empire and the Soviet Union. She concentrated on irrigation and water infrastructure in the countryside and in cities and had a regional focus on Central Asia. She is the author of *Imperial Desert Dreams. Cotton Growing and Irrigation in Central Asia, 1860–1991* (V&R unipress, 2017). She has co-edited, together with Christine Bichsel, Peter Mollinga und Timothy Moss, *Water, Infrastructure and Political Rule, Water Alternatives* 9 (2016) 2. Together with Olga Malinova-Tziafeta, she has co-authored an article in Russian entitled 'Istoriia Gorodov i Vodnye Infrastruktury v Rossiiskoi imperii i SSSR' [Urban History and Water Infrastructure in the Russian Empire and the USSR], *Noveishaia istoriia Rossii. Mezhdistsiplinarnyi nauchno-teoreticheskii zhurnal* 9 (2019) 1: 173–201. Most recently, she co-edited, together with Jonas van der Straeten, a special issue entitled *Technology, Temporality and the Study of Central Asia, Central Asian Survey* 41 (2022) 2. She was an editorial board member of the academic journals *Environmental History* and *Slavic Review*.

Omprasad

Omprasad holds a PhD in history of science from Jawaharlal Nehru University, New Delhi. He was a recipient of an archival research grant from the Charles Wallace Trust and a Doctoral Fellowship from the Indian Council of Social Science Research. His research interests include the history of social movements of scientists and the relationship between science and politics.

Arnaud Passalacqua

Arnaud Passalacqua is Professor in Urban Planning (Université Paris Est Créteil, Lab'Urba, F-94010 Créteil, France). His work focusses on the history of mobility in Europe, and he has been particularly interested in various issues such as the public space in cities, energy choices, the imaginations associated with mobility, and, more recently, the international circulation of transport systems. He is an active member of the major French and international networks in this field (including T2M, P2M, and Rails et histoire).

David Pretel

David Pretel is Professor of Economic History at Autonomous University of Madrid and has been visiting scholar at the universities of Harvard, Cambridge, UCLA, and the Max Planck Institute. He was educated in Madrid and Cambridge and has held teaching and research positions at the European University Institute, the University of London, El Colegio de México, and Pompeu Fabra University. His first book, *Institutionalising Patents in Nineteenth-Century Spain* (Palgrave, 2015), examined the development of the Spanish patent system, providing a fundamental reassessment of its evolution in an international and imperial context. He is co-editor of the volumes *The Caribbean and the Atlantic World Economy, 1650-1914* (Palgrave, 2015) and *Technology and Globalisation: Networks of Experts in World History* (Palgrave, 2018). His recent publications include articles in the journals *History of Science, Global Environment, Business History* and *Technology & Culture* as well as chapters in several edited volumes published in Europe, the United States and Mexico, such as the *Oxford Handbook of Commodity History*. His works, CV, and other details can be found on his HYPERLINK "http://www.davidpretel.com/".

Laure Quennouëlle-Corre

Laure Quennouëlle-Corre is Director of Research at the French National Centre for Scientific Research (CNRS). She received her PhD in 2000 (École des hautes études en sciences sociales) and published *La Direction du Trésor. L'État-Banquier et la croissance 1947–1967* (CHEFF Comité pour l'histoire économique et financière de la France) in the same year. Her fields of research relate to economic and financial modern history seen from an international perspective. Her current research focusses on the state, enterprises, and markets in monetary and financial areas including work on public debt, capital markets, financial elites, and women in the banking sector. She recently published *La place financière de Paris au xxe siècle. Des ambitions contrariées* (CHEFF/IGPDE Institut de la gestion publique et du développement économique, 2015); *Politique et finance dans l'Europe du xxe siècle. Entretiens avec Robert Jablon* (with Robert Jablon and André Straus, Peter Lang, 2009); and 'French Bankers and the Transformation of the Financial System in the Second Half of the Twentieth Century', in *Financial Elites and European Banking. Historical Perspectives*, ed. by Youssef Cassis and Giuseppe Telesca (Oxford University Press, 2018).

Dhruv Raina

Dhruv Raina is Professor at the Jawaharlal Nehru University, New Delhi. He studied physics at the Indian Institute of Technology, Mumbai and received his PhD in the philosophy of science from Göteborg University. His research has focussed on the politics and cultures of scientific knowledge in South Asia, as well as the history and historiography of mathematics. Among his authored works are *Images and Contexts* (Oxford University Press, 2003); *Needham's Indian Network* (Yoda Press, 2015), and the co-authored with S. Irfan Habib *Domesticating Modern Science* (Tulika Books, 2004). His co-edited works include *Situating the History of Science* (Oxford University Press, 1999), *Social History of Sciences in Colonial India* (2007), and *Science Between Europe and Asia* (Springer, 2010). His more recent work addresses the contemporary concerns of the emergence of inter- and transdisciplinarity, with a special focus on fields across the natural and social sciences.

Arnaud Saint-Martin

Arnaud Saint-Martin is a sociologist and CNRS research fellow. His research alternates between the history of science and technology, particularly astronomy, and the study of transformations in astronautics, from the Cold War to the advent of "New Space". He is currently completing research on the recent history of the privatization of space activities in the United States and the emergence of astrocapitalism.

Nayani Sarma

Nayani Sarma is currently a PhD Research Scholar of Zakir Husain Centre for Educational Studies in Jawaharlal Nehru University, New Delhi. Her research focusses on the history of the university and higher education in contemporary India.

Caterina Sganga

Caterina Sganga is Associate Professor of Comparative Private Law at Scuola Superiore Sant'Anna (Pisa). She holds a PhD in comparative private law from Sant'Anna, an LLM from Yale Law School, and an LLB and JD from University of Pisa. Her key research area is international and comparative intellectual property law, with a focus on European copyright law, IP and new technologies, the interplay between IP and human rights, and data ownership and management in the era of big data. She is a member of several international IP and property law associations (EPIP, ATRIP, ALPS) and a fellow of the European Law Institute. She published *Propertizing European Copyright. History, Challenges and Opportunities* (Edward Elgar Publishing, 2018). She is *avvocato* at the Italian State Bar and attorney-at-law at the New York State Bar.

Kevin Troch

Kevin Troch is curator of collections at the University of Mons. He defended his PhD in history in 2018 at the University of Lille and the University of Namur ('Ne pas grever l'avenir au bénéfice du présent. Une histoire environnementale de l'extraction du charbon de la fin du 18e siècle à l'Entre-deux-guerres: un développement non soutenable. L'exemple du Couchant de Mons et du Valenciennois'), which focussed on the environmental history of coal mining in Belgium and France from the 1870s to the 1940s. His current research focusses on the history of mineralogy and scientific instruments.

Jonas van der Straeten

Jonas van der Straeten is Assistant Professor at the Technology, Innovation and Society Group (Eindhoven University of Technology). He has been a postdoctoral researcher in the project 'A Global History of Technology, 1850–2000' at the Technical University of Darmstadt, funded by the European Research Council. His postdoctoral research project focussed on the temporality of technology in Central Asia. Based on archival and field research in Samarkand and Tashkent, Uzbekistan, he currently works on the history of building and housing in the Uzbek SSR. He is author of the forthcoming book *Capital Grids: A Global History of Electricity in East Africa*, to be published with Palgrave Macmillan.

Erik van der Vleuten

Erik van der Vleuten serves as Professor and Chair of History of Technology at Eindhoven University of Technology, and as scientific director of the Foundation for the History of Technology SHT (both since 2015). He was a co-founding member and Chair of the pan-European history of technology community *Tensions of Europe*, and co-chaired *the European Social Science History Conference's* history of technology program. His teaching and research revolve around pluralizing linear and universalistic discourses on technology and socio-ecological change. His dissertation (1998) explored a 'symmetrical history' of co-existing, conflicting, and entwined systems for electricity supply and use. Later he studied diverse, contested and connected histories of (trans)national infrastructure and socio-ecological change in modern and contemporary Europe. His current focus is on the plural-yet-connected (un)sustainability histories and futures of distant regions across the globe. In 2023-28 he co-leads the Dutch Research Council Open Competition program *Soy Stories. Connected 'Sustainability' Histories and Futures of the global Soyacene.*

Heike Weber

Heike Weber is Professor for History of Technology at the Technische Universität Berlin. Her primary research is situated at the intersection of consumption history, environmental history, and the history of technology. She has worked on twentieth-century everyday technologies and the cross-roads of media and mobility history and currently studies the history of waste, recycling, and repair. Recent publications include *Provokationen der Technikgeschichte. Zum Reflexionszwang historischer Forschung* (co-edited with Martina Heßler, Schöningh, 2019); and *Histories of Technology's Persistence: Repair, Reuse and Disposal* (co-edited with Stefan Krebs, Transcript, 2020).